U0276505

国外优秀数学著作
原 版 系 列

离散与计算几何手册

Handbook of Discrete and Computational Geometry, 3e

［美］ 雅各布·E. 古德曼（Jacob E. Goodman）

［美］ 约瑟夫·奥罗克（Joseph O'Rourke）

［美］ 乔鲍·D. 托特（Csaba D. Tóth）

主编

（下册·第三版）

（英文）

哈尔滨工业大学出版社

HARBIN INSTITUTE OF TECHNOLOGY PRESS

黑版贸审字 08－2020－202 号

Handbook of Discrete and Computational Geometry, 3e/by Jacob E. Goodman, Joseph O'Rourke, Csaba D. Tóth/ISBN:978－1－4987－1139－5

Copyright © 2018 by CRC Press.

Authorized translation from English reprint edition published by CRC Press, part of Taylor & Francis Group LLC；All rights reserved；本书原版由 Tadylor & Francis 出版集团旗下，CRC 出版公司出版，并经其授权出版影印版. 版权所有，侵权必究.

Harbin Institute of Technology Press Ltd is authorized to publish and distribute exclusively the English reprint edition. This edition is authorized for sale throughout Mainland of China. No Part of the publication may be reproduced or distributed by any means, or stored in a database or retrieval system, without the prior written permission of the publisher. 本书英文影印版授权由哈尔滨工业大学出版社独家出版并仅限在中国大陆地区销售. 未经出版者书面许可，不得以任何方式复制或发行本书的任何部分.

Copies of this book sold without a Taylor & Francis sticker on the cover are unauthorized and illegal. 本书封面贴有 Taylor & Francis 公司防伪标签，无标签者不得销售.

图书在版编目(CIP)数据

离散与计算几何手册：第三版＝Handbook of Discrete and Computational Geometry, 3e. 下册：英文/(美)雅各布·E. 古德曼(Jacob E. Goodman)，(美)约瑟夫·奥罗克(Joseph O'Rourke)，(美)乔鲍·D. 托特(Csaba D. Tóth)主编. —哈尔滨：哈尔滨工业大学出版社，2023.1

ISBN 978-7-5767-0652-9

Ⅰ.①离… Ⅱ.①雅… ②约… ③乔… Ⅲ.①离散数学－计算几何－英文 Ⅳ.①O18

中国国家版本馆 CIP 数据核字(2023)第 030334 号

LISAN YU JISUAN JIHE SHOUCE：DI-SAN BAN(XIACE)

策划编辑	刘培杰　杜莹雪
责任编辑	刘立娟
封面设计	孙茵艾
出版发行	哈尔滨工业大学出版社
社　　址	哈尔滨市南岗区复华四道街 10 号　邮编 150006
传　　真	0451－86414749
网　　址	http://hitpress.hit.edu.cn
印　　刷	哈尔滨市石桥印务有限公司
开　　本	787 mm×1 092 mm　1/16　印张 123　字数 2 371 千字
版　　次	2023 年 1 月第 1 版　2023 年 1 月第 1 次印刷
书　　号	ISBN 978-7-5767-0652-9
定　　价	248.00 元(全 3 册)

(如因印装质量问题影响阅读，我社负责调换)

Part VII

APPLICATIONS OF DISCRETE AND COMPUTATIONAL GEOMETRY

49 LINEAR PROGRAMMING

Martin Dyer, Bernd Gärtner, Nimrod Megiddo, and Emo Welzl

INTRODUCTION

Linear programming has many important practical applications, and has also given rise to a wide body of theory. See Section 49.9 for recommended sources. Here we consider the linear programming problem in the form of maximizing a linear function of d variables subject to n linear inequalities. We focus on the relationship of the problem to computational geometry, i.e., we consider the problem in small dimension. More precisely, we concentrate on the case where $d \ll n$, i.e., $d = d(n)$ is a function that grows very slowly with n. By linear programming duality, this also includes the case $n \ll d$. This has been called *fixed-dimensional* linear programming, though our viewpoint here will not treat d as constant. In this case there are strongly polynomial algorithms, provided the rate of growth of d with n is small enough.

The plan of the chapter is as follows. In Section 49.2 we consider the simplex method, in Section 49.3 we review deterministic linear time algorithms, in Section 49.4 randomized algorithms, and in Section 49.5 we consider the derandomization of the latter. Section 49.6 discusses the combinatorial framework of LP-type problems, which underlie most current combinatorial algorithms and allows their application to a host of optimization problems. We briefly describe the more recent combinatorial framework of unique sink orientations, in the context of striving for algorithms with a milder dependence on d. In Section 49.7 we examine parallel algorithms for this problem, and finally in Section 49.8 we briefly discuss related issues. The emphasis throughout is on complexity-theoretic bounds for the linear programming problem in the form (49.1.1).

49.1 THE BASIC PROBLEM

Any linear program (LP) may be expressed in the inequality form

$$\begin{aligned} \text{maximize} \quad & z = c \cdot x \\ \text{subject to} \quad & Ax \leq b, \end{aligned} \tag{49.1.1}$$

where $c \in \mathbb{R}^d$, $b \in \mathbb{R}^n$, and $A \in \mathbb{R}^{n \times d}$ are the input data and $x \in \mathbb{R}^d$ the variables. Without loss of generality, the columns of A are assumed to be linearly independent. The vector inequality in (49.1.1) is with respect to the componentwise partial order on $\mathbb{R}^n$. We will write a_i for the ith row of A, so the constraint may also be expressed in the form

$$a_i \cdot x = \sum_{j=1}^{d} a_{ij} x_j \leq b_i \quad (i = 1, \ldots, n). \tag{49.1.2}$$

GLOSSARY

Constraint: A condition that must be satisfied by a solution.

Objective function: The linear function to be maximized over the set of solutions.

Inequality form: The formulation of the linear programming problem where all the constraints are weak inequalities $a_i \cdot x \le b_i$.

Feasible set: The set of points that satisfy all the constraints. In the case of linear programming, it is a convex polyhedron in $\mathbb{R}^d$.

Defining hyperplanes: The hyperplanes described by the equalities $a_i \cdot x = b_i$.

Tight constraint: An inequality constraint is tight at a certain point if the point lies on the corresponding hyperplane.

Infeasible problem: A problem with an empty feasible set.

Unbounded problem: A problem with no finite maximum.

Vertex: A feasible point where at least d linearly independent constraints are tight.

Nondegenerate problem: A problem where at each vertex precisely d constraints are tight.

Strongly polynomial-time algorithm: An algorithm for which the total number of arithmetic operations and comparisons (on numbers whose size is polynomial in the input length) is bounded by a polynomial in n and d alone.

We observe that (49.1.1) may be infeasible or unbounded, or have multiple optima. A complete algorithm for linear programming must take account of these possibilities. In the case of multiple optima, we assume that we have merely to identify *some* optimum solution. (The task of identifying *all* optima is considerably more complex; see [Dye83, AF92].) An optimum of (49.1.1) will be denoted by x^0. At least one such solution (assuming one exists) is known to lie at a vertex of the feasible set. There is little loss in assuming nondegeneracy for theoretical purposes, since we may "infinitesimally perturb" the problem to ensure this using well known methods [Sch86]. However, a complete algorithm must recognize and deal with this possibility.

It is well known that linear programs can be solved in time polynomial in the total length of the input data [GLS93]. However, it is not known in general if there is a *strongly* polynomial-time algorithm. This is true even if randomization is permitted. (Algorithms mentioned below may be assumed deterministic unless otherwise stated.) The "weakly" polynomial-time algorithms make crucial use of the size of the numbers, so seem unlikely to lead to strongly polynomial-time methods. However, strong bounds are known in some special cases. Here are two examples. If all a_{ij} are bounded by a constant, then É. Tardos [Tar86] has given a strongly polynomial-time algorithm. If every row a_i has at most two nonzero entries, Megiddo [Meg83a] has shown how to find a feasible solution in strongly polynomial time.

49.2 THE SIMPLEX METHOD

GLOSSARY

Simplex method: For a nondegenerate problem in inequality form, this method seeks an optimal vertex by iteratively moving from one vertex to a neighboring vertex of higher objective function value.

Pivot rule: The rule by which a neighboring vertex is chosen.

Random-edge simplex algorithm: A randomized variant of the simplex method where the neighboring vertex is chosen uniformly at random.

Dantzig's simplex method is probably still the most commonly used method for solving large linear programs in practice, but (with Dantzig's original pivot rule) Klee and Minty showed that the algorithm may require an exponential number of iterations in the worst case. For example, it may require $2^d - 1$ iterations when $n = 2d$. Other variants were subsequently shown to have similar behavior. While it is not known for certain that all suggested variants of the simplex method have this bad worst case, there seems to be no reason to believe otherwise. In fact, two long-standing candidates for a polynomial-time variant (Zadeh's and Cunningham's rule) have recently been shown to have bad worst-case behavior as well [Fri11, AF16]. This breakthrough is due to a new technique by Friedmann who constructed lower bound instances in a combinatorial way from certain games on graphs. Earlier lower bound constructions were typically in the form of "raw" linear programs, with carefully crafted coordinates. But this direct approach (pioneered by Klee and Minty [KM72]) could so far not be applied to history-dependent rules such as Zadeh's and Cunningham's.

When we assume $d \ll n$, the simplex method may require $\Omega(n^{\lfloor d/2 \rfloor})$ iterations [KM72, AZ99], and thus it is polynomial only for $d = O(1)$. This is asymptotically no better than enumerating all vertices of the feasible region.

By contrast, Kalai [Kal92] gave a randomized simplex-like algorithm that requires only $2^{O(\sqrt{d \log n})}$ iterations. (An identical bound was also given by Matoušek, Sharir, and Welzl [MSW96] for a closely related algorithm; see Section 49.4.) Combined with Clarkson's methods [Cla95], this results in a bound of $O(d^2 n) + e^{O(\sqrt{d \log d})}$ (cf. [MSW96]). This is the best "strong" bound known, other than for various special cases, and it is evidently polynomial-time provided $d = O(\log^2 n / \log \log n)$. No complete derandomization of these algorithms is known, and it is possible that randomization may genuinely help here. However, Friedmann, Hansen and Zwick recently showed that the complexity of the random-edge simplex algorithm (where the pivot is chosen uniformly at random) is actually superpolynomial, defeating the most promising candidate for an (expected) polynomial-time pivot rule [FHZ11]. By giving a lower bound also for the random-facet rule, they actually prove that the analysis underlying the best strong bound above is essentially tight. These are the first superpolynomial lower bound constructions for natural randomized simplex variants, again exploiting the new game-based technique by Friedmann. Some less natural randomized variants have not been defeated yet, and it remains to be seen whether this would also be possible through Friedmann's approach.

49.3 LINEAR-TIME LINEAR PROGRAMMING

The study of linear programming within computational geometry was initiated by Shamos [Sha78] as an application of an $O(n \log n)$ convex hull algorithm for the intersection of halfplanes. Muller and Preparata [MP78] gave an $O(n \log n)$ algorithm for the intersection of halfspaces in $\mathbb{R}^3$. Dyer [Dye84] and Megiddo [Meg83b] found, independently, an $O(n)$ time algorithm for the linear programming problem in the cases $d = 2, 3$.

Megiddo [Meg84] generalized the approach of these algorithms to arbitrary d, arriving at an algorithm of time complexity $O(2^{2^d} n)$, which is polynomial for $d \leq \log \log n + O(1)$. This was subsequently improved by Clarkson [Cla86b] and Dyer [Dye86] to $O(3^{d^2} n)$, which is polynomial for $d = O(\sqrt{\log n})$. Megiddo [Meg84, Meg89] and Dyer [Dye86, Dye92] showed that Megiddo's idea could be used for many related problems: Euclidean one-center, minimum ball containing balls, minimum volume ellipsoid, etc.; see also the derandomized methods and LP-type problems in the sections below.

GLOSSARY

Multidimensional search: Given a set of hyperplanes and an oracle for locating a point relative to any hyperplane, locate the point relative to all the input hyperplanes.

MEGIDDO'S ALGORITHMS

The basic idea in these algorithms is as follows. It follows from convexity considerations that either the constraints in a linear program are tight (i.e., satisfied with equality) at x^0, or they are irrelevant. We need identify only d linearly independent tight constraints to identify x^0. We do this by discarding a fixed proportion of the irrelevant constraints at each iteration. Determining whether the ith constraint is tight amounts to determining which case holds in $a_i \cdot x^0 \gtreqless b_i$. This is embedded in a multidimensional search problem. Given *any* hyperplane $\alpha \cdot x = \beta$, we can determine which case of $\alpha \cdot x^0 \gtreqless \beta$ holds by (recursively) solving three linear programs in $d - 1$ variables. These are (49.1.1) plus $\alpha \cdot x = \gamma$, where $\gamma \in \{\beta - \epsilon, \beta, \beta + \epsilon\}$ for "small" $\epsilon > 0$. (We need not define ϵ explicitly; it can be handled symbolically.) In each of the three linear programs we eliminate one variable to get $d - 1$. The largest of the three objective functions tells us where x^0 lies with respect to the hyperplane. We call this an *inquiry* about $\alpha \cdot x = \beta$. The problem now reduces to locating x^0 with respect to a proportion $P(d)$ of the n hyperplanes using only $N(d)$ inquiries.

The method recursively uses the following observation in $\mathbb{R}^2$. Given two lines through the origin with slopes of opposite sign, knowing which quadrant a point lies in allows us to locate it with respect to *at least one* of the lines (see Figure 49.3.1). We use this on the first two coordinates of the problem in $\mathbb{R}^d$. First rotate until $\frac{1}{2}n$ defining hyperplanes have positive and $\frac{1}{2}n$ negative "slopes" on these

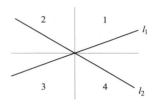

FIGURE 49.3.1
Quadrants $1, 3$ locate for l_2; quadrants $2, 4$ locate for l_1.

coordinates. This can be done in $O(n)$ time using median-finding. Then arbitrarily pair a positive with a negative to get $\frac{1}{2}n$ pairs of the form

$$ax_1 + bx_2 + \cdots \;=\; \cdots$$
$$cx_1 - dx_2 + \cdots \;=\; \cdots,$$

where a, b, c, and d represent nonnegative numbers, and the $\cdots$ represent linear functions of $x_3, \ldots, x_d$ on the left and arbitrary numbers on the right. Eliminating x_2 and x_1 in each pair gives two families S_1, S_2 of $\frac{1}{2}n$ hyperplanes each in $d-1$ dimensions of the form

$$S_1: \quad x_1 \qquad + \cdots \;=\; \cdots$$
$$S_2: \qquad\quad x_2 \;+ \cdots \;=\; \cdots.$$

We recursively locate with respect to $\frac{1}{2}P(d-1)n$ hyperplanes with $N(d-1)$ inquiries in S_1, and then locate with respect to a $P(d-1)$-fraction of the corresponding paired hyperplanes in S_2. We have then located $\frac{1}{2}P(d-1)^2n$ pairs with $2N(d-1)$ inquiries. Using the observation above, each pair gives us location with respect to at least one hyperplane in d dimensions, i.e.,

$$P(d) = \tfrac{1}{2}P(d-1)^2, \qquad N(d) = 2N(d-1). \tag{49.3.1}$$

Since $P(1) = \frac{1}{2}, N(1) = 1$ (by locating with respect to the median in $\mathbb{R}^1$), (49.3.1) yields

$$P(d) = 2^{-(2^d-1)}, \qquad N(d) = 2^{d-1},$$

giving the following time bound $T(n, d)$ for solving (49.1.1).

$$T(n, d) \le 3 \cdot 2^{d-1}T(n, d-1) + T((1 - 2^{-(2^d-1)})n, d) + O(nd),$$

with solution $T(n, d) < 2^{2^{d+2}}n$.

THE CLARKSON-DYER IMPROVEMENT

The Clarkson/Dyer improvement comes from repeatedly locating in S_1 and S_2 to increase $P(d)$ at the expense of $N(d)$.

49.4 RANDOMIZED ALGORITHMS

Dyer and Frieze [DF89] showed that, by applying an idea of Clarkson [Cla86a] to give a *randomized* solution of the multidimensional search in Megiddo's algorithm [Meg84], an algorithm of complexity $O(d^{3d+o(d)}n)$ was possible. Clarkson [Cla88, Cla95] improved this dramatically. We describe this below, but first outline a simpler algorithm subsequently given by Seidel [Sei91].

Suppose we order the constraints randomly. At stage k, we have solved the linear program subject to constraints $i = 1, \ldots, k - 1$. We now wish to add constraint k. If it is satisfied by the current optimum we finish stage k and move to $k+1$. Otherwise, the new constraint is clearly tight at the optimum over constraints $i = 1, \ldots, k - 1$. Thus, recursively solve the linear program subject to this equality (i.e., in dimension $d-1$) to get the optimum over constraints $i = 1, \ldots, k$, and move on to $k + 1$. Repeat until $k = n$.

The analysis hinges on the following observation. When constraint k is added, the probability it is not satisfied is exactly d/k (assuming, without loss, nondegeneracy). This is because only d constraints are tight at the optimum and this is the probability of writing one of these *last* in a random ordering of $1, 2, \ldots, k$. This leads to an expected time of $O(d!n)$ for (49.1.1). Welzl [Wel91] extended Seidel's algorithm to solve other problems such as smallest enclosing ball or ellipsoid, and described variants that perform favorably in practice.

Sharir and Welzl [SW92] modified Seidel's algorithm, resulting in an improved running time of $O(d^3 2^d n)$. They put their algorithm in a general framework of solving "LP-type" problems (see Section 49.6 below). Matoušek, Sharir, and Welzl [MSW96] improved the analysis further, essentially obtaining the same bound as for Kalai's "primal simplex" algorithm. The algorithm was extended to LP-type problems by Gärtner [Gär95], with a similar time bound.

CLARKSON'S ALGORITHM

The basic idea is to choose a random set of r constraints, and solve the linear program subject to these. The solution will violate "few" constraints among the remaining $n - r$, and, moreover, one of these must be tight at x^0. We solve a new linear program subject to the violated constraints and a new random subset of the remainder. We repeat this procedure (aggregating the old violated constraints) until there are no new violated constraints, in which case we have found x^0. Each repetition gives an extra tight constraint for x^0, so we cannot perform more than d iterations.

Clarkson [Cla88] gave a different analysis, but using Seidel's idea we can easily bound the expected number of violated constraints (see also [GW01] for further simplifications of the algorithm). Imagine all the constraints ordered randomly, our sample consisting of the first r. For $i > r$, let $I_i = 1$ if constraint i is violated, $I_i = 0$ otherwise. Now $\Pr(I_i = 1) = \Pr(I_{r+1} = 1)$ for all $i > r$ by symmetry, and $\Pr(I_{r+1} = 1) = d/(r + 1)$ from above. Thus the expected number of violated constraints is

$$\mathbb{E} \left(\sum_{i=r+1}^{n} I_i \right) = \sum_{i=r+1}^{n} \Pr(I_i = 1) = (n - r)d/(r + 1) < nd/r.$$

(In the case of degeneracy, this will be an upper bound by a simple perturbation argument.)

Thus, if $r = \sqrt{n}$, say, there will be at most $d\sqrt{n}$ violated constraints in expectation. Hence, by Markov's inequality, with probability $\frac{1}{2}$ there will be at most $2d\sqrt{n}$ violated constraints in actuality. We must therefore recursively solve about $2(d + 1)$ linear programs with at most $(2d^2 + 1)\sqrt{n}$ constraints. The "small" base cases can be solved by the simplex method in $d^{O(d)}$ time. This can now be applied recursively, as in [Cla88], to give a bound for (49.1.1) of

$$O(d^2 n) + (\log n)^{\log d + 2} d^{O(d)}.$$

Clarkson [Cla95] subsequently modified his algorithm using a different "iterative reweighting" algorithm to solve the $d + 1$ small linear programs, obtaining a better bound on the execution time.

Each constraint receives an initial weight of 1. Random samples of total weight $10d^2$ (say) are chosen at each iteration, and solved by the simplex method. If W is the current total weight of all constraints, and W' the weight of the unsatisfied constraints, then $W' \leq 2Wd/10d^2 = W/5d$ with probability at least $\frac{1}{2}$ by the discussion above, regarding the weighted constraints as a multiset. We now double the weights of all violated constraints and repeat until there are no violated constraints (see [BG11] for a further simplification of this algorithm). This terminates in $O(d \log n)$ iterations by the following argument. After k iterations we have

$$W \leq \left(1 + \frac{1}{5d}\right)^k n \leq n e^{k/5d},$$

and W^*, the total weight of the d optimal constraints, satisfies $W^* \geq 2^{k/d}$, since at least one is violated at each iteration. Now it is clear that $W^* < W$ only while $k < Cd \ln n$, for some constant C. Applying this to the $d + 1$ small linear programs gives overall complexity

$$O(d^2 n + d^4 \sqrt{n} \log n) + d^{O(d)} \log n.$$

This is almost the best expected time bound known for linear programming, except that Kalai's algorithm (or [MSW96]) can be used to solve the base cases rather than the simplex method. Then we get the improved bound (cf. [GW96])

$$O(d^2 n) + e^{O(\sqrt{d \log d})}.$$

This is polynomial for $d = O(\log^2 n / \log \log n)$, and is the best expected time bound to date.

49.5 DERANDOMIZED METHODS

Somewhat surprisingly, the randomized methods of Section 49.4 can also lead to the best *deterministic* algorithms for (49.1.1). Chazelle and Matoušek [CM96] produced a first derandomized version of Clarkson's algorithm.

The idea, which has wider application, is based on finding (in linear time) *approximations* to the constraint set. If N is a constraint set, then for each $x \in \mathbb{R}^d$ let $V(x, N)$ be the set of constraints violated at x. A set $S \subseteq N$ is an ϵ-approximation to N if, for all x,

$$\left| \frac{|V(x, S)|}{|S|} - \frac{|V(x, N)|}{|N|} \right| < \epsilon.$$

(See also Chapters 40 and 47.) Since $n = |N|$ hyperplanes partition $\mathbb{R}^d$ into only $O(n^d)$ regions, there is essentially only this number of possible cases for x, i.e., only this number of different sets $V(x, N)$. It follows from the work of Vapnik and Chervonenkis that a (d/r)-approximation of size $O(r^2 \log r)$ always exists, since a random subset of this size has the property with nonzero probability. If we can find such an approximation deterministically, then we can use it in Clarkson's algorithm in place of random sampling. If we use a (d/r)-approximation, then, if x^* is the linear programming optimum for the subset S, $|V(x^*, S)| = 0$, so that

$$|V(x^*, N)| < |N|d/r = nd/r,$$

as occurs in expectation in the randomized version. The implementation involves a refinement based on two elegant observations about approximations, which both follow directly from the definition.

(i) An ϵ-approximation of a δ-approximation is an $(\epsilon + \delta)$-approximation of the original set.

(ii) If we partition N into q equal sized subsets $N_1, \ldots, N_q$ and take an (equal sized) ϵ-approximation S_i in each N_i $(i = 1, \ldots, q)$, then $S_1 \cup \ldots \cup S_q$ is an ϵ-approximation of N.

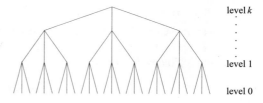

level k

level 1

FIGURE 49.5.1
A partition tree of height k, with $q = 3$.

level 0

We then *recursively* partition N into q equal sized subsets, to give a "partition tree" of height k, say, as in Figure 49.5.1 (cf. Section 40.2). The sets at level 0 in the partition tree are "small." We calculate an ϵ_0-approximation in each. We now take the union of these approximations at level 1 and calculate an ϵ_1-approximation of this union. This is an $(\epsilon_0 + \epsilon_1)$-approximation of the whole level 1 set, by the above observations. Continuing up the tree, we obtain an overall $(\sum_{i=0}^{k} \epsilon_i)$-approximation of the entire set. At each stage, the sets on which we have to find the approximations remain "small" if the ϵ_i are suitably chosen. Therefore we can use a relatively inefficient method of finding an approximation. A suitable method is the *method of conditional probabilities* due to Raghavan and Spencer. It is (relatively) straightforward to implement this on a set of size m to run in $O(m^{d+1})$ time. However, since this has to be applied only to small sets (in comparison with n), the total work can be bounded by a linear function of n. Chazelle and Matoušek [CM96] used $q = 2$, and an ϵ_i that corresponds to roughly halving the union at each level $i = 1, \ldots, k$.

The algorithm cannot completely mimic Clarkson's, however, since we can no longer use $r = \sqrt{n}$. Such a large approximation cannot be determined in linear time by the above methods. But much smaller values of r suffice (e.g., $r = 10d^3$) simply to get *linear-time* behavior in the recursive version of Clarkson's algorithm. Using this observation, Chazelle and Matoušek [CM96] obtained a deterministic algorithm with time-complexity $d^{O(d)}n$. As far as the asymptotics of the exponent is concerned, this is still the best deterministic time bound known for solving (49.1.1), and it remains polynomial for $d = O(\log n/\log \log n)$. Later research has improved the constant in the exponent. The currently best (and most simple) approach is due to Chan [Cha16].

Recall that with $r = 10d^3$, Chazelle and Matoušek's algorithm as sketched above computes an $O(1/d^2)$-approximation S of the constraint set. Actually, an $O(1/d^2)$-net S suffices. An ϵ-net has the ϵ-approximation property only for vectors x satisfying $V(x, S) = \emptyset$. But this is enough, as (the derandomized version of) Clarkson's algorithm solves the problem subject to the constraints in S first, and then only keeps adding constraints. Hence, all solutions x coming up during the algorithm satisfy $V(x, S) = \emptyset$.

Chan's observation is that the $O(1/d^2)$-net does not have to be of optimal size, hence one might be able to employ a simpler and faster deterministic construction algorithm. Indeed, as long as the subproblems solved in the recursive calls of Clarkson's algorithm are by a factor of $2d$, say, smaller than the original problem, the linear-time analysis still works.

Chan's main contribution is a simple and fast deterministic method to compute an $O(1/d^2)$-net of size roughly $n/(2d)$. For this, he first provides a very simple and fast greedy method to compute an ϵ-net for a set of k constraints, of size $O((d/\epsilon) \log k)$. This is clearly suboptimal, as there are ϵ-nets whose size is independent of k. The trick now is to subdivide the constraint set arbitrarily into $n/f(d)$ groups of size $f(d)$, for a suitable polynomial f. For each group, the fast algorithm is applied with $k = f(d)$, meaning that the size penalty is only logarithmic in d, per group. The final ϵ-net is simply the union of the ϵ-nets for the groups.

The resulting algorithm has runtime $O(d^{3d}n)$, up to $(\log d)^{O(d)}$ factors. A further reduction to $O(d^{d/2}n)$, again up to $(\log d)^{O(d)}$ factors, is obtained through a derandomization of a new variant of Clarkson's second algorithm, and improved ϵ-net constructions. Moreover, this seems to reach a natural barrier in the sense that an exponent of $d/2$ is already incurred by the simplex algorithm that is used as a subroutine for linear programs with $O(d^2)$ constraints.

49.6   LP-TYPE PROBLEMS

The randomized algorithms above by Clarkson and in [MSW96, Gär95] can be formulated in an abstract framework called *LP-type problems*. With an extra condition (involving VC-dimension of certain set-systems) this extends to the derandomization in [CM96]. In this way, the algorithms are applicable to a host of problems including smallest enclosing ball, polytope distance, smallest enclosing ellipsoid, largest ellipsoid in polytope, smallest ball intersecting a set of convex objects, angle-optimal placement in polygon, rectilinear 3-centers in the plane, spherical separability, width of thin point sets in the plane, and integer linear programming (see [MSW96, GW96] for descriptions of these problems and of the reductions needed). A different abstraction called *abstract objective functions* is described by Kalai in [Kal97], and for the even more general setting of *abstract optimization problems* see [Gär95].

For the definitions below, the reader should think of optimization problems in which we are given some set H of constraints and we want to *minimize* some given function under those constraints. For every subset G of H, let $w(G)$ denote the optimum value of this function when all constraints in G are satisfied. The function w is only given implicitly via some basic operations to be specified below. The goal is to compute an inclusion-minimal subset B_H of H with the same value as H (from which, in general, the value is easy to determine).

GLOSSARY

LP-type problem: A pair (H, w), where H is a finite set and $w : 2^H \to \mathcal{W}$ for a linearly ordered set $(\mathcal{W}, \leq)$ with a minimal element $-\infty$, so that the monotonicity and locality axioms below are satisfied.

Monotonicity axiom: For any F, G with $F \subseteq G \subseteq H$, we have $w(F) \leq w(G)$.

Locality axiom (for LP-type problems):   For any $F \subseteq G \subseteq H$ with $-\infty \neq w(F) = w(G)$ and for any $h \in H$, $w(G) < w(G \cup \{h\})$ implies $w(F) < w(F \cup \{h\})$.

Constraints of LP-type problem: Given an LP-type problem (H, w), the elements of H are called constraints.

Basis: A set B of constraints is called a basis if $w(B') < w(B)$ for every proper subset of B.

Basis of set of constraints: Given a set G of constraints, a subset $B \subseteq G$ is called a basis of G if it is a basis and $w(B) = w(G)$ (i.e., an inclusion-minimal subset of G with equal w-value).

Combinatorial dimension: The maximum cardinality of any basis in an LP-type problem (H, w), denoted by $\delta = \delta_{(H,w)}$.

Basis regularity: An LP-type problem (H, w) is basis-regular if, for every basis B with $|B| = \delta$ and for every constraint h, all bases of $B \cup \{h\}$ have exactly δ elements.

Violation test: Decides whether or not $w(B) < w(B \cup \{h\})$, for a basis B and a constraint h.

Basis computation: Delivers a basis of $B \cup \{h\}$, for a basis B and a constraint h.

Violator space: A pair $(H, \mathcal{V})$, where H is a finite set and $\mathcal{V} : 2^H \to 2^H$ such that the consistency and locality axioms below are satisfied.

Consistency axiom: For any $G \subseteq H$, we have $G \cap \mathcal{V}(G) = \emptyset$.

Locality axiom (for violator spaces):   For any $F \subseteq G \subseteq H$, $G \cap \mathcal{V}(F) = \emptyset$ implies $\mathcal{V}(F) = \mathcal{V}(G)$.

Unique Sink Orientation: Orientation of the n-dimensional hypercube graph such that every subgraph induced by a nonempty face has a unique sink.

A simple example of an LP-type problem is the smallest enclosing ball problem (this problem traces back to J.J. Sylvester [Syl57]): Let S be a finite set of points in $\mathbb{R}^d$, and for $G \subseteq S$, let $\rho(G)$ be the radius of the ball of smallest volume containing G (with $\rho(\emptyset) = -\infty$). Then (S, ρ) is an LP-type problem with combinatorial dimension at most $d + 1$. A violation test amounts to a test deciding whether a point lies in a given ball, while an efficient implementation of basis computations is not obvious (cf. [Gär95]).

Many more examples have been indicated above. As the name suggests, linear programming can be formulated as an LP-type problem, although some care is needed in the presence of degeneracies. Let us assume that we want to maximize the objective function $-x_d$ in (49.1.1), i.e., we are looking for a point in $\mathbb{R}^d$ of smallest x_d-coordinate. In the underlying LP-type problem, the set H of constraints is given by the halfspaces as defined by (49.1.2). For a subset G of these constraints, let $v(G)$ be the backwards lexicographically smallest point satisfying these constraints, with $v(G) := -\infty$ if G gives rise to an unbounded problem, and with $v(G) := \infty$ in case of infeasibility. We assume the backwards lexicographical ordering on $\mathbb{R}^d$ to be extended to $\mathbb{R}^d \cup \{-\infty, \infty\}$ by letting $-\infty$ and ∞ be the minimal and maximal element, resp. The resulting pair (H, v) is LP-type of combinatorial dimension at most $d + 1$. In fact, if the problem is feasible and bounded, then the LP-type problem is basis-regular of combinatorial dimension d. The violation test and basis computation (this amounts to a dual pivot step) are easy to implement.

Matoušek, Sharir, and Welzl [MSW96] showed that a basis-regular LP-type problem (H, w) of combinatorial dimension δ with n constraints can be solved (i.e., a basis of H can be determined) with an expected number of at most

$$\min\{e^{2\sqrt{\delta \ln((n-\delta)/\sqrt{\delta})}+O(\sqrt{\delta}+\ln n)}, 2^{\delta+2}(n-\delta)\} \tag{49.6.1}$$

violations tests and basis computations, provided an initial basis B_0 with $|B_0| = \delta$ is available. (For linear programming one can easily generate such an initial basis by adding d symbolic constraints at "infinity.") Then Gärtner [Gär95] generalized this bound to all LP-type problems. Combining this with Clarkson's methods, one gets a bound (cf. [GW96]) of

$$O(\delta n) + e^{O(\sqrt{\delta \log \delta})}$$

for the expected number of violation tests and basis computations, the best bound known up to now. In the important special case where $n = O(\delta)$, the resulting bound of $e^{O(\sqrt{\delta \log \delta})}$ can be improved to $e^{O(\sqrt{\delta})}$ by using a dedicated algorithm for this case [Gär95]. Recently, Hansen and Zwick [HZ15] have developed and analyzed a new variant of the general algorithm, essentially replacing the denominator of $\sqrt{\delta}$ under the root in (49.6.1) with δ. For $n = O(\delta)$, this also results in an improved bound of $e^{O(\sqrt{\delta})}$.

Matoušek [Mat94] provided a family of LP-type problems, for which the bound (49.6.1) is almost tight for the algorithm provided in [MSW96], for the case $n = 2d$. It was an open problem, though, whether the algorithm performs faster when applied to linear programming instances. In fact, Gärtner [Gär02] showed that the algorithm is quadratic on the instances in Matousek's lower bound family which are realizable as linear programming problems as in (49.1.1). Only the recent subexponential lower bound for the random facet simplex algorithm due to Friedmann, Hansen and Zwick [FHZ11] implies that for $n = 2d$, the bound (49.6.1) is tight up to a possible replacement of the square root by a different root.

Amenta [Ame94] considers the following extension of the abstract framework: Suppose we are given a family of LP-type problems (H, w_λ), parameterized by a real parameter λ; the underlying ordered value set $\mathcal{W}$ has a maximum element ∞ representing *infeasibility*. The goal is to find the smallest λ for which (H, w_λ) is feasible, i.e., $w_\lambda(H) < \infty$. [Ame94] provides conditions under which such a problem can be transformed into a single LP-type problem, and she gives bounds on the resulting combinatorial dimension. This work exhibits interesting relations between LP-type problems and Helly-type theorems (see also [Ame96]).

An interesting combinatorial generalization of LP-type problems that removes the objective function w has been introduced by Škovroň [Ško07]. A *violator space* is a pair $(H, \mathcal{V})$, where H is a finite set that we again think of as being constraints.

$\mathcal{V} : 2^H \to 2^H$ is a function that assigns to each subset G of constraints a set $\mathcal{V}(G)$ of constraints violated by (the solution subject to the constraints in) G. $\mathcal{V}$ is required to satisfy *consistency* ($G \cap \mathcal{V}(G) = \emptyset$ for all G) as well as *locality* (if $F \subseteq G$ and $G \cap \mathcal{V}(F) = \emptyset$, then $\mathcal{V}(F) = \mathcal{V}(G)$). If (H, w) is an LP-type problem (without a set of value $-\infty$) and $\mathcal{V}(G) := \{h \in H \setminus G : w(G \cup \{h\}) > w(G)\}$, then $(H, \mathcal{V})$ is a violator space. A variant of the definition that covers $-\infty$ also exists [Ško07].

It turns out that violator spaces form in a well-defined sense the most general framework in which Clarkson's random-sampling based methods still work [GMRS08, BG11]. In this sense, violator spaces form a "combinatorial core" of linear programming, useful for theoretical considerations. The concept properly generalizes LP-type problems: in the absence of an objective function, simplex-type algorithms may cycle even in nondegenerate situations [GMRS08]. In the applications, however, most violator spaces can actually be cast as LP-type problems.

UNIQUE SINK ORIENTATIONS

As seen in Section 49.4, the currently best "strong" bound for linear programming is
$$O(d^2 n) + e^{O(\sqrt{d \log d})}$$
in expectation. In order to further improve the (subexponential) dependence on d, one would have to make progress on "small" problems, most notably the case $n = O(d^2)$.

A natural and important but poorly understood special case is that of linear programs whose feasible set is a (deformed) d-dimensional cube. Here, $n = 2d$.

The combinatorial framework of unique sink orientations (USO) has been introduced to deal with such small problems; in fact, there is only one parameter, namely n. A USO is an orientation of the n-dimensional hypercube graph, with the property that every subgraph induced by a nonempty face of the cube has a unique sink [SW01].

The algorithmic problem is the following: given a *vertex evaluation* oracle that for a given vertex returns the orientations of the n incident edges, how many vertex evaluations need to be performed in order to find the unique global sink?

Under suitable nondegeneracy assumptions, a linear program over a (deformed) cube directly yields an acyclic USO; from its global sink, one can read off the solution to the linear program. But in fact, any linear program with n inequality constraints reduces to sink-finding in a (possibly cyclic) n-cube USO [GS06].

Simplex-type methods (following a directed path in the USO) are the most natural sink-finding strategies; in fact, the idea behind the subexponential algorithm for LP-type problems [MSW96] can be understood in its purest form on acyclic USO [Gär02]. This, however, does not imply any new strong bounds for linear programming.

But the cube structure also allows for algorithms that "jump," meaning that they do not necessarily follow a path in the cube graph but visit cube vertices in a random access fashion. The Fibonacci Seesaw [SW01] is a deterministic jumping algorithm that is able to find the sink of any n-cube USO with $O(1.61^n)$ vertex evaluations. Via the reduction to USO, this yields the best known deterministic strong bound for linear programs with $n = 2d$ constraints [GS06]. Most notably, this bound is smaller than the worst-case number of vertices, all of which the known deterministic simplex algorithms might have to visit.

49.7 PARALLEL ALGORITHMS

GLOSSARY

PRAM: Parallel Random Access Machine. (See Section 46.1 for more information on this and the next two terms.)

EREW: Exclusive Read Exclusive Write.

CRCW: Concurrent Read Concurrent Write.

P: The class of polynomial time problems.

NC: The class of problems that have poly-logarithmic parallel time algorithms running a polynomial number of processors.

P-*complete problem:* A problem in P whose membership in NC implies $P = NC$.

Expander: A graph in which, for every set of nodes, the set of the neighbors of the nodes is relatively large.

We will consider only PRAM algorithms. (See also Section 46.2.)

The general linear programming problem has long been known to be P-complete, so there is little hope of very fast parallel algorithms. However, the situation is different in the case $d \ll n$, where the problem is in NC if d grows slowly enough.

First, we note that there is a straightforward parallel implementation of Megiddo's algorithm [Meg83b] that runs in $O((\log n)^d)$ time on an EREW PRAM. However, this algorithm is rather inefficient in terms of processor utilization, since at the later stages, when there are few constraints remaining, most processors are idle. However, Deng [Den90] gave an "optimal" $O(n)$ work implementation in the plane running in $O(\log n)$ time on a CRCW PRAM with $O(n/\log n)$ processors. Deng's method does not seem to generalize to higher dimensions.

Alon and Megiddo [AM94] gave a *randomized* parallel version of Clarkson's algorithm which, with high probability, runs in constant time on a CREW PRAM in fixed dimension. Here the "constant" is a function of dimension only, and the probability of failure to meet the time bound is small for $n \gg d$.

Ajtai and Megiddo [AM96] attempted to improve the processor utilization in parallelizing Megiddo's algorithm for general d. They gave an intricate algorithm based on using an expander graph to select more nondisjoint pairs so as to utilize all the processors and obtain more rapid elimination. The resulting algorithm for (49.1.1) runs in $O((\log \log n)^d)$ time, but in a nonuniform model of parallel computation based on Valiant's comparison model. The model, which is stronger than the CRCW PRAM, requires $O(\log \log n)$ time median selection from n numbers using n processors, and employs an $O(\log \log n)$ time scheme for compacting the data after deletions, again based on a nonuniform use of expander graphs. A lower bound of $\Omega(\log n/\log \log n)$ time for median-finding on the CRCW PRAM follows from results of Beame and Håstad. Thus Ajtai and Megiddo's algorithm could not be implemented directly on the CRCW PRAM. Within Ajtai and Megiddo's model there is a lower bound $\Omega(\log \log n)$ for the case $d = 1$ implied by results of Valiant. This extends to the CRCW PRAM, and is the only lower bound known for solving (49.1.1) in this model.

Dyer [Dye95] gave a different parallelization of Megiddo's algorithm, which avoids the use of expanders. The method is based on forming groups of size $r \geq 2$, rather than simple pairs. As constraints are eliminated, the size of the groups is gradually increased to utilize the extra processors. Using this, Dyer [Dye95] establishes an $O(\log n (\log \log n)^{d-1})$ bound in the EREW model. It is easy to show that there is an $\Omega(\log n)$ lower bound for solving (49.1.1) on the EREW PRAM, even with $d = 1$. (See [KR90].) Thus improvements on Dyer's bound for the EREW model can only be made in the $\log \log n$ term. However, there was still an open question in the CRCW model, since exact median-finding and data compaction cannot be performed in time polynomial in $\log \log n$.

Goodrich [Goo93] solved these problems for the CRCW model by giving fast implementations of derandomization techniques similar to those outlined in Section 49.5. However, the randomized algorithm that underlies the method is not a parallelization of Clarkson's algorithm, but is similar to a parallelized version of that of Dyer and Frieze [DF89]. He achieves a work-optimal (i.e., $O(n)$ work) algorithm running in $O(\log \log n)^d$ time on the CRCW PRAM. The methods also imply a work-optimal EREW algorithm, but only with the same time bound as Dyer's. Neither Dyer nor Goodrich is explicit about the dependence on d of the execution time of their algorithms.

Independently of Goodrich's work, Sen [Sen95] has shown how to directly modify Dyer's algorithm to give a work-optimal algorithm with $O((\log \log n)^{d+1})$ execution time in the CRCW model. The "constant" in the running time is shown to be $2^{O(d^2)}$. To achieve this, he uses approximate median-finding and approximate data compaction operations, both of which can be done in time polynomial in $\log \log n$ on the common CRCW PRAM. These additional techniques are, in fact, both examples of derandomized methods and similar to those Goodrich uses for the same purpose. Note that this places linear programming in NC provided $d = O(\sqrt{\log n})$. This is the best result known, although Goodrich's algorithm may give a better behavior once the "constant" has been explicitly evaluated. We may also observe that the Goodrich/Sen algorithms improve on Deng's result in $\mathbb{R}^2$.

There is still room for some improvements in this area, but there now seems to be a greater need for sharper lower bounds, particularly in the CRCW case.

49.8 RELATED ISSUES

GLOSSARY

Integer programming problem: A linear programming problem with the additional constraint that the solution must be integral.

k-Violation linear programming: A problem as in (49.1.1), except that we want to maximize the linear objective function subject to all but at most k of the given linear constraints.

Average case analysis: Expected performance of an algorithm for random input (under certain distributions).

Smoothed analysis: Expected performance of an algorithm under small random perturbations of the input.

Linear programming is a problem of interest in its own right, but it is also representative of a class of geometric problems to which similar methods can be applied. Many of the references given below discuss closely related problems, and we have mentioned them in passing above.

An important related area is integer programming. Here the size of the numbers cannot be relegated to a secondary consideration. In general this problem is NP-hard, but in fixed dimension is polynomial-time solvable. See [Sch86] for further information. It may be noted that Clarkson's methods and the LP-type framework are applicable in this situation; some care with the interpretation of the primitive operations is in place, though.

We have considered only the solution of a single linear program. However, there are some situations where one might wish to solve a sequence of closely related linear programs. In this case, it may be worth the computational investment of building a data structure to facilitate fast solution of the linear programs. For results of this type see, for example, [Epp90, Mat93, Cha96, Cha98].

Finally there has been some work about optimization, where we are asked to satisfy all but at most k of the given constraints, see, e.g., [RW94, ESZ94, Mat95b, DLSS95, Cha99]. In particular, Matoušek [Mat95a] has investigated this question in the general setting of LP-type problems. Chan [Cha02] solved this problem in $\mathbb{R}^2$ with a randomized algorithm in expected time $O((n + k^2) \log k)$ (see this paper for the best bounds known for $d = 3, 4$.)

A direction we did not touch upon here is *average-case analysis*, where we analyze a deterministic algorithm for random inputs [Bor87, Sma83, AM85, Tod86, AKS87]. In the latter three, bounds are proven with respect to classes of probability distributions, where the numerical entries of A, b, and c are fixed in a nondegenerate way and the directions of inequalities are picked at random. More recently, there has been an interesting new direction, smoothed analysis, where a simplex algorithm is analyzed for small random perturbations of the input. The main result of Spielmann and Teng [ST04] is that (superpolynomial) worst-case instances for the shadow vertex simplex algorithm are "isolated." This means that after small random perturbations of the input, the algorithm will be fast in expectation. Indeed, if one looks at the geometry of the known (contrived) worst-case instances, one sees 2-dimensional projections ("shadows") with exponentially many vertices, tightly packed along the boundary of a convex polygon. It is not surprising that such "bad" projections disappear even under small random perturbations of the input. What is more surprising—and this is the main contribution of Spielmann and Teng—is that not only these bad instances, but *all* instances have small shadows after random perturbations.

49.9 SOURCES AND RELATED MATERIAL

BOOKS AND SURVEYS

Linear programming is a mature topic, and its foundations are well-covered in classic textbooks. A good general introduction to linear programming may be found in Chvátal's book [Chv83]. A theoretical treatment is given in Schrijver's book [Sch86]. The latter is a very good source of additional references. The book by Matoušek and Gärtner [MG07] is a concise introduction from a computer science perspective. Karp and Ramachandran [KR90] is a good source of information on models of parallel computation. See [Mat96] for a survey of derandomization techniques for computational geometry.

RELATED CHAPTERS

Chapter 15: Basic properties of convex polytopes
Chapter 19: Polytope skeletons and paths
Chapter 36: Computational convexity
Chapter 46: Parallel algorithms in geometry
Chapter 67: Software

REFERENCES

[AF92]   D. Avis and K. Fukuda. A pivoting algorithm for convex hulls and vertex enumeration of arrangements and polyhedra. *Discrete Comput. Geom.*, 8:295–313, 1992.

[AF16]   D. Avis and O. Friedmann. An exponential lower bound for Cunningham's rule. *Math. Program. Ser. A*, to appear.

[AKS87] I. Adler, R.M. Karp, and R. Shamir. A simplex variant solving an $(m \times d)$ linear program in $O(\min(m^2, d^2))$ expected number of pivot steps. *J. Complexity*, 3:372–387, 1987.

[AM85]   I. Adler and N. Megiddo. A simplex algorithm whose average number of steps is bounded between two quadratic functions of the smaller dimension. *J. ACM*, 32:871–895, 1985.

[AM94]   N. Alon and N. Megiddo. Parallel linear programming in fixed dimension almost surely in constant time. *J. ACM*, 41:422–434, 1994.

[AM96]   M. Ajtai and N. Megiddo. A deterministic poly(log log n)-time n-processor algorithm for linear programming in fixed dimension. *SIAM J. Comput.*, 25:1171–1195, 1996.

[Ame94] N. Amenta. Helly-type theorems and generalized linear programming. *Discrete Comput. Geom.*, 12:241–261, 1994.

[Ame96] N. Amenta. A new proof of an interesting Helly-type theorem. *Discrete Comput. Geom.*, 15:423–427, 1996.

[AZ99]   N. Amenta and G.M. Ziegler. Deformed products and maximal shadows. In B. Chazelle, J.E. Goodman, and R. Pollack, editors, *Advances in Discrete and Computational Geometry*, vol. 223 of *Contemp. Math.*, pages 57–90, AMS, Providence, 1998.

[BG11]   Y. Brise and B. Gärtner. Clarkson's algorithm for violator spaces. *Comput. Geom.*, 44:70–81, 2011.

[Bor87] K.H. Borgwardt. *The Simplex Method: A Probabilistic Analysis*. Vol. 1 of *Algorithms Combin.*, Springer-Verlag, Berlin, 1987.

[Cha96] T.M. Chan. Fixed-dimensional linear programming queries made easy. In *Proc. 12th Sympos. Comput. Geom.*, pages 284–290, ACM Press, 1996.

[Cha98] T.M. Chan. Deterministic algorithms for 2-d convex programming and 3-d online linear programming. *J. Algorithms*, 27:147–166, 1998.

[Cha99] T.M. Chan. Geometric applications of a randomized optimization technique. *Discrete Comput. Geom.*, 22:547–567, 1999.

[Cha02]   T.M. Chan. Low-dimensional linear programming with violations. In *Proc. 43rd IEEE Sympos. Found. Comput. Sci.*, pages 570–579, 2002.

[Cha16]   T.M. Chan. Improved deterministic algorithms for linear programming in low dimensions. In *Proc. 27th ACM-SIAM Sympos. Discrete Algorithms*, pages 1213–1219, 2016.

[Chv83]   V. Chvátal. *Linear Programming*. Freeman, New York, 1983.

[Cla86a]   K.L. Clarkson. Further applications of random sampling to computational geometry. In *Proc. 18th ACM Sympos. Theory Comput.*, pages 414–423, 1986.

[Cla86b]   K.L. Clarkson. Linear programming in $O(n3^{d^2})$ time. *Inform. Process. Lett.*, 22:21–24, 1986.

[Cla88]   K.L. Clarkson. Las Vegas algorithms for linear and integer programming when the dimension is small. In *Proc. 29th IEEE Sympos. Found. Comput. Sci.*, pages 452–456, 1988.

[Cla95]   K.L. Clarkson. Las Vegas algorithms for linear and integer programming when the dimension is small. *J. ACM*, 42:488–499, 1995. (Improved version of [Cla88].)

[CM96]   B. Chazelle and J. Matoušek. On linear-time deterministic algorithms for optimization in fixed dimension. *J. Algorithms* 21:579–597, 1996.

[Den90]   X. Deng. An optimal parallel algorithm for linear programming in the plane. *Inform. Process. Lett.*, 35:213–217, 1990.

[DF89]   M.E. Dyer and A.M. Frieze. A randomized algorithm for fixed-dimensional linear programming. *Math. Program.*, 44:203–212, 1989.

[DLSS95]   A. Datta, H.-P. Lenhof, C. Schwarz, and M. Smid. Static and dynamic algorithms for k-point clustering problems. *J. Algorithms*, 19:474–503, 1995.

[Dye83]   M.E. Dyer. The complexity of vertex enumeration methods. *Math. Oper. Res.*, 8:381–402, 1983.

[Dye84]   M.E. Dyer. Linear time algorithms for two- and three-variable linear programs. *SIAM J. Comput.*, 13:31–45, 1984.

[Dye86]   M.E. Dyer. On a multidimensional search problem and its application to the Euclidean one-centre problem. *SIAM J. Comput.*, 15:725–738, 1986.

[Dye92]   M.E. Dyer. A class of convex programs with applications to computational geometry. In *Proc. 8th Sympos. Comput. Geom.*, pages 9–15, ACM Press, 1992.

[Dye95]   M.E. Dyer. A parallel algorithm for linear programming in fixed dimension. In *Proc. 11th Sympos. Comput. Geom.*, pages 345–349, ACM Press, 1995.

[Epp90]   D. Eppstein. Dynamic three-dimensional linear programming. *ORSA J. Comput.*, 4:360–368, 1990.

[ESZ94]   A. Efrat, M. Sharir, and A. Ziv. Computing the smallest k-enclosing circle and related problems. *Comput. Geom.*, 4:119–136, 1994.

[FHZ11]   O. Friedmann, T.D. Hansen, and U. Zwick. Subexponential lower bounds for randomized pivoting rules for the simplex algorithm. In *Proc. 43rd ACM Sympos. Theory Comput.*, pages 283–292, 2011.

[Fri11]   O. Friedmann. A subexponential lower bound for Zadehs pivoting rule for solving linear programs and games. In *Proc. 15th Conf. Integer Program. Combin. Opt.*, pages 192–206, vol. 6655 of *LNCS*, Springer, Berlin, 2011.

[Gär95]   B. Gärtner. A subexponential algorithm for abstract optimization problems. *SIAM J. Comput.*, 24:1018–1035, 1995.

[Gär02]   B. Gärtner. The random-facet simplex algorithm on combinatorial cubes. *Random Structures Algorithms*, 20:353–381, 2002.

[GLS93]   M. Grötschel, L. Lovász, and A. Schrijver. *Geometric Algorithms and Combinatorial Optimization*, 2nd corrected edition. Vol. 2 of *Algorithms and Combinatorics*, Springer, Berlin, 1993

[GMRS08]   B. Gärtner, J. Matoušek, L. Rüst, and P. Škovroň. Violator spaces: Structure and algorithms. *Discrete Appl. Math.*, 156:2124–2141, 2008.

[Goo93]   M.T. Goodrich. Geometric partitioning made easier, even in parallel. In *Proc. 9th Sympos. Comput. Geom.*, pages 73–82, ACM Press, 1993.

[GS06]   B. Gärtner and I. Schurr. Linear programming and unique sink orientations. In *Proc. 17th ACM-SIAM Sympos. Discrete Algorithms*, pages 749–757, 2006

[GW01]   B. Gärtner and E. Welzl. A simple sampling lemma: Analysis and applications in geometric optimization. *Discrete Comput. Geom.*, 25:569–590, 2001.

[GW96]   B. Gärtner and E. Welzl. Linear programming – randomization and abstract frameworks. In *Proc. 13th Sympos. Theoret. Aspects Comp. Sci.*, vol. 1046 of *LNCS*, pages 669–687, Springer, Berlin, 1996.

[HZ15]   T.D. Hansen and U. Zwick. An improved version of the random-facet pivoting rule for the simplex algorithm. In *Proc. 47th ACM Sympos. Theory Comput.*, pages 209–218, 2015.

[Kal92]   G. Kalai. A subexponential randomized simplex algorithm. In *Proc. 24th ACM Sympos. Theory Comput.*, pages 475–482, 1992.

[Kal97]   G. Kalai. Linear programming, the simplex algorithm and simple polytopes. *Math. Program.*, 79:217–233, 1997.

[KM72]   V. Klee and G.J. Minty. How good is the simplex algorithm? In O. Shisha, editor, *Inequalities III*, pages 159–175, Academic Press, New York, 1972.

[KR90]   R.M. Karp and V. Ramachandran. Parallel algorithms for shared-memory machines. In J. van Leeuwen, editor, *Handbook of Theoretical Computer Science, Vol. A: Algorithms and Complexity*, pages 869–941, Elsevier, Amsterdam, 1990.

[Mat93]   J. Matoušek. Linear optimization queries. *J. Algorithms*, 14:432–448, 1993.

[Mat94]   J. Matoušek. Lower bounds for a subexponential optimization algorithm. *Random Structures Algorithms*, 5:591–607, 1994.

[Mat95a]   J. Matoušek. On geometric optimization queries with few violated constraints. *Discrete Comput. Geom.*, 14:365–384, 1995.

[Mat95b]   J. Matoušek. On enclosing k points by a circle. *Inform. Process. Lett.*, 53:217–221, 1995.

[Mat96]   J. Matoušek. Derandomization in computational geometry. *J. Algorithms*, 20:545–580, 1996.

[Meg83a]   N. Megiddo. Towards a genuinely polynomial algorithm for linear programming. *SIAM J. Comput.*, 12:347–353, 1983.

[Meg83b]   N. Megiddo. Linear time algorithms for linear programming in $\mathbb{R}^3$ and related problems. *SIAM J. Comput.*, 12:759–776, 1983.

[Meg84]   N. Megiddo. Linear programming in linear time when dimension is fixed. *J. ACM*, 31:114–127, 1984.

[Meg89]   N. Megiddo. On the ball spanned by balls. *Discrete Comput. Geom.*, 4:605–610, 1989.

[MG07] J. Matoušek and B. Gärtner. *Understanding and Using Linear Programming*. Springer, Berlin, 2007.

[MP78] D.E. Muller and F.P. Preparata. Finding the intersection of two convex polyhedra. *Theoret. Comput. Sci.*, 7:217–236, 1978.

[MSW96] J. Matoušek, M. Sharir, and E. Welzl. A subexponential bound for linear programming. *Algorithmica*, 16:498–516, 1996.

[RW94] T. Roos and P. Widmayer. k-violation linear programming. *Inform. Process. Lett.*, 52:109–114, 1994.

[Sch86] A. Schrijver. *Introduction to the Theory of Linear and Integer Programming*. Wiley, Chichester, 1986.

[Sei91] R. Seidel. Low dimensional linear programming and convex hulls made easy. *Discrete Comput. Geom.*, 6:423–434, 1991.

[Sen95] S. Sen. A deterministic poly($\log \log n$) time optimal CRCW PRAM algorithm for linear programming in fixed dimension. Technical Report 95-08, Dept. of Comp. Sci., Univ. of Newcastle, Australia, 1995.

[Sha78] M.I. Shamos. *Computational Geometry*. PhD thesis, Yale Univ., New Haven, 1978.

[Ško07] P. Škovroň. *Abstract Models of Optimization Problems*. PhD thesis, Charles University, Prague, 2007.

[Sma83] S. Smale. On the average number of steps in the simplex method of linear programming. *Math. Program.*, 27:241–262, 1983.

[ST04] D.A. Spielman and S.-H. Teng. Smoothed analysis of algorithms: Why the simplex algorithm usually takes polynomial time. *J. ACM*, 51:385–463, 2004.

[SW92] M. Sharir and E. Welzl. A combinatorial bound for linear programming and related problems. In *Proc. 9th Sympos. Theoret. Aspects Comput. Sci.*, vol. 577 of *LNCS*, pages 569–579, Springer, Berlin, 1992.

[SW01] T. Szabó and E. Welzl. Unique sink orientations of cubes. In *Proc. 42nd IEEE Sympos. Found. Comp. Sci.*, pages 547–555, 2001.

[Syl57] J.J. Sylvester. A question in the geometry of situation. *Quart. J. Math.*, 1:79, 1857.

[Tar86] É. Tardos. A strongly polynomial algorithm to solve combinatorial linear programs. *Oper. Res.*, 34:250–256, 1986.

[Tod86] M.J. Todd. Polynomial expected behavior of a pivoting algorithm for linear complementarity and linear programming problems. *Math. Program.*, 35:173–192, 1986.

[Wel91] E. Welzl. Smallest enclosing disks (balls and ellipsoids). In H. Maurer, editor, *New Results and New Trends in Computer Science*, vol. 555 of *LNCS*, pages 359–370, Springer, Berlin, 1991.

50 ALGORITHMIC MOTION PLANNING
Dan Halperin, Oren Salzman, and Micha Sharir

INTRODUCTION

Motion planning is a fundamental problem in robotics. It comes in a variety of forms, but the simplest version is as follows. We are given a robot system B, which may consist of several rigid objects attached to each other through various joints, hinges, and links, or moving independently, and a 2D or 3D environment V cluttered with obstacles. We assume that the shape and location of the obstacles and the shape of B are known to the planning system. Given an initial placement Z_1 and a final placement Z_2 of B, we wish to determine whether there exists a collision-avoiding motion of B from Z_1 to Z_2, and, if so, to plan such a motion. In this simplified and purely geometric setup, we ignore issues such as incomplete information, nonholonomic constraints, control issues related to inaccuracies in sensing and motion, nonstationary obstacles, optimality of the planned motion, and so on.

Since the early 1980s, motion planning has been an intensive area of study in robotics and computational geometry. In this chapter we will focus on *algorithmic motion planning*, emphasizing theoretical algorithmic analysis of the problem and seeking worst-case asymptotic bounds, and only mention briefly practical heuristic approaches to the problem. The majority of this chapter is devoted to the simplified version of motion planning, as stated above. Section 50.1 presents general techniques and lower bounds. Section 50.2 considers efficient solutions to a variety of specific moving systems with a small number of degrees of freedom. These efficient solutions exploit various sophisticated methods in computational and combinatorial geometry related to arrangements of curves and surfaces (Chapter 28). Section 50.3 then briefly discusses various extensions of the motion planning problem such as computing *optimal* paths with respect to various quality measures, computing the path of a tethered robot, incorporating uncertainty, moving obstacles, and more.

50.1 GENERAL TECHNIQUES AND LOWER BOUNDS

GLOSSARY

Some of the terms defined here are also defined in Chapter 51.

Robot B: A mechanical system consisting of one or more rigid bodies, possibly connected by various joints and hinges.

Physical space (workspace): The 2D or 3D environment in which the robot moves.

Placement: The portion of physical space occupied by the robot at some instant.

Degrees of freedom k: The number of real parameters that determine the robot B's placements. Each placement can be represented as a point in $\mathbb{R}^k$.

Free placement: A placement at which the robot is disjoint from the obstacles.

Semifree placement: A placement at which the robot does not meet the interior of any obstacle (but may be in contact with some obstacles).

Configuration space $\mathcal{C}$:   A portion of k-space (where k is the number of degrees of freedom of B) that represents all possible robot placements; the coordinates of any point in this space specify the corresponding placement.

Expanded obstacle / C-obstacle / forbidden region: For an obstacle O, this is the portion O^* of configuration space consisting of placements at which the robot intersects (collides with) O.

Free configuration space $\mathcal{F}$:   The subset of configuration space consisting of free placements of the robot: $\mathcal{F} = \mathcal{C} \setminus \bigcup_O O^*$. (In the literature, this usually also includes semifree placements. In that case, $\mathcal{F}$ is the complement of the union of the *interiors* of the expanded obstacles.)

Contact surface: For an obstacle feature a (corner, edge, face, etc.) and for a feature b of the robot, this is the locus in $\mathcal{C}$ of placements at which a and b are in contact with each other. In most applications, these surfaces are semialgebraic sets of constant description complexity (see definitions below).

Collision-free motion of B: A path contained in $\mathcal{F}$. Any two placements of B that can be reached from each other via a collision-free path must lie in the same (arcwise-)connected component of $\mathcal{F}$.

Arrangement $\mathcal{A}(\Sigma)$:   The decomposition of k-space into cells of various dimensions, induced by a collection Σ of surfaces in $\mathbb{R}^k$. Each cell is a maximal connected portion of the intersection of some fixed subcollection of surfaces that does not meet any other surface. See Chapter 28. Since a collision-free motion should not cross any contact surface, $\mathcal{F}$ is the union of some of the cells of $\mathcal{A}(\Sigma)$, where Σ is the collection of contact surfaces.

Semialgebraic set: A subset of $\mathbb{R}^k$ defined by a Boolean combination of polynomial equalities and inequalities in the k coordinates. See Chapter 37.

Constant description complexity: Said of a semialgebraic set if it is defined by a constant number of polynomial equalities and inequalities of constant maximum degree (where the number of variables is also assumed to be constant).

EXAMPLE

Let B be a rigid polygon with k edges, moving in a planar polygonal environment V with n edges. The system has three degrees of freedom, (x, y, θ), where (x, y) are the coordinates of some reference point on B, and θ is the orientation of B. Each contact surface is the locus of placements where some vertex of B touches some edge of V, or some edge of B touches some vertex of V. There are $2kn$ contact surfaces, and if we replace θ by $\tan \frac{\theta}{2}$, then each contact surface becomes a portion of some algebraic surface of degree at most 4, bounded by a constant number of algebraic arcs, each of degree at most 2.

50.1.1 GENERAL SOLUTIONS

GLOSSARY

Cylindrical algebraic decomposition of $\mathcal{F}$:   A recursive decomposition of $\mathcal{C}$ into cylindrical-like cells originally proposed by Collins [Col75]. Over each cell of the decomposition, each of the polynomials involved in the definition of $\mathcal{F}$ has a fixed sign (positive, negative, or zero), implying that $\mathcal{F}$ is the union of some of the cells of this decomposition. See Chapter 37 for further details.

Connectivity graph: A graph whose nodes are the (free) cells of a decomposition of $\mathcal{F}$ and whose arcs connect pairs of adjacent cells.

Roadmap $\mathcal{R}$:   A network of one-dimensional curves within $\mathcal{F}$, having the properties that (i) it *preserves the connectivity* of $\mathcal{F}$, in the sense that the portion of $\mathcal{R}$ within each connected component of $\mathcal{F}$ is (nonempty and) connected; and (ii) it is **reachable**, in the sense that there is a simple procedure to move from any free placement of the robot to a placement on $\mathcal{R}$; we denote the mapping resulting from this procedure by $\phi_{\mathcal{R}}$.

Retraction of $\mathcal{F}$ onto $\mathcal{R}$:   A continuous mapping of $\mathcal{F}$ onto $\mathcal{R}$ that is the identity on $\mathcal{R}$. The roadmap mapping $\phi_{\mathcal{R}}$ is usually a retraction. When this is the case, we note that for any path ψ within $\mathcal{F}$, represented as a continuous mapping $\psi : [0, 1] \mapsto \mathcal{F}$, $\phi_{\mathcal{R}} \circ \psi$ is a path within $\mathcal{R}$, and, concatenating to it the motions from $\psi(0)$ and $\psi(1)$ to $\mathcal{R}$, we see that there is a collision-free motion of B between two placements Z_1, Z_2 iff there is a path within $\mathcal{R}$ between $\phi_{\mathcal{R}}(Z_1)$ and $\phi_{\mathcal{R}}(Z_2)$.

Silhouette: The set of critical points of a mapping; see Chapter 37.

CELL DECOMPOSITION

$\mathcal{F}$ is a semialgebraic set in $\mathbb{R}^k$. Applying Collins's cylindrical algebraic decomposition results in a collection of cells whose total complexity is $O((nd)^{3^k})$, where d is the maximum algebraic degree of the polynomials defining the contact surfaces; the decomposition can be constructed within a similar time bound. If the coordinate axes are generic, then we can also compute all pairs of cells of $\mathcal{F}$ that are **adjacent** to each other (i.e., cells whose closures (within $\mathcal{F}$) overlap), and store this information in the form of a connectivity graph. It is then easy to search for a collision-free path through this graph, if one exists, between the (cell containing the) initial robot placement and the (cell containing the) final placement. This leads to a doubly-exponential general solution for the motion planning problem:

THEOREM 50.1.1   *Cylindrical Cell Decomposition* [SS83]

Any motion planning problem, with k degrees of freedom, for which the contact surfaces are defined by a total of n polynomials of maximum degree d, can be solved by Collins's cylindrical algebraic decomposition, in randomized expected time $O((nd)^{3^k})$.

Remarks. (1) The randomization is needed only to choose a generic direction for the coordinate axes. (2) Here and throughout the chapter we adhere to the *real RAM* model of computation, which is standard in computational geometry [PS85].

ROADMAPS

An improved solution is given in [Can87, BPR00] based on the notion of a **roadmap** $\mathcal{R}$, a network of one-dimensional curves within (the closure of) $\mathcal{F}$, having properties defined in the glossary above. Once such a roadmap $\mathcal{R}$ has been constructed, any motion planning instance reduces to path searching within $\mathcal{R}$, which is easy to do. $\mathcal{R}$ is constructed recursively, as follows. One projects $\mathcal{F}$ onto some generic 2-plane, and computes the silhouette of $\mathcal{F}$ under this projection. Next, the critical values of the projection of the silhouette on some line are found, and a roadmap is constructed recursively within each slice of $\mathcal{F}$ at each of these critical values. The resulting "sub-roadmaps" are then merged with the silhouette, to obtain the desired $\mathcal{R}$.

The original algorithm of Canny [Can87] relies heavily on the polynomials defining $\mathcal{F}$ being in general position, and on the availability of a generic plane of projection. This algorithm runs in $n^k (\log n) d^{O(k^4)}$ deterministic time, and in $n^k (\log n) d^{O(k^2)}$ expected randomized time. Later work [BPR00] addresses and overcomes the general position issue, and produces a roadmap for any semialgebraic set; the running time of this solution is $n^{k+1} d^{O(k^2)}$.

If we ignore the dependence on the degree d, the algorithm of Canny is close to optimal in the worst case, assuming that some representation of the entire $\mathcal{F}$ has to be output, since there are easy examples where the free configuration space consists of $\Omega(n^k)$ connected components.

THEOREM 50.1.2   *Roadmap Algorithm* [Can87, BPR00]

Any motion planning problem, as in the preceding theorem, in general position can be solved by the roadmap technique in $n^k (\log n) d^{O(k^4)}$ deterministic time, and in $n^k (\log n) d^{O(k^2)}$ expected randomized time.

50.1.2 LOWER BOUNDS

The upper bounds for both general solutions are (at least) exponential in k (but are polynomial in the other parameters when k is fixed). This raises the issue of calibrating the complexity of the problem when k can be arbitrarily large.

THEOREM 50.1.3   *Lower Bounds*

The motion planning problem, with arbitrarily many degrees of freedom, is PSPACE-hard for the instances of: (a) coordinated motion of many rectangular boxes along a rectangular floor [HSS84]; the problem remains PSPACE-hard even if only two types of rectangles are used [HD05] or if only unit squares are used [SH15b] (b) motion planning of a planar mechanical linkage with many links [HJW84]; and (c) motion planning for a multi-arm robot in a 3-dimensional polyhedral environment [Rei87].

All early results can also be found in the collection [HSS87]. There are NP-hardness results for other systems; see, e.g., [HJW85] and [SY84].   Hearn and Demaine [HD09] introduced a general tool called the *nondeterministic constraint logic* (NCL) model of computation that facilitates derivation of hardness results (in particular, the results in [HD05] and [SH15b]). Using the NCL model, Hearn and Demaine also derive the PSPACE-hardness of a variety of motion-related puzzle-like problems that consist of sliding game pieces. In particular, they applied their technique to the SOKOBAN puzzle, where multiple "crates" need to be pushed to target locations, and the Rush Hour game, where a parking attendant has to evacuate a car from a parking lot, by clearing a route blocked by other cars.

Facing the aforementioned hardness results, we consider specific problems with small values of k, with the goal of obtaining solutions better than those yielded by the general techniques. Alternatively, we can approach the general problem with heuristic or approximate schemes. We will mostly survey here the former approach, which allows for efficient computation for a restricted set of problems. However, as the general motion-planning problem is of practical interest, considerable research effort has been devoted to practical solutions to this problem. Noteworthy is the *sampling-based* approach in which $\mathcal{F}$ is approximated via a roadmap constructed by randomly sampling the configuration space. We will review this practical approach (as well as some alternative approaches) briefly and refer the reader to Chapter 51 for an in-depth discussion of sampling-based algorithms.

50.2 MOTION PLANNING WITH A SMALL NUMBER OF DEGREES OF FREEDOM

In this main section of the chapter, we review solutions to a variety of specific motion-planning problems, most of which have 2 or 3 degrees of freedom. Exploiting the special structure of these problems leads to solutions that are more efficient than the general methods described above.

GLOSSARY

Jordan arc/curve: The image of the closed unit interval under a continuous bijective mapping into the plane. A closed Jordan curve is the image of the unit circle under a similar mapping, and an unbounded Jordan curve is an image of the open unit interval (or of the entire real line) that separates the plane.

Randomized algorithm: An algorithm that applies internal randomization ("coin-flips"). We consider here "Las Vegas" algorithms that always terminate, and produce the correct output, but whose running time is a random variable that depends on the internal coin-flips. We will state upper bounds on the expectation of the running time (the **randomized expected time**) of such an algorithm, which hold for any input. See Chapter 44.

General position: The input to a geometric problem is said to be in general position if no nontrivial algebraic identity with integer coefficients holds among the parameters that specify the input (assuming the input is not overspecified). For example: no three input points should be collinear, no four points cocircular, no three lines concurrent, etc. (In general, this requirement is too restrictive and many instances explicitly specify which identities are not supposed to hold.)

Minkowski sum: For two planar (or spatial) sets A and B, their Minkowski sum, or pointwise vector addition, is the set $A \oplus B = \{x + y \mid x \in A, y \in B\}$.

Convex distance function: A convex region B that contains the origin in its interior induces a convex distance function d_B defined by

$$d_B(p, q) = \min \{\lambda \mid q \in p \oplus \lambda B\}.$$

If B is centrally symmetric with respect to the origin then d_B is a metric whose unit ball is B.

B-Voronoi diagram: For a set S of sites, and a convex region B as above, the B-Voronoi diagram $\mathrm{Vor}_B(S)$ of S is a decomposition of space into Voronoi cells $V(s)$, for $s \in S$, such that

$$V(s) = \{p \mid d_B(p, s) \leq d_B(p, s') \text{ for all } s' \in S\}.$$

Here $d_B(p, s) = \min_{q \in s} d_B(p, q)$.

$\alpha(n)$:   The extremely slowly-growing inverse Ackermann function; see Chapter 28.

Contact segment: The locus of (not necessarily free) placements of a polygon B translating in a planar polygonal workspace, at each of which either some specific vertex of B touches some specific obstacle edge, or some specific edge of B touches some specific obstacle vertex.

Contact curve: A generalization of "contact segment" to the locus of (not necessarily free) placements of a more general robot system B, assuming that B has only two degrees of freedom, where some specific feature of B makes contact with some specific obstacle feature.

50.2.1 TWO DEGREES OF FREEDOM

A TRANSLATING POLYGON IN 2D

This is a system with two degrees of freedom (translations in the x and y directions).

A CONVEX POLYGON

Suppose first the translating polygon B is a *convex* k-gon, and there are m convex polygonal obstacles, $A_1, \ldots, A_m$, with pairwise disjoint interiors, having a total of n edges. The region of configuration space where B collides with A_i is the *Minkowski sum*

$$K_i = A_i \oplus (-B) = \{x - y \mid x \in A_i,\ y \in B\}.$$

The free configuration space is the complement of $\bigcup_{i=1}^m K_i$. Assuming general position, one can show:

THEOREM 50.2.1   [KLPS86]

(a) *Each K_i is a convex polygon, with $n_i + k$ edges, where n_i is the number of edges of A_i.*

(b) *For each $i \neq j$, the boundaries of K_i and K_j intersect in at most two points. (This also holds when the A_i's and B are not polygons but are still convex.)*

(c) *Given a collection of planar regions $K_1, \ldots, K_m$, each enclosed by a closed Jordan curve, such that any pair of the bounding curves intersects at most twice, then the boundary of the union $\bigcup_{i=1}^m K_i$ consists of at most $6m - 12$ maximal connected portions of the boundaries of the K_i's, provided $m \geq 3$, and this bound is tight in the worst case.*

Such a collection $K_1, \ldots, K_m$ is called a collection of *pseudo-disks*. Now, these properties, combined with several algorithmic techniques [KLPS86, MMP+91, BDS95], imply:

THEOREM 50.2.2

(a) *The free configuration space for a translating convex polygon, as above, is a polygonal region with at most $6m - 12$ convex vertices and $N = \sum_{i=1}^{m}(n_i + k) = n + km$ nonconvex vertices.*

(b) *$\mathcal{F}$ can be computed in deterministic time $O(N \log n \log m)$ or in randomized expected time $O(N \log n)$.*

If the robot is translating in a convex room with n walls, then the complexity of the free space is $O(n)$ and it can be computed in $O(n + k)$ time.

AN ARBITRARY POLYGON

Suppose next that B is an arbitrary polygonal region with k edges. Let A be the union of all obstacles, which is another polygonal region with n edges. As above, the free configuration space is the complement of the Minkowski sum

$$K = A \oplus (-B) = \{x - y \mid x \in A, \, y \in B\}.$$

K is again a polygonal region, but, in this case, its maximum possible complexity is $\Theta(k^2 n^2)$ (see, e.g., [AFH02]), so computing it might be considerably more expensive than in the convex case. Efficient practical algorithms for the exact computation of Minkowski sums in this case (together with their implementation) are described in [AFH02, Wei06, BFH+15].

A single face suffices. If the initial placement Z of B is given, then we do not have to compute the entire (complement of) K; it suffices to compute the connected component f of the complement of K that contains Z, because no other placement is reachable from Z via a collision-free motion.

Let Σ be the collection of all contact segments; there are $2kn$ such segments. The desired component f is the face of $\mathcal{A}(\Sigma)$ that contains Z. Using the theory of *Davenport-Schinzel sequences* (Chapter 28), one can show that the maximum possible combinatorial complexity of a single face in a two-dimensional arrangement of N segments is $\Theta(N\alpha(N))$. A more careful analysis [HCA+95], combined with the algorithmic techniques of [CEG+93, GSS89], shows:

THEOREM 50.2.3

(a) *The maximum combinatorial complexity of a single face in the arrangement of contact segments for the case of an arbitrary translating polygon is $\Theta(kn\alpha(k))$ (this improvement is significant only when $k \ll n$).*

(b) *Such a face can be computed in deterministic time $O(kn\alpha(k) \log^2 n)$ [GSS89], or in randomized expected time $O(kn\alpha(k) \log n)$ [CEG+93].*

VORONOI DIAGRAMS

Another approach to motion planning for a translating *convex* object B is via generalized *Voronoi diagrams* (see Chapter 27), based on the convex distance function

$d_B(p, q)$. This function effectively places B centered at p and expands it until it hits q. The scaling factor at this moment is the d_B-distance from p to q (if B is a unit disk, d_B is the Euclidean distance). d_B satisfies the triangle inequality, and is thus "almost" a metric, except that it is not symmetric in general; it is symmetric iff B is centrally symmetric with respect to the point of reference.

Using this distance function d_B, a *B-Voronoi diagram* $\text{Vor}_B(\mathcal{S})$ of $\mathcal{S}$ may be defined for a set $\mathcal{S}$ of m pairwise disjoint obstacles. See [LS87a, Yap87].

THEOREM 50.2.4

Assuming that each of B and the obstacles in $\mathcal{S}$ has constant-description complexity, and that they are in general position, the B-Voronoi diagram has $O(m)$ complexity, and can be computed in $O(m \log m)$ time. If B and the obstacles are convex polygons, as above, then the complexity of $\text{Vor}_B(\mathcal{S})$ is $O(N)$ and it can be computed in time $O(N \log m)$, where $N = n + km$.

One can show that if Z_1 and Z_2 are two free placements of B, then there exists a collision-free motion from Z_1 to Z_2 if and only if there exists a collision-free motion of B where its center moves only along the edges of $\text{Vor}_B(\mathcal{S})$, between two corresponding placements W_1, W_2, where W_i, for $i = 1, 2$, is the placement obtained by pushing B from the placement Z_i away from its d_B-nearest obstacle, until it becomes equally nearest to two or more obstacles (so that its center lies on an edge of $\text{Vor}_B(\mathcal{S})$).

Thus motion planning of B reduces to path-searching in the one-dimensional network of edges of $\text{Vor}_B(\mathcal{S})$. This technique is called the ***retraction technique***, and can be regarded as a special case of the general roadmap algorithm. The resulting motions have "high clearance," and so are safer than arbitrary motions, because they stay equally nearest to at least two obstacles.

THEOREM 50.2.5

The motion-planning problem for a convex object B translating amidst m convex and pairwise disjoint obstacles can be solved in $O(m \log m)$ time, by constructing and searching in the B-Voronoi diagram of the obstacles, assuming that B and the obstacles have constant description complexity each. If B and the obstacles are convex polygons, then the same technique yields an $O(N \log m)$ solution, where $N = n + km$ is as above.

THE GENERAL MOTION-PLANNING PROBLEM WITH TWO DE-GREES OF FREEDOM

If B is any system with two degrees of freedom, its configuration space is 2D, and, for simplicity, let us think of it as the plane (spaces that are topologically more complex can be decomposed into a constant number of "planar" patches). We construct a collection Σ of contact curves, which, under reasonable assumptions concerning B and the obstacles, are each an algebraic Jordan arc or curve of some fixed maximum degree b. In particular, each pair of contact curves will intersect in at most some constant number, $s \leq b^2$, of points.

As above, it suffices to compute the single face of $\mathcal{A}(\Sigma)$ that contains the initial placement of B. The theory of Davenport-Schinzel sequences implies that the complexity of such a face is $O(\lambda_{s+2}(n))$, where $\lambda_{s+2}(n)$ is the maximum length of an $(n, s + 2)$-Davenport-Schinzel sequence (Chapter 28), which is slightly super-linear in n when s is fixed.

The face in question can be computed in deterministic time $O(\lambda_{s+2}(n) \log^2 n)$, using a fairly involved divide-and-conquer technique based on line-sweeping; see [GSS89] and Chapter 28. (Some slight improvements in the running time have subsequently been obtained.) Using randomized incremental (or divide-and-conquer) techniques, the face can be computed in randomized expected $O(\lambda_{s+2}(n) \log n)$ time [CEG+93, SA95].

THEOREM 50.2.6 [GSS89, CEG+93, BDS95]

Under the above assumptions, the general motion-planning problem for systems with two degrees of freedom can be solved in deterministic time $O(\lambda_{s+2}(n) \log^2 n)$, or in $O(\lambda_{s+2}(n) \log n)$ randomized expected time.

50.2.2 THREE DEGREES OF FREEDOM

A ROD IN A PLANAR POLYGONAL ENVIRONMENT

We next pass to systems with three degrees of freedom. Perhaps the simplest instance of such a system is the case of a line segment B ("rod," "ladder," "pipe") moving (translating and rotating) in a planar polygonal environment with n edges. The maximum combinatorial complexity of the free configuration space $\mathcal{F}$ of B is $\Theta(n^2)$ (recall that the naive bound for systems with three degrees of freedom is $O(n^3)$). A cell-decomposition representation of $\mathcal{F}$ can be constructed in (deterministic) $O(n^2 \log n)$ time [LS87b]. Several alternative near-quadratic algorithms have also been developed, including one based on constructing a Voronoi diagram in $\mathcal{F}$ [OSY87]. A worst-case optimal algorithm, with running time $O(n^2)$, has been given in [Veg90].

An $\Omega(n^2)$ lower bound for this problem has been established in [KO88]. It exhibits a polygonal environment with n edges and two free placements of B that are reachable from each other. However, any free motion between them requires $\Omega(n^2)$ "elementary moves," that is, the specification of any such motion requires $\Omega(n^2)$ complexity. This is a fairly strong lower bound, since it does not rely on lower bounding the complexity of the free configuration space (or of a single connected component thereof); after all, it is not clear why a motion planning algorithm should have to produce a full description of the whole free space (or of a single component).

THEOREM 50.2.7

Motion planning for a rod moving in a polygonal environment bounded by n edges can be performed in $O(n^2)$ time. There are instances where any collision-free motion of the rod between two specified placements requires $\Omega(n^2)$ "elementary moves."

A CONVEX POLYGON IN A PLANAR POLYGONAL ENVIRONMENT

Here B is a convex k-gon, free to move (translate and rotate) in an arbitrary polygonal environment bounded by n edges. The free configuration space is 3D, and there are at most $2kn$ contact surfaces, of maximum degree 4. The naive bound on the complexity of $\mathcal{F}$ is $O((kn)^3)$ (attained if B is nonconvex), but, using Davenport-Schinzel sequences, one can show that the complexity of $\mathcal{F}$ is only $O(kn\lambda_6(kn))$. Geometrically, a vertex of $\mathcal{F}$ is a semifree placement of B at which it makes simultaneously three obstacle contacts. The above bound implies that the number of such *critical placements* is only slightly super-quadratic (and not cubic) in kn.

Computing $\mathcal{F}$ in time close to this bound has proven more difficult, and only in the late 1990s has a complete solution, running in $O(kn\lambda_6(kn)\log kn)$ time and constructing the entire $\mathcal{F}$, been attained [AAS99]. Previous solutions, that were either incomplete with the same time bound, or complete and somewhat more expensive, are given in [KS90, HS96, KST97].

Another approach was given in [CK93]. It computes the Delaunay triangulation of the obstacles under the distance function d_B, when the orientation of B is fixed, and then traces the discrete combinatorial changes in the diagram as the orientation varies. The number of changes was shown to be $O(k^4n\lambda_3(n))$. Using this structure, the algorithm of [CK93] produces a high-clearance motion of B between any two specified placements, in time $O(k^4n\lambda_3(n)\log n)$.

Since all these algorithms are fairly complicated, one might consider in practice an alternative approximate scheme, proposed in [AFK$^+$92]. This scheme, originally formulated for a rectangle, discretizes the orientation of B, solves the translational motion planning for B at each of the discrete orientations, and finds those placements of B at which it can rotate (without translating) between two successive orientations. This scheme works very well in practice.

THEOREM 50.2.8

Motion planning for a k-sided convex polygon, translating and rotating in a planar polygonal environment bounded by n edges, can be performed in $O(kn\lambda_6(kn)\log kn)$ or $O(k^4n\lambda_3(n)\log n)$ time.

EXTREMAL PLACEMENTS

A related problem is to find the free placement of the largest scaled copy of B in the given polygonal environment. This has applications in manufacturing, where one wants to cut out copies of B that are as large as possible from a sheet of some material.

If only translations are allowed, the B-Voronoi diagram can be used to find the largest free homothetic copy of B. If general rigid motions are allowed, the technique of [CK93] computes the largest free similar copy of B in time $O(k^4n\lambda_3(n)\log n)$. An alternative technique is given in [AAS98], with randomized expected running time $O(kn\lambda_6(kn)\log^4 kn)$. Both bounds are nearly quadratic in n. See also earlier work on this problem in [ST94].

Finally, we mention the special case where the polygonal environment is the interior of a convex n-gon. This is simpler to analyze. The number of free critical placements of (similar copies of) B, at which B makes simultaneously four obstacle contacts, is $O(kn^2)$ [AAS98], and they can all be computed in $O(kn^2\log n)$ time. If only translations are allowed, this problem can easily be expressed as a linear program, and can be solved in $O(n+k)$ time [ST94].

THEOREM 50.2.9

The largest similar placement of a k-sided convex polygon in a planar polygonal environment bounded by n edges can be computed in randomized expected time $O(kn\lambda_6(kn)\log^4 kn)$ or in deterministic time $O(k^4 n\lambda_3(n)\log n)$. When the environment is the interior of an n-sided convex polygon, the running time improves to $O(kn^2 \log n)$, and to $O(n + k)$ if only translations are allowed.

A NONCONVEX POLYGON

Next we consider the case where B is an arbitrary polygonal region (not necessarily connected), translating and rotating in a polygonal environment bounded by n edges, as above. Here one can show that the maximum complexity of $\mathcal{F}$ is $\Theta((kn)^3)$. Using standard techniques, $\mathcal{F}$ can be constructed in $\Theta((kn)^3 \log kn)$ time, and algorithms with this running time bound have been implemented; see, e.g., [ABF89]. However, as in the purely translational case, it usually suffices to construct the connected component of $\mathcal{F}$ containing the initial placement of B. The general result, stated below, for systems with three degrees of freedom, implies that the complexity of such a component is only near-quadratic in kn. A special-purpose algorithm that computes the component in time $O((kn)^{2+\epsilon})$ is given in [HS96], where the constant of proportionality depends on ϵ. A more general algorithm with a similar running time bound is reported below. An earlier work considered the case where B is an L-shaped object moving amid n point obstacles [HOS92]. Motion planning can be performed in this case in time $O(n^2 \log^2 n)$.

THEOREM 50.2.10

Motion planning for an arbitrary k-sided polygon, translating and rotating in a planar polygonal environment bounded by n edges, can be performed in time $O((kn)^{2+\epsilon})$, for any $\epsilon > 0$. If the polygon is L-shaped and the obstacles are points, the running time improves to $O(n^2 \log^2 n)$.

A TRANSLATING POLYTOPE IN A 3D POLYHEDRAL ENVIRONMENT

Another interesting motion planning problem with three degrees of freedom involves a polytope B, with a total of k vertices, edges, and facets, translating amidst polyhedral obstacles in $\mathbb{R}^3$, with a total of n vertices, edges, and faces. The contact surfaces in this case are planar polygons, composed of a total of $O(kn)$ triangles in 3-space. Without additional assumptions, the complexity of $\mathcal{F}$ can be $\Theta((kn)^3)$ in the worst case. However, the complexity of a single component is only $O((kn)^2 \log kn)$. Such a component can be constructed in $O((kn)^{2+\epsilon})$ time, for any $\epsilon > 0$ [AS94].

If B is a convex polytope, and the obstacles consist of m convex polyhedra, with pairwise disjoint interiors and with a total of n faces, the complexity of the entire $\mathcal{F}$ is $O(kmn \log m)$ and it can be constructed in $O(kmn \log^2 m)$ time [AS97]. (Note that, in analogy with the two-dimensional case, $\mathcal{F}$ is the complement of the union of the Minkowski sums $A_i \oplus (-B)$, where A_i are the given obstacles. The above-cited bound is about the complexity and construction of such a union.) An earlier study [HY98] considered the case where B is a box, and obtained an $O(n^2\alpha(n))$ bound for the complexity of $\mathcal{F}$. Efficient practical algorithms for the exact computation of Minkowski sums for convex polyhedra are described in [FH07], and for general polyhedra in [Hac09].

THEOREM 50.2.11

Translational motion planning for an arbitrary polytope with k facets, in an arbitrary 3D polyhedral environment bounded by n facets, can be performed in time $O((kn)^{2+\epsilon})$, for any $\epsilon > 0$. If B is a convex polytope, and there are m convex pairwise disjoint obstacles with a total of n facets, then the motion planning can be performed in $O(kmn \log^2 m)$ time.

A BALL IN A 3D POLYHEDRAL ENVIRONMENT

Let B be a ball moving in 3D amidst polyhedral obstacles with a total of n vertices, edges, and faces. The complexity of the entire $\mathcal{F}$ is $O(n^{2+\epsilon})$, for any $\epsilon > 0$ [AS00a]. Note that, for the special case of line obstacles, the expanded obstacles are congruent (infinite) cylinders, and $\mathcal{F}$ is the complement of their union.

THEOREM 50.2.12

Motion planning for a ball in an arbitrary 3D polyhedral environment bounded by n facets can be performed in time $O(n^{2+\epsilon})$, for any $\epsilon > 0$.

It is worth mentioning that the combinatorial complexity of the union of n infinite cylinders in $\mathbb{R}^3$, having arbitrary radii, is $O(n^2 + \epsilon)$, for any $\epsilon > 0$ where the bound is almost tight in the worst case [Ezr11].

3D B-VORONOI DIAGRAMS

A more powerful approach to translational motion planning in three dimensions is via B-Voronoi diagrams, defined in three dimensions in full analogy to the two-dimensional case mentioned above. The goal is to establish a near-quadratic bound for the complexity of such a diagram. This would yield near-quadratic algorithms for planning the motion of the moving body B, for planning a high-clearance motion, and for finding largest homothetic free placements of B. The analysis of B-Voronoi diagrams is considerably more difficult in 3-space, and there are only a few instances where a near-quadratic complexity bound is known. One instance is for the case where B is a translating convex polytope with $O(1)$ facets in a 3D polyhedral environment [KS04]; the complexity of the diagram in this case is $O(n^{2+\epsilon})$. If the obstacles are lines or line segments, the complexity is $O(n^2\alpha(n)\log n)$ [CKS+98, KS04].

The case where B is a ball appears to be more challenging. Even for the special case where the obstacles are lines, no near-quadratic bounds are known. However, if the obstacles are n lines with a constant number of orientations, the B-diagram has complexity $O(n^{2+\epsilon})$ [KS03].

THE GENERAL MOTION PLANNING PROBLEM WITH THREE DEGREES OF FREEDOM

The last several instances were special cases of the general motion planning problem with three degrees of freedom. In abstract terms, we have a collection Σ of N contact surfaces in $\mathbb{R}^3$, where these surfaces are assumed to be (semi-algebraic patches of) algebraic surfaces of constant maximum degree. The free configuration space consists of some cells of the arrangement $\mathcal{A}(\Sigma)$, and a single connected component of $\mathcal{F}$ is just a single cell in that arrangement.

Inspecting the preceding cases, a unifying observation is that while the maximum complexity of the entire $\mathcal{F}$ can be $\Theta(N^3)$, the complexity of a single component is invariably only near-quadratic in N. This was shown in [HS95a] to hold in general: the combinatorial complexity of a single cell of $\mathcal{A}(\Sigma)$ is $O(N^{2+\epsilon})$, for any $\epsilon > 0$, where the constant of proportionality depends on ϵ and on the maximum degree of the surfaces; see Chapter 28.

A general-purpose algorithm for computing a single cell in such an arrangement was given in [SS97]. It runs in randomized expected time $O(N^{2+\epsilon})$, for any $\epsilon > 0$, and is based on *vertical decompositions* in such arrangements (see Chapter 28).

THEOREM 50.2.13

An arbitrary motion planning problem with three degrees of freedom, involving N contact surface patches, each of constant description complexity, can be solved in time $O(N^{2+\epsilon})$, for any $\epsilon > 0$.

50.2.3 OTHER PROBLEMS WITH FEW DEGREES OF FREEDOM

MORE DEGREES OF FREEDOM

The general motion planning problem for systems with d degrees of freedom, for $d \geq 4$, calls for estimating the complexity of a single cell in the d-dimensional arrangement of the appropriate contact surfaces, and for efficient algorithms for constructing such a cell. Basu [Bas03] shows that the complexity of such a cell in a d-dimensional arrangement of n surfaces of constant description complexity is $O(n^{d-1+\epsilon})$, for any $\epsilon > 0$, where the constant of proportionality depends on d, ϵ, and the maximum degree of the polynomials defining the surfaces.

In contrast, computing such a cell within a comparable time bound remains an open problem.

COORDINATED MOTION PLANNING

Another class of motion-planning problems involves coordinated motion planning of several independently moving systems. Conceptually, this situation can be handled as just another special case of the general problem: Consider all the moving objects as a single system, with $k = \sum_{i=1}^{t} k_i$ degrees of freedom, where t is the number of moving objects, and k_i is the number of degrees of freedom of the ith object. However, k will generally be too large, and the problem then will be more difficult to tackle (see, e.g., Section 50.1.2).

A better approach is as follows [SS91]. Let $B_1, \ldots, B_t$ be the given independent objects. For each $i = 1, \ldots, t$, construct the free configuration space $\mathcal{F}^{(i)}$ for B_i alone (ignoring the presence of all other moving objects). The actual free configuration space $\mathcal{F}$ is a subset of $\prod_{i=1}^{t} \mathcal{F}^{(i)}$. Suppose we have managed to decompose each $\mathcal{F}^{(i)}$ into subcells of constant description complexity. Then $\mathcal{F}$ is a subset of the union of Cartesian products of the form $c_1 \times c_2 \times \cdots \times c_t$, where c_i is a subcell of $\mathcal{F}^{(i)}$.

We next compute the portion of $\mathcal{F}$ within each such product. Each such sub-problem can be intuitively interpreted as the coordinated motion planning of our objects, where each moves within a small portion of space, amidst only a constant number of nearby obstacles; so these subproblems are much easier to solve. More-over, in typical cases, for most products $P = c_1 \times c_2 \times \cdots \times c_t$ the problem is trivial, because P represents situations where the moving objects are far from one another, and so cannot interact at all, meaning that $\mathcal{F} \cap P = P$. The number of subproblems that really need to be solved will be relatively small.

The connectivity graph that represents $\mathcal{F}$ is also relatively easy to construct. Its nodes are the connected components of the intersections of $\mathcal{F}$ with each of the above cell products P, and two nodes are connected to each other if they are adjacent in the overall $\mathcal{F}$. In many typical cases, determining this adjacency is easy.

As an example, one can apply this technique to the coordinated motion plan-ning of k disks moving in a planar polygonal environment bounded by n edges, to get a solution with $O(n^k)$ running time [SS91]. Since this problem has $2k$ degrees of freedom, this is a significant improvement over the bound $O(n^{2k} \log n)$ yielded by Canny's general algorithm.

See [ABS+99] for another treatment of coordinated motion planning, for two or three general independently moving robots, where algorithms that are also faster than Canny's general technique are developed.

UNLABELED MOTION PLANNING

A recent variant of the coordinated motion-planning problem considers a collection of m identical and indistinguishable moving objects. Given a set of m initial and m final placements for the objects, the goal is to find a collision-free path for the objects. In contrast to the standard (labeled) formulation of the problem, here we are not interested in a specific assignment between robots and final placements as long as each final placement is occupied by some robot at the end of the motion.

In general, the unlabeled problem is PSPACE-hard [SH15b] (as well as several simplified variants of the problem) for the case of unit squares. In contrast, the set-ting with unit disks can be solved efficiently, if one makes simplifying assumptions regarding the separation among the initial and final placements of the robots, and sometimes also between these placements and the obstacles. Adler et al. [ABHS15] give an $O(m^2 + mn)$-time algorithm that solves the problem for the case of simple polygonal environments. Turpin et al. [TuMK14] present an $O(m^4 + m^2 n^2)$-time algorithm (with some additional poly-logarithmic factors), which also guarantees to return a solution that minimizes the length of the longest path of an individual robot. More recently, Solovey et al. [SYZH15] devised an algorithm with similar running time, which minimizes the sum of lengths of the individual paths. In partic-ular, their algorithm returns a near-optimal solution whose additive approximation factor is $O(m)$.

A ROD IN A 3D POLYHEDRAL ENVIRONMENT

This problem has five degrees of freedom (three of translation and two of rotation). The complexity of $\mathcal{F}$ can be $\Omega(n^4)$ [KO87] in such a setting. This bound has almost been matched by Koltun [Kol05] who gave an $O(n^{4+\epsilon})$-time algorithm to solve this problem.

MOTION PLANNING AND ARRANGEMENTS

As can be seen from the preceding subsections, motion planning is closely related to the study of arrangements of surfaces in higher dimensions. Motion planning has motivated many problems in arrangements, such as the problem of bounding the complexity of, and designing efficient algorithms for, computing a single cell in an arrangement of n low-degree algebraic surface patches in d dimensions, the problem of computing the union of geometric objects (the expanded obstacles), and the problem of decomposing higher-dimensional arrangements into subcells of constant description complexity. These problems are only partially solved and present major challenges in the study of arrangements. See Chapter 28 and [SA95] for further details.

IMPLEMENTATION OF COMPLETE SOLUTIONS

Previously, complete solutions have rarely been implemented, mainly due to lack of the nontrivial infrastructure that is needed for such tasks. With the recent advancement in the laying out of such infrastructure, and in particular with tools now available in the software libraries LEDA [MN99] and CGAL [CGAL] (cf. Chapter 68), implementing complete solutions to motion planning has become feasible. A summary of progress and prospects in this domain can be found in [Hal02, FHW12].

SUMMARY

Some of the above results are summarized in Table 50.2.1. For each specific system, only one or two algorithms are listed.

TABLE 50.2.1   Summary of motion planning algorithms.

SYSTEM	MOTION	ENVIRONMENT	df	RUNNING TIME
Convex k-gon	translation	planar polygonal	2	$O(N \log m)$
Arbitrary k-gon	translation	planar polygonal	2	$O(kn \log^2 n)$
General			2	$O(\lambda_{s+2}(n) \log^2 n)$
Line segment	trans & rot	planar polygonal	3	$O(n^2 \log n)$
Convex k-gon	trans & rot	planar polygonal	3	$O(k^4 n \lambda_3(n) \log n)$
				$O(kn\lambda_6(kn) \log n)$
Arbitrary k-gon	trans & rot	planar polygonal	3	$O((kn)^{2+\epsilon})$
Convex polytope	translation	3D polyhedral	3	$O(kmn \log^2 m)$
Arbitrary polytope	translation	3D polyhedral	3	$O((kn)^{2+\epsilon})$
Ball		3D polyhedral	3	$O(n^{2+\epsilon})$
General 3 D.O.F.			3	$O(N^{2+\epsilon})$

50.2.4 PRACTICAL APPROACHES TO MOTION PLANNING

When the number of degrees of freedom is even moderately large, exact and complete solutions of the motion planning problem are very inefficient in practice, so one seeks heuristic or other incomplete but practical solutions. Several such techniques have been developed.

Potential field. The first heuristic regards the robot as moving in a potential field [Kha86] induced by the obstacles and by the target placement, where the obstacles act as repulsive barriers, and the target as a strongly attracting source. By letting the robot follow the gradient of such a potential field, we obtain a motion that avoids the obstacles and that can be expected to reach the goal. An attractive feature of this technique is that planning and executing the desired motion are done in a single stage. Another important feature is the generality of the approach; it can easily be applied to systems with many degrees of freedom.

This technique, however, may lead to a motion where the robot gets stuck at a local minimum of the potential field, leaving no guarantee that the goal will be reached (see [KB91] and references within). To overcome this problem, several solutions have been proposed. One is to try to escape from such a "potential well" by making a few small random moves, in the hope that one of them will put the robot in a position from which the field leads it away from this well. Another approach is to use the potential field only for subproblems where the initial and final placements are close to each other, so the chance to get stuck at a local minimum is small.

Probabilistic roadmaps. Over the past two decades, this method has picked up momentum, and has become the method of choice in many practical motion-planning systems [BKL+97, KSLO96, Lat91, CBH+05, Lav06, KL08, MLL08].

The general approach is to generate many random free placements throughout the workspace, and to apply any "local" simple-minded planner to plan a motion between pairs of these placements; one may use for this purpose the potential field approach, or simply attempt to connect the two placements by a straight line segment in configuration space. If the configuration space is sufficiently densely sampled, enough local free paths will be generated, and they will form a roadmap, in the sense of Section 50.1.1, which can then be used to perform motion planning between any pair of input placements. These algorithms are categorized as *single-query* algorithms such as the Rapidly Exploring Random Tree (RRT) [LK99], or *multi-query* algorithms such as the Probabilistic Roadmap Method [KSLO96]. Many of these algorithms are *probabilistically complete*. Namely, the probability that they will return a solution (if one exists) approaches one as the number of samples tends to infinity. By now, both *asymptotically optimal* and *asymptotically near-optimal* variants were devised (see e.g., [KF11, JSCP15, SH16, DB14] for a partial list). We say that an algorithm is asymptotically (near)-optimal if the cost of the solution returned by the algorithm (nearly) approaches the cost of the optimal solution as the number of samples tends to infinity.

Interestingly, sampling-based motion-planning algorithms have been applied to molecular simulations. Specifically, these algorithms have been used for the analysis of conformational transitions, protein folding and unfolding, and protein-ligand interactions [ASC12].

A significant problem that arises is how to sample well the free configuration space; informally, the goal is to detect all "tight" passages within $\mathcal{F}$, which will be missed unless some placements are generated near them. See [ABD+98, BKL+97, HLM99, KSLO96, KL01] and Chapter 51 for more details concerning this technique, its extensions and variants.

The geometric methods for exact and complete analysis of low-dimensional configuration spaces as described so far in this chapter are combined in [SHRH13, SHH15] with the practical, considerably simpler sampling-based approaches that are appropriate for higher dimensions. This is done by taking samples that are entire low-dimensional manifolds of the configuration space and that capture the connectivity of the configuration space much better than isolated point samples. To do so, on each low-dimensional manifold an arrangement is computed, which subdivides the manifold into free and forbidden regions. Subsequently, geometric algorithms for analysis of low-dimensional manifolds provide powerful primitive operations to construct a roadmap-like data structure. Experiments show that although this hybrid approach uses heavy machinery of exact algebraic computing, it significantly outperforms the sampling-based algorithms in tight settings, where the robots need to move in densely cluttered environments.

Fat obstacles.   Another technique exploits the fact that, in typical layouts, the obstacles can be expected to be "fat" (this has several definitions; intuitively, they do not have long and skinny parts). Also, the obstacles tend not to be too clustered, in the sense that each placement of the robot can interact with only a constant number of obstacles. These facts tend to make the problem easier to solve in such so-called *realistic* input scenes. See [SHO93] for the case of fat obstacles, [SOBV98] for the case of environments with low obstacle density, and [BKO+02] for two other models of realistic input scenes.

50.3 VARIANTS OF THE MOTION PLANNING PROBLEM

We now briefly review several variants of the basic motion planning problem, in which additional constraints are imposed on the problem. Further material on many of these problems can be found in Chapter 51.

OPTIMAL MOTION PLANNING

The preceding section described techniques for determining the existence of a collision-free motion between two given placements of some moving system. It paid no attention to the optimality of the motion, which is an important consideration in practice. There are several problems involved in optimal motion planning. First, optimality is a notion that can be defined in many ways, each of which leads to different algorithmic considerations. Second, optimal motion planning is usually much harder than motion planning per se.

SHORTEST PATHS

The simplest case is when the moving system B is a single point. In this case the cost of the motion is simply the length of the path traversed by the point (normally, we use the Euclidean distance, but other metrics have been considered as well). We thus face the problem of computing **shortest paths** amidst obstacles in a 2D or 3D environment.

The planar case.   Let V be a closed planar polygonal environment bounded by n edges, and let s (the "source") be a point in V. For any other point $t \in V$, let $\pi(s,t)$ denote the (Euclidean) shortest path from s to t within V. Finding $\pi(s,t)$ for any t is facilitated by construction of the *shortest path map* $SPM(s,V)$ from s

in V, a decomposition of V into regions detailed in Chapter 31. Computing the map can be done in optimal $O(n \log n)$ time [HS99].

The same problem may be considered in other metrics. For example, it is easier to give an $O(n \log n)$ algorithm for the shortest path problem under the L_1 or L_∞ metric. See Chapter 31. Another issue that arises in this context is the *clearance* of the path (namely, the minimal distance to the closest obstacle). The Euclidean shortest path may touch obstacle boundaries and therefore its clearance at certain points may be zero. Conversely, if maximizing the distance from the obstacles is the main optimization criterion, then the path can be computed by constructing a minimum spanning tree in the Voronoi diagram of the obstacles [OY85] in $O(n \log n)$ time. Wein et al. [WBH07] considered the problem of computing shortest paths that have a minimal given clearance. Specifically, they precompute a data structure called the *visibility Voronoi complex* in time $O(n^2 \log n)$ which allows to compute shortest paths that have clearance at least δ, for any specified parameter δ. An alternative measure to quantify the tradeoff between the length and the clearance was suggested by Wein et al. [WBH08] where the optimization criterion is minimizing the reciprocal of the clearance, integrated over the length of the path. While it is still not known whether the problem of computing the optimal path in this measure is NP-hard, only recently, the first polynomial-time approximation algorithm for this problem was proposed [AFS16]; it produces a $(1+\epsilon)$-approximation in time $O(\frac{n^2}{\epsilon^2} \log \frac{n}{\epsilon})$.

The three-dimensional case. Let V be a closed polyhedral environment bounded by a total of n faces, edges, and vertices. Again, given two points $s, t \in V$, we wish to compute the shortest path $\pi(s,t)$ within V from s to t. Here $\pi(s,t)$ is a polygonal path, bending at *edges* (sometimes also at vertices) of V. To compute $\pi(s,t)$, we need to solve two subproblems: to find the sequence of edges (and vertices) of V visited by $\pi(s,t)$ (the *shortest-path sequence* from s to t), and to compute the actual points of contact of $\pi(s,t)$ with these edges. These points obey the rule that the incoming angle of $\pi(s,t)$ with an edge is equal to the outgoing angle. Hence, given the shortest-path sequence of length m, we need to solve a system of m polynomial equations in m variables in order to find the contact points; each equation turns out to be quadratic. This can be solved either approximately, using an iterative scheme, or exactly, using techniques of computational real algebraic geometry; the latter method requires exponential time. Even the first, more "combinatorial," problem of computing the shortest-path sequence is NP-hard [CR87], so the general shortest-path problem is certainly much harder in three dimensions.

Many special cases of this problem, with more efficient solutions, have been studied. The simplest instance is the problem of computing shortest paths on a convex polytope. Schreiber and Sharir [SS08] present an optimal $O(n \log n)$ algorithm, following earlier near-quadratic solutions [MMP87, CH96] and a simple linear-time approximation [AHSV97]. A related problem involves shortest paths on a polyhedral terrain, where near-quadratic exact algorithms, as well as more efficient approximation algorithms are known [MMP87, VA01, LMS97]. Approximation algorithms were also developed for computing shortest paths in weighted polyhedral surfaces [ADG+10] and in polyhedral domains [ADMS13]. See also Chapter 31.

VARIOUS OPTIMAL MOTION PLANNING PROBLEMS

Suppose next that the moving system B is a rigid body free only to translate in two or three dimensions. Then the notion of optimality is still well defined—it is the total distance traveled by (any reference point attached to) B. One can then apply the same techniques as above, after replacing the obstacles by their expanded versions. For example, if B is a convex polygon in the plane, and the obstacles are m pairwise openly-disjoint convex polygons $A_1, \ldots, A_m$, then we form the Minkowski sums $K_i = A_i \oplus (-B)$, for $i = 1, \ldots, m$, and compute a shortest path in the complement of their union. Since the K_i's may overlap, we first need to compute the complement of their union, as above. A similar approach can be used for planning shortest motion of a polyhedron translating amidst polyhedra in 3-space, etc.

If B admits more complex motions, then the notion of optimality begins to be fuzzy. For example, consider the case of a line segment ("rod") translating and rotating in a planar polygonal environment. One could measure the cost of a motion by the total distance traveled by a designated endpoint (or the centerpoint) of B, or by a weighted average between such a distance and the total turning angle of B, etc. A version of this problem has been shown to be NP-hard [AKY96]. See Chapter 31.

The notion of optimality gets even more complicated when one introduces kinematic constraints on the motion of B (for example, bounds on the radius of the curvature of the path [AW01, ABL$^+$02]). It is then often challenging even without obstacles; see Chapter 51.

MOTION PLANNING FOR A TETHERED ROBOT

An interesting family of motion-planning problems occurs when the robot is anchored to a fixed base point by a finite tether. In the basic form of this problem, the tether is not an obstacle (the robot may drive over it), and the additional constraint is to ensure that the robot does not get too far from the base. The objective is to compute the shortest path of the robot between any two given points while satisfying the constraint that the distance between the base and the robot following the tether does not exceed the tether's length. Xu et al. [XBV15] study the problem in the plane and give an algorithm that runs in time $O(kn^2 \log n)$ where n is the number of obstacle vertices and k is the number of segments in the polyline defining the initial tether configuration. Salzman and Halperin [SH15a] considered the problem of preprocessing a planar workspace to efficiently answer multiple queries. Their work relies on an extension of the *visibility graph* which encodes for each vertex of the graph, all homotopy classes that can be used to reach that vertex using a tether of predefined length.

A different variant was considered by Hert and Lumelsky [HL99] who study the motion of multiple tethered point robots in a workspace with no obstacles. This problem focuses on allowing each robot to reach a goal without undue tangling. They devised algorithms which take start and goal configurations for the robots and produce an ordering of the robots. In the planar case [HL96], the tethers are tangled, but in a prespecified way, and the problem is to find an ordering for the robots' motions such that the goal is reached with the specified tether locations. This is done in $O(n^3 \log n)$ time where n is the number of robots. In the spatial case [HL99] (applicable to multiple underwater vehicles), the tethers remain untangled, and the problem is to find an ordering in which to move the robots such that tangling does not occur. This algorithm runs in $O(n^4)$ time.

EXPLORATORY MOTION PLANNING

If the environment in which the robot moves is not known to the system a priori, but the system is equipped with sensory devices, motion planning assumes a more "exploratory" character. If only tactile (or proximity) sensing is available, then a plausible strategy might be to move along a straight line (in physical or configuration space) directly to the target position, and when an obstacle is reached, to follow its boundary until the original straight line of motion is reached again. This technique has been developed and refined for arbitrary systems with two degrees of freedom (see, e.g., [LS87]). It can be shown that this strategy provably reaches the goal, if at all possible, with a reasonable bound on the length of the motion. This technique has been implemented on several real and simulated systems, and has applications to maze-searching problems.

One attempt to extend this technique to a system with three degrees of freedom is given in [CY91]. This technique computes within $\mathcal{F}$ a certain one-dimensional skeleton (roadmap) $\mathcal{R}$ which captures the connectivity of $\mathcal{F}$. The twist here is that $\mathcal{F}$ is not known in advance, so the construction of $\mathcal{R}$ has to be done in an incremental, exploratory manner. This exploration can be implemented in a controlled manner that does not require too many "probing" steps, and which enables the system to recognize when the construction of $\mathcal{R}$ has been completed (if the goal has not been reached beforehand).

If vision is also available, then other possibilities need to be considered, e.g., the system can obtain partial information about its environment by viewing it from the present placement, and then "explore" it to gain progressively more information until the desired motion can be fully planned. Results that involve such *model-building* tasks can be found in [GMR97, ZF96]. Online algorithms for mobile robots that use vision for searching a target and for exploring a region in the plane are surveyed in [GK10]. Variants of this basic problem have been introduced which include the use of only a discrete number of visibility queries [FS10, FMS12] or minimizing the number of turns that the robot performs while exploring [DFG06].

This problem is closely related to the problem of *coverage* in which the robot is equipped with the task of determining a path that passes over all points of an area or volume of interest while avoiding obstacles. For surveys on the subject see [Cho01, GC13].

This problem becomes substantially harder when errors in localization and in mapping exist. However, by combining localization and mapping into one process, the error converges. This paradigm, called *Simultaneous Localization and Mapping*, or SLAM, is the computational problem of constructing or updating a map of an unknown environment while simultaneously keeping track of the robot's location within it. Pioneering work in this field include the work by Smith et al. [SC86] on representing and estimating spatial uncertainty. This has become a thriving area of research; see, e.g., the surveys [DB06a, DB06b] and Chapter 34.

TIME-VARYING ENVIRONMENTS

Interesting generalizations of the motion planning problem arise when some of the obstacles in the robot's environment are assumed to be moving along known trajectories. In this case the robot's goal will be to "dodge" the moving obstacles while moving to its target placement. In this "kinetic" motion planning problem, it is reasonable to assume some limit on the robot's velocity and/or acceleration. Two studies of this problem are [SM88, RS94]. They show that the problem of avoiding moving obstacles is substantially harder than the corresponding static problem. By using time-related configuration changes to encode Turing machine states, they show that the problem is PSPACE-hard even for systems with a small and fixed number of degrees of freedom. However, polynomial-time algorithms are available in a few particularly simple special cases. Another variant of this problem involves movable obstacles, which the robot B can, say, push aside to clear its passage. Again, it can be shown that the general problem of this kind is PSPACE-hard, some special instances are NP-hard, and polynomial-time algorithms are available in certain other special cases [Wil91, DZ99]. There exist sampling-based planners (see., e.g., [NSO06]) that solve this problem successfully using heuristics to provide efficient solutions in time-varying environments encountered in practical situations.

COMPLIANT MOTION PLANNING

In realistic situations, the moving system has only approximate knowledge of the geometry of the obstacles and/or of its current position and velocity, and it has an inherent amount of error in controlling its motion. The objective is to devise a strategy that will guarantee that the system reaches its goal, where such a strategy usually proceeds through a sequence of free motions (until an obstacle is hit) intermixed with *compliant motions* (sliding along surfaces of contacted obstacles) until it can be ascertained that the goal has been reached.

A standard approach to this problem is through the construction of pre-images (or back projections) [LPMT84]. Specific algorithms that solve various special cases of the problem can be found in [Bri89, Don90, FHS96]. See Chapter 51.

NONHOLONOMIC MOTION PLANNING

Another realistic constraint on the possible motions of a given system is kinematic (or *kinodynamic*). For example, the moving object B might be constrained not to exceed certain velocity or acceleration thresholds, or has only limited steering capability. Even without any obstacles, such problems are usually quite hard, and the presence of (stationary or moving) obstacles makes them extremely complicated to solve. These so-called *nonholonomic motion planning* problems are usually handled using tools from control theory. A relatively simple special case is that of a car-like robot in a planar workspace, with a bound on the radius of curvature of its motion. Issues like reachability between two given placements (even in the absence of obstacles) raise interesting geometric considerations, where one of the goals is to identify canonical motions that always suffice to get to any reachable placement. See [Lat91, LC92, Lau98] for several books that cover this topic, and Chapter 51. Kinodynamic motion planning is treated in [CDRX88, CRR91]. The problem of finding a shortest curvature-constrained path in a polygonal domain with holes was recently shown to be NP-Hard [KKP11]. Simplified cases of this problem as well as approximation algorithms are treated in [AW01, RW98, ABL+02, ACMV12, BK05, BB07, KC13].

GENERAL TASK AND ASSEMBLY PLANNING

In task planning problems, the system is given a complex task to perform, such as assembling a part from several components or restructuring its workcell into a new layout, but the precise sequence of substeps needed to attain the final goal is not specified and must be inferred by the system.

Suppose we want to manufacture a product consisting of several parts. Let S be the set of parts in their final assembled form. The first question is whether the product can be disassembled by translating in some fixed direction one part after the other, so that no collision occurs. An order of the parts that satisfies this property is called a **depth order**. It need not always exist, but when it does, the product can be assembled by translating the constituent parts one after another, in the reverse of the depth order, to their target positions. Products that can be assembled in this manner are called **stack products** [WL94]. The simplicity of the assembly process makes stack products attractive to manufacture. Computing a depth order in a given direction (or deciding that no such order exists) can be done in $O(m^{4/3+\epsilon})$ time, for any $\epsilon > 0$, for a set of polygons in 3-space with m vertices in total [BOS94]. Faster algorithms are known for the special cases of axis-parallel polygons, c-oriented polygons, and "fat" objects.

Many products, however, are not stack products, that is, a single direction in which the parts must be moved is not sufficient to (dis)assemble the product. One solution is to search for an assembly sequence that allows a subcollection of parts to be moved as a rigid body in *some* direction. This can be accomplished in polynomial time, though the running time is rather high in the worst case: it may require $\Omega(m^4)$ time for a collection of m tetrahedra in 3-space [WL94]. A more modest, but considerably more efficient, solution allows each disassembly step to proceed in one of a few given directions [ABHS96]. It has running time $O(m^{4/3+\epsilon})$, for any $\epsilon > 0$.

A general approach to assembly planning, based on the concept of a *nondirectional blocking graph* [WL94], is proposed in [HLW00]. It is called the *motion space approach*, where the motion space plays a role parallel to configuration space in motion planning. Every point in the motion space represents a possible (dis)assembly sequence motion, all having the same number of degrees of freedom. The motion space is decomposed into an arrangement of cells where in each cell the blocking relations among the parts are invariant, namely, for a every pair of parts P, Q, P will either hit Q for all the possible motions of a cell, or avoid it. It thus suffices to check one specific motion sequence from each cell, leading to a finite complete solution.

Often we restrict ourselves to two-handed partitioning steps, meaning that we partition the given assembly into two complementing subsets each treated as a rigid body. Even for two-handed partitioning, if we allow arbitrary translational motions the problem becomes NP-hard [KK95]. When we restrict ourselves to infinite translations, efficient algorithms together with exact implementations [FH13] exist. For a recent survey on assembly planning, see [Jim13]. Sampling-based motion-planning algorithms (see Section 50.2.4) have been used to (heuristically) overcome the hardness of assembly planning (see, e.g., [CJS08, SHH15]).

See Chapter 51 and [HML91] for further details on assembly sequencing, and Chapter 57 for related problems.

ON-LINE MOTION PLANNING

Consider the problem of a point robot moving through a planar environment filled with polygonal obstacles, where the robot has no a priori information about the obstacles that lie ahead. One models this situation by assuming that the robot knows the location of the target position and of its own absolute position, but that it only acquires knowledge about the obstacles as it contacts them. The goal is to minimize the distance that the robot travels. See also the discussion on exploratory motion planning above.

Because the robot must make decisions without knowing what lies ahead, it is natural to use the ***competitive ratio*** to evaluate the performance of a strategy. In particular, one would like to minimize the ratio between the distance traveled by the robot and the length of the shortest start-to-target path in that scene. The competitive ratio is the worst-case ratio achieved over all scenes having a given source-target distance. A special case of interest is when all obstacles are axis-parallel rectangles of width at least 1 located in Euclidean plane. Natural greedy strategies yield a competitive ratio of $\Theta(n)$, where n is the Euclidean source-target distance. More sophisticated algorithms obtain competitive ratios of $\Theta(\sqrt{n})$ [BRS97]. Randomized algorithms can do much better [BBF+96]. Through the use of randomization, one can transform the case of arbitrary convex obstacles [BRS97] to rectilinearly-aligned rectangles, at the cost of some increase in the competitive ratio. If the scene is not on an infinite plane but rather within some finite rectangular "warehouse," and the start location is one of the warehouse corners, then the competitive ratio drops to $\log n$ [BBFY94].

COLLISION DETECTION

Although not a motion planning problem per se, collision detection is a closely related problem in robotics [LG98]. It arises, for example, when one tries to use some heuristic approach to motion planning, where the planned path is not guaranteed apriori to be collision-free. In such cases, one wishes to test whether collisions occur during the proposed motion. Collision detection is also used as a primitive operation in sampling-based algorithms (see Section 50.2.4 and Chapter 51). Several methods have been developed, including: (a) Keeping track of the closest pair of features between two objects, at least one of which is moving, and updating the closest pair, either at discrete time steps, or using *kinetic data structures* (Chapter 53). (b) Using a hierarchical representation of more complex moving systems, by means of bounding boxes or spheres, and testing for collision recursively through the hierarchical representation (see, e.g., [LGLM00, TaMK14] and references therein). See Chapter 39 for more details.

50.4 SOURCES AND RELATED MATERIAL

SURVEYS

The results not given an explicit reference above, and additional material on motion planning and related problems may be traced in these surveys:

[Lat91, CBH+05, Lav06, KL08, MLL08]: Books or chapters of books devoted to robot motion planning.

[HSS87]: A collection of early papers on motion planning.

[SA95]: A book on Davenport-Schinzel sequences and their geometric applications; contains a section on motion planning.

[HS95b]: A review on arrangements and their applications to motion planning.

[SS88, SS90, Sha89, Sha95, AY90]: Several survey papers on algorithmic motion planning.

[AS00b, AS00c]: Surveys on Davenport-Schinzel sequences and on higher-dimensional arrangements.

RELATED CHAPTERS

REFERENCES

[AAS98]   P.K. Agarwal, N. Amenta, and M. Sharir. Largest placement of one convex polygon inside another. *Discrete Comput. Geom.*, 19:95–104, 1998.

[AAS99]   P.K. Agarwal, B. Aronov, and M. Sharir. Motion planning for a convex polygon in a polygonal environment. *Discrete Comput. Geom.*, 22:201–221, 1999,.

[ABD+98]   N.M. Amato, B. Bayazit, L. Dale, C. Jones, and D. Vallejo. OBPRM: An obstacle-based PRM for 3D workspaces. In *Robotics: The Algorithmic Perspective (WAFR'98)*, pages 155–168, A.K. Peters, Wellesley, 1998.

[ABF89]   F. Avnaim, J.-D. Boissonnat, and B. Faverjon. A practical exact motion planning algorithm for polygonal objects amidst polygonal obstacles. In *Geometry and Robotics*, vol. 391 of *LNCS*, pages 67–86, Springer, Berlin, 1989.

[ABHS96]   P.K. Agarwal, M. de Berg, D. Halperin, and M. Sharir. Efficient generation of k-directional assembly sequences. In *Proc. 7th ACM-SIAM Sympos. Discrete Algorithms*, pages 122–131, 1996.

[ABHS15]   A. Adler, M. de Berg, D. Halperin, and K. Solovey. Efficient multi-robot motion planning for unlabeled discs in simple polygons. *IEEE Trans. Autom. Sci. Eng.*, 12:1309–1317, 2015.

[ABL+02]   P.K. Agarwal, T.C. Biedl, S. Lazard, S. Robbins, S. Suri, and S. Whitesides. Curvature-constrained shortest paths in a convex polygon. *SIAM J. Comput.*, 31:1814–1851, 2002.

[ABS+99]   B. Aronov, M. de Berg, A.F. van der Stappen, P. Švestka, and J. Vleugels. Motion planning for multiple robots. *Discrete Comput. Geom.*, 22:505–525, 1999.

[ACMV12]   H.-K. Ahn, O. Cheong, J. Matoušek, and A. Vigneron. Reachability by paths of bounded curvature in a convex polygon. *Comput. Geom.*, 45:21–32, 2012.

[ADG+10]   L. Aleksandrov, H.N. Djidjev, H. Guo, A. Maheshwari, D. Nussbaum, and J.-R. Sack. Algorithms for approximate shortest path queries on weighted polyhedral surfaces. *Discrete Comput. Geom.*, 44:762–801, 2010.

[ADMS13]   L. Aleksandrov, H. Djidjev, A. Maheshwari, and J.-R. Sack.An approximation algorithm for computing shortest paths in weighted 3-d domains. *Discrete Comput. Geom.*, 50:124–184, 2013.

[AFH02]   P.K. Agarwal, E. Flato, and D. Halperin. Polygon decomposition for efficient construction of Minkowski sums. *Comput. Geom.*, 21:39–61, 2002.

[AFK+92]   H. Alt, R. Fleischer, M. Kaufmann, K. Mehlhorn, S. Näher, S. Schirra, and C. Uhrig. Approximate motion planning and the complexity of the boundary of the union of simple geometric figures. *Algorithmica*, 8:391–406, 1992.

[AFS16]   P.K. Agarwal, K. Fox, and O. Salzman. An efficient algorithm for computing high-quality paths amid polygonal obstacles. In *Proc. 27th ACM-SIAM Sympos. Discrete Algorithms*, pages 1179–1192, 2016.

[AHSV97]   P.K. Agarwal, S. Har-Peled, M. Sharir, and K.R. Varadarajan. Approximate shortest paths on a convex polytope in three dimensions. *J. ACM*, 44:567–584, 1997.

[AKY96]   T. Asano, D.G. Kirkpatrick, and C.K. Yap. d_1-optimal motion for a rod. In *Proc. 12th Sympos. Comput. Geom.*, pages 252–263, ACM Press, 1996.

[AS94]   B. Aronov and M. Sharir. Castles in the air revisited. *Discrete Comput. Geom.*, 12:119–150, 1994.

[AS97]   B. Aronov and M. Sharir. On translational motion planning of a convex polyhedron in 3-space. *SIAM J. Comput.*, 26:1785–1803, 1997.

[AS00a]   P.K. Agarwal and M. Sharir. Pipes, cigars, and kreplach: The union of Minkowski sums in three dimensions. *Discrete Comput. Geom.*, 24:645–685, 2000.

[AS00b]   P.K. Agarwal and M. Sharir. Davenport-Schinzel sequences and their geometric applications. In J.-R. Sack and J. Urrutia, editors, *Handbook of Computational Geometry*, pages 1–47, Elsevier, Amsterdam, 2000.

[AS00c]   P.K. Agarwal and M. Sharir. Arrangements of surfaces in higher dimensions. in J.-R. Sack and J. Urrutia, editors, *Handbook of Computational Geometry*, pages 49–119, Elsevier, Amsterdam, 2000.

[ASC12]   I. Al-Bluwi, T. Siméon, and J. Cortés. Motion planning algorithms for molecular simulations: A survey. *Comput. Sci. Rev.*, 6:125–143, 2012.

[AW01]   P.K. Agarwal and H. Wang. Approximation algorithms for shortest paths with bounded curvature. *SIAM J. Comput.*, 30:1739–1772, 2001.

[AY90]   H. Alt and C.K. Yap. Algorithmic aspects of motion planning: A tutorial, Parts 1 and 2. *Algorithms Rev.*, 1:43–60 and 61–77, 1990.

[Bas03]   S. Basu. On the combinatorial and topological complexity of a single cell. *Discrete Comput. Geom.*, 29:41–59, 2003.

[BB07]   J. Backer and D.G. Kirkpatrick. Finding curvature-constrained paths that avoid polygonal obstacles. In *Proc. 23rd Sympos. Comput. Geom.*, pages 66–73, ACM Press, 2007.

[BBF⁺96]   P. Berman, A. Blum, A. Fiat, H. Karloff, A. Rosen, and M. Saks. Randomized robot navigation algorithms. In *Proc. 7th ACM-SIAM Sympos. Discrete Algorithms*, pages 75–84, 1996.

[BBFY94]   E. Bar-Eli, P. Berman, A. Fiat, and P. Yan. On-line navigation in a room. *J. Algorithms*, 17:319–341, 1994.

[BDS95]   M. de Berg, K. Dobrindt, and O. Schwarzkopf. On lazy randomized incremental construction. *Discrete Comput. Geom.*, 14:261–286, 1995.

[BFH⁺15]   A. Baram, E. Fogel, D. Halperin, M. Hemmer, and S. Morr. Exact Minkowski sums of polygons with holes. In *Proc. 23rd European Sympos. Algorithms*, vol. 9294 of *LNCS*, pages 71–82, Springer, Berlin, 2015.

[BK05]   S. Bereg and D.G. Kirkpatrick. Curvature-bounded traversals of narrow corridors. In *Proc. 21st Sympos. Comput. Geom.*, pages 278–287, ACM Press, 2005.

[BKL⁺97]   J. Barraquand, L.E. Kavraki, J.-C. Latombe, T.-Y. Li, R. Motwani, and P. Raghavan. A random sampling framework for path planning in large-dimensional configuration spaces. *I. J. Robot. Res.*, 16:759–774, 1997.

[BKO⁺02]   M. de Berg, M.J. Katz, M.H. Overmars, A.F. van der Stappen, and J. Vleugels. Models and motion planning. *Comput. Geom.*, 23:53–68, 2002.

[BOS94]   M. de Berg, M.H. Overmars, and O. Schwarzkopf. Computing and verifying depth orders. *SIAM J. Comput.*, 23:437–446, 1994.

[BPR00]   S. Basu, R. Pollack, and M.-F. Roy. Computing roadmaps of semi-algebraic sets on a variety. *J. Amer. Math. Soc.*, 13:55–82, 2000.

[Bri89]   A.J. Briggs. An efficient algorithm for one-step planar compliant motion planning with uncertainty. In *Proc. 5th Sympos. Comput. Geom.*, pages 187–196, ACM Press, 1989.

[BRS97]   A. Blum, P. Raghavan, and B. Schieber. Navigating in unfamiliar geometric terrain. *SIAM J. Comput.*, 26:110–137, 1997.

[Can87]   J.F. Canny. *The Complexity of Robot Motion Planning*. MIT Press, Cambridge, 1987. See also: Computing roadmaps in general semi-algebraic sets. *Comput. J.*, 36:504–514, 1993.

[CBH⁺05]   H. Choset, W. Burgard, S. Hutchinson, G.A. Kantor, L.E. Kavraki, K. Lynch, and S. Thrun. *Principles of Robot Motion: Theory, Algorithms, and Implementation*. MIT Press, Cambridge, 2005.

[CDRX88]   J.F. Canny, B.R. Donald, J.H. Reif, and P. Xavier. On the complexity of kinodynamic planning. In *Proc. 29th IEEE Sympos. Found. Comp. Sci.*, pages 306–316, 1988.

[CEG⁺93]   B. Chazelle, H. Edelsbrunner, L.J. Guibas, M. Sharir, and J. Snoeyink. Computing a face in an arrangement of line segments and related problems. *SIAM J. Comput.*, 22:1286–1302, 1993.

[CGAL]   CGAL, The Computational Geometry Algorithms Library. http://www.cgal.org.

[CH96]   J. Chen and Y. Han. Shortest paths on a polyhedron. *Internat. J. Comput. Geom. Appl.*, 6:127–144, 1996.

[Cho01]   H. Choset. Coverage for robotics – A survey of recent results. *Ann. Math. Artif. Intell.*, 31:113–126, 2001.

[CJS08]   J. Cortés, L. Jaillet, and T. Siméon Disassembly path planning for complex articulated objects. *IEEE Trans. Robot.*, 24:475–481, 2008

[CK93]   L.P. Chew and K. Kedem. A convex polygon among polygonal obstacles: Placement and high-clearance motion. *Comput. Geom.*, 3:59–89, 1993.

[CKS⁺98] L.P. Chew, K. Kedem, M. Sharir, B. Tagansky, and E. Welzl. Voronoi diagrams of lines in three dimensions under polyhedral convex distance functions. *J. Algorithms*, 29:238–255, 1998.

[Col75]   G.E. Collins. Quantifier elimination for real closed fields by cylindrical algebraic decomposition. In *Proc. 2nd GI Conf. Automata Theory Formal Languages*, vol. 33 of *LNCS*, pages 134–183, Springer, Berlin, 1975.

[CR87]   J.F. Canny and J.H. Reif. New lower bound techniques for robot motion planning problems. In *Proc. 28th IEEE Sympos. Found. Comp. Sci.*, pages 49–60, 1987.

[CRR91]   J.F. Canny, A. Rege, and J.H. Reif. An exact algorithm for kinodynamic planning in the plane. *Discrete Comput. Geom.*, 6:461–484, 1991.

[CY91]   J. Cox and C.K. Yap. On-line motion planning: Case of a planar rod. *Ann. Math. Artif. Intell.*, 3:1–20, 1991.

[DB06a]   H.F. Durrant-Whyte and T. Bailey. Simultaneous localization and mapping: Part I. *IEEE Robot. Automat. Mag.*, 13:99–110, 2006.

[DB06b]   H.F. Durrant-Whyte and T. Bailey. Simultaneous localization and mapping: Part II. *IEEE Robot. Automat. Mag.*, 13:108–117, 2006.

[DB14]   A. Dobson and K.E. Bekris. Sparse roadmap spanners for asymptotically near-optimal motion planning. *I. J. Robot. Res.*, 33:18–47, 2014.

[DFG06]   E.D. Demaine, S. Fekete, and S. Gal. Online searching with turn cost. *Theoret. Comput. Sci.*, 361:342–355, 2006.

[Don90]   B.R. Donald. The complexity of planar compliant motion planning under uncertainty. *Algorithmica*, 5:353–382, 1990.

[DZ99]   D. Dor and U. Zwick. SOKOBAN and other motion planning problems. *Comput. Geom.*, 13:215–228, 1999.

[Ezr11]   E. Ezra. On the union of cylinders in three dimensions. *Discrete Comput. Geom.*, 45:45–46, 2011.

[FH07]   E. Fogel and D. Halperin. Exact and efficient construction of Minkowski sums of convex polyhedra with applications. *Comput.-Aided Des.*, 39:929–940, 2007.

[FH13]   E. Fogel and D. Halperin. Polyhedral assembly partitioning with infinite translations or the importance of being exact. *IEEE Trans. Autom. Sci. Eng.*, 10:227–247, 2013.

[FHS96]   J. Friedman, J. Hershberger, and J. Snoeyink. Efficiently planning compliant motion in the plane. *SIAM J. Comput.*, 25:562–599, 1996.

[FHW12]   E. Fogel, D. Halperin, and R. Wein. *CGAL Arrangements and Their Applications: A Step-by-step Guide.* vol. 7 of *Geometry and Computing*, Springer, Berlin, 2012.

[FMS12]   S. Fekete, J.S.B. Mitchell, and C. Schmidt. Minimum covering with travel cost. *J. Comb. Optim.*, 24:32–51, 2012.

[FS10]   S. Fekete and C. Schmidt. Polygon exploration with time-discrete vision. *Comput. Geom.*, 43:148–168, 2010.

[GC13]   E. Galceran and M. Carreras. A survey on coverage path planning for robotics. *Robotics Autonomous Systems*, 61:1258–1276, 2013.

[GK10]   S.K. Ghosh and R. Klein. Online algorithms for searching and exploration in the plane. *Computer Science Review*, 4:189–201, 2010.

[GMR97]   L.J. Guibas, R. Motwani, and P. Raghavan. The robot localization problem. *SIAM J. Comput.*, 26:1120–1138, 1997.

[GSS89]   L.J. Guibas, M. Sharir, and S. Sifrony. On the general motion planning problem with two degrees of freedom. *Discrete Comput. Geom.*, 4:491–521, 1989.

[Hac09]   P. Hachenberger. Exact Minkowksi sums of polyhedra and exact and efficient decomposition of polyhedra into convex pieces. *Algorithmica*, 55:329–345, 2009.

[Hal02]   D. Halperin. Robust geometric computing in motion. *I. J. Robot. Res.*, 21:219–232, 2002.

[HCA⁺95]   S. Har-Peled, T.M. Chan, B. Aronov, D. Halperin, and J. Snoeyink. The complexity of a single face of a Minkowski sum. In *Proc. 7th Canad. Conf. Comput. Geom.*, Québec City, pages 91–96, 1995.

[HD05]   R.A. Hearn and E.D. Demaine. PSPACE-completeness of sliding-block puzzles and other problems through the nondeterministic constraint logic model of computation. *Theor. Comput. Sci.*, 343:72–96, 2005.

[HD09]   R.A. Hearn and E.D. Demaine. *Games, Puzzles and Computation.* A. K. Peters, Boston, 2009.

[HJW84]   J. Hopcroft, D. Joseph, and S. Whitesides. Movement problems for 2-dimensional linkages. *SIAM J. Comput.*, 13:610–629, 1984.

[HJW85]   J. Hopcroft, D. Joseph, and S. Whitesides. On the movement of robot arms in 2-dimensional bounded regions. *SIAM J. Comput.* 14:315–333, 1985.

[HL96]   S. Hert and V.J. Lumelsky. The ties that bind: Motion planning for multiple tethered robots. *Robotics Autonomous Systems*, 17:187–215, 1996.

[HL99]   S. Hert and V.J. Lumelsky. Motion planning in $\mathbb{R}^3$ for multiple tethered robots. *IEEE Trans. Robot. Autom.*, 15:623–639, 1999.

[HLM99]   D. Hsu, J.-C. Latombe, and R. Motwani. Path planning in expansive configuration spaces. *Internat. J. Comput. Geom. Appl.*, 9:495–512, 1999.

[HLW00]   D. Halperin, J.-C. Latombe, and R.H. Wilson. A general framework for assembly planning: The motion space approach. *Algorithmica*, 26:577–601, 2000.

[HML91]   L.S. Homem de Mello and S. Lee, editors. *Computer-Aided Mechanical Assembly Planning.* Kluwer Academic Publishers, Boston, 1991.

[HOS92]   D. Halperin, M.H. Overmars, and M. Sharir. Efficient motion planning for an L-shaped object in the plane. *SIAM J. Comput.* 21:1–23, 1992.

[HS95a]   D. Halperin and M. Sharir. Almost tight upper bounds for the single cell and zone problems in three dimensions. *Discrete Comput. Geom.*, 14:385–410, 1995.

[HS95b]   D. Halperin and M. Sharir. Arrangements and their applications in robotics: Recent developments. In K. Goldberg, D. Halperin, J.-C. Latombe, and R. Wilson, editors, *The Algorithmic Foundations of Robotics*, pages 495–511, A.K. Peters, Boston, 1995.

[HS96]   D. Halperin and M. Sharir. A near-quadratic algorithm for planning the motion of a polygon in a polygonal environment. *Discrete Comput. Geom.*, 16:121–134, 1996.

[HS99]   J. Hershberger and S. Suri. An optimal algorithm for Euclidean shortest paths in the plane. *SIAM J. Comput.*, 28:2215–2256, 1999.

[HSS84]   J.E. Hopcroft, J.T. Schwartz, and M. Sharir. On the complexity of motion planning for multiple independent objects: P-space hardness of the "Warehouseman's Problem." *I. J. Robot. Res.*, 3:76–88, 1984.

[HSS87]   J.E. Hopcroft, J.T. Schwartz, and M. Sharir, editors. *Planning, Geometry, and Complexity of Robot Motion.* Ablex, Norwood, 1987.

[HY98]   D. Halperin and C.K. Yap. Combinatorial complexity of translating a box in polyhedral 3-space. *Comput. Geom.*, 9:181–196, 1998.

[Jim13]   P. Jiménez. Survey on assembly sequencing: A combinatorial and geometrical perspective *J. Intelligent Manufacturing*, 24:235–250, 2013.

[JSCP15]   L. Janson, E. Schmerling, A.A. Clark, and M. Pavone. Fast marching tree: A fast marching sampling-based method for optimal motion planning in many dimensions. *I. J. Robot. Res.*, 34:883–921, 2015.

[KB91]   Y. Koren and J. Borenstein. Potential field methods and their inherent limitations for mobile robot navigation. In *Proc. IEEE Int. Conf. Robot. Autom.*, pages 1398–1404, 1991.

[KC13]   H.-S. Kim and O. Cheong. The cost of bounded curvature. *Comput. Geom.*, 46:648–672, 2013.

[KF11]   S. Karaman and E. Frazzoli. Sampling-based algorithms for optimal motion planning. *I. J. Robot. Res.*, 30:846–894, 2011.

[Kha86]   O. Khatib. Real-time obstacle avoidance for manipulators and mobile robots *I. J. Robot. Res.*, 5:90–98, 1986.

[KK95]   L.E. Kavraki and M.N. Kolountzakis. Partitioning a planar assembly into two connected parts is NP-complete. *Inform. Process. Lett.*, 55:159–165, 1995.

[KKP11]   D.G. Kirkpatrick, I. Kostitsyna, and V. Polishchuk. Hardness results for two-dimensional curvature-constrained motion planning. In *Proc. 23rd Canad. Conf. Comput. Geom.*, pages 27–32, 2011.

[KL01]   J.J. Kuffner and S.M. LaValle. Rapidly exploring random trees: Progress and prospects. In *Algorithmic and Computational Robotics: New Dimensions (WAFR'00)*, pages 293–308, A.K. Peters, Wellesley, 2001.

[KL08]   L.E. Kavraki and S.M. LaValle. Motion planning. In *Springer Handbook of Robotics*, pages 109–131, 2008.

[KLPS86]   K. Kedem, R. Livne, J. Pach, and M. Sharir. On the union of Jordan regions and collision-free translational motion amidst polygonal obstacles. *Discrete Comput. Geom.*, 1:59–71, 1986.

[KO87]   Y. Ke and J. O'Rourke. Moving a ladder in three dimensions: upper and lower bounds. In *Proc. 3rd Sympos. Comput. Geom.*, pages 136–145, ACM Press, 1987.

[KO88]   Y. Ke and J. O'Rourke. Lower bounds on moving a ladder in two and three dimensions. *Discrete Comput. Geom.*, 3:197–217, 1988.

[Kol05]   V. Koltun. Pianos are not flat: Rigid motion planning in three dimensions. In *Proc. 16th ACM-SIAM Sympos. Discrete Algorithms*, pages 505–514, 2005.

[KS90]   K. Kedem and M. Sharir. An efficient motion planning algorithm for a convex rigid polygonal object in 2-dimensional polygonal space. *Discrete Comput. Geom.*, 5:43–75, 1990.

[KS03]   V. Koltun and M. Sharir. Three-dimensional Euclidean Voronoi diagrams of lines with a fixed number of orientations. *SIAM J. Comput.*, 32:616–642, 2003.

[KS04]   V. Koltun and M. Sharir. Polyhedral Voronoi diagrams of polyhedral sites in three dimensions. *Discrete Comput. Geom.*, 31:93–124, 2004.

[KSLO96]   L.E. Kavraki, P. Švestka, J.-C. Latombe, and M.H. Overmars. Probabilistic roadmaps for fast path planning in high dimensional configuration spaces. *IEEE Trans. Robot. Autom.*, 12:566–580, 1996.

[KST97]   K. Kedem, M. Sharir, and S. Toledo. On critical orientations in the Kedem-Sharir motion planning algorithm. *Discrete Comput. Geom.*, 17:227–239, 1997.

[Lat91]   J.-C. Latombe. *Robot Motion Planning*. Kluwer Academic, Boston, 1991.

[Lau98]   J.-P. Laumond, editor. *Robot Motion Planning and Control*. Vol. 229 of *Lectures Notes Control Inform. Sci.*, Springer-Verlag, Berlin, 1998.

[Lav06]   S.M. LaValle. *Planning Algorithms*. Cambridge University Press, 2006.

[LC92]   Z. Li and J.F. Canny, editors. *Nonholonomic Motion Planning*. Kluwer Academic, Norwell, 1992.

[LG98]   M.C. Lin and S. Gottschalk. Collision detection between geometric models: A survey. In *Proc. IMA Conf. Math. Surfaces*, pages 37–56, 1998.

[LGLM00]   E. Larsen, S. Gottschalk, M.C. Lin, and D. Manocha. Fast distance queries using rectangular swept sphere volume. In *Proc. IEEE Int. Conf. Robotics Autom.*, pages 3719–3726, 2000.

[LK99]   S.M. LaValle and J.J. Kuffner. Randomized kinodynamic planning. In *Proc. IEEE Int. Conf. Robot. Autom.*, pages 473–479, 1999.

[LMS97]   M. Lanthier, A. Maheshwari, and J.-R. Sack. Approximating weighted shortest paths on polyhedral surfaces. In *Proc. 13th Sympos. Comput. Geom.*, pages 274–283, ACM Press, 1997.

[LPMT84]   T. Lozano-Pérez, M.T. Mason, and R.H. Taylor. Automatic synthesis of fine-motion strategies for robots. *I. J. Robot. Res.*, 3:3–24, 1984.

[LS87a]   D. Leven and M. Sharir. Planning a purely translational motion for a convex object in two-dimensional space using generalized Voronoi diagrams. *Discrete Comput. Geom.*, 2:9–31, 1987.

[LS87b]   D. Leven and M. Sharir. An efficient and simple motion planning algorithm for a ladder moving in 2-dimensional space amidst polygonal barriers. *J. Algorithms*, 8:192–215, 1987.

[LS87]   V.J. Lumelsky and A.A. Stepanov. Path-planning strategies for a point mobile automaton moving amidst unknown obstacles of arbitrary shape. *Algorithmica*, 2:403–430, 1987.

[MLL08]   J. Minguez, F. Lamiraux, and J.-P. Laumond. Motion planning and obstacle avoidance. In *Springer Handbook of Robotics*, pages 827–852, 2008.

[MMP87]   J.S.B. Mitchell, D.M. Mount, and C.H. Papadimitriou. The discrete geodesic problem. *SIAM J. Comput.*, 16:647–668, 1987.

[MMP+91]   J. Matoušek, N. Miller, J. Pach, M. Sharir, S. Sifrony, and E. Welzl. Fat triangles determine linearly many holes. In *Proc. 32nd IEEE Sympos. Found. Comput. Sci.*, pages 49–58, 1991.

[MN99]   K. Mehlhorn and S. Näher. *The LEDA Platform of Combinatorial and Geometric Computing*, Cambridge University Press, 1999.

[NSO06]   D. Nieuwenhuisen, A.F. van der Stappen, and M.H. Overmars. An effective framework for path planning amidst movable obstacles. In *Algorithmic Foundation of Robotics VII (WAFR'06)*, pages 87–102, Springer, Berlin, 2006.

[OSY87]   C. Ó'Dúnlaing, M. Sharir, and C.K. Yap. Generalized Voronoi diagrams for a ladder: II. Efficient construction of the diagram. *Algorithmica*, 2:27–59, 1987.

[OY85]   C. Ó'Dúnlaing and C.K. Yap. A "retraction" method for planning the motion of a disc. *J. Algorithms*, 6:104–111, 1985.

[PS85]   F.P. Preparata and M.I. Shamos. *Computational Geometry: An Introduction*, Springer-Verlag, New York, 1985

[Rei87]   J.H. Reif. Complexity of the generalized mover's problem. In J.E. Hopcroft, J.T. Schwartz, and M. Sharir, editors, *Planning, Geometry, and Complexity of Robot Motion*, pages 267–281, Ablex, Norwood, 1987.

[RS94]   J.H. Reif and M. Sharir. Motion planning in the presence of moving obstacles. *J. ACM*, 41:764–790, 1994.

[RW98]   J.H. Reif and H. Wang. The complexity of the two-dimensional curvature-constrained shortest-path problem. In *Proc. 3rd Workshop the Algo. Found. Robotics (WAFR'98)*, pages 49–58, A.K. Peters, Natick, 1998.

[SA95]   M. Sharir and P.K. Agarwal. *Davenport-Schinzel Sequences and Their Geometric Applications*. Cambridge University Press, 1995.

[SC86]   R.C. Smith and P. Cheeseman. On the representation and estimation of spatial uncertainty. *I. J. Robot. Res.*, 5:56–68, 1986.

[SH15a]   O. Salzman and D. Halperin. Optimal motion planning for a tethered robot: Efficient preprocessing for fast shortest paths queries. In *Proc. IEEE Int. Conf. Robotics Autom.*, pages 4161–4166, 2015.

[SH15b]   K. Solovey and D. Halperin. On the hardness of unlabeled multi-robot motion planning. In *Robotics: Science and Systems*, 2015.

[SH16]   O. Salzman and D. Halperin. Asymptotically near-optimal RRT for fast, high-quality motion planning. *IEEE Trans. Robot.*, 32:473–483, 2016.

[Sha89]   M. Sharir. Algorithmic motion planning in robotics. *Computer*, 22:9–20, 1989.

[Sha95]   M. Sharir. Robot motion planning. *Comm. Pure Appl. Math.*, 48:1173–1186, 1995. Also in E. Schonberg, editor, *The Houses That Jack Built*. Courant Institute, New York, 1995, 287–300.

[SHH15]   O. Salzman, M. Hemmer, and D. Halperin. On the power of manifold samples in exploring configuration spaces and the dimensionality of narrow passages. *IEEE Trans. Autom. Sci. Eng.*, 12:529–538, 2015.

[SHRH13]   O. Salzman, M. Hemmer, B. Raveh, and D. Halperin. Motion planning via manifold samples. *Algorithmica*, 67:547–565, 2013.

[SHO93]   A.F. van der Stappen, D. Halperin, and M.H. Overmars. The complexity of the free space for a robot moving amidst fat obstacles. *Comput. Geom.*, 3:353–373, 1993.

[SM88]   K. Sutner and W. Maass. Motion planning among time-dependent obstacles. *Acta Inform.*, 26:93–122, 1988.

[SOBV98]   A.F. van der Stappen, M.H. Overmars, M. de Berg, and J. Vleugels. Motion planning in environments with low obstacle density. *Discrete Comput. Geom.*, 20:561–587, 1998.

[SS83]   J.T. Schwartz and M. Sharir. On the piano movers' problem: II. General techniques for computing topological properties of real algebraic manifolds. *Adv. Appl. Math.*, 4:298–351, 1983.

[SS87]   S. Sifrony and M. Sharir. A new efficient motion planning algorithm for a rod in two-dimensional polygonal space. *Algorithmica*, 2:367–402, 1987.

[SS88]   J.T. Schwartz and M. Sharir. A survey of motion planning and related geometric algorithms. *Artif. Intell.*, 37:157–169, 1988. Also in D. Kapur and J. Mundy, editors, *Geometric Reasoning*, pages 157–169. MIT Press, Cambridge, 1989. And in S.S. Iyengar and A. Elfes, editors, *Autonomous Mobile Robots*, volume I, pages 365–374. IEEE Computer Society Press, Los Alamitos, 1991.

[SS90]   J.T. Schwartz and M. Sharir. Algorithmic motion planning in robotics. In J. van Leeuwen, editor, *Handbook of Theoret. Comput. Sci., Volume A: Algorithms and Complexity*, pages 391–430, Elsevier, Amsterdam, 1990.

[SS91]   M. Sharir and S. Sifrony. Coordinated motion planning for two independent robots. *Ann. Math. Artif. Intell.*, 3:107–130, 1991.

[SS97]   O. Schwarzkopf and M. Sharir. Vertical decomposition of a single cell in a 3-dimensional arrangement of surfaces. *Discrete Comput. Geom.*, 18:269–288, 1997.

[SS08]   Y. Schreiber and M. Sharir. An optimal-time algorithm for shortest paths on a convex polytope in three dimensions. *Discrete Comput. Geom.*, 39:500–579, 2008.

[ST94]   M. Sharir and S. Toledo. Extremal polygon containment problems. *Comput. Geom.*, 4:99–118, 1994.

[SY84]   P.G. Spirakis and C.-K. Yap. Strong NP-Hardness of moving many discs. *Inform. Process. Lett.*, 19:55–59, 1984.

[SYZH15]   K. Solovey, J. Yu, O. Zamir, and D. Halperin. Motion planning for unlabeled discs with optimality guarantees. In *Robotics: Science and Systems*, 2015.

[TaMK14]   M. Tang, D. Manocha, and Y.J. Kim. Hierarchical and controlled advancement for continuous collision detection of rigid and articulated models. *IEEE Trans. Vis. Comput. Graph*, 20:755–766, 2014.

[TuMK14]   M. Turpin, N. Michael, and V. Kumar. Capt: Concurrent assignment and planning of trajectories for multiple robots. *I. J. Robot. Res.*, 33:98–112, 2014.

[VA01]   K.R. Varadarajan and P.K. Agarwal. Approximate shortest paths on a nonconvex polyhedron. *SIAM J. Comput.*, 30:1321–1340, 2001.

[Veg90]   G. Vegter. The visibility diagram: A data structure for visibility problems and motion planning. In *Proc. 2nd Scand. Workshop Algorithm Theory*, vol. 447 of *LNCS*, pages 97–110, Springer, Berlin, 1990.

[WBH07]   R. Wein, J.P. van den Berg, and D. Halperin. The visibility-Voronoi complex and its applications. *Comput. Geom.*, 36:66–87, 2007.

[WBH08]   R. Wein, J.P. van den Berg, and D. Halperin. Planning high-quality paths and corridors amidst obstacles. *I. J. Robot. Res.*, 27:1213–1231, 2008.

[Wei06]   R. Wein. Exact and efficient construction of planar Minkowski sums using the convolution method. In *Proc. 14th European Sympos. Algorithms*, vol. 4168 of *LNCS*, pages 829–840, Springer, Berlin, 2006.

[Wil91]   G. Wilfong. Motion planning in the presence of movable obstacles. *Ann. Math. Artif. Intell.*, 3:131–150, 1991.

[WL94]   R.H. Wilson and J.-C. Latombe. Geometric reasoning about mechanical assembly. *Artif. Intell.*, 71:371–396, 1994.

[XBV15]   N. Xu, P. Brass, and I. Vigan. An improved algorithm in shortest path planning for a tethered robot. *Comput. Geom.*, 48:732–742, 2015.

[Yap87]   C.K. Yap. An $O(n \log n)$ algorithm for the Voronoi diagram of a set of simple curve segments. *Discrete Comput. Geom.*, 2:365–393, 1987.

[ZF96]   Z. Zhang and O. Faugeras. A 3D world model builder with a mobile robot. *I. J. Robot. Res.*, 11:269–285, 1996.

51 ROBOTICS

Dan Halperin, Lydia E. Kavraki, and Kiril Solovey

INTRODUCTION

Robotics is concerned with the generation of computer-controlled motions of physical objects in a wide variety of settings. Because physical objects define spatial distributions in 3-space, geometric representations and computations play an important role in robotics. As a result the field is a significant source of practical problems for computational geometry. There are substantial differences, however, in the ways researchers in robotics and in computational geometry address related problems. Robotics researchers are primarily interested in developing methods that work well in practice and can be combined into integrated systems. They often pay less attention than researchers in computational geometry to the underlying combinatorial and complexity issues (the focus of Chapter 50). This difference in approach will become clear in the present chapter.

We start this chapter with *part manipulation* (Section 51.1), which is a frequently-performed operation in industrial robotics concerned with the physical rearrangement of multiple parts, in preparation for further automated steps. Next, we consider the problem of *assembly sequencing* (Section 51.2), where multiple parts need to be assembled into a target shape. A significant part of this chapter is devoted to *motion planning* (Section 51.3), which is concerned with finding a collision-free path for a robot operating in an environment cluttered with obstacles. We put special emphasis on *sampling-based* techniques for dealing with the problem. We then review various extensions of this problem in Section 51.4. The three final and brief sections are dedicated to software for the implementation of sampling-based motion-planning algorithms, applications of motion planning beyond robotics and bibliographic sources.

GLOSSARY

Workspace W: A subset of 2D or 3D physical space: $W \subset \mathbb{R}^k$ ($k = 2$ or 3).

Body: Rigid physical object modeled as a compact manifold with boundary $B \subset \mathbb{R}^k$ ($k = 2$ or 3). B's boundary is assumed piecewise-smooth. We will use the terms "body," "physical object," and "part" interchangeably.

Robot: A collection of bodies capable of generating their own motions.

Configuration: Any mathematical specification of the position and orientation of every body composing a robot, relative to a fixed coordinate system. The configuration of a single body is also called a **placement** or a **pose**.

Configuration space C: Set of all configurations of a robot. For almost any robot, this set is a smooth manifold. We will always denote the configuration space of a robot by C and its dimension by m. Given a robot A, we will let $A(q)$ denote the subset of the workspace occupied by A at configuration q.

Number of degrees of freedom: The dimension m of C. In the following we will abbreviate "degree of freedom" by **dof**.

51.1 PART MANIPULATION

Part manipulation is one of the most frequently performed operations in industrial robotics: parts are grasped from conveyor belts, they are oriented prior to feeding assembly workcells, and they are immobilized for machining operations.

GLOSSARY

Wrench: A pair $[\boldsymbol{f}, \boldsymbol{p} \times \boldsymbol{f}]$, where $\boldsymbol{p}$ denotes a point in the boundary of a body B, represented by its coordinate vector in a frame attached to B, $\boldsymbol{f}$ designates a force applied to B at $\boldsymbol{p}$, and $\times$ is the vector cross-product. If $\boldsymbol{f}$ is a unit vector, the wrench is said to be a **unit** wrench.

Finger: A tool that can apply a wrench.

Grasp: A set of unit wrenches $\boldsymbol{w}_i = [\boldsymbol{f}_i, \boldsymbol{p}_i \times \boldsymbol{f}_i]$, $i = 1, \ldots, p$, defined on a body B, each created by a finger in contact with the boundary, ∂B, of B. For each $\boldsymbol{w}_i$, if the contact is frictionless, $\boldsymbol{f}_i$ is normal to ∂B at $\boldsymbol{p}_i$; otherwise, it can span the friction cone defined by the Coulomb law.

Force-closure grasp: A grasp $\{\boldsymbol{w}_i\}_{i=1,\ldots,p}$ such that, for any arbitrary wrench $\boldsymbol{w}$, there exists a set of real values $\{f_1, \ldots, f_p\}$ achieving $\Sigma_{i=1}^{p} f_i \boldsymbol{w}_i = -\boldsymbol{w}$. In other words, a force-closure grasp can resist any external wrenches applied to B. If contacts are nonsticky, we require that $f_i \geq 0$ for all $i = 1, \ldots, p$, and the grasp is called **positive**. In this section we only consider positive grasps.

Form-closure grasp: A positive force-closure grasp in which all finger-body contacts are frictionless.

51.1.1 GRASPING

Grasp analysis and synthesis has been an active research area over the last decade and has contributed to the development of robotic hands and grasping mechanisms. A comprehensive review of robot grasping foundations can be found at [LMS14, Chapter 2].

SIZE OF A FORM/FORCE CLOSURE GRASP

The following results are shown in [MLP90, MSS87]:

- Bodies with rotational symmetry (e.g., disks in 2-space, spheres and cylinders in 3-space) admit no form-closure grasps.

- All other bodies admit a form-closure grasp with at most four fingers in 2-space and twelve fingers in 3-space.

- All polyhedral bodies have a form-closure grasp with seven fingers.

- With frictional finger-body contacts, all bodies admit a force-closure grasp that consists of three fingers in 2-space and four fingers in 3-space.

TESTING FOR FORM/FORCE CLOSURE

A necessary and sufficient condition for force closure in 2-space (resp. 3-space) is that the finger wrenches span three (resp. six) dimensions and that a strictly positive linear combination of them be zero. Said otherwise, the null wrench (the origin) should lie in the interior of the convex hull H of the finger wrenches [MSS87]. This condition provides an effective test for deciding in constant time whether a given grasp achieves force closure. A related quantitative measure of the quality of a grasp (one among several metrics proposed) is the radius of the maximum ball centered at the origin and contained in the convex hull H [KMY92].

SYNTHESIZING FORM/FORCE CLOSURE GRASPS

Most research has concentrated on computing grasps with two to four nonsticky fingers. Optimization techniques and elementary Euclidean geometry are used in [MLP90] to derive an algorithm computing a single force-closure grasp of a polygonal or polyhedral part. This algorithm is linear in the part complexity. Other linear-time techniques using results from combinatorial geometry (Steinitz's theorem) are presented in [MSS87, Mis95]. Optimal force-closure grasps are synthesized in [FC92] by maximizing the set of external wrenches that can be balanced by the contact wrenches.

Finding the maximal regions on a body where fingers can be positioned independently while achieving force closure makes it possible to accommodate errors in finger placement. Geometric algorithms for constructing such regions are proposed in [Ngu88] for grasping polygons with two fingers (with friction) and four fingers (without friction), and for grasping polyhedra with three fingers (with frictional contact capable of generating torques) and seven fingers (without friction). Curved obstacles have also been studied [PSS+97]. The latter paper contains a good overview of work on the effect of curvature at contact points on grasp planning.

DEXTROUS GRASPING

Reorienting a part by moving fingers on the part's surface is often considered to lie in the broader realm of grasping. Finger gait algorithms [Rus99] and nonholonomic rolling contacts [HGL+97] for fingertips have been explored. A more general approach relying on extrinsic resources to the hand is considered in [DRP+14].

51.1.2 CAGING

The problem of *caging* is concerned with surrounding an object by the fingers of a robotic hand such that the object may be able to move locally, but cannot escape the cage. This problem was first introduced in [RB99] with the motivation of simplifying the challenging task of grasping an object by first caging it. The relation between caging and grasping is studied in [RMF12]. This work characterizes caging configurations from which the object can be grasped without breaking the cage.

An important parameter of caging is the number of fingers involved in the construction of the cage—caging of complex objects typically requires more fingers than simple ones, and is also more complex to compute. Two-finger cagings were studied in [RB99, PS06, PS11, ABR12] for a polygonal object, and in [PS11, ARB14] for a polytope; three-finger cagings were the concern of [DB98, ETRP07, VS08, ARB15];

more complex settings are studied in [PVS08, RMF12]. Caging was also used in the context of manipulation (see Section 51.4.2), where several robots cage the object in order to manipulate it [SRP02, FHK08].

51.1.3 FIXTURING

Most manufacturing operations require fixtures to hold parts. To avoid the custom design of fixtures for each part, modular reconfigurable fixtures are often used. A typical modular fixture consists of a workholding surface, usually a plane, that has a lattice of holes where locators, clamps, and edge fixtures can be placed. Locators are simple round pins, while clamps apply pressure on the part.

Contacts between fixture elements and parts are generally assumed to be frictionless. In modular fixturing, contact locations are restricted by the lattice of holes, and form closure cannot always be achieved. In particular, when three locators and one clamp are used on a workholding plane, there exist polygons of arbitrary size for which no fixture design can be achieved [ZG96]. But if parts are restricted to be rectilinear, a fixture can always be found as long as all edges have length at least four lattice units [Mis91]. Algorithms for computing all placements of (frictionless) point fingers that put a polygonal part in form closure and all placements of point fingers that achieve "2nd-order immobility" [RB98] of a polygonal part are presented in [SWO00].

When the fixturing kit consists of a latticed workholding plane, three locators, and one clamp, the algorithm in [BG96] finds all possible placements of a given part on the workholding surface where form closure can be achieved, along with the corresponding positions of the locators and the clamp. The algorithm in [ORSW95] computes the form-closure fixtures of input polygonal parts using a kit containing one edge fixture, one locator, and one clamp.

An algorithm for fixturing an assembly of parts that are not rigidly fastened together is proposed in [Mat95]. A large number of fixturing contacts are first scattered at random on the external boundary of the assembly. Redundant contacts are then pruned until the stability of the assembly is no longer guaranteed.

Fixturing is also studied for more complex parts such as sheet metal with holes [GGB+04], deformable parts that are modeled as linearly-elastic polygons [GG05] and polygonal chains [CSG+07, RS12].

51.1.4 PART FEEDING

Part feeders account for a large fraction of the cost of a robotic assembly workcell. A typical feeder must bring parts at subsecond rates with high reliability. The problem of part-feeder design is formalized in [Nat89] in terms of a set of functions—called *transfer functions*—which map configurations to configurations. The goal is then to find a composition of these functions that maps each configuration to a unique final configuration (or a small set of final configurations). Given k transfer functions and n possible configurations, the shortest composition that will result in the smallest number of final configurations can be found in $O(kn^4)$ [Nat89]. If the transfer functions are all monotone, the complexity is reduced to $O(kn^2)$ [Epp90].

Part feeding often relies on *nonprehensile manipulation*. Nonprehensile manipulation exploits task mechanics to achieve a goal state without grasping and frequently allows accomplishing complex feeding tasks with few dofs. It may also enable a robot to move parts that are too large or heavy to be grasped and lifted.

Pushing is one form of nonprehensile manipulation. Work on pushing originated in [Mas82] where a simple rule is established to qualitatively determine the motion of a pushed object. This rule makes use of the position of the center of friction of the object on the supporting surface. Given a part we can compute its *push* transfer function. The push function, $p_\alpha : S_1 \to S_1$, when given an orientation θ returns the orientation of the part $p_\alpha(\theta)$ after it has been pushed from direction α by a fence orthogonal to the push direction. With a sequence of different push operations it is possible to uniquely orient a part. The push function has been used in several nonprehensile manipulation algorithms:

- A planning algorithm for a robot that tilts a tray containing a planar part of known shape to orient it to a desired orientation [EM88]. This algorithm was extended to the polyhedral case in [EMJ93].

- An algorithm to compute the design of a sequence of curved fences along a conveyor belt to reorient a given polygonal part [WGPB96]. See also [BGOS98].

- An algorithm that computes a sequence of motions of a single articulated fence on a conveyor belt that achieves a goal orientation of an object [AHLM00].

A frictionless parallel-jaw gripper was used in [Gol93] to orient polygonal parts. For any part P having an n-sided convex hull, there exists a sequence of $2n - 1$ squeezes achieving a single orientation of P (up to symmetries of the convex hull). This sequence is computed in $O(n^2)$ time [CI95]. The result has been generalized to planar parts having a piecewise algebraic convex hull [RG95]. It was shown [SG00] that one could design plans whose length depends on a parameter that describes the part's shape (called *geometric eccentricity* in [SG00]) rather than on the description of the combinatorial complexity of the part. For the parallel-jaw gripper we can define the *squeeze* transfer function. In [MGEF02] another transfer function is defined: the *roll* function. With this function a part is rolled between the jaws by making one jaw slide in the tangential direction. Using a combination of squeeze and roll primitives a polygonal part can be oriented without changing the orientation of the gripper.

Distributed manipulation systems provide another form of nonprehensile manipulation. These systems induce motions on objects through the application of many external forces. The part-orienting algorithm for the parallel-jaw gripper has been adapted for arrays of microelectromechanical actuators which—due to their tiny size—can generate almost continuous fields [BDM99]. Algorithms that position and orient parts based on identifying a finite number (depending on the number of vertices of the part) of distinct equilibrium configurations were also given in [BDM99, BDH99]. Subsequent work showed that using a carefully selected actuators field, it is possible to position and orient parts in two stable equilibrium configurations [Kav97]. Finally, a long standing conjecture was proved, namely that there exists a field that can uniquely position and orient parts in a single step [BDKL00]. In fact, two different such fields were fully analyzed in [LK01b, SK01]. On the macroscopic scale it was shown that in-plane vibration can be used for closed-loop manipulation of objects using vision systems for feedback [RMC00], that arrays of controllable airjets can manipulate paper [YRB00], and that foot-sized discrete actuator arrays can handle heavier objects under various manipulation strategies [LMC01]. The work [BBDG00] describes a sensorless technique for manipulating objects using vibrations.

Some research is devoted to the design of a *vibratory bowl feeder*, which consists of a bowl filled with parts surrounded by a helical metal track. Vibrations cause the parts to advance along the track, where they encounter a sequence of mechanisms, such as traps [ACH01, BGOS01] or blades [GGS06], which are designed to discard parts in improper orientations. Eventually, only parts that are oriented in the desired position reach to the top of the bowl.

OPEN PROBLEMS

A major open practical problem is to predict feeder throughputs to evaluate alternative feeder designs, given the geometry of the parts to be manipulated. In relation to this problem, simulation algorithms have been proposed to predict the pose of a part dropped on a surface [MZG+96, ME02, PS15, Vár14]. In distributed manipulation, an open problem is to analyze the effect of discrete arrays of actuators on the positioning and orientation of parts [LMC01, LK01b].

51.2 ASSEMBLY SEQUENCING

Most mechanical products consist of multiple parts. The goal of assembly sequencing is to compute both an order in which parts can be assembled and the corresponding required movements of the parts.

GLOSSARY

Assembly: Collection of bodies in some given relative placements.

Subassembly: Subset of the bodies composing an assembly A in their relative positions and orientations in A.

Separated subassemblies: Subassemblies that are arbitrarily far apart from one another.

Hand: A tool that can hold an arbitrary number of bodies in fixed relative placements.

Assembly operation: A motion that merges s pairwise-separated subassemblies ($s \geq 2$) into a new subassembly; each subassembly moves as a single body. No overlapping between bodies is allowed during the operation. The parameter s is called the **number of hands of the operation**. We call the reverse of an assembly operation **assembly partitioning**.

Assembly sequence: A total ordering on assembly operations that merge the separated parts composing an assembly into this assembly. The maximum, over all the operations in the sequence, of the number of hands required by an operation is called the **number of hands of the sequence**.

Monotone assembly sequence: A sequence in which no operation brings a body to an intermediate placement (relative to other bodies), before another operation transfers it to its final placement.

NUMBER OF HANDS IN ASSEMBLY

Every assembly of convex polygons in the plane has a two-handed assembly sequence of translations. In the worst case, s hands are necessary and sufficient for assemblies of s star-shaped polygons/polyhedra [Nat88].

There exists an assembly of six tetrahedra without a two-handed assembly sequence of translations, but with a three-handed sequence of translations. Every assembly of five or fewer convex polyhedra admits a two-handed assembly sequence of translations. There exists an assembly of thirty convex polyhedra that cannot be assembled with two hands [SS94].

COMPLEXITY OF ASSEMBLY SEQUENCING

When arbitrary sequences are allowed, assembly sequencing is PSPACE-hard. The problem remains PSPACE-hard even when the bodies are polygons, each with a constant number of vertices [Nat88].

When only two-handed monotone sequences are permitted, deciding if an assembly A can be partitioned into two subassemblies S and $A \backslash S$ such that they can be separated by an arbitrary motion is NP-complete [WKLL95]. The problem remains NP-complete when both S and $A \backslash S$ are required to be connected and motions are restricted to translations [KK95].

MONOTONE TWO-HANDED ASSEMBLY SEQUENCING

A popular approach to assembly sequencing is disassembly sequencing [HdMS91]. A sequence that separates an assembly into its individual components is first generated and next reversed. Most existing assembly sequencers can only generate two-handed monotone sequences. Such a sequence is computed by partitioning the assembly and, recursively, the resulting subassemblies into two separated subassemblies.

The **nondirectional blocking graph** (NDBG) is proposed in [WL94] to represent all the blocking relations in an assembly. It is a subdivision of the space of all allowable motions of separation into a finite number of cells such that within each cell the set of blocking relations between all pairs of parts remains fixed. Within each cell this set is represented in the form of a directed graph, called the directional blocking graph (DBG). The NDBG is the collection of the DBGs over all the cells in the subdivision.

We illustrate this approach for polyhedral assemblies when the allowable motions are infinite translations. The partitioning of an assembly consisting of polyhedral parts into two subassemblies is performed as follows. For an ordered pair of parts P_i, P_j, the 3-vector $\vec{d}$ is a **blocking direction** if translating P_i to infinity in direction $\vec{d}$ will cause P_i to collide with P_j. For each ordered pair of parts, the set of blocking directions is constructed on the unit sphere $\mathcal{S}^2$ by drawing the boundary arcs of the union of the blocking directions (each arc is a portion of a great circle). The resulting collection of arcs partitions $\mathcal{S}^2$ into maximal regions such that the blocking relation among the parts is the same for any direction inside such a region.

Next, the blocking graph is computed for one such maximal region. The algorithm then moves to an adjacent region and updates the DBG by the blocking relations that change at the boundary between the regions, and so on. After each time the construction of a DBG is completed, this graph is checked for strong connectivity in time linear in its number of edges. The algorithm stops the first time it encounters a DBG that is not strongly connected and it outputs the two subassemblies of the partitioning. The overall sequencing algorithm continues recursively with the resulting subassemblies. If all the DBGs that are produced during a partitioning step are strongly connected, the algorithm reports that the assembly does not admit a two-handed monotone assembly sequence with infinite translations.

Polynomial-time algorithms are proposed in [WL94] to compute and exploit NDBGs for restricted families of motions. In particular, the case of partitioning a polyhedral assembly by a single translation to infinity is analyzed in detail, and it is shown that partitioning an assembly of m polyhedra with a total of v vertices takes $O(m^2 v^4)$ time. Another case studied in [WL94] is where the separating motions are infinitesimal rigid motions. Then partitioning the polyhedral assembly takes $O(mc^5)$ time, where m is the number of pairs of parts in contact and c is the number of independent point-plane contact constraints. (This result is improved in [GHH+98] by using the concept of maximally covered cells; see Section 28.6.) Using these algorithms, every feasible disassembly sequence can be generated in polynomial time.

In [WL94], NDBG's are defined only for simple families of separating motions (infinitesimal rigid motions and infinite translations). An extension, called the *interference diagram*, is proposed in [WKLL95] for more complex motions. In the worst case, however, this diagram yields a partitioning algorithm that is exponential in the number of surfaces describing the assembly. When each separating motion is restricted to be a short sequence of concatenated translations (for example, a finite translation followed by an infinite translation), rather efficient partitioning algorithms are available [HW96]. A unified and general framework for assembly planning, based on the NDBG, called the *motion space approach* is presented in [HLW00]. On the practical side, an exact and robust implementation of this framework for infinite-translation motions is discussed in [FH13]. We mention the works [CJS08, LCS09], which employ sampling-based techniques (see Section 51.3) in order to produce assembly sequences. The reader is referred to a recent review on the subject of assembly sequencing in [Jim13].

OPEN PROBLEM

1. The complexity of the NDBG grows exponentially with the number of parameters that control the allowable motions, making this approach highly time consuming for assembly sequencing with compound motions. For the case of infinitesimal rigid motion it has been observed that only a (relatively small) subset of the NDBG needs to be constructed [GHH+98]. Are there additional types of motion where similar gains can be made? Are there situations where the full NDBG (or a structure of comparable size) must be constructed?

2. Most efficient algorithms for assembly planning deal with the two-handed case. Devise efficient algorithms for 3-(and higher)handed assembly planning.

51.3 MOTION PLANNING

Motion planning is aimed to provide robots with the capability of deciding automatically which motions to execute in order to achieve goals specified by spatial
arrangements of physical objects. It arises in a variety of forms. The simplest
form—the *basic path-planning problem*—requires finding a geometric collision-free
path for a single robot in a known static workspace. The path is represented by a
curve connecting two points in the robot's configuration space [LP83]. This curve
must not intersect a forbidden region, the *C-obstacle region*, which is the image
of the workspace obstacles. Other motion planning problems require dealing with
moving obstacles, multiple robots, movable objects, uncertainty, etc.

   In this section we consider the basic motion planning. In the next one we review
other motion planning problems. Most of our presentation focuses on practical
methods, and in particular, sampling-based algorithms. See Chapter 50 for a more
extensive review of theoretical motion planning.

GLOSSARY

Forbidden space: The set of configurations $\mathcal{B} \subset \mathcal{C}$ in which the robot collides
   with obstacles or violate the mechanical limits of its joints.

Free space: The complement of the forbidden space in region in $\mathcal{C}$, $\mathcal{F} = \mathcal{C} \backslash \mathcal{B}$.

Path: A continuous map $\tau : [0, 1] \to \mathcal{C}$. A path is also *free* if it lies entirely in $\mathcal{F}$.

Motion-planning problem: Compute a free path between two input configurations.

Path planning query: Given two points in configuration space find a free path
   between them. The term is often used in connection with algorithms that pre-
   process the configuration space in preparation for many queries.

Complete algorithm: A motion planning algorithm is complete if it is guaranteed to find a free path between two given configurations whenever such a path
   exists, and report that there is no free path otherwise. Complete algorithms
   are sometimes referred to as *exact* algorithms. There are weaker variants of
   completeness, for example, *probabilistic completeness*.

SAMPLING-BASED PLANNERS

The complexity of motion planning for robots with many dofs (more than 4 or 5)
has led to the development of sampling-based algorithms that trade off completeness against applicability in practical settings. Such techniques avoid computing an
explicit geometric representation of the free space. Instead, in their simplest form,
they generate a large set of configuration samples and discard samples that represent forbidden configurations of the robot. Then, an attempt is made to connect
nearby configurations, given a metric which is defined over the configuration space.
This construction induces a graph structure, termed a *roadmap*, with the property
that a path connecting two vertices in the roadmap represents a free path in the
configuration space which connects the corresponding configurations. Although
such techniques are inherently incomplete, in many cases they can be shown to be

probabilistically complete—guaranteed to find a solution with high probability if one exists, given a sufficient number of samples.

The implementation of sampling-based techniques relies on two main geometric components: A *collision detector* determines whether a given configuration is free or forbidden. Collision-detection calls typically dominate the running time of sampling-based planners. A *nearest-neighbor search* structure processes a collection of points to efficiently answer queries returning the set of nearest neighbors of a given configuration, or the set of configurations that lie within a certain distance from the given configuration. For further information on these components see Chapter 39 and Chapter 43. We provide references for software implementation of collision detection and nearest-neighbor search in Section 51.5 below.

A prime example of a sampling-based algorithm is the *probabilistic roadmap method* (PRM) [KŠLO96], which proceeds as follows. A collection of randomly-sampled configurations is generated, and the non-free configurations are identified using the collision detector, and consequently discarded. The remaining configurations represent the vertex set of the roadmap. Then, for every sampled configuration a collection of nearby configurations are identified using nearest-neighbor search. Finally, a simple (usually, straight-line) path is generated between pairs of nearby configurations, and the path is checked for collision by dense sampling along the path (or by more costly continuous collision-detection techniques). Paths that are identified as collision free are added as edges to the roadmap. Once a roadmap has been computed, it is used to process an arbitrary number of path-planning queries. The density of sampling can be selectively increased to speed up the roadmap connection.

Other sampling strategies, such as *rapidly-exploring random trees* (RRT) [LK01c] and *expansive-space trees* (EST) [HLM99] assume that the initial and goal configurations are given, and incrementally build a tree structure until these two configurations are connected. In the frequently used RRT, for example, in every step a random configuration is generated and the configuration that is the closest to the sample in the current tree is identified. In the most simplified version of the algorithm, the path between the two configurations is checked for collision. If it is indeed free, the sampled configuration is added as a vertex to the tree, along with the new edge connecting it to the tree. It should also be noted that RRT, EST, and several other sampling-based algorithms have found applications in cases where non-holonomic, dynamic, and other constraints are considered (see [KL16] and Section 51.4.4).

The results reported in [KLMR98, HLM99, LK01c, KKL98, LK04b] provide a probabilistic-completeness analysis of the aforementioned planners.

Some research has focused on designing efficient sampling and connection strategies. For instance, the Gaussian sampling strategy produces a greater density of milestones near the boundary of the free space $\mathcal{F}$, whose connectivity is usually more difficult to capture by a roadmap than wide-open areas of $\mathcal{F}$ [BOS99]. Different methods to create milestones near the boundary of $\mathcal{F}$ were obtained in [YTEA12]. A lazy-evaluation of the roadmap has been suggested in [BK00, SL03, SL02] while visibility has been exploited in [SLN00]. Sampling and connection strategies are reviewed in [SL02].

Standard sampling-based planners are unable to determine that a given problem has no solution. Several works deal with *disconnection proofs*, which are able to determine, in certain settings, that two given configurations reside in two distinct components of the free space [BGHN01, ZKM08, MBH12].

There is a large variety of applications beyond robot motion where sampling-based planners are useful. For a survey of few such applications, see Section 51.6.

OPTIMALITY IN SAMPLING-BASED PLANNING

So far we have been concerned with finding *a* solution to a given motion-planning problem. In practical settings one is usually interested in finding a *high-quality* path, in terms of, for example, the path's length, clearance from obstacles, energy consumption, and so on. Initial efforts in this area include applying shortcuts on the returned solution to improve its quality [GO07], merging several solutions to produce an improved one [REH11]. Some works fine-tune existing sampling-based planners by modifying the sampling strategy [LTA03, US03, SLN00], or the connection scheme to new samples [US03, SLN00]. Other approaches include supplementing edges for shortcuts in existing roadmaps [NO04] and random restarts [WB08]. On the negative side, it was shown that existing methods can produce arbitrarily bad solutions [NRH10, KF11].

The breakthrough came in [KF11]. This work introduced PRM* and RRT*, variants of PRM and RRT, and proved that they are *asymptotically optimal*—the solution returned by these algorithms converges to the optimal solution, as the number of samples tends to infinity. This work establishes a bound on the connection radius of PRM, which ensures asymptotic optimality. In particular, the connection radius is a function of the total number of samples n and results in a roadmap of size $O(n \log n)$. The adaptation of RRT to an optimal planner requires to consider additional neighbors for connection and a rewiring scheme, which ensures that the resulting roadmap has a tree structure.

Since then, other asymptotically-optimal planners have been introduced. RRT# [AT13] extends RRT, but through a different technique than RRT*, and has (empirically) better convergence rates than the latter. *Fast marching trees* (FMT*) [JSCP15] is another single-query planner which was shown to have faster convergence rates than RRT*. An improvement of FMT* is introduced in [SH15a]. A lazy version of PRM* is described in [Hau15]. Finally, we note that by using nearest-neighbor data structures tailored for these algorithms, additional speedup in the running times can be obtained [KSH15].

The guarantee of optimality comes at the price of increased space and time consumption of the planners. As a result, several techniques were developed to lower these attributes, by relaxing the optimality guarantees to *asymptotic near optimality*. This property implies that the cost of the returned solution converges to within a factor of $(1 + \varepsilon)$ of the cost of the optimal solution. Several methods were introduced to reduce the size of a probabilistic roadmaps via *spanners* [DB14, WBC15] or *edge contractions* [SSAH14]. The planner described in [SH14a] allows for continuous interpolation between the fast RRT algorithm and the asymptotically-optimal RRT*. A different approach is taken in [LLB14], where an asymptotically near-optimal algorithm is described, which is also suitable for systems with dynamics.

A recent work [SSH16b] describes a general framework for asymptotic analysis of sampling-based planners. The framework exploits a relation between such planners and standard models of *random geometric graphs*, which have been extensively studied for several decades and many of their aspects are by now well understood.

Most of the analysis for sampling-based algorithms is concerned with *asymptotic* properties, i.e., those that hold with high probability as the number of samples tends to infinity. A recent work [DMB15] provides bounds on the probability of finding a near-optimal solution with PRM* after a finite number of iterations. We also mention that the work on FMT* [JSCP15] studies a similar setting for this algorithm, albeit for a limited case of an obstacle-free workspace.

OTHER ALGORITHMS

Several non-sampling based methods have been proposed for path planning. Some of them work well in practice, but they usually offer no performance guarantee.

Heuristic algorithms often search a regular grid defined over the configuration space and generate a path as a sequence of adjacent grid points [Don87]. Early heuristics employed *potential fields* to guide the grid search. A potential field is a function over the free space that (ideally) has a global minimum at the goal configuration. This function may be constructed as the sum of an attractive and a repulsive field [Kha86]. More modern approaches use A*-flavored graph-search techniques that drastically reduce the amount of explored grid vertices, and in some cases can guarantee convergence to an optimal solution (see, e.g., [LF09, ASN$^+$16]). Other approaches combine discrete search and sampling-based planners. For example, SyCloP [PKV10] combines A*-flavored searched over a grid imposed on the workspace, and the result of the search is used to guide sampling-based planners.

One may also construct grids at variable resolution. Hierarchical space decomposition techniques such as octrees and boxtrees have been used for that purpose [BH95]. At any decomposition level, each grid cell is labeled EMPTY, FULL, or MIXED depending on whether it lies entirely in the free space, lies in the C-obstacle region, or overlaps both. Only the MIXED cells are decomposed further, until a search algorithm finds a sequence of adjacent FREE cells connecting the initial and goal configurations. A recent subdivision-based planner [WCY15] reduces the computation time by incorporating inexact predicates for checking the property of cells. This planner accepts a resolution parameter and can determine whether a solution determined by this parameter exists or not.

We note that several optimizing planners that are based on optimization methods have been recently developed (e.g., [Kob12, SDH$^+$14]).

OPEN PROBLEMS

1. Analyze convergence rates for sampling-based algorithms (see [DMB15, JSCP15]).

2. Given a fixed amount of time, can one decide if it is best to use an optimizing sampling-based planner, or a non-optimizing sampling-based planner and then optimization methods to improve the resulting path?

51.4 EXTENSIONS OF THE BASIC MOTION PLANNING PROBLEM

There are many useful extensions of the basic motion-planning problem. Several are surveyed in Chapter 50, e.g., shortest paths, time-varying workspaces (moving obstacles), and exploratory motion planning. Below we focus on the following extensions: multi-robot motion planning, manipulation planning, task planning, and planning with differential constraints.

GLOSSARY

Movable object: Body that can be grasped, pushed, pulled, and moved by a robot.

Manipulation planning: Motion planning with movable objects.

Trajectory: Path parameterized by time.

Tangent space: Given a smooth manifold M and a point $p \in M$, the vector space $T_p(M)$ spanned by the tangents at p to all smooth curves passing through p and contained in M. The tangent space has the same dimension as M.

Nonholonomic robot: Robot whose permissible velocities at every configuration q span a subset $\Omega(q)$ of the tangent space $T_q(\mathcal{C})$ of lower dimension. Ω is called the *set of controls* of the robot.

Feasible path: A piecewise differentiable path of a nonholonomic robot whose tangent at every point belongs to the robot's set of controls, i.e., satisfies the nonholonomic velocity constraints.

Locally controllable robot: A nonholonomic robot is locally controllable if for every configuration q_0 and any configuration q_1 in a neighborhood U of q_0, there exists a feasible path connecting q_0 to q_1 which is entirely contained in U.

Kinodynamic planning: Find a minimal-time trajectory between two given configurations of a robot, given the robot's dynamic equation of motion.

51.4.1 MULTI-ROBOT MOTION PLANNING

In *multi-robot motion planning* several robots operate in a shared workspace. The goal is to move the robots to their assigned target positions, while avoiding collisions with obstacles, as well as with each other. The problem can be viewed as an instance of the single-robot case in which the robot consists of several independent parts (robots). In particular, suppose the problem consists of planning the motion of m robots, and denote by C_i the configuration space of the ith robot. Then it can be restated as planning the motion of a single robot whose configuration space is defined to be $C_1 \times \ldots \times C_m$. This observation allows to apply single-robot tools directly to the multi-robot case. However, such a naive approach overlooks the special characteristics of the problem at hand and tends to be inefficient both in theory and practice. Various techniques were specifically designed for this problem. Complete algorithms are covered in Chapter 50.2.3 ("coordinated motion planning"). Here we focus on heuristics and sampling-based techniques.

We start with the *decentralized* approach, which asserts that conflicts between robots should be resolved locally and without central coordination. In the *reciprocal* method [SBGM11, BB15], the individual robots take into consideration the current position and the velocity of other robots to compute their future trajectories in order to avoid collisions. The work in [KEK10] takes a game-theoretic approach and develops a reward function that reduces time and resources which are spent on coordination, and maximizes the time between conflicts. Several works [KR12, GGGDC12] develop collision-avoidance heuristics based on the sociology of pedestrian interaction.

Centralized planners produce a global plan, which the individual robots need to execute. Such planners fall into the two categories of *coupled* and *decoupled* planners, which indicate whether the planners account for all the possible robot-robot interactions or only partially do so, respectively. Coupled planners typically operate in the combined configuration space of the problem, and as a result can usually guarantee *probabilistic completeness*. Due to the high dimensionality of the combined configuration space, *decoupled* planners typically solve separate problems for the individual robots and combine the sub-solutions into one for the entire problem [LLS99, BBT02, BO05, BSLM09, BGMK12].

We proceed to describe coupled planners. In [SL02] an evaluation of single-robot sampling-based algorithms is given, when directly applied to the multi-robot case. The works [HH02, SHH15] exploit the geometric properties of the two-robot setting and develop techniques that combine sampling-based methods and geometric tools. The work in [ŠO98] introduces a sampling-based technique, which was exploited in two recent works to solve complex settings of the multi-robot problem involving many more robots than the original method could. The algorithm described in [ŠO98] produces probabilistic roadmaps for the individual robots and combines them into a *composite* roadmap embedded in the joint configuration space. Due to the size of this roadmap it can be explicitly constructed only for settings in which the number of robots is very small. In particular, the number of its vertices and the number of neighbors each vertex has are exponential in the number of robots. In [WC15] a search-based method, termed M*, is developed to traverse this enormous roadmap while representing it implicitly. M* can efficiently solve problems involving multiple robots, as long as the solution requires only mild coordination between the robots. Otherwise, it is forced to consider all the neighbors of explored vertices, which is prohibitively costly. In [SSH16a] a different traversal technique, termed *discrete-RRT* (dRRT) was introduced. It is an adaptation of RRT for the exploration of a geometrically-embedded graphs; it can cope with more tight settings of the problem as it does not require to consider all the neighbors of a given vertex in order to make progress.

In [SH14b] a centralized algorithm is described for a generalized problem termed *k-color multi-robot motion planning*—the robots are partitioned into k groups (colors) such that within each group the robots are interchangeable. Every robot is required to move to one of the target positions that are assigned to its group, such that at the end of the motion each target position is occupied by exactly one robot. This work exploits a connection between the *continuous* multi-robot problem and a discrete variant of it called *pebble motion on graphs* (see, e.g., [KMS84, GH10, KLB13, YL13]), which consists of moving a collection of pebbles from one set of vertices to another while abiding by a certain set of rules. The algorithm reduces the k-color problem into several pebble problems such that the solution to the latter can be transformed into a solution to the continuous problem.

The special case of the k-color problem with $k = 1$ is usually termed *unlabeled* multi-robot motion planning. It was first studied in [KH06], where a sampling-based algorithm was introduced. Very recently, efficient and complete algorithms for the unlabeled problem were introduced [TMMK14, ABHS15, SYZH15]. These algorithms make several simplifying assumptions concerning the separation between start and target positions, without which the problem was shown to be PSPACE-hard [SH15b]. See additional information in Chapter 50.2.3.

51.4.2 MANIPULATION PLANNING

Many robot tasks consist of achieving arrangements of physical objects. Such objects, called movable objects, cannot move autonomously; they must be grasped by a robot. Planning with movable objects is called manipulation planning. The problem is considered to be very challenging as one has to reason in a prohibitively large search space, which encompasses the various positions of the robot and the objects. In contrast with Section 51.1, here we take the viewpoint of the manipulator, rather than that of the manipulated objects.

Different instances of the problem that consist of a single robot and a single movable object with relatively simple geometry can be solved rather efficiently in a complete manner [Wil91, ALS95, BG13]. However, the problem becomes computationally intractable when multiple movable objects come into play [Wil91]. The problem of manipulation planning of multiple objects also relates to the popular *SOKOBAN* game, which was shown to be hard to solve on several occasions (see, e.g., [HD05]). Navigation Among Moveable Obstacles (NAMO) considers the path planning problem where obstacles can be moved out of the way [SK05]. Sampling-based techniques were developed to tackle more general and challenging instances of manipulation planning involving several manipulators and objects [NSO07, NSO08, BSK$^+$08].

With the growing demand for robotic manipulators that can perform tasks in cluttered human environments an additional level of complexity was added. In *rearrangement planning*, a robotic manipulator with complex kinodynamic constraints is required to rearrange a set of objects. In [SSKA07] a technique for *monotone* instances of the problem, in which it is assumed that each object can move at most once, is described. A more recent work [KSD$^+$14] which can cope also with non-monotone instances draws upon a relation between the problem at hand and multi-robot motion planning. The technique described in [KB15] extends previous work for monotone settings to the non-monotone case.

Finally, we mention that recent advances in the area of machine learning have led to the development of highly effective techniques for dealing with various complex manipulation tasks. See, e.g., [LWA15, LLG$^+$15, SEBK15].

51.4.3 TASK PLANNING

Thus far the focus was on tasks with relatively simple specifications for the robot to follow. For instance, in motion planning the robot was required to "move from s to t while avoiding collision with obstacles." Real-world problems usually require to perform much more complex operations that force the robot to make decisions while performing a task. In some cases, a richer task specification can assist in performing a given task. Consider for example a robotic arm that needs to collect several books from a desk and place them on a shelf. A human operator can assist by specifying that the manipulator should first remove objects placed on top of the books before proceeding to the actual books.

Such complex problems, which are often grouped under the umbrella term *task planning*, usually require the combination of tools from motion planning, AI, control, and model checking. The predominant methods for task planning are heuristic search [HN01] and constraint satisfaction [KS92]. We mention that a recent workshop was devoted to the subject[1].

A prevalent approach for task planning in the context of motion planning is to supplement the basic input with a Boolean logical expression, usually in the form of linear temporal logic [Pnu77], which has to be satisfied in order to achieve the goal. The workspace is decomposed into regions and the expression describes the conditions on the order of visitation of regions. For instance, "the robot has to first visit A, then B, and C afterwards; it cannot pass through D, unless B has been reached." Many of the approaches in this area implement a two layer architecture: a discrete planner generates a high-level plan that the robot has to fulfill ("the robot should move from B to D"), whereas a low-level continuous planner needs to plan a path that executes the high-level plan. Earlier works integrated potential-field methods for the low-level plans (see, e.g., [KB08, FGKP09]). More recent efforts rely on sampling-based techniques for the continuous planning (see, e.g., [BKV10, BMKV11, KF12, VB13]). A recent work [LAF+15] explores a setting in which every satisfying assignment is associated with a value and the goal is to find the "best" satisfying plan. Besides using logics, the combination of Satisfiability Modulo Theories (SMT) solvers and sampling-based planners for problems involving tasks that include manipulation is presented in [DKCK16].

Task planning has also earned some attention in the context of multi-robot systems [KDB11, WUB+13, USD+13, IS15] and manipulation [SK11, SFR+14].

51.4.4 MOTION PLANNING WITH CONSTRAINTS

In robotics, there is a need to satisfy both global constraints and local constraints. One could consider the avoidance of obstacles as a global constraint. Often local constraints are modeled with differential equations and are called differential constraints. Some of the first such constraints considered in the robotics community were nonholonomic constraints, which raise very interesting geometric problems. They are discussed in this section. However planning under constraints includes planning with velocity and acceleration considerations (the problem is often called kinodynamic planning) and other variations that are also discussed below.

PLANNING FOR NONHOLONOMIC ROBOTS

The trajectories of a nonholonomic robot are constrained by $p \geq 1$ nonintegrable scalar equality constraints:

$$G(\boldsymbol{q}(t), \dot{\boldsymbol{q}}(t)) = (G^1(\boldsymbol{q}(t), \dot{\boldsymbol{q}}(t)), \cdots, G^p(\boldsymbol{q}(t), \dot{\boldsymbol{q}}(t))) = (0, \ldots, 0),$$

where $\dot{\boldsymbol{q}}(t) \in T_{\boldsymbol{q}(t)}(\mathcal{C})$ designates the velocity vector along the trajectory $\boldsymbol{q}(t)$. At every $\boldsymbol{q}$, the function $G_{\boldsymbol{q}} = G(\boldsymbol{q}, .)$ maps the tangent space $T_{\boldsymbol{q}}(\mathcal{C})$ into $\mathbb{R}^p$. If $G_{\boldsymbol{q}}$

---

[1]See http://www.kavrakilab.org/2016-rss-workshop/

is smooth and its Jacobian has full rank (two conditions that are often satisfied), the constraint $G_q(\dot{q}) = (0, \ldots, 0)$ constrains $\dot{q}$ to be in a linear subspace of $T_q(\mathcal{C})$ of dimension $m - p$. The nonholonomic robot may also be subject to scalar inequality constraints of the form $H^j(q, \dot{q}) > 0$. The subset of $T_q(\mathcal{C})$ that satisfies all the constraints on $\dot{q}$ is called the set $\Omega(q)$ of controls at q. A feasible path is a piecewise differentiable path whose tangent lies everywhere in the control set.

A car-like robot is a classical example of a nonholonomic robot. It is constrained by one equality constraint (the linear velocity points along the car's main axis). Limits on the steering angle impose two inequality constraints. Other nonholonomic robots include tractor-trailers, airplanes, and satellites.

Given an arbitrary subset $U \subset \mathcal{C}$, the configuration $q_1 \in U$ is said to be U-*accessible* from $q_0 \in U$ if there exists a piecewise constant control $\dot{q}(t)$ in the control set whose integral is a trajectory joining q_0 to q_1 that lies fully in U. Let $A_U(q_0)$ be the set of configurations U-accessible from q_0. The robot is said to be **locally controllable** at q_0 iff for every neighborhood U of q_0, $A_U(q_0)$ is also a neighborhood of q_0. It is locally controllable iff this is true for all $q_0 \in \mathcal{C}$. Car-like robots and tractor-trailers that can go forward and backward are locally controllable [BL93].

Let X and Y be two smooth vector fields on $\mathcal{C}$. The Lie bracket of X and Y, denoted by $[X, Y]$, is the smooth vector field on $\mathcal{C}$ defined by $[X, Y] = dY \cdot X - dX \cdot Y$, where dX and dY, respectively, denote the $m \times m$ matrices of the partial derivatives of the components of X and Y w.r.t. the configuration coordinates in a chart placed on $\mathcal{C}$. To get a better intuition of the Lie bracket, imagine a trajectory starting at an arbitrary configuration q_s and obtained by concatenating four subtrajectories: the first is the integral curve of X during time δt; the second, third, and fourth are the integral curves of Y, $-X$, and $-Y$, respectively, each during the same δt. Let q_f be the final configuration reached. A Taylor expansion yields:

$$\lim_{\delta t \to 0} \frac{q_f - q_s}{\delta t^2} = [X, Y].$$

Hence, if $[X, Y]$ is not a linear combination of X and Y, the above trajectory allows the robot to move away from q_s in a direction that is not contained in the vector subspace defined by $X(q_s)$ and $Y(q_s)$. But the motion along this new direction is an order of magnitude slower than along any direction $\alpha X(q_s) + \beta Y(q_s)$.

The **control Lie algebra** associated with the control set Ω, denoted by $L(\Omega)$, is the space of all linear combinations of vector fields in Ω closed by the Lie bracket operation. The following result derives from the Controllability Rank Condition Theorem [BL93]:

> *A robot is locally controllable if, for every $q \in \mathcal{C}$, $\Omega(q)$ is symmetric with respect to the origin of $T_q(\mathcal{C})$ and the set $\{X(q) \mid X \in L(\Omega(q))\}$ has dimension m.*

The minimal length of the Lie brackets required to construct $L(\Omega)$, when these brackets are expressed with vectors in Ω, is called the **degree of nonholonomy** of the robot. The degree of nonholonomy of a car-like robot is 2. Except at some singular configurations, the degree of nonholonomy of a tractor towing a chain of s trailers is $2 + s$ [LR96]. Intuitively, the higher the degree of nonholonomy, the more complex (and the slower) the robot's maneuvers to perform some motions.

Nonholonomic motion planning was also studied in a purely-geometric setting, where one is interested in finding a path of *bounded curvature* (see e.g., [KKP11, KC13]). The reader is referred to Chapter 50 for further information.

PLANNING FOR CONTROLLABLE NONHOLONOMIC ROBOTS

Let A be a locally controllable nonholonomic robot. A necessary and sufficient condition for the existence of a feasible free path of A between two given configurations is that they lie in the same connected component of the *open* free space. Indeed, local controllability guarantees that a possibly nonfeasible path can be decomposed into a finite number of subpaths, each short enough to be replaced by a feasible free subpath. Hence, deciding if there exists a free path for a locally controllable nonholonomic robot has the same complexity as deciding if there exists a path for the holonomic robot having the same geometry.

Transforming a nonfeasible free path τ into a feasible one can be done by recursively decomposing τ into subpaths. The recursion halts at every subpath that can be replaced by a feasible free subpath. Specific substitution rules (e.g., Reeds and Shepp curves) have been defined for car-like robots [LJTM94]. The complexity of transforming a nonfeasible free path τ into a feasible one is of the form $O(\epsilon^d)$, where ϵ is the smallest clearance between the robot and the obstacles along τ and d is the degree of nonholonomy of the robot (see [LJTM94] for the case $d = 2$).

The algorithm in [BL93] directly constructs a nonholonomic path for a car-like or a tractor-trailer robot by searching a tree obtained by concatenating short feasible paths, starting at the robot's initial configuration. The planner is ***asymptotically complete***, i.e., it is guaranteed to find a path if one exists, provided that the lengths of the short feasible paths are small enough. It can also find paths that minimize the number of cusps (changes of sign of the linear velocity).

PLANNING FOR NONCONTROLLABLE ROBOTS

Motion planning for nonholonomic robots that are not locally controllable is much less understood. Research has almost exclusively focused on car-like robots that can only move forward. Results include:

- No obstacles: A complete synthesis of the shortest, no-cusp path for a point moving with a lower-bounded turning radius [BSBL94].

- Polygonal obstacles: An algorithm to decide whether there exists such a path between two configurations; it runs in time exponential in obstacle complexity [FW91].

- Convex obstacles: The algorithm in [ART95] computes a path in polynomial time under the assumptions that all obstacles are convex and their boundaries have a curvature radius greater than the minimum turning radius of the point.

- Other polynomial algorithms (e.g., [BL93]) require some sort of discretization and are only asymptotically complete.

PLANNING UNDER DIFFERENTIAL CONSTRAINTS

If a planning problem involves constraints on at least velocity and acceleration the problem is often referred to as kinodynamic planning. Typically nonholonomic constraints are kinematic and arise from wheels in contact. But nonholonomic constraints may arise from dynamics. Trajectory planning refers to the problem of planning for both a path and a velocity function. Recently, the term planning under differential constraints is used as a catch-all term that includes nonholonomic, kinodynamic and sometimes trajectory planning. Planning under differential constraints may involve discretization of constraints and decoupled approaches where some constraints are ignored at the beginning but are then gradually introduced (see [KL16]).

Sampling-based planners were initially developed for the "geometric," i.e., holonomic, setting of the problem, but it was not long before similar techniques for planning under differential constraints started to emerge. However, even in the randomized setting of sampling-based planners, planning under differential constraints remains to be very challenging and the design of such planners usually involves much more than straightforward extensions of existing geometric planners.

Sampling-based planners for problems with differential constraints operate in the *state space* $\mathcal{X}$. Typically, every *state* represents a pairing of a configuration and a velocity vector. An inherent difficulty of the problem involves the connection of two given states in the absence of obstacles—an operation called *steering*, or the *two-point boundary-value problem* (BVP).

For robotic systems in which steering can be quickly performed (see, e.g., [ŠO97, LSL99, KF10, WB13, PPK+12, XBPA15]) holonomic sampling-based planners, or their extensions, can be applied.

Sampling-based algorithms that build trees avoid steering. They simply attempt to move in the direction of sampled states and do not require to precisely reach them (see, e.g., [HKLR02, LK01c, PKV10, SK12, LLB14] and many others). As one of many examples, the planner described in [LK04a] samples robot actions, which consist of a velocity vector and time duration, in order to avoid steering. The idea of avoiding steering has been proven very powerful in the context of sampling-based planners and has led to the solution of complex problems with differential constraints.

Finally, a number of related works exist that augment sampling-based algorithms for instances of whole-body manipulation for humanoids [KKN+02, HBL+08, YPL+10, BSK11, BHB13, Hau14]. The primary concern in these works is the generation of dynamically-stable motions for high-dimensional bipedal robots.

PLANNING UNDER OTHER TYPES OF CONSTRAINTS

Some robotic systems are subjected to constraints for particular tasks, restricting the set of valid configurations to a manifold, which has measure zero with respect to the ambient space. Naturally, sampling in the ambient space can no longer effectively produce valid configurations. Some approaches target specific kinds of constraint problems, such as open- or closed-loop kinematic linkage constraints [TTCA10, CS05], end-effector pose constraints [BS10], or n-point contact constraints [Hau14]. Some approaches project invalid samples through gradient descent [YLK01] or by more robust means [KBV+12]. Recent successful algorithms for planning under abstract constraints are CBiRRT2 [BSK11]—which like RRT steps toward sampled points, but after every step it projects the current point back to the manifold—as well as AtlasRRT [JP13] and [KTSP16]. AtlasRRT constructs a piecewise-linear approximation of the manifold by computing tangent spaces. This approximation allows the planner to reason about a difficult, implicitly-defined space by projecting regions onto simpler spaces.

51.5   SOFTWARE FOR SAMPLING-BASED PLANNERS

Robot motion planning requires the integration of numerous software components, and several existing packages address this goal. The Open Motion Planning Library (OMPL) [SMK12] includes abstract implementations of many planning algorithms discussed in this chapter. The robot modeling—loading 3D meshes, kinematics, visualization, etc.—necessary to apply OMPL is handled by external packages such as MoveIt! [SC15], which integrates with the ROS framework [QGC+09], and Amino [Dan16] which provides capabilities to support task planning and real-time control.

Software packages performing nearest-neighbor search include E2LSH [AI06], ANN [MA10], FLANN [ML10] and RTG [KSH15]. Collision checking is typically handled by specialized packages such as the Flexible Collision Library [PCM12] and libccd [Fis16]. The Orocos Kinematics and Dynamics Library (KDL) [Smi] is widely used for basic kinematics and dynamics computations. Several dedicated packages provide more advanced capabilities for dynamic simulation [S+05, LGS+16, C+16, SSD11]. Other all-in-one robotics packages include OpenRAVE [Dia10] and Klamp't [Hau16].

51.6   APPLICATIONS BEYOND ROBOTICS

Although the focus of this chapter was restricted to applications of motion planning in robotics, we mention that the problem naturally arises in many other settings [Lat99]. For instance, in computer games[2] motion planning is used for determining the motion of single or multiple agents. In structural bioinformatics motion-planning tools are employed in order to study the motion of molecules such a proteins, which can be modeled as highly-complex robots. This is one application domain where sampling-based tools enable motion planning with hundreds of degrees of freedom [LK01a, RESH09, GMK13, ETA15, ADS03, ABG+03]. In Computer-Aided-Design (CAD) motion planning has been used to find intricate separation paths to take out parts for maintenance out of a complex and cluttered environment [CL95, FL04].

51.7   SOURCES AND RELATED MATERIAL

Craig's book [Cra05] provides an introduction to robot arm kinematics, dynamics, and control. The mechanics of robotic manipulation is covered in Mason's book [Mas01]. The book by Siciliano and Khatib [SK16] provides a comprehensive overview of the field of robotics.

Robot motion planning and its variants are discussed in Latombe's book [Lat91]. This book takes an algorithmic approach to a variety of advanced issues in robotics. The books by LaValle [LaV06] and Choset et al. [CLS+05] provide a more current overview of motion planning, and also include detailed description

---

[2]See the *Motion in Games* conference: http://www.motioningames.org/.

of several sampling-based approaches.

Algorithmic issues in robotics are covered in several conferences including the Workshop on Algorithmic Foundations of Robotics (WAFR), Robotics: Science and Systems (RSS), and the broader IEEE International Conference on Robotics and Automation (ICRA) and the IEEE/RSJ International Conference on Intelligent Robots and Systems (IROS).

Several computational-geometry books contain sections on robotics or motion planning [O'R98, SA95, BKOC08]. The book by Fogel et al. [FHW12] includes implementation details for basic geometric constructions useful in motion planning, such as Minkowski sums.

RELATED CHAPTERS

Chapter 28: Arrangements
Chapter 31: Shortest paths and networks
Chapter 33: Visibility
Chapter 34: Geometric reconstruction problems
Chapter 37: Computational and quantitative real algebraic geometry
Chapter 39: Collision and proximity queries
Chapter 50: Algorithmic motion planning
Chapter 53: Modeling motion
Chapter 60: Geometric applications of the Grassmann-Cayley algebra
Chapter 68: Two computational geometry libraries: LEDA and CGAL

REFERENCES

[ABG+03]   M.S. Apaydin, D.L. Brutlag, C. Guestrin, D. Hsu, J.-C. Latombe, and C. Varma. Stochastic roadmap simulation: An efficient representation and algorithm for analyzing molecular motion. *J. Comput. Biol.*, 10:257–281, 2003.

[ABHS15]   A. Adler, M. de Berg, D. Halperin, and K. Solovey. Efficient multi-robot motion planning for unlabeled discs in simple polygons. *IEEE Trans. Autom. Sci. Eng.*, 12:1309–1317, 2015.

[ABR12]   T.F. Allen, J.W. Burdick, and E. Rimon. Two-fingered caging of polygons via contact-space graph search. In *Proc. IEEE Int. Conf. Robot. Autom.*, pages 4183–4189, 2012.

[ACH01]   P.K. Agarwal, A.D. Collins, and J.L. Harer. Minimal trap design. In *Proc. IEEE Int. Conf. Robot. Autom.*, pages 2243–2248, 2001.

[ADS03]   N.M. Amato, K.A. Dill, and G. Song. Using motion planning to map protein folding landscapes and analyze folding kinetics of known native structures. *J. Comput. Biol.*, 10:239–255, 2003.

[AHLM00]   S. Akella, W.H. Huang, K.M. Lynch, and M.T. Mason. Parts feeding on a conveyor with a one joint robot. *Algorithmica*, 26:313–344, 2000.

[AI06]   A. Andoni and P. Indyk. LSH algorithm and implementation. **http://web.mit.edu/andoni/www/LSH/**, 2006.

[ALS95]   R. Alami, J.-P. Laumond, and T. Siméon. Two manipulation planning algorithms. In *Algorithmic Foundations of Robotics*, pages 109–125, A.K. Peters, Natick, 1995.

[ARB14]   T.F. Allen, E. Rimon, and J.W. Burdick. Two-finger caging of 3D polyhedra using contact space search. In *Proc. IEEE Int. Conf. Robot. Autom.*, pages 2005–2012, 2014.

[ARB15]   T.F. Allen, E. Rimon, and J.W. Burdick. Robust three-finger three-parameter caging of convex polygons. In *Proc. IEEE Int. Conf. Robot. Autom.*, pages 4318–4325, 2015.

[ART95]   P.K. Agarwal, P. Raghavan, and H. Tamaki. Motion planning for a steering-constrained robot through moderate obstacles. In *27th ACM Sympos. Theory Comput.*, pages 343–352, 1995.

[ASN$^+$16]   S. Aine, S. Swaminathan, V. Narayanan, V. Hwang, and M. Likhachev. Multi-heuristic A*. *I. J. Robotics Res.*, 35:224–243, 2016.

[AT13]   O. Arslan and P. Tsiotras. Use of relaxation methods in sampling-based algorithms for optimal motion planning. In *Proc. IEEE Int. Conf. Robot. Autom.*, pages 2421–2428, 2013.

[BB15]   D. Bareiss and J.P. van den Berg. Generalized reciprocal collision avoidance. *I. J. Robotics Res.*, 34:1501–1514, 2015.

[BBDG00]   K.-F. Böhringer, V. Bhatt, B.R. Donald, and K.Y. Goldberg. Algorithms for sensorless manipulation using a vibrating surface. *Algorithmica*, 26:389–429, 2000.

[BBT02]   M. Bennewitz, W. Burgard, and S. Thrun. Finding and optimizing solvable priority schemes for decoupled path planning techniques for teams of mobile robots. *Robotics Autonomous Systems*, 41:89–99, 2002.

[BDH99]   K.-F. Böhringer, B.R. Donald, and D. Halperin. On the area bisectors of a polygon. *Discrete Comput. Geom.*, 22:269–285, 1999.

[BDKL00]   K.-F. Böhringer, B.R. Donald, L.E. Kavraki, and F. Lamiraux. Part orientation with one or two stable equilibria using programmable force fields. *IEEE Trans. Robot. Autom.*, 16:157–170, 2000.

[BDM99]   K.F. Böhringer, B.R. Donald, and N.C. MacDonald. Programmable force fields for distributed manipulation, with applications to MEMS actuator arrays and vibratory parts feeders. *I. J. Robotics Res.*, 18:168–200, 1999.

[BG96]   R.C. Brost and K.Y. Goldberg. A complete algorithm for designing planar fixtures using modular components. *IEEE Trans. Robot. Autom.*, 12:31–46, 1996.

[BG13]   M. de Berg and D.H.P. Gerrits. Computing push plans for disk-shaped robots. *Internat. J. Comput. Geom. Appl.*, 23:29–48, 2013.

[BGHN01]   J. Basch, L.J. Guibas, D. Hsu, and A.T. Nguyen. Disconnection proofs for motion planning. In *Proc. IEEE Int. Conf. Robot. Autom.*, pages 1765–1772, 2001.

[BGMK12]   K.E. Bekris, D.K. Grady, M. Moll, and L.E. Kavraki. Safe distributed motion coordination for second-order systems with different planning cycles. *I. J. Robotics Res.*, 31:129–150, 2012.

[BGOS98]   R.-P. Berretty, K.Y. Goldberg, M.H. Overmars, and A.F. van der Stappen. Computing fence designs for orienting parts. *Comput. Geom.*, 10:249–262, 1998.

[BGOS01]   R.-P. Berretty, K.Y. Goldberg, M.H. Overmars, and A.F. van der Stappen. Trap design for vibratory bowl feeders. *I. J. Robotics Res.*, 20:891–908, 2001.

[BH95]   M. Barbehenn and S. Hutchinson. Efficient search and hierarchical motion planning by dynamically maintaining single-source shortest paths trees. *IEEE Trans. Robot. Autom.*, 11:198–214, 1995.

[BHB13]   F. Burget, A. Hornung, and M. Bennewitz. Whole-body motion planning for manipulation of articulated objects. In *Proc. IEEE Int. Conf. Robot. Autom.*, pages 1656–1662, 2013.

[BK00]   R. Bohlin and L.E. Kavraki. Path planning using lazy PRM. In *Proc. IEEE Int. Conf. Robot. Autom.*, pages 521–528, 2000.

[BKOC08]   M. de Berg, M. van Kreveld, M. Overmars, and O. Cheong. *Computational Geometry: Algorithms and Applications*, 3rd edition. Springer-Verlag, Berlin, 2008.

[BKV10]   A. Bhatia, L.E. Kavraki, and M.Y. Vardi. Motion planning with hybrid dynamics and temporal goals. In *Proc. IEEE Conf. Decision and Control*, pages 1108–1115, 2010.

[BL93]   J. Barraquand and J.-C. Latombe. Nonholonomic multibody mobile robots: Controllability and motion planning in the presence of obstacles. *Algorithmica*, 10:121–155, 1993.

[BMKV11]   A. Bhatia, M.R. Maly, L.E. Kavraki, and M.Y. Vardi. Motion planning with complex goals. *IEEE Robot. Autom. Mag.*, 18:55–64, 2011.

[BO05]   J.P. van den Berg and M.H. Overmars. Prioritized motion planning for multiple robots. In *Proc. IEEE/RSJ Int. Conf. Intelligent Robots and Systems*, pages 430–435, 2005.

[BOS99]   V. Boor, M.H. Overmars, and A.F. van der Stappen. The Gaussian sampling strategy for probabilistic roadmap planners. In *Proc. IEEE Int. Conf. Robot. Autom.*, pages 1018–1023, 1999.

[BS10]   D. Berenson and S. Srinivasa. Probabilistically complete planning with end-effector pose constraints. In *Proc. IEEE Int. Conf. Robot. Autom.*, pages 2724–2730, 2010.

[BSBL94]   X.N. Bui, P. Souères, J.-D. Boissonnat, and J.-P. Laumond. Shortest path synthesis for dubins non-holonomic robot. In *Proc. IEEE Int. Conf. Robot. Autom.*, pages 2–7, 1994.

[BSK+08]   J. van den Berg, M. Stilman, J. Kuffner, M.C. Lin, and D. Manocha. Path planning among movable obstacles: A probabilistically complete approach. In *Algorithmic Foundations of Robotics VIII*, pages 599–614, Springer, Berlin, 2008.

[BSK11]   D. Berenson, S. Srinivasa, and J.J. Kuffner. Task space regions: A framework for pose-constrained manipulation planning. *I. J. Robotics Res.*, 30:1435–1460, 2011.

[BSLM09]   J.P. van den Berg, J. Snoeyink, M.C. Lin, and D. Manocha. Centralized path planning for multiple robots: Optimal decoupling into sequential plans. In *Robotics: Science and Systems*, 2009.

[C+16]   E. Coumans et al. Bullet physics library. `http://bulletphysics.org/`, 2016.

[CI95]   Y.B. Chen and D.J. Ierardi. The complexity of oblivious plans for orienting and distinguishing polygonal parts. *Algorithmica*, 14:367–397, 1995.

[CJS08]   J. Cortés, L. Jaillet, and T. Siméon. Disassembly path planning for complex articulated objects. *IEEE Trans. Robot.*, 24:475–481, 2008.

[CL95]   H. Chang and T.-Y. Li. Assembly maintainability study with motion planning. In *Proc. IEEE Int. Conf. Robot. Autom.*, pages 1012–1019, 1995.

[CLS+05]   H. Choset, K.M. Lynch, S.Hutchinson, G.A. Kantor, W. Burgard, L.E. Kavraki, and S. Thrun. *Principles of Robot Motion: Theory, Algorithms, and Implementations*. MIT Press, Cambridge, 2005.

[Cra05]   J.J. Craig. *Introduction to Robotics: Mechanics and Control*, 3rd edition. Pearson Prentice Hall, Upper Saddle River, 2005.

[CS05]   J. Cortés and T. Siméon. Sampling-based motion planning under kinematic loop-closure constraints. In *Algorithmic Foundations of Robotics VI*, pages 75–90, Springer, Berlin, 2005.

[CSG+07]   J.-S. Cheong, A.F. van der Stappen, K.Y. Goldberg, M.H. Overmars, and E. Rimon. Immobilizing hinged polygons. *Internat. J. Comput. Geom. Appl.*, 17:45–70, 2007.

[Dan16]   N.T. Dantam. Amino. http://amino.golems.org, 2016.

[DB98]   C. Davidson and A. Blake. Caging planar objects with a three-finger one-parameter gripper. In *Proc. IEEE Int. Conf. Robot. Autom.*, vol. 3, pages 2722–2727, 1998.

[DB14]   A. Dobson and K.E. Bekris. Sparse roadmap spanners for asymptotically near-optimal motion planning. *I. J. Robotics Res.*, 33:18–47, 2014.

[Dia10]   R. Diankov. *Automated Construction of Robotic Manipulation Programs*. PhD thesis, Carnegie Mellon University, Robotics Institute, August 2010.

[DKCK16]   N.T. Dantam, Z.K. Kingston, S. Chaudhuri, and L.E. Kavraki. Incremental task and motion planning: A constraint-based approach. In *Robotics: Science and Systems*, 2016.

[DMB15]   A. Dobson, G.V. Moustakides, and K.E. Bekris. Geometric probability results for bounding path quality in sampling-based roadmaps after finite computation. In *Proc. IEEE Int. Conf. Robot. Autom.*, pages 4180–4186, 2015.

[Don87]   B.R. Donald. A search algorithm for motion planning with six degrees of freedom. *Artif. Intell.*, 31(3):295–353, 1987.

[DRP+14]   N.C. Dafle, A. Rodriguez, R. Paolini, B. Tang, S.S. Srinivasa, M. Erdmann, M.T. Mason, I. Lundberg, H. Staab, and T. Fuhlbrigge. Extrinsic dexterity: In-hand manipulation with external forces. In *Proc. IEEE Int. Conf. Robot. Autom.*, pages 1578–1585, 2014.

[EM88]   M.A. Erdmann and M.T. Mason. An exploration of sensorless manipulation. *IEEE J. Robot. Autom.*, 4:369–379, 1988.

[EMJ93]   M.A. Erdmann, M.T. Mason, and G. Vaněček Jr. Mechanical parts orienting: The case of a polyhedron on a table. *Algorithmica*, 10:226–247, 1993.

[Epp90]   D. Eppstein. Reset sequences for monotonic automata. *SIAM J. Comput.*, 19:500–510, 1990.

[ETA15]   C. Ekenna, S. Thomas, and N.M. Amato. Adaptive local learning in sampling based motion planning for protein folding. In *Proc. IEEE Intl. Conf. Bioinf. Biomed.*, pages 61–68, 2015.

[ETRP07]   J. Erickson, S. Thite, F. Rothganger, and J. Ponce. Capturing a convex object with three discs. *IEEE Trans. Robot.*, 23:1133–1140, 2007.

[FC92]   C. Ferrari and J. Canny. Planning optimal grasps. In *Proc. IEEE Int. Conf. Robot. Autom.*, vol. 3, pages 2290–2295, 1992.

[FGKP09]   G.E. Fainekos, A. Girard, H. Kress-Gazit, and G.J. Pappas. Temporal logic motion planning for dynamic robots. *Automatica*, 45:343–352, 2009.

[FH13]   E. Fogel and D. Halperin. Polyhedral assembly partitioning with infinite translations or the importance of being exact. *IEEE Trans. Autom. Sci. Eng.*, 10:227–241, 2013.

[FHK08]   J. Fink, M.A. Hsieh, and V. Kumar. Multi-robot manipulation via caging in environments with obstacles. In *Proc. IEEE Int. Conf. Robot. Autom.*, pages 1471–1476, 2008.

[FHW12]   E. Fogel, D. Halperin, and R. Wein. *CGAL Arrangements and Their Applications: A Step-by-Step Guide*. Vol 7 of *Geometry and Computing*, Springer, Berlin, 2012.

[Fis16]   D. Fiser. `libccd`. `http://libccd.danfis.cz/`, 2016.

[FL04]   E. Ferre and J.-P. Laumond. An iterative diffusion algorithm for part disassembly. In *Proc. IEEE Int. Conf. Robot. Autom.*, pages 3149–3154, 2004.

[FW91]   S. Fortune and G. Wilfong. Planning constrained motion. *Ann. Math. Artif. Intell.*, 3:21–82, 1991.

[GG05]   K. Gopalakrishnan and K.Y. Goldberg. D-space and deform closure grasps of deformable parts. *I. J. Robotics Res.*, 24:899–910, 2005.

[GGB+04]   K. Gopalakrishnan, K.Y. Goldberg, G.M. Bone, M. Zaluzec, R. Koganti, R. Pearson, and P. Deneszczuk. Unilateral fixtures for sheet-metal parts with holes. *IEEE Trans. Autom. Sci. Eng.*, 1:110–120, 2004.

[GGGDC12]   J. Guzzi, A. Giusti, L.M. Gambardella, and G.A. Di Caro. Bioinspired obstacle avoidance algorithms for robot swarms. In *Bio-Inspired Models of Network, Information, and Computing Systems*, pages 120–134, Springer, Berlin, 2012.

[GGS06]   O.C. Goemans, K.Y. Goldberg, and A.F. van der Stappen. Blades: A new class of geometric primitives for feeding 3D parts on vibratory tracks. In *Proc. IEEE Int. Conf. Robot. Autom.*, pages 1730–1736, 2006.

[GH10]   G. Goraly and R. Hassin. Multi-color pebble motion on graphs. *Algorithmica*, 58:610–636, 2010.

[GHH+98]   L.J. Guibas, D. Halperin, H. Hirukawa, J.-C. Latombe, and R.H. Wilson. Polyhedral assembly partitioning using maximally covered cells in arrangements of convex polytopes. *Internat. J. Comput. Geom. Appl.*, 8:179–200, 1998.

[GMK13]   B. Gipson, M. Moll, and L.E. Kavraki. SIMS: A hybrid method for rapid conformational analysis. *PLoS ONE*, 8:e68826, 2013.

[GO07]   R. Geraerts and M.H. Overmars. Creating high-quality paths for motion planning. *I. J. Robotics Res.*, 26:845–863, 2007.

[Gol93]   K.Y. Goldberg. Orienting polygonal parts without sensors. *Algorithmica*, 10:210–225, 1993.

[Hau14]   K. Hauser. Fast interpolation and time-optimization with contact. *I. J. Robotics Res.*, 33:1231–1250, 2014.

[Hau15]   K. Hauser. Lazy collision checking in asymptotically-optimal motion planning. In *proc. IEEE Int. Conf. Robot. Autom.*, pages 2951–2957, 2015.

[Hau16]   K. Hauser. Robust contact generation for robot simulation with unstructured meshes. In *Robotics Research*, pages 357–373, Springer, Berlin, 2016. `http://motion.pratt.duke.edu/klampt/`.

[HBL+08]   K. Hauser, T. Bretl, J.-C. Latombe, K. Harada, and B. Wilcox. Motion planning for legged robots on varied terrain. *I. J. Robotics Res.*, 27:1325–1349, 2008.

[HD05]   R.A. Hearn and E.D. Demaine. PSPACE-completeness of sliding-block puzzles and other problems through the nondeterministic constraint logic model of computation. *Theor. Comput. Sci.*, 343:72–96, 2005.

[HdMS91]   L.S. Homem de Mello and A.C. Sanderson. A correct and complete algorithm for the generation of mechanical assembly sequences. *IEEE Trans. Robot. Autom.*, 7:228–240, 1991.

[HGL+97]   L. Han, Y.S. Guan, Z.X. Li, Q. Shi, and J.C. Trinkle. Dextrous manipulation with rolling contacts. In *Proc. IEEE Int. Conf. Robot. Autom.*, vol. 2, pages 992–997, 1997.

[HH02]   S. Hirsch and D. Halperin. Hybrid motion planning: Coordinating two discs moving among polygonal obstacles in the plane. In *Algorithmic Foundations of Robotics V*, pages 239–256, Springer, Berlin, 2002.

[HKLR02]   D. Hsu, R. Kindel, J.-C. Latombe, and S.M. Rock. Randomized kinodynamic motion planning with moving obstacles. *I. J. Robotics Res.*, 21:233–256, 2002.

[HLM99]   D. Hsu, J.-C. Latombe, and R. Motwani. Path planning in expansive configuration spaces. *Internat. J. Comput. Geom. Appl.*, 9:495–512, 1999.

[HLW00]   D. Halperin, J.-C. Latombe, and R.H. Wilson. A general framework for assembly planning: The motion space approach. *Algorithmica*, 26:577–601, 2000.

[HN01]   J. Hoffmann and B. Nebel. The FF planning system: Fast plan generation through heuristic search. *J. Artif. Int. Res.*, 14:253–302, 2001.

[HW96]   D. Halperin and R.H. Wilson. Assembly partitioning along simple paths: The case of multiple translations. *Advanced Robotics*, 11:127–145, 1996.

[IS15]   F. Imeson and S.L. Smith. Multi-robot task planning and sequencing using the SAT-TSP language. In *Proc. IEEE Int. Conf. Robot. Autom.*, pages 5397–5402, 2015.

[Jim13]   P. Jiménez. Survey on assembly sequencing: A combinatorial and geometrical perspective. *J. Intell. Manufacturing*, 24:235–250, 2013.

[JP13]   L. Jaillet and J.M. Porta. Path planning under kinematic constraints by rapidly exploring manifolds. *IEEE Trans. Robot.*, 29:105–117, 2013.

[JSCP15]   L. Janson, E. Schmerling, A.A. Clark, and M. Pavone. Fast marching tree: A fast marching sampling-based method for optimal motion planning in many dimensions. *I. J. Robotics Res.*, 34:883–921, 2015.

[Kav97]   L.E. Kavraki. Part orientation with programmable vector fields: Two stable equilibria for most parts. In *Proc. IEEE Int. Conf. Robot. Autom.*, pages 2446–2451, 1997.

[KB08]   M. Kloetzer and C. Belta. A fully automated framework for control of linear systems from temporal logic specifications. *IEEE Trans. Autom. Control*, 53:287–297, 2008.

[KB15]   A. Krontiris and K.E. Bekris. Dealing with difficult instances of object rearrangement. In *Proc. Robotics: Science and Systems*, 2015.

[KBV+12]   P. Kaiser, D. Berenson, N. Vahrenkamp, T. Asfour, R. Dillmann, and S. Srinivasa. Constellation – An algorithm for finding robot configurations that satisfy multiple constraints. In *Proc. IEEE Int. Conf. Robot. Autom.*, pages 436–443, 2012.

[KC13]   H.-S. Kim and O. Cheong. The cost of bounded curvature. *Comput. Geom.*, 46:648–672, 2013.

[KDB11]   M. Kloetzer, X.C. Ding, and C. Belta. Multi-robot deployment from LTL specifications with reduced communication. In *Proc. IEEE Conf. Decision and Control*, pages 4867–4872, 2011.

[KEK10]   G.A. Kaminka, D. Erusalimchik, and S. Kraus. Adaptive multi-robot coordination: A game-theoretic perspective. In *Proc. IEEE Int. Conf. Robot. Autom.*, pages 328–334, 2010.

[KF10]   S. Karaman and E. Frazzoli. Optimal kinodynamic motion planning using incremental sampling-based methods. In *Proc. IEEE Conf. Decision and Control*, pages 7681–7687, 2010.

[KF11]   S. Karaman and E. Frazzoli. Sampling-based algorithms for optimal motion planning. *I. J. Robotics Res.*, 30:846–894, 2011.

[KF12]   S. Karaman and E. Frazzoli. Sampling-based algorithms for optimal motion planning with deterministic μ-calculus specifications. In *Proc. American Control Conference*, pages 735–742, 2012.

[KH06]   S. Kloder and S. Hutchinson. Path planning for permutation-invariant multirobot formations. *IEEE Trans. Robot.*, 22:650–665, 2006.

[Kha86]   O. Khatib. Real-time obstacle avoidance for manipulators and mobile robots. *I. J. Robotics Res.*, 5:90–98, 1986.

[KK95]   L.E. Kavraki and M.N. Kolountzakis. Partitioning a planar assembly into two connected parts is NP-complete. *Inform. Process. Lett.*, 55:159–165, 1995.

[KKL98]   L.E. Kavraki, M.N. Kolountzakis, and J.-C. Latombe. Analysis of probabilistic roadmaps for path planning. *IEEE Trans. Robot. Autom.*, 14:166–171, 1998.

[KKN$^+$02] J.J. Kuffner, S. Kagami, K. Nishiwaki, M. Inaba, and H. Inoue. Dynamically-stable motion planning for humanoid robots. *Autonomous Robots*, 12:105–118, 2002.

[KKP11]   D. Kirkpatrick, I. Kostitsyna, and V. Polishchuk. Hardness results for two-dimensional curvature-constrained motion planning. In *Proc. 23rd Canad. Conf. Comput. Geom.*, Toronto, 2011.

[KL16]   L.E. Kavraki and S.M. LaValle. Motion planning. Chap. 7 in B. Siciliano and O. Khatib, editors, *Springer Handbook of Robotics*, 2nd edition, pages 139–158, Springer, Berlin, 2016.

[KLB13]   A. Krontiris, R. Luna, and K.E. Bekris. From feasibility tests to path planners for multi-agent pathfinding. In *Proc. 6th AAAI Int. Sympos. Combin. Search*, 2013.

[KLMR98]   L.E. Kavraki, J.-C. Latombe, R. Motwani, and P. Raghavan. Randomized query processing in robot path planning. *J. Comput. Syst. Sci.*, 57:50–66, 1998.

[KMS84]   D. Kornhauser, G. Miller, and P. Spirakis. Coordinating pebble motion on graphs, the diameter of permutation groups, and applications. In *Proc. 54th IEEE Sympos. Found. Comp. Sci.*, pages 241–250, 1984.

[KMY92]   D. Kirkpatrick, B. Mishra, and C.K. Yap. Quantitative Steinitz's theorems applications to multifingered grasping. *Discrete Comput. Geom.*, 7:295–318, 1992.

[Kob12]   M. Kobilarov. Cross-entropy motion planning. *I. J. Robotics Res.*, 31:855–871, 2012.

[KR12]   R.A. Knepper and D. Rus. Pedestrian-inspired sampling-based multi-robot collision avoidance. In *Proc. IEEE Int. Sympos. Robot Human Interactive Comm.*, pages 94–100, 2012.

[KS92]   H.A. Kautz and B. Selman. Planning as satisfiability. *proc. 10th European Conf. Artif. Intell.*, pages 359–363, Wiley, New York, 1992.

[KSD$^+$14] A. Krontiris, R. Shome, A. Dobson, A. Kimmel, and K.E. Bekris. Rearranging similar objects with a manipulator using pebble graphs. In *Proc. IEEE/RAS Int. Conf. Humanoid Robots*, pages 1081–1087, 2014.

[KSH15]   M. Kleinbort, O. Salzman, and D. Halperin. Efficient high-quality motion planning by fast all-pairs r-nearest-neighbors. In *IEEE Int. Conf. Robot. Autom.*, pages 2985–2990, 2015.

[KŠLO96]   L.E. Kavraki, P. Švestka, J.-C. Latombe, and M.H. Overmars. Probabilistic roadmaps for path planning in high-dimensional configuration spaces. *IEEE Trans. Robot. Autom.*, 12:566–580, 1996.

[KTSP16]   B. Kim, T.U. Taewoong, C. Suh, and F.C. Park. Tangent bundle RRT: A randomized algorithm for constrained motion planning. *Robotica*, 34:202–225, 2016.

[LAF+15]   M. Lahijanian, S. Almagor, D. Fried, L.E. Kavraki, and M.Y. Vardi. This time the robot settles for a cost: A quantitative approach to temporal logic planning with partial satisfaction. In *AAAI Conf. Artif. Intell.*, pages 3664–3671, 2015.

[Lat91]   J.-C. Latombe. *Robot Motion Planning*. Volume 124 of *Int. Ser. Engin. Comp. Sci.*, Kluwer, Boston, 1991.

[Lat99]   J.-C. Latombe. Motion planning: A journey of robots, molecules, digital actors, and other artifacts. *I. J. Robotics Res.*, 18:1119–1128, 1999.

[LaV06]   S.M. LaValle. *Planning Algorithms*. Cambridge University Press, 2006.

[LCS09]   D.T. Le, J. Cortés, and T. Siméon. A path planning approach to (dis)assembly sequencing. In *Proc. IEEE Int. Conf. Autom. Sci. Eng.*, pages 286–291, 2009.

[LF09]   M. Likhachev and D. Ferguson. Planning long dynamically feasible maneuvers for autonomous vehicles. *I. J. Robotics Res.*, 28:933–945, 2009.

[LGS+16]   J. Lee, M.X. Grey, M. Stilman, K. Liu, et al. Dart simulator. `http://dartsim.github.io/`, 2016.

[LJTM94]   J.-P. Laumond, P.E. Jacobs, M. Taïx, and R.M. Murray. A motion planner for nonholonomic mobile robots. *IEEE Trans. Robot. Autom.*, 10:577–593, 1994.

[LK01a]   F. Lamiraux and L.E. Kavraki. Planning paths for elastic objects under manipulation constraints. *I. J. Robotics Res.*, 20:188–208, 2001.

[LK01b]   F. Lamiraux and L.E. Kavraki. Positioning of symmetric and nonsymmetric parts using radial and constant fields: Computation of all equilibrium configurations. *I. J. Robotics Res.*, 20:635–659, 2001.

[LK01c]   S.M. LaValle and J.J. Kuffner. Randomized kinodynamic planning. *I. J. Robotics Res.*, 20:378–400, 2001.

[LK04a]   A.M. Ladd and L.E. Kavraki. Fast tree-based exploration of state space for robots with dynamics. In *Algorithmic Foundations of Robotics VI*, pages 297–312, Springer, Berlin, 2004.

[LK04b]   A.M. Ladd and L.E. Kavraki. Measure theoretic analysis of probabilistic path planning. *IEEE Trans. Robot. Autom.*, 20:229–242, 2004.

[LLB14]   Y. Li, Z. Littlefield, and K.E. Bekris. Sparse methods for efficient asymptotically optimal kinodynamic planning. In *Algorithmic Foundations of Robotics XI*, pages 263–282, Springer, Berlin, 2014.

[LLG+15]   A.X. Lee, H. Lu, A. Gupta, S. Levine, and P. Abbeel. Learning force-based manipulation of deformable objects from multiple demonstrations. In *Proc. IEEE Int. Conf. Robot. Autom.*, pages 177–184, 2015.

[LLS99]   S. Leroy, J.-P. Laumond, and T. Siméon. Multiple path coordination for mobile robots: A geometric algorithm. In *Proc. 16th Int. Joint Conf. Artif. Intell.*, pages 1118–1123, Morgan Kaufmann, San Francisco, 1999.

[LMC01]   J.E. Luntz, W.C. Messner, and H. Choset. Distributed manipulation using discrete actuator arrays. *I. J. Robotics Res.*, 20:553–583, 2001.

[LMS14]   B. León, A. Morales, and J. Sancho-Bru. *From Robot to Human Grasping Simulation*. Vol. 19 of *Cognitive Systems Monographs*, Springer, Berlin, 2014.

[LP83]   T. Lozano-Pérez. Spatial planning: A configuration space approach. *IEEE Trans. Comput.*, C-32:108–120, 1983.

[LR96]   J.-P. Laumond and J.-J. Risler. Nonholonomic systems: Controllability and complexity. *Theoret. Comput. Sci.*, 157:101–114, 1996.

[LSL99]   F. Lamiraux, S. Sekhavat, and J.-P. Laumond. Motion planning and control for hilare pulling a trailer. *IEEE Trans. Robot. Autom.*, 15:640–652, 1999.

[LTA03]   J.-M. Lien, S.L. Thomas, and N.M. Amato. A general framework for sampling on the medial axis of the free space. In *Proc. IEEE Int. Conf. Robot. Autom.*, pages 4439–4444, 2003.

[LWA15]   S. Levine, N. Wagener, and P. Abbeel. Learning contact-rich manipulation skills with guided policy search. In *Proc. IEEE Int. Conf. Robot. Autom.*, pages 156–163, 2015.

[MA10]   D.M. Mount and S. Arya. ANN: A library for approximate nearest neighbor searching. `https://www.cs.umd.edu/~mount/ANN/`, 2010.

[Mas82]   M.T. Mason. *Manipulation by Grasping and Pushing Operations*. PhD. thesis, MIT, Artificial Intelligence Lab., Cambridge, 1982.

[Mas01]   M.T. Mason. *Mechanics of Robotic Manipulation*. MIT Press, Cambridge, 2001.

[Mat95]   R.S. Mattikalli. *Mechanics Based Assembly Planning*. PhD. thesis, Carnegie Mellon Univ., Pittsburgh, 1995.

[MBH12]   Z. McCarthy, T. Bretl, and S. Hutchinson. Proving path non-existence using sampling and alpha shapes. In *Proc. IEEE Int. Conf. Robot. Autom.*, pages 2563–2569, 2012.

[ME02]   M. Moll and M.A. Erdmann. Manipulation of pose distributions. *I. J. Robotics Res.*, 21:277–292, 2002.

[MGEF02]   M. Moll, K. Goldberg, M.A. Erdmann, and R. Fearing. Aligning parts for micro assemblies. *Assembly Automation*, 22:46–54, 2002.

[Mis91]   B. Mishra. Workholding: analysis and planning. In *Proc. IEEE/RSJ Int. Workshop Intell. Robots and Systems*, pages 53–57, 1991.

[Mis95]   B. Mishra. Grasp metrics: Optimality and complexity. In *Proc. Algorithmic Foundations of Robotics I*, pages 137–165, A.K. Peters, Natick, 1995.

[ML10]   M. Muja and D.G. Lowe. FLANN - Fast library for approximate nearest neighbors. `http://www.cs.ubc.ca/research/flann/`, 2010.

[MLP90]   X. Markenscoff, N. Luqun, and C.H. Papadimitriou. The geometry of grasping. *I. J. Robotics Res.*, 9:61–74, 1990.

[MSS87]   B. Mishra, J.T. Schwartz, and M. Sharir. On the existence and synthesis of multi-finger positive grips. *Algorithmica*, 2:541–558, 1987.

[MZG⁺96]   B. Mirtich, Y. Zhuang, K.Y. Goldberg, J. Craig, R. Zanutta, B. Carlisle, and J.F. Canny. Estimating pose statistics for robotic part feeders. In *Proc. IEEE Int. Conf. Robot. Autom.*, pages 1140–1146, 1996.

[Nat88]   B.K. Natarajan. On planning assemblies. In *Proc. 4th Sympos. Comput. Geom.*, pages 299–308, ACM Press, 1988.

[Nat89]   B.K. Natarajan. Some paradigms for the automated design of parts feeders. *I. J. Robotics Res.*, 8:98–109, 1989.

[Ngu88]   V.-D. Nguyen. Constructing force-closure grasps. *I. J. Robotics Res.*, 7:3–16, 1988.

[NO04]   D. Nieuwenhuisen and M.H. Overmars. Useful cycles in probabilistic roadmap graphs. In *Proc. IEEE Int. Conf. Robot. Autom.*, pages 446–452, 2004.

[NRH10]   O. Nechushtan, B. Raveh, and D. Halperin. Sampling-diagram automata: A tool for analyzing path quality in tree planners. In *Algorithmic Foundations of Robotics IX*, pages 285–301, Springer, Berlin, 2010.

[NSO08]   D. Nieuwenhuisen, A.F. van der Stappen, and M.H. Overmars. An effective frame-work for path planning amidst movable obstacles. In *Algorithmic Foundations of Robotics VIII*, pages 87–102, Springer, Berlin, 2008.

[NSO07]   D. Nieuwenhuisen, A.F. van der Stappen, and M.H. Overmars. Pushing a disk using compliance. *IEEE Trans. Robot.*, 23:431–442, 2007.

[O'R98]   J. O'Rourke. *Computational Geometry in C.* Cambridge University Press, 1998.

[ORSW95]   M.H. Overmars, A. Rao, O. Schwarzkopf, and C. Wentink. Immobilizing polygons against a wall. In *Proc. 11th Sympos. Comput. Geom.*, pages 29–38, ACM Press, 1995.

[PCM12]   J. Pan, S. Chitta, and D. Manocha. FCL: A general purpose library for collision and proximity queries. In *Proc. IEEE Int. Conf. Robot. Autom.*, pages 3859–3866, 2012.

[PKV10]   E. Plaku, L.E. Kavraki, and M.Y. Vardi. Motion planning with dynamics by a synergistic combination of layers of planning. *IEEE Trans. Robot.*, 26:469–482, jun. 2010.

[Pnu77]   A. Pnueli. The temporal logic of programs. In *Proc. 18th IEEE Sympos. Found. Comp. Sci.*, pages 46–57, 1977.

[PPK+12]   A. Perez, R. Platt, G. Konidaris, L.P. Kaelbling, and T Lozano-Pérez. LQR-RRT*: Optimal sampling-based motion planning with automatically derived extension heuristics. In *Proc. IEEE Int. Conf. Robot. Autom.*, pages 2537–2542, 2012.

[PS06]   P. Pipattanasomporn and A. Sudsang. Two-finger caging of concave polygon. In *Proc. IEEE Int. Conf. Robot. Autom.*, pages 2137–2142, 2006.

[PS11]   P. Pipattanasomporn and A. Sudsang. Two-finger caging of nonconvex polytopes. *IEEE Trans. Robot.*, 27:324–333, 2011.

[PS15]   F. Panahi and A.F. van der Stappen. Reprint of: Bounding the locus of the center of mass for a part with shape variation. *Comput. Geom.*, 48:398–406, 2015.

[PSS+97]   J. Ponce, S. Sullivan, A. Sudsang, J.-D. Boissonnat, and J.-P. Merlet. On computing four-finger equilibrium and force-closure grasps of polyhedral objects. *I. J. Robotics Res.*, 16:11–35, 1997.

[PVS08]   P. Pipattanasomporn, P. Vongmasa, and A. Sudsang. Caging rigid polytopes via finger dispersion control. In *Proc. IEEE Int. Conf. Robot. Autom.*, pages 1181–1186, 2008.

[QGC+09]   M. Quigley, B. Gerkey, K. Conley, J. Faust, T. Foote, J. Leibs, E. Berger, R. Wheeler, and A. Ng. ROS: An open-source robot operating system. In *IEEE ICRA Workshop on Open Source Robotics*, 2009.

[RB98]   E. Rimon and J.W. Burdick. Mobility of bodies in contact. i. A 2nd-order mobility index for multiple-finger grasps. *IEEE Trans. Robot. Autom.*, 14:696–708, 1998.

[RB99]   E. Rimon and A. Blake. Caging planar bodies by one-parameter two-fingered gripping systems. *I. J. Robotics Res.*, 18:299–318, 1999.

[REH11]   B. Raveh, A. Enosh, and D. Halperin. A little more, a lot better: Improving path quality by a path-merging algorithm. *IEEE Trans. Robot.*, 27:365–371, 2011.

[RESH09]   B. Raveh, A. Enosh, O. Schueler-Furman, and D. Halperin. Rapid sampling of molecular motions with prior information constraints. *PLoS Comput. Biol.*, 5:e1000295, 2009.

[RG95]   A.S. Rao and K.Y. Goldberg. Manipulating algebraic parts in the plane. *IEEE Trans. Robot. Autom.*, 11:598–602, 1995.

[RMC00]   D. Reznik, E. Moshkovich, and J.F. Canny. Building a universal planar manipulator. In K.F. Böhringer and H. Choset, editors, *Distributed Manipulation*, pages 147–171, Springer, New York, 2000.

[RMF12]   A. Rodriguez, M.T. Mason, and S. Ferry. From caging to grasping. *I. J. Robotics Res.*, 31(7):886–900, 2012.

[Rus99]   D. Rus. In-hand dexterous manipulation of piecewise-smooth 3-d objects. *I. J. Robotics Res.*, 18:355–381, 1999.

[RS12]   E. Rimon and A.F. van der Stappen. Immobilizing 2-d serial chains in form-closure grasps. *IEEE Trans. Robot*, 28:32–43, 2012.

[S$^+$05]   R. Smith et al. Open dynamics engine. http://www.ode.org/, 2005.

[SA95]   M. Sharir and P.K. Agarwal. *Davenport-Schinzel Sequences and Their Geometric Applications*. Cambridge University Press, 1995.

[SBGM11] J. Snape, J.P. van den Berg, S.J. Guy, and D. Manocha. The hybrid reciprocal velocity obstacle. *IEEE Trans. Robot.*, 27:696–706, 2011.

[SC15]   I. Sucan and S. Chitta. MoveIt! http://moveit.ros.org, 2015.

[SDH$^+$14] J. Schulman, Y. Duan, J. Ho, A. Lee, I. Awwal, H. Bradlow, J. Pan, S. Patil, K. Goldberg, and P. Abbeel. Motion planning with sequential convex optimization and convex collision checking. *I. J. Robotics Res.*, 33:1251–1270, 2014.

[SEBK15] J.A. Stork, C.H. Ek, Y. Bekiroglu, and D. Kragic. Learning predictive state representation for in-hand manipulation. In *Proc. IEEE Int. Conf. Robot. Autom.*, pages 3207–3214, 2015.

[SFR$^+$14] S. Srivastava, E. Fang, L. Riano, R. Chitnis, S.J. Russell, and P. Abbeel. Combined task and motion planning through an extensible planner-independent interface layer. In *Proc. IEEE Int. Conf. Robot. Autom.*, pages 639–646, 2014.

[SG00]   A.F. van der Stappen and K.Y. Goldberg. Geometric eccentricity and the complexity of manipulation plans. *Algorithmica*, 26:494–514, 2000.

[SH14a]   O. Salzman and D. Halperin. Asymptotically near-optimal RRT for fast, high-quality, motion planning. In *Proc. IEEE Int. Conf. Robot. Autom.*, pages 4680–4685, 2014.

[SH14b]   K. Solovey and D. Halperin. *k*-Color multi-robot motion planning. *I. J. Robotics Res.*, 33:82–97, 2014.

[SH15a]   O. Salzman and D. Halperin. Asymptotically-optimal motion planning using lower bounds on cost. In *Proc. IEEE Int. Conf. Robot. Autom.*, pages 4167–4172, 2015.

[SH15b]   K. Solovey and D. Halperin. On the hardness of unlabeled multi-robot motion planning. In *Proc. Robotics: Science and Systems*, 2015.

[SHH15]   O. Salzman, M. Hemmer, and D. Halperin. On the power of manifold samples in exploring configuration spaces and the dimensionality of narrow passages. *IEEE Trans. Autom. Sci. Eng.*, 12:529–538, 2015.

[SK01]   A. Sudsang and L.E. Kavraki. A geometric approach to designing a programmable force field with a unique stable equilibrium for parts in the plane. In *Proc. IEEE Int. Conf. Robot. Autom.*, pages 1079–1085, 2001.

[SK05]   M. Stilman and J.J Kuffner. Navigation among movable obstacles: Real-time reasoning in complex environments. *I. J. Humanoid Robot.*, 2:479–503, 2005.

[SK11]   I.A. Sucan and L.E. Kavraki. Mobile manipulation: Encoding motion planning options using task motion multigraphs. In *Proc. IEEE Int. Conf. Robot. Autom.*, pages 5492–5498, 2011.

[SK12]   I. Sucan and L.E. Kavraki. A sampling-based tree planner for systems with complex dynamics. *IEEE Trans. Robot.*, 28:116–131, 2012.

[SK16]   B. Siciliano and O. Khatib, editors. *Springer Handbook of Robotics*. Springer, Berlin, 2016.

[SL03]   G. Sánchez and J.-C. Latombe. A single-query bi-directional probabilistic roadmap planner with lazy collision checking. In *Robotics Research*, vol. 6 of *Tracts in Advanced Robotics*, pages 403–417, Springer, Berlin, 2003.

[SL02]   G. Sánchez and J.-C. Latombe. On delaying collision checking in PRM planning. *I. J. Robotics Res.*, 21:5–26, 2002.

[SLN00]   T. Siméon, J.-P. Laumond, and C. Nissoux. Visibility-based probabilistic roadmaps for motion planning. *Advanced Robotics*, 14:477–493, 2000.

[Smi]   R. Smits. KDL: Kinematics and Dynamics Library. http://www.orocos.org/kdl.

[SMK12]   I. Sucan, M. Moll, and L.E. Kavraki. The open motion planning library. *Robotics & Automation Magazine*, 19(4):72–82, 2012.

[ŠO97]   P. Švestka and M.H. Overmars. Motion planning for carlike robots using a probabilistic learning approach. *I. J. Robotics Res.*, 16:119–143, 1997.

[ŠO98]   P. Švestka and M.H. Overmars. Coordinated pat planning for multiple robots. *Robotics Autonom. Syst.*, 23:125–152, 1998.

[SRP02]   A. Sudsang, F. Rothganger, and J. Ponce. Motion planning for disc-shaped robots pushing a polygonal object in the plane. *IEEE Trans. Robot. Autom.*, 18:550–562, 2002.

[SS94]   J. Snoeyink and J. Stolfi. Objects that cannot be taken apart with two hands. *Discrete Comput. Geom.*, 12:367–384, 1994.

[SSAH14] O. Salzman, D. Shaharabani, P.K. Agarwal, and D. Halperin. Sparsification of motion-planning roadmaps by edge contraction. *I. J. Robotics Res.*, 33:1711–1725, 2014.

[SSD11]   M.A. Sherman, A. Seth, and S.L. Delp. Simbody: multibody dynamics for biomedical research. *Procedia IUTAM*, 2:241–261, 2011.

[SSH16a] K. Solovey, O. Salzman, and D. Halperin. Finding a needle in an exponential haystack: Discrete RRT for exploration of implicit roadmaps in multi-robot motion planning. *I. J. Robotics Res.*, 35:501–513, 2016.

[SSH16b] K. Solovey, O. Salzman, and D. Halperin. New perspective on sampling-based motion planning via random geometric graph. In *Proc. Robotics: Science and Systems*, 2016.

[SSKA07] M. Stilman, J.U. Schamburek, J.J. Kuffner, and T. Asfour. Manipulation planning among movable obstacles. In *Proc. IEEE Int. Conf. Robot. Autom.*, pages 3327–3332, 2007.

[SWO00]   A.F. van der Stappen, C. Wentink, and M.H. Overmars. Computing immobilizing grasps of polygonal parts. *I. J. Robotics Res.*, 19:467–479, 2000.

[SYZH15] K. Solovey, J. Yu, O. Zamir, and D. Halperin. Motion planning for unlabeled discs with optimality guarantees. In *Proc. Robotics: Science and Systems*, 2015.

[TMMK14] M. Turpin, K. Mohta, N. Michael, and V. Kumar. Goal assignment and trajectory planning for large teams of interchangeable robots. *Auton. Robots*, 37:401–415, 2014.

[TTCA10] X. Tang, S. Thomas, P. Coleman, and N.M. Amato. Reachable distance space: Efficient sampling-based planning for spatially constrained systems. *I. J. Robotics Res.*, 29:916–934, 2010.

[US03]   C. Urmson and R.G. Simmons. Approaches for heuristically biasing RRT growth. In *Proc. IEEE/RSJ Int. Conf. Intell. Robots and Systems*, pages 1178–1183, 2003.

[USD+13]   A. Ulusoy, S.L. Smith, X.C. Ding, C. Belta, and D. Rus. Optimality and robustness in multi-robot path planning with temporal logic constraints. *I. J. Robotics Res.*, 32:889–911, 2013.

[Vár14]   P.L. Várkonyi. Estimating part pose statistics with application to industrial parts feeding and shape design: New metrics, algorithms, simulation experiments and datasets. *IEEE Trans. Autom. Sci. Eng*, 11:658–667, 2014.

[VB13]   C.I. Vasile and C. Belta. Sampling-based temporal logic path planning. In *Proc. IEEE/RSJ Int. Conf. Intell. Robots and Systems*, pages 4817–4822, 2013.

[VS08]   M. Vahedi and A.F. van der Stappen. Caging polygons with two and three fingers. *I. J. Robotics Res.*, 27:1308–1324, 2008.

[WB13]   D.J. Webb and J.P. van den Berg. Kinodynamic RRT*: Asymptotically optimal motion planning for robots with linear dynamics. In *Proc. IEEE Int. Conf. Robot. Autom.*, pages 5054–5061, 2013.

[WB08]   N.A. Wedge and M.S. Branicky. On heavy-tailed runtimes and restarts in rapidly-exploring random trees. In *Proc. AAAI Conf. Artif. Intell.*, pages 127–133, 2008.

[WBC15]   W. Wang, D. Balkcom, and A. Chakrabarti. A fast online spanner for roadmap construction. *I. J. Robotics Res.*, 34:1418–1432, 2015.

[WC15]   G. Wagner and H. Choset. Subdimensional expansion for multirobot path planning. *Artif. Intell.*, 219:1–24, 2015.

[WCY15]   C. Wang, Y.-J. Chiang, and C. Yap. On soft predicates in subdivision motion planning. *Comput. Geom.*, 48:589–605, 2015.

[WGPB96]   J. Wiegley, K.Y. Goldberg, M.A. Peshkin, and M. Brokowski. A complete algorithm for designing passive fences to orient parts. In *Proc. IEEE Int. Conf. Robot. Autom.*, pages 1133–1139, 1996.

[Wil91]   G.T. Wilfong. Motion planning in the presence of movable obstacles. *Ann. Math. Artif. Intell.*, 3:131–150, 1991.

[WKLL95]   R.H. Wilson, L.E. Kavraki, T. Lozano-Pérez, and J.-C. Latombe. Two-handed assembly sequencing. *I. J. Robotics Res.*, 14:335–350, 1995.

[WL94]   R.H. Wilson and J.-C. Latombe. Geometric reasoning about mechanical assembly. *Artif. Intell.*, 71:371–396, 1994.

[WUB+13]   T. Wongpiromsarn, A. Ulusoy, C. Belta, E. Frazzoli, and D. Rus. Incremental synthesis of control policies for heterogeneous multi-agent systems with linear temporal logic specifications. In *Proc. IEEE Int. Conf. Robot. Autom.*, pages 5011–5018, 2013.

[XBPA15]   C. Xie, J.P. van den Berg, S. Patil, and P. Abbeel. Toward asymptotically optimal motion planning for kinodynamic systems using a two-point boundary value problem solver. In *Proc. IEEE Int. Conf. Robot. Autom.*, pages 4187–4194, 2015.

[YL13]   J. Yu and S.M. LaValle. Structure and intractability of optimal multi-robot path planning on graphs. In *Proc. AAAI Conf. Artif. Intell.*, 2013.

[YLK01]   J.H. Yakey, S.M. LaValle, and L.E. Kavraki. Randomized path planning for linkages with closed kinematic chains. *IEEE Trans. Robot. Autom.*, 17:951–958, 2001.

[YPL+10]   E. Yoshida, M. Poirier, J.-P. Laumond, O. Kanoun, F. Lamiraux, R. Alami, and K. Yokoi. Pivoting based manipulation by a humanoid robot. *Auton. Robots*, 28:77–88, 2010.

[YRB00]   M. Yim, J. Reich, and A.A. Berlin. Two approaches to distributed manipulation. In K.F. Böhringer and H. Choset, editors, *Distributed Manipulation*, pages 237–261, Springer, New York, 2000.

[YTEA12]   H.-Y. Yeh, S.L. Thomas, D. Eppstein, and N.M. Amato. UOBPRM: A uniformly distributed obstacle-based PRM. In *Proc. IEEE/RSJ Int. Conf. Intell. Robots and Systems*, pages 2655–2662, 2012.

[ZG96]   Y. Zhuang and K.Y. Goldberg. On the existence of solutions in modular fixturing. *I. J. Robotics Res.*, 15:646–656, 1996.

[ZKM08]   L. Zhang, Y.J. Kim, and D. Manocha. Efficient cell labelling and path non-existence computation using C-obstacle query. *I. J. Robotics Res.*, 27:1246–1257, 2008.

52 COMPUTER GRAPHICS
David Dobkin and Seth Teller

INTRODUCTION

Computer graphics has often been cited as a prime application area for the techniques of computational geometry. The histories of the two fields have a great deal of overlap, with similar methods (e.g., sweep-line and area subdivision algorithms) arising independently in each. Both fields have often focused on similar problems, although with different computational models. For example, hidden surface removal (visible surface identification) is a fundamental problem in both fields. At the same time, as the fields have matured, they have brought different requirements to similar problems. The body of this chapter contains an updated version of our chapter in an earlier edition of this Handbook [DT04] in which we aimed to highlight both similarities and differences between the fields.

Computational geometry is fundamentally concerned with the efficient quantitative representation and manipulation of ideal geometric entities to produce exact results. Computer graphics shares these goals, in part. However, graphics practitioners also model the interaction of objects with light and with each other, and the media through which these effects propagate. Moreover, graphics researchers and practitioners: (1) typically use finite precision (rather than exact) representations for geometry; (2) rarely formulate closed-form solutions to problems, instead employing sampling strategies and numerical methods; (3) often design into their algorithms explicit tradeoffs between running time and solution quality; (4) often analyze algorithm performance by defining as primitive operations those that have been implemented in particular hardware architectures rather than by measuring asymptotic behavior; and (5) implement most algorithms they propose.

The relationship described above has faded over the years. At the present time, the overlap between computer graphics and computational geometry has retreated from many of the synergies of the past. The availability of fast hardware driven by the introduction of graphics processing units (GPUs) that provide high speed computation at low cost has encouraged computer graphics practitioners to seek solutions that are engineering based rather than more scientifically driven. At the same time, the basic science underlying computational geometry continues to grow providing a firm basis for understanding basic processes. Adding to this divide is the unfortunate split between theoretical and practical considerations that arise when handling the very large data sets that are becoming common in the computer graphics arena. In the past, it was hoped that the asymptotic nature of many algorithms in computational geometry would bear fruit as the number of elements being processed in a computer graphics scene grew large. Unfortunately, other considerations have been more significant in the handling of large scenes. The gain in asymptotic behavior of many algorithms is overwhelmed by the additional cost in data movement required for their implementation.

Recognizing that the divide between computational geometry and computer graphics is wide at the moment, this chapter has been updated from the earlier edition but new material has not been added. This decision was made in the hope that as processing power and memory sizes of more conventional CPUs become dominant in the future, there will arise a time when computer graphics practitioners will again need a firmer scientific basis on which to develop future results. When this day arrives, the ideas covered in this chapter could provide a productive landscape from which this new collaboration can develop.

52.1 RELATIONSHIP TO COMPUTATIONAL GEOMETRY

In this section we elaborate these five contacts and contrasts.

GEOMETRY VS. RADIOMETRY AND PSYCHOPHYSICS

One fundamental computational process in graphics is *rendering*: the synthesis of realistic images of physical objects. This is done through the application of a simulation process to quantitative models of light, materials, and transmission media to predict (i.e., *synthesize*) appearance. Of course, this process must account for the shapes of and spatial relationships among objects and the viewer, as must computational geometric algorithms. In graphics, however, objects are imbued further with material properties, such as *reflectance* (in its simplest form, color), *refractive index*, *opacity*, and (for light sources) *emissivity*. Moreover, physically justifiable graphics algorithms must model *radiometry*: quantitative representations of light sources and the electromagnetic radiation they emit (with associated attributes of intensity, wavelength, polarization, phase, etc.), and the psychophysical aspects of the human visual system. Thus rendering is a kind of radiometrically and psychophysically "weighted" counterpart to computational geometry problems involving interactions among objects.

CONTINUOUS IDEAL VS. DISCRETE REPRESENTATIONS

Computational geometry is largely concerned with ideal objects (points, lines, circles, spheres, hyperplanes, polyhedra), continuous representations (effectively infinite precision arithmetic), and exact combinatorial and algebraic results. Graphics algorithms (and their implementations) model such objects as well, but do so in a discrete, finite-precision computational model. For example, most graphics algorithms use a floating-point or fixed-point coordinate representation. Thus, one can think of many computer graphics computations as occurring on a (2D or 3D) sample grid. However, a practical difficulty is that the grid spacing is not constant, causing certain geometric predicates (e.g., sidedness) to change under simple transformations such as scaling or translation (see Chapter 45).

An analogy can be made between this distinct choice of coordinates, and the way in which geometric objects—infinite collections of points—are represented by geometers and graphics researchers. Both might represent a sphere similarly—say, by a center and radius. However, an algorithm to render the sphere must select a finite set of sample points on its surface. These sample points typically arise from the placement of a synthetic camera and from the locations of display elements on a two-dimensional display device, for example pixels on a computer monitor or ink dots on a page in a computer printer. The colors computed at these (zero-area) sample points, through some radiometric computation, then serve as an assignment to the discrete value of each (finite-area) display element.

CLOSED-FORM VS. NUMERICAL SOLUTION METHODS

Rarely does a problem in graphics demand a closed-form solution. Instead, graphicists typically rely on numerical algorithms to estimate solution values in an iterative fashion. Numerical algorithms are chosen by reason of efficiency, or of simplicity. Often, these are antagonistic goals. Aside from the usual dangers of quantization into finite-precision arithmetic (Chapter 45), other types of error may arise from numerical algorithms. First, using a point-sampled value to represent a finite-area function's value leads to discretization errors—differences between the reconstructed (interpolated) function, which may be piecewise-constant, piecewise-linear, piecewise-polynomial, etc., and the piecewise-continuous (but unknown) true function. These errors may be exacerbated by a poor choice of sampling points, by a poor piecewise function representation or basis, or by neglect of boundaries along which the true function or its derivative have strong discontinuities. Also, numerical algorithms may suffer bias and converge to incorrect solutions (e.g., due to the misweighting, or omission, of significant contributions).

TRADING SOLUTION QUALITY FOR COMPUTATION TIME

Many graphics algorithms recognize sources of error and seek to bound them by various means. Moreover, for efficiency's sake an algorithm might deliberately introduce error. For example, during rendering, objects might be crudely approximated to speed the geometric computations involved. Alternatively, in a more general illumination computation, many instances of combinatorial interactions (e.g., reflections) between scene elements might be ignored except when they have a significant effect on the computed image or radiometric values. Graphics practitioners have long sought to exploit this intuitive tradeoff between solution quality and computation time.

THEORY VS. PRACTICE

Graphics algorithms, while often designed with theoretical concerns in mind, are typically intended to be of practical use. Thus, while computational geometers and computer graphicists have a substantial overlap of interest in geometry, graphicists develop computational strategies that can feasibly be implemented on modern machines. These strategies change as the nature of the machines changes. For example, the rise of high speed specialized hardware has caused graphics processes to change overnight. Within computational geometry, it is rare to see such drastic changes in the set of primitive operations against which an algorithm is evaluated. Also, while computational geometric algorithms often assume "generic" inputs, in practice geometric degeneracies do occur, and inputs to graphics algorithms are at times highly degenerate (for example, comprised entirely of isothetic rectangles).

Thus, algorithmic strategies are shaped not only by challenging inputs that arise in practice, but also by the technologies available at the time the algorithm is proposed. The relative bandwidths of CPU, bus, memory, network connections, and tertiary storage have major implications for graphics algorithms involving interaction or large amounts of simulation data, or both. For example, in the 1980s the decreasing cost of memory, and the need for robust processing of general datasets, brought about a fundamental shift in most practitioners' choice of computational techniques for resolving visibility (from combinatorial, object-space algorithms to brute force, screen-space algorithms). The increasing power of general-purpose processors, the emergence of sophisticated, robust visibility algorithms, and the wide availability of dedicated, programmable low-level graphics hardware have brought about yet another fundamental shift in recent decades.

TOWARD A MORE FRUITFUL OVERLAP

Given such substantial overlap, there is potential for fruitful collaboration between geometers and graphicists [CAA+96]. One mechanism for spurring such collaboration is the careful posing of models and open problems to both communities. To that end, these are interspersed throughout the remainder of this chapter.

52.2   GRAPHICS AS A COMPUTATIONAL PROCESS

This section gives an overview of three fundamental graphics operations: *acquisition* of some *representation* of model data, its associated attributes and illumination sources; *rendering*, or simulating the appearance of static scenes; and simulating the *behavior* of dynamic scenes either in isolation or under the influence of user *interaction*.

GLOSSARY

Rendering problem: Given quantitative descriptions of surfaces and their properties, light sources, and the media in which all these are embedded, rendering is the application of a computational model to predict appearance; that is, rendering is the synthesis of images from simulation data. Rendering typically involves for each surface a *visibility* computation followed by a *shading* computation.

Visibility: Determining which pairs of a set of objects in a scene share an unobstructed line of sight.

Shading: The determination of radiometric values on a surface (eventually interpreted as colors) as viewed by the observer.

Simulation: The representation of a natural process by a computation.

Psychophysics: The study of the human visual system's response to electromagnetic stimuli.

REPRESENTATION: GEOMETRY, LIGHT, AND FORCES

Every computational process requires some representation in a form amenable to simulation. In graphics, the quantities to be represented span shape, appearance, and illumination. In simulation or interactive settings, forces must also be represented; these may arise from the environment, from interactions among objects, or from the user's actions.

The graphics practitioner's choice of representation has significant implications. For example, how is the data comprising the representation to be acquired? For efficient manipulation or increased spatial or temporal coherence, the representation might have to include, or be amenable to, spatial indexing. A number of intrinsic (winged-edge, quad-edge, facet-edge, etc.) and extrinsic (quadtree, octree, k-d tree, BSP tree, B-rep, CSG, etc.) data structures have been developed to represent geometric data. Continuous, implicit functions have been used to model shape, as have discretized volumetric representations, in which data types or densities are associated with spatial "voxels." A subfield of modeling, Solid Modeling (Chapter 57), represents shape, mass, material, and connectivity properties of objects, so that, for example, complex object assemblies may be defined for use in Computer-Aided Machining environments. Some of these data structures can be adaptively subdivided, and made persistent (that is, made to exist in memory and in nonvolatile storage; see Chapter 38), so that models with wide-scale variations, or simply enormous data size, may be handled. None of these data structures is universal; each has been brought to bear in specific circumstances, depending on the nature of the data (manifold vs. nonmanifold, polyhedral vs. curved, etc.) and the problem at hand. We forego here a detailed discussion of representational issues; see Chapters 56 and 57.

The data structures alluded to above represent "macroscopic" properties of scene geometry—shape, gross structure, etc. Representing material properties, including reflectance over each surface, and possibly surface microstructure (such as roughness) and substructure (as with layers of skin or other tissue), is another fundamental concern of graphics. For each material, computer graphics researchers craft and employ quantitative descriptions of the interaction of radiant energy and/or physical forces with objects having these properties. Examples include human-made objects such as machine parts, furniture, and buildings; organic objects such as flora and fauna; naturally occurring objects such as molecules, terrains, and galaxies; and wholly synthetic objects and materials. Analogously, suitable representations of radiant energy and physical forces also must be crafted in order that the simulation process can model such effects as erosion [DPH96].

ACQUISITION

In practice, algorithms require input. Realistic scene generation can demand complex geometric and radiometric models—for example, of scene geometry and reflectance properties, respectively. Nongeometric scene generation methods can use sparse or dense collections of images of real scenes. Geometric and image inputs must arise from some source; this ***model acquisition*** problem is a core problem in graphics. Models may be generated by a human designer (for example, using Computer-Aided Design packages), generated procedurally (for example, by applying recursive rules), or constructed by machine-aided manipulation of image data (for example, generating 3D topographical maps of terrestrial or extra-terrestrial terrain from multiple photographs), or other machine-sensing methods (e.g., [CL96]). Methods for largely automatic (i.e., minimally human-assisted) acquisition of large-scale geometric models have arisen in the past decade. For example, Google StreetView [GSV16] acquisition combined with methods of image stitching (e.g., [Sze10]) developed in the computer vision community have largely automated the acquisition task.

RENDERING

We partition the simulation process of *rendering* into *visibility* and *shading* sub-components, which are treated in separate subsections below.

For static scenes, and with more difficulty when conditions change with time, rendering can be factored into geometrically and radiometrically view-independent tasks (such as spatial partitioning for surface intervisibility, and the computation of diffuse illumination) and their view-dependent counterparts (culling and specular illumination, respectively). View-independent tasks can be cast as precomputations, while at least some view-dependent tasks cannot occur until the instantaneous viewpoint is known. These distinctions have been blurred by the development of data structures that organize lazily-computed, view-dependent information for use in interactive settings [TBD96].

INTERACTION (SIMULATION OF DYNAMICS)

Graphics brings to bear a wide variety of simulation processes to predict behavior. For example, one might detect collisions to simulate a pair of tumbling dice, or simulate frictional forces in order to provide haptic (touch) feedback through a mechanical device to a researcher manipulating a virtual object [LMC94]. Increasingly, graphics researchers are incorporating spatialized sounds into simulations as well. These physically-based simulations are integral to many graphics applications. However, the generation of synthetic imagery is the most fundamental operation in graphics. The next section describes this "rendering" problem.

When datasets become extremely large, some kind of hierarchical, persistent spatial database is required for efficient storage and access to the data [FKST96], and simplification algorithms are necessary to store and display complex objects with varying fidelity (see, e.g., [CVM+96, HDD+92]).

We first discuss algorithmic aspects of model acquisition, a fundamental first step in graphics (Section 52.3). We next introduce rendering, with its intertwined operations of visibility determination, shading, and sampling (Section 52.4). We then pose several challenges for the future, listing problems of current or future interest in computer graphics on which computational geometry may have a substantial impact (Section 52.5). Finally, we list further references (Section 52.6).

52.3 ACQUISITION

Model acquisition is fundamental in achieving realistic, complex simulations. In some cases, the required model information may be "authored" manually, for example by a human user operating a computer-aided design application. Clearly manual authoring can produce only a limited amount of data. For more complex inputs, simulation designers have crafted "procedural" models, in which code is written to generate model geometry and attributes. However, such models often have limited expressiveness. To achieve both complexity and expressiveness, practitioners employ sensors such as cameras and range scanners to "capture" representations of real-world objects.

GLOSSARY

Model capture: Acquiring a data representation of a real-world object's shape, appearance, or other properties.

GEOMETRY CAPTURE

In crafting a geometry capture method, the graphics practitioner must choose a sensor, for example a (passive) camera or multi-baseline stereo camera configuration, or an (active) laser range-finder. Regardless of sensor choice, data fusion from several sensors requires intrinsic and extrinsic sensor *calibration* and *registration* of multiple sensor observations. The fundamental algorithmic challenges here include handling noisy data, and solving the *data association* problem, i.e., determining which features match or correspond across sensor observations. When the device output (e.g., a point cloud) is not immediately useful as a geometric model, an intermediate step is required to infer geometric structure from the unorganized input [HDD⁺92, AB99]. These problems are particularly challenging in an interactive context, for example when merging range scans acquired at video rate [RHHL02]. A decade ago, the data size became enormous in the "Digital Michelangelo" project [LPC⁺00] or in GIS (geographical information systems) applied over large land areas (see Chapter 59). More recently, the scanning provided by projects such as Google Earth (estimated at over 20 petabytes of information) [Mas12] has added many orders of magnitude to the complexity of problems that arise in the capture of geometry. At the same time, the availability of (relatively) high speed processing in handheld devices coupled with bandwidth that allows these devices to download from servers with little latency has facilitated the rapid display of captured images in a variety of contexts.

One thrust common to both computer graphics and computer vision includes attempts to recover 3D geometry from many cameras situated outside or within the object or scene of interest. These "volumetric stereo" algorithms must face representational issues: a voxel data structure grows in size as the cube of the scene's linear dimension, whereas a boundary representation is more efficient but requires additional a priori information.

Another class of challenges arises from hybrids of procedural and data-driven methods. For example, there exist powerful "grammars" that produce complex synthetic flora using recursive elaboration of simple shapes [MP96]. These methods have a high "amplification factor" in the sense that they can produce complex geometry from a small number of parameters. However, they are notoriously difficult to invert; that is, given a set of observations of a tree, it is apparently difficult to recover an L-system (a particular string rewriting system) that reproduces the tree.

APPEARANCE CAPTURE

Another aspect of capture arises in the process of acquiring texture properties or other "appearance" attributes of geometric models. A number of powerful procedural methods exist for texture generation [Per85] and 3D volumetric effects such as smoke, fire, and clouds [SF95]. Researchers are challenged to make these methods data-driven, i.e., to find the procedural parameters that reproduce observations.

Recently, appearance capture approaches have emerged that attempt to avoid explicit geometry capture. These "image-based" modeling techniques [MB95] gather typically dense collections of images of the object or scene of interest, then use the acquired data to reconstruct images from novel viewpoints (i.e., viewpoints not occupied by the camera). Outstanding challenges for developers of these methods include: crafting effective sampling and reconstruction strategies; achieving effective storage and compression of the input images, which are often highly redundant; and achieving classical graphics effects such as re-illumination under novel lighting conditions when the underlying object geometry is unknown or only approximately known. Acquisition strategies are also needed when capturing materials with complex appearance due to, for example, subsurface effects (e.g., veined marble) [LPC$^+$00].

MOTION CAPTURE

Capturing geometry and appearance of static scenes populated by rigid bodies is challenging. Yet this problem can itself be generalized in two ways. First, scenes may be dynamic, i.e., dependent on the passage of time. Second, scene objects may be articulated, i.e., composed of a number of rigid or deformable subobjects, linked through a series of geometric transformations. Although the dimensionality of the observed data may be immense, the actual number of degrees of freedom can be significantly lower; the computational challenge lies in discovering and representing the reduced dimensions efficiently and without an unacceptable loss of fidelity to the original motion. Thus motion capture yields a host of problems: segmenting objects from one another and from outlier data; inference of object substructure and degrees of freedom; and scaling up to complex articulated assemblies. Some of these problems have been addressed in Computer Vision (see also Chapter 54), although in graphics the same problems arise when processing 3D range (in contrast to 2D image) data.

OPEN PROBLEMS

Given the existence of the ultra-large data sets that have been developed in the computer graphics and computer vision communities, find the proper mathematical framework in the computational geometry community through which to evaluate the relative performance of competing algorithms for a variety of tasks. Even a partial solution to this problem will require a thorough understanding of current hardware as well as a reasonable projection of future hardwares.

52.4 RENDERING

Rendering is the process through which a computer image of a model (acquired or otherwise) is created. To render an image that is perceived by the human visual system as being accurate is often considered to be the fundamental problem of computer graphics (*photorealistic rendering*). To do so requires visibility computations to determine which portions of objects are not obscured. Also required are shading computations to model the photometry of the situation. Because the resultant image will be sampled on a discrete grid, we must also consider techniques for minimizing sampling artifacts from the resultant image.

GLOSSARY

Visibility computation: The determination of whether some set of surfaces, or sample points, is visible to a synthetic observer.

Shading computation: The determination of radiometric values on the surface (eventually interpreted as colors) as viewed by the observer.

Pixel: A picture element, for example on a raster display.

Viewport: A 2D array of pixels, typically comprising a rectangular region on a computer display.

View frustum: A truncated rectangular pyramid, representing the synthetic observer's field of view, with the synthetic eyepoint at the apex of the pyramid. The truncation is typically accomplished using *near* and *far* clipping planes, analogous to the "left, right, top, and bottom" planes that define the rectangular field of view. (If the synthetic eyepoint is placed at infinity, the frustum becomes a rectangular parallelepiped.) Only those portions of the scene geometry that fall inside the view frustum are rendered.

Rasterization: The transformation of a continuous scene description, through discretization and sampling, into a discrete set of pixels on a display device.

Ray casting: A hidden-surface algorithm in which, for each pixel of an image, a ray is cast from the synthetic eyepoint through the center of the pixel [App68]. The ray is parameterized by a variable t such that $t = 0$ is the eyepoint, and $t > 0$ indexes points along the ray increasingly distant from the eye. The first intersection found with a surface in the scene (i.e., the intersection with minimum positive t) locates the visible surface along the ray. The corresponding pixel is assigned the intrinsic color of the surface, or some computed value.

Depth-buffering: (also *z-buffering*) An algorithm that resolves visibility by storing a discrete depth (initialized to some large value) at each pixel [Cat74]. Only when a rendered surface fragment's depth is less than that stored at the pixel can the fragment's color replace that currently stored at the pixel.

Irradiance: Total power per unit area impinging on a surface element. Units: POWER PER RECEIVER AREA.

BRDF: The Bidirectional Reflectance Distribution Function, which maps incident radiation (at general position and angle of incidence) to reflected exiting radiation (at general position and angle of exiting). Unitless, in $[0, 1]$.

BTDF: The Bidirectional Transmission Distribution Function, which maps incident radiation (at general position and angle of incidence) to transmitted exiting radiation (at general position and angle of exiting). Analogous to the BRDF.

Radiance: The fundamental quantity in image synthesis, which is conserved along a ray traveling through a nondispersive medium, and is therefore "the quantity that should be associated with a ray in ray tracing" [CW93]. Units: POWER PER SOURCE AREA PER RECEIVER STERADIAN.

Radiosity: A global illumination algorithm for ideal diffuse environments. Radiosity algorithms compute shading estimates that depend only on the surface normal and the size and position of all other surfaces and light sources, and that are independent of view direction. Also: a physical quantity, with units POWER PER SOURCE AREA.

Ray tracing: An image synthesis algorithm in which ray casting is followed, at each surface, by a recursive shading operation involving a spherical/hemispherical integral of irradiance at each surface point. Ray tracing algorithms are best suited to scenes with small light sources and specular surfaces.

Hybrid algorithm: A global illumination algorithm that models both diffuse and specular interactions (e.g., [SP89]).

VISIBILITY

LOCAL VISIBILITY COMPUTATIONS

Given a scene composed of modeling primitives (e.g., polygons, or spheres), and a viewing frustum defining an eyepoint, a view direction, and field of view, the visibility operation determines which scene points or fragments are *visible*—connected to the eyepoint by a line segment that meets the closure of no other primitive. The visibility computation is global in nature, in the sense that the determination of visibility along a single ray may involve all primitives in the scene. Typically, however, visibility computations can be organized to involve coherent subsets of the model geometry.

In practice, algorithms for visible surface identification operate under severe constraints. First, available memory may be limited. Second, the computation time allowed may be a fraction of a second—short enough to achieve interactive refresh rates under changes in viewing parameters (for example, the location or viewing direction of the observer). Third, visibility algorithms must be simple enough to be practical, but robust enough to apply to highly degenerate scenes that arise in practice.

The advent of machine rendering techniques brought about a cascade of screen-space and object-space combinatorial hidden-surface algorithms, famously surveyed and synthesized in [SSS74]. However, a memory-intensive screen-space technique—*depth-buffering*—soon won out due to its simplicity and the decreasing cost of memory. In depth-buffering, specialized hardware performs visible surface determination independently at each pixel. Each polygon to be rendered is rasterized, producing a collection of pixel coordinates and an associated depth for each. A polygon fragment is allowed to "write" its color into a pixel only if the depth of the fragment at hand is less than the depth stored at the pixel (all pixel depths are initialized to some large value). Thus, in a complex scene each pixel might be written many times to produce the final image, wasting computation and memory bandwidth. This is known as the **overdraw** problem.

Four decades of spectacular improvement in graphics hardware have ensued, and high-end graphics workstations now contain hundreds of increasingly complex processors that clip, illuminate, rasterize, and texture billions of polygons per second. This capability increase has naturally led users to produce ever more complex geometric models, which suffer from increasing overdraw. Object simplification algorithms, which represent complex geometric assemblages with simpler shapes, do little to reduce overdraw. Thus, visible-surface identification (hidden-surface elimination) algorithms have again come to the fore (Section 33.8.1).

GLOBAL VISIBILITY COMPUTATIONS

Real-time systems perform visibility computations from an instantaneous synthetic viewpoint along rays associated with one or more samples at each pixel of some viewport. However, visibility computations also arise in the context of global illumination algorithms, which attempt to identify *all* significant light transport among point and area emitters and reflectors, in order to simulate realistic visual effects such as diffuse and specular interreflection and refraction. A class of *global* visibility algorithms has arisen for these problems. For example, in radiosity computations, a fundamental operation is determining **area-area** visibility in the presence of *blockers*; that is, the identification of those (area) surface elements visible to a given element, and for those partially visible, all tertiary elements causing (or potentially causing) occlusion [HW91, HSA91].

CONSERVATIVE ALGORITHMS

Graphics algorithms often employ *quadrature* techniques in their innermost loops— for example, estimating the energy arriving at one surface from another by casting multiple rays and determining an energy contribution along each. Thus, any efficiency gains in this frequent process (e.g., omission of energy sources known not to contribute energy at the receiver, or omission of objects known not to be blockers) will significantly improve overall system performance. Similarly, occlusion culling algorithms (omission of objects known not to contribute pixels to the rendered image) can significantly reduce overdraw. Both techniques are examples of **conservative** algorithms, which overestimate some geometric set by combinatorial means, then perform a final sampling-based operation that produces a (discrete) solution or quadrature. Of course, the success of conservative algorithms in practice depends on two assumptions: first, that through a relatively simple computation, a usefully tight bound can be attained on whatever set would have been computed by a more sophisticated (e.g., exact) algorithm; and second, that the aggregate time of the conservative algorithm and the sampling pass is less than that of an exact algorithm for input sizes encountered in practice.

This idea can be illustrated as follows. Suppose the task is to render a scene of n polygons. If visible fragments must be rendered *exactly*, any correct algorithm must expend at least kn^2 time, since n polygons (e.g., two slightly misaligned combs, each with $n/2$ teeth) can cause $O(n^2)$ visible fragments to arise. But a conservative algorithm might simply render all n polygons, incurring some overdraw (to be resolved by a depth-buffer) at each pixel, but expending only time linear in the size of the input.

This highlights an important difference between computational geometry and computer graphics. Standard computational geometry cost measures would show that the $O(n^2)$ algorithm is optimal in an output-sensitive model (Section 33.8.1). In computer graphics, hardware considerations motivate a fundamentally different approach: rendering a (judiciously chosen) superset of those polygons that will contribute to the final image. A major open problem is to unify these approaches by finding a cost function that effectively models such considerations (see below).

HARDWARE TRENDS

In previous decades, several hybrid object-space/screen-space visibility algorithms emerged (e.g., [GKM93]). As general-purpose processors became faster, such hybrid algorithms became more widely used. In certain situations, these algorithms operated entirely in object space, without relying on special-purpose graphics hardware [CT96]. In the past decade, the arrival of high powered graphics processing units (GPUs) from companies such as nVidia have allowed programming to occur at the level of programmable shaders at the vertex, geometry and pixel level. These shaders tend to be largely driven by hardware considerations but give hybrid solutions. Vertex shaders operate at the level of the vertex and can do the calculations necessary to properly light the vertex and, as necessary transform the neighborhood of the vertex. Geometry shaders operate at the level of the geometric primitive, typically the triangle, and can perform calculations at this level. Vertices and geometries are then scan-converted to provide information for the pixel shader which operates solely in image space and provides a rendering of the scene.

SHADING

Through sampling and visibility operations, a visible surface point or fragment is identified. This point or fragment is then **shaded** according to a *local* or *global* illumination algorithm. Given scene light sources and material reflection and transmission properties, and the propagative media comprising and surrounding the scene objects, the shading operation determines the color and intensity of the incident and exiting radiation at the point to be shaded. Shading computations can be characterized further as **view-independent** (modeling only purely diffuse interactions, or directional interactions with no dependence on the instantaneous eye position) or **view-dependent**.

Most graphics workstations perform a local shading operation in hardware, which, given a point light source, a surface point, and an eye position, evaluates the energy reaching the eye via a single reflection from the surface. This local operation is implemented in the software and hardware offered by most workstations. However, this simple model cannot produce realistic lighting cues such as shadows, reflection, and refraction. These require more extensive, global computations as described below.

SHADING AS RECURSIVE WEIGHTED INTEGRATION

Most generally, the shading operation computes the energy leaving a differential surface element in a specified differential direction. This energy depends on the surface's emittance and on the product of the surface's reflectance with the total energy incident from all other surfaces. This relation is known as the *Rendering Equation* [Kaj86], which states intuitively that each surface fragment's appearance, as viewed from a given direction, depends on any light it emits, plus any light (gathered from other objects in the scene) that it reflects in the direction of the observer. Thus, shading can be cast as a recursive integration; to shade a surface fragment F, shade all fragments visible to F, then sum those fragments' illumination upon F (appropriately weighted by the BRDF or BTDF) with any direct illumination of F. Effects such as diffuse illumination, motion blur, Fresnel effects, etc., can be simulated by supersampling in space, time, and wavelength, respectively, and then averaging [CPC84].

Of course, a base case for the recursion must be defined. Classical ray tracers truncate the integration when a certain recursion depth is reached. If this maximum depth is set to 1, ray casting (the determination of visibility for eye rays only) results. More common is to use a small constant greater than one, which leads to "Whitted" or "classical" ray tracing [Whi80]. For efficiency, practitioners also employ a thresholding technique: when multiple reflections cause the weight with which a particular contribution will contribute to the shading at the root to drop below a specified threshold, the recursion ceases. These termination conditions can, under some conditions, cause important energy-bearing paths to be overlooked. For example, a bright light source (such as the sun) filtering through many parts of a house to reach an interior space may be incorrectly discounted by this condition.

In recent years, a hardware trend has developed in support of "programmable shading," in which a (typically short, straight-line) program can be downloaded into graphics hardware for application to every vertex, polygon or pixel processed.[1] This trend has spurred research into, for example, ways to "factor" complex shading calculations into suitable components for mapping to hardware.

ALIASING

From a purely physical standpoint, the amount of energy leaving a surface in a particular direction is the product of the spherical integral of incoming energy and the bidirectional reflectance (and transmittance, as appropriate) in the exiting direction. From a psychophysical standpoint, the perceived color is an inner product of the energy distribution incident on the retina with the retina's spectral response function. We do not explore psychophysical considerations further here.

Global illumination algorithms perform an integration of irradiance at each point to be shaded. Ray tracing and radiosity are examples of global illumination algorithms. Since no closed-form solutions for global illumination are known for general scenes, practitioners employ sampling strategies. Graphics algorithms typically attempt "reconstruction" of some illumination function (e.g., irradiance, or radiance), given some set of samples of the function's values and possibly other information, for example about light source positions, etc. However, such reconstruction is subject to error for two reasons.

First, the well-known phenomenon of *aliasing* occurs when insufficient samples are taken to find all high-frequency terms in a sampled signal. In image processing, samples arise from measurements, and reconstruction error arises from samples that are too widely spaced. However, in graphics, the sample values arise from a simulation process, for example, the evaluation of a local illumination equation, or the numerical integration of irradiance. Thus, reconstruction error can arise from simulation errors in generating the samples. This second type of error is called *biasing*.

---

[1] Current manufacturers include NVIDIA `http://nvidia.com/` and AMD `http://amd.com/`.

For example, classical ray tracers [Whi80] may suffer from biasing in three ways. First, at each shaded point, they compute irradiance only: from direct illumination by point lights; along the reflected direction; and along the refracted direction. Significant "indirect" illumination that occurs along any direction other than these is not accounted for. Thus, indirect reflection and focusing effects are missed. Classical ray tracers also suffer biasing by truncating the depth of the recursive ray tree at some finite depth d; thus, they cannot find significant paths of energy from light source to eye of length greater than d. Third, classical ray tracers truncate ray trees when their weight falls below some threshold. This can fail to account for large radiance contributions due to bright sources illuminating surfaces of low reflectance.

SAMPLING

Sampling patterns can arise from a regular grid (e.g., pixels in a viewport) but these lead to a stair-stepping kind of aliasing. One solution is to *supersample* (i.e., take multiple samples per pixel) and average the results. However, one must take care to supersample in a way that does not align with the scene geometry or some underlying attribute (e.g., texture) in a periodic, spatially varying fashion; otherwise aliasing (including Moiré patterns) will result.

DISCREPANCY

The quality of sampling patterns can be evaluated with a measure known as *discrepancy* (Chapter 47). For example, if we are sampling in a pixel, features interacting with the pixel can be modeled by line segments (representing parts of edges of features) crossing the pixel. These segments divide the pixel into two regions. A good sampling strategy will ensure that the proportion of sample points in each region approximates the proportion of pixel area in that region. The difference between these quantities is the discrepancy of the point set with respect to the line segment. We define the discrepancy of a set of samples (in this case) as the maximum discrepancy with respect to all line segments. Other measures of discrepancy are possible, as described below. See also Chapter 13.

Sampling patterns are used to solve integral equations. The advantage of using a low-discrepancy set is that the solution will be more accurately approximated, resulting in a better image. These differences are expressed in solution convergence rates as a function of the number of samples. For example, truly random sampling has a discrepancy that grows as $O(N^{-\frac{1}{2}})$ where N is the number of samples. There are other sampling patterns (e.g., the *Hammersley points*) that have discrepancies growing as $O(N^{-1} \log^{k-1} N)$. Sometimes one wishes to combine values obtained by different sampling methods [VG95]. The search for good sampling patterns, given a fixed number of samples, is often done by running an optimization process which aims to find sets of ever-decreasing discrepancy. A crucial part of any such process is the ability to quickly compute the discrepancy of a set of samples.

COMPUTING THE DISCREPANCY

There are two common questions that arise in the study of discrepancy: first, given fixed N, how to construct a good sampling pattern in the model described above; second, how to construct a good sampling pattern in an alternative model.

For concreteness, consider the problem of finding low discrepancy patterns in the unit square, modeling an individual pixel. As stated above, the geometry of objects is modeled by edges that intersect the pixel dividing it into two regions, one where the object exists and one where it does not. An ideal sampling method would sample the regions in proportion to their relative areas.

We model this as a discrepancy problem as follows. Let S be a sample set of points in the unit square. For a line l (actually, a segment arising from a polygon boundary in the scene being rendered), define the two regions S^+ and S^- into which l divides S. Ideally, we want a sampling pattern that has the same fraction of samples in the region S^+ as the area of S^+. Thus, in the region S^+, the discrepancy with respect to l is

$$|\sharp(S \cap S^+)/\sharp(S) - \text{Area}(S^+)|,$$

where $\sharp(\cdot)$ denotes the cardinality operator. The discrepancy of the sample set S with respect to a line l is defined as the larger of the discrepancies in the two regions. The discrepancy of set S is then the maximum, over *all* lines l, of the discrepancy of S with respect to l.

Finding the discrepancy in this setting is an interesting computational geometry problem. First, we observe that we do not need to consider all lines. Rather, we need consider only those lines that pass through two points of S, plus a few lines derived from boundary conditions. This suggests the $O(n^3)$ algorithm of computing the discrepancy of each of the $O(n^2)$ lines separately. This can be improved to $O(n^2 \log n)$ by considering the fan of lines with a common vertex (i.e., one of the sample points) together. This can be further improved by appealing to duality. The traversal of this fan of lines is merely a walk in the arrangement of lines in dual space that are the duals of the sample points. This observation allows us to use techniques similar to those in Chapter 28 to derive an algorithm that runs asymptotically as $O(n^2)$. Full details are given in [DEM93].

There are other discrepancy models that arise naturally. A second obvious candidate is to measure the discrepancy of sample sets in the unit square with respect to axis-oriented rectangles. Here we can achieve a discrepancy of $O(n^2 \log n)$, again using geometric methods. We use a combination of techniques, appealing to the incremental construction of 2D convex hulls to solve a basic problem, then using the sweep paradigm to extend this incrementally to a solution of the more general problem. The sweep is easier in the case in which the rectangle is anchored with one vertex at the origin, yielding an algorithm with running time $O(n \log^2 n)$.

The model given above can be generalized to compute **bichromatic discrepancy**. In this case, we have sample points that are colored either black or red. We can now define the discrepancy of a region as the difference between its number of red and black points. Alternatively, we can look for regions (of the allowable type) that are most nearly monochromatic in red while their complements are nearly monochromatic in black. This latter model has application in computational learning theory. For example, red points may represent situations in which a concept is true, black situations where it is false. The minimum discrepancy rectangle is now a classifier of the concept. This is a popular technique for computer-assisted medical diagnosis.

The relevance of these algorithms to computational geometry is that they will lead to faster algorithms for testing the "goodness" of sampling patterns, and thus eventually more efficient algorithms with bounded sampling error. Also, algorithms for computing the discrepancy relative to a particular set system are directly related to the system's VC-dimension (see Section 47.1).

OPEN PROBLEMS

An enormous literature of adaptive, backward, forward, distribution, etc. ray tracers has evolved to address sampling and bias errors. However, the fundamental issues can be stated simply. (Each of the problems below assumes a geometric model consisting of n polygons.)

A related *inverse* problem arises in machine vision, now being adopted by computer graphics practitioners as a method for acquiring large-scale geometric models from imagery.

The problems below are open for both the unit cube and unit ball in all dimensions.

1. The set of visible fragments can have complexity $\Omega(n^2)$ in the worst case. However, the complexity is lower for many scenes. If k is the number of edge incidences (vertices) in the projected visible scene, the set of visible fragments can be computed in optimal output-sensitive $O(nk^{1/2}\log n)$ time [SO92]. Although specialized results have been obtained, optimality has not been reached in many cases. See Table 33.8.1.

2. Give a spatial partitioning and ray casting algorithm that runs in amortized nearly-constant time (that is, has only a weak asymptotic dependence on total scene complexity). Identify a useful "density" parameter of the scene (e.g., the largest number of simultaneously visible polygons), and express the amortized cost of a ray cast in terms of this parameter.

3. Give an output-sensitive algorithm which, for specified viewing parameters, determines the set of "contributing" polygons—i.e., those which contribute their color to at least one viewport pixel.

4. Give an output-sensitive algorithm which, for specified viewing parameters, approximates the visible set to within ϵ. That is, produce a superset of the visible polygons of size (alternatively, total projected area) at most $(1 + \epsilon)$ times the size (resp., projected area) of the true set. Is the lower bound for this problem asymptotically smaller than that for the exact visibility problem?

5. For machine-dependent parameters A and B describing the transform (per-vertex) and fill (per-pixel) costs of some rendering architecture, give an algorithm to compute a superset S of the visible polygon set minimizing the rendering cost on the specified architecture.

6. In a local illumination computation, identify those polygons (or a superset) visible from the synthetic observer, and construct, for each visible polygon P, an efficient function $V(p)$ that returns 1 iff point $p \in P$ is visible from the viewpoint.

7. In a global illumination computation, identify all pairs (or a superset) of intervisible polygons, and for each such pair P, Q, construct an efficient function $V(p, q)$ that returns 1 iff point $p \in P$ is visible from point $q \in Q$.

8. *Image-based rendering* [MB95]: Given a 3D model, generate a minimal set of images of the model such that for all subsequent query viewpoints, the correct image can be recovered by combination of the sample images.

9. Given a geometric model M, a collection of light sources L, a synthetic viewpoint E, and a threshold ϵ, identify all optical paths to E bearing radiance greater than ϵ.

10. Given a geometric model M, a collection of light sources L, and a threshold ϵ, identify *all* optical paths bearing radiance greater than ϵ.

11. An observation of a real object comprises the *product* of irradiance and reflection (BRDF). How can one deduce the BRDF from such observations?

12. Given N, generate a minimum-discrepancy pattern of N samples.

13. Given a low-discrepancy pattern of K points, generate a low (or lower) discrepancy pattern of $K + 1$ points.

52.5 FURTHER CHALLENGES

We have described several core problems of computer graphics and illustrated the impact of computational geometry. We have only scratched the surface of a highly fruitful interaction; the possibilities are expanding, as we describe below. These computer graphics problems all build on the combinatorial framework of computational geometry and so have been, and continue to be, ripe candidates for application of computational geometry techniques. Numerous other problems remain whose combinatorial aspects are perhaps less obvious, but for which interaction may be equally fruitful.

INDEX AND SEARCH

The proliferation of geometric models leads to a problem analogous to that in document storage: how to index models so that they can be efficiently found later. In particular, we might wish to define the Google of 3D models. Searching by name is of limited utility, since in many cases a model's author may not have named it suggestively, or as expected by the seeker. Searching by attributes or appearance is likely to be more fruitful or at the least, a necessary adjunct to searching by name. Perhaps the most successful search mechanisms to date are those relying on geometric "shape signatures" of objects, along with name and attribute metadata where available [FMK+03]. One promising class of signatures related to the medial axis transform is the "shock graph" [LK01]. A first step toward building such a system appears in [OFCD02].

TRANSMISSION AND LEVEL OF DETAIL

Fast network connectivity is not yet universally deployed, and the number and size of available models is growing inexorably with time. Thus in many contexts it is important to store, transmit, and display geometric information efficiently. A variety of techniques have been developed for "progressive" [Hop97] or "multi-resolution" geometry representation [GSS99], as well as for automated level-of-detail generation from source objects [GH97]. For specific model classes, e.g., terrain, efficient algorithms have been developed for varying the fidelity of the display across the field of view [BD98]. Finally, some practitioners have proposed techniques to choose levels of detail, within some time rendering budget, to optimize some image quality criterion [FKST96].

OPEN PROBLEM

Robust simplification. Cheng et al. [CDP04] recently gave a method for computing levels of details that preserve the genus of the original surface. Combine their techniques with techniques for robust computation to derive a robust and efficient scheme for simplification that can be easily implemented. See Chapter 45.

INTERACTION

In addition to off-line or batch computations, graphics practitioners develop on-line computations which involve a user in an interactive loop with input and output (display) stages, such as scientific visualization. For responsiveness, such applications may have to produce many outputs per second: rendering applications typically must maintain 30Hz or faster, whereas haptic or force-feedback applications may operate at 10KHz. Modern applications must also cope with large datasets, only parts of which may be memory-resident at any moment. Thus effective techniques for indexing, searching, and transmitting model data are required. For out-of-core data, predictive fetching strategies are required to avoid high-latency "hiccups" in the user's display.

Beyond seeing and feeling virtual representations of an object, new "3D printing" techniques have emerged for rapid prototyping applications that create real, physical models of objects. Computational geometry algorithms are required to plan the slicing or deposition steps needed. Also, "augmented reality" (AR) methods attempt to provide synthetically generated image overlays onto real scenes, for example using head-mounted displays or hand-held projectors. AR methods require good, low-latency 6-DOF tracking of the user's head or device position and orientation in extended environments.

An exciting new class of "pervasive computing" and "mobile computing" applications attempts to move computation away from the desktop and out into the extended work, home, or outdoor environment. These applications are by nature integrative, encompassing geometric and functional models, position and orientation tracking, proximity data structures, ad hoc networks, and distributed self-calibration algorithms [PMBT01].

OPEN PROBLEM

Collision detection and force feedback. Imagine that every object has an associated motion, and that some objects (e.g., virtual probes) are interactively controlled. Suppose further that when pairs of objects intersect, there is a reaction (due, e.g., to conservation of momentum). Here we wish to render frames and generate haptic feedback while accounting for such physical considerations. Are there suitable data structures and algorithms within computational geometry to model and solve this problem (e.g., [LMC94, MC95])?

DYNAMICS

When simulations include objects that affect each other through force exchange or collision, they must efficiently identify the actual interactions. Usually there is significant temporal coherence, i.e., the set of objects near a given object changes slowly over time. A number of techniques have been proposed to track moving objects in a spatial index or closest-pair geometric data structure in order to detect collisions efficiently [MC95, LMC94, BGH99]. The "object" of interest may be the geometric representation of a user, for example of a finger or hand probing a virtual scene. Recently, some authors have proposed synthesizing sound information to accompany the visual simulation outputs [OSG02].

We have focused this chapter on problems in which the parameters are static; that is, the geometry is unchanging, and nothing is moving (except perhaps the synthetic viewpoint). Now, we briefly describe situations where this is not the case and deeper analysis is required. In these situations it is likely that computational geometry can have a tremendous impact; we sketch some possibilities here.

Each of the static assumptions above may be relaxed, either alone or in combination. For example, objects may evolve with time; we may be interested in transient rather than steady-state solutions; material properties may change over time; object motions may have to be computed and resolved; etc. It is a challenge to determine how techniques of computational geometry can be modified to address state-of-the-art and future computer graphics tasks in dynamic environments.

Among the issues we have not addressed where these considerations are important are the following.

MODEL CHANGES OVER TIME

In a realistic model, even unmoving objects change over time, for example becoming dirty or scratched. In some environments, objects rust or suffer other corrosive effects. Sophisticated geometric representations and algorithms are necessary to capture and model such phenomena [DPH96]. See Chapter 53.

INVERSE PROCESSES

Much of what we have described is a feed-forward process in which one specifies a model and a simulation process and computes a result. Of equal importance in design contexts is to specify a result and a simulation process, and compute a set of initial conditions that would produce the desired result. For example, one might wish to specify the appearance of a stage, and deduce the intensities of scores or hundreds of illuminating light sources that would result in this appearance [SDSA93]. Or, one might wish to solve an inverse kinematics problem in which an object with multiple parts and numerous degrees of freedom is specified. Given initial and final states, one must compute a smooth, minimal energy path between the states, typically in an underconstrained framework. This is a common problem in robotics (see Section 50.1). However, the configurations encountered in graphics tend to have very high complexity. For example, convincingly simulating the motion of a human figure requires processing kinematic models with hundreds of degrees of freedom.

EXTERNAL MEMORY ALGORITHMS

Computational geometry assumes a realm in which all data can be stored in RAM and accessed at no cost (or unit cost per word). Increasingly often, this is not the case in practice. For example, many large databases cannot be stored in main memory. Only a small subset of the model contributes to each generated image, and algorithms for efficiently identifying this subset, and maintaining it under small changes of the viewpoint or model, form an active research area in computer graphics. Given that motion in virtual environments is usually smooth, and that hard real-time constraints preclude the use of purely reactive, synchronous techniques, such algorithms must be *predictive* and *asynchronous* in nature [FKST96]. Achieving efficient algorithms for appropriately shuttling data between secondary (and tertiary) storage and main memory is an interesting challenge for computational geometry.

52.6 SOURCES AND RELATED MATERIAL

SURVEYS

All results not given an explicit reference above may be traced in these surveys:

[Dob92]: A survey article on computational geometry and computer graphics.

[Dor94]: Survey of object-space hidden-surface removal algorithms.

[Yao92, LP84]: Surveys of computational geometry.

[CCSD03]: Survey of visibility for walkthroughs.

RELATED CHAPTERS

Chapter 13: Geometric discrepancy theory and uniform distribution
Chapter 16: Subdivisions and triangulations of polytopes
Chapter 30: Polygons
Chapter 33: Visibility
Chapter 39: Collision and proximity queries
Chapter 41: Ray shooting and lines in space
Chapter 50: Algorithmic motion planning
Chapter 56: Splines and geometric modeling
Chapter 45: Robust geometric computation
Chapter 57: Solid modeling

REFERENCES

[AB99]   N. Amenta and M. Bern. Surface reconstruction by Voronoi filtering. *Discrete Comput. Geom.*, 22:481–504, 1999.

[App68]   A. Appel. Some techniques for shading machine renderings of solids. In *Proc. Spring Joint Computer Conference*, pages 37–45, ACM Press, 1968.

[BD98]   M. de Berg and K. Dobrindt. On levels of detail in terrains. *Graphical Models Image Proc.*, 60:1–12, 1998.

[BGH99]   J. Basch, L.J. Guibas, and J. Hershberger. Data structures for mobile data. *J. Algorithms*, 31:1–28, 1999.

[CAA⁺96]   B. Chazelle, N. Amenta, Te. Asano, G. Barequet, M. Bern, J.-D. Boissonnat, J.F. Canny, K.L. Clarkson, D.P. Dobkin, B.R. Donald, S. Drysdale, H. Edelsbrunner, D. Eppstein, A.R. Forrest, S.J. Fortune, K.Y. Goldberg, M.T. Goodrich, L.J. Guibas, P. Hanrahan, C.M. Hoffmann, D.P. Huttenlocher, H. Imai, D.G. Kirkpatrick, D.T. Lee, K. Mehlhorn, V.J. Milenkovic, J.S.B. Mitchell, M.H. Overmars, R. Pollack, R. Seidel, M. Sharir, J. Snoeyink, G.T. Toussaint, S. Teller, H. Voelcker, E. Welzl, and C.K. Yap. Application Challenges to Computational Geometry: CG Impact Task Force Report. Tech. Rep. TR-521-96, Princeton CS Dept., 1996. `https://www.cs.princeton.edu/research/techreps/TR-521-96`

[Cat74]   E.E. Catmull. *A Subdivision Algorithm for Computer Display of Curved Surfaces.* Ph.D. thesis, Univ. Utah, TR UTEC-CSc-74-133, 1974.

[CCSD03]   D. Cohen-Or, Y. Chrysanthou, C. Silva, and F. Durand. A survey of visibility for walkthrough applications. *IEEE. Trans. Visualization Comput. Graphics*, 9:412–431, 2003.

[CDP04]   S.-W. Cheng, T.K. Dey, and S.-H. Poon. Hierarchy of surface models and irreducible triangulation. *Comput. Geom.*, 27:135–150, 2004.

[CL96]   B. Curless and M. Levoy. A volumetric method for building complex models from range images. In *Proc. 23rd ACM Conf. SIGGRAPH*, pages 303–312, 1996.

[CPC84]   R.L. Cook, T. Porter, and L. Carpenter. Distributed ray tracing. *SIGGRAPH Comput. Graph.*, 18:137–145, 1984.

[CT96]   S. Coorg and S. Teller. Temporally coherent conservative visibility. *Comput. Geom.* 12,105–124, 1999.

[CVM⁺96]   J. Cohen, A. Varshney, D. Manocha, G. Turk, H. Weber, P.K. Agarwal, F.P. Brooks, Jr., and W.V. Wright. Simplification envelopes. In *Proc. 23rd ACM Conf. SIGGRAPH*, pages 119–128, 1996.

[CW93]   M.F. Cohen and J.R. Wallace. *Radiosity and Realistic Image Synthesis.* Academic Press, Cambridge, 1993.

[DEM93]   D.P. Dobkin, D. Eppstein, and D.P. Mitchell. Computing the discrepancy with applications to supersampling patterns. In *Proc. 9th Sympos. Comput. Geom.*, pages 47–52, ACM Press, 1993.

[Dob92]   D.P. Dobkin. Computational geometry and computer graphics. *Proc. IEEE*, 80:1400–1411, 1992.

[Dor94]   S.E. Dorward. A survey of object-space hidden surface removal. *Internat. J. Comput. Geom. Appl.*, 4:325–362, 1994.

[DPH96]   J. Dorsey, H. Pedersen, and P. Hanrahan. Flow and changes in appearance. In *Proc. 23rd ACM Conf. SIGGRAPH*, pages 411–420, 1996.

[DT04]   D.P. Dobkin and S. Teller. Computer graphics. In J.E. Goodman and J. O'Rourke, editors, *CRC Handbook of Discrete and Computational Geometry*, 2nd edition, CRC Press, Boca Raton, 2004.

[FKST96]   T. Funkhouser, D. Khorramabadi, C. Séquin, and S. Teller. The UCB system for interactive visualization of large architectural models. *Presence*, 5:13–44, 1996.

[FMK⁺03]   T. Funkhouser, P. Min, M. Kazhdan, J. Chen, A. Halderman, D.P. Dobkin, and D. Jacobs. A search engine for 3D models. *ACM Trans. Graph.*, 22:83–105, 2003.

[GH97]   M. Garland and P.S. Heckbert. Surface simplification using quadric error metrics. In *Proc. 24th ACM Conf. SIGGRAPH*, pages 209–216, 1997.

[GKM93]   N. Greene, M. Kass, and G.L. Miller. Hierarchical Z-buffer visibility. In *Proc. 20th ACM Conf. SIGGRAPH*, pages 231–238, 1993.

[GSS99]   I. Guskov, W. Sweldens, and P. Schröder. Multiresolution signal processing for meshes. In *Proc. 26th ACM Conf. SIGGRAPH*, pages 325–334, 1999.

[GSV16]   Understand Google Street View, `https://www.google.com/maps/streetview/understand/`, retrieved February 7, 2016.

[HDD⁺92]   H. Hoppe, T.D. DeRose, T. Duchamp, J. McDonald, and W. Stuetzle. Surface reconstruction from unorganized points. In *Proc. 19th ACM Conf. SIGGRAPH*, pages 71–78, 1992.

[Hop97]   H. Hoppe. View-dependent refinement of progressive meshes. In *Proc. 24th ACM Conf. SIGGRAPH*, pages 189–198, 1997.

[HSA91]   P. Hanrahan, D. Salzman, and L.J. Aupperle. A rapid hierarchical radiosity algorithm. In *Proc. 18th ACM Conf. SIGGRAPH*, pages 197–206, 1991.

[HW91]   E. Haines and J.R. Wallace. Shaft culling for efficient ray-traced radiosity. In *Proc. 2nd Eurographics Workshop Rendering*, pages 122–138, 1991.

[Kaj86]   J.T. Kajiya. The rendering equation. *SIGGRAPH Comput. Graph.*, 20:143–150, 1986.

[LK01]   F.F. Leymarie and B.B. Kimia. The shock scaffold for representing 3D shape. In *Proc. 4th Internat. Workshop Visual Form*, pages 216–229. Springer-Verlag, Berlin, 2001.

[LMC94]   M.C. Lin, D. Manocha, and J.F. Canny. Fast contact determination in dynamic environments. In *Proc. Internat. Conf. Robot. Autom.*, pages 602–609, 1994.

[LP84]   D.T. Lee and F.P. Preparata. Computational geometry: A survey. *IEEE Trans. Comput.*, 33:1072–1101, 1984.

[LPC⁺00]   M. Levoy, K. Pulli, B. Curless, S. Rusinkiewicz, D. Koller, L. Pereira, M. Ginzton, S. Anderson, J. Davis, J. Ginsberg, J. Shade, and D. Fulk. The digital Michelangelo project: 3D scanning of large statues. In *Proc. 27th ACM Conf. SIGGRAPH*, pages 131–144, 2000.

[Mas12]   Machable. 11 Fascinating Facts About Google Maps, `http://mashable.com/2012/08/22/google-maps-facts/#0Czys4bUBkq0`, retrieved February 3, 2016.

[MB95]   L. McMillan and G. Bishop. Plenoptic modeling: An image-based rendering system. In *Proc. 22nd ACM Conf. SIGGRAPH 95*, pages 39–46, 1995.

[MC95]   B. Mirtich and J.F. Canny. Impulse-based simulation of rigid bodies. In *1995 Sympos. Interactive 3D Graphics*, pages 181–188, 1995.

[MP96]   R. Mech and P. Prusinkiewicz. Visual models of plants interacting with their environment. In *Proc. 23rd ACM Conf. SIGGRAPH*, pages 397–410, 1996.

[OFCD02]   R. Osada, T. Funkhouser, B. Chazelle, and D.P. Dobkin. Shape distributions. *ACM Trans. Graph.*, 21:807–832, 2002.

[OSG02]   J. O'Brien, C. Shen, and C. Gatchalian. Natural phenomena: Synthesizing sounds from rigid-body simulations. *Proc. ACM SIGGRAPH Sympos. Computer Animation*, 2002.

[Per85]   K. Perlin. An image synthesizer. *SIGGRAPH Comput. Graph.*, 19:287–296, 1985.

[PMBT01] N. Priyantha, A. Miu, H. Balakrishnan, and S. Teller. The cricket compass for context-aware mobile applications. In *Proc. 7th ACM Conf. Mobile Comput. Network*, pages 1–14, 2001.

[RHHL02] S. Rusinkiewicz, O. Hall-Holt, and M. Levoy. Real-time 3D model acquisition. In *Proc. 29th ACM Conf. SIGGRAPH*, pages 438–446, 2002.

[SDSA93] C. Schoeneman, J. Dorsey, B. Smits, and J. Arvo. Painting with light. *Proc. 20th ACM Conf. SIGGRAPH*, pages 143–146, 1993.

[SF95] J. Stam and E. Fiume. Depicting fire and other gaseous phenomena using diffusion processes. In *Proc. 22nd ACM Conf. SIGGRAPH*, pages 129–136, 1995.

[SO92] M. Sharir and M.H. Overmars. A simple output-sensitive algorithm for hidden surface removal. *ACM Trans. Graph.*, 11:1–11, 1992.

[SP89] F.X. Sillion and C. Puech. A general two-pass method integrating specular and diffuse reflection. *SIGGRAPH Comput. Graph.*, 23:335–344, 1989.

[SSS74] I.E. Sutherland, R.F. Sproull, and R.A. Schumacker. A characterization of ten hidden-surface algorithms. *ACM Comput. Surv.*, 6:1–55, 1974.

[Sze10] R. Szeliski. *Computer Vision: Algorithms and Applications*. Springer, Berlin, 2010.

[TBD96] S. Teller, K. Bala, and J. Dorsey. Conservative radiance envelopes for ray tracing. In *Proc. 7th Eurographics Workshop Rendering*, pages 105–114, 1996.

[VG95] E. Veach and L.J. Guibas. Optimally combining sampling techniques for Monte Carlo rendering. In *Proc. 22nd ACM Conf. SIGGRAPH*, pages 419–428, 1995.

[Whi80] T. Whitted. An improved illumination model for shading display. *Commun. ACM*, 23:343–349, 1980.

[Yao92] F.F. Yao. Computational geometry. In *Algorithms and Complexity, Handbook of Theoretical Computer Science*, volume A, Elsevier, Amsterdam, pages 343–389, 1992.

53 MODELING MOTION

Leonidas J. Guibas and Marcel Roeloffzen

INTRODUCTION

Motion is ubiquitous in the physical world, yet its study is much less developed than that of another common physical modality, namely shape. While we have several standardized mathematical shape descriptions, and even entire disciplines devoted to that area—such as *Computer-Aided Geometric Design* (CAGD)—the state of formal motion descriptions is still in flux. This in part because motion descriptions span many levels of detail; they also tend to be intimately coupled to an underlying physical process generating the motion (dynamics). Thus, there is a wide variety of work on algorithm descriptions of motion and their associated complexity measure. This chapter aims to show how an algorithmic study of motion is intimately tied to discrete and computational geometry. We first discuss some earlier work and various motion models. In Section 53.1 we then go into more detail on models that consider so-called *incremental motion*. We then devote the bulk of this chapter to discussing the framework of *Kinetic Data Structures* (Section 53.2) [Gui98, BGH99].

MOTION IN COMPUTATIONAL GEOMETRY

Dynamic computational geometry refers to the study of combinatorial changes in a geometric structure, as its defining objects undergo prescribed motions. For example, we may have n points moving linearly with constant velocities in $\mathbb{R}^2$, and may want to know the time intervals during which a particular point appears on their convex hull, the steady-state form of the hull (after all changes have occurred), or get an upper bound on how many times the convex hull changes during this motion. Such problems were introduced and studied in [Ata85].

A number of other authors have dealt with geometric problems arising from motion, such as collision detection (Chapter 39) or minimum separation determination [GJS96, ST95b, ST95a]. For instance, [ST95a] shows how to check in subquadratic time whether two collections of simple geometric objects (spheres, triangles) collide with each other under specified polynomial motions.

MOTION MODELS

An issue in the above research is that object motion(s) are assumed to be known in advance, sometimes in explicit form. A common assumption is that the coordinates of each point are bounded-degree polynomial functions of time and that these functions are known in advance. In essence, the proposed methods reduce questions about moving objects to other questions about derived static objects.

While most evolving physical systems follow known physical laws, it is also frequently the case that discrete events occur (such as collisions) that alter the motion law of one or more of the objects. Thus motion may be predictable in the short term, but becomes less so further into the future. Because of such discrete events, algorithms for modeling motion must be able to adapt in a dynamic way to motion model modifications. Furthermore, the occurrence of these events must be either predicted or detected, incurring further computational costs. Nevertheless, any truly useful model of motion must accommodate this *on-line* aspect of the temporal dimension, differentiating it from spatial dimensions, where all information is typically given at once.

Here, we distinguish two general motion models. The first considers unknown movement, where we don't have direct knowledge of motions. Instead, we rely on external updates or explicitly requesting new locations of the objects. However, most of the algorithms in this setting, as discussed in Section 53.1, make some assumptions on the motions, such as a bounded movement speed.

The second model is more similar to that of previous work and assumes that trajectories are known at least on the short-term. More precisely, the model assumes that point trajectories are known exactly and any changes in this trajectory are explicitly reported. This model is used by data structures in the KDS-framework discussed in Section 53.1. In most cases the motions are assumed to be polynomial (i.e., coordinates defined by bounded degree polynomials) or pseudo-algebraic, though this is mainly needed to prove efficiency, not correctness of the data structures. (The definition of pseudo-algebraic motion is given in Section 53.2 as it depends on the data-structure, however, polynomial movement is generally also pseudo-algebraic.)

53.1 INCREMENTAL MOTION

In real-world settings, the motion of objects may be imperfectly known and better information may only be obtainable at considerable expense. There have been several studies that incorporate this concept into their motion models. Most of these models are based on updating motions or locations of objects at discrete times. These updates may be obtained either by *pushing* or *pulling*, where in the former case an outside source or component provides new information and in the latter case the data structure must query for new location information when needed.

One way to deal with this more unpredictable motion is to separate the continuous tracking of motion from the more discrete maintenance of the data structure. Several works consider an observer component whose task is to observe motion and provide updated location or motion predictions to a builder component. The builder is then responsible for maintaining the data structure itself. The observer always has access to the exact location of the objects and the goal is to minimize communication between the two components. Communication can happen through *pushing* or *pulling*.

In case of pushing the observer is tasked with ensuring that the builder always has a sufficiently accurate location [YZ12]. Efficiency can then be measured in terms of competitive ratio, that is, by comparing the number of updates sent by the observer with those by an optimal update schedule that knows the full motions, but must adhere to the same accuracy bounds.

Information can also be pulled from the observer by the builder. In this case the builder queries the observer for the location of specific objects when more precise information is needed to ensure a correct data structure [Kah91]. This is convenient when there are strict bounds on the movement rates or directions of objects that enable to builder to compute its own guarantees on the objects locations (e.g., an object must be within a certain region). A compromise between the pulling and pushing mechanism can be found by allowing the builder to provide triggers to the observer [MNP⁺04, CMP09]. A trigger is a simple assumption on the location of the objects, for example that it remains within a certain region or follow a given trajectory. When such an assumption is no longer valid the observer can then inform the builder, which can then update its internal structures and provide new triggers. Triggers are similar to *certificates* that form the basis of the kinetic data structures discussed in Section 53.2.

In practice, the observer and builder may not be part of the same system and communication is not instantaneous. That is, we cannot neglect the fact that time passes and objects move while a location update is send from the observer to the builder. This can be modeled by maintaining regions in which each object resides, as time passes these regions grow. A location update then leads to one region shrinking to a point while every other region grows. When considering objects with a bounded movement rate, but unknown motions, the basic problem of minimizing the number of regions that overlap—that is, the maximum number of regions containing the same point—is hard, even when we aim to minimize ply only at a given time in the future. However, a strategy that approximates this number exists under certain conditions [EKLS16].

A much simpler motion model is that of *stepwise incremental motion*, where objects cannot be continuously observed. Instead, their locations are updated at discrete points in time [BRS12a, BRS12b, BRS13]. At each point the data structure must be repaired. When considering basic constraints on movement and distribution of the objects, this allows for updating strategies that are faster than rebuilding the whole structure from scratch with each location update.

53.2 KINETIC DATA STRUCTURES

Suppose we are interested in tracking high-level attributes of a geometric system of objects in motion such as, for example, the convex hull of a set on n points moving in $\mathbb{R}^2$. Note that as the points move continuously, their convex hull will be a continuously evolving convex polygon. At certain discrete moments, however, the combinatorial structure of the convex hull will change (that is, the circular sequence of a subset of the points that appear on the hull will change). In between such moments, tracking the hull is straightforward: its geometry is determined by the positions of the sequence of points forming the hull. How can we know when the combinatorial structure of the hull changes? The idea is that we can focus on certain elementary geometric relations among the n points, a set of cached assertions we call *certificates*, which altogether certify the correctness of the current combinatorial structure of the hull. Furthermore, we can hope to choose these relations in such a way so that when one of them fails because of point motion, both the hull and its set of certifying relations can be updated locally and incrementally, so that the whole process can continue.

GLOSSARY

Kinetic data structure: A kinetic data structure (KDS) for a geometric attribute is a collection of simple geometric relations that certifies the combinatorial structure of the attribute, as well as a set of rules for repairing the attribute and its certifying relations when one relation fails.

Certificate: A certificate is an elementary geometric relation used in a KDS.

Event: An event is the failure of a KDS certificate during motion. Events are classified as *external* when the combinatorial structure of the attribute changes, and *internal*, when the structure of the attribute remains the same, but its certification needs to change.

Event queue: In a KDS, all certificates are placed in an event queue, according to their earliest failure time.

Pseudo-algebraic motion: The class of allowed motions is usually specified as the class of pseudo-algebraic motions, in which each KDS certificate can flip between true and false at most a bounded number of times. Note that this class is specific for each KDS, but points whose coordinates are bounded-degree polynomials of time generally satisfy this condition.

The inner loop of a KDS consists of repeated certificate failures and certification repairs, as depicted in Figure 53.2.1.

FIGURE 53.2.1
The inner loop of a kinetic data structure.

We remark that in the KDS framework, objects are allowed to change their motions at will, with appropriate notification to the data structure. When this happens all certificates involving the object whose motion has changed must reevaluate their failure times.

CONVEX HULL EXAMPLE

Suppose we have four points a, b, c, and d in $\mathbb{R}^2$, and wish to track their convex hull. For the convex hull problem, the most important geometric relation is the CCW predicate: $\mathrm{CCW}(a, b, c)$ asserts that the triangle abc is oriented counterclockwise. Figure 53.2.2 shows a configuration of four points and four CCW relations that hold among them. It turns out that these four relations are sufficient to prove that the convex hull of the four points is the triangle abc. Indeed the points can move and form different configurations, but as long as the four certificates shown remain valid, the convex hull must be *abc*.

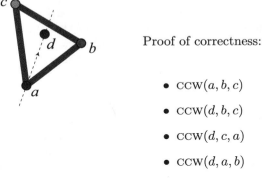

Proof of correctness:

- CCW(a, b, c)
- CCW(d, b, c)
- CCW(d, c, a)
- CCW(d, a, b)

FIGURE 53.2.2
Determining the convex hull of the points.

Now suppose that points a, b, and c are stationary and only point d is moving, as shown in Figure 53.2.3. At some time t_1 the certificate CCW(d, b, c) will fail, and at a later time t_2 CCW(d, a, b) will also fail. Note that the certificate CCW(d, c, a) will never fail in the configuration shown even though d is moving. So the certificates CCW(d, b, c) and CCW(d, a, b) schedule events that go into the event queue. At time t_1, CCW(d, b, c) ceases to be true and its negation, CCW(c, b, d), becomes true. In this simple case the three old certificates, plus the new certificate CCW(c, b, d), allow us to conclude that the convex hull has now changed to *abdc*.

Old proof	New proof
CCW(a, b, c)	CCW(a, b, c)
CCW(d, b, c)	CCW(c, b, d)
CCW(d, c, a)	CCW(d, c, a)
CCW(d, a, b)	CCW(d, a, b)

FIGURE 53.2.3
Updating the convex hull of the points.

If the certificate set is chosen judiciously, the KDS repair can be a local, incremental process—a small number of certificates may leave the cache, a small number may be added, and the new attribute certification will be closely related to the old one. A good KDS exploits the continuity or coherence of motion and change in the world to maintain certifications that themselves change only incrementally and locally as assertions in the cache fail.

PERFORMANCE MEASURES FOR KDS

Because a KDS is not intended to facilitate a terminating computation but rather an on-going process, we need to use somewhat different measures to assess its complexity. In classical data structures there is usually a tradeoff between operations that interrogate a set of data and operations that update the data. We commonly seek a compromise by building indices that make queries fast, but such that updates to the set of indexed data are not that costly as well. Similarly in the KDS setting, we must at the same time have access to information that facilitates or trivializes the computation of the attribute of interest, yet we want information that is relatively stable and not so costly to maintain. Thus, in the same way that classical data structures need to balance the efficiency of access to the data with the ease of its update, kinetic data structures must tread a delicate path between "knowing too little" and "knowing too much" about the world. A good KDS will select a certificate set that is at once economical and stable, but also allows a quick repair of itself and the attribute computation when one of its certificates fails. To measure these concerns, four quality criteria are introduced below. To measure efficiency, we need to define a class of motions that the KDS accepts. Generally this is the class of *pseudo-algebraic* motions. That is, the class of all motions where each certificate can flip between true and false at most a bounded number of times. Although this definition depends on the data-structure, the class of polynomial motions (motions defined by bounded degree polynomial functions) are also pseudo-algebraic in mostly all KDSs.

GLOSSARY

Responsiveness: A KDS is *responsive* if the cost, when a certificate fails, of repairing the certificate set and updating the attribute computation is small. By "small" we mean polylogarithmic in the problem size—in general we consider small quantities that are polylogarithmic or $O(n^\epsilon)$ in the problem size.

Efficiency: A KDS is *efficient* if the number of certificate failures (total number of events) it needs to process is comparable to the number of required changes in the combinatorial attribute description (external events), over some class of allowed motions. Technically, we require that the ratio of total events to external events is small.

Compactness: A KDS is *compact* if the size of the certificate set it needs is close to linear in the degrees of freedom of the moving system.

Locality: A KDS is *local* if no object participates in too many certificates; this condition makes it easier to re-estimate certificate failure times when an object changes its motion law. (The existence of local KDSs is an intriguing theoretical question for several geometric attribute functions.)

CONVEX HULL, REVISITED

We now briefly describe a KDS for maintaining the convex hull of n points moving around in the plane [BGH99].

The key goal in designing a KDS is to produce a ***repairable certification*** of the geometric object we want to track. In the convex hull case it turns out that it is a bit more intuitive to look at the dual problem, that of maintaining the upper (and lower) envelope of a set of moving lines in the plane, instead of the convex hull of the primal points. For simplicity we focus only on the upper envelope part from now on; the lower envelope case is entirely symmetric. Using a standard divide-and-conquer approach, we partition our lines into two groups of size roughly $n/2$ each, and assume that recursive invocations of the algorithm maintain the upper envelopes of these groups. For convenience, call the groups red and blue.

In order to produce the upper envelope of all the lines, we have to merge the upper envelopes of the red and blue groups and also certify this merge, so we can detect when it ceases to be valid as the lines move; see Figure 53.2.4.

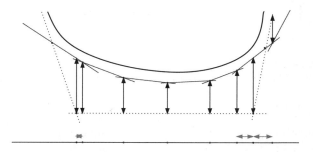

FIGURE 53.2.4
Merging the red and blue upper envelopes. In this example, the red envelope (solid line) is above the blue (dotted line), except at the extreme left and right areas. Vertical double-ended arrows represent y-certificates and horizontal double-ended arrows represent x-certificates, as described below.

Conceptually, we can approach this problem by sweeping the envelopes with a vertical line from left to right. We advance to the next red (blue) vertex and determine if it is above or below the corresponding blue (red) edge, and so on. In this process we determine when red is above blue or vice versa, as well as when the two envelopes cross. By stitching together all the upper pieces, whether red or blue, we get a representation of the upper envelope of all the lines.

The certificates used in certifying the above merge are of three flavors:

- x-certificates $(<_x)$ are used to certify x-ordering among the red and blue vertices; these involve four original lines.

- y-certificates $(<_y)$ are used to certify that a vertex is above or below an edge of the opposite color; these involve three original lines and are exactly the duals of the CCW certificates discussed earlier.

- s-certificates $(<_s)$ are slope comparisons between pairs of original lines; though these did not arise in our sweep description above, they are needed to make the KDS local [BGH99].

Figure 53.2.5 shows examples of how these types of certificates can be used to specify x-ordering constraints and to establish intersection or nonintersection of the envelopes. A total of $O(n)$ such certificates suffice to verify the correctness of the upper envelope merge.

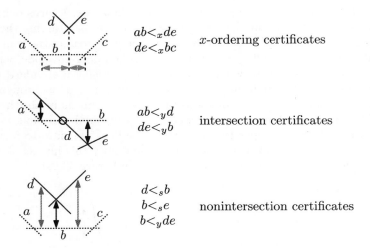

$ab<_x de$
$de<_x bc$ x-ordering certificates

$ab<_y d$
$de<_y b$ intersection certificates

$d<_s b$
$b<_s e$ nonintersection certificates
$b<_y de$

FIGURE 53.2.5
Using the different types of certificates to certify the red-blue envelope merge.

Whenever the motion of the lines causes one of these certificates to fail, a local, constant-time process suffices to update the envelope and repair the certification. Figure 53.2.6 shows an example where a y-certificate fails, allowing the blue envelope to poke up above the red.

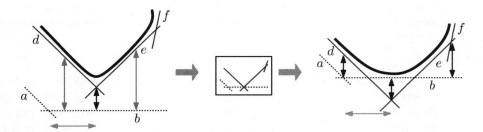

FIGURE 53.2.6
Envelope repair after a certificate failure. In the event shown, lines b, d, and e become concurrent, producing a red-blue envelope intersection.

It is straightforward to prove that this kinetic upper envelope algorithm is responsive, local, and compact, using the logarithmic depth of the hierarchical structure of the certification. In order to bound the number of events processed, however, we must assume that the line motions are polynomial or at least pseudo-algebraic. A proof of efficiency can be developed by extruding the moving lines into space-time surfaces. Using certain well-known theorems about the complexity of upper envelopes of surfaces [Sha94] and the overlays of such envelopes [ASS96] (cf. Chapter 50), it can be shown that in the worst case the number of events processed by this algorithm is near quadratic ($O(n^{2+\epsilon})$, for any constant $\epsilon > 0$, where the constant of proportionality depends on ϵ). Since the convex hull of even linearly moving points can change $\Omega(n^2)$ times [AGHV01], the efficiency result follows. No comparable structure is known for the convex hull of points in dimensions $d \geq 3$.

EXTENT PROBLEMS

A number of problems for which kinetic data structures were developed are aimed at different measures of how "spread out" a moving set of points in $\mathbb{R}^2$ is—one example is the convex hull, whose maintenance was discussed in the previous subsection. Other measures of interest include the diameter, width, and smallest area or perimeter bounding rectangle for a moving set S of n points. All these problems can be solved using the kinetic convex hull algorithm above; the efficiency of the algorithms is $O(n^{2+\epsilon})$, for any $\epsilon > 0$. There are also corresponding $\Omega(n^2)$ lower bounds for the number of combinatorial changes in these measures. Surprisingly, the best upper bound known for maintaining the smallest enclosing disk containing S is still near-cubic. Extensions of these results to dimensions higher than two are also lacking.

These costs can be dramatically reduced if we consider approximate extent measures. If we are content with $(1 + \epsilon)$-approximations to the measures, then an approximate smallest axis-aligned rectangle, diameter, and smallest enclosing disk can be maintained with a number of events that is a function ϵ only and not of n [AHP01]. For example, the bound of the number of approximate diameter updates in $\mathbb{R}^2$ under linear motion of the points is $O(1/\epsilon)$.

PROXIMITY PROBLEMS

The fundamental proximity structures in computational geometry are the Voronoi diagram and the Delaunay triangulation (Chapter 27). The edges of the Delaunay triangulation contain the closest pair of points, the closest neighbor to each point, as well as a wealth of other proximity information among the points. From the kinetic point of view, these are nice structures, because they admit completely local certifications. Delaunay's 1934 theorem [Del34] states that if a local empty sphere condition is valid for each $(d-1)$-simplex in a triangulation of points in $\mathbb{R}^d$, then that triangulation must be Delaunay. This makes it simple to maintain a Delaunay triangulation under point motion: an update is necessary only when one of these empty sphere conditions fails. Furthermore, whenever that happens, a local retiling of space (of which the classic "edge-flip" in $\mathbb{R}^2$ is a special case; cf. Section 29.3) easily restores Delaunayhood. Thus the KDS for Delaunay (and Voronoi) that follows from this theorem is both responsive and efficient—in fact, each KDS event is an external event in which the structure changes. Tight bounds on the number of events were a long-standing open problem with only a quadratic lower bound and near-cubic upper bound being known [AGMR98]. Here, near-cubic (or near-quadratic) indicates at most a factor n^ϵ difference with cubic (or quadratic) for an arbitrarily small constant $\epsilon > 0$. Progress towards closing this gap has been made in several different ways. Firstly, by using a randomized triangulation an expected near-quadratic number of events occur [KRS11]. Secondly, the Voronoi diagram for a convex polygonal distance measure—a distance measure where the unit ball is a convex polygon—can also be maintained with near-quadratic events [AKRS15]. Thirdly, better bounds on the number of events in the Euclidean setting can be proven when motions are more restricted. When the points move along straight-line trajectories with constant speed, then only near-quadratic events occur [Rub15]. Lastly, one could consider not maintaining all edges. Specifically the *stable* edges of a Delaunay triangulation can be maintained with a near-quadratic number of events [AGG+15]. Here a Delaunay edge is stable if the angle from either of its endpoints to the endpoints of the corresponding Voronoi edge is large enough. Intuitively, stable edges are those that are not close to flipping as their Voronoi edges are far from collapsing.

A set of easily checked local conditions that implies a global property has also been used in kinetizing other proximity structures. For instance, in the *power diagram* [Aur87] of a set of disjoint balls, the two closest balls must be neighbors [GZ98]—and this diagram can be kinetized by a similar approach. Voronoi diagrams of more general objects, such as convex polytopes, have also been investigated. For example, in $\mathbb{R}^2$ [GSZ00] shows how to maintain a compact Voronoi-like diagram among moving disjoint convex polygons; again, a set of local conditions is derived which implies the global correctness of this diagram. As the polygons move, the structure of this diagram allows one to know the nearest pair of polygons at all times.

In many applications we do not need the full Delaunay triangulation and instead maintaining the nearest neighbor for each vertex or the *minimum spanning tree* may be sufficient. Instead of maintaining the Delaunay graph we can instead maintain the *Theta-* or *Yao*-graph. These are defined as follows. For each point p, the plane is divided into cones, then for each nonempty cone we create an edge from p to the nearest point in that cone, where nearest for the Theta-graph is nearest in the direction of the cone, whereas for the Yao graph it is the nearest in the Euclidean metric. The Theta-graph based on six cones can be used to maintain the global nearest neighbor pair and the graph encounters $\Theta(n^2)$ events as the number of events is bounded by the number of changes in the sorted orders along the directions of the cones, which are fixed [BGZ97]. Using variations of the Yao graph it is also possible to maintain the minimum spanning tree and nearest neighbors for all vertices, and, although the analysis is more complicated, this also yields a near-quadratic number of events [RAK+15].

TRIANGULATIONS AND TILINGS

Many areas in scientific computation and physical modeling require the maintenance of a triangulation (or more generally a simplicial complex) that approximates a manifold undergoing deformation. The problem of maintaining the Delaunay triangulation of moving points in the plane mentioned above is a special case. More generally, local re-triangulations are necessitated by collapsing triangles, and sometimes required in order to avoid undesirably "thin" triangles. In certain cases the number of nodes (points) may also have to change in order to stay sufficiently faithful to the underlying physical process; see, for example, [CDES01]. Because in general a triangulation meeting certain criteria is not unique or canonical, it becomes more difficult to assess the efficiency of kinetic algorithms for solving such problems. The lower-bounds in [ABB+00] indicate that one cannot hope for a subquadratic bound on the number of events in the worst case for the maintenance of *any* triangulation, even if a linear number of additional Steiner points is allowed.

When considering deterministic algorithms the best-known result for maintaining a triangulation uses $O(n^{7/3})$ events [ABG+02]. However, when allowing randomized triangulation it was shown that indeed $O(n^{2+\epsilon})$ events is possible with high probability. The first near-quadratic triangulation uses a hierarchical scheme where first a subset of points is triangulated, splitting the pointset into subsets that are then further triangulated [AWY06]. A more recent result slightly improves the bounds on number of events, but is also simpler [KRS11]. In fact, the approach is closer to the much earlier deterministic result; in both cases, first a pseudotriangulation is created, which is then triangulated further.

COLLISION DETECTION

Collision detection is the problem of determining whether there are insections between objects of a given input set. The more interesting version of this problem is of course when the objects move and collisions have to be detected as soon as they occur, see Chapter 39 for more background on collision detection. Kinetic methods are naturally applicable to the problem of collision detection between moving geometric objects. Typically collisions occur at irregular intervals, so that fixed-time stepping methods have difficulty selecting an appropriate sampling rate to fit both the numerical requirements of the integrator as well as those of collision detection. A kinetic method based on the discrete events that are the failures of relevant geometric conditions can avoid the pitfalls of both oversampling and undersampling the system. For two moving convex polygons in the plane, a kinetic algorithm where the number of events is a function of the relative separation of the two polygons is given in [EGSZ99]. The algorithm is based on constructing certain outer hierarchies on the two polygons. Analogous methods for 3D polytopes were presented in [GXZ01], together with implementation data. Such methods however do not easily extend to situations with many objects, hence a different approach is needed.

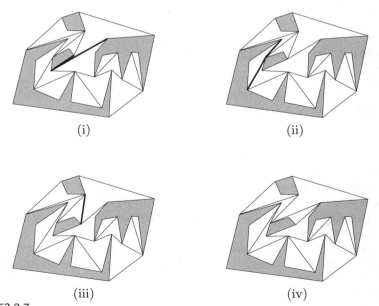

(i)   (ii)

(iii)   (iv)

FIGURE 53.2.7

Snapshots of the mixed pseudotriangulation of [ABG+02]. As the center trapezoid-like polygon moves to the right, the edges corresponding to the next about-to-fail certificate are highlighted.

One option is to tile the free space around objects. Such a tiling can serve as a proof of nonintersection of the objects. If such a tiling can be efficiently maintained under object motion, then it can be the basis of a kinetic algorithm for collision detection. Several papers have developed techniques along these lines, including the case of two moving simple polygons in the plane [BEG+04], or multiple moving polygons [ABG+02, KSS02]. These developments all exploit deformable pseudotriangulations of the free space—tilings which undergo fewer combinatorial changes than, for example, triangulations. An example from [ABG+02] is shown in Figure 53.2.7. The figure shows how the pseudotriangulation adjusts by local retiling to the motion of the inner quadrilateral. An advantage of all these methods is that the number of certificates needed is close to the size of the min-link separating subdivision of the objects, and thus sensitive to how intertwined the objects are.

Deformable and nonpolygonal objects are more challenging to handle. Static methods such as bounding volume hierarchies [GLM96] deal with more complex objects by using two phases. In the *broad phase* a selection of pairs of objects is found that may be colliding. In the *narrow phase* these pairs of objects are explicitly tested for intersection. Efficiency of this method depends heavily on the number of pairs produced in the broad phase, so also broad-phase algorithms have been studied within the KDS framework. In a kinetic setting the pairs produced by the broad phase can be used as certificates of nonintersection. Based on this idea efficient KDSs have been produced for deformable [AGN+04] and constant-complexity convex objects in 3D [ABPS09]. In both results the number of potentially intersecting pairs, and hence, the number of certificates is close to linear. For unit balls of similar size a kinetic data structure has also been developed based on dividing space into cells and maintaining which cells the spheres intersect [KGS98]. This method was also used in a more general setting where motions are not known exactly and experimentally shown to work well under reasonable motions [KGS05].

CONNECTIVITY AND CLUSTERING

Closely related to proximity problems is the issue of maintaining structures encoding connectivity among moving geometric objects. Connectivity problems arise frequently in *ad hoc* mobile communication and sensor networks, where the viability of links may depend on proximity or direct line-of-sight visibility among the stations desiring to communicate. With some assumptions, the communication range of each station can be modeled by a geometric region, so that two stations can establish a link if and only if their respective regions overlap. There has been work on kinetically maintaining the connected components of the union of a set of moving geometric regions for the case of rectangles [HS99] and unit disks [GHSZ01].

Clustering mobile nodes is an essential step in many algorithms for establishing communication hierarchies, or otherwise structuring *ad hoc* networks. Nodes in close proximity can communicate directly, using simpler protocols; correspondingly, well-separated clusters can reuse scarce resources, such as the same frequency or time-division multiplexing communication scheme, without interference. Maintaining clusters of mobile nodes requires a tradeoff between the tightness, or optimality of the clustering, and its stability under motion. A basic clustering scheme is to cover the points with a minimum number of unit boxes. A randomized scheme that maintains such a clustering was given by Gao *et al.* [GGH+03]. This scheme maintains a number of boxes that is within a (large) constant factor of the minimum number needed. Later, a deterministic solution was found that works in higher dimensions with an approximation ratio of 3^d in d dimensions [Her05]. Both schemes have the nice property that the clustering is smooth, that is, each event affects only a small number of boxes.

A next step in establishing a connected network between mobile nodes is to create a connectivity graph that can be used for routing. Graphs that work well for this purpose have a bounded *spanning ratio*. That is, for any two nodes, the ratio between their Euclidean distance and their shortest distance along the graph is bounded. Here, the distance along the graph is the sum of Euclidean lengths of all edges along a path between the two nodes. Graph for which this spanning ratio is bounded are generally referred to as *spanners*. One of the earliest kinetic spanners was based on the randomized clustering scheme from [GGH+03] and had a constant spanning ratio. Using a different approach Gao *et al.* [GGN06] proposed a spanner with a spanning ratio of $(1 + \epsilon)$ for any constant $\epsilon > 0$. This spanner also allows logarithmic time insertion and deletion of points. However, the number of events and query time depend on the spread of the input set. That is, the ratio between the shortest distance and longest distance between any pair of points. The dependency on the spread was removed by Abam *et al.* who present two spanners with spanning ratio $(1 + \epsilon)$. The first works for any constant number of dimensions [AB11], but is somewhat complex, whereas the other works only in two dimensions, but is much simpler [ABG10].

VISIBILITY

The problem of maintaining the visible parts of the environment when an observer is moving is one of the classic questions in computer graphics and has motivated significant developments, such as binary space partition trees, the hardware depth buffer, etc. The difficulty of the question increases significantly when the environment itself includes moving objects; whatever visibility structures accelerate occlusion culling for the moving observer must now themselves be maintained under object motion.

Binary space partitions (BSP) are hierarchical partitions of space into convex tiles obtained by performing planar cuts (Section 33.8.2). Tiles are refined by further cuts until the interior of each tile is free of objects or contains geometry of limited complexity. Once a BSP tree is available, a correct visibility ordering for all geometry fragments in the tiles can be easily determined and incrementally maintained as the observer moves. A kinetic algorithm for visibility can be devised by maintaining a BSP tree as the objects move. The key insight is to certify the correctness of the BSP tree through certain combinatorial conditions, whose failure triggers localized tree rearrangements—most of the classical BSP construction algorithms do not have this property. In $\mathbb{R}^2$, a randomized algorithm for maintaining a BSP of moving disjoint line segments is given in [AGMV00]. The algorithm processes $O(n^2)$ events, the expected cost per tree update is $O(\log n)$, and the expected tree size is $O(n \log n)$. The maintenance cost increases to $O(n\lambda_{s+2}(n)\log^2 n)$ [AEG98] for disjoint moving triangles in $\mathbb{R}^3$ (s is a constant depending on the triangle motion). Both of these algorithms are based on variants of vertical decompositions (many of the cuts are parallel to a given direction). It turns out that in practice these generate "sliver-like" BSP tiles that lead to robustness issues [Com99].

As the pioneering work on the visibility complex has shown [PV96], another structure that is well suited to visibility queries in $\mathbb{R}^2$ is an appropriate pseudo-triangulation. Given a moving observer and convex moving obstacles, a full radial decomposition of the free space around the observer is quite expensive to maintain. One can build pseudotriangulations of the free space that become more and more like the radial decomposition as we get closer to the observer. Thus one can have a structure that compactly encodes the changing visibility polygon around the observer, while being quite stable in regions of the free space well occluded from the observer [HH02].

FACILITY LOCATION

A common networking strategy is to use hierarchies, where a small set of the nodes is selected as hubs that serve as communication gateways for the nodes around them. We can then establish a network between these hubs either inductively, by sampling superhubs, or through another paradigm. These networks have the advantage that routing is generally very simple as each node just needs to communicate to its nearest hub and the hubs can be connected with much simpler protocols, because there are only few. Selecting hubs from a set of nodes can be seen as the classic facility location problem, originally studied in the context of a facility (store, distribution centers, post office) serving a number of clients (customers, stores, addresses). Here the facilities are the network hubs and the clients the other nodes. Many variations of this problem exist, which consider the maintenance cost of operating a facility, the distance of clients to the facility, and the number of clients a facility can serve.

With the increased number of mobile networks it becomes interesting to study this problem in a mobile setting, where clients and facilities move. Clients may then be promoted to servers and servers demoted to clients. A recently studied variant considers a set of moving points that may be used as client or as facility moving (as before) along pseudo-algebraic trajectories. Each point has a maintenance cost when it is used as a facility and a demand when it is a client [DGL10]. The quality measure here is the total maintenance cost of all facilities plus the distances of each client to its nearest facility multiplied by their demand. In this setting a constant factor approximation of the optimal solution can be maintained with a quadratic number of events and a logarithmic number of status changes (changing a client to a facility of vice versa) per event. Note that the number of events and number of changes per event also depend on the ratio between the largest and smallest maintenance cost as well as the ratio between the largest and smallest demand of any point.

Unfortunately these status changes may be quite expensive in practice as they necessitate changes in network structure, which often results in loss of packages. An open problem is whether an efficient strategy exists that takes the cost of a status change into account explicitly. Such an algorithm would have to be analyzed in a competitive way.

QUERYING MOVING OBJECTS

Continuous tracking of a geometric attribute may be more than is needed for some applications. There may be time intervals during which the value of the attribute is of no interest; in other scenarios we may be just interested to know the attribute value at certain discrete query times. For example, given n moving points in $\mathbb{R}^2$, we may want to pose queries asking for all points inside a rectangle R at time t, for various values of R and t, or for an interval of time Δt, etc. A number of other classical range-searching structures, such as k-d-trees and R-trees been investigated for moving objects [PAHP02, ABS09].

Another interesting observation is that maintenance of a kinetic data structure may be made more efficient if we allow queries to spend more time. Such trade-offs were investigated for kinetic dictionaries [Ber03, AAE03] and for sorting and convex hulls [AB07].

OPEN PROBLEMS

As mentioned above, we still lack efficient kinetic data structures for many fundamental geometric questions. Here is a short list of such open problems:

1. Find an efficient (and responsive, local, and compact) KDS for maintaining the convex hull of points moving in dimensions $d \geq 3$.

2. Find an efficient KDS for maintaining the smallest enclosing disk in $d \geq 2$. For $d = 2$, a goal would be an $O(n^{2+\epsilon})$ algorithm.

3. Find a KDS to maintain the MST of moving points under the Euclidean metric achieving subquadratic bounds.

4. Maintain mobile facilities with a guaranteed small number of facility changes while maintaining low cost in terms of facility maintenance and travel distance of clients.

Beyond specific problems, there are also several important structural issues that require further research in the KDS framework. These include:

Recovery after multiple certificate failures. We have assumed up to now that the KDS assertion cache can detect certificate failures exactly. In real-world applications this may be impossible due to inexact knowledge of the trajectories or finite precision computations. Multiple certificates may fail at the same time or may be treated in the wrong order. Although for some structures work has been done to create robust KDSs this is still ongoing work and a key step in making KDSs more applicable in practice [AABY11].

There is also a related subtlety in the way that a KDS assertion cache can certify the value, or a computation yielding the value, of the attribute of interest. Suppose our goal is to certify that a set of moving points in the plane, in a given circular order, always forms a convex polygon. A plausible certificate set for convexity is that all interior angles of the polygon are convex. See Figure 53.2.8. In the normal KDS setting where we can always predict accurately the next certificate failure, it turns out that the above certificate set is sufficient, *as long as at the beginning of the motion the polygon was convex.* One can draw, however, nonconvex self-intersecting polygons all of whose interior angles are convex, as shown in the same figure. The point here is that a standard KDS can offer a *historical* proof of the convexity of the polygon by relying on the fact that the certificate set is valid *and* that the polygon was convex during the prior history of the motion. Indeed the counterexample shown cannot arise under continuous motion without one of the angle certificates failing first. On the other hand, if an oracle can move the points when "we are not looking," we can wake up and find all the angle certificates to be valid, yet our polygon need not be convex. Thus in this oracle setting, since we cannot be sure that no certificates failed during the time step, we must insist on *absolute* proofs—certificate sets that in any state of the world fully validate the attribute computation or value.

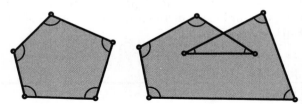

FIGURE 53.2.8
Certifying the convexity of a polygon.

Hierarchical motion descriptions. Objects in the world are often organized into groups and hierarchies and the motions of objects in the same group are highly correlated. For example, though not all points in an elastic bouncing ball follow exactly the same rigid motion, the trajectories of nearby points are very similar and the overall motion is best described as the superposition of a global rigid motion with a small local deformation. Similarly, the motion of an articulated figure, such as a man walking, is most succinctly described as a set of relative motions, say that of the upper right arm relative to the torso, rather than by giving the trajectory of each body part in world coordinates.

What both of these examples suggest is that there can be economies in motion description, if the motion of objects in the environment can be described as a superposition of terms, some of which can be shared among several objects. Such hierarchical motion descriptions can simplify certificate evaluations, as certificates are often local assertions concerning nearby objects, and nearby objects tend to share motion components. For example, in a simple articulated figure, we may wish to assert $\mathrm{CCW}(A, B, C)$ to indicate that an arm is not fully extended, where $\overline{AB}$ and $\overline{BC}$ are the upper and lower parts of the arm, respectively. Evaluating this certificate is clearly better done in the local coordinate frame of the upper arm than in a world frame—the redundant motions of the legs and torso have already been factored out.

Motion sensitivity. As already mentioned, the motions of objects in the world are often highly correlated and it behooves us to find representations and data structures that exploit such motion coherence. It is also important to find mathematical measures that capture the degree of coherence of a motion and then use this as a parameter to quantify the performance of motion algorithms. If we do not do this, our algorithm design may be aimed at unrealistic worst-case behavior, without capturing solutions that exploit the special structure of the motion data that actually arise in practice—as already discussed in a related setting in [BSVK02]. Thus it is important to develop a class of kinetic *motion-sensitive* algorithms, whose performance can be expressed as a function of how coherent the motions of the underlying objects are.

Noncanonical structures. The complexity measures for KDSs mentioned earlier are more suitable for maintaining *canonical* geometric structures, which are uniquely defined by the position of the data, e.g., convex hull, closest pair, and Delaunay triangulation. In these cases the notion of external events is well defined and is independent of the algorithm used to maintain the structure. On the other hand, as we already discussed, suppose we want to maintain a triangulation of a moving point set. Since the triangulation of a point set is not unique, the external events depend on the triangulation being maintained, and thus depend on the algorithm. This makes it difficult to analyze the efficiency of a kinetic triangulation algorithm. Most of the current approaches for maintaining noncanonical structures artificially impose canonicality and maintain the resulting canonical structure. But this typically increases the number of events. So it is entirely possible that methods in which the current form of the structure may depend on its past history can be more efficient. Unfortunately, we lack mathematical techniques for analyzing such history-dependent structures.

53.3 SOURCES AND RELATED MATERIALS

SURVEYS

Results not given an explicit reference above may be traced in these surveys.

[Gui98]: An early, and by now somewhat dated, survey of KDS work.

[AGE⁺02]: A report based on an NSF-ARO workshop, addressing several issues on modeling motion from the perspective of a variety of disciplines.

[Gui02]: A "popular-science" type article containing material related to the costs of sensing and communication for tracking motion in the real world.

RELATED CHAPTERS

Chapter   9: Geometry and topology of polygonal linkages
Chapter 26: Convex hull computations
Chapter 27: Voronoi diagrams and Delaunay triangulations
Chapter 29: Triangulations and mesh generation
Chapter 39: Collision and proximity queries

REFERENCES

[AABY11]   M.A. Abam, P.K. Agarwal, M. de Berg, and H. Yu. Out-of-order event processing in kinetic data structures. *Algorithmica*, 60:250–273, 2011.

[AAE03]   P.K. Agarwal, L. Arge, and J. Erickson. Indexing moving points. *J. Comp. Syst. Sci.*, 66:207–243, 2003.

[AB07]   M.A. Abam and M. de Berg. Kinetic sorting and kinetic convex hulls. *Comput. Geom.*, 37:16–26, 2007.

[AB11]   M.A. Abam and M. de Berg. Kinetic spanners in $\mathbb{R}^d$. *Discrete Comput. Geom.*,

45:723–736, 2011.

[ABB+00]   P.K. Agarwal, J. Basch, M. de Berg, L.J. Guibas, and J. Hershberger. Lower bounds for kinetic planar subdivisions. *Discrete Comput. Geom.*, 24:721–733, 2000.

[ABG+02]   P.K. Agarwal, J. Basch, L.J. Guibas, J. Hershberger, and L. Zhang. Deformable free-space tilings for kinetic collision detection. *I. J. Robotic Res.*, 21:179–198, 2002.

[ABG10]   M.A. Abam, M. de Berg, and J. Gudmundsson. A simple and efficient kinetic spanner. *Comput. Geom.*, 43:251–256, 2010.

[ABPS09]   M.A. Abam, M. de Berg, S.-H. Poon, and B. Speckmann. Kinetic collision detection for convex fat objects. *Algorithmica*, 53:457–473, 2009.

[ABS09]   M.A. Abam, M. de Berg, and B. Speckmann. Kinetic kd-trees and longest-side kd-trees. *SIAM J. Comput.*, 39:1219–1232, 2009.

[AEG98]   P.K. Agarwal, J. Erickson, and L.J. Guibas. Kinetic binary space partitions for intersecting segments and disjoint triangles (extended abstract). In *Proc. 9th ACM-SIAM Sympos. Discrete Algorithms*, pages 107–116, 1998.

[AGG+15]   P.K. Agarwal, J. Gao, L.J. Guibas, H. Kaplan, N. Rubin, and M. Sharir. Stable Delaunay graphs. *Discrete Comput. Geom.*, 54:905–929, 2015.

[AGHV01]   P.K. Agarwal, L.J. Guibas, J. Hershberger, and E. Veach. Maintaining the extent of a moving point set. *Discrete Comput. Geom.*, 26:353–374, 2001.

[AGMR98]   G. Albers, L.J. Guibas, J.S.B. Mitchell, and T. Roos. Voronoi diagrams of moving points. *Internat. J. Comput. Geom. Appl.*, 8:365–380, 1998.

[AGMV00]   P.K. Agarwal, L.J. Guibas, T.M. Murali, and J.S. Vitter. Cylindrical static and kinetic binary space partitions. *Comput. Geom.*, 16:103–127, 2000.

[AGE+02]   P.K. Agarwal, L. Guibas, H. Edelsbrunner, et al. Algorithmic issues in modeling motion. *ACM Computing Surveys*, 34:550-572, 2002.

[AGN+04]   P.K. Agarwal, L.J. Guibas, A. Nguyen, D. Russel, and L. Zhang. Collision detection for deforming necklaces. *Comput. Geom.*, 28:137–163, 2004.

[AHP01]   P.K. Agarwal and S. Har-Peled. Maintaining approximate extent measures of moving points. In *Proc. 12th ACM-SIAM Sympos. Discrete Algorithms*, pages 148–157, 2001.

[AKRS15]   P.K. Agarwal, H. Kaplan, N. Rubin, and M. Sharir. Kinetic Voronoi diagrams and Delaunay triangulations under polygonal distance functions. *Discrete Comput. Geom.*, 54:871–904, 2015.

[ASS96]   P.K. Agarwal, O. Schwarzkopf, and M. Sharir. The overlay of lower envelopes and its applications. *Discrete Comput. Geom.*, 15:1–13, 1996.

[Ata85]   M.J. Atallah. Some dynamic computational geometry problems. *Comput. Math. Appl.*, 11:1171–1181, 1985.

[Aur87]   F. Aurenhammer. Power diagrams: Properties, algorithms and applications. *SIAM J. Comput.*, 16:78–96, 1987.

[AWY06]   P.K. Agarwal, Y. Wang, and H. Yu. A two-dimensional kinetic triangulation with near-quadratic topological changes. *Discrete Comput. Geom.*, 36:573–592, 2006.

[BEG+04]   J. Basch, J. Erickson, L.J. Guibas, J. Hershberger, and L. Zhang. Kinetic collision detection between two simple polygons. *Comput. Geom.*, 27:211–235, 2004.

[Ber03]   M. de Berg. Kinetic dictionaries: How to shoot a moving target. In *Proc. 11th European Sympos. Algorithms*, pages 172–183, vol. 2832 of *LNCS*, Springer, Berlin, 2003.

[BGH99]   J. Basch, L.J. Guibas, and J. Hershberger. Data structures for mobile data. *J. Algorithms*, 31:1–28, 1999.

[BGZ97]   J. Basch, L.J. Guibas, and L. Zhang. Proximity problems on moving points. In *Proc. 13th Sympos. Comput. Geom.*, pages 344–351, ACM Press, 1997.

[BRS12a]   M. de Berg, M. Roeloffzen, and B. Speckmann. Kinetic compressed quadtrees in the black-box model with applications to collision detection for low-density scenes. In *Proc. 20th European Sympos. Algorithms*, pages 383–394, vol. 7501 of *LNCS*, Springer, Berlin, 2012.

[BRS12b]   M. de Berg, M. Roeloffzen, and B. Speckmann. Kinetic convex hulls, Delaunay triangulations and connectivity structures in the black-box model. *J. Comput. Geom.*, 3:222–249, 2012.

[BRS13]   M. de Berg, M. Roeloffzen, and B. Speckmann. Kinetic 2-centers in the black-box model. In *Proc. 29th Sympos. Comput. Geom.*, pages 145–154, ACM Press, 2013.

[BSVK02]   M. de Berg, A.F. van der Stappen, J. Vleugels, and M.J. Katz. Realistic input models for geometric algorithms. *Algorithmica*, 34:81–97, 2002.

[CDES01]   H.-L. Cheng, T.K. Dey, H. Edelsbrunner, and J. Sullivan. Dynamic skin triangulation. *Discrete Comput. Geom.*, 25:525–568, 2001.

[CMP09]   M. Cho, D.M. Mount, and E. Park. Maintaining nets and net trees under incremental motion. In *Proc. 20th Sympos. Internat. Sympos. Algorithms Computation*, pages 1134–1143, vol. 5878 of *LNCS*, Springer, Berlin, 2009.

[Com99]   J. Comba. *Kinetic Vertical Decomposition Trees*. PhD thesis, Stanford University, 1999.

[Del34]   B.N. Delaunay. Sur la sphére vide: à la mémoire de Georges Voronoi. *Izv. Akad. Nauk SSSR, Otdelenie Matematicheskih i Estestvennyh Nauk*, 7:793–800, 1934.

[DGL10]   B. Degener, J. Gehweiler, and C. Lammersen. Kinetic facility location. *Algorithmica*, 57:562–584, 2010.

[EGSZ99]   J. Erickson, L.J. Guibas, J. Stolfi, and L. Zhang. Separation-sensitive collision detection for convex objects. In *Proc. 10th ACM-SIAM Sympos. Discrete Algorithms*, pages 102–111, 1999.

[EKLS16]   W. Evans, D. Kirkpatrick, M. Löffler, and F. Staals. Minimizing Co-location potential of moving entities. *SIAM J. Comput.*, 45:1870–1893, 2016.

[GGH⁺03]   J. Gao, L. Guibas, J. Hershberger, L. Zhang, and A. Zhu. Discrete mobile centers. *Discrete Comput. Geom.*, 30:45–63, 2003.

[GGN06]   J. Gao, L.J. Guibas, and A.T. Nguyen. Deformable spanners and applications. *Comput. Geom.*, 35:2–19, 2006.

[GHSZ01]   L.J. Guibas, J. Hershberger, S. Suri, and L. Zhang. Kinetic connectivity for unit disks. *Discrete Comput. Geom.*, 25:591–610, 2001.

[GJS96]   P. Gupta, R. Janardan, and M. Smid. Fast algorithms for collision and proximity problems involving moving geometric objects. *Comput. Geom.*, 6:371–391, 1996.

[GLM96]   S. Gottschalk, M.C. Lin, and D. Manocha. OBB-Tree: A hierarchical structure for rapid interference detection. In *Proc. 23rd Conf. Comp. Graphics Interactive Techniques*, pages 171–180, ACM Press, 1996.

[GSZ00]   L. Guibas, J. Snoeyink, and L. Zhang. Compact Voronoi diagrams for moving convex polygons. In *Proc. Scand. Workshop Algorithms Data Structures*, vol. 1851 of *LNCS*, pages 339–352, Springer, Berlin, 2000.

[Gui98]   L. Guibas. Kinetic data structures: A state of the art report. In *Proc. 3rd Workshop Algorithmic Found. Robot.*, pages 191–209, A.K. Peters, Natick, 1998.

[Gui02]   L.J. Guibas. Sensing, tracking, and reasoning with relations. *IEEE Signal Proc. Magazine*, 19:73–85, 2002.

[GXZ01]   L.J. Guibas, F. Xie, and L. Zhang. Kinetic collision detection: Algorithms and experiments. In *Proc. IEEE Internat. Conf. Robotics and Automation*, pages 2903–2910, 2001.

[GZ98]   L. Guibas and L. Zhang. Euclidean proximity and power diagrams. In *Proc. 10th Canad. Conf. Comput. Geom.*, pages 90–91, Montréal, 1998.

[Her05]   J. Hershberger. Smooth kinetic maintenance of clusters. *Comput. Geom.*, 31:3–30, 2005.

[HH02]   O. Hall-Holt. *Kinetic Visibility*. PhD thesis, Stanford University, 2002.

[HS99]   J. Hershberger and S. Suri. Kinetic connectivity of rectangles. In *Proc. 15th Sympos. Comput. Geom.*, pages 237–246, ACM Press, 1999.

[Kah91]   S. Kahan. A model for data in motion. In *23rd ACM Sympos. Theory of Comput.*, pages 267–277, 1991.

[KGS98]   D.-J. Kim, L.J. Guibas, and S.-Y. Shin. Fast collision detection among multiple moving spheres. *IEEE Trans. Vis. Comput. Graph.*, 4:230–242, 1998.

[KGS05]   H.K. Kim, L.J. Guibas, and S.Y. Shin. Efficient collision detection among moving spheres with unknown trajectories. *Algorithmica*, 43:195–210, 2005.

[KRS11]   H. Kaplan, N. Rubin, and M. Sharir. A kinetic triangulation scheme for moving points in the plane. *Comput. Geom.*, 44:191–205, 2011.

[KSS02]   D.G. Kirkpatrick, J. Snoeyink, and B. Speckmann. Kinetic collision detection for simple polygons. *Int. J. Comput. Geom. Appl.*, 12:3–27, 2002.

[MNP+04]   D.M. Mount, N.S. Netanyahu, C.D. Piatko, R. Silverman, and A.Y. Wu. A computational framework for incremental motion. In *Proc. 20th Sympos. Comput. Geom.*, pages 200–209, ACM Press, 2004.

[PAHP02]   C.M. Procopiuc, P.K. Agarwal, and S. Har-Peled. STAR-tree: An efficient self-adjusting index for moving points. In *Proc. 4th Workshop on Algorithms Engineering Experiments*, vol. 2409 of *LNCS*, pages 178–193, Springer, Berlin, 2002.

[PV96]   M. Pocchiola and G. Vegter. The visibility complex. *Internat. J. Comput. Geom. Appl.*, 6:279–308, 1996.

[RAK+15]   Z. Rahmati, M.A. Abam, V. King, S. Whitesides, and A. Zarei. A simple, faster method for kinetic proximity problems. *Comput. Geom.*, 48:342–359, 2015.

[Rub15]   N. Rubin. On kinetic Delaunay triangulations: A near-quadratic bound for unit speed motions. *J. ACM*, 62:25, 2015.

[Sha94]   M. Sharir. Almost tight upper bounds for lower envelopes in higher dimensions. *Discrete Comput. Geom.*, 12:327–345, 1994.

[ST95a]   E. Schömer and C. Thiel. Subquadratic algorithms for the general collision detection problem. In *Abstracts 12th European Workshop Comput. Geom.*, pages 95–101, Linz, 1995.

[ST95b]   E. Schömer and C. Thiel. Efficient collision detection for moving polyhedra. In *Proc. 11th Sympos. Comput. Geom.*, pages 51–60, ACM Press, 1995.

[YZ12]   K. Yi and Q. Zhang. Multidimensional online tracking. *ACM Trans. Algorithms*, 8:12, 2012.

54 PATTERN RECOGNITION
Joseph O'Rourke and Godfried T. Toussaint

INTRODUCTION

The two fundamental problems in a pattern recognition system are feature extraction (shape measurement) and classification. The problem of extracting a vector of shape measurements from a digital image can be further decomposed into three subproblems. The first is the image segmentation problem, i.e., the separation of objects of interest from their background. The cluster analysis methods discussed in Section 54.1 are useful here. The second subproblem is that of finding the objects in the segmented image. An example is the location of text lines in a document as illustrated in Section 54.2. The final subproblem is extracting the shape information from the objects detected. Here there are many tools available depending on the properties of the objects that are to be classified. The Hough transform (Section 54.2), polygonal approximation (Section 54.3), shape measurement (Section 54.4), and polygon decomposition (Section 54.6), are some of the favorite tools used here. Important to many of these tasks is finding a nice viewpoint from which extraction is robust and efficient (Section 54.5). Proximity graphs, discussed in Section 54.2, are used extensively for both cluster analysis and shape measurement.

The classification problem involves the design of efficient decision rules with which to classify the feature vector. The most powerful decision rules are the nonparametric rules which make no assumptions about the underlying distributions of the feature vectors. Of these the nearest-neighbor (NN) rule, treated in Section 54.7, is the most well known. This section covers the three main issues concerning NN-rules: how to edit the data set so that little storage space is used, how to search for the nearest neighbor of a vector efficiently, and how to estimate the future performance of a rule both reliably and efficiently.

54.1 CLUSTER ANALYSIS AND CLASSIFICATION

GLOSSARY

Cluster analysis problem: Partitioning a collection of n points in some fixed-dimensional space into $m < n$ groups that are "natural" in some sense. Here m is usually much smaller than n.

Image segmentation problem: Partitioning the pixels in an image into "meaningful" regions, usually such that each region is associated with one physical object.

Block-based segmentation: Image segmentation based on characteristics of rectangular blocks of pixels. Often contrasted with layer-based segmentation.

Dendrogram: A tree representing a hierarchy of categories or clusters.

Hierarchical clustering algorithms: Those that produce a dendrogram whose root is the whole set and whose leaves are the individual points.

Graph-theoretic clustering: Clustering based on deleting edges from a proximity graph.

K-means clustering: Tracking clusters over time by comparing new data with old means.

Data mining: Algorithmic processing of large (and often unstructured) data sets to uncover previously unknown, informative patterns.

Classical cluster analysis requires partitioning points into natural clumps. "Natural" may mean that the clustering agrees with human perception, or it may simply optimize some natural mathematical measure of similarity or distance so that points that belong to one cluster are similar to each other and points far away from each other are assigned to different clusters. It is not surprising that such a general and fundamental tool has been applied to widely different subproblems in pattern recognition. One obvious application is to the determination of the number and description of classes in a pattern recognition problem where the classes are not known a priori, such as disease classification in particular or taxonomy in general. In this case m is not known beforehand and the cluster analysis reveals it.

A fundamental problem in pattern recognition of images is the segmentation problem: distinguishing the figure from the background. Clustering is one of the most powerful approaches to image segmentation, applicable even to complicated images such as those of outdoor scenes. In this approach each pixel in the $N \times N$ image is treated as a complicated object by associating it with a local neighborhood. For example, we may define the 5×5 neighborhood of pixel p_{ij}, denoted by $N_5[p_{ij}]$, as $\{p_{mn} \mid i - 2 \leq m \leq i + 2, j - 2 \leq n \leq j + 2\}$. We next measure k properties of p_{ij} by making k measurements in $N_5[p_{ij}]$. Such measurements may include various moments of the intensity values (grey levels) found in $N_5[p_{ij}]$, etc. Thus each pixel is mapped into a point in k-dimensional pixel-space. Performing a cluster analysis of all the resulting $N \times N$ points in pixel-space yields the desired partitioning of the pixels into categories. Labeling each category of pixels with a different color then produces the segmentation. See [BR14] for a survey of traditional and graph-theoretical techniques for image segmentation, as well as combinations of both approaches.

See [Gor96] for a survey of clustering methods that include constraints beyond similarity, or constraints on the topology of the resulting dendrograms; and see [ZA15] for a survey of image segmentation techniques, with an emphasis on block-based methods. See also Section 47.7 of this Handbook.

HIERARCHICAL CLUSTERING

In taxonomy there is no special number m that we want to discover; rather the goal is the production of a *dendrogram* (tree) that grows all the way from one cluster to n clusters and shows us at once how good a partitioning is obtained for any number of clusters between one and n. Such methods are referred to as *hierarchical* methods. They fall into two groups: **agglomerative** (bottom-up, merging) and **divisive** (top-down, splitting). Furthermore, each of these methods can be used with a variety of distance measures between subsets to determine when to merge or split. Two popular agglomerative clustering algorithms are the *single-link* and the *complete-linkage* (or farthest-neighbor) algorithm. In the former, cluster similarity is measured by the minimum of the distances between all pairs of elements, one in each subset, whereas in the latter similarity is measured by the maximum pairwise distance. The complete-linkage criterion tends to produce more compact clusters, while the single-link criterion can suffer from chaining [JMF99]. Krznaric and Levcopoulos [KL02] show that a complete-linkage hierarchy can be computed in optimal $O(n \log n)$ time and $O(n)$ space.

GRAPH-THEORETIC CLUSTERING

The most powerful methods of clustering in difficult problems, which give results having the best agreement with human performance, are the graph-theoretic methods [JT92]. The idea is simple: Compute some proximity graph (such as the minimum spanning tree) of the original points. Then delete (in parallel) any edge in the graph that is much longer (according to some criterion) than its neighbors. The resulting forest is the clustering (each tree in this forest is a cluster). Proximity graphs have also been used effectively to design cluster validity tests [PB97], [YL04], and for various specific problems that occur in pattern recognition such as measuring association, outlier detection, and dimensionality reduction [Mar04].

A useful generalization is the *graph-cut* method, which cuts edges to optimize some energy function. The max-cut/min-flow theorem permits energy minimization via maximizing the flow. See [PZZ13] for a survey.

K-MEANS TYPE CLUSTERING

There are many applications where we know that there are exactly k clusters, for example in character recognition. However, because of external factors such as the variations in people's hand-printing over time, or a change in the parameters of a machine due to wear or weather conditions, the clusters must be "tracked" over time. One of the most popular methods for doing this is the *k-means algorithm*. The k-means algorithm searches for k cluster centroids in $\mathbb{R}^d$ with the property that the mean squared (Euclidean) distance between each of the n points and its nearest centroid ("mean") is minimized [Mac67]. A determining characteristic of this approach is that the number of clusters k is fixed. A typical heuristic starts with an initial partition, computes centers, assigns data points to their nearest center, recomputes the centroids, and iterates until convergence is achieved according to some criterion. Unfortunately, this attractively simple algorithm's performance depends upon the initial partitioning, and in fact can be forced into a suboptimal solution of arbitrarily high approximation ratio. Moreover, it can take exponential (in n) time to converge, even in the plane [Vat11]. Finding the optimal solution is NP-hard in the plane [MNV12]. This led to developing algorithms with performance guarantees. Matoušek achieved an $O(n \log^k n)$ ϵ-approximation algorithm under the assumption that k and d are fixed [Mat00]. This was improved to an $(9 + \epsilon)$-approximation algorithm via a center-swap heuristic with this provable upper bound [KMN⁺04]. On the other hand, there is some evidence that the exact k-means algorithm, even though exponential, can be implemented to work well in practice for small d [PM99].

A variation on the k-means algorithm permits splitting and merging of clusters. This technique is employed by the ISODATA algorithm (Interactive Self-Organizing Data Analysis Technique) [Jen96].

DISTANCES BETWEEN SETS

A fundamental computational primitive of almost all clustering algorithms is the frequent computation of some distance measure between two sets (subsets) of points. This is especially so in the popular hierarchical methods discussed above. There exists a large variety of distance and more general similarity measures for this purpose. Here we mention a few. Most efficient algorithms for distance between sets apply only in $\mathbb{R}^2$ but some methods extend to higher dimensions; see [Smi00]. Let P and Q be two convex polygons of n sides each. Two distance measures should be distinguished: the minimum *element* distance, the smallest distance between a vertex or edge of P and a vertex or edge of Q, and the minimum *vertex* distance, the minimum distance between a vertex of P and a vertex of Q. The minimum element distance can be computed in $O(\log n)$ time [Ede85]. On the other hand, computation of the minimum vertex distance between P and Q has a linear lower bound. For the case of two nonintersecting convex polygons several different $O(n)$ time algorithms are available, and the same bound can be achieved for crossing convex polygons [Tou84]. For the case when the two polygons do not intersect, there exist additional distance measures that apply to specific applications areas. One such measure is the Grenander distance [Gre76], [Tou14c] defined as the sum of the lengths of the two chords determined by the critical separating supporting lines, less the sum of the lengths of all the edges of the polygons that belong to the inner chains determined by the supporting lines.

Let P and Q be two x-monotonic rectilinear (isothetic) polygonal functions in two dimensions. This setting occurs in several applications such as measuring music similarity in which case the functions act as models of melodies, and the similarity between two melodies P and Q is measured by the area that lies in between the two curves [ÓMa98]. If this distance is minimized under translations of P (with Q fixed) in the x or y directions then the problem falls under the topic of polygonal matching problems [AFL06], [ACH91].

Let R be a set of n red points and B a set of n blue points in the plane. Both the minimum distance and the maximum distance between R and B can be computed in $O(n \log n)$ time. For the latter problem, two algorithms are available. The first [TM82] is simple but does not appear to generalize to higher dimensions. The second [BT83] works by reducing the maximum distance problem between R and B to computing the diameter of 81 convex polygons. These are obtained by computing the convex hulls of the unions of 81 carefully selected subsets of R and B, and then reporting the maximum of these 81 diameters as the maximum distance. These ideas can be extended to obtain efficient algorithms for all dimensions [Rob93]. Therefore, any improvement in high-dimensional diameter algorithms automatically improves maximum-distance algorithms.

Another distance that has applications to machine learning in general, and support vector machines (SVMs) in particular, is the width of the widest empty separating strip between the two sets of points if they are linearly separable (*strong separators*), and otherwise the thinnest strip that minimizes a given number of classification errors (*weak separators*) [Hou93].

DATA MINING

The explosion of the Web has given new impetus to automated methods to uncover informative patterns from huge, often unstructured, data repositories ("big data"). This activity has become known as *data mining.* Clustering to discover structures in the data set is a key component of data mining; see [Ber06] for a survey of clustering in data mining. The k-means algorithm and its variants remain popular, not only in geometric domains (e.g., in geological databases [JMF99], celestial databases [PM99], image databases, and so on), but even for text-document clustering [SKK00]. There is some movement in the literature away from point distances for categorical attributes, for which the iterative centroid-based clustering algorithms are often inappropriate. For example, "links" [GRS00] or context-based measures [DM00] are frequently employed, which places this work close to graph-theoretic clustering. A related direction is finding "unusual" strings of ACTG characters within the human genome [ABL02] [BC15].

54.2 EXTRACTING SHAPE FROM DOT PATTERNS

HOUGH TRANSFORMS

The *Hough transform* was originally proposed (and patented) as an algorithm to detect straight lines in digital images [Lea92]. The method may be used to detect any parametrizable pattern, and has been generalized to locate arbitrary shapes in images. The basic idea is to let each above-threshold pixel in the image vote for the points in the parameter space that could generate it. The votes are accumulated in a quantized version of the parameter space, with high vote tallies indicating detection.

EXAMPLES

1. *Lines.* Let the lines in the (x, y) image plane be parametrized as $y = mx + b$. Then a pixel at (x_0, y_0) is a witness to a line passing through it, that is, an (m, b) pair satisfying $y_0 = mx_0 + b$. Thus, (x_0, y_0) votes for all those (m, b) pairs: the line in parameter space dual to the pixel.

2. *Circles.* Parametrize the circles by their center and radius, (x_c, y_c, r). Then a pixel (x_0, y_0) gives evidence for all the parameter triples on the circular cone in the 3-parameter space with apex at $(x_0, y_0, 0)$ and axis parallel to the r-axis.

3. *Object location*. Suppose a known but arbitrary shape S is expected to be found in an image, and its most likely location is sought. For translation-only, the parameter space represents the location of some fixed point of S. Each pixel in the image of the right shading or color votes for all those translations that cover it appropriately.

The above approaches are not necessarily optimal for the tasks listed. For example, it was shown in [CT77] that nonuniform (maximum entropy) quantization with ρ-θ (normal-form) parametrization for lines is superior to uniform quantization with m-b (slope-intercept) parametrization. Since earning a patent in 1962 the Hough transform has been generalized in a variety of ways in order to overcome several of its limitations [MC15].

The demands of high-dimensional vote accumulators have engendered the study of *dynamic quantization* of the parameter space, and **geometric hashing**. This latter technique has features in the image vote for each member of a library of shapes by hashing into a table using the feature coordinates as a key. Each table entry may correspond to several shapes at several displacements, but all receive votes. Geometric hashing has been applied with some success to the *molecular docking* problem [LW88]; see also [MSW02].

Variants of the Hough transform inspired by results in computational geometry have also appeared. In [AK96] two such algorithms are presented and studied with respect to the tradeoff that exists between computational complexity and effectiveness of line detection. They obtain efficient implementations by using the plane-sweep paradigm.

A significant generalization of the technique is proposed in [FO12], which seeks to detect subspace alignments in noisy multidimensional data sets.

More recent applications of the Hough transform and its variants include the detection of lane lines in traffic and transportation science [BK13], [SSSS10], as well as detecting land mines in mine fields [LSC97]. Present challenges in this area include extending the existing methods that have concentrated on binary images, to also work well on color images, and to reduce the computational complexity of the algorithms by developing parallel implementations.

TEXT LINE ORIENTATION INFERENCE

In an automated document analysis system, given a block (paragraph) of text, the text line orientation inference problem consists of determining the location and direction of the lines of text on the page. Almost always these lines are either horizontal (e.g., English) or vertical (e.g., Chinese) The fundamental geometric property that allows this problem to be solved is the fact that according to a universal typesetting convention guided by ease of reading, characters are printed closer together within text lines than between text lines.

One of the most successful, robust, skew-tolerant, simple, and elegant techniques for text line orientation inference was proposed by Ittner [Itt93], and later improved by Bose et al. [BCG98]. Assume that the given text block B consists of n black connected components (characters). The three key steps in Ittner's procedure are: (1) idealize each character by a point, thus obtaining a set S of n points in the plane; (2) construct the Euclidean minimum spanning tree MST(S) of the n points obtained in (1); and (3) determine the text line orientation by analysis of the distribution of the orientations of the edges in MST(S). Step (1) is done by computing the center of the bounding box of each character. Cheriton and Tarjan proposed a simple algorithm for computing the MST of a graph in $O(E)$ time where E is the number of edges in the graph [CT76]. Fortunately there are many graphs defined on S (usually belonging to the class of proximity graphs [JT92]) that have the property that they contain the MST(S) and also have $O(n)$ edges. For these graphs the Cheriton-Tarjan algorithm runs in $O(n)$ time.

The traditional approaches to text detection described above were focused on two-dimensional inputs such as newspapers, magazines, and books, in which the input consisted of binary (black-white) images. More recently attention has shifted to more difficult problems of detecting text in color images of complex three-dimensional scenes. Ye and Doermann [YD15] provide a survey of new approaches to these more challenging problems.

TEXT BLOCK ISOLATION

The text-block isolation problem consists of extracting blocks of text (paragraphs) from a digitized document. By finding the enclosing rectangles around each connected component (character) and around the entire set of characters we obtain a well structured geometric object, namely, a rectangle with n rectangular "holes." This problem is ideally suited to a computational geometric treatment. Here we mention an elegant method that analyzes the empty (white) spaces in the document [BJF90]. This approach enumerates all maximal white rectangles implied by the black rectangles. A white rectangle is called maximal if it cannot be enlarged while remaining outside the black rectangles. Their enumeration algorithm takes quadratic time in the worst case, but a clever heuristic exploiting properties of layouts leads to $O(n)$ expected time.

RELATIVE NEIGHBORHOOD GRAPHS

Relative neighborhood graphs (RNG's), introduced in [Tou80], capture proximity between points by connecting nearby points with a graph edge. The many possible notions of "nearby" (in several metrics) lead to a variety of related graphs, including the *sphere of influence graph* (SIG). We defer to Chapter 32 for definitions and details, and only mention here several applications. The RNG was proposed as an approach to Marr's "primal sketch" in computer vision, and it continues to have application in low-level vision, extracting perceptually meaningful features from images. The RNG, the SIG, and related proximity graphs find use in nearest-neighbor searching, cluster analysis, data mining, and outlier detection. See [Tou05], [Tou14a], and [Tou14b] for further details.

54.3 POLYGONAL APPROXIMATION

Let $P = (p_1, p_2, \ldots, p_n)$ be a polygonal curve or chain in the plane, consisting of n points p_i joined by line segments $p_i p_{i+1}$. In general P may be closed and self-intersecting. Polygonal curves occur frequently in pattern recognition either as representations of the boundaries of figures or as functions of time representing, e.g., speech. In order to reduce the complexity of costly processing operations, it is often desirable to approximate a curve P with one that is composed of far fewer segments, yet is a close enough replica of P for the intended application. Some methods of reduction attempt smoothing as well. An important instance of the problem is to determine a new curve $Q = (q_1, q_2, \ldots, q_m)$ such that (1) m is significantly smaller than n; (2) the q_j are selected from among the p_i; and (3) any segment $q_j q_{j+1}$ that replaces the chain $q_j = p_r, \ldots, p_s = q_{j+1}$ is such that the distance between $q_j q_{j+1}$ and each p_k, $r \leq k \leq s$, is less than some predetermined error tolerance ω. Different notions of distance, or error criteria, lead to different algorithmic issues. Moreover, for each distance definition, there are two constrained optimization problems that are of interest, Min-# and Min-ϵ.

GLOSSARY

Distance from point p to segment s: Minimum distance from p to any point of s.

Parallel-strip criterion: All the vertices $p_i, \ldots, p_t$ lie in a parallel strip of width 2ϵ whose center line is collinear with $q_j q_{j+1}$ [ET94].

Segment criterion, or Hausdorff measure: For each p_k, $r \leq k \leq s$, the distance from p_k to $q_j q_{j+1}$ is less than ϵ [MO88, CC96].

Min-# problem: Given the error tolerance ϵ, find a curve $Q = (q_1, \ldots, q_m)$ satisfying the constraint such that m is minimum.

Min-ϵ problem: Given m, find a curve $Q = (q_1, \ldots, q_m)$ satisfying the constraint such that the error tolerance is minimized.

Fréchet distance curves: The smallest "leash length" for a master to walk along one curve while a leashed dog walks along the other.

The main research focus has been the Min-# problem under the Hausdorff error measure, also called ϵ-simplification. Near quadratic-time, $O(n^2 \log n)$, algorithms have been achieved for arbitrary polygonal curves [MO88]. The quadratic-time barrier seems difficult to break, so approximation algorithms have been explored, which achieve an error no more than ϵ but compromise on m, the min-#. A popular algorithm is the Douglas-Peucker algorithm, *iterative endpoint fitting*, which, although easily implemented, does not guarantee performance and could be quadratic in the worst case. Now near-linear performance with a guarantee has been achieved for simplifying monotone curves [AHM+05].

At least three other simplification directions have been pursued:

1. Rather than using the Hausdorff distance, a large amount of recent work has explored guaranteeing a maximum Fréchet distance between the two polygonal curves. Again near-linear approximations are available [AHM+05]. For

polygonal chains in dimensions $d \geq 3$, and an $O(n \log n)$ exact algorithm has been achieved [BJW+08].

2. Tracking objects over long periods has led to exploring streaming algorithms, with limited storage. Here $O(1)$ competitive ratios have been achieved for several cases under both the Hausdorff and Fréchet distance measures [ADB+10].

3. The task of polygonal approximation has been given new significance in three dimensions for its importance in simplifying polyhedral models in computer graphics. There is now a module within CGAL for "Triangulated Surface Mesh Simplification." This topic is covered in detail in Chapter 56.

54.4 SHAPE MEASUREMENT AND REPRESENTATION

MEDIAL AXIS

GLOSSARY

Medial axis: The set of points of a region P with more than one closest point among the boundary points ∂P of the region. Equivalently, it is the set of centers of maximal balls, i.e., of balls in P that are themselves not enclosed in another ball in P.

Voronoi diagram: The partition of a polygonal region P into cells each consisting of the points closer to a particular open edge or vertex than to any other.

The *medial* or *symmetric axis* was introduced by Blum [Blu67] to capture biological shape, and it has since found many other applications, for example, to geometric modeling (*offset* computations; see Section 50.2) and to mesh generation [SNT+92] (Section 29.4). It provides a central "skeleton" for an object that has found many uses. It connects to several other mathematical concepts, including the *cut locus* and most importantly, the Voronoi diagram (Chapter 27).

The medial axis of a convex polygon is a tree with each vertex a leaf. For a nonconvex polygon, the medial axis may have a parabolic arc associated with each reflex vertex (Figure 50.1.5). The basic properties of the medial axis were detailed by Lee [Lee82], who showed that the medial axis of a polygon P is just the Voronoi diagram minus the Voronoi edges incident to reflex vertices, and provided an $O(n \log n)$ algorithm for constructing it. After a long search by the community, an $O(n)$ algorithm was obtained [CSW99]. The simplest implementations are, however, quadratic [YR91].

The medial axis has also found much use in image processing, where its digital computation is via *thinning algorithms*. Pioneered by Rosenfeld, these algorithms are very simple and easily parallelized [Cyc94].

The definition of medial axis extends to $\mathbb{R}^d$, and has found application in $\mathbb{R}^3$. Although an exact algorithm is available for the medial axis of (perhaps nonconvex) polyhedra in $\mathbb{R}^3$ [Cul00], there is no implementation that constructs the exact medial axis of an arbitrary semi-algebraic set [ABE09]. For applications, exact medial axes are rarely needed. An approximate medial axis of a set of points in $\mathbb{R}^3$ can be computed directly from the Voronoi diagram, with guaranteed convergence: [DZ04]. See [TDS+16] for an overview of skeletons in $\mathbb{R}^3$, and [AW13] for closely related work on the *straight skeleton* in $\mathbb{R}^3$.

POINT PATTERN MATCHING

Exact point pattern matching is an interesting algorithmic question related to string matching, but pattern recognition applications usually require some type of approximate matching. Two types may be distinguished [AG00]: one-to-one matching, and Hausdorff matching.

GLOSSARY

One-to-one approximate matching: Let two finite sets of points A and B have the same cardinality. One-to-one matching requires finding a transformation (of a given type) of B such that each point of B is mapped to within a distance of ϵ of a matched point of A. The matches are either determined by **labels** on the points, or the points are **unlabeled** and the match is to be discovered.

Decision problem: Given ϵ, is there such a matching?

Optimization problem: Find the minimum ϵ for which an approximate matching exists.

Hausdorff distance: For two finite sets A and B, perhaps of different cardinalities, the largest of the between-sets nearest-neighbor distances.

Hausdorff matching: Find a transformation of B that minimizes the Hausdorff distance from A.

The most combinatorially interesting point matching (unrealistically) demands exact matching. One version of this is the **congruent subset detection problem**: Given a pattern set A of m points, find all subsets of a background set B of n points that are congruent to A. Solving this in the plane relies on the unsolved Erdős problem of bounding the number of unit-distance pairs among n points, whose best upper bound is $O(n^{4/3})$ (Chapter 10). Important variations are obtained by acting on the pattern by some group, e.g., translations. Results here are surveyed in [Bra02], from which the results shown in Table 54.4.1 are gathered ($\alpha()$ is the near-constant inverse Ackermann function; cf. Chapter 28).

A window-restricted version of the problem led Brass to pose the following interesting conjecture:

• Any set of n points in the plane contains only $O(n)$ empty congruent triangles.

There are sets with $\binom{n}{3}$ empty triangles.

Results on **one-to-one approximate matching** algorithms obtained for a variety of permissible transformations in [AMW+88] are shown in Table 54.4.2.

Hausdorff matching leads to analysis of envelopes of *Voronoi surfaces*. Typical results are shown in Table 54.4.3. Here we show the complexities when $|A| = |B| = n$, although the algorithms work for sets of different cardinalities.

TABLE 54.4.1   Subset detection of m points among n points.

GROUP	DIM	MATCHES	ALGORITHM
Congruence	2	$O(n^{4/3})$	$O(mn^{4/3}\log n)$
Congruence	3	$\Omega(n^{4/3})$	$O(mn^{5/3}\log n 2^{O(\alpha(n)^2)}))$
Translation	d	$n - \Theta(n^{1-1/k}),\ k \le d$	$O(mn\log n)$
Homothets	d	$O(n^{1+1/k}),\ k \le d$	$O(mn^{1+1/d}\log n)$
Similarity	d	$O(n^d)$	$O(mn^d\log n)$
Affine	d	$O(n^{d+1})$	$O(mn^{d+1}\log n)$

TABLE 54.4.2   One-to-one point matching in two dimensions.

MOVEMENTS	LABELED	ϵ	COMPLEXITY
Translation	labeled	dec, opt	$O(n)$
Translation	unlabeled	decision	$O(n^6)$
Translation	unlabeled	optimization	$O(n^6\log n)$
Rotation	labeled	decision	$O(n\log n)$
Rotation	labeled	optimization	$O(n^2)$
Trans+rot+refl	labeled	decision	$O(n^3\log n)$
Trans+rot+refl	unlabeled	decision	$O(n^8)$

TABLE 54.4.3   Hausdorff matching in the L_2 metric.

MOVEMENTS	DIM	COMPLEXITY
Translation	2	$O(n^3\log n)$
Translation + rotation	2	$O(n^6\log n)$
Translation	3	$O(n^{5+\epsilon})$

Another type of matching is *order type matching* (cf. Section 5.2). In [GP83], an $O(n^3)$ algorithm is given for finding all matchings between two planar point configurations in which their order types agree.

SYMMETRY DETECTION

Symmetry is an important feature in the analysis and synthesis of shape and form and has received considerable attention in the pattern recognition and computer graphics literatures. In [WWV85] an $O(n\log n)$ algorithm is presented for computing the rotational symmetries of polygons and polyhedra of n vertices, but the constant in $\mathbb{R}^3$ is very large. Jiang and Bunke [JB91] give a simple and practical $O(n^2)$ time algorithm for polyhedra. One of the earliest applications of computational geometry to symmetry detection was the algorithm of Akl and Toussaint [AT79] to check for polygon similarity. For a survey of the early work on detecting symmetry, see [Ead88]. Since then attention has been given to other aspects of symmetry and for objects other than polygons. Sugihara [Sug84] shows how a modification of the planar graph-isomorphism algorithm of Hopcroft and Tarjan can be used to find all symmetries of a wide class of polyhedra in $O(n \log n)$ time.

Most previous methods for measuring the symmetries of polygons and polyhedra have focused on a binary output: an object either has or does not have some symmetries. However, for many problems in pattern recognition a more useful measure is a graded (or continuous) descriptor that yields an indication of the amount of symmetry contained in an object. Kazhdan et al. [KCD+04] present an algorithm that computes a reflective symmetry descriptor in $O(n^4 \log n)$ time for an $n \times n \times n$ grid of voxels in $3D$. The approach in [KCD+04] is to measure the amount of reflective symmetry along an axis, through the center of gravity of the pattern, by the amount of overlap between the original pattern and the pattern reflected about that axis. This provides a *global* graded symmetry descriptor. An alternate approach is to measure all *local* reflection symmetries (sub-symmetries) rather than only global ones, and do so with respect to all locations in the object rather than just the center of mass. Such symmetries correlate better with human judgments of complexity than do global symmetries [TOV15].

A related topic is centers of symmetry. Given a convex polygon P, associate with each point p in P the minimum area of the polygon to the left of any chord through p. The maximum over all points in P is known as **Winternitz's measure of symmetry** and the point p^* that achieves this maximum is called the **center of area**. Diaz and O'Rourke [DO94] show that p^* is unique and propose an algorithm for computing p^* in time $O(n^6 \log^2 n)$. See [BT10] for an improved implementation and analysis of the Diaz-O'Rourke algorithm for finding the Simpson point of a convex polygon.

THE ALPHA HULL

The α-shape $\mathcal{S}_\alpha$ of a set S of n points in $\mathbb{R}^3$ is a polyhedral surface whose boundary is a particular collection of triangles, edges, and vertices determined by the points of S [EM94]. It is similar in spirit to the β-skeleton of Section 54.2 in that it is a parametrized collection of shapes determined by an empty balls condition, but it emphasizes the external rather than the internal structure of the set. Let T be a subset of S of 1, 2, or 3 points. Then the convex hull of T, conv(T), is part of the boundary $\partial\mathcal{S}_\alpha$ of the α-shape iff the surface of some ball of radius α includes exactly the points of T while its interior is empty of points of S. Thus a triangle conv(T) is part of $\partial\mathcal{S}_\alpha$ iff there is an open α-ball that can "erase" all of the triangle but leave its vertices. $\mathcal{S}_\alpha$ is defined for all $0 \leq \alpha \leq \infty$, with $\mathcal{S}_0 = S$ and $\mathcal{S}_\infty = \text{conv}(S)$.

Every edge and triangle of $\mathcal{S}_\alpha$ is present in the Delaunay triangulation DT of S, and every edge and triangle in DT is present in some $\mathcal{S}_\alpha$. If α is varied continuously over its full range starting from ∞, the convex hull of S is gradually "eaten away" by smaller and smaller α-ball erasers, eventually exposing the original set of points. In between, the α-shape bounds a subcomplex of DT that represents the shape of S.

The alpha shape has been used for cluster analysis, molecular modeling, and the analysis of medical data, among other applications. High-quality code is available: CGAL includes alpha shapes in its basic library (Chapter 68), and a package in R is also available [LPPD16].

54.5 NICE VIEWPOINTS AND PROJECTIONS

A robot navigating in 3D space faces a variety of pattern recognition problems that involve classifying objects modeled as polyhedra. A polyhedral object in 3D space is often well represented by a set of points (vertices) and line segments (edges) that act as its features. The feature extraction process involves obtaining *nice* viewpoints of the polyhedron. By a nice viewpoint of an object we mean a projective view in which all (or most) of the features of the object, relevant for some task, are clearly visible. Such a view is often called a nondegenerate view or projection. A recent survey of this topic can be found in [Tou00].

GLOSSARY

Nice viewpoint: A projection of a 3D object (set of points, etc.) onto a plane such that it has some desirable special property.

Knot diagram: A regular projection of a polygon in 3-dimensions onto a plane.

Degeneracies: Properties of objects such as three points collinear.

General position: A configuration of an object such that some specified degeneracies are absent.

Orthogonal projection: A projection from a point at infinity.

Perspective projection: A projection from a point not at infinity.

Robust algorithm: One that works correctly even in the presence of degeneracies.

Regular projection: An orthogonal projection of S such that no three points of S project to the same point on H, and no vertex of S projects to the same point on H as any other point of S.

Wirtinger projections: Regular projections in which no two consecutive edges of the 3D polygon have collinear projections.

Robust nondegenerate projection: A projection that remains nondegenerate even if the object is slightly perturbed.

Decision problem: Given an object and a projection of it, does the projection contain a degeneracy?

Computation problem: Given an object, compute a projection that does not contain a specified degeneracy.

Optimization problem: Given an object, compute the most robust nondegenerate projection.

REGULAR PROJECTIONS

The earliest work on nondegenerate orthogonal projections appears to be in the area of knot theory. Let S be a set of n disjoint line segments in $\mathbb{R}^3$ specified by the Cartesian coordinates of their endpoints (vertices of S) and let H be a plane. Let

SH be the orthogonal projection of S onto H. An orthogonal projection of S is said to be *regular* if no three points of S project to the same point on H and no vertex of S projects to the same point on H as any other point of S [Liv93]. This definition implies that for disjoint line segments (1) no point of SH corresponds to more than one vertex of S, (2) no point of SH corresponds to a vertex of S and an interior point of an edge of S, and (3) no point of SH corresponds to more than two interior points of edges of S. Therefore the only crossing points (intersections) allowed in a regular projection are those points that belong to the interiors of precisely two edges of S. This condition is crucial for the successful visualization and manipulation of knots [Liv93].

Regular projections of 3D polygons were first studied by the knot theorist K. Reidemeister [Rei32] in 1932 who showed that all 3D polygons (knots) admit a regular projection, and in fact almost all projections of polygons are regular. Reidemeister however was not concerned with computing regular projections. The computational aspects of regular projections of knots were first investigated by Bose et al., [BGRT99] under the real RAM model of computation. Given a polygonal object (geometric graph, wire-frame or skeleton) in $\mathbb{R}^3$ (such as a simple polygon, knot, skeleton of a Voronoi diagram or solid model mesh), they consider the problem of computing several "nice" orthogonal projections of the object. One such projection, well known in the graph-drawing literature, is a projection with few crossings. They consider the most general polygonal object, i.e., a set of n disjoint line segments, and show that deciding whether it admits a crossing-free projection can be done in $O(n^2 \log n + k)$ time and $O(n^2 + k)$ space, where k is the number of intersections among a set of "forbidden" quadrilaterals on the direction sphere, and $k = O(n^4)$. This implies for example that, given a knot, one can determine if there exists a plane on which its projection is a simple polygon, within the same complexity. Furthermore, if such a projection does not exist, a minimum-crossing projection can be found in $O(n^4)$ time and $O(n^2)$ space. They showed (independently of Reidemeister) that a set of line segments in space (which includes polygonal objects as special cases) always admits a regular projection, and that such a projection can be obtained in $O(n^3)$ time. A description of the set of all directions which yield regular projections can be computed in $O(n^3 \log n + k \log n)$ time, where k is the number of intersections of a set of quadratic arcs on the direction sphere and $k = O(n^6)$. Finally, when the objects are polygons and trees in space, they consider monotonic projections, i.e., projections such that every path from the root of the tree to every leaf is monotonic in some common direction on the projection plane. For example, given a polygonal chain P, they can determine in $O(n)$ time if P is monotonic on the projection plane, and in $O(n \log n)$ time they can find all the viewing directions with respect to which P is monotonic. In addition, in $O(n^2)$ time, they can determine all directions for which a given tree or a given simple polygon is monotonic.

COMPUTER VISION

In the computer vision field there is both a theoretical [BWR93] interest in nondegenerate projections and a practical one [DWT99]. The theoretical work resembles the work described in the previous section in that it is assumed that the object consists of idealized points and line segments or polygons and polyhedra. A tool used for computing viewpoints from which the maximum number of faces of a solid polyhedron is visible, is the *aspect graph* [PD90] (Chapter 33).

WIRTINGER PROJECTIONS

That certain types of nondegenerate orthogonal projections of 3D polygons always exist for some directions of projection was rediscovered by Bhattacharya and Rosenfeld [BR94] for a restricted class of regular projections. Those regular projections, in which it is also required that no two consecutive edges of the 3D polygon have collinear projections, are known as *Wirtinger projections*. Bose et al. [BGRT99] study the complexity of computing a single Wirtinger projection as well as constructing a description of all such projections for the more general input consisting of disjoint line segments. These results include therefore results for 3D chains, polygons, trees and geometric graphs in general. The description of all projections allows one to obtain Wirtinger projections that optimize additional properties. For example, one may be interested in obtaining the most robust projection in the sense that it maximizes the deviation of the viewpoint required to violate the Wirtinger property.

VISUALIZATION

In computer graphics one is interested in visualizing objects well, and therefore *nice* views and nondegenerate views are major concerns. For example, Kamada and Kawai [KK88] proposed a method to obtain nice projections by making sure that in the projection, parallel line segments on a plane in 3D project as far away from each other as possible. Intuitively, the viewer should be as orthogonal as possible to every face of the 3D object. Of course this is not possible and therefore they suggest minimizing (over all faces) the maximum angle deviation between a normal to the face and the line of sight from the viewer. They then propose an algorithm to solve this problem in $O(n^6 \log n)$ time, where n is the number of edges in the polyhedral object in 3D. Gómez et al. [GRT01] reduce this complexity to $O(n^4)$ time. Furthermore, they show that if one is restricted to viewing an object from only a hemisphere, as is the case with a building on top of flat ground, then a further reduction in complexity is possible to $O(n^2)$ time.

A rather different optimization problem regarding projections arises in visualization applications in which it is desired to display visual content on three-dimensional real-world objects rather than on a traditional two-dimensional displays, possibly using more than one projector. The problems here consist of determining optimal placements of the projectors to optimize a variety of objective functions under possible constraints [LAM10].

REMOVING DEGENERACIES

Algorithms in computational geometry are usually designed for the real RAM (random access machine) assuming that the input is in *general position*. More specifically, the general position assumption implies that the input to an algorithm for solving a specific problem is free of certain degeneracies. Yap [Yap90] has distinguished between intrinsic or *problem-induced* and extrinsic or *algorithm-induced* degeneracies (see also Chapter 45). For example, in computing the convex hull of a set of points in the plane, where "left" turns and "right" turns are fundamental primitives, three collinear points constitute a problem-induced degeneracy. On the other hand, for certain vertical line-sweep algorithms, two points with the same x-coordinate constitute an algorithm-induced degeneracy. Computational geometers make these assumptions because doing so makes it not only much easier to design algorithms but often yields algorithms with reduced worst-case complexities. On the other hand, to the implementers of geometric algorithms these assumptions are frustrating. Programmers would like the algorithms to work for any input that they may encounter in practice, regardless of any degeneracies contained in the input.

Often a typical computational geometry paper will make a nondegeneracy assumption that can in fact be removed (*without* perturbing the input) by a global rigid transformation of the input (such as a rotation, for example). Once the solution is obtained on the transformed nondegenerate input, the solution can be transformed back trivially (by an inverse rotation) to yield the solution to the original problem. In these situations, by applying suitable *pre-* and *post*-processing steps, one obtains the *exact* solution to the *original* problem using an algorithm that assumes a nondegenerate input, even when that input is in fact degenerate. This approach not only handles algorithm-induced degeneracies via orthogonal projections but some problem-induced degeneracies as well with the aid of perspective projections.

Gómez et al. [GRT01] consider several nondegeneracy assumptions that are typically made in the literature, propose efficient algorithms for performing the suitable rotations that remove these degeneracies, analyze their complexity in the real RAM model of computation and, for some of these problems, give lower bounds on their worst-case complexity. The assumptions considered in [GRT01] are summarized in Tables 54.5.1 and reftab:degen.segs $\lambda(\cdot)$ is nearly linear; cf. Section 28.10).

PERSPECTIVE PROJECTIONS AND INTRINSIC DEGENERACIES

Intrinsic degeneracies cannot be removed by rotations of the input. If a set of points S in 3D contains three collinear points then so does every orthogonal projection of S. This is where *perspective* projections come to the rescue. However, not all intrinsic degeneracies can be removed with perspective projections. Intrinsic degeneracies that can be removed via perspective projections are called *quasi-intrinsic degeneracies* [HS97, GHS+01].

Gómez et al. [GHS+01] consider computing nondegenerate *perspective* projections of sets of points and line segments in 3D space. For sets of points they give algorithms for computing perspective projections such that (1) all points have distinct x-coordinates, (2) all points have both distinct x- and y-coordinates, (3) no three points in the projection are collinear, and (4) no four points in the projection are cocircular. For sets of line segments they present an algorithm for computing a perspective projection with no two segments parallel. All their algorithms have time and space complexities bounded by low degree polynomials.

TABLE 54.5.1   Removing degeneracies: Point sets.

PROBLEM	DECISION	COMPUTATION	OPTIMIZATION
2D No two on vertical line	$\Theta(n \log n)$	$O(n \log n)$	$O(n^2 \log n)$ time, $O(n^2)$ space
			$O(n^2)$ time, space with floor functions
3D No two on vertical line	$\Theta(n \log n)$	$O(n \log n)$	$O(n^2 \log n)$ time, $O(n^2)$ space
No two with same x-coordinate	$\Theta(n \log n)$	$O(n \log n)$	$O(n^4)$ time, space
			$O(n^2 \lambda_6(n^2) \log n)$ time, $O(n^2)$ space
No two with same x, y or z-coord	$\Theta(n \log n)$	$O(n \log n)$	OPEN
No three on vertical plane	(3SUM-hard) $O(n^2)$ time, space	(3SUM-hard) $O(n^2)$ time, space $O(n^3)$ time, $O(n)$ space	$O(n^6)$ time and space

TABLE 54.5.2   Removing degeneracies: Line segments and faces.

PROBLEM	DECISION	COMPUTATION	OPTIMIZATION
LINE SEGMENTS			
2D No vertical	$\Theta(n)$	$\Theta(n)$	$O(n \log n)$ time, $O(n)$ space
3D No vertical	$\Theta(n)$	$\Theta(n)$	$O(n \log n)$ time, $O(n)$ space
No two on vertical plane	$O(n \log n)$	$O(n^2)$ time, $O(n)$ space	$O(n^4)$ time, space
			$O(n^2 \lambda_6(n^2) \log n)$ time, $O(n^2)$ space
FACES			
No face of poly-hedron vertical	$\Theta(n)$	$\Theta(n)$	$O(n^2)$ time, space
			$O(n \lambda_6(n) \log n)$ time, $O(n)$ space

FINITE-RESOLUTION MODELS OF VIEW DEGENERACY

View degeneracy is a central concern in robotics where a robot must navigate and recognize objects based on views of the scene at hand [DPR92a, DPR92b]. In the idealized world assumed in the previous sections, degenerate views are not much of a problem if a viewpoint is chosen at random, since almost all projections are not degenerate. On the other hand, real-world digital cameras have a finite resolution and therefore view degeneracy can no longer be ignored [KF87].

OPEN PROBLEMS

A more practical approach would give some thickness to the objects, i.e., consider the points as little balls and the edges of the polyhedra as thin cylinders, and then to redesign the algorithms accordingly. This may turn out to be rather expensive. In practice it may be much more efficient to perform a half-dozen *random* rotations to obtain a nice projection. After all, for many problems in the idealized infinite precision model, a single random rotation yields a nice projection with probability one. Computing optimal projections on the other hand is another matter. Here approximate algorithms may yield efficient solutions that are near-optimal, but these are open problems.

54.6 NEAREST-NEIGHBOR DECISION RULES

GLOSSARY

Nearest-neighbor decision rule: Classifies a feature vector with the closest sample point in parameter space.

In the typical nonparametric classification problem (see Devroye, Gyorfy and Lugosi [DGL96]) we have available a set of d measurements or observations (also called a feature vector) taken from each member of a data set of n objects (patterns) denoted by $\{X, Y\} = \{(X_1, Y_1), (X_2, Y_2), \ldots, (X_n, Y_n)\}$, where X_i and Y_i denote, respectively, the feature vector on the ith object and the class label of that object. One of the most attractive decision procedures is the nearest-neighbor rule (1-*NN*-rule) [FH51]. Let Z be a new pattern (feature vector) to be classified and let X_j be the feature vector in $\{X, Y\} = \{(X_1, Y_1), (X_2, Y_2), \ldots, (X_n, Y_n)\}$ closest to Z. The nearest neighbor decision rule classifies the unknown pattern Z into class Y_j. In the 1960s and 1970s many pattern recognition practitioners resisted using the 1-*NN*-rule on the grounds of the mistaken assumptions that (1) all the data $\{X, Y\}$ must be stored in order to implement such a rule, (2) to determine Y_j, distances must be computed between the unknown vector Z and all the members of $\{X, Y\}$, and (3) such a rule is difficult to implement in parallel using a neural network. Computational geometry research in the 1980s and 1990s along with faster and cheaper hardware has made the *NN*-rules a practical reality [Tou02].

MINIMAL-SIZE TRAINING-SET CONSISTENT SUBSETS

A question that has received a lot of attention in the past fifty years concerns the problem of reducing the number of patterns in the training set $\{X, Y\}$ without degrading the performance of the decision rule. In 1968 Hart was the first to propose an algorithm for reducing the size of the stored data for the nearest neighbor decision rule [Har68]. Hart defined a *consistent* subset of the data as one that classified the entire set correctly with the nearest neighbor rule. He then proposed an $O(n^3)$ time algorithm that he called *CNN* (Condensed Nearest Neighbor) for selecting a consistent subset by heuristically searching for data that were near the decision boundary. However, the method does not in general yield a minimal-size consistent subset.

The first researchers to deal with computing a *minimal-size* consistent subset were Ritter et al. [RWLI75]. They proposed a procedure they called a *selective* nearest neighbor rule (*SNN*) to obtain a minimal-size consistent subset of $\{X, Y\}$, call it S, with one additional property that Hart's *CNN* does not have. Any consistent subset C obtained by *CNN* has the property that every element of $\{X, Y\}$ is nearer to an element in C of the same class than to any element in C of a different class. On the other hand, the consistent subset S of Ritter et al. [RWLI75] has the additional property that every element of $\{X, Y\}$ is nearer to an element in S of the same class than to any element, in the *complete* set, $\{X, Y\}$ of a different class. This additional property of *SNN* tends to keep points closer to the decision boundary than does *CNN*, and allows Ritter et al. [RWLI75] to compute the selected subset S without testing all possible subsets of $\{X, Y\}$. Nevertheless, their algorithm still runs in time exponential in n (Wilfong [Wil91]) in the worst case. However, Wilson and Martinez [WM97] and Wilson [WM00] claim that the average running time of *SNN* is $O(n^3)$. In 1994 Dasarathy [Das94] proposed a complicated algorithm intended to compute a *minimal-size* consistent subset but did not provide a proof of optimality. However, counter-examples to this claim were found by Kuncheva and Bezdek [KB98], Cerverón and Fuertes [CF98] and Zhang and Sun [ZS02]. Wilfong [Wil91] showed in 1991 that the problem of finding the smallest size training-set consistent subset is NP-complete when there are three or more classes. The problem is still open for two classes. Furthermore, he showed that even for only two classes the problem of finding the smallest size consistent *selective* subset (Ritter et al. [RWLI75]) is also NP-complete.

Given the computational difficulty of computing minimum-size training-set consistent subsets, subsequent research has focused on computing small-size training-set consistent subsets. The main challenges regarding the computation of small-size training-set consistent subsets are the scalability for very large datasets as well as for high-dimensional data. An example of this approach is the Fast Condensed Nearest Neighbor rule, which appears to be a promising approach to tackle both problems [Ang07]. Furthermore, since the introduction of the nearest neighbor decision rules more than half a century ago, the Support Vector Machine (SVM) classifier has emerged as a very popular and powerful classifier that rivals the nearest neighbor rules in terms of classification accuracy. Hence one of the main areas of research today is the empirical comparison of the nearest neighbor methods with SVMs in terms of both classification accuracy and computation time, using a wide variety of datasets [GGG+15].

TRAINING-DATA EDITING RULES

Methods have been developed [TBP85] to edit (delete) "redundant" members of $\{X, Y\}$ in order to obtain a subset of $\{X, Y\}$ that implements exactly the same decision boundary as would be obtained using all of $\{X, Y\}$. Such methods depend on the computation of Voronoi diagrams and of other proximity graphs that are subgraphs of the Delaunay triangulation, such as the Gabriel graph. Furthermore, the fraction of data discarded in such a method is a useful measure of the resulting reliability of the rule. If few vectors are discarded the feature space is relatively empty and more training data are needed. During the past thirty-five years proximity graphs have proven to be very useful both in theory and in practice for solving many of the problems encountered with *NN*-rules. A description of many of these graphs along with related computational geometry problems can be found in [Tou02].

NEAREST-NEIGHBOR SEARCHING

Another important issue in the implementation of nearest-neighbor decision rules, whether editing has or has not been performed, concerns the efficient search for the nearest neighbor of an unknown vector in order to classify it. Various methods exist for computing a nearest neighbor without computing distances to all the candidates. The problem is in general quite difficult when the dimension is high, which it is for most pattern recognition tasks. Simple brute-force search yields $O(dn)$ query time. To improve upon this, one builds a data structure for the points that supports more efficient queries, often at the expense of space for the data structure. For a set of n points in $\mathbb{R}^d$, one could construct a Voronoi diagram for the points of size $O(n^{\lceil d/2 \rceil})$ (Chapter 26), and respond to a query in $O(\log n)$ time. But the exponential space makes this impractical beyond $d \leq 3$. Range searching (Chapter 40) supports structures with linear space and achieving slightly sublinear time. But all constants are exponential in d. This has led to intensive work on approximate nearest-neighbor search, where one seeks a point within $(1 + \epsilon)$ of the distance to the true nearest neighbor. An example of an important early milestone along these lines is an algorithm by Arya et al. [AMN+98], which constructs a data structure of size $O(dn)$ that can report approximate nearest neighbors in $O(c \log n)$ time, with $c = O(d(1 + d/\epsilon)^d)$. The algorithm traverses down a balanced box-tree decomposition (BDD) of $O(\log n)$ height and stops when the approximation criterion is satisfied. The query time is logarithmic in n but still the constant is exponential in d. The many advances beyond this and similar algorithms with exponential query time or space requirements are described in Chapters 43 (high dimensions) and 32 (low dimensions).

ESTIMATION OF MISCLASSIFICATION

A very important problem in pattern recognition is the estimation of the performance of a decision rule [McL92]. Many geometric problems occur here also, for which computational geometry offers elegant and efficient solutions. For example, a good method of estimating the performance of the NN-rule is to delete each member of $\{X, Y\} = \{(X_1, Y_1), (X_2, Y_2), \ldots, (X_n, Y_n)\}$ in turn and classify it with the remaining set. Geometrically this problem reduces to computing for a given set of points in d-space the nearest neighbor of each (the **all-nearest neighbors problem**). Vaidya [Vai89] gives an $O(n \log n)$ time algorithm to solve this problem.

54.7 PATTERN COMPLEXITY

GLOSSARY

Kolmogorov complexity: The length of the shortest algorithm that generates the pattern.

Papentin complexity: The length of the shortest description of the pattern by means of a specified hierarchy of description languages.

Normalized pairwise variability index (nPVI): The average variation of a set of distances that are obtained from successive adjacent ordered pairs of elements in the pattern.

Sub-symmetries: The contiguous subsets of a pattern that possess reflection symmetries.

Perimetric complexity: The sum of the exterior perimeter of the pattern and the perimeter of the holes in the pattern (interior perimeter), squared, divided by the pattern area, and divided by 4π.

The study of complexity in pattern recognition is one of the more recent developments in the field, and it concerns two rather different problems, the solutions to which require different approaches: the measurement of the complexity of a single pattern, and the complexity of a dataset (training set) of patterns. A single pattern can be one-dimensional, e.g., temporal data such as speech and music, two-dimensional as in character recognition and image processing, or three or higher dimensional. In addition to being of theoretical interest, in practice the complexity of a pattern serves as a feature (shape measurement) of the pattern. For example, in musical rhythm the complexity of a rhythm is akin to the amount of syncopation the rhythm contains. The complexity of a collection of patterns (the training set) is referred to as *data complexity*, where data consist of usually multidimensional feature vectors (points in some space) [HBL06].

The investigation of data complexity via geometrical characteristics of training data is motivated by the desire to determine the existence of independent factors that contribute to the difficulty of pattern classification, and hence guide the selection of an appropriate classification algorithm for a specific application [Can13]. The measures usually considered include measures of overlaps in feature values, measures of the separability of pattern classes, and measures of the shape, density, and interconnectedness of the data manifolds. The emphasis in this research has been on the relative effectiveness of predicting the pattern classification accuracy of several popular decision rules, such as nearest neighbor rules [HBL06] [CRV12]. However, investigation of the relative computational complexity of these geometric measures remains mostly open, apart from a few exceptions such as measures of class separability [Hou93].

Definitions of measures of the complexity of individual patterns vary depending on the definitions of "patterns," which depends in turn on the particular domain that investigates patterns [TT14]. For one-dimensional patterns such as symbol sequences, an appropriate definition for the complexity of an individual pattern (rather than a collection of patterns), is the Kolmogorov complexity [Kol65], which is defined as the length of the shortest algorithm that generates the pattern. Although this measure is theoretically fruitful, it is not computable in practice, and hence there is a need for suitable approximations to the Kolmogorov complexity. One attractive approximation (lower bound) of the Kolmogorov complexity was proposed by Papentin [Pap80] in terms of the shortest description of the pattern computed in a well defined efficiently computable hierarchy of specific description languages.

A measure of one-dimensional pattern complexity that originated in computational linguistics, but has made a significant entry into the study of musical rhythm, is the *normalized pairwise variability index* (*nPVI*) [RAR10]. The *nPVI* is defined as the average variation of a set of distances that are obtained from successive adjacent ordered pairs of elements in the pattern. For a comparison of the *nPVI* to other measures of one-dimensional pattern complexity in the domain of musical rhythm see [Tou13].

Another easily computable measure of the simplicity of one-dimensional patterns is the number of *sub-symmetries* that the pattern contains. A sub-symmetry of a sequence is defined as a contiguous (connected) subsequence of the sequence (pattern) that possesses mirror symmetry. For example, in the pattern BBWWWBB there are four sub-symmetries of length two (BB, WW, WW, and BB), one of length three (WWW), one of length five (BWWWB), and one of length seven (BBWWWBB), for a total of seven. A pattern that contains many sub-symmetries is considered to be simple, and one with few sub-symmetries is complex. This measure was originally shown to predict human judgments of the complexity of visual one-dimensional patterns. However, it was recently discovered that it also predicts human judgments of the complexity of auditory patterns (musical rhythms) [TB13]. These measures can be generalized to apply to two-dimensional patterns. The simplest extension is simply to compute the one-dimensional versions for every row, column and diagonal of the two-dimensional pattern and add up all the counts obtained [TOV15].

A popular measure of pattern complexity with a variety of applications to character recognition is the *perimetric complexity*. For continuous plane shapes the perimetric complexity of a binary pattern with holes is defined as the sum the exterior perimeter of the pattern and the perimeter of the holes (interior perimeter), squared, divided by the pattern area, and divided by 4pi. Several variants for the case of digital images, and for taking blurring into account, are proposed by Watson [Wat12], who also cites many applications of this measure.

54.8 SOURCES AND RELATED MATERIAL

SURVEYS

[BR14]: A survey of traditional and graph theoretical techniques for image segmentation.

[JMF99]: A survey of clustering from the pattern recognition point of view.

[MQ94]: A survey of properties and types of sphere of influence graphs.

[MPW13]: An exhaustive survey of the state-of-the-art in symmetry detection methods for geometric data sets.

[MC15]: An extensive survey of the Hough transform and its variants.

[ABE09]: A survey of algorithms for computing the medial axis.

[Tou91]: A survey of computer vision problems where computational geometry may be applied. This survey references several others; the entire collection is of interest as well.

[Tou00]: A survey on computing nice viewpoints of objects in space.

[Tou14a]: A survey on applications of the relative neighborhood graph to pattern recognition problems.

[Tou14b]: A survey on theoretical results and applications of the sphere of influence graph to pattern recognition problems.

[Tou14c]: A survey on pattern recognition problems, including several distance measures between sets, that may be computed with the *rotating calipers*.

RELATED CHAPTERS

Chapter 1: Finite point configurations
Chapter 10: Geometric graph theory
Chapter 27: Voronoi diagrams and Delaunay triangulations
Chapter 30: Polygons
Chapter 32: Proximity algorithms
Chapter 38: Point location
Chapter 40: Range searching
Chapter 43: Nearest-neighbors in high-dimensional spaces
Chapter 45: Robust geometric computation

REFERENCES

[ABE09]   D. Attali, J.-D. Boissonnat, and H. Edelsbrunner. Stability and computation of medial axes-a state-of-the-art report. In *Mathematical Foundations of Scientific Visualization, Computer Graphics, and Massive Data Exploration*, pages 109–125, Springer, Berlin, 2009.

[ABL02]   A. Apostolico, M.E. Bock, and S. Lonardi. Monotony of surprise and large-scale quest for unusual words. *J. Comput. Biology* 10.3-4: 283-311, 2003.

[ACH91]   E.M. Arkin, L.P. Chew, D.P. Huttenlocher, K. Kedem, and J.S.B. Mitchell. An efficiently computable metric for comparing polygonal shapes. *IEEE Trans. Pattern Anal. Mach. Intell.*, 13:209–206, 1991.

[ADB$^+$10]   M.A. Abam, M. de Berg, P. Hachenberger, and A. Zarei. Streaming algorithms for line simplification. *Discrete Comput. Geom.*, 43:497–515, 2010.

[AFL06]   G. Aloupis, T. Fevens, S. Langerman, T. Matsui, A. Mesa, Y. Nuñez, D. Rappaport, and G. Toussaint. Algorithms for computing geometric measures of melodic similarity. *Computer Music Journal*, 30:67–76, 2006.

[AG00]   H. Alt and L.J. Guibas. Discrete geometric shapes: Matching, interpolation, and approximation. In J.-R. Sack and J. Urrutia, editors, *Handbook of Computational Geometry*, pages 121–153, Elsevier, Amsterdam, 2000.

[AHM+05]   P.K. Agarwal, S. Har-Peled, N.H. Mustafa, and Y. Wang. Near-linear time approximation algorithms for curve simplification. *Algorithmica*, 42:203–219, 2005.

[AK96]   T. Asano and N. Katoh. Variants for the Hough transform for line direction. *Comput. Geom.*, 6:231–252, 1996.

[AMN+98]   S. Arya, D.M. Mount, N.S. Netanyahu, R. Silverman, and A.Y. Wu. An optimal algorithm for approximate nearest neighbor searching in fixed dimensions. *J. ACM*, 45:891–923, 1998.

[AMW+88]   H. Alt, K. Mehlhorn, H. Wagener, and E. Welzl. Congruence, similarity and symmetries of geometric objects. *Discrete Comput. Geom.*, 3:237–256, 1988.

[Ang07]   F. Angiulli. Fast nearest neighbor condensation for large data sets classification. *IEEE Trans. Knowl. Data Eng.*, 19:1450–1464, 2007.

[AT79]   S.G. Akl and G.T. Toussaint. Addendum to "An improved algorithm to check for polygon similarity." *Inform. Process. Lett.*, 8:157–158, 1979.

[AW13]   F. Aurenhammer and G. Walzl. Three-dimensional straight skeletons from bisector graphs. In *Proc. 5th Internat. Conf. Analytic Number Theory and Spatial Tessellations*, pages 58–59, 2013.

[BC15]   D. Belazzougui and F. Cunial. Space-efficient detection of unusual words. In *Proc. 22nd Sympos. String Process. Info. Retrieval*, vol. 9309 of *LNCS*, pages 222–233, Springer, Berlin, 2015.

[BCG98]   P. Bose, J. Caron, and K. Ghoudi. Detection of text-line orientation. In *10th Canadian Conf. Comput. Geom.*, Québec, 1998.

[Ber06]   P. Berkhin. A survey of clustering data mining techniques. In *Grouping Multidimensional Data*, pages 25–71, Springer, Berlin, 2006.

[BGRT99]   P. Bose, F. Gómez, P. Ramos, and G.T. Toussaint. Drawing nice projections of objects in space. *J. Visual Commun. Image Rep.*, 10:155–172, 1999.

[BJF90]   H.S. Baird, S.E. Jones, and S.J. Fortune. Image segmentation by shape-directed covers. In *Proc. 10th IEEE Conf. Pattern Recogn.*, pages 820–825, 1990.

[BJW+08]   S. Bereg, M. Jiang, W. Wang, B. Yang, and B. Zhu. Simplifying 3D polygonal chains under the discrete Fréchet distance. In *Proc. 8th Latin American Sympos. Theoret. Informatics*, vol. 4957 of *LNCS*, pages 630–641, Springer, Berlin, 2008.

[BK13]   A.K. Bajwa, and R. Kaur. Fast lane detection using improved Hough transform. *J. Computing Technologies*, 2:10–13, 2013.

[Blu67]   H. Blum. A transformation for extracting new descriptors of shape. In W. Wathen-Dunn, editor, *Models for the Perception of Speech and Visual Form*, pages 362–380, MIT Press, Cambridge, 1967.

[BR94]   P. Bhattacharya and A. Rosenfeld. Polygons in three dimensions. *J. Visual Communication and Image Representation*, 5:139–147, 1994.

[BR14]   B. Basavaprasad and S.H. Ravindra. A survey on traditional and graph theoretical techniques for image segmentation. In *Proc. National Conf. Recent Advances in Information Technology*, pages 38–46, 2014.

[Bra02]   P. Brass. Combinatorial geometry problems in pattern recognition. *Discrete Comput. Geom.*, 28:495–510, 2002.

[BT83]   B.K. Bhattacharya and G.T. Toussaint. Efficient algorithms for computing the maximum distance between two finite planar sets. *J. Algorithms*, 4:121–136, 1983.

[BT10]   J. Bhadury, and C.A. Tovey. An improved implementation and analysis of the Diaz and O'Rourke algorithm for finding the Simpson point of a convex polygon. *Int. J. Computer Math.*, 87:244–259, 2010.

[BWR93]   J.B. Burns, R.S. Weiss, and E.M. Riseman. View variation of point-set and line-segment features. *IEEE Trans. Pattern Anal. Mach. Intell.*, 15:51–68, 1993.

[Can13]   J.-R. Cano. Analysis of data complexity measures for classification. *Expert Syst. Appl.*, 40:4820–4831, 2013.

[CC96]   W.S. Chan and F.Y.L. Chin. Approximation of polygonal curves with minimum number of line segments or minimum error. *Internat. J. Comput. Geom. Appl*, 6:59–77, 1996.

[CF98]   V. Cerverón and A. Fuertes. Parallel random search and Tabu search for the minimum consistent subset selection problem. In *Randomization and Approximation Techniques in Computer Science*, vol. 1518 of *LNCS*, pages 248–259, Springer, Berlin, 1998.

[CRV12]   G.D.C. Cavalcanti, I.R. Tsang, and B.A. Vale. Data complexity measures and nearest neighbor classifiers: A practical analysis for meta-learning. In *Proc. 24th IEEE Conf. Tools with Artificial Intelligence*, volume 1, pages 1065–1069, 2012.

[CSW99]   F. Chin, J. Snoeyink, and C.A. Wang. Finding the medial axis of a simple polygon in linear time. *Discrete Comput. Geom.*, 21:405–420, 1999.

[CT76]   D. Cheriton and R.E. Tarjan. Finding minimum spanning trees. *SIAM J. Comput.*, 5:724–742, 1976.

[CT77]   M. Cohen and G.T. Toussaint. On the detection of structures in noisy pictures. *Pattern Recogn.*, 9:95–98, 1977.

[Cul00]   T. Culver. Computing the Medial Axis of a Polyhedron Reliably and Efficiently. PhD thesis, University of North Carolina, Chapel Hill, 2000.

[Cyc94]   J.M. Cychosz. Efficient binary image thinning using neighborhood maps. In P. Heckbert, editor, *Graphics Gems IV*, pages 465–473. Academic Press, Boston, 1994.

[Das94]   B.V. Dasarathy. Minimal consistent set (MCS) identification for optimal nearest neighbor decision system design. *IEEE Trans. Syst. Man Cybern.*, 24:511–517, 1994.

[DGL96]   L. Devroye, L. Györfi, and G. Lugosi. *A Probabilistic Theory of Pattern Recognition*. Springer, New York, 1996.

[DM00]   G. Das and H. Mannila. Context-based similarity measures for categorical databases. In *Principles of Data Mining and Knowledge Discovery*, pages 201–210, 2000.

[DO94]   M. Díaz and J. O'Rourke. Algorithms for computing the center of area of a convex polygon. *Visual Comput.*, 10:432–442, 1994.

[DPR92a]   S.J. Dickinson, A. Pentland, and A. Rosenfeld. 3D shape recovery using distributed aspect matching. *IEEE Trans. Pattern Anal. Mach. Intell.*, 14:174–197, 1992.

[DPR92b]   S.J. Dickinson, A. Pentland, and A. Rosenfeld. From volumes to views: An approach to 3D object recognition. *CVGIP: Image Understanding*, 55:130–154, 1992.

[DWT99]   S.J. Dickinson, D. Wilkes, and J.K. Tsotsos. A computational model of view degeneracy. *IEEE Trans. Pattern Anal. Mach. Intell.*, 21:673–689, 1999.

[DZ04]   T.K. Dey and W. Zhao. Approximating the medial axis from the Voronoi diagram with a convergence guarantee. *Algorithmica*, 38:179–200, 2004.

[Ead88]   P. Eades. Symmetry finding algorithms. In G.T. Toussaint, editor, *Computational Morphology*, pages 41–51, North-Holland, Amsterdam, 1988.

[Ede85]   H. Edelsbrunner. Computing the extreme distances between two convex polygons. *J. Algorithms*, 6:213–224, 1985.

[EM94]   H. Edelsbrunner and E.P. Mücke. Three-dimensional alpha shapes. *ACM Trans. Graph.*, 13:43–72, 1994.

[ET94]   D. Eu and G.T. Toussaint. On approximating polygonal curves in two and three dimensions. *CVGIP: Graph. Models Image Process.*, 56:231–246, 1994.

[FH51]   E. Fix and J. Hodges. Discriminatory analysis. Nonparametric discrimination: Consistency properties. Tech. Report 4, USAF School of Aviation Medicine, Randolph Field, Texas, 1951.

[FO12]   L.A.F. Fernandes and M.M. Oliveira. A general framework for subspace detection in unordered multidimensional data. *Pattern Recogn.*, 45:3566–3579, 2012.

[GGG+15]   M. Gamboni, A. Garg, O. Grishin, S. Man Oh, F. Sowani, A. Spalvieri-Kruse, G.T. Toussaint, and L. Zhang. An empirical comparison of support vector machines versus nearest neighbour methods for machine learning applications. In *Proc. 3rd Conf. Pattern Recogn. Appl. Methods*, vol. 9443 of *LNCS*, pages 110–129, Springer, Berlin, 2015.

[GHS+01]   F. Gómez, F. Hurtado, J.A. Sellarès, and G.T. Toussaint. On degeneracies removable by perspective projection. *Internat. J. Math. Alg.*, 2:227–248, 2001.

[Gor96]   A.D. Gordon. A survey of constrained classification. *Comput. Stat. Data Anal.*, 21:17–29, 1996.

[GP83]   J.E. Goodman and R. Pollack. Multidimensional sorting. *SIAM J. Comput.*, 12:484–507, 1983.

[Gre76]   U. Grenander. *Pattern Synthesis*. Springer, New York, 1976.

[GRS00]   S. Guha, R. Rastogi, and K. Shim. ROCK: A robust clustering algorithm for categorical attributes. *Info. Sys.*, 25:345–366, 2000.

[GRT01]   F. Gómez, S. Ramaswami, and G.T. Toussaint. On computing general position views of data in three dimensions. *J. Visual Commun. Image Rep.*, 12:387–400, 2001.

[Har68]   P.E. Hart. The condensed nearest neighbor rule. *IEEE Trans. Inform. Theory*, 14:515–516, 1968.

[HBL06]   T.K. Ho, M. Basu, and M.H.C. Lau. Measures of geometrical complexity in classification problems. In M. Basu and T. Kam Ho, editors, *Data Complexity in Pattern Recognition*, pages 3–23, Springer, London, 2006.

[Hou93]   M.E. Houle. Algorithms for weak and wide separation of sets. *Discrete Appl. Math.*, 45:139–159, 1993.

[HS97]   F. Hurtado and A.A. Sellarès. Proyecciones perspectivas regulares: Correspondencia por proyeccion perspectiva entre configuraciones planas de puntos. In *Actas VII Encuentros de Geometria Computacional*, pages 57–70, 1997.

[Itt93]   D.J. Ittner. Automatic inference of textline orientation. In *Proc. 2nd Sympos. Document Anal. Info. Retrieval*, pages 123–133, 1993.

[JB91]   X.-Y. Jiang and H. Bunke. Determination of the symmetries of polyhedra and an application to object recognition. In *Proc. Comput. Geom.: Methods, Algorithms, Appl.*, vol. 553 of *LNCS*, pages 113–121, Springer, Berlin, 1991.

[Jen96]   J.R. Jensen. *Introductory Digital Image Processing: A Remote Sensing Perspective*, 2nd edition. Prentice-Hall, Englewood Cliffs, 1996.

[JMF99]   A.K. Jain, M.N. Murty, and P.J. Flynn. Data clustering: A review. *ACM Comput. Surv.*, 31:264–323, 1999.

[JT92]   J.W. Jaromczyk and G.T. Toussaint. Relative neighborhood graphs and their relatives. *Proc. IEEE*, 80:1502–1517, 1992.

[KB98]   L.I. Kuncheva and J.C. Bezdek. Nearest prototype classification: Clustering, genetic algorithms, or random search. *IEEE Trans. Syst. Man Cybern.*, 28:160–164, 1998.

[KCD+04]   M. Kazhdan, B. Chazelle, D. Dobkin, T. Funkhouser, and S. Rusinkiewicz. A reflective symmetry descriptor for 3D models. *Algorithmica*, 38:201–225, 2004.

[KF87]   J. Kender and D. Freudenstein. What is a degenerate view? In *Proc. 10th Joint Conf. Artif. Intell.*, pages 801–804, 1987.

[KK88]   T. Kamada and S. Kawai. A simple method for computing general position in displaying three-dimensional objects. *Comput. Vision Graph. Image Process.*, 41:43–56, 1988.

[KL02]   D. Krznaric and C. Levcopoulos. Optimal algorithms for complete linkage clustering in d dimensions. *Theoret. Comput. Sci.*, 286:1–149, 2002.

[KMN+04]   T. Kanungo, D.M. Mount, N.S. Netanyahu, C.D. Piatko, R. Silverman, and A.Y. Wu. A local search approximation algorithm for k-means clustering. *Comput. Geom.*, 28:89–112, 2004.

[Kol65]   A.N. Kolmogorov. Three approaches to the quantitative definition of information. *Problems in Information Transmission*, 1:1–7, 1965.

[LAM10]   A.J. Law, D.G. Aliaga, and A. Majumder. Projector placement planning for high quality visualizations on real-world colored objects. *IEEE Trans. Vis. Comput. Graphics*, 16:1633–1641, 2010.

[Lea92]   V.F. Leavers. *Shape Detection in Computer Vision Using the Hough Transform*. Springer, Berlin, 1992.

[Lee82]   D.T. Lee. Medial axis transformation of a planar shape. *IEEE Trans. Pattern Anal. Mach. Intell.*, PAMI-4:363–369, 1982.

[Liv93]   C. Livingston. *Knot Theory*. Math. Assoc. Amer., Washington, 1993.

[LPPD16]   T. Lafarge, B. Pateiro-Lopez, A. Possolo, and J.P. Dunkers. R implementation of a polyhedral approximation to a 3D set of points using the α-shape. American Statistical Association, 2014. https://cran.r-project.org/web/packages/alphashape3d/.

[LSC97]   D.E. Lake, B.M. Sadler, and S.D. Casey. Detecting regularity in minefields using collinearity and a modified Euclidean algorithm. In *Proc. SPIE 3079, Detection and Remediation Technologies for Mines and Minelike Targets II*, 1997.

[LW88]   Y. Lamdan and H.J. Wolfson. Geometric hashing: A general and efficient model-based recognition scheme. In *Proc. 2nd Internat. Conf. on Computer Vision*, pages 238–249, 1988.

[Mac67]   J. MacQueen. Some methods for classification and analysis of multivariate observations. In *Proc. 5th Berkeley Sympos. Math. Stat. Probab.*, pages 281–296, 1967.

[Mar04]   D.J. Marchette. *Random Graphs for Statistical Pattern Recognition*. John Wiley & Sons, Hoboken, 2004.

[Mat00]   J. Matoušek. On approximate geometric k-clustering. *Discrete Comput. Geom.*, 24:61–84, 2000.

[McL92]   G.J. McLachlan. *Discriminant Analysis and Statistical Pattern Recognition*. John Wiley & Sons, Hoboken, 1992.

[MNV12]   M. Mahajan, P. Nimbhorkar, and K. Varadarajan. The planar k-means problem is NP-hard. *Theoret. Comput. Sci.*, 442:13–21, 2012.

[MO88]   A. Melkman and J. O'Rourke. On polygonal chain approximation. In G.T. Toussaint, editor, *Computational Morphology*, pages 87–95, North-Holland, Amsterdam, 1988.

[MPW13]   N.J. Mitra, M. Pauli, M. Wand, and D. Ceylan. Symmetry in 3D geometry: Extraction and applications. *Computer Graphics Forum*, 22:1–23, 2013.

[MQ94]   T.S. Michael and T. Quint. Sphere of influence graphs: A survey. *Congressus Numerantium*, 105:153–160, 1994.

[MSW02]   R. Nussinov, M. Shatsky, and H.J. Wolfson. Flexible protein alignment and hinge detection. *Proteins*, 48:242–256, 2002.

[MC15]   P. Mukhopadhyay and B.B. Chaudhuri. A survey of Hough transform. *Pattern Recogn.*, 48:993–1010, 2015.

[ÓMa98]   D.S. Ó Maidín. A geometrical algorithm for melodic difference. *Computing in Musicology*, 11:65–72, 1998.

[Pap80]   F. Papentin. On order and complexity. I. General considerations. *J. Theoret. Biology*, 87:421–456, 1980.

[PB97]   N.R. Pal and J. Biswas. Cluster validation using graph theoretic concepts. *Pattern Recogn.*, 30:847–857, 1997.

[PD90]   H. Plantinga and C.R. Dyer. Visibility, occlusion, and the aspect graph. *Internat. J. Comput. Vision*, 5:137–160, 1990.

[PM99]   D. Pelleg and A. Moore. Accelerating exact k-means algorithms with geometric reasoning. In *Knowledge Discovery and Data Mining*, pages 277–281, AAAI Press, New York, 1999.

[PZZ13]   B. Peng, L. Zhang, and D. Zhang. A survey of graph theoretical approaches to image segmentation *Pattern Recogn.*, 46:1020–1038, 2013.

[RAR10]   M. Raju, E.L. Asu, and J. Ross. Comparison of rhythm in musical scores and performances as measured with the pairwise variability index. *Musicae Scientiae*, 14:51–71, 2010.

[Rei32]   K. Reidemeister. *Knotentheorie*. Springer, Berlin, 1932. L.F. Boron, C.O. Christenson and B.A. Smith (English translation), *Knot Theory*, BSC Associates, Moscow, Idaho, USA, 1983.

[Rob93]   J.-M. Robert. Maximum distance between two sets of points in E^d. *Pattern Recogn. Lett.*, 14:733–735, 1993.

[RWLI75]   G.L. Ritter, H.B. Woodruff, S.R. Lowry, and T.L. Isenhour. An algorithm for a selective nearest neighbor decision rule. *IEEE Trans. Inform. Theory*, 21:665–669, 1975.

[SKK00]   M. Steinbach, G. Karypis, and V. Kumar. A comparison of document clustering techniques. In *KDD Workshop on Text Mining*, vol. 400, pages 525–526, 2000.

[Smi00]   M. Smid. Closest point problems in computational geometry. In J.-R. Sack and J. Urrutia, editors, *Handbook of Computational Geometry*, pages 877–935, Elsevier, Amsterdam, 2000.

[SNT$^+$92]   V. Srinivasan, L.R. Nackman, J.-M. Tang, and S.N. Meshkat. Automatic mesh generation using the symmetric axis transform of polygonal domains. *Proc. IEEE*, 80:1485–1501, 1992.

[SSSS10]   R.K. Satzoda, S. Sathyanarayana, T. Srikanthan, and S. Sathyanarayana. Hierarchical additive Hough transform for lane detection. *IEEE Embedded Syst. Lett.*, 2:23–26, 2010.

[Sug84]   K. Sugihara. An $n \log n$ algorithm for determining the congruity of polyhedra. *J. Comput. Syst. Sci.*, 29:36–47, 1984.

[TB13]   G.T. Toussaint and J.F. Beltran. Subsymmetries predict auditory and visual pattern complexity. *Perception*, 42:1095–1100, 2013.

[TBP85]   G.T. Toussaint, B.K. Bhattacharya, and R.S. Poulsen. The application of Voronoi diagrams to nonparametric decision rules. In *Computer Science and Statistics: The Interface*, pages 97–108, 1985.

[TDS$^+$16]   A. Tagliasacchi, T. Delame, M. Spagnuolo, N. Amenta, and A. Telea. 3D skeletons: A state-of-the-art report. *Computer Graphics Forum*, 35:573–597, 2016.

[TM82]   G.T. Toussaint and M.A. McAlear. A simple $O(n \log n)$ algorithm for finding the maximum distance between two finite planar sets. *Pattern Recogn. Lett.*, 1:21–24, 1982.

[Tou80]   G.T. Toussaint. The relative neighbourhood graph of a finite planar set. *Pattern Recogn.*, 12:261–268, 1980.

[Tou84]   G.T. Toussaint. An optimal algorithm for computing the minimum vertex distance between two crossing convex polygons. *Computing*, 32:357–364, 1984.

[Tou91]   G.T. Toussaint. Computational geometry and computer vision. In B. Melter, A. Rosenfeld, and P. Bhattacharya, editors, *Vision Geometry*, pages 213–224, AMS, Providence, 1991.

[Tou00]   G.T. Toussaint. The complexity of computing nice viewpoints of objects in space. In *Proc. Vision Geometry IX, SPIE Internat. Sympos. Optical Sci. Tech.*, pages 1–11, 2000.

[Tou02]   G.T. Toussaint. Proximity graphs for nearest neighbor decision rules: Recent progress. In *Proc. 34th Sympos. Comput. Statist. (Geoscience and Remote Sensing)*, Montreal, 2002.

[Tou05]   G.T. Toussaint. Geometric proximity graphs for improving nearest neighbor methods in instance-based learning and data mining. *Internat. J. Comput. Geom. Appl.*, 15:101–150, 2005.

[Tou13]   G.T. Toussaint. The pairwise variability index as a measure of rhythm complexity. *Analytical Approaches to World Music*, 2:1–42, 2013.

[Tou14a]   G.T. Toussaint. Applications of the relative neighborhood graph. *Int. J. Advances Comp. Sci. Appl.*, 4:77–85, 2014.

[Tou14b]   G.T. Toussaint. The sphere of influence graph: Theory and applications. *Int. J. Inform. Tech. Comp. Sci.*, 14:37–42, 2014.

[Tou14c]   G.T. Toussaint. Applications of the rotating calipers to geometric problems in two and three dimensions. *Int. J. Digital Information and Wireless Communications*, 4:108–122, 2014.

[TOV15]   G.T. Toussaint, N. Onea, and Q. Vuong. Measuring the complexity of two-dimensional patterns: Sub-symmetries versus Papentin complexity. In *Proc. 14th IAPR Conf. Machine Vision Appl.*, pages 80–83, 2015.

[TT14]   E.R. Toussaint and G.T. Toussaint. What is a pattern? *Proc. Bridges: Mathematics, Music, Art, Architecture, Culture*, pages 293–300, 2014.

[Vai89]   P.M. Vaidya. An $O(n \log n)$ algorithm for the all-nearest-neighbors problem. *Discrete Comput. Geom.*, 4:101–115, 1989.

[Vat11]   A. Vattani. k-means requires exponentially many iterations even in the plane. *Discrete Comput. Geom.*, 45: 596–616, 2011.

[Wat12]   A.B. Watson. Perimetric complexity of binary digital images. *The Mathematica Journal*, 14:1–40, 2012.

[Wil91]   G. Wilfong. Nearest neighbor problems. In *Proc. 7th Sympos. Comput. Geom.*, pages 224–233, ACM Press, 1991.

[WM97]   D.R. Wilson and T.R. Martinez. Instance pruning techniques. In D. Fisher, editor, *Proc. 14th Internat. Conf. Machine Learning*, pages 404–411, Morgan Kaufmann Publishers, San Francisco, 1997.

[WM00]   D.R. Wilson and T.R. Martinez. Reduction techniques for instance-based learning algorithms. *Mach. Learning*, 38:257–286, 2000.

[WWV85]   J.D. Wolter, T. Woo, and R.A. Volz. Optimal algorithms for symmetry detection in two and three dimensions. *Visual Comput.*, 1:37–48, 1985.

[Yap90]   C.K. Yap. Symbolic treatment of geometric degeneracies. *J. Symb. Comput.*, 10:349–370, 1990.

[YD15]   Q. Ye and D.S. Doermann. Text detection and recognition in imagery: A survey. *IEEE Trans. Pattern Anal. Mach. Intell.*, 37:1480–1500, 2015.

[YL04]   J. Yang and I. Lee. Cluster validity through graph-based boundary analysis. In *Proc. Int. Conf. Information and Knowledge Engineering*, pages 204–210, 2004.

[YR91]   C. Yao and J.G. Rokne. A straightforward algorithm for computing the medial axis of a simple polygon. *Internat. J. Comput. Math.*, 39:51–60, 1991.

[ZA15]   N.M. Zaitoun and M.J. Aqel. Survey on image segmentation techniques. *Procedia Comput. Sci.*, 65:797–806, 2015.

[ZS02]   H. Zhang and G. Sun. Optimal reference subset selection for nearest neighbor classification by tabu search. *Pattern Recogn.*, 35:1481–1490, 2002.

55 GRAPH DRAWING

Emilio Di Giacomo, Giuseppe Liotta, and Roberto Tamassia

INTRODUCTION

Graph drawing addresses the problem of constructing geometric representations of graphs, and has important applications to key computer technologies such as software engineering, database systems, visual interfaces, and computer-aided design. Research on graph drawing has been conducted within several diverse areas, including discrete mathematics (topological graph theory, geometric graph theory, order theory), algorithmics (graph algorithms, data structures, computational geometry, VLSI), and human-computer interaction (visual languages, graphical user interfaces, information visualization). This chapter overviews aspects of graph drawing that are especially relevant to computational geometry. Basic definitions on drawings and their properties are given in Section 55.1. Bounds on geometric and topological properties of drawings (e.g., area and crossings) are presented in Section 55.2. Section 55.3 deals with the time complexity of fundamental graph drawing problems. An example of a drawing algorithm is given in Section 55.4. Techniques for drawing general graphs are surveyed in Section 55.5.

55.1 DRAWINGS AND THEIR PROPERTIES

TYPES OF GRAPHS

First, we define some terminology on graphs pertinent to graph drawing. Throughout this chapter let n and m be the number of graph vertices and edges respectively, and d the maximum vertex degree (i.e., number of edges incident to a vertex).

GLOSSARY

Degree-k graph: Graph with maximum degree $d \leq k$.

Digraph: Directed graph, i.e., graph with directed edges.

Acyclic digraph: Digraph without directed cycles.

Transitive edge: Edge (u, v) of a digraph is transitive if there is a directed path from u to v not containing edge (u, v).

Reduced digraph: Digraphs without transitive edges.

Source: Vertex of a digraph without incoming edges.

Sink: Vertex of a digraph without outgoing edges.

st-digraph: Acyclic digraph with exactly one source and one sink, which are joined by an edge (also called **bipolar digraph**).

Connected graph: Graph in which any two vertices are joined by a path.

Biconnected graph: Graph in which any two vertices are joined by two vertex-disjoint paths.

Triconnected graph: Graph in which any two vertices are joined by three (pairwise) vertex-disjoint paths.

Layered (di)graph: (Di)graph whose vertices are partitioned into sets, called layers, such that no two vertices in the same layer are adjacent.

k-layered (di)graph: Layered (di)graph with k layers.

Tree: Connected graph without cycles.

Directed Tree: Digraph whose underlying undirected graph is a tree.

Rooted tree: Directed tree with a distinguished vertex, the **root**, such that each vertex lies on a directed path to the root. A rooted tree is also viewed as a layered digraph where the layers are sets of vertices at the same distance from the root.

Binary tree: Rooted tree where each vertex has at most two incoming edges.

Ternary tree: Rooted tree where each vertex has at most three incoming edges.

Series-parallel digraph (SP digraph): A digraph with a single source s and a single sink t recursively defined as follows: (i) a single edge (s, t) is a series-parallel digraph. Given two series-parallel digraphs G' and G'' with sources s' and s'', respectively and sinks t' and t'', respectively, (ii) the digraph obtained by identifying t' with s'' is a series-parallel digraph; (iii) the digraph obtained by identifying s' with s'' and t' with t'' is a series-parallel digraph. The series-parallel digraphs defined above are often called **two-terminal series parallel digraphs**. Throughout this section series-parallel digraphs have no multiple edges.

Series-parallel graph (SP graph): The underlying undirected graph of a series-parallel digraph.

Bipartite (di)graph: (Di)graph whose vertices are partitioned into two sets and each edge connects vertices in different sets. A bipartite (di)graph is also viewed as a 2-layered (di)graph.

TYPES OF DRAWINGS

In a drawing of a graph one has to geometrically represent the vertices and their adjacencies (edges). This can be done in several different ways. In the most common types of drawing, vertices are represented by points (or by geometric figures such as circles or rectangles) and edges are represented by curves such that any two edges intersect at most in a finite number of points. In other types of drawings vertices can be represented by various geometric objects (segments, curves, polygons) while adjacencies can be represented by intersections, contacts, or visibility of the objects representing the vertices.

GLOSSARY

Polyline drawing: Each edge is a polygonal chain (Figure 55.1.1(a)).

Straight-line drawing: Each edge is a straight-line segment (Figure 55.1.1(b)).

Orthogonal drawing: Each edge is a chain of horizontal and vertical segments (Figure 55.1.1(c)).

Bend: In a polyline drawing, point where two segments belonging to the same edge meet (Figure 55.1.1(a)).

Orthogonal Representation: Description of an orthogonal drawing in terms of bends along each edge and angles around each vertex with no information about the length of the segments that connect vertices and bends.

Crossing: Intersection point of two edges that is not a common vertex nor a touching (tangential) point (Figure 55.1.1(b)).

Grid drawing: Polyline drawing such that vertices and bends have integer coordinates.

Planar drawing: Drawing where no two edges cross (see Figure 55.1.1(d)).

Planar (di)graph: (Di)graph that admits a planar drawing.

Face: A connected region of the plane defined by a planar drawing, where the unbounded region is called the ***external face***.

Embedded (di)graph: Planar (di)graph with a prespecified topological embedding (i.e., set of faces), which must be preserved in the drawing.

Outerplanar (di)graph: A planar (di)graph that admits a planar drawing with all vertices on the boundary of the external face.

Convex drawing: Planar straight-line drawing such that the boundary of each face is a convex polygon.

Upward drawing: Drawing of a digraph where each edge is monotonically non-decreasing in the vertical direction (see Figure 55.1.1(d)).

Upward planar digraph: Digraph that admits an upward planar drawing.

Layered drawing: Drawing of a layered graph such that vertices in the same layer lie on the same horizontal line (also called ***hierarchical drawing***).

Dominance drawing: Upward drawing of an acyclic digraph such that there exists a directed path from vertex u to vertex v if and only if $x(u) \leq x(v)$ and $y(u) \leq y(v)$, where $x(\cdot)$ and $y(\cdot)$ denote the coordinates of a vertex.

hv-drawing: Upward orthogonal straight-line drawing of a binary tree such that the drawings of the subtrees of each node are separated by a horizontal or vertical line.

FIGURE 55.1.1
Types of drawings: (a) *polyline drawing of* $K_{3,3}$; (b) *straight-line drawing of* $K_{3,3}$; (c) *orthogonal drawing of* $K_{3,3}$; (d) *planar upward drawing of an acyclic digraph.*

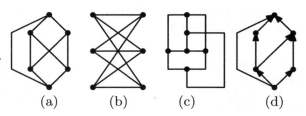

(a)   (b)   (c)   (d)

Straight-line and orthogonal drawings are special cases of polyline drawings. Polyline drawings provide great flexibility since they can approximate drawings with curved edges. However, edges with more than two or three bends may be difficult to "follow" for the eye. Also, a system that supports editing of polyline drawings is more complicated than one limited to straight-line drawings. Hence, depending on the application, polyline or straight-line drawings may be preferred. If vertices are represented by points, orthogonal drawings are possible only for graphs of maximum vertex degree 4.

PROPERTIES OF DRAWINGS

GLOSSARY

Area: Area of the smallest axis-aligned rectangle (*bounding box*) containing the drawing. This definition assumes that the drawing is constrained by some resolution rule that prevents it from being reduced by an arbitrary scaling (e.g., requiring a grid drawing, or stipulating a minimum unit distance between any two vertices).

Total edge length: Sum of the lengths of the edges.

Maximum edge length: Maximum length of an edge.

Curve complexity: Maximum number of bends along an edge of a polyline drawing.

Angular resolution: Smallest angle formed by two edges, or segments of edges, incident on the same vertex or bend, in a polyline drawing.

Perfect angular resolution: A drawing has perfect angular resolution if for every vertex v the angle formed by any two consecutive edges around v is $\frac{2\pi}{d(v)}$, where $d(v)$ is the degree of v.

Aspect ratio: Ratio of the longest to the shortest side of the smallest rectangle with horizontal and vertical sides covering the drawing.

There are infinitely many drawings for a graph. In drawing a graph, we would like to take into account a variety of properties. For example, planarity and the display of symmetries are highly desirable in visualization applications. Also, it is customary to display trees and acyclic digraphs with upward drawings. In general, to avoid wasting valuable space on a page or a computer screen, it is important to keep the area of the drawing small. In this scenario, many graph drawing problems can be formalized as multi-objective optimization problems (e.g., construct a drawing with minimum area and minimum number of crossings), so that tradeoffs are inherent in solving them. Typically, it is desirable to maximize the angular resolution and to minimize the other measures.

FIGURE 55.1.2
(a–b) *Tradeoff between planarity and symmetry in drawing K_4.* (c–d) *Tradeoff between planarity and upwardness in drawing an acyclic digraph G.*

   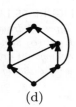

(a)   (b)   (c)   (d)

The following examples illustrate two typical tradeoffs in graph drawing problems. Figure 55.1.2(a–b) shows two drawings of K_4, the complete graph on four vertices. The drawing of part (a) is planar, while the drawing of part (b) "maximizes symmetries." It can be shown that no drawing of K_4 is optimal with respect to both criteria, i.e., the maximum number of symmetries cannot be achieved by a planar drawing. Figure 55.1.2(c–d), shows two drawing of the same acyclic digraph G. The drawing of part (c) is upward, while the drawing of part (d) is planar. It can be shown that there is no drawing of G which is both planar and upward.

55.2 BOUNDS ON DRAWING PROPERTIES

For various classes of graphs and drawing types, many universal/existential upper and lower bounds for specific drawing properties have been discovered. Such bounds typically exhibit tradeoffs between drawing properties. A universal bound applies to all the graphs of a given class. An existential bound applies to infinitely many graphs of the class. In the following tables, the abbreviations PSL, PSLO, PO, and PPL are used for planar straight-line, planar straight-line orthogonal, planar orthogonal, and planar polyline, respectively.

BOUNDS ON THE AREA

Tables 55.2.1 and 55.2.2 summarize selected universal upper bounds and existential lower bounds on the area of drawings of trees and graphs, respectively. The following comments apply to Tables 55.2.1 and 55.2.2, where specific rows of the table are indicated within parentheses. Linear or almost-linear bounds on the area can be achieved for several families of trees (1–7, 10–14, and 17–27); typically, superlinear lower bounds are associated with order preserving drawings (6, 9, 12, 14, 16, 19, 26). For directed trees, if a given embedding must be preserved, exponential area is required (28). See Table 55.2.5 for tradeoffs between area and aspect ratio in drawings of trees. Planar graphs admit planar drawings with quadratic area both in the straight-line, polyline and orthogonal model (10–12). For planar straight-line drawings, outerplanar graphs are the only class of graphs for which a sublinear upper bound is known (4). For polyline drawings this is true also for series-parallel graphs (5 and 7). Series-parallel graphs are the only subclass of planar graphs for which a superlinear lower bound is known both for straight-line and polyline drawings (6–7). If drawings are not required to be planar, linear area can be achieved for planar graphs (13). If, however, the drawing is required to be orthogonal, then superlinear lower bounds exists both for planar and non-planar graphs (2 and 3). In this case, almost linear area can be achieved for planar graphs (2), while linear area is possible for outerplanar graphs (1). Studies about the nature of the crossings in compact straight-line drawings of planar graphs are presented in [DDLM12, DDLM13]. Upward planar drawings provide an interesting tradeoff between area and having straight-line edges or not (15–24). Indeed, if a straight-line drawing is required, the area can become exponential even for subclasses of upward planar digraphs like outerplanar or bipartite DAGs (15, 18, 20, 22). In these cases a quadratic area bound is achieved only at the expense of a linear number of bends (16, 19, 21, 24). Other cases for which a polynomial bound is known are series-parallel graphs, when one is allowed to change the embedding (17), and upward dominance drawings of reduced planar st-graphs (23).

TABLE 55.2.1   Universal upper and existential lower bounds on the area of drawings of trees.

	CLASS OF GRAPHS	DRAWING TYPE	AREA	
1	Fibonacci tree	strict upw PSL	$\Omega(n)$ trivial	$O(n)$ [Tre96]
2	AVL tree	strict upw PSL	$\Omega(n)$ trivial	$O(n)$ [CPP98]
3	Balanced binary tree	strict upw PSL	$\Omega(n)$ trivial	$O(n)$ [CP98]
4	Binary tree	PSL	$\Omega(n)$ trivial	$O(n)$ [GR04]
5	Binary tree	ord pres PSL	$\Omega(n)$ trivial	$O(n \log \log n)$ [GR03a]
6	Binary tree	strict upw ord pres PSL	$\Omega(n \log n)$ [CDP92]	$O(n \log n)$ [GR03a]
7	Binary tree	PSLO	$\Omega(n)$ trivial	$O(n \log \log n)$ [SKC00]
8	Binary tree	ord pres PSLO	$\Omega(n)$ trivial	$O(n^{1.5})$ [Fra08b]
9	Binary tree	upw ord pres PSLO	$\Omega(n^2)$ [Fra08b]	$O(n^2)$ [Fra08b]
10	Binary tree	ord pres PO	$\Omega(n)$ trivial	$O(n)$ [DT81]
11	Binary tree	upw PO	$\Omega(n \log \log n)$ [GGT96]	$O(n \log \log n)$ [GGT96, Kim95, SKC00]
12	Binary tree	upw ord pres PO	$\Omega(n \log n)$ [GGT96]	$O(n \log n)$ [Kim95]
13	Binary tree	upw PPL	$\Omega(n)$ trivial	$O(n)$ [GGT96]
14	Binary tree	upw ord pres PPL	$\Omega(n \log n)$ [GGT96]	$O(n \log n)$ [GGT96]
15	Ternary tree	PSLO	$\Omega(n)$ trivial	$O(n^{1.631})$ [Fra08b]
16	Ternary tree	ord pres PSLO	$\Omega(n^2)$ [Fra08b]	$O(n^2)$ [Fra08b]
17	Ternary tree	PO	$\Omega(n)$ trivial	$O(n)$ [Val81]
18	Ternary tree	ord pres PO	$\Omega(n)$ trivial	$O(n)$ [DT81]
19	Ternary tree	upw ord pres PO	$\Omega(n \log n)$ [GGT96]	$O(n \log n)$ [Kim95]
20	deg-$O(1)$ rooted tree	upw PSL	$\Omega(n)$ trivial	$O(n \log \log n)$ [SKC00]
21	deg-$O(n^{\frac{a}{2}})$ rooted tree	PSL	$\Omega(n)$ trivial	$O(n)$ [GR03b]
22	deg-$O(n^a)$ rooted tree	upw PPL	$\Omega(n)$ trivial	$O(n)$ [GGT96]
23	Rooted tree	ord pres PSL	$\Omega(n)$ trivial	$O(n \log n)$ [GR03a]
24	Rooted tree	upw PSL	$\Omega(n)$ trivial	$O(n \log n)$ [CDP92]
25	Rooted tree	strict upw PSL	$\Omega(n \log n)$ [CDP92]	$O(n \log n)$ [CDP92]
26	Rooted tree	strict upw ord pres PSL	$\Omega(n \log n)$ [CDP92]	$O(n 4^{\sqrt{2 \log_2 n}})$ [Cha02]
27	Directed trees	upw PSL	$\Omega(n \log n)$ [CDP92]	$O(n \log n)$ [Fra08a]
28	Directed trees	upw ord pres PSL	$\Omega(b^n)$ [Fra08a]	$O(c^n)$ [GT93]

Note: n is the number of vertices, a, b, and c are constants such that $0 \le a < 1$, $1 < b < c$.
All bounds assume grid drawings.

TABLE 55.2.2   Universal upper and existential lower bounds on the area of drawings of graphs.

	CLASS OF GRAPHS	DRAWING TYPE	AREA	
1	Outerpl deg-4 graph	orthogonal	$\Omega(n)$ trivial	$O(n)$ [Lei80]
2	Planar deg-4 graph	orthogonal	$\Omega(n \log n)$ [Lei84]	$O(n \log^2 n)$ [Lei80, Val81]
3	Deg-4 graph	orthogonal	$\Omega(n^2)$ [Val81]	$O(n^2)$ [BK98a, BS15, PT98, Sch95, Val81]
4	Outerplanar graph	PSL	$\Omega(n)$ trivial	$O(n^{1.48})$ [DF09]
5	Outerplanar graph	PPL	$\Omega(n)$ trivial	$O(n \log n)$ [Bie11]
6	SP graph	PSL	$\Omega(n2^{\sqrt{\log_2 n}})$ [Fra10]	$O(n^2)$ [FPP90, Sch90]
7	SP graph	PPL	$\Omega(n2^{\sqrt{\log_2 n}})$ [Fra10]	$O(n^{1.5})$ [Bie11]
8	Triconn pl graph	convex PSL	$\Omega(n^2)$ [FPP90, FP08, MNRA11, Val81]	$O(n^2)$ [BFM07, CK97, DTV99, ST92]
9	Triconn pl graph	strict convex PSL	$\Omega(n^3)$ [And63, BP92, BT04, Rab93]	$O(n^4)$ [BR06]
10	Planar graph	PSL	$\Omega(n^2)$ [FPP90]	$O(n^2)$ [FPP90, Sch90]
11	Planar graph	PPL	$\Omega(n^2)$ [FPP90]	$O(n^2)$ [DT88, DTT92, Kan96]
12	Planar graph	PO	$\Omega(n^2)$ [FPP90]	$O(n^2)$ [BK98a, Kan96, Tam87, TT89]
13	Planar graph	straight-line	$\Omega(n)$ trivial	$O(n)$ [Woo05]
14	General graph	polyline	$\Omega(n)$ trivial	$O((n + \chi)^2)$ [BK98a, Kan96, Tam87, TT89]
15	Outerplanar DAG	upw PSL	$\Omega(b^n)$ [Fra08a]	$O(c^n)$ [GT93]
16	Outerplanar DAG	upw PPL	$\Omega(n^2)$ [Fra08a]	$O(n^2)$ [DT88, DTT92]
17	SP digraph	upw PSL	$\Omega(n^2)$ trivial	$O(n^2)$ [BCD$^+$94]
18	Embed SP digraph	upw PSL	$\Omega(b^n)$ [BCD$^+$94]	$O(c^n)$ [GT93]
19	Embed SP digraph	upw PPL	$\Omega(n^2)$ trivial	$O(n^2)$ [DT88, DTT92]
20	Bipartite DAG	upw PSL	$\Omega(b^n)$ [Fra08a]	$O(c^n)$ [GT93]
21	Bipartite DAG	upw PPL	$\Omega(n^2)$ [Fra08a]	$O(n^2)$ [DT88, DTT92]
22	Upward pl digraph	upw PSL	$\Omega(b^n)$ [DTT92]	$O(c^n)$ [GT93]
23	Reduced pl *st*-digraph	upw PSL dominance	$\Omega(n^2)$ [FPP90]	$O(n^2)$ [DTT92]
24	Upward pl digraph	upw PPL	$\Omega(n^2)$ [FPP90]	$O(n^2)$ [DT88, DTT92]

Note: n is the number of vertices, χ is the number of crossings in the drawing, a, b, and c are constants such that $0 \leq a < 1$, $1 < b < c$. All bounds assume grid drawings.

BOUNDS ON THE ANGULAR RESOLUTION

Table 55.2.3 summarizes selected universal lower bounds and existential upper bounds on the angular resolution of drawings of graphs. The bounds of the first row are stated for $n \geq 5$ because any planar straight-line drawing of K_4 has angular resolution lower than $\frac{\pi}{4}$.

TABLE 55.2.3 Universal lower and existential upper bounds on angular resolution.

CLASS OF GRAPHS	DRAWING TYPE	ANGULAR RESOLUTION	
deg-3 plan graph †	PSL	$\geq \frac{\pi}{4}$ [DLM14]	$\leq \frac{\pi}{4}$ [DLM14]
SP graph	PSL	$\Omega\left(\frac{1}{d}\right)$ [LLMN13]	$O\left(\frac{1}{d}\right)$ trivial
General graph	straight-line	$\Omega\left(\frac{1}{d^2}\right)$ [FHH+93]	$O\left(\frac{\log d}{d^2}\right)$ [FHH+93]
Planar graph	straight-line	$\Omega\left(\frac{1}{d}\right)$ [FHH+93]	$O\left(\frac{1}{d}\right)$ trivial
Planar graph	PSL	$\Omega\left(\frac{1}{c^d}\right)$ [MP94]	$O\left(\sqrt{\frac{\log d}{d^3}}\right)$ [GT94]
Planar graph	PSL	$\Omega\left(\frac{1}{n^2}\right)$ [FPP90, Sch90]	$O\left(\frac{1}{n}\right)$ trivial
Planar graph	PPL	$\Omega\left(\frac{1}{d}\right)$ [Kan96]	$O\left(\frac{1}{d}\right)$ trivial
Note: n is the number of vertices, d is the maximum vertex degree c is a constant such that $c > 1.$ † $n \geq 5$;			

BOUNDS ON THE NUMBER OF BENDS

Table 55.2.4 summarizes selected universal upper bounds and existential lower bounds on the total number of bends and on the curve complexity of orthogonal drawings. Some bounds are stated for $n \geq 5$ or $n \geq 7$ because the maximum number of bends is at least two for K_4 and at least three for the skeleton graph of an octahedron, in any planar orthogonal drawing.

TABLE 55.2.4 Orthogonal drawings: universal upper and existential lower bounds on the number of bends.

CLASS OF GRAPHS	DRAWING TYPE	TOTAL NUM. OF BENDS		CC	REF
deg-4 †	orthog	$\geq n$	$\leq 2n+2$	2	[BK98a]
Planar deg-4 †	orthog planar	$\geq 2n-2$	$\leq 2n+2$	2	[BK98a, TTV91]
Embed deg-4	orthog planar	$\geq 2n-2$	$\leq \frac{12}{5}n+2$	3	[EG95, LMS91, TT89, TTV91]
Biconn embed deg-4	orthog planar	$\geq 2n-2$	$\leq 2n+2$	3	[EG95, LMS91, TT89, TTV91]
Triconn embed deg-4	orthog planar	$\geq \frac{4}{3}(n-1)+2$	$\leq \frac{3}{2}n+4$	2	[Kan96]
Embed deg-3 ‡	orthog planar	$\geq \frac{1}{2}n+1$	$\leq \frac{1}{2}n+1$	1	[Kan96, LMPS92]
Note: CC stands for curve complexity, while n is the number of vertices. † $n \geq 7$; ‡ $n \geq 5$.					

TRADEOFF BETWEEN AREA AND ASPECT RATIO

The ability to construct area-efficient drawings is essential in practical visualization applications, where screen space is at a premium. However, achieving small area is not enough, e.g., a drawing with high aspect ratio may not be conveniently placed on a workstation screen, even if it has modest area. Hence, it is important to keep the aspect ratio small. Ideally, one would like to obtain small area for any given aspect ratio in a wide range. This would provide graphical user interfaces with the flexibility of fitting drawings into arbitrarily shaped windows. A variety of tradeoffs for the area and aspect ratio arise even when drawing graphs with a simple structure, such as trees. Table 55.2.5 summarizes selected universal bounds that can be simultaneously achieved on the area and the aspect ratio of various types of drawings of trees. Only for a few cases there exist algorithms that guarantee efficient area performance and that can accept any user-specified aspect ratio in a given range. For such cases, the aspect ratio in Table 55.2.5 is given as an interval.

TABLE 55.2.5   Trees: Universal upper bounds simultaneously achievable for area and aspect ratio.

CLASS OF GRAPHS	DRAWING TYPE	AREA	ASPECT RATIO	REF
Binary tree	PSL	$O(n)$	$[O(1), O(n^b)]$	[GR04]
Binary tree	ord pres PSL	$O(n \log n)$	$\left[O(1), O\left(\frac{n}{\log n}\right)\right]$	[GR04]
Binary tree	ord pres PSL	$O(n \log \log n)$	$O\left(\frac{n \log \log n}{\log^2 n}\right)$	[GR04]
Binary tree	upw ord pres PSL	$O(n \log n)$	$O\left(\frac{n}{\log n}\right)$	[GR04]
Binary tree	PSLO	$O(n \log \log n)$	$\left[O(1), O\left(\frac{n \log \log n}{\log^2 n}\right)\right]$	[SKC00]
Binary tree	upward PO	$O(n \log \log n)$	$O\left(\frac{n \log \log n}{\log^2 n}\right)$	[GGT96]
Binary tree	upward PSLO	$O(n \log n)$	$\left[O(1), O\left(\frac{n}{\log n}\right)\right]$	[CGKT02]
deg-4 tree	orthogonal	$O(n)$	$O(1)$	[Lei80, Val81]
deg-4 tree	orthogonal, leaves on hull	$O(n \log n)$	$O(1)$	[BK80]
Rooted deg-$O(n^a)$ tree	upward PPL	$O(n)$	$[O(1), O(n^b)]$	[GGT96]
Rooted tree	upward PSL layered	$O(n^2)$	$O(1)$	[RT81]
Rooted tree	upward PSL	$O(n \log n)$	$O\left(\frac{n}{\log n}\right)$	[CDP92]

Note: n is the number of vertices, a and b are constants such that $0 \le a, b < 1$.

All bounds assume grid drawings.

While upward planar straight-line drawings are the most natural way of visualizing rooted trees, the existing drawing techniques are unsatisfactory with respect to either the area requirement or the aspect ratio. Regarding polyline drawings, linear area can be achieved with a prescribed aspect ratio. However, for rooted trees, straight-line drawing remains by far the most used convention. For non-upward drawings of trees, linear area and optimal aspect ratio are possible for planar orthogonal drawings, and a small (logarithmic) amount of extra area is needed if the leaves are constrained to be on the convex hull of the drawing (e.g., pins on the boundary of a VLSI circuit). However, the non-upward drawing methods for rooted trees are better suited for VLSI layout than for visualization applications.

TRADEOFF BETWEEN AREA AND ANGULAR RESOLUTION

Table 55.2.6 summarizes selected universal bounds that can be simultaneously achieved on the area and the angular resolution of drawings of graphs. Universal lower bounds on the angular resolution exist that depend only on the degree of the graph. Also, substantially better bounds can be achieved by drawing a planar graph with bends or in a nonplanar way. Concerning trade-offs between area and angular resolution, Garg and Tamassia [GT94] proved that for any chosen angular resolution ρ, there exists a planar graph such that any planar straight-line drawing with angular resolution ρ has area $\Omega(a^{\rho n})$, for a constant $a > 1$. Duncan et al. [DEG+13] proved that there are trees that require exponential area for any order preserving planar straight-line drawing having perfect angular resolution. Duncan et al. also proved that perfect angular resolution and polynomial area can be simultaneously achieved for trees if order is not preserved or if the edges are drawn as circular arcs.

TABLE 55.2.6   Universal upper bounds for area and lower bounds for angular resolution, simultaneously achievable.

CLASS OF GRAPHS	DRAWING TYPE	AREA	ANGULAR RESOLUTION	REF
Tree	PSL	$O(n^8)$	$\Omega\left(\frac{1}{d}\right)$	[DEG+13]
Planar graph	SL grid	$O(d^6 n)$	$\Omega\left(\frac{1}{d^2}\right)$	[FHH+93]
Planar graph	SL grid	$O(d^3 n)$	$\Omega\left(\frac{1}{d}\right)$	[FHH+93]
Planar graph	PSL grid	$O(n^2)$	$\Omega\left(\frac{1}{n^2}\right)$	[FPP90, Sch90]
Planar graph	PSL grid	$O(b^n)$	$\Omega\left(\frac{1}{c^d}\right)$	[MP94]
Planar graph	PPL grid	$O(n^2)$	$\Omega\left(\frac{1}{d}\right)$	[Kan96]

Note: n is the number of vertices, d is the maximum vertex degree, b and c are constants such that $b > 1$ and $c > 1$.

OPEN PROBLEMS

1. Determine the area requirement of planar straight-line orthogonal drawings of binary and ternary trees. There are currently wide gaps between the known upper and lower bounds (Table 55.2.1, rows 8 and 15).

2. Determine the area requirement of (upward) planar straight-line drawings of trees. There is currently an $O(\log n)$ gap between the known upper and lower bounds (Table 55.2.1, row 24).

3. Determine the area requirement of (outer)planar straight-line grid drawings of outerplanar graphs. There is currently an $O(n^{0.48})$ gap between the known upper and lower bounds (Table 55.2.2, row 4).

4. Determine the area requirement of planar straight-line grid drawing of series-parallel graphs. In particular it would be interesting to prove a subquadratic upper bound (Table 55.2.2, row 6).

5. Determine the area requirement of orthogonal (or, more generally, polyline) nonplanar drawings of planar graphs. There is currently an $O(\log n)$ gap between the known upper and lower bounds (Table 55.2.2, row 2).

6. Close the gap between the $\Omega(\frac{1}{d^2})$ universal lower bound and the $O(\frac{\log d}{d^2})$ existential upper bound on the angular resolution of straight-line drawings of general graphs (Table 55.2.3).

7. Close the gap between the $\Omega(\frac{1}{c^d})$ universal lower bound and the $O(\sqrt{\frac{\log d}{d^3}})$ existential upper bound on the angular resolution of planar straight-line drawings of planar graphs (Table 55.2.3).

8. Determine the best possible aspect ratio and area that can be simultaneously achieved for (upward) planar straight-line drawings of trees (Table 55.2.5).

55.3 COMPLEXITY OF GRAPH DRAWING PROBLEMS

Tables 55.3.1–55.3.4 summarize selected results on the time complexity of some fundamental graph drawing problems.

It is interesting that apparently similar problems exhibit very different time complexities. For example, while planarity testing can be done in linear time, upward planarity testing is NP-hard. Note that, as illustrated in Figure 55.1.2 (c–d), planarity and acyclicity are necessary but not sufficient conditions for upward planarity. While many efficient algorithms exist for constructing drawings of trees and planar graphs with good universal area bounds, exact area minimization for most types of drawings is NP-hard, even for trees.

OPEN PROBLEMS

1. Reduce the time complexity of upward planarity testing for embedded digraphs (which is currently $O(n^2)$), biconnected series-parallel digraphs (currently $O(n^4)$), and biconnected outerplanar digraphs (currently $O(n^2)$), or prove a superlinear lower bound (Table 55.3.1).

2. Reduce the time complexity of computing a planar straight-line drawing of an outerplanar graph such that the vertices are represented by a set of given points in general position (currently $O(n \log^3 n)$) or prove an $\omega(n \log n)$ lower bound (Table 55.3.2).

TABLE 55.3.1   Time complexity of some fundamental graph drawing problems: general graphs and digraphs.

CLASS OF GRAPHS	PROBLEM	TIME COMPLEXITY	
General graph	minimize crossings	NP-hard [GJ83]	
2-layered graph	minimize crossings in layered drawing with preassigned order on one layer	NP-hard [EW94]	
General graph	planarity testing and computing a planar embedding	$\Omega(n)$ trivial	$O(n)$ [BL76, CNAO85, FR82, ET76, HT74, LEC67]
General graph	maximum planar subgraph	NP-hard [GJ79]	
General graph	maximal planar subgraph	$\Omega(n+m)$ trivial	$O(n+m)$ [CHT93, Dji95, DT89, La94]
General graph	test the existence of a drawing where each edge is crossed at most once	NP-hard [GB07, KM09]	
General graph with $m = 4n - 8$	test the existence of a drawing where each edge is crossed at most once	$\Omega(n)$ trivial	$O(n^3)$ † [CGP06]
General graph	test the existence of a straight-line drawing where edges cross forming right angles	NP-hard [ABS12]	
2-layered graph	test the existence of a straight-line layered drawing where edges cross forming right angles	$\Omega(n)$ trivial	$O(n)$ [DDEL14]
General digraph	upward planarity testing	NP-hard [GT95]	
Embedded digraph	upward planarity testing	$\Omega(n)$ trivial	$O(n^2)$ [BDLM94]
Biconnected series-parallel digraphs	upward planarity testing	$\Omega(n)$ trivial	$O(n^4)$ [DGL10]
Biconnected outerplanar digraphs	upward planarity testing	$\Omega(n)$ trivial	$O(n^2)$ [Pap95]
Biconnected bipartite digraphs	upward planarity testing	$\Omega(n)$ trivial	$O(n)$ [DLR90]
Single-source digraph	upward planarity testing	$\Omega(n)$ trivial	$O(n)$ [BDMT98, HL96]
General graph	draw as the intersection graph of a set of unit diameter disks in the plane	NP-hard [BK98b]	

Note: n is the number of vertices, m is the number of edges.

†Brandenburg [Bra15] recently announced an $O(n)$ time algorithm for this problem.

TABLE 55.3.2   Time complexity of some fundamental graph drawing problems: Planar graphs and digraphs.

CLASS OF GRAPHS	PROBLEM	TIME COMPLEXITY	
Planar graph	planar straight-line drawing with prescribed edge lengths	NP-hard [EW90]	
Planar graph	planar straight-line drawing with maximum angular resolution	NP-hard [Gar95]	
Embedded graph	test the existence of a planar straight-line drawing with prescribed angles between pairs of consecutive edges incident on a vertex	NP-hard [Gar95]	
Maximal planar graph	test the existence of a planar straight-line drawing with prescribed angles between pairs of consecutive edges incident on a vertex	$\Omega(n)$ trivial	$O(n)$ [DV96]
Planar graph	planar straight-line grid drawing with $O(n^2)$ area and $O(1/n^2)$ angular resolution	$\Omega(n)$ trivial	$O(n)$ [FPP90, Sch90]
Planar graph	planar polyline drawing with $O(n^2)$ area, $O(n)$ bends, and $O(1/d)$ angular resolutions	$\Omega(n)$ trivial	$O(n)$ [Kan96]
Triconn planar graph	planar straight-line convex grid drawing with $O(n^2)$ area and $O(1/n^2)$ angular resolution	$\Omega(n)$ trivial	$O(n)$ [Kan96]
Triconn planar graph	planar straight-line strictly convex drawing	$\Omega(n)$ trivial	$O(n)$ [CON85, Tut60, Tut63]
Reduced planar st-digraph	upward planar grid straight-line dominance drawing with minimum area	$\Omega(n)$ trivial	$O(n)$ [DTT92]
Upward planar digraph	upward planar polyline grid drawing with $O(n^2)$ area and $O(n)$ bends	$\Omega(n)$ trivial	$O(n)$ [DT88, DTT92]
Planar graph	planar straight-line drawing such that the vertices are represented by a set of given points	NP-hard [Cab06]	
Outerplanar graph	planar straight-line drawing such that the vertices are represented by a set of given points in general position	$\Omega(n \log n)$ [BMS97]	$O(n \log^3 n)$ [Bos02]
Planar graph	planar drawing such that the vertices are collinear and each edge has at most one bend	NP-hard [BK79]	
Series-parallel (di)graph	(upward) planar drawing such that the vertices are collinear and each edge has at most one bends	$\Omega(n)$ trivial	$O(n)$ [DDLW06]
Planar graph	planar drawing such that the vertices are collinear and each edge has at most two bends	$\Omega(n)$ trivial	$O(n)$ [DDLW05]

Note: n is the number of vertices.

TABLE 55.3.3   Time complexity of some fundamental graph drawing problems: Planar graphs and digraphs.

CLASS OF GRAPHS	PROBLEM	TIME COMPLEXITY	
Planar deg-4 graph	planar orthogonal grid drawing with minimum number of bends	NP-hard [GT95]	
Planar biconnected deg-3 graph	planar orthogonal grid drawing with minimum number of bends and $O(n^2)$ area	$\Omega(n)$ trivial	$O(n^5 \log n)$ [DLV98]
Embedded deg-3 graph	planar orthogonal grid drawing with minimum number of bends (and $O(n^2)$ area)	$\Omega(n)$ trivial	$O(n)$ [RN02]
Planar biconnected deg-4 series-parallel graph	planar orthogonal grid drawing with minimum number of bends and $O(n^2)$ area	$\Omega(n)$ trivial	$O(n^4)$ [DLV98]
Planar biconnected deg-3 series-parallel graph	planar orthogonal grid drawing with minimum number of bends and $O(n^2)$ area	$\Omega(n)$ trivial	$O(n^3)$ [DLV98]
Embedded deg-4 graph	planar orthogonal grid drawing with minimum number of bends and $O(n^2)$ area	$\Omega(n)$ trivial	$O(n^{3/2})$ [CK12]
Planar deg-4 graph	planar orthogonal grid drawing with $O(n^2)$ area and $O(n)$ bends	$\Omega(n)$ trivial	$O(n)$ [BK98a, Kan96, TT89]
Embedded deg-4 graph	test the existence of a PSLOg drawing with rectangular faces	$\Omega(n)$ trivial	$O\left(\frac{n^{1.5}}{\log n}\right)$ [MHN06]
Planar deg-3 graph	test the existence of a PSLOg drawing with rectangular faces	$\Omega(n)$ trivial	$O(n)$ [RNG04]
Planar deg-3 graph	test the existence of a PSLOg drawing	$\Omega(n)$ trivial	$O(n)$ [RNN03]
Deg-3 series-parallel graph	test the existence of a planar orthogonal grid with no bends	$\Omega(n)$ trivial	$O(n)$ [REN06]
Planar orthog rep	planar orthogonal grid drawing with minimum area	NP-hard [Pat01]	

Note: n is the number of vertices.

3. Reduce the time complexity of bend minimization for planar orthogonal drawings of degree-3 graphs and degree-3 and degree-4 series-parallel graphs (Table 55.3.3).

4. Reduce the time complexity of bend minimization for planar orthogonal drawings of embedded graphs (currently $O(n^{3/2})$), or prove a superlinear lower bound (Table 55.3.3).

5. Reduce the time complexity of testing the existence of a planar straight-line orthogonal drawing with rectangular faces (currently $O(n^{1.5}/\log n)$), or prove a superlinear lower bound (Table 55.3.3).

6. Reduce the time complexity of area minimization of hv-drawings of binary trees (from $O(n\sqrt{n}\log n)$), or prove a superlinear lower bound (Table 55.3.4).

TABLE 55.3.4   Time complexity of some fundamental graph drawing problems: trees.

CLASS OF GRAPHS	PROBLEM	TIME COMPLEXITY	
Tree	draw as the Euclidean minimum spanning tree of a set of points in the plane	NP-hard [EW96]	
degree-4 tree	minimize area in planar orthogonal grid drawing	NP-hard [Bra90, DLT85, KL85, Sto84]	
degree-4 tree	minimize total/maximum edge length in planar orthogonal grid drawing	NP-hard [BC87, Bra90, Gre89]	
Rooted tree	minimize area in a planar straight-line upward layered grid drawing that displays symmetries and isomorphisms of subtrees	NP-hard [SR83]	
Rooted tree	minimize area in a planar straight-line upward layered drawing that displays symmetries and isomorphisms of subtrees	$\Omega(n)$ trivial	$O(n^k)$, $k \geq 1$ [SR83]
Binary tree	minimize area in hv-drawing	$\Omega(n)$ trivial	$O(n\sqrt{n}\log n)$ [ELL92]
Rooted tree	planar straight-line upward layered grid drawing with $O(n^2)$ area	$\Omega(n)$ trivial	$O(n)$ [RT81]
Rooted tree	planar polyline upward grid drawing with $O(n)$ area	$\Omega(n)$ trivial	$O(n)$ [GGT93]
Tree	planar straight-line drawing such that the vertices are represented by a set of given points in general position	$\Omega(n \log n)$ [BMS97]	$O(n \log n)$ [BMS97]

Note: n is the number of vertices, m is the number of edges.

55.4   EXAMPLE OF A GRAPH DRAWING ALGORITHM

In this section we outline the algorithm in [Tam87] for computing, for an embedded degree-4 graph G, a planar orthogonal grid drawing with the minimum number of bends and using $O(n^2)$ area (see Table 55.3.2). This algorithm is the core of a practical drawing algorithm for general graphs (see Section 55.5 and Figure 55.4.1 (d)). The algorithm consists of two main phases:

1. Computation of an orthogonal representation for G, where only the bends and the angles of the orthogonal drawing are defined.

2. Assignment of integer lengths to the segments of the orthogonal representation.

Phase 1 uses a transformation into a network flow problem (Figure 55.4.1 (a–c)), where each unit of flow is associated with a right angle in the orthogonal drawing. Hence, angles are viewed as a commodity that is produced by the vertices, transported across faces by the edges through their bends, and eventually consumed by the faces. From the embedded graph G we construct a flow network N as follows. The nodes of network N are the vertices and faces of G. Let $\deg(f)$ denote the number of edges of the circuit bounding face f. Each vertex v supplies $\sigma(v) = 4$ units of flow, and each face f consumes $\tau(f)$ units of flow, where

$$\tau(f) = \begin{cases} 2\deg(f) - 4 & \text{if } f \text{ is an internal face} \\ 2\deg(f) + 4 & \text{if } f \text{ is the external face.} \end{cases}$$

By Euler's formula, $\sum_v \sigma(v) = \sum_f \tau(f)$, i.e., the total supply is equal to the total consumption.

Network N has two types of arcs:

- arcs of the type (v, f), where f is a face incident on vertex v; the flow in (v, f) represents the angle at vertex v in face f, and has lower bound 1, upper bound 4, and cost 0;

- arcs of the type (f, g), where face f shares an edge e with face g; the flow in (f, g) represents the number of bends along edge e with the right angle inside face f, and has lower bound 0, upper bound $+\infty$, and cost 1.

The conservation of flow at the vertices expresses the fact that the sum of the angles around a vertex is equal to 2π. The conservation of flow at the faces expresses the fact that the sum of the angles at the vertices and bends of an internal face is equal to $\pi(p - 2)$, where p is the number of such angles. For the external face, the above sum is equal to $\pi(p + 2)$. It can be shown that every feasible flow ϕ in network N corresponds to an admissible orthogonal representation for graph G, whose number of bends is equal to the cost of flow ϕ. Hence, an orthogonal representation for G with the minimum number of bends can be computed from a minimum-cost flow in G. This flow can be computed in $O(n^{1.5})$ time [CK12]. Phase 2 uses a simple compaction strategy derived from VLSI layout, where the lengths of the horizontal and vertical segments are computed independently after a preliminary refinement of the orthogonal representation that decomposes each face into rectangles. The resulting drawing is shown in Figure 55.4.1 (d).

55.5   TECHNIQUES FOR DRAWING GRAPHS

In this section we outline some of the most successful techniques that have been devised for drawing general graphs.

PLANARIZATION

The planarization approach is motivated by the availability of many efficient and well-analyzed drawing algorithms for planar graphs (see Table 55.3.2). If the graph is nonplanar, it is transformed into a planar graph by means of a preliminary planarization step that replaces each crossing with a fictitious vertex. The planarization approach consists of two main steps: in the first step a maximal planar subgraph G' of the input graph G is computed; in the second step, all the edges of G that are not in G' are added to G' and the crossings formed by each added edge are replaced with dummy vertices. Clearly when adding an edge one wants to produce as few crossings as possible. The two optimization problems arising in the two steps of the planarization approach, i.e., the maximum planar subgraph problem and the

FIGURE 55.4.1

(a) *Embedded graph G.* (b) *Minimum cost flow in network N: the flow is shown next to each arc; arcs with zero flow are omitted; arcs with unit cost are drawn with thick lines; a face f is represented by a box labeled with $\tau(f)$.* (c) *Planar orthogonal grid drawing of G with minimum number of bends.* (d) *Orthogonal grid drawing of a nonplanar graph produced by a drawing method for general graphs based on the algorithm of Section 55.4.*

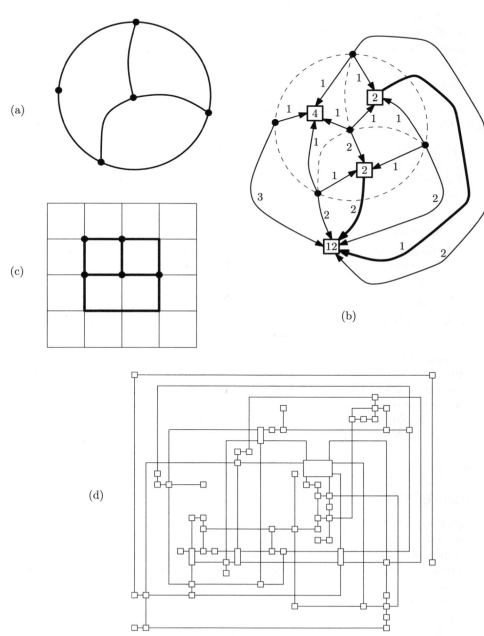

edge insertion problem, are NP-hard. Hence, existing planarization algorithms use heuristics. The best available heuristic for the maximum planar subgraph problem is described in [JM96]. This method has a solid theoretical foundation in polyhedral combinatorics, and achieves good results in practice. A sophisticated algorithm for edge insertion (that inserts each edge minimizing the number of crossings over all possible embeddings of the planar subgraph) is described in [GMW05]. See also [BCG+13] for more references.

A successful drawing algorithm based on the planarization approach and a bend-minimization method [Tam87] is described in [TDB88] (Figure 55.4.1(d) was generated by this algorithm). It has been widely used in software visualization systems.

LAYERING

The layering approach for constructing polyline drawings of directed graphs transforms the digraph into a layered digraph and then constructs a layered drawing. A typical algorithm based on the layering approach consists of the following main steps:

1. Assign each vertex to a layer, with the goal of maximizing the number of edges oriented upward.

2. Insert fictitious vertices along the edges that cross layers, so that each edge in the resulting digraph connects vertices in consecutive layers. (The fictitious vertices will be displayed as bends in the final drawing.)

3. Permute the vertices on each layer with the goal of minimizing crossings.

4. Adjust the positions of the vertices in each layer with the goal of distributing the vertices uniformly and minimizing the number of bends.

Most of the subproblems involved in the various steps are NP-hard, hence heuristics must be used. The layering approach was pioneered by Sugiyama et al. [STT81] and since then a lot of research has been devoted to all optimization problems in each of the four steps above (see, e.g., [BBBH10, BK02, BWZ10, CGMW10, CGMW11, EK86, ELS93, EW94, GKNV93, HN02, JM97, MSM99, Nag05, NY04, TNB04]). See also [HN13] for more references.

FORCE DIRECTED

This approach uses a physical model where the vertices and edges of the graph are viewed as objects subject to various forces. Starting from an initial configuration (which can be randomly defined or suitably chosen), the physical system evolves into a final configuration of minimum energy, which yields the drawing. Rather than solving a system of differential equations, the evolution of the system is usually simulated using numerical methods (e.g., at each step, the forces are computed and corresponding incremental displacements of the vertices are performed).

Drawing algorithms based on the physical simulation approach are often able to detect and display symmetries in the graph. However, their running time is typically high. The physical simulation approach was pioneered in [Ead84, KS80].

Sophisticated developments and applications include [BP07, DH96, DM14, EH00, FR91, GGK04, GKN05, HJ05, HK02, KK89]. See also [Kob13] for additional references.

55.6 SOURCES AND RELATED MATERIAL

Several books devoted to graph drawing are published [DETT99, JM03, Kam89, KW01, NR04, Sug02, Tam13]. Among the early books, [Kam89] describes declarative approaches to graph drawing;[Sug02], motivated by software engineering applications, mostly focuses on layered drawings. [DETT99] is the first book that collects different techniques for graph drawing; [NR04] focuses on planar graphs. [KW01], [JM03], and [Tam13] are collections of surveys by different authors; [JM03] is devoted to graph drawing software and libraries while [Tam13] is the most recent handbook on Graph Drawing and Network Visualization. Sites with pointers to graph drawing resources and tools include the Web site `http://graphdrawing.org`, the Graph drawing e-print archive (`http://gdea.informatik.uni-koeln.de/`), and the Graph-Archive (`http://www.graph-archive.org/doku.php`).

RELATED CHAPTERS

Chapter 1: Finite point configurations
Chapter 10: Geometric graph theory
Chapter 23: Computational topology of graphs on surfaces
Chapter 27: Voronoi diagrams and Delaunay triangulations
Chapter 29: Triangulations and mesh generation

REFERENCES

[ABS12]   E.N. Argyriou, M.A. Bekos, and A. Symvonis. The straight-line RAC drawing problem is NP-hard. *J. Graph Algorithms Appl.*, 16:569–597, 2012.

[And63]   G.E. Andrews. A lower bound for the volume of strictly convex bodies with many boundary lattice points. *Trans. Amer. Math. Soc.*, 106:270–279, 1963.

[BBBH10]   C. Bachmaier, F.J. Brandenburg, W. Brunner, and F. Hübner. A global k-level crossing reduction algorithm. In *Proc. 4th Workshop Algorithms Comput.*, vol. 5942 of *LNCS*, pages 70–81, Springer, Berlin, 2010.

[BC87]   S.N. Bhatt and S.S. Cosmadakis. The complexity of minimizing wire lengths in VLSI layouts. *Inform. Process. Lett.*, 25:263–267, 1987.

[BCD⁺94]   P. Bertolazzi, R.F. Cohen, G. Di Battista, R. Tamassia, and I.G. Tollis. How to draw a series-parallel digraph. *Internat. J. Comput. Geom. Appl.*, 4:385–402, 1994.

[BCG⁺13]   C. Buchheim, M. Chimani, C. Gutwenger, M. Jünger, and P. Mutzel. Crossings and planarization. In R. Tamassia, editor, *Handbook of Graph Drawing and Visualization*, pages 43–85, CRC Press, Boca Raton, 2013.

[BDLM94]   P. Bertolazzi, G. Di Battista, G. Liotta, and C. Mannino. Upward drawings of triconnected digraphs. *Algorithmica*, 12:476–497, 1994.

[BDMT98]   P. Bertolazzi, G. Di Battista, C. Mannino, and R. Tamassia. Optimal upward planarity testing of single-source digraphs. *SIAM J. Comput.*, 27:132–169, 1998.

[BFM07]   N. Bonichon, S. Felsner, and M. Mosbah. Convex drawings of 3-connected plane graphs. *Algorithmica*, 47:399–420, 2007.

[Bie11]   T. Biedl. Small drawings of outerplanar graphs, series-parallel graphs, and other planar graphs. *Discrete Comput. Geom.*, 45:141–160, 2011.

[BK79]   F. Bernhart and P.C. Kainen. The book thickness of a graph. *J. Combin. Theory Ser. B*, 27:320–331, 1979.

[BK80]   R.P. Brent and H. Kung. On the area of binary tree layouts. *Inform. Process. Lett.*, 11:46–48, 1980.

[BK98a]   T. Biedl and G. Kant. A better heuristic for orthogonal graph drawings. *Comput. Geom.*, 9:159–180, 1998.

[BK98b]   H. Breu and D.G. Kirkpatrick. Unit disk graph recognition is NP-hard. *Comput. Geom.*, 9:3–24, 1998.

[BK02]   U. Brandes and B. Köpf. Fast and simple horizontal coordinate assignment. In *Proc. 9th Sympos. Graph Drawing*, vol. 2265 of *LNCS*, pages 33–36, Springer, Berlin, 2002.

[BL76]   K.S. Booth and G.S. Lueker. Testing for the consecutive ones property, interval graphs, and graph planarity using pq-tree algorithms. *J. Comp. Syst. Sci.*, 13:335–379, 1976.

[BMS97]   P. Bose, M. McAllister, and J. Snoeyink. Optimal algorithms to embed trees in a point set. *J. Graph Algorithms Appl.*, 1:1–15, 1997.

[Bos02]   P. Bose. On embedding an outer-planar graph in a point set. *Comput. Geom.*, 23:303–312, 2002.

[BP92]   I. Bárány and J. Pach. On the number of convex lattice polygons. *Combin. Probab. Comput.*, 1:295–302, 1992.

[BP07]   U. Brandes and C. Pich. Eigensolver methods for progressive multidimensional scaling of large data. In *Proc. 14th Sympos. Graph Drawing*, vol. 4372 of *LNCS*, pages 42–53, Springer, Berlin, 2007.

[BR06]   I. Bárány and G. Rote. Strictly convex drawings of planar graphs. *Doc. Math.*, 11:369–391, 2006.

[Bra90]   F.J. Brandenburg. Nice drawings of graphs are computationally hard. In P. Gorny and M.J. Tauber, editors, *Visualization in Human-Computer Interaction*, vol. 439 of *LNCS*, pages 1–15, Springer, Berlin, 1990.

[Bra15]   F.J. Brandenburg. On 4-map graphs and 1-planar graphs and their recognition problem. Preprint, `arXiv:1509.03447`, 2015.

[BS15]   T. Biedl and J.M. Schmidt. Small-area orthogonal drawings of 3-connected graphs. In *Proc. 23rd Sympos. Graph Drawing Network Vis.*, vol. 9411 of *LNCS*, pages 153–165, Springer, Berlin, 2015.

[BT04]   I. Bárány and N. Tokushige. The minimum area of convex lattice *n*-gons. *Combinatorica*, 24:171–185, 2004.

[BWZ10]   C. Buchheim, A. Wiegele, and L. Zheng. Exact algorithms for the quadratic linear ordering problem. *INFORMS J. Computing*, 22:168–177, 2010.

[Cab06]   S. Cabello. Planar embeddability of the vertices of a graph using a fixed point set is NP-hard. *J. Graph Algorithms Appl.*, 10:353–363, 2006.

[CDP92]   P. Crescenzi, G. Di Battista, and A. Piperno. A note on optimal area algorithms for upward drawings of binary trees. *Comput. Geom.*, 2:187–200, 1992.

[CGKT02]   T.M. Chan, M.T. Goodrich, S.R. Kosaraju, and R. Tamassia. Optimizing area and aspect ratio in straight-line orthogonal tree drawings. *Comput. Geom.*, 23:153–162, 2002.

[CGMW10]   M. Chimani, C. Gutwenger, P. Mutzel, and H.-M. Wong. Layer-free upward crossing minimization. *J. Exper. Algorithmics*, 15:#2.2, 2010.

[CGMW11]   M. Chimani, C. Gutwenger, P. Mutzel, and H.-M. Wong. Upward planarization layout. *J. Graph Algorithms Appl.*, 15:127–155, 2011.

[CGP06]   Z.-Z. Chen, M. Grigni, and C.H. Papadimitriou. Recognizing hole-free 4-map graphs in cubic time. *Algorithmica*, 45:227–262, 2006.

[Cha02]   T.M. Chan. A near-linear area bound for drawing binary trees. *Algorithmica*, 34:1–13, 2002.

[CHT93]   J. Cai, X. Han, and R.E. Tarjan. An $O(m \log n)$-time algorithm for the maximal planar subgraph problem. *SIAM J. Comput.*, 22:1142–1162, 1993.

[CK97]   M. Chrobak and G. Kant. Convex grid drawings of 3-connected planar graphs. *Internat. J. Comput. Geom. Appl.*, 7:211–223, 1997.

[CK12]   S. Cornelsen and A. Karrenbauer. Accelerated bend minimization. *J. Graph Algorithms Appl.*, 16:635–650, 2012.

[CNAO85]   N. Chiba, T. Nishizeki, S. Abe, and T. Ozawa. A linear algorithm for embedding planar graphs using pq-trees. *J. Comp. Syst. Sci.*, 30:54–76, 1985.

[CON85]   N. Chiba, K. Onoguchi, and T. Nishizeki. Drawing plane graphs nicely. *Acta Informatica*, 22:187–201, 1985.

[CP98]   P. Crescenzi and P. Penna. Strictly-upward drawings of ordered search trees. *Theoret. Comput. Sci.*, 203:51–67, 1998.

[CPP98]   P. Crescenzi, P. Penna, and A. Piperno. Linear area upward drawings of AVL trees. *Comput. Geom.*, 9:25–42, 1998.

[DDEL14]   E. Di Giacomo, W. Didimo, P. Eades, and G. Liotta. 2-layer right angle crossing drawings. *Algorithmica*, 68:954–997, 2014.

[DDLM12]   E. Di Giacomo, W. Didimo, G. Liotta, and F. Montecchiani. h-quasi planar drawings of bounded treewidth graphs in linear area. In *Proc. 38th Workshop Graph-Theoret. Concepts Comp. Sci.*, vol. 7551 of *LNCS*, pages 91–102, Springer, Berlin, 2012.

[DDLM13]   E. Di Giacomo, W. Didimo, G. Liotta, and F. Montecchiani. Area requirement of graph drawings with few crossings per edge. *Comput. Geom.*, 46:909–916, 2013.

[DDLW05]   E. Di Giacomo, W. Didimo, G. Liotta, and S.K. Wismath. Curve-constrained drawings of planar graphs. *Comput. Geom.*, 30:1–23, 2005.

[DDLW06]   E. Di Giacomo, W. Didimo, G. Liotta, and S.K. Wismath. Book embeddability of seriesparallel digraphs. *Algorithmica*, 45:531–547, 2006.

[DEG+13]   C.A. Duncan, D. Eppstein, M.T. Goodrich, S.G. Kobourov, and M. Nöllenburg. Drawing trees with perfect angular resolution and polynomial area. *Discrete Comput. Geom.*, 49:157–182, 2013.

[DETT99]   G. Di Battista, P. Eades, R. Tamassia, and I.G. Tollis. *Graph Drawing: Algorithms for the Visualization of Graphs*. Prentice Hall, Upper Saddle River, 1999.

[DF09]   G. Di Battista and F. Frati. Small area drawings of outerplanar graphs. *Algorithmica*, 54:25–53, 2009.

[DGL10]   W. Didimo, F. Giordano, and G. Liotta. Upward spirality and upward planarity testing. *SIAM J. Discrete Math.*, 23:1842–1899, 2010.

[DH96]   R. Davidson and D. Harel. Drawing graphs nicely using simulated annealing. *ACM Trans. Graph.*, 15:301–331, 1996.

[Dji95]   H.N. Djidjev. A linear algorithm for the maximal planar subgraph problem. In *Proc. 4th Workshop Algorithms and Data Structures*, vol. 955 of *LNCS*, pages 369–380, Springer, Berlin, 1995.

[DLM14]   E. Di Giacomo, G. Liotta, and F. Montecchiani. The planar slope number of subcubic graphs. In *Proc. 11th Latin American Sympos. Theoret. Informatics*, vol. 8392 of *LNCS*, pages 132–143, Springer, Berlin, 2014.

[DLR90]   G. Di Battista, W.-P. Liu, and I. Rival. Bipartite graphs, upward drawings, and planarity. *Inform. Process. Lett.*, 36:317–322, 1990.

[DLT85]   D. Dolev, T. Leighton, and H. Trickey. Planar embedding of planar graphs. In F.P. Preparata, editor, *VLSI Theory*, vol. 2 of *Advances in Computing Research*, pages 147–161, JAI Press, Greenwich, 1985.

[DLV98]   G. Di Battista, G. Liotta, and F. Vargiu. Spirality and optimal orthogonal drawings. *SIAM J. Comput.*, 27:1764–1811, 1998.

[DM14]   W. Didimo and F. Montecchiani. Fast layout computation of clustered networks: Algorithmic advances and experimental analysis. *Information Sciences*, 260:185–199, 2014.

[DT81]   D. Dolev and H. Trickey. On linear area embedding of planar graphs. Tech. report STAN-CS-81-876, Stanford University, 1981.

[DT88]   G. Di Battista and R. Tamassia. Algorithms for plane representations of acyclic digraphs. *Theoret. Comput. Sci.*, 61:175–198, 1988.

[DT89]   G. Di Battista and R. Tamassia. Incremental planarity testing. In *Proc. 30th IEEE Sympos. Found. Comp. Sci.*, pages 436–441, 1989.

[DTT92]   G. Di Battista, R. Tamassia, and I.G. Tollis. Area requirement and symmetry display of planar upward drawings. *Discrete Comput. Geom.*, 7:381–401, 1992.

[DTV99]   G. Di Battista, R. Tamassia, and L. Vismara. Output-sensitive reporting of disjoint paths. *Algorithmica*, 23:302–340, 1999.

[DV96]   G. Di Battista and L. Vismara. Angles of planar triangular graphs. *SIAM J. Discrete Math.*, 9:349–359, 1996.

[Ead84]   P. Eades. A heuristic for graph drawing. *Congressus Numerantium*, 42:149–160, 1984.

[EG95]   S. Even and G. Granot. Grid layouts of block diagrams—bounding the number of bends in each connection (extended abstract). In *Proc. Graph Drawing*, vol. 894 of *LNCS*, pages 64–75, Springer, Berlin, 1995.

[EH00]   P. Eades and M. Huang. Navigating clustered graphs using force-directed methods. *J. Graph Algorithms Appl.*, 4:157–181, 2000.

[EK86]   P. Eades and D. Kelly. Heuristics for drawing 2-layered networks. *Ars Combin.*, 21:89–98, 1986.

[ELL92]   P. Eades, T. Lin, and X. Lin. Minimum size h-v drawings. In *Advanced Visual Interfaces*, vol. 36 of *World Scientific Series in Computer Science*, pages 386–394, 1992.

[ELS93]   P. Eades, X. Lin, and W.F. Smyth. A fast and effective heuristic for the feedback arc set problem. *Inform. Process. Lett.*, 47:319–323, 1993.

[ET76]   S. Even and R.E. Tarjan. Computing an *st*-numbering. *Theoret. Comput. Sci.*, 2:339–344, 1976.

[EW90]   P. Eades and N.C. Wormald. Fixed edge-length graph drawing is NP-hard. *Discrete Appl. Math.*, 28:111–134, 1990.

[EW94]   P. Eades and N.C. Wormald. Edge crossings in drawings of bipartite graphs. *Algorithmica*, 11:379–403, 1994.

[EW96]   P. Eades and S. Whitesides. The realization problem for Euclidean minimum spanning trees is NP-hard. *Algorithmica*, 16:60–82, 1996.

[FHH⁺93] M. Formann, T. Hagerup, J. Haralambides, M. Kaufmann, F.T. Leighton, A. Symvonis, E. Welzl, and G.J. Woeginger. Drawing graphs in the plane with high resolution. *SIAM J. Comput.*, 22:1035–1052, 1993.

[FP08]   F. Frati and M. Patrignani. A note on minimum-area straight-line drawings of planar graphs. In *Proc. 15th Sympos. Graph Drawing*, vol. 4875 of *LNCS*, pages 339–344, Springer, Berlin, 2008.

[FPP90]   H. de Frayesseix, J. Pach, and R. Pollack. How to draw a planar graph on a grid. *Combinatorica*, 10:41–51, 1990.

[FR82]   H. de Fraysseix and P. Rosenstiehl. A depth-first-search characterization of planarity. *Ann. Discrete Math.*, 13:75–80, 1982.

[FR91]   T.M.J. Fruchterman and E.M. Reingold. Graph drawing by force-directed placement. *Software–Practice Exper.*, 21:1129–1164, 1991.

[Fra08a] F. Frati. On minimum area planar upward drawings of directed trees and other families of directed acyclic graphs. *Internat. J. Comput. Geom. Appl.*, 18:251–271, 2008.

[Fra08b] F. Frati. Straight-line orthogonal drawings of binary and ternary trees. In *Proc. 15th Sympos. Graph Drawing*, vol. 4875 of *LNCS*, pages 76–87, Springer, Berlin, 2008.

[Fra10]   F. Frati. Lower bounds on the area requirements of series-parallel graphs. *Discrete Math. Theor. Comp. Sci.*, 12:139–174, 2010.

[Gar95]   A. Garg. On drawing angle graphs. In *Proc. Graph Drawing*, vol. 894 of *LNCS*, pages 84–95, Springer, Berlin, 1995.

[GB07]   A. Grigoriev and H.L. Bodlaender. Algorithms for graphs embeddable with few crossings per edge. *Algorithmica*, 49:1–11, 2007.

[GGK04]   P. Gajer, M.T. Goodrich, and S.G. Kobourov. A multi-dimensional approach to force-directed layouts of large graphs. *Comput. Geom.*, 29:3–18, 2004.

[GGT93]   A. Garg, M.T. Goodrich, and R. Tamassia. Area-efficient upward tree drawings. In *Proc. 9th Sympos. Comput. Geom.*, pages 359–368, ACM Press, 1993.

[GGT96]   A. Garg, M.T. Goodrich, and R. Tamassia. Planar upward tree drawings with optimal area. *Internat. J. Comput. Geom. Appl.*, 6:333–356, 1996.

[GJ79]   M.R. Garey and D.S. Johnson. *Computers and Intractability: A Guide to the Theory of NP-Completeness*. W.H. Freeman, New York, 1979.

[GJ83]   M.R. Garey and D.S. Johnson. Crossing number is NP-complete. *SIAM J. Algebraic Discrete Methods*, 4:312–316, 1983.

[GKN05]   E.R. Gansner, Y. Koren, and S. North. Graph drawing by stress majorization. In *Proc. 12th Sympos. Graph Drawing*, vol. 3383 of *LNCS*, pages 239–250, Springer, Berlin, 2005.

[GKNV93] E.R. Gansner, E. Koutsofios, S.C. North, and K.-P. Vo. A technique for drawing directed graphs. *IEEE Trans. Software Eng.*, 19:214–230, 1993.

[GMW05]   C. Gutwenger, P. Mutzel, and R. Weiskircher. Inserting an edge into a planar graph. *Algorithmica*, 41:289–308, 2005.

[GR03a]   A. Garg and A. Rusu. Area-efficient order-preserving planar straight-line drawings of ordered trees. *Internat. J. Comput. Geom. Appl.*, 13:487–505, 2003.

[GR03b]   A. Garg and A. Rusu. Straight-line drawings of general trees with linear area and arbitrary aspect ratio. In *Proc. Conf. Comput. Sci. Appl., Part III*, vol 2669 of *LNCS*, pages 876–885, Springer, Berlin, 2003.

[GR04]   A. Garg and A. Rusu. Straight-line drawings of binary trees with linear area and arbitrary aspect ratio. *J. Graph Algorithms Appl.*, 8:135–160, 2004.

[Gre89]   A. Gregori. Unit-length embedding of binary trees on a square grid. *Inform. Process. Lett.*, 31:167–173, 1989.

[GT93]   A. Garg and R. Tamassia. Efficient computation of planar straight-line upward drawings. In *Proc. ALCOM Workshop on Graph Drawing)*, 1993.

[GT94]   A. Garg and R. Tamassia. Planar drawings and angular resolution: Algorithms and bounds. In *Proc. 2nd European Sympos. Algorithms*, vol. 855 of *LNCS*, pages 12–23. Springer, Berlin, 1994.

[GT95]   A. Garg and R. Tamassia. On the computational complexity of upward and rectilinear planarity testing. In *Proc. Graph Drawing*, vol. 894 of *LNCS*, pages 286–297, Springer, Berlin, 1995.

[HJ05]   S. Hachul and M. Jünger. Drawing large graphs with a potential-field-based multilevel algorithm. In *Proc. 12th Graph Drawing*, vol. 3383 of *LNCS*, pages 285–295, Springer, Berlin, 2005.

[HK02]   D. Harel and Y. Koren. A fast multi-scale method for drawing large graphs. *J. Graph Algorithms Appl.*, 6:179–202, 2002.

[HL96]   M.D. Hutton and A. Lubiw. Upward planar drawing of single-source acyclic digraphs. *SIAM J. Comput.*, 25:291–311, 1996.

[HN02]   P. Healy and N.S. Nikolov. A branch-and-cut approach to the directed acyclic graph layering problem. In *Proc. 10th Graph Drawing*, vol. 2528 of *LNCS*, pages 98–109, Springer, Berlin, 2002.

[HN13]   P. Healy and N.S. Nikolov. Hierarchical drawing algorithms. In R. Tamassia, editor, *Handbook of Graph Drawing and Visualization*, pages 409–453, CRC Press, Boca Raton, 2013.

[HT74]   J. Hopcroft and R.E. Tarjan. Efficient planarity testing. *J. ACM*, 21:549–568, 1974.

[JM96]   M. Jünger and P. Mutzel. Maximum planar subgraphs and nice embeddings: Practical layout tools. *Algorithmica*, 16:33–59, 1996.

[JM97]   M. Jünger and P. Mutzel. 2-layer straightline crossing minimization: Performance of exact and heuristic algorithms. *J. Graph Algorithms Appl.*, 1:1–25, 1997.

[JM03]   M. Jünger and P. Mutzel, editors. *Graph Drawing Software*. Springer, Berlin, 2003.

[Kam89]   T. Kamada. *Visualizing Abstract Objects and Relations*. World Scientific, Singapore, 1989.

[Kan96]   G. Kant. Drawing planar graphs using the canonical ordering. *Algorithmica*, 16:4–32, 1996.

[Kim95]   S.K. Kim. Simple algorithms for orthogonal upward drawings of binary and ternary trees. In *Proc. 7th Canad. Conf. Comput. Geom.*, pages 115–120, 1995.

[KK89]   T. Kamada and S. Kawai. An algorithm for drawing general undirected graphs. *Inform. Process. Lett.*, 31:7–15, 1989.

[KL85]   M.R. Kramer and J. van Leeuwen. The complexity of wire-routing and finding minimum area layouts for arbitrary VLSI circuits. In F.P. Preparata, editor, *VLSI Theory*, vol. 2 of *Advances in Computing Research*, pages 129–146, JAI Press, Greenwich, 1985.

[KM09]   V.P. Korzhik and B. Mohar. Minimal obstructions for 1-immersions and hardness of 1-planarity testing. In *Proc. 16th Sympos. Graph Drawing*, vol. 5417 of *LNCS*, pages 302–312, Springer, Berlin, 2009.

[Kob13]   S.G. Kobourov. Force-directed drawing algorithms. In R. Tamassia, editor, *Handbook of Graph Drawing and Visualization*, pages 383–408, CRC Press, Boca Raton, 2013.

[KS80]   J.B. Kruskal and J.B. Seery. Designing network diagrams. In *Proc. 1st General Conf. Social Graphics*, pages 22–50, U.S. Department of the Census, 1980.

[KW01]   M. Kaufmann and D. Wagner, editors. *Drawing Graphs, Methods and Models*. Vol. 2025 of *LNCS*, Springer, Berlin, 2001.

[La94]   J.A. La Poutré. Alpha-algorithms for incremental planarity testing (preliminary version). In *Proc. 26th ACM Sympos. Theory Comput.*, pages 706–715, 1994.

[LEC67]   A. Lempel, S. Even, and I. Cederbaum. An algorithm for planarity testing of graphs. In *Proc. Internat. Sympos. Theory of Graphs*, pages 215–232. Gordon and Breach, New York, 1967.

[Lei80]   C.E. Leiserson. Area-efficient graph layouts. In *21st IEEE Sympos. Found. Comp. Sci.*, pages 270–281, 1980.

[Lei84]   F.T. Leighton. New lower bound techniques for VLSI. *Mathematical Systems Theory*, 17:47–70, 1984.

[LLMN13]   W. Lenhart, G. Liotta, D. Mondal, and R.I. Nishat. Planar and plane slope number of partial 2-trees. In *Proc. 21st Sympos. Graph Drawing*, vol. 8242 of *LNCS*, pages 412–423, Springer, Berlin, 2013.

[LMPS92]   Y. Liu, P. Marchioro, R. Petreschi, and B. Simeone. Theoretical results on at most 1-bend embeddability of graphs. *Acta Math. Appl. Sin.*, 8:188–192, 1992.

[LMS91]   Y. Liu, A. Morgana, and B. Simeone. General theoretical results on rectilinear embedability of graphs. *Acta Math. Appl. Sin.*, 7:187–192, 1991.

[MHN06]   K. Miura, H. Haga, and T. Nishizeki. Inner rectangular drawings of plane graphs. *Internat. J. Comput. Geom. Appl.*, 16:249–270, 2006.

[MNRA11]   D. Mondal, R.I. Nishat, M.S. Rahman, and M.J. Alam. Minimum-area drawings of plane 3-trees. *J. Graph Algorithms Appl.*, 15:177–204, 2011.

[MP94]   S. Malitz and A. Papakostas. On the angular resolution of planar graphs. *SIAM J. Discrete Math.*, 7:172–183, 1994.

[MSM99]   C. Matuszewski, R. Schönfeld, and P. Molitor. Using sifting for k-layer straightline crossing minimization. In *Proc. 7th Sympos. Graph Drawing*, vol. 1731 of *LNCS*, pages 217–224, Springer, Berlin, 1999.

[Nag05]   H. Nagamochi. On the one-sided crossing minimization in a bipartite graph with large degrees. *Theoret. Comput. Sci.*, 332:417 – 446, 2005.

[NR04]   T. Nishizeki and M.S. Rahman. *Planar Graph Drawing*. World Scientific, Singapore, 2004.

[NY04]   H. Nagamochi and N. Yamada. Counting edge crossings in a 2-layered drawing. *Inform. Process. Lett.*, 91:221–225, 2004.

[Pap95]   A. Papakostas. Upward planarity testing of outerplanar dags. In *Proc. Sympos. Graph Drawing*, vol. 894 of *LNCS*, pages 298–306, Springer, Berlin, 1995.

[Pat01]   M. Patrignani. On the complexity of orthogonal compaction. *Comput. Geom.*, 19:47–67, 2001.

[PT98]   A. Papakostas and I.G. Tollis. Algorithms for area-efficient orthogonal drawings. *Comput. Geom.*, 9:83–110, 1998.

[Rab93]   S. Rabinowitz. $O(n^3)$ bounds for the area of a convex lattice n-gon. *Geombinatorics*, 2:85–88, 1993.

[REN06]   M.S. Rahman, N. Egi, and T. Nishizeki. No-bend orthogonal drawings of series-parallel graphs. In *Proc. 13th Sympos. Graph Drawing*, vol. 3843 of *LNCS*, pages 409–420, Springer, Berlin, 2006.

[RN02]   M. Rahman and T. Nishizeki. Bend-minimum orthogonal drawings of plane 3-graphs. In *Proc. 28th Workshop on Graph-Theoretic Concepts Comp. Sci.*, vol. 2573 of *LNCS*, pages 367–378, Springer, Berlin, 2002.

[RNG04]   M. Rahman, T. Nishizeki, and S. Ghosh. Rectangular drawings of planar graphs. *J. Algorithms*, 50:62–78, 2004.

[RNN03]   M.S. Rahman, T. Nishizeki, and M. Naznin. Orthogonal drawings of plane graphs without bends. *J. Graph Algorithms Appl.*, 7:335–362, 2003.

[RT81]   E.M. Reingold and J.S. Tilford. Tidier drawings of trees. *IEEE Trans. Software Eng.*, SE-7:223–228, 1981.

[Sch90]   W. Schnyder. Embedding planar graphs on the grid. In *Proc. 1st ACM-SIAM Sympos. Discrete Algorithms*, pages 138–148, 1990.

[Sch95]   M. Schäffter. Drawing graphs on rectangular grids. *Discrete Appl. Math.*, 63:75–89, 1995.

[SKC00]   C.-S. Shin, S.K. Kim, and K.-Y. Chwa. Area-efficient algorithms for straight-line tree drawings. *Comput. Geom.*, 15:175–202, 2000.

[SR83]   K.J. Supowit and E.M. Reingold. The complexity of drawing trees nicely. *Acta Inform.*, 18:377–392, 1983.

[ST92]   W. Schnyder and W.T. Trotter. Convex embeddings of 3-connected plane graphs. *Abstracts Amer. Math. Soc.*, 13:502, 1992.

[Sto84]   J.A. Storer. On minimal node-cost planar embeddings. *Networks*, 14:181–212, 1984.

[STT81]   K. Sugiyama, S. Tagawa, and M. Toda. Methods for visual understanding of hierarchical system structures. *IEEE Trans. Systems, Man and Cybernetics*, 11:109–125, 1981.

[Sug02]   K. Sugiyama. *Graph Drawing and Applications for Software and Knowledge Engineers*. World Scientific, Singapore, 2002.

[Tam87]   R. Tamassia. On embedding a graph in the grid with the minimum number of bends. *SIAM J. Comput.*, 16:421–444, 1987.

[Tam13]   R. Tamassia, editor. *Handbook of Graph Drawing and Visualization*. CRC Press, Boca Raton, 2013.

[TDB88]   R. Tamassia, G. Di Battista, and C. Batini. Automatic graph drawing and readability of diagrams. *IEEE Trans. Systems, Man and Cybernetics*, 18:61–79, 1988.

[TNB04]   A. Tarassov, N.S. Nikolov, and J. Branke. A heuristic for minimum-width graph layering with consideration of dummy nodes. In *In Proc. 3rd Workshop on Experimental and Efficient Algorithms*, vol. 3059 of *LNCS*, pages 570–583, Springer, Berlin, 2004.

[Tre96]   L. Trevisan. A note on minimum-area upward drawing of complete and Fibonacci trees. *Inform. Process. Lett.*, 57:231–236, 1996.

[TT89]   R. Tamassia and I.G. Tollis. Planar grid embedding in linear time. *IEEE Trans. Circuits Syst.*, 36:1230–1234, 1989.

[TTV91]   R. Tamassia, I.G. Tollis, and J.S. Vitter. Lower bounds for planar orthogonal drawings of graphs. *Inform. Process. Lett.*, 39:35–40, 1991.

[Tut60]   W.T. Tutte. Convex representations of graphs. *Proc. London Math. Soc.*, 10:304–320, 1960.

[Tut63]   W.T. Tutte. How to draw a graph. *Proc. London Math. Soc.*, 13:743–768, 1963.

[Val81]   L.G. Valiant. Universality considerations in VLSI circuits. *IEEE Trans. Comput.*, 30:135–140, 1981.

[Woo05]   D.R. Wood. Grid drawings of k-colourable graphs. *Comput. Geom.*, 30:25–28, 2005.

56 SPLINES AND GEOMETRIC MODELING
Chandrajit L. Bajaj

INTRODUCTION

Piecewise polynomials of fixed degree and continuously differentiable up to some order are known as *splines* or *finite elements*. Splines are used in applications ranging from computer-aided design, computer graphics, data visualization, geometric modeling, and image processing to the solution of partial differential equations via finite element analysis. The spline-fitting problem of constructing a mesh of finite elements that interpolate or approximate data is by far the primary research problem in geometric modeling. *Parametric splines* are vectors of a set of multivariate polynomial (or rational) functions while *implicit splines* are zero contours of collections of multivariate polynomials. This chapter dwells mainly on spline surface fitting methods in real Euclidean space. We first discuss tensor product surfaces (Section 56.1), perhaps the most popular. The next sections cover generalized spline surfaces (Section 56.2), free-form surfaces (Section 56.3), and subdivision surfaces (Section 56.4). This classification is not strict, and some overlap exists. Interactive editing of surfaces is discussed in the final section (Section 56.5).

The various spline methods may be distinguished by several criteria:

- Implicit or parametric representations.
- Algebraic and geometric degree of the spline basis.
- Adaptivity and number of surface patches.
- Computation (time) and memory (space) required.
- Stability of fitting algorithms.
- Local or nonlocal interpolation.
- Spline Support and splitting of input mesh.
- Convexity of the input and solution.
- Fairness of the solution (first- and second-order variation).

These distinctions will guide the discussions throughout the chapter.

56.1 TENSOR PRODUCT SURFACES

Tensor product B-splines have emerged as the polynomial basis of choice for working with parametric surfaces [Boo68, CLR80]. The theory of tensor product splines or surface patches requires that the data have a rectangular geometry and that the parametrizations of opposite boundary curves be similar. It is based on the concept of bilinear interpolation. B-splines are generated by a rectangular (tensor product) mesh of control points. A reduced or decimated version of rectangular control points with T-junctions yields a T-spline tensor product [SZB+03, DCL+08,

DLP13, Mou10] The most general results obtained to date are further summarized in Table 56.1.1, and will be discussed below.

GLOSSARY

Affine invariance: A property of a curve or surface generation scheme, implying invariance with respect to whether computation of a point on a curve or surface occurs before or after an affine map is applied to the input data.

A-spline: Collection of bivariate Bernstein-Bézier polynomials, each over a triangle and with prescribed geometric continuity, such that the zero contour of each polynomial defines a smooth and single-sheeted real algebraic curve segment. ("A" stands for "algebraic.")

A-patch: Smooth and "functional" zero contour of a Bernstein-Bézier polynomial over a tetrahedron.

Barycentric combination: A weighted average where the sum of the weights equals one.

Barycentric coordinates: A point in $\mathbb{R}^2$ may be written as a unique barycentric combination of three points. The coefficients in this combination are its barycentric coordinates. Similarly, a point in $\mathbb{R}^3$ may be written as a unique barycentric combination of four or more points. These latter are often referred to as generalized barycentric coordinates.

Basis function: Functions form linear spaces, which have bases. The elements of these bases are the basis functions.

Bernstein-Bézier form: Let p_1, p_2, p_3, $p_4 \in \mathbb{R}^3$ be affinely independent. Then the tetrahedron with these points as vertices is $V = [p_1 p_2 p_3 p_4]$. Any polynomial $f(p)$ of degree n can be expressed in the Bernstein-Bézier (BB) form over V as

$$f(p) = \sum_{|\lambda|=n} b_\lambda \, B_\lambda^n(\alpha), \ \lambda \in \mathcal{Z}_+^4 \,, \tag{56.1.1}$$

where

$$B_\lambda^n(\alpha) = \frac{n!}{\lambda_1! \lambda_2! \lambda_3! \lambda_4!} \, \alpha_1^{\lambda_1} \alpha_2^{\lambda_2} \alpha_3^{\lambda_3} \alpha_4^{\lambda_4}$$

are Bernstein polynomials, $|\lambda| = \sum_{i=1}^4 \lambda_i$ with $\lambda = (\lambda_1, \lambda_2, \lambda_3, \lambda_4)^T$, the barycentric coordinates of p are $\alpha = (\alpha_1, \alpha_2, \alpha_3, \alpha_4)^T$, $b_\lambda = b_{\lambda_1 \lambda_2 \lambda_3 \lambda_4}$ are the *control points*, and $\mathcal{Z}_+^4$ is the set of all four-dimensional vectors with nonnegative integer components.

Bernstein polynomials: The basis functions for Bézier curves and surfaces.

Bézier curve: A curve whose points are determined by the parameter u in the equation $\sum_{i=0}^n B_i^n(u) P_i$, where the $B_i^n(u)$ are basis functions, and the P_i control points.

Bilinear interpolation: A tensor product of two orthogonal linear interpolants and the "simplest" surface defined by values at four points on a rectangle.

Blending functions: The basis functions used by interpolation schemes such as Gordon surfaces.

B-spline surface: Traditionally, a tensor product of curves defined using piecewise basis polynomials (B-spline basis). Any B-spline can be written in piecewise Bézier form. ("B" stands for "basis.")

C^k *continuity:*   Smoothness defined in terms of matching of up to kth order derivatives along patch boundaries.

Control point: The coefficients in the expansion of a Bézier curve in terms of Bernstein polynomials.

Convex hull: The smallest convex set that contains a given set.

Convex set: A set such that the straight line segment connecting any two points of the set is completely contained within the set.

G^k *continuity:*   Geometric continuity with smoothness defined in terms of matching of up to kth order derivatives allowing for reparametrization. For example, G^1 smoothness is defined in terms of matching tangent planes along patch boundaries.

Knots: A spline curve is defined over a partition of an interval of the real line. The points that define the partition are called knots.

Manifold Spline: A spline surface defined over a planar domain which can be extended to 2-manifolds in space, with arbitrary topology, and with or without boundaries.

Mesh: A decomposition of a geometric domain into finite elements; see Chapter 29.

Radial basis function: A real-valued function with value dependent on the distance from a point. Euclidean distance norm is typical. Quadric, multi-quadric, poly-harmonic and thin-plate splines are based on radial basis functions.

Ruled (lofted) surface: A surface that interpolates two given curves using linear interpolation.

Subdivision surface: A surface that is iteratively refined from a surface mesh using splitting and averaging (linear) operations.

Tensor product surfaces: A surface represented with basis functions that are constructed as products of univariate basis functions. A tensor product Bézier surface is given by the equation $\sum_{i=0}^{n} \sum_{j=0}^{m} B_i^n(u) B_j^m(v) P_{ij}$, where the $B_i^n(u)$ and $B_j^m(v)$ are the univariate Bernstein polynomial basis functions, and the P_{ij} are control points.

T-spline surface: A reduced or decimated version of rectangular control points of B-splines with T-junctions produces or a reduced representation of the same tensor product B-spline.

Transfinite interpolation: Interpolating entire curves as opposed to values at discrete points.

Variation diminishing: A curve or surface scheme has this property if its output "wiggles less" than the control points from which it is constructed.

PARAMETRIC BÉZIER AND B-SPLINES

Tensor product Bézier surfaces are obtained by repeated applications of bilinear interpolation. Properties of tensor product Bézier patches include affine invariance, the "convex hull property," and the variation diminishing property. The boundary curves of a patch are polynomial curves that have their Bézier polygon given by the boundary polygons of the control net of the patch. Hence the four corners of the control net lie on the patch.

TABLE 56.1.1   Tensor product surfaces.

TYPE	INPUT	PROPERTIES
Piecewise Bézier and Hermite	rectangular grid of points, corner twists	C^1, initial global data survey data to determine the tangent and cross-derivative vectors at patch corners
Bicubic B-spline	rectangular grid of points	C^1
Coons patches	4 boundary curves	C^1
Gordon surfaces	rectangular network of curves	C^1, Gregory square
Biquadratic B-spline	limit of Doo-Sabin subdivision of rectangular faces	C^1
Bicubic B-spline	limit of Catmull-Clark subdivision of rectangular faces	C^1
Biquadratic splines	control points on mesh with arbitrary topology	G^1, system of linear equations for smoothness conditions around singular vertices
Biquartic splines	cubic curve mesh	C^1, interpolate second-order data at mesh points
Bisextic B-spline	rectangular network of cubic curves	C^1
Triquadratic/tricubic A-patches	rectilinear 3D grid points	C^1, local calculation of first-order cross derivatives
Triple products of B-splines	rectangular boxes	mixed orders possible
T-spline surface	reduced rectangular mesh with some T-junctions	C^1

Piecewise bicubic Bézier patches may be used to fit a C^1 surface through a rectangular grid of points. After the rectangular network of curves has been created, there are four coefficients left to determine the corner twists of each patch. These four corner twists cannot be specified independently and must satisfy a "compatibility constraint." Common twist estimation methods include zero twists, Adini's twist, Bessel twist, and Brunet's twist [Far98]. To obtain C^1 continuity between two patches the directions and lengths of the polyhedron edges must be matched across the common polyhedron boundary defining the common boundary curve. Piecewise bicubic Hermite patches are similar to the piecewise bicubic Bézier patches, but take points, partials, and mixed partials as input. The mixed partials affect only the interior shape of the patch, and are also called *twist vectors*.

It is not possible to model a general closed surface or a surface with handles as a single non-degenerate B-spline. To represent free-form surfaces a significant amount of recent work has been done in the areas of geometric continuity, non-tensor product patches, and generalizing B-splines [CF83, Pet90a, Pet90b, GW91, DM83, GH87]. Common schemes include splitting, convex combinations of blending functions, subdivision, and local interpolation by construction [For95, HF84, MLL+92, Pet93, Pet02, PR08].

IMPLICIT BÉZIER AND B-SPLINES

Patrikalakis and Kriezis [PK89] demonstrate how implicit algebraic surfaces can be manipulated in rectangular boxes as functions in a tensor product B-spline basis. This work, however, leaves open the problem of selecting weights or specifying knot sequences for C^1 meshes of tensor product implicit algebraic surface patches that fit given spatial data. Moore and Warren [MW91] extend the "marching cubes" scheme to compute a C^1 piecewise tensor product triquadratic approximation to scattered data using a Powell-Sabin-like split over subcubes. In [BBC$^+$99] an incremental and adaptive approach is used to construct C^1 spline functions defined over an octree subdivision that approximate a dense set of multiple volumetric scattered scalar values. Further details are provided in subsequent sub-sections on A-patches and implicit free-form surfaces.

COONS PATCHES AND GORDON SURFACES

Coons patches interpolate four boundary curves. They are constructed by composing two ruled, or lofted, surfaces and one bilinear surface, and hence are called *bilinearly blended surfaces*. A Coons patch has four blending functions $f_i(u)$, $g_i(v)$, $i = 1, 2$. There are only two restrictions on the f_i and g_i: each pair must sum to one, and we must have $f_1(0) = g_1(0) = 1$ and $f_2(1) = g_2(1) = 0$ in order to interpolate.

A network of curves may be filled in with a C^1 surface using bicubically blended Coons patches. For this, the four twists at the data points and the four cross boundary derivatives must be computed. Compatibility problems may arise in computing the twists. If $\mathbf{x}(u, v)$ is twice differentiable, we have $\mathbf{x}_{uv} = \mathbf{x}_{vu}$, but this simplification does not apply here. One approach is to adjust the given data so that the incompatibilities disappear. Or if the data cannot be changed one can use a method known as *Gregory's square* that replaces the constant twist terms by variable twists that are computed from the cross boundary derivatives. The resulting surface does not have continuous twists at the corners and is rational parametric, which may not be acceptable geometry for certain geometric modeling systems.

Gordon surfaces are a generalization of Coons patches used to construct a surface that interpolates a rectangular network of curves. The idea is to take a univariate interpolation scheme, apply it to all curves, add the resulting surfaces, and subtract the tensor product interpolant that is defined by the univariate scheme. Polynomial interpolation or spline interpolation schemes may be used. Methods for Coons patches and Gordon surfaces can be formulated in terms of Boolean sums and projectors. This has also been generalized to create triangular Coons patches.

56.2 GENERALIZED SPLINE SURFACES

B-PATCHES

The B-patches developed by Seidel [Sei89, DMS92] are based on the study of symmetric recursive evaluation algorithms, and are defined by generalizing the deBoor algorithm for the evaluation of a B-spline segment from curves to surfaces. A polynomial surface that has a symmetric recursive evaluation algorithm is called a *B-patch*. B-patches generalize Bézier patches over triangles, and are characterized by control points and a three-parameter family of knots. Every bivariate polynomial $F : \mathbb{R}^2 \to \mathbb{R}^d$ of degree n has a unique representation

$$F(U) = \sum_{|\vec{i}|=n} N_{\vec{i}}^n(U)P_{\vec{i}}, \qquad P_{\vec{i}} \in \mathbb{R}^d$$

as a B-patch, with parameters $\mathcal{K} = R_0, \ldots, R_{n-1}, S_0, \ldots, S_{n-1}, T_0, \ldots, T_{n-1}$ in $\mathbb{R}^2$, if the parameters (R_i, S_j, T_k) are affinely independent for $0 \leq |\vec{i}| \leq n - 1$. The real-valued polynomials $N_{\vec{i}}^n(U)$ are called the **normalized B-weights** of degree n over $\mathcal{K}$.

MULTISIDED PATCHES

Multisided patches can be generated in basically two ways. Either the polygonal domain which is to be mapped into $\mathbb{R}^3$ is subdivided in the parametric plane, or one uniform equation is used as a combination of equations. In the former case, triangular or rectangular elements are put together or recursive subdivision is applied. In the latter case, either the known control point methods are generalized, or a weighted sum of interpolants is used. With constrained domain mapping, a domain point for an n-sided patch is represented by n dependent parameters. If the remainder of the parameters can be computed when any two parameters are independently chosen, it is called a *symmetric system of parameters*. The main results from multisided patch schemes obtained to date are summarized in Table 56.2.1.

TABLE 56.2.1   Multisided schemes.

TYPE	LIMITATIONS	PROPERTIES	DOMAIN POINTS
Sabin	n=3,5	C^1	constrained domain mapping, symmetric system of parameters
Gregory/Charrot	n=3,5	C^1	barycentric coordinates
Hosaka/Kimura	$n \leq 6$	C^1	constrained domain mapping, symmetric system of parameters
Varady		VC^1	$2n$ variables constrained along polygon sides
Base points	$n = 4, 5, 6$	rational Bézier surfaces	base points in the parametric domain map to rational curves in $\mathbb{R}^3$
S-patches	multisided	G^1 rational bi-quadratic and bicubic B-splines	embed n-sided domain polygon into simplex of dimension $n - 1$
Multisided A-patches	"polynomial surfaces" boundary curves	C^1, C^2 implicit Bezier surfaces	Hermite interpolation of boundary curves
Generalized Barycentric Finite Elements	"functional" multisided	G^1 rational bi-quadratic and bicubic B-splines	defined on n-sided domain polygons

TRIANGULAR RATIONAL PATCHES WITH BASE POINTS

Another approach to creating multisided patches is to introduce base points into rational parametric functions. Base points are parameter values for which the homogeneous coordinates (x, y, z, w) are mapped to $(0, 0, 0, 0)$ by the rational parametrization. Gregory's patch [Gre83] is defined using a special collection of rational basis functions that evaluate to $0/0$ at vertices of the parametric domain, and thus introduce base points in the resulting parametrization. Warren [War92] uses base points to create parametrizations of four-, five-, and six-sided surface patches using rational Bézier surfaces defined over triangular domains. Setting a triangle of weights to zero at one corner of the domain triangle produces a four-sided patch that is the image of the domain triangle.

S-PATCHES

Loop and DeRose [LD89, LD90] present generalizations of biquadratic and bicubic B-spline surfaces that are capable of representing surfaces of arbitrary topology by placing restrictions on the connectivity of the control mesh, relaxing C^1 continuity to G^1 (geometric) continuity, and allowing n-sided finite elements. This generalized view considers the spline surface to be a collection of possibly rational polynomial maps from independent n-sided polygonal domains, whose union possesses continuity of some number of geometric invariants, such as tangent planes. This more general view allows patches to be sewn together to describe free-form surfaces in more complex ways.

An n-sided S-patch S is constructed by embedding its n-sided domain polygon P into a simplex $\triangle$ whose dimension is one less than the number of sides of the polygon. The edges of the polygon map to edges of the simplex. A Bézier simplex **B** is then constructed using $\triangle$ as a domain. The patch representation S is obtained by restricting the Bézier simplex to the embedded domain polygon.

A-PATCHES

The A-patch technique provides simple ways to guarantee that a constructed implicit surface is single-sheeted and free of undesirable singularities. The technique uses the zero contouring surfaces of trivariate Bernstein-Bézier polynomials to construct a piecewise smooth surface. We call such iso-surfaces *A-patches*. Algorithms to fill an n-sided hole, using either a single multisided A-patch or a network of A-patches, are given in [BE95]. The blends may be C^0, C^1, or C^2 exact fits (interpolation), as well as C^1 or C^2 least squares fits (interpolation and approximation).

For degree-bounded patches, a triangular network of A-patches for the hole may be generated in two ways. First, the n-sided hole is projected onto a plane and the result of a planar triangulation is projected back onto the hole. Second, an initial multisided A-patch is created for the hole and then a coarse triangulation for the patch is generated using a rational spline approximation [BX94].

MULTIVARIATE SPLINES, SIMPLEX AND BOX SPLINES

Multivariate splines are a generalization of univariate B-splines to a multivariate setting [Dah80, DM83, Boo88, Hol82]. Multivariate splines have applications in data fitting, computer-aided design, the finite element method, and image analysis. Work on splines has traditionally been for a given planar triangulation using a polynomial function basis. Box splines are multivariate generalizations of B-splines with uniform knots. Many of the basis functions used in finite element calculations on uniform triangles occur as special instances of box splines. In general a box spline is a locally supported piecewise polynomial. One can define translates of box splines that form a negative partition of unity.

In the bivariate case, box splines correspond to surfaces defined over a regular tessellation of the plane. If the tessellation is composed of triangles, it is possible to represent the surface as a collection of Bernstein-Bézier patches. The two most commonly used special tessellations arise from a rectangular grid by drawing in lines in north-easterly diagonals in each subrectangle or by drawing in both diagonals for each subrectangle. For these special triangulations there is an elegant way to construct locally supported splines.

Multivariate splines defined as projections of simplices are called *simplex splines*. Auerbach [AGM+91] constructs approximations with simplex splines over irregular triangles. Bivariate quadratic simplicial B-splines defined by their corresponding sets of knots derived from a (suboptimal) constrained Delaunayi triangulation of the domain are employed to obtain a C^1 surface. This approach is well suited for scattered data.

Fong and Seidel [FS86, FS92] construct multivariate B-splines for quadratics and cubics by matching B-patches with simplex splines. The surface scheme is an approximation scheme based on blending functions and control points and allows the modeling of C^{k-1} continuous piecewise polynomial surfaces of degree k over arbitrary triangulations of the parameter plane. The resulting surfaces are defined as linear combinations of the blending functions, and are parametric piecewise polynomials over a triangulation of the parameter plane whose shape is determined by their control points.

SPHERICAL SPLINES

Spherical splines are piecewise representations of functions on the sphere. These can be applied to data fitting problems: (a) the input data is scattered on a unit sphere, (b) a boundary value approximation where the boundary is a unit sphere, (c) solving spherical partial differential equations. The splines can be defined using BB-basis polynomials on spherical triangles, or spherical quads, and are linearly independent and form a basis for functions on the sphere [LS07].

56.3 FREE-FORM SURFACES

The representation of free-form surfaces is one of the major issues in geometric modeling. These surfaces are generally defined in a piecewise manner by smoothly joining several, mostly four-sided, patches. Common approaches to constructing surfaces over irregular meshes are local construction, blending polynomial pieces, and splitting.

GLOSSARY

Blending polynomial pieces: Constructing k pieces for a k-sided mesh facet such that each piece matches a part of the facet data, and a convex combination of the pieces matches the whole.

Vertex enclosure constraint: Not every mesh of polynomial curves with a well-defined tangent plane at the mesh points can be interpolated by a smooth regularly parametrized surface with one polynomial piece per facet. This constraint on the mesh is a necessary and sufficient condition to guarantee the existence of such an interpolant [Pet91]. Rational patches, singular parametrizations, and the splitting of patches are techniques to enforce the vertex enclosure constraint.

MAIN RESULTS

Blending approaches prescribe a mesh of boundary curves and their normal derivatives. For this approach, however, the existence of a well-defined tangent plane at the data points is not sufficient to guarantee the existence of a C^1 mesh interpolant, because the mixed derivatives p_{uv} and p_{vu} are given independently at any point p. Splitting approaches, on the other hand, expect to be given at least tangent vectors at the data points, and sometimes the complete boundary. Mann et al. [MLL$^+$92] conclude that local polynomial interpolants generally produce unsatisfactory shapes.

With splitting schemes, every triangle in the triangulation of the data points (also called a *macro-triangle*) is split into several *mini-triangles*. Split-triangle interpolants do not require derivative information of higher order than the continuity of the desired interpolant. The simplest of the split-triangle interpolants is the C^1 Clough-Tocher interpolant. Each vertex is joined to the centroid, and the macro-triangle is split into three mini-triangles. The first-order data that this interpolant requires are position and gradient value at the macro-triangle vertices, plus some cross-boundary derivative at the midpoint of each edge. There are twelve data per macro-triangle, and cubic polynomials are used over each mini-triangle. The C^1 Powell-Sabin interpolants produce C^1 piecewise quadratic interpolants to C^1 data at the vertices of a triangulated data set. Each macro-triangle is split into six or twelve mini-triangles.

PARAMETRIC PATCH SCHEMES

These patches are given in vector-valued parametric form, generally mapping a rectangular or triangular parametric domain into $\mathbb{R}^3$. Parametric free-form surface patch schemes are summarized in Table 56.3.1.

TABLE 56.3.1   Free-form parametric schemes.

DEGREE	SCHEME	INPUT	PROPERTIES
Piecewise biquartic	local interpolation	cubic curve mesh	C^1, interpolate second-order data at mesh points
Piecewise biquadratic	G-edges	control points on a mesh with arbitrary topology	G^1, system of linear eqns for smoothness conditions around singular vertices
Sextic triangular pieces	approximation, no local splitting	triangular control mesh	G^1
Quadratic/cubic triangular pieces	splitting, subdivision	irregular mesh of points	C^1, refine mesh by Doo-Sabin to isolate regions of irregular points

IMPLICIT PATCH SCHEMES

While it is possible to model a general closed surface of arbitrary genus as a single implicit surface patch, the geometry of such a global surface is difficult to specify, interactively control, and polygonize. The main difficulties stem from the fact that implicit representations are iso-contours which generally have multiple real sheets, self-intersections, and several other undesirable singularities. Looking on the bright side, implicit polynomial splines of the same geometric degree have more degrees of freedom compared with parametric splines, and hence potentially are more flexible for approximating a complicated surface with fewer pieces and for achieving a higher order of smoothness. The potential of implicits remains largely latent: virtually all commercial and many research modeling systems are based on the parametric representation. An exception is SHASTRA, which allows modeling with both implicit and parametric splines [Baj93]. Implicit free-form surface schemes are summarized in Table 56.3.2.

TABLE 56.3.2   Free-form implicit schemes.

DEGREE	SCHEME	INPUT	PROPERTIES
5, 7	local interpolation, no splitting	curve mesh from spatial triangulation	C^1 interpolate or approximate
2	simplicial hull construction	spatial triangulation	
3	simplicial hull construction, Clough–Tocher split	spatial triangulation	C^1
3	simplicial hull construction, Clough–Tocher split of coplanar faces	spatial triangulation	C^1 A-patches, 3 or 4 sides
5	simplicial hull construction, Clough–Tocher split of coplanar faces	spatial triangulation	C^2 A-patches

A-SPLINES

An *A-spline* is a piecewise G^k-continuous chain of real algebraic curve segments, such that each curve segment is a smooth and single-sheeted zero contour of a bivariate Bernstein-Bézier polynomial (called a regular curve segment). A-splines are a suitable polynomial form for working with piecewise implicit polynomial curves. A characterization of A-splines defined over triangles or quadrilaterals is available [BX99, XB00], as is a detailing of their applications in curve design and fitting [BX01a].

CURVILINEAR MESH SCHEMES

Bajaj and Ihm [BI92a] construct implicit surfaces to solve the scattered data-fitting problem. The resulting surfaces approximate or contain with C^1 continuity any collection of points and algebraic space curves with derivative information. Their Hermite interpolation algorithm solves a homogeneous linear system of equations to compute the coefficients of the polynomial defining the algebraic surface. This idea has been extended to C^k (rescaling continuity) interpolate or least squares approximate implicit or parametric curves in space [BIW93]. This problem is formulated as a constrained quadratic minimization problem, where the algebraic distance is minimized instead of the geometric distance.

In a *curvilinear-mesh-based* scheme, Bajaj and Ihm [BI92b] construct low-degree implicit polynomial spline surfaces by interpolating a mesh of curves in space using the techniques of [BI92a, BIW93]. They consider an arbitrary spatial triangulation $\mathcal{T}$ consisting of vertices in $\mathbb{R}^3$ (or more generally, a simplicial polyhedron $\mathcal{P}$ when the triangulation is closed), with possibly normal vectors at the vertex points. Their algorithm constructs a C^1 mesh of real implicit algebraic surface patches over $\mathcal{T}$ or $\mathcal{P}$. The scheme is local (each patch has independent free parameters) and there is no local splitting. The algorithm first converts the given triangulation or polyhedron into a curvilinear wire frame, with at most cubic parametric curves which C^1 interpolate all the vertices. The curvilinear wire frame is then fleshed to produce a single implicit surface patch of degree at most 7 for each triangular face $\mathcal{T}$ of $\mathcal{P}$. If the triangulation is convex then the degree is at most 5. Similar techniques exist for parametrics [Pet91, Sar87]; however, the geometric degrees of the solution surfaces tend to be prohibitively high.

SIMPLEX- AND BOX-BASED SCHEMES

In a ***simplex-based*** approach, one first constructs a tetrahedral mesh (called the simplicial hull) conforming to a surface triangulation $\mathcal{T}$ of a polyhedron $\mathcal{P}$. The implicit piecewise polynomial surface consists of the zero set of a Bernstein-Bézier polynomial, defined within each tetrahedron (simplex) of the simplicial hull. A simplex-based approach enforces continuity between adjacent patches by enforcing that vertex/edge/face-adjacent trivariate polynomials are continuous with one another.

Similar to the trivariate interpolation case, Powell-Sabin or Clough-Tocher splits are used to introduce degree-bounded vertices to prevent the continuity system from propagating globally. Such splitting, however, could result in a large number of patches. However, as only the zero set of the polynomial is of interest, one does not need a complete mesh covering the entire space.

Sederberg [Sed85] showed how various smooth implicit algebraic surfaces, represented in trivariate Bernstein basis form, can be manipulated as functions in Bézier control tetrahedra with finite weights. He showed that if the coefficients of the Bernstein-Bézier form of the trivariate polynomial on the lines that parallel one edge, say L, of the tetrahedron all increase (or decrease) monotonically in the same direction, then any line parallel to L will intersect the zero contour algebraic surface patch at most once.

Guo [Guo91] used cubics to create free-form geometric models and enforced monotonicity conditions on a cubic polynomial along the direction from one vertex to a point of the face opposite the vertex. A Clough-Tocher split is used to subdivide each tetrahedron of the simplicial hull. Dahmen and Thamm-Scharr [DTS93] utilize a single cubic patch per tetrahedron, except for tetrahedra on coplanar faces.

Lodha [Lod92] constructed low degree surfaces with both parametric and implicit representations and investigated their properties. A method is described for creating quadratic triangular Bézier surface patches that lie on implicit quadric surfaces. Another method is described for creating biquadratic tensor product Bézier surface patches that lie on implicit cubic surfaces. The resulting patches satisfy all the standard properties of parametric Bézier surfaces, including interpolation of the corners of the control polyhedron and the convex hull property.

Bajaj and Ihm, Guo, and Dahmen [BI92b, Guo91, Guo93, Dah89] provide heuristics based on monotonicity and least square approximation to circumvent the multiple-sheeted and singularity problems of implicit patches.

Bajaj, Chen, and Xu [BCX95] construct 3- and 4-sided A-patches that are implicit surfaces in Bernstein-Bézier (BB) form and that are smooth and single-sheeted. They give sufficiency conditions for the BB form of a trivariate polynomial within a tetrahedron, such that the zero contour of the polynomial is a single-sheeted nonsingular surface within the tetrahedron, and its cubic-mesh complex for the polyhedron $\mathcal{P}$ is guaranteed to be both nonsingular and single-sheeted. They distinguish between convex and non-convex facets and edges of the triangulation. A double-sided tetrahedron is built for nonconvex facets and edges, and single-sided tetrahedra are built for convex facets and edges. A generalization of Sederberg's condition is given for a three-sided j-patch where any line segment passing through the jth vertex of the tetrahedron and its opposite face intersects the patch only once. Instead of having coefficients be monotonically increasing or decreasing there is a single sign change condition. There are also free parameters for both local and global shape control.

Reconstructing surfaces and scalar fields defined over the surface from scattered data using implicit Bézier splines is described in [BBX95, BX01b]. See also Chapter 35.

GENERALIZED BARYCENTRIC FINITE ELEMENTS

A *generalized-barycentric* scheme for the construction of free-form curve or surface elements, is an extension of the afore-mentioned simplex and box based schemes to meshes of convex polytopes, especially in dimensions 2 and 3 [War96, FHK06, GRB16]. It is well known that on simplicial and tensor product meshes, standard barycentric coordinates provide a local basis for scalar-valued finite element spaces, commonly called the Lagrange interpolation elements. Using generalized barycentric coordinates on can construct local interpolation bases with the same global continuity and polynomial reproduction properties as their simplicial counterparts. Further, local bases for the lowest-order vector-valued Brezzi-Douglas-Marini [BDM85], Raviart-Thomas [RT77], and Nedelec [BDD+87, Néd80, Néd86] finite element spaces on simplices can also be defined in a canonical fashion from an associated set of standard barycentric functions.

In [GRB12, RGB13] linear order, scalar-valued interpolants on polygonal meshes are considered use four different types of generalized barycentric coordinates: Wachspress [Wac11], Sibson [Far90, Sib80], harmonic [Chr08, JMD+07, MKB+08], and mean value [Flo03, FHK06, FKR05]. The analysis was extended by Gillette, Floater, and Sukumar in the case of Wachspress coordinates to convex polytopes in any dimension [FGS14], based on work by Warren and colleagues [JSW+05, War96, WSH+07]. In a related vein, it has also been shown how taking pairwise products of generalized barycentric coordinates can be used to construct quadratic order methods on polygons [RGB14]. Applications of generalized barycentric coordinates to finite element methods have primarily focused on scalar-valued PDE problems [MP08, RS06, SM06, ST04, WBG07] though extensions to vector-valued interpolants on dual meshes is the method of choice for solution of mixed or coupled PDEs [AFW09].

56.4 MULTIRESOLUTION SPLINE SURFACES

SUBDIVISION SURFACES

Subdivision techniques can be used to produce generally pleasing surfaces from arbitrary control meshes. The faces of the mesh need not be planar, nor need the vertices lie on a topologically regular mesh. Subdivision consists of splitting and averaging. Each edge or face is split, and each new vertex introduced by the splitting is positioned at a fixed affine combination of its neighbor's weights. Subdivision schemes are summarized in Table 56.4.1.

TABLE 56.4.1   Subdivision schemes.

TYPE	PROPERTIES
Doo-Sabin; Catmull-Clark	C^1, interpolate centroids of all faces at each step
Nasri	interpolate points/normals on irregular networks
Loop	C^1, split each triangle of a triangular mesh into 4 triangles
Hoppe et al.	extends Loop's method to incorporate shape edges in limit surfaces; initial vertices belong to vertex, edge, or face of limit surface
Storry and Ball	C^1 n-sided B-spline patch to fit in bicubic surface, one dof
Dyn, Levin, and Gregory	interpolatory butterfly subdivision, modify set of deterministic rules for subdivision
Bajaj, Chen, and Xu	approximation, one step subdivision to build simplicial hull, C^1 cubic and C^2 quintic A-patches
Reif	regularity conditions

MAIN ALGORITHMS

Subdivision algorithms start with a polyhedral configuration of points, edges, and faces. The control mesh will in general consist of large regular regions and isolated singular regions. Subdivision enlarges the regular regions of the control net and shrinks the singular regions. Each application of the subdivision algorithm constructs a refined polyhedron, consisting of more points and smaller faces, tending in the limit to a smooth surface. In general the new control points are computed as linear combinations of old control points. The associated matrix is called the ***subdivision matrix***. Except for some special cases, the limiting surface does not have an explicit analytic representation. If each face of the polyhedron is a rectangle, the Doo-Sabin subdivision rules generate biquadratic tensor product B-splines, and the Catmull-Clark subdivision rules generate bicubic tensor product B-splines. Also, the subdivision technique of Loop generates three-direction box splines.

Reif [Rei92] presents a unified approach to subdivision algorithms for meshes with arbitrary topology and gives a sufficient condition for the regularity of the surface. The existence of a smooth regular parametrization for the generated surface near the point is determined from the leading eigenvalues of the subdivision matrix and an associated characteristic map. Details and further discussion of recent subdivision schemes are available from [WW02].

APPROXIMATING SCHEMES

Bajaj, Chen, and Xu [BCX94] construct an "inner" simplicial hull after one step of subdivision of the input polyhedron $\mathcal{P}$. As in traditional subdivision schemes, $\mathcal{P}$ is used as a control mesh for free-form modeling, while an inner surface triangulation $\mathcal{T}$ of the hull can be considered as the second-level mesh. Both a C^1 mesh with cubic A-patches and a C^2 mesh with quintic patches can be constructed to approximate the polyhedron $\mathcal{P}$ [XBE01].

INTERPOLATING SCHEMES

There are two key approaches to constructing interpolating subdivision surfaces. One approach is to first compute a new configuration of vertices, edges, and faces with the same topology such that the vertices of the new configuration converge to the given vertices in the limit. The subdivision technique is then applied to this new configuration The other approach is to modify the deterministic subdivision rules so that the limiting surface interpolates the vertices.

HIERARCHICAL SPLINES

Hierarchical splines are a multiresolution approach to the representation and manipulation of free-form surfaces. A hierarchical B-spline is constructed from a base surface (level 0) and a series of overlays are derived from the immediate parent in the hierarchy. Forsey and Bartels [FB88] present a refinement scheme that uses a hierarchy of rectangular B-spline overlays to produce C^2 surfaces. Overlays can be added manually to add detail to the surface, and local or global changes to the surface can be made by manipulating control points at different levels.

Forsey and Wang [FW93] create hierarchical bicubic B-spline approximations to scanned cylindrical data. The resulting hierarchical spline surface is interactively modifiable using editing capabilities of the hierarchical surface representation, allowing either local or global changes to surface shape while retaining the details of the scanned data. Oscillations occur, however, when the data have high-amplitude or high-frequency regions. Forsey and Bartels use a hierarchical wavelet-based representation for fitting tensor product parametric spline surfaces to gridded data in [FB95]. The multiresolution representation is extend to include arbitrary meshes in [EDD+95]. The method is based on approximating an arbitrary mesh by a special type of mesh and using a continuous parametrization of the arbitrary mesh over a simple domain mesh.

Multiresolution representations for spherical splines or spherical wavelets is given in [SS95] Further discussion of wavelet based multiresolution schemes and some of their applications is available from [SDS96].

56.5   PHYSICALLY BASED APPROACHES TO SURFACE MODELING

ENERGY-BASED SPLINES

A group of researchers [TF88, PB88, TPB⁺87, WFB87, BHN99] have presented discrete models which are based extensively on the theory of elasticity and plasticity, using energy fields to define and enforce constraints [AFW06, AFW10]. Haumann [Hau87] used the same approach but used a triangulated model and a simpler physical model based on points, springs, and hinges. Thingvold and Cohen [TC90] defined a model of elastic and plastic B-spline surfaces which supports both animation and design operations. The basis for the physical model is a generalized point-mass/spring/hinge model that has been adapted into a simultaneous refinement of the geometric/physical model. Always having a sculptured surface representation as well as the physical hinge/spring/mesh model allows the user to intertwine physical-based operations, such as force application, with geometrical modeling. Refinement operations for spring and hinge B-spline models are compatible with the physics and mathematics of B-spline models. The models of elasticity and plasticity are written in terms of springs and hinges, and can be implemented with standard integration techniques to model realistic motions of elastic and plastic surfaces. These motions are controlled by the physical properties assigned and by kinematic constraints on various portions of the surface. Terzopoulos and Qin [TQ94] develop a dynamic generalization of the nonuniform rational B-spline (NURBS) model. They present a physics-based model that incorporates mass distributions, internal deformation energies, and other physical quantities into the NURBS geometric substrate. These dynamic NURBS can be used in applications such as rounding of solids, optimal surface fitting to unstructured data, surface design from cross-sections, and free-form deformations.

DIFFERENTIAL EQUATIONS AND SURFACE SPLINES

Early research on using partial differential equations (PDEs) to handle surface modeling problems traces back to Bloor et al.'s work at the end of the 1980s ([BW89a, BW89b, BW90]). The basic idea of these papers is the use of biharmonic equations on a rectangular domain to solve blending and hole filling problems. One of the advantages of using the biharmonic equation is that it is linear, and therefore easier to solve. However, the solution of the equation depends on the surface parametrization.

The evolution technique, based on the heat equation $\partial_t x - \Delta x = 0$, has been extensively used in the area of image processing (see [PM87, PR99, Wei98], where Δ is a $2D$ Laplace operator. This was extended later to smoothing or fairing noisy surfaces (see [CDR00, DMS⁺99, DMS⁺03]). For a surface M, the counterpart of the Laplacian Δ is the Laplace-Beltrami operator Δ_M (see [Car92]). One then obtains the geometric diffusion equation $\partial_t x - \Delta_M x = 0$ for a surface point $x(t)$ on the surface $M(t)$. Taubin [Tau95] discusses the discretized operator of the Laplacian and related approaches in the context of generalized frequencies on meshes. Kobbelt

[Kob96] considers discrete approximations of the Laplacian in the construction of fair interpolatory subdivision schemes. This work was extended in [KCV+98] to arbitrary connectivity for purposes of multi-resolution interactive editing. Desbrun et al. [DMS+99] used an implicit discretization of geometric diffusion to obtain a strongly stable numerical smoothing scheme. The same strategy of discretization is also adopted and analyzed by Deckelnick and Dziuk [DD02] with the conclusion that this scheme is unconditionally stable. Clarenz et al. [CDR00] introduced anisotropic geometric diffusion to enhance features while smoothing. Ohtake et al. [OBB00] combined an inner fairness mechanism in their fairing process to increase the mesh regularity. Bajaj and Xu [BX03] smooth both surfaces and functions on surfaces, in a C^2 smooth function space defined by the limit of triangular subdivision surfaces (quartic Box splines).

Similar to surface diffusion using the Laplacian, a more general class of PDE based methods called *flow surface techniques* have been developed which simulate different kinds of flows on surfaces (see [WJE00] for references) using the equation $\partial_t x - v(x,t) = 0$, where $v(x,t)$ represents the instantaneous stationary velocity field.

Level set methods were also used in surface fairing and surface reconstruction; see [BCO+01, BSC+00, CS99, MBW+02, OF03, WB98, ZOF01, ZOM+00]. In these methods, surfaces are formulated as iso-surfaces (level surfaces) of 3D functions, which are usually defined from the signed distance over Cartesian grids of a volume. An evolution PDE on the volume governs the behavior of the level surface. These level-set methods have several attractive features including, ease of implementation, arbitrary topology [BW01] and a growing body of theoretical results. Often, fine surface structures are not captured by level sets, although it is possible to use adaptive [PR99] and triangulated grids as well as Hermite data [JLS+02, KBS+01]. To reduce the computationally complexity, Bertalmio et al. [BCO+01, BSC+00] solve the PDE in a narrow band for deforming vectorial functions on surfaces (with a fixed surface represented by the level surface).

Recently, surface diffusion flow has been used to solve the surface blending problem and free-form surface design problem. In [SK00], fair meshes with G^1 conditions are created in the special case where the meshes are assumed to have subdivision connectivity. In this work, local surface parametrization is still used to estimate the surface curvatures. A later paper [SK01] uses the same equation for smoothing meshes while satisfying G^1 boundary conditions. Outer fairness (the smoothness in the classical sense) and inner fairness (the regularity of the vertex distribution) criteria are used in their fairing process.

Another category of surface fairing research is based on utilizing optimization techniques. In this category, one constructs an optimization problem that minimizes certain objective functions [Gre94, GL10, HG00, MS92, Sap94, Wah90, WW92], such as thin plate energy, membrane energy [KCV+98], total curvature [KHP+97, WW94], or sum of distances [Mal92]. Using local interpolation or fitting, or replacing differential operators with divided difference operators, the optimization problems are discretized to arrive at finite dimensional linear or nonlinear systems. Approximate solutions are then obtained by solving the constructed systems. In general, such an approach is quite computationally intensive.

56.6 SOURCES AND RELATED MATERIAL

SURVEYS

All results not given an explicit reference above may be traced in these surveys.

[Alf89]: Scattered data fitting and multivariate splines.

[Baj97]: Summary of data fitting with implicit algebraic splines.

[BBB87]: Application of B-splines.

[Dau92]: An introduction to wavelets.

[BHR93]: An introduction to Box splines.

[DM83]: Scattered data fitting and multivariate splines.

[Far98]: Summary of the history of triangular Bernstein-Bézier patches.

[GL93]: An introduction to Knot manipulation techniques in splines.

[Hol82]: Scattered data fitting and multivariate splines.

[SDS96]: Application of wavelet representations.

[Wah90]: Penalized regression splines.

[WW02, PR08]: Subdivision techniques.

RELATED CHAPTERS

Chapter 29: Triangulations and mesh generation
Chapter 37: Computational and quantitative real algebraic geometry
Chapter 52: Computer graphics
Chapter 57: Solid modeling

REFERENCES

[AFW06]   D. Arnold, R. Falk, and R. Winther. Finite element exterior calculus, homological techniques, and applications. *Acta Numer.*, pages 1–155, 2006.

[AFW09]   D. Arnold, R. Falk, and R. Winther. Geometric decompositions and local bases for spaces of finite element differential forms. *Comput. Methods Appl. Mech. Engrg.*, 198:1660–1672, 2009.

[AFW10]   D. Arnold, R. Falk, and R. Winther. Finite element exterior calculus: from Hodge theory to numerical stability. *Bull. Amer. Math. Soc.*, 47:281–354, 2010.

[AGM$^+$91]   S. Auerbach, R.H.J. Gmelig Meyling, M. Neamtu, and H. Schaeben. Approximation and geometric modeling with simplex B-splines associated with irregular triangles. *Comput. Aided Geom. Design*, 8:67–87, 1991.

[Alf89]   P. Alfeld. Scattered data interpolation in three or more variables. In T. Lyche and L.L. Schumaker, editors, *Mathematical Methods in Computer Aided Geometric Design*, pages 1–34, Academic Press, San Diego, 1989.

[Baj93]   C.L. Bajaj. The emergence of algebraic curves and surfaces in geometric design. In R. Martin, editor, *Directions in Geometric Computing*, pages 1–29, Information Geometers, Winchester, 1993.

[Baj97]   C.L. Bajaj. Implicit surface patches. In J. Bloomenthal, editor, *Introduction to Implicit Surfaces*, pages 98–125, Morgan Kaufman, San Francisco, 1997.

[BBB87]   R. Bartels, J. Beatty, and B. Barsky. *An Introduction to Splines for Use in Computer Graphics and Geometric Modeling*. Morgan Kaufmann, San Francisco, 1987.

[BBC+99]   F. Bernardini, C.L. Bajaj, J. Chen, and D. Schikore. Automatic reconstruction of 3D CAD models from digital scans. *Internat. J. Comput. Geom. Appl.*, 9:327–369, 1999.

[BBX95]   C.L. Bajaj, F. Bernardini, and G. Xu. Automatic reconstruction of surfaces and scalar fields from 3D scans. *Proc. 22nd Conf. Comput. Graph. Interactive Tech.*, pages 109–118, ACM Press, 1995.

[BCO+01]   M. Bertalmio, L.-T. Cheng, S. Osher, and G. Sapiro. Variational problems and partial differential equations on implicit surfaces. *J. Comput. Physics*, 174:759–780, 2001.

[BCX94]   C.L. Bajaj, J. Chen, and G. Xu. Smooth low degree approximations of polyhedra. Comput. Sci. Tech. Rep., CSD-TR-94-002, Purdue Univ., 1994.

[BCX95]   C.L. Bajaj, J. Chen, and G. Xu. Modeling with cubic A-patches. *ACM Trans. Graph.*, 14:103–133, 1995.

[BDD+87]   F. Brezzi, J. Douglas Jr., R. Durán, and M. Fortin. Mixed finite elements for second order elliptic problems in three variables. *Numer. Math.*, 51:237–250, 1987.

[BDM85]   F. Brezzi, J. Douglas Jr., and L.D. Marini. Two families of mixed finite elements for second order elliptic problems. *Numer. Math.*, 47:217–235, 1985.

[BE95]   C.L. Bajaj and S.B. Evans. Smooth multi-sided blends with A-patches. Presented at *4th SIAM Conf. Geom. Design*, 1995.

[BHN99]   C.L. Bajaj, R. Holt, and A. Netravali. Energy formulations for A-splines. *Comput. Aided Geom. Design*, 16:39–59, 1999.

[BHR93]   C. de Boor, K. Hollig, and S. Riemenschneider. *Box Splines*. Springer-Verlag, New York, 1993.

[BI92a]   C.L. Bajaj and I. Ihm. Algebraic surface design with Hermite interpolation. *ACM Trans. Graph.*, 11:61–91, 1992.

[BI92b]   C.L. Bajaj and I. Ihm. C^1 Smoothing of polyhedra with implicit algebraic splines. *SIGGRAPH Comput. Graph.*, 26:79–88, 1992.

[BIW93]   C.L. Bajaj, I. Ihm, and J. Warren. Higher-order interpolation and least-squares approximation using implicit algebraic surfaces. *ACM Trans. Graph.*, 12:327–347, 1993.

[Boo68]   C. de Boor. On calculating with B-splines. *J. Approx. Theory*, 1:219–235, 1968.

[Boo88]   C. de Boor. What is a multivariate spline? In *Proc 1st Internat. Conf. Industrial Appl. Math.*, pages 91–101, SIAM, Philadelphia, 1988.

[BSC+00]   M. Bertalmio, G. Sapiro, L.T. Cheng, and S. Osher. A framework for solving surface partial differential equations for computer graphics applications. CAM Report 00-43, UCLA, Math. Dept., 2000.

[BW89a]   M.I.G. Bloor and M.J. Wilson. Generating blend surfaces using partial differential equations. *Comput. Aided Design*, 21:165–171, 1989.

[BW89b]   M.I.G. Bloor and M.J. Wilson. Generating N-sided patches with partial differential equations. In *Advances in Comput. Graph.*, pages 129–145. Springer-Verlag, Berlin, 1989.

[BW90]   M.I.G. Bloor and M.J. Wilson. Using partial differential equations to generate free-form surfaces. *Comput. Aided Design*, 22:221–234, 1990.

[BW01]   D. Breen and R. Whitaker. A level-set approach for the metamorphosis of solid models. *IEEE Trans. Vis. Comput. Graph.*, 7:173–192, 2001.

[BX94]   C.L. Bajaj and G. Xu. Rational spline approximations of real algebraic curves and surfaces. In H.P. Dikshit and C. Micchelli, editors, *Advances in Computational Mathematics*, pages 73–85, World Scientific, Singapore, 1994.

[BX99]   C.L. Bajaj and G. Xu. A-splines: Local interpolation and approximation using G^k-continuous piecewise real algebraic curves. *Comput. Aided Geom. Design*, 16:557–578, 1999.

[BX01a]   C.L. Bajaj and G. Xu. Regular algebraic curve segments (III)—applications in interactive design and data fitting. *Comput. Aided Geom. Design*, 18:149–173, 2001.

[BX01b]   C.L. Bajaj and G. Xu. Smooth shell construction with mixed prism fat surfaces. In *Geometric Modeling*, pages 19–36, Springer, Berlin, 2001.

[BX03]   C.L. Bajaj and G. Xu. Anisotropic diffusion of surfaces and functions on surfaces. *ACM Trans. Graph.*, 22:4–32, 2003.

[Car92]   M. do Carmo. *Riemannian Geometry*. Birkhäuser, Boston, 1992.

[CDR00]   U. Clarenz, U. Diewald, and M. Rumpf. Anisotropic geometric diffusion in surface processing. In *Proc. IEEE Visualization*, pages 397–505, 2000.

[CF83]   H. Chiyokura and K. Fumihiko. Design of solids with free form surfaces. *SIGGRAPH Comput. Graph.*, 17:289–298, 1983.

[Chr08]   S.H. Christiansen. A construction of spaces of compatible differential forms on cellular complexes. *Math. Models Methods Appl. Sci.*, 18:739–757, 2008.

[CLR80]   E. Cohen, T. Lyche, and R. Riesenfeld. Discrete *B*-splines and subdivision techniques in computer aided geometric design and computer graphics. *Comput. Graph. Image Process.*, 14:87–111, 1980.

[CS99]   D.L. Chopp and J.A. Sethian. Motion by intrinsic Laplacian of curvature. *Interfaces and Free Boundaries*, 1:1–18, 1999.

[Dah80]   W. Dahmen. On multivariate *B*-splines. *SIAM J. Numer. Anal.*, 17:179–191, 1980.

[Dah89]   W. Dahmen. Smooth piecewise quadratic surfaces. In T. Lyche and L.L. Schumaker, editors, *Mathematical Methods in Computer Aided Geometric Design*, pages 181–193, Academic Press, Boston, 1989.

[Dau92]   I. Daubechies. *Ten Lectures on Wavelets*. SIAM, Philadelphia, 1992.

[DCL+08] J. Deng, F. Chen, X. Li, C. Hu, W. Tong, Z. Yang, and Y. Feng. Polynomial splines over hierarchical T-meshes. *Graphical Models*, 70:76–86, 2008.

[DD02]   K. Deckelnick and G. Dziuk. A fully discrete numerical scheme for weighted mean curvature flow. *Numer. Math.*, 91:423–452, 2002.

[DLP13]   T. Dokken, T. Lyche, and K. Pettersen. Polynomial splines over locally refined box partitions. *Comput. Aided Geom. Design*, 30:331–356, 2013.

[DM83]   W. Dahmen and C. Micchelli. Recent progress in multivariate splines. In L. Shumaker, C. Chui, and J. Word, editors, *Approximation Theory IV*, pages 27–121, Academic Press, 1983.

[DMS92]   W. Dahmen, C. Micchelli, and H.-P. Seidel. Blossoming begets B-spline bases built better by B-patches. *Mathematics of Computation*, 59:97–115, 1992.

[DMS⁺99]   M. Desbrun, M. Meyer, P. Schröder, and A.H. Barr. Implicit fairing of irregular meshes using diffusion and curvature flow. In *Proc. 26th Conf. Comput. Graph. Interactive Tech*, pages 317–324, ACM Press, 1999.

[DMS⁺03]   M. Meyer, M. Desbrun, P. Schröder, A.H. Barr. Discrete differential-geometry operators for triangulated 2-manifolds. In H.-C. Hege and K. Polthier, editors, *Visualization and Mathematics III*, pages 35–57, Springer, Berlin, 2003.

[DTS93]   W. Dahmen and T-M. Thamm-Schaar. Cubicoids: Modeling and visualization. *Comput. Aided Geom. Design*, 10:89–108, 1993.

[EDD⁺95]   M. Eck, T. DeRose, T. Duchamp, H. Hoppe, M. Lounsbery, and W. Stuetzle. Multiresolution analysis of arbitrary meshes. In *Proc. 22nd Conf. Comput. Graph. Interactive Tech*, pages 173–180, ACM Press, 1995.

[Far90]   G. Farin. Surfaces over Dirichlet tessellations. *Comput. Aided Geom. Design*, 7:281–292, 1990.

[Far98]   G. Farin. *Curves and Surfaces for Computer Aided Geometric Design: A Practical Guide*. Academic Press, Boston, 1998.

[FB88]   D. Forsey and R. Bartels. Hierarchical B-spline refinement. *SIGGRAPH Comput. Graph.*, 22:205–212, 1988.

[FB95]   D. Forsey and R. Bartels. Surface fitting with hierarchical splines. *ACM Trans. Graph.*, 14:134–161, 1995.

[FGS14]   M.S. Floater, A. Gillette, and N. Sukumar. Gradient bounds for Wachspress coordinates on polytopes. *SIAM J. Numer. Anal.*, 52:515–532, 2014.

[FHK06]   M. Floater, K. Hormann, and G. Kós. A general construction of barycentric coordinates over convex polygons. *Adv. Comput. Math.*, 24:311–331, 2006.

[FKR05]   M. Floater, G. Kós, and M. Reimers. Mean value coordinates in 3D. *Comput. Aided Geom. Design*, 22:623–631, 2005.

[Flo03]   M. Floater. Mean value coordinates. *Comput. Aided Geom. Design*, 20:19–27, 2003.

[For95]   D. Forsey. Surface fitting with hierarchical splines. *ACM Trans. Graph.*, 14:134–161, 1995.

[FS86]   P. Fong and H.-P. Seidel. Control points for multivariate B-spline surfaces over arbitrary triangulations. *Comput. Graphics Forum*, 10:309–317, 1986.

[FS92]   P. Fong and H.-P. Seidel. An implementation of multivariate B-spline surfaces over arbitrary triangulations. In *Proc. Graphics Interface*, pages 1–10, Vancouver, 1992.

[FW93]   D. Forsey and L. Wang. Multi-resolution surface approximation for animation. In *Proc. Graphics Interface*, pages 192–199, Toronto, 1993.

[GH87]   J.A. Gregory and J. Hahn. Geometric continuity and convex combination patches. *Comput. Aided Geom. Design*, 4:79–89, 1987.

[GL93]   R.N. Goldman and T. Lyche. *Knot Insertion and Deletion Algorithms for B-spline Curves and Surfaces*. SIAM, Philadelphia, 1993.

[GL10]   S. Guillas and M.-J. Lai. Bivariate splines for spatial functional regression models. *J. Nonparametr. Stat.*, 22:477–497, 2010.

[GRB12]   A. Gillette, A. Rand, and C. Bajaj. Error estimates for generalized barycentric coordinates. *Adv. Comput. Math.*, 37:417–439, 2012.

[GRB16]   A. Gillette, A. Rand, and C. Bajaj. Construction of scalar and vector finite element families on polygonal and polyhedral meshes. *Comput. Methods Appl. Math.*, 2016.

[Gre83]   J.A. Gregory. C^1 rectangular and non-rectangular surface patches. In R.E. Barnhill and W. Boehm, editors, *Surfaces in Computer Aided Geometric Design*, pages 25–33, North–Holland, Amsterdam, 1983.

[Gre94]   G. Greiner. Variational design and fairing of spline surface. *Comput. Graph. Forum*, 13:143–154, 1994.

[Guo91]   B. Guo. Surface generation using implicit cubics. In N.M. Patrikalakis, editor, *Scientific Visualization of Physical Phenomena*, pages 485–530, Springer, Tokyo, 1991.

[Guo93]   B. Guo. Non-splitting macro patches for implicit cubic spline surfaces. *Comput. Graph. Forum*, 12:434–445, 1993.

[GW91]   T. Garrity and J. Warren. Geometric continuity. *Comput. Aided Geom. Design*, 8:51–65, 1991.

[Hau87]   D. Haumann. Modeling the physical behavior of flexible objects. ACM SIGGRAPH Course Notes #17, 1987.

[HF84]   M. Hosaka and K. Fumihiko. Non-four-sided patch expressions with control points. *Comput. Aided Geom. Design*, 1:75–86, 1984.

[HG00]   A. Hubeli and M. Gross. Fairing of non-manifolds for visualization. In *Proc. IEEE Visualization*, pages 407–414, 2000.

[Hol82]   K. Hollig. Multivariate splines. *SIAM J. Numer. Anal.*, 19:1013–1031, 1982.

[JLS+02]   T. Ju, F. Losasso, S. Schaefer, and J. Warren. Dual contouring of Hermite data. *ACM Trans. Graph.*, 21:339–346, 2002.

[JMD+07]   P. Joshi, M. Meyer, T. DeRose, B. Green, and T. Sanocki. Harmonic coordinates for character articulation. *ACM Trans. Graph.*, 26:71, 2007.

[JSW+05]   T. Ju, S. Schaefer, J, Warren, and M. Desbrun. A geometric construction of coordinates for convex polyhedra using polar duals. In *Proc. 3rd Eurographics Sympos. Geom. Process.*, pages 181–186, 2005.

[KBS+01]   L.P. Kobbelt, M. Botsch, U. Schwanecke, and H.-P. Seidel. Feature sensitive surface extraction from volume data. In *Proc. 28th Conf. Comput. Graph. Interactive Tech.*, pages 57–66, ACM Press, 2001.

[KCV+98]   L. Kobbelt, S. Campagna, J. Vorsatz, and H.-P. Seidel. Interactive muti-resolution modeling on arbitrary meshes. In *Proc. 25th Conf. Comput. Graph. Interactive Tech.*, pages 105–114, ACM Press, 1998.

[KHP+97]   L. Kobbelt, T. Hesse, H. Prautzsch, and K. Schweizerhof. Iterative mesh generation for FE-computation on free form surfaces. *Engng. Comput.*, 14:806–820, 1997.

[Kob96]   L. Kobbelt. Discrete fairing. In T. Goodman and R. Martin, editors, *The Mathematics of Surfaces VII*, pages 101–129, Information Geometers, 1996.

[LD89]   C.T. Loop and T.D. DeRose. A multisided generalization of Bézier surfaces. *ACM Trans. Graph.*, 8:205–234, 1989.

[LD90]   C. Loop and T.D. DeRose. Generalized B-spline surfaces of arbitrary topology. *Comput. Graph.*, 24:347–356, 1990.

[Lod92]   S. Lodha. *Surface Approximation by Low Degree Patches with Multiple Representations*. Ph.D. thesis, Purdue Univ., West Lafayette, 1992.

[LS07]   M.-J. Lai and L.L. Schumaker. *Spline Functions on Triangulations*. Vol. 110 of *Encyclopedia of Mathematics and its Applications*, Cambridge University Press, 2007.

[Mal92]   J.L. Mallet. Discrete smooth interpolation in geometric modelling. *Comput. Aided Design*, 24:178–191, 1992.

[MBW⁺02]   K. Museth, D. Breen, R. Whitaker, and A.H. Barr. Level set surface editing operators. *ACM Trans. Graph.*, 21:330–338, 2002.

[MKB⁺08]   S. Martin, P. Kaufmann, M. Botsch, M. Wicke, and M. Gross. Polyhedral finite elements using harmonic basis functions. In *Proc. Eurographics Sympos. Geom. Process.*, pages 1521–1529, 2008.

[MLL⁺92]   S. Mann, C. Loop, M. Lounsbery, D. Meyers, J. Painter, T.D. DeRose, and K. Sloan. A survey of parametric scattered data fitting using triangular interpolants. In H. Hagen, editor, *Curve and Surface Modeling*, pages 145–172, SIAM, Philadelphia, 1992.

[Mou10]   B. Mourrain. On the dimension of spline spaces on planar t-meshes. *Math. Comput.*, 83:847–871, 2010.

[MP08]   P. Milbradt and T. Pick. Polytope finite elements. *Int. J. Numer. Methods Eng.*, 73:1811–1835, 2008.

[MS92]   H.P. Moreton and C.H. Séquin. Functional optimization for fair surface design. *ACM Comput. Graph.*, 26:167–176 , 1992.

[MW91]   D. Moore and J. Warren. Approximation of dense scattered data using algebraic surfaces. In *Proc. 24th Hawaii Int. Conf. System Sci.*, pages 681–690, IEEE, 1991.

[Néd80]   J.-C. Nédélec. Mixed finite elements in $\mathbf{R}^3$. *Numer. Math.*, 35:315–341, 1980.

[Néd86]   J.-C. Nédélec. A new family of mixed finite elements in $\mathbb{R}^3$. *Numer. Math.*, 50:57–81, 1986.

[OBB00]   Y. Ohtake, A.G. Belyaev, and I.A. Bogaevski. Polyhedral surface smoothing with simultaneous mesh regularization. In *Proc. Geom. Modeling Process.*, pages 229–237, 2000.

[OF03]   S.J. Osher and R. Fedkiw. *Level Set Methods and Dynamic Implicit Surfaces*. Vol. 153 of *Applied Mathematical Sciences*, Springer-Verlag, New York, 2003.

[PB88]   J. Platt and A.H. Barr. Constraint methods for flexible models. *SIGGRAPH Comput. Graph.*, 22:279–288, 1988.

[Pet90a]   J. Peters. Local cubic and bicubic C^1 surface interpolation with linearly varying boundary normal. *Comput. Aided Geom. Design*, 7:499–516, 1990.

[Pet90b]   J. Peters. Smooth mesh interpolation with cubic patches. *Comput. Aided Design*, 22:109–120, 1990.

[Pet91]   J. Peters. Smooth interpolation of a mesh of curves. *Constr. Approx.*, 7:221–246, 1991.

[Pet93]   J. Peters. Smooth free-form surfaces over irregular meshes generalizing quadratic splines. *Comput. Aided Geom. Design*, 10:347–361, 1993.

[Pet02]   J. Peters. C^2 free-form surfaces of degree (3,5). *Comput. Aided Geom. Design*, 19:113–126, 2002.

[PK89]   N.M. Patrikalakis and G.A. Kriezis. Representation of piecewise continuous algebraic surfaces in terms of b-splines. *Visual Comput.*, 5:360–374, 1989.

[PM87]   P. Perona and J. Malik. Scale space and edge detection using anisotropic diffusion. In *IEEE Comput. Soc. Workshop Comput. Vision*, 1987.

[PR99]   T. Preußer and M. Rumpf. An adaptive finite element method for large scale image processing. In *Scale-Space Theories in Computer Vision*, pages 232–234, 1999.

[PR08]   J. Peters and U. Reif. *Subdivision Surfaces*. Springer, Berlin, 2008.

[Rei92]   U. Reif. A unified approach to subdivision algorithms. Tech. Rep. 92-16, Mathematisches Institut A, Universität Stuttgart, 1992.

[RGB13]   A. Rand, A. Gillette, and C. Bajaj. Interpolation error estimates for mean value coordinates. *Adv. Comput. Math.*, 39:327–347, 2013.

[RGB14]   A. Rand, A. Gillette, and C. Bajaj. Quadratic serendipity finite elements on polygons using generalized barycentric coordinates. *Math. Comput.*, 83:2691–2716, 2014.

[RS06]   M. Rashid and M. Selimotic. A three-dimensional finite element method with arbitrary polyhedral elements. *Int. J. Numer. Methods Eng.*, 67:226–252, 2006.

[RT77]   P.-A. Raviart and J.M. Thomas. A mixed finite element method for 2nd order elliptic problems. In *Mathematical Aspects of Finite Element Methods*, pages 292–315, vol. 606 of *Lecture Notes in Math.*, Springer, Berlin, 1977.

[Sap94]   N. Sapidis. *Designing Fair Curves and Surfaces*. SIAM, Philadelphia, 1994.

[Sar87]   R.F. Sarraga. G^1 interpolation of generally unrestricted cubic Bézier curves. *Comput. Aided Geom. Design*, 4:23–39, 1987.

[Sch94]   L.L. Schumaker. Applications of multivariate splines. In *Math. Comput., 1943–1993: A Half-century of Computations Mathematics*, vol. 48 of *Proc. Symposia Appl. Math.*, AMS, Providence, 1994.

[SDS96]   E.J. Stollnitz, T.D. DeRose, and D. Salesin. *Wavelets for Computer Graphics: Theory and Applications*. Morgan Kaufmann, San Francisco, 1996.

[Sed85]   T.W. Sederberg. Piecewise algebraic patches. *Comput. Aided Geom. Design*, 2:53–59, 1985.

[Sei89]   H.-P. Seidel. A new multiaffine approach to B-splines. *Comput. Aided Geom. Design*, 6:23–32, 1989.

[Sib80]   R. Sibson. A vector identity for the Dirichlet tessellation. *Math. Proc. Cambridge Philos. Soc.*, 87:151–155, 1980.

[SK00]   R. Schneider and L. Kobbelt. Generating fair meshes with G^1 boundary conditions. In *Geometric Modeling Processing*, pages 251–261, 2000.

[SK01]   R. Schneider and L. Kobbelt. Geometric fairing of triangular meshes for free-form surface design. *Comput. Aided Geom. Design*, 18:359–379, 2001.

[SM06]   N. Sukumar and E.A. Malsch. Recent advances in the construction of polygonal finite element interpolants. *Archives Comput. Methods. Eng.*, 13:129–163, 2006.

[SS95]   P. Schröder and W. Sweldens. Spherical wavelets: Efficiently representing functions on the sphere. *Comput. Graphics*, 29:161–172, 1995.

[ST04]   N. Sukumar and A. Tabarraei. Conforming polygonal finite elements. *Int. J. Numer. Methods Eng.*, 61:2045–2066, 2004.

[SZB+03]   T.W. Sederberg, J. Zheng, A. Bakenov, and A. Nasri. T-splines and T-NURCCs. *ACM Trans. Graph.*, 22:477–484, 2003.

[Tau95]   G. Taubin. A signal processing approach to fair surface design. In *Proc. 22nd Conf. Comput. Graph. Interactive Tech.*, pages 351–358, ACM Press, 1995.

[TC90]   J. Thingvold and E. Cohen. Physical modeling with B-spline surfaces for interactive design and animation. *Comput. Graphics*, 24:129–137, 1990.

[TF88]   D. Terzopoulos and K. Fleischer. Modeling inelastic deformation: Visoelasticity, plasticity, fracture. In *SIGGRAPH Comput. Graph.*, 22:269–278, 1988.

[TPB+87]   D. Terzopoulos, J. Platt, A. Barr, and K. Fleischer. Elastically deformable models. *Comput. Graphics*, 21:205–214, 1987.

[TQ94]   D. Terzopoulos and H. Qin. Dynamic nurbs with geometric constraints for interactive sculpting. *ACM Trans. Graph.*, 13:103–136, 1994.

[Wac11]   E.L. Wachspress. Barycentric coordinates for polytopes. *Comput. Math. Appl.*, 61:3319–3321, 2011.

[Wah90]   G. Wahba. *Spline Models for Observational Data.* CBMS-NSF in Applied Mathematics, SIAM, Philadelphia, 1990.

[War92]   J. Warren. Creating multisided rational Bézier surfaces using base points. *ACM Trans. Graph.*, 11:127–139, 1992.

[War96]   J. Warren. Barycentric coordinates for convex polytopes. *Adv. Comput. Math.*, 6:97–108, 1996.

[WB98]   R. Whitaker and D. Breen. Level set models for the deformation of solid objects. In *Proc. 3rd Int. Eurographics Workshop Implicit Surfaces*, pages 19–35, 1998.

[WBG07]   M. Wicke, M. Botsch, and M. Gross. A finite element method on convex polyhedra. *Comput. Graphics Forum*, 26:355–364, 2007.

[Wei98]   J. Weickert. *Anisotropic Diffusion in Image Processing.* B.G. Teubner, Stuttgart, 1998.

[WFB87]   A. Witkin, K. Fleischer, and A.H. Barr. Energy constraints on parameterized models. In *SIGGRAPH Comput. Graph.*, 21:225–232, 1987.

[WJE00]   R. Westermann, C. Johnson, and T. Ertl. A level-set method for flow visualization. In *Proc. IEEE Visualization*, pages 147–154, 2000.

[WSH+07] J. Warren, S. Schaefer, A.N. Hirani, and M. Desbrun. Barycentric coordinates for convex sets. *Adv. Comput. Math.*, 27:319–338, 2007.

[WW92]   W. Welch and A. Witkin. Variational surface modeling. *Comput. Graph.*, 26:157–166, 1992.

[WW94]   W. Welch and A. Witkin. Free-form shape design using triangulated surfaces. In *Proc. 21st Conf. Comput. Graph. Interactive Tech.*, pages 247–256, 1994.

[WW02]   J. Warren and H. Weimer. *Subdivision Methods for Geometric Design: A Constructive Approach.* Morgan Kaufmann, San Francisco, 2002.

[XB00]   G. Xu and C.L. Bajaj. Regular algebraic curve segments (I)—Definitions and characteristics. *Comput. Aided Geom. Des.*, 17:485–501, 2000.

[XBE01]   G. Xu, C.L. Bajaj, and S. Evans. C^1 modeling with hybrid multiple-sided A-patches. *Internat. J. Found. Comput. Sci.*, 13:261–284, 2001.

[ZOF01]   H.-K. Zhao, S. Osher, and R. Fedkiw. Fast surface reconstruction using the level set method. In *Proc. IEEE Workshop Variational Level Set Methods Comput. Vis.*, 2001.

[ZOM+00] H.K. Zhao, S.J. Osher, B. Merriman, and M. Kang. Implicit and non-parametric shape reconstruction from unorganized points using variational level set method. *Comput. Vision Graph. Image Understanding*, 80:295–319, 2000.

57 SOLID MODELING

Christoph M. Hoffmann and Vadim Shapiro

INTRODUCTION

The objective of solid modeling is to represent, manipulate, and reason about the 3D shape of solid physical objects, by computer.

As an application-oriented field, the scope of solid modeling, as well as the relative emphasis on its parts, change over time. Since its inception, [RV82, Bra75], in the 1970s, the focus on modeling shape alone has continued to expand to integrate widening domains of physical properties. Monographs of the subject appeared in the late 1980s and include [Chi88, Hof89, Män88]. The expansion was and continues to be driven by major applications that include manufacturing, architecture, construction, computer vision, materials science, medicine, biological engineering, graphics, and virtual reality. With the integration of new applications, and with the availability of inexpensive, powerful computing platforms, the relative importance of specific shape representations can change. Along with such shifts, new modeling and analysis techniques are added. As a result, the field draws on very diverse technologies, including numerical analysis, symbolic algebraic computation, approximation theory, point set and algebraic topology, differential geometry, algebraic geometry, and computational geometry.

In this chapter, we first review the major representations of solids in Section 57.1. They include constructive solid geometry, boundary representation, spatial subdivisions of various types, and medial surface representations. With changing scope and focus, different representations enter and exit center stage. For instance, polygonal meshes and voxel-based representations became relatively marginal in the late 1990s, only to gain renewed interest and importance recently with the advent of 3D printing, because they allow representing a structured interior of modeled shapes. Procedural and declarative representations are becoming more popular as geometric programming methods are increasingly used for creating complex solid models and assemblies in architecture and additive manufacturing.

For decades modeled solids were generally assumed to have a homogeneous interior. This fact reflected common manufacturing applications and manufacturing processes that separated the production of raw materials from subsequent processing of those materials into parts, assemblies, etc. With the advent of additive manufacturing, that is, machinery that builds physical objects by additive processes, laying down material layer by layer, there is growing interest in building solids that have a complex interior structure. An efficient and comprehensive representation of heterogeneous interior has not yet emerged. Along with explorations of printer technologies and of what can or should be built there is a wide-ranging, diverse body of research, that we will comment on in Section 57.1.7, tracing this development.

Next, major layers of abstraction in a typical solid modeling system are characterized in Section 57.2. The lowest level of abstraction comprises a substratum of basic service algorithms. At an intermediate level of abstraction, there are algorithms for larger, more conceptual operations. Many functions and operations have been devised over the years and at varying levels of abstraction. Note that this layer can be further sub-structured into a variety of levels. Finally, a yet higher level of abstraction is offered by constraint- and feature-based designs: the constraints and the feature parameters, index instance designs. The resulting generic design, prior to valuating parameters and constraints, defines families of shape instances, and ventures into territory that is only partially mastered, mathematically and computationally.

The rich infrastructure of solid modeling representations and operations can be accessed in several ways:

1. A graphical user interface (GUI) presents the capabilities, usually targeting specific application areas. This is by far the most common type of access.

2. An application programming interface (API) presents the infrastructure much like any other software library. Most systems have an API, but the vendor API does not necessarily expose every system functionality.

3. A current trend in system architecture is a shift toward modularized confederations of plug-compatible functional components. Using a plug-and-play perspective, components with a specific functionality collaborate in the system. This can be done in two ways: (a) design and analysis data is exchanged and/or shared by the system components, or (b) components query each other, so accumulating needed information to do their work.

The first style of plug-and-play requires that the data is understood by the cooperating components. If the components come from different vendors, understanding the exchanged data has to overcome the absence of a mathematically sound semantics of the data. Approach (3a) thus favors systems implemented by the same vendor who may use an idiosyncratic data format but can deliver a consistent interpretation. Approach (3b) has an object-oriented character. Moreover, it takes a page from *Geometric Dimensioning and Tolerancing*; e.g., [Sri03]. This recent work seeks to abstract the role of a solid model in applications, reducing the interface to a set of queries so as to avoid representation translation/interpretation altogether. All this will be discussed in Section 57.3. Open problems are gathered in Section 57.4.

57.1 MAJOR REPRESENTATION SCHEMATA

GLOSSARY

Solid representation: Any representation allowing a deterministic, algorithmic point membership test.

Constructive solid geometry (CSG): The solid is represented as union, intersection, and difference of primitive solids that are positioned in space by rigid-body transformations.

Boundary representation (Brep): The solid surface is represented as a quilt of vertices, edges, and faces.

Mesh representation: A boundary representation whose faces are planar polygons. Topological information may be reduced or implicit.

Spatial subdivision: The solid is decomposed into a set of primitive volumes with nonintersecting interior, for instance voxels.

A solid representation must allow the unambiguous, algorithmic determination of point membership: given any point $p = (x, y, z)$, there must be an algorithm that determines whether the point is inside, outside, or on the surface of the solid; [Req77]. Moreover, restrictions are placed on the topology of the solid and its embedding, excluding, for example, fractal solids.

These restrictions are eminently reasonable. Increasingly, however, solid modeling systems depart from this strict notion of solid and permit representing a mixture of solids, surfaces, curves, and points, for example, in surface modeling in graphics via "particle systems." The additional geometric structures are useful for certain design processes, for interfacing with applications such as meshing solid volumes, and for abstracting solid features, to name a few. More than that, solid models used to partition space into three point sets: points exterior to the solid, points contained in the solid, and points on the boundary, separating inside from outside. This is changing with a growing interest in representing structured solid interiors.

GEOMETRIC COVERAGE

The range and geometric representation of solid surfaces is referred to as *geometric coverage*. Polyhedral modeling restricts to planes. Classical CSG allows only planes, cones, cylinders, spheres, and tori. Experimental modelers have been built allowing arbitrary algebraic halfspaces. SGDL used implicit algebraic surfaces of degree up to 4.

Most commercial and many research modelers use B-splines (uniform or nonuniform, integral or rational) or Bézier surfaces. The properties and algorithmic treatment of these surfaces are studied by computer-aided geometric design. See Chapter 56, as well as the monographs and surveys [Far88, Hos92, HL93].

57.1.1 CONSTRUCTIVE SOLID GEOMETRY

GLOSSARY

Primitive solids: Traditionally: block, sphere, cylinder, cone, and torus. More general primitives are possible.

Sweep: Volume covered by moving a solid or a closed contour in space.

Extrusion: Sweep along a straight line segment.

Revolution: Circular sweep.

Regularized Boolean operation: The closure of the interior of a set-theoretic union, intersection, or difference.

Algebraic halfspace: Points such that $f(x, y, z) \leq 0$ where f is an irreducible polynomial.

Irreducible polynomial: Polynomial that cannot be factored over the complex numbers.

Constructive Solid Geometry is a special case because it is not only a particular representation of solid shapes, but is also a methodology for composing primitive solids to represent complex solid shapes. Visual representation of a CSG solid, moreover, soon was done converting the CSG representation into boundary representation for faster rendering. In this section, we restrict to the representational aspect of CSG.

Classical Constructive Solid Geometry (CSG) represents a solid as a set-theoretic Boolean expression of *primitive* solid objects, of a simpler structure [RV77]. The traditional CSG primitives are block, sphere, cylinder, cone, and torus. The traditional operations are regularized union, intersection, and difference. A regularized set operation is obtained by taking the closure of the interior of the set-theoretic result.

Each solid has a default coordinate system that can be changed with a rigid body transformation. A Boolean operation identifies the two coordinate systems of the solids to be combined and makes it the default coordinate system of the resulting solid.

Primitives, and solids obtained from them with the CSG operations, are represented by expressions that can be conceptualized as expression trees. The leaves are the primitives, the internal tree nodes are rigid-body transformations and regularized Boolean operations.

Both the surface and the interior of the final solid are thereby defined, albeit implicitly. The CSG representation is valid if the primitives are valid.

FIGURE 57.1.1
Left and middle: CSG primitives block(w, d, h) *and* cylinder(r, h) *with default coordinate systems. Right: T-bracket as union of two blocks minus a cylinder.*

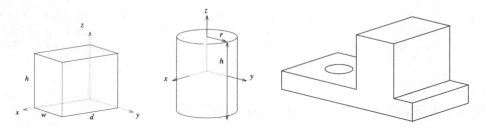

As an example, consider Figure 57.1.1. Using the coordinate system conventions shown, the CSG representation of the bracket is the expression

$$\text{block}(8, 3, 1) \cup^* \text{move}(\text{block}(2, 2.5, 3), (0, 4.5, 1))$$
$$-^* \text{move}(\text{cylinder}(0.75, 1), (1.5, 1.5, -0.5))$$

where the * indicates a regularized operation. (See also Figure 42.4.1.)

Algorithmic infrastructure operations the system can perform include classifying points, curves, and surfaces with respect to a solid; eliminating redundancies in the CSG expressions; and rendering solids visually. A CSG solid can be rendered by ray tracing directly, or by converting the solid to a boundary representation and rendering the resulting data structure.

More general primitives may be added if they support the CSG operations. Examples include the volume covered by sweeping a solid along a space curve, or by sweeping a planar contour bounding an area. Defining a sweep is delicate, requiring many parameters to be exactly defined; for instance consider the operation of blending discussed in Section 57.2.2. However, simple cases are widely used. They include extrusion, i.e., sweep along a straight line; and revolution, i.e., a sweep about an axis. The evaluation of general sweeps can be accomplished by a number of methods.

CSG representations define unambiguously the surface and homogeneous interior of solids, provided the primitives do, a fact that is straightforward. In contrast, boundary representations must satisfy both local and global properties if they are to partition space in a like manner [Wei86].

57.1.2 BOUNDARY REPRESENTATION

In boundary representation (Brep), the surface of a solid is explicitly represented, and the interior is implicitly represented. As before, the interior is assumed to be homogeneous. The solid surface is represented as a quilt of faces, edges, and vertices; [Bra75]. A distinction is drawn between the topological entities, vertex, edge, and face, related to each other by incidence and adjacency, and the geometric location and shape of these entities. See also Figure 57.1.2. For example, when polyhedra are represented, the faces are polygons described geometrically by a face equation plus a description of the polygon boundary.

Face equations are often parametric, with nonuniform rational B-splines the dominant form. They can also be implicit algebraic, perhaps of limited maximum degree. The SGDL modeler used implicit algebraic surface patches of degree 4 or less.

Geometrically, the entities in a Brep are not permitted to intersect anywhere except in edges and vertices that are explicitly represented in the topology data structure. In addition to the classification operations mentioned for CSG, Boolean union, intersection, and difference operations are usually implemented for Brep systems. Both regularized and nonregularized Boolean operations may occur. Early systems maintained Brep and CSG representations in parallel; [RS00].

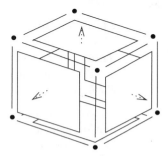

FIGURE 57.1.2
Topological entities of a box. Adjacency and incidence are recorded in Brep. Dotted arrows indicate face orientation.

Different Brep schemata appear in the literature, divided into two major families. One family restricts the solid surfaces to oriented manifolds. Here, every edge is incident to two faces, and every vertex is the apex of a single cone of incident edges and faces. The second family of Brep schemata allows oriented nonmanifolds in which edges are adjacent to an even number of faces. When these faces are ordered radially around the common edge, consecutive face pairs alternatingly bound solid interior and exterior. See Figure 57.1.3 for examples.

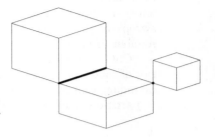

FIGURE 57.1.3
A nonmanifold solid without dangling or interior faces, edges, and vertices; the nonmanifold edges and vertices are drawn with a thicker pen.

More general nonmanifold Breps are used in systems that combine surface modeling with solid modeling [Tak92]. In such representation schemata, a solid may have interior (two-sided) faces, dangling edges, and so on. Solid modelers often integrate surface modeling capabilities.

The topology may be restricted in other ways. For instance, the interior of a face may be required to be homeomorphic to a disk, and edges to have two distinct vertices. In that case, the Brep of a cylinder would have at least four faces, two planar and two curved. This may be desirable because of the geometric surface representation, or may be intended to simplify the algorithms operating on solids.

All such topological representations may be viewed as instances of a chain complex, a concept commonly used in algebraic topology that involves representing a sequence of boundary relationships between the spaces of k-chains constructed over cellular decomposition of a solid. When the boundary operators are represented by sparse incidence matrices, many boundary representations and algorithms reduce to problems in linear algebra [DPS14].

57.1.3 MESH REPRESENTATIONS

GLOSSARY

Mesh boundary representation: A simplified boundary representation in which faces are planar, often polygons with few vertices.

Triangulated surface representation: A boundary mesh representation where all faces are triangles.

STL file format: A triangle mesh specification widely used in 3D printing.

Mesh representations of solids restrict to planar faces. Some topological information is usually given, for instance by using vertex identifiers when defining faces. Which side of a face is to the outside of the solid could be decided using a right-hand rule and restricting to convex polygons, or be based on locally outward pointing vertex normals; see, e.g., [CMS98]. Mesh and octree representations are treated in [BN90, Hof95, Sam89a, Sam89b, TWM85], including the associated conversion problems.

The STL file format, used in 3D printing, restricts to triangles and uses vertex normals; e.g., [RW91]. Vertex coordinates are local to the triangle defined. Consequently no adjacencies are explicit and all topology has to be reconstructed/inferred.

57.1.4 SPATIAL SUBDIVISION REPRESENTATIONS

GLOSSARY

Boundary conforming subdivision: Spatial subdivision of a solid that represents the boundary of the solid exactly or to within a given tolerance.

Boundary approximating subdivision: Spatial subdivision that represents the boundary of the solid only approximately to within the size of the subdivision.

Regular subdivision: A subdivision whose cells are congruent. Grids are regular subdivisions.

Voxel/voxelized subdivision: A regular subdivision whose cells are cubes.

Irregular subdivision: A subdivision with noncongruent cells.

Octree: Recursive selective subdivision of a cuboid volume into eight subcuboids.

Binary space partition (BSP) tree: Recursive irregular subdivision of space, traditionally by halfplanes. See also Sections 33.8.2 and 42.5.

Spatial subdivision decomposes a solid into cells, each with a simple topological structure and often also with a simple geometric structure; [TN87, Mea82]. Subdivision representations are divided into *boundary conforming* and *boundary approximating*.

Important boundary conforming subdivision schemata are finite-element meshes and the *BSP tree*. Mesh representations are used in finite element analysis, a method for solving continuous physical problems. The mesh elements can be geometric tetrahedra, hexahedra, or other simple polyhedra, or they can be deformations of topological polyhedra so that curved boundaries can be approximated exactly. See Sections 29.4–5.

FIGURE 57.1.4
A polygon and a representing BSP tree.

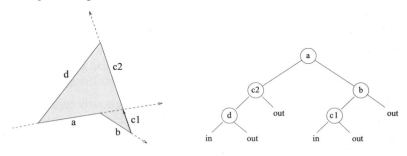

Binary space partition trees are recursive subdivisions of 3-space. Each interior node of the tree separates space into two disjoint point sets. In the simplest case, the root denotes a separator plane. All points of $\mathbb{R}^3$ below or on the plane are represented by one subtree, all points above the plane are represented by the other subtree. The two point sets are recursively subdivided by halfplanes at the subtree nodes. The leaves of the tree represent cells that are labeled IN or OUT. The (half) planes are usually face planes of a polyhedron, and the union of all cells labeled IN is the polyhedron. For an example in $\mathbb{R}^2$ see Figure 57.1.4. Note that algebraic halfspaces can be used as separators, so that curved solids can be represented exactly.

Boundary approximating representations are *grids*, *voxels*, and *octrees*. In grids, space is subdivided in conformity with a coordinate system. For Cartesian coordinates, the division is into hexahedra whose sides are parallel to the coordinate planes. In cylindrical coordinate systems, the division is into concentric sectors, and so on. The grids may be regular or adaptive, and may be used to solve continuous physical problems by differencing schemes. Rectilinear grids that are geometrically deformed can be boundary-conforming. Otherwise, they approximate curved boundaries. Voxels are rectilinear grids, each hexahedron/voxel of equal size.

An octree divides a cube into eight subcubes. Each subcube may be further subdivided recursively. Cubes and their subdivision cubes are labeled white, black, or grey. A grey cube is one that has been subdivided and contains both white and black subcubes. A subcube is black if it is inside the solid to be represented, white if it is outside. In some variants, grey cells describe the contained boundary surface, leading to a boundary-conforming representation [BN90]. Quadtrees, the two-dimensional analogue of octrees, are used in many geographical information systems. See Figure 42.5.1. Some 3D printers rasterize the slices to be printed, using nonconforming spatial subdivisions.

The conversion between boundary representation and CSG can be considered a generalization of the binary space partition tree and is explored in [Hof93b, Nay90, NR95, Sha91b, SV93].

57.1.5 MEDIAL SURFACE REPRESENTATIONS

GLOSSARY

Maximal inscribed ball: Ball inscribed in a domain and not properly contained in another inscribed ball.

Medial surface transformation: Closure of the locus of centers of maximal inscribed spheres, and a function giving the minimum distance to the solid boundary. Usually called the *MAT* for "medial axis transformation."

Procedural representation: The solid is described by a scripting language or a notational schema that is declarative and must be evaluated deductively. Examples include: level sets of a scalar field, R-functions, subdivision surfaces.

The medial axis and medial surface can unambiguously represent two-dimensional domains and 3D solids, respectively [Blu73]. The representations are not widely used for this purpose at this time; more frequently they are used for shape recognition (see Section 54.4). However, as explained below, some meshing algorithms are based on the medial axis and the medial surface [FAR16].

The medial axis of a two-dimensional domain is defined as the closure of the locus of centers of maximal disks inscribed within the domain. A disk is maximal if no other disk properly contains it. An example is shown in Figure 57.1.5 along with some maximal disks.

The medial surface of a solid is the closure of the locus of centers of maximal inscribed spheres. When we know the radius (the limit radius in case of closure points) of the corresponding sphere for each point on the medial surface, then an

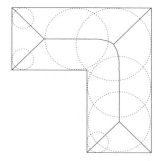

FIGURE 57.1.5
L-shaped domain and associated medial axis. Some maximal in-scribed circles contributing to the medial axis are shown.

unambiguous solid representation is obtained that is sometimes called the medial axis transform (MAT). The MAT is the deformation retract of a solid and has a number of intriguing mathematical properties. For example, by enlarging the radius values by a constant, the MAT of a dilatation of the solid is obtained.

Inverting, we can construct tubular surfaces as envelope of a family of spheres centered on (a segment of) a space curve. The torus is a simple example, the Dupin cyclides a more complex one. Generalized cylinders have been used in so-called skeleton models. They also play a role in geometric dimensioning and tolerancing (GD&T).

Early on, solid modeling has investigated the MAT for the purpose of constructing shell solids (obtained by subtracting a small inset), for organizing finite element meshing algorithms, and for recognizing form features; e.g., [Hof92, SAR95, Ver94]. More recently, the role of the MAT in surface reconstruction has begun to impact solid modeling; see Chapter 35. Surface reconstruction arises in solid modeling for its application in reverse engineering where a model is to be constructed from a physical object by an automated measuring strategy.

57.1.6 CONSTRUCTIVE REPRESENTATIONS

Early on, solid models were constructed by extending a programming language with special operations for creating primitive solids, arranging them with respect to a coordinate system, and combining them. The latter used, for the most part, regularized Boolean operations. This approach created a scripting language that is effectively a constructive representation. For example, the PADL system and its variants used Fortran as script language to specify solids. CSG expressions and directives were embedded into the Fortran program. The solid was then evaluated into an internal format, usually CSG expressions plus a Brep evaluation for efficient rendering. Other modeling systems at the time used other programming languages. Since such scripting languages were based on general, procedural programming languages, the solid evaluation can be highly complex and may include any deterministic, step-by-step computation. Procedural script languages include Fortran for PADL [Bro82], Lisp for Alpha_1 [GDC94], and Scheme for SGDL [Sys01].

But the same constructive representation can also be declarative, describing requirements and properties of the complex solid model, without committing to a specific evaluation sequence. For example, instead of procedurally executing CSG operations, CSG expressions may be used as implicit representations [BB97] specifying which subsets of space (points, line segments, faces, voxels, etc.) are contained within the represented set. Specific algorithms for evaluating such sub-sets vary widely with applications and include polygonization, rendering, and mass property computations [Rot82, Til80]. Another example of declarative modeling is *variational geometric constraint solving* discussed in Section 57.2.5.

Constructive representations (both procedural and declarative) have experi-enced a resurgence recently, fueled by the need and desire to create complex solid models in architecture and in additive manufacturing. This resurgence was made possible by a rich repertoire of advanced constructions, and their compositions, including sweeps, offsets, projections, Minkowski operations, blends, free-form de-formations, and interpolations [PASS95, WGG99, PPV+03, MUSA15]. The se-mantics of these constructions is usually hidden from the user (see Section 57.3.1) and may be specified in terms of set-theoretic operations, affine and smooth trans-formations, polynomial constraints, R-functions [Sha07], and B-splines to name a few. As described in Section 57.3.4, these constructions are usually evaluated into Breps, mesh representations, or spatial discretizations for further processing and downstream applications.

57.1.7 MODELS OF INTERIOR

With few exceptions, solid models historically described a division of space into three point sets: the interior of the solid, its surface, and the exterior of the solid. Neither exterior nor interior were considered to have further structure. This con-ceptualization reflects the early focus of solid modeling on engineering applications for rigid objects with homogeneous interior such as those that were manufactured by extant manufacturing techniques of that time, including CNC machining and other unit manufacturing processes [UMP95]. The interior's homogeneous material properties were abstracted by appropriate label and material constants describing how the material behaves under various physical conditions (structural, thermal, electrical, etc.)

Industry's adaptation of solid modeling progressed quickly to more complex ap-plications requiring models and representations of solids with *discrete,* piecewise ho-mogeneous interiors partitioned by "interior boundaries," for example in modeling VLSI layouts and Micro-Electro-Mechanical Systems (MEMS). Rapid generaliza-tion of traditional solid modeling representations ensued, as surveyed, for example in [Tak92]. Notably, Brisson [Bri93] observed that most of such representations are essentially spatial subdivision representations based on cellular stratifications of d-dimensional manifolds. Generalizations of both boundary representations [RO89] and CSG representations [RR91] have been proposed for modeling of such cellular structures.

However, many natural and engineered objects exhibit *continuous* spatial vari-ability of material properties. Such properties include density of bone and other natural tissues, porosity of mineral materials, volume fraction and anisotropy in functionally graded materials (FGM), and many other properties of natural and engineered materials. The continuous properties are naturally abstracted by fields: functions of spatial coordinates defined over a spatial decomposition of a solid model. Such fields may be represented in a piecewise continuous fashion over de-compositions of solids [KD97] or as discrete distributions (algebraic topological chains defined over cell complexes) [PS93]. The need to model complex spatially varying fields over decomposed solids gave prominence to a subfield of *heterogeneous object modeling,* which conceptually is formalized in terms of fiber bundles where the spatial decomposition of the solid serves as atlas and collections of material attributes are represented by sections of the bundles [Zag97, KBDH99].

A key technical issue that dominates heterogeneous object modeling is conceptualization, representation, and control of continuous property fields, as surveyed in [KT07]. In a typical modeling scenario, values of material property of interest are specified by a user or application at certain locations in the solid considered *material features*. From these features, the field values at any location within the solid must be automatically determined, subject to gradient constraints, experimental data, as well as empirical and physical laws. As discussed in [BST04], solving such problems requires combining interpolation methods (most commonly weighted by powers of distances to material features) and numerical solvers of boundary value problems. Hence heterogeneous modeling methods may be categorized based on the methods used for interpolating property values and gradients, and based on the spatial decomposition used to evaluate the solution. Popular choices for the latter include cell complexes [KD97, AKK+02], conforming and nonconforming finite element meshes [JLP+99, BST04] and voxelizations of solid [DKP+92, CT00].

With the advent of *additive manufacturing,* also referred to as *3D printing,* the need for modeling solids with heterogeneous interior has become obvious and growing, because heterogeneity is capable of describing composites, active structures, engineered materials, and so on. Proposed representations vary widely and are, in many cases, closely focused on specific applications, with voxel representations gaining in popularity as modern 3D printers are capable of depositing unique material at spatial resolutions of a few microns. Project Maxwell, e.g., [DKP+92], included efforts to develop heterogeneous manufacturing paradigms. This, and more recent projects, proposed to employ shape and topology optimization techniques on the voxel-represented objects to be printed in order to realize functional properties via geometrical representations.

However mechanical and physical functions of solids are usually designed and controlled at a much coarser scale, suggesting that it may be beneficial to aggregate the individual voxels into local *unit cells* serving particular purpose. Such unit cells may be selected from a great variety of periodic and stochastic tessellations, or produced automatically by shape and topology optimization algorithms [Ros07]. Other benefits of using unit cells in design of materials and tissue engineering [Hol05, SSND05] include increased material and manufacturing efficiency, the ability to mimic natural and synthetic materials (for example, based on Voronoi diagrams), as well as creating and fine tuning new custom meta-materials with unusual properties, such as auxetic materials, medical scaffolds, and lightweight composite structures.

The great variety of unit cells and their combinations lead to the need for modeling and representing graded, anisotropic, periodic and stochastic lattice structures that can fill the interior of a solid. This is an active area of research with competing proposals to extend Voronoi and mesh representations [YBSH16], boundary representations [WCR05], trivariate parameterizations [Elb15], implicit periodic representations [FVP13], and stochastically generated structures [LS15], as well as their combinations [KT10, LS16]. Such lattices are not, however, boundary-conforming and additional techniques are needed to blend them with classical representations of the boundary. In medical imaging applications, organ boundaries may be approximated, resulting in a nonmanifold boundary representation whose cells represent specific organs or parts thereof. Finally, we note that cells in a lattice themselves could be further structured, leading to the notion of multi-scale solid modeling.

The extraordinary diversity of research into the technologies and applications of 3D printing presages that there will be a great variety of representations of the interior, and that calls for universal standards [KT07] may not come to pass. Nevertheless, a cell-structured interior, based on Voronoi tessellations, is a popular choice for material performance models, since it can generate realistic homogeneous and heterogeneous structures in both 2D and 3D; e.g., material performance sound absorption [JDKK07], polycrystalline materials [GLM96], thermal insulation [MY14], and thermo-elastic behavior of functionally graded materials [Bin01]. Other representations of cell structures employ medial surfaces to model organ boundaries; e.g., [NSK+97], or derive smooth boundaries by approximating (polyhedral) boundaries of cells or aggregations of cells with B-splines, [KT07]. The representation of fruits in [MVH+08, AVH+13] uses Voronoi cells whose vertices have been snubbed to model internal passages for respiration.

Voxel-based representations are basically a sampling of the internal structure. Uniform sampling has to contend with unfavorable scaling behavior [CMP95]. This applies as well to procedural voxel representations where the voxels are constructed on the fly; e.g., [GQD13, Oxm11, DTD+15]. Nonuniform samplings, on the other hand, argue for an explicit cell representation where different cells would be sampled at different densities. Khoda and Koc [KK13] propose a layered tissue representation where each slice to be printed has a preferred grain direction. Slices vary angle and density of the extruded filament.

57.2 LEVELS OF ABSTRACTION

GLOSSARY

Substratum: Basic computational primitives of a solid modeler, such as incidence tests, matrix algorithms, etc.

Algorithmic infrastructure: Major algorithms implementing conceptual operations, such as surface intersection, edge blending, etc.

Graphical user interface (GUI):   Visual presentation of the functionality of the system.

Application programming interface (API):   Presentation of system functionality in terms of methods and routines that can be included in user programs.

Substratum problem: Unreliability of logical decisions based on floating-point computations.

Large software systems should be structured into layers of abstraction. Doing so simplifies the implementation effort because the higher levels of abstraction can be compactly programmed in terms of the functionality of the lower levels. Thereby, the complexity of the system is reduced. A solid modeling system spans several levels of abstraction:

1. On the lowest level, there is the substratum of arithmetic and symbolic computations that are used as primitives by the algorithmic infrastructure. This level contains point and vector manipulation routines, incidence tests, and so on.

2. Next, there is an intermediate level comprising the algorithmic infrastructure. This level implements the conceptual operations available in the user interface, as well as a wide range of auxiliary tools needed by these operations. There is often an application programming interface (API) available with which programs can be written that use the algorithmic infrastructure of the modeling system.

3. Designing families of shapes is at a higher level yet, through the use of features and constraints. Features and constraints are defined using feature parameters and geometric and dimensional constraints. An instance design so prepared can be changed to another instance design by changing parameters and/or constraints — which triggers a re-evaluation of the design by the system in accordance with these changes. This style of shape design can be conceptualized as a generic design of a family of specific designs or instance designs.

Ideally, the levels of abstraction should be kept separate, with the higher levels leveraging the functionality of the lower levels. However, this separation is fundamentally limited by the interaction of numeric and symbolic computation. As it were, the substratum is not fully dependable.

More than that, there is no complete semantic characterization of the generic design process at the third level, since the re-evaluation of the changed parameters and constraints requires an understanding of how specific shape elements (vertices, edges, faces) correspond, between different family members, and what it means when such a correspondence does not exist or has ambiguities.

The hierarchy of tools and concepts can be used in a variety of ways, creating different views of the capabilities:

1. Traditionally, a graphical user interface (GUI) presents to the user a view of the functional capabilities of the system. Interaction with the GUI exercises these functions, for instance, for solid design. Tools for editing and archiving solids are included.

2. Another way to access is by an application programming interface (API). An API exposes functionality of the system much as a software library would. Most common is access to algorithmic infrastructure, but APIs can also expose GUI functionality, giving users the tools to customize the GUI for their purposes.

3. A third style of accessing system capabilities is through a query interface. Here, a client–server architecture can be constructed that allows various systems to collaborate using a message-passing paradigm.

57.2.1 THE SUBSTRATUM

The substratum consists of many low-level computations and tests; for example, matrix computations, simple incidence tests, and computations for ordering points along a simple curve in space. Ideally, these operations create an abstract machine whose functionality simplifies the algorithms at the intermediate level of abstraction. But it turns out that this abstract machine is unreliable in a subtle way when implemented using floating-point arithmetic. Exact arithmetic would remedy this unreliability, but is held by many to be unacceptably inefficient when dealing with solids that have curved boundaries. See Section 45.4. Problems include input accuracy as well as tolerances used in deciding incidence and other numerical predicates.

A root problem is the dichotomy between approximate numerical floating-point computation and exact topological incidence decisions based on it; [Hof89][Chap. 4]. It is therefore no surprise that systems built on a substratum that only does exact computation avoid this unreliability and do not experience consequent failures. However, exact arithmetic is not efficient unless geometric coverage is very limited. An interesting variant was proposed in [For97], where for polyhedral solids the needed predicates are evaluated exactly only when the floating-point computation cannot be guaranteed to be correct. Although the problem was observed early-on, its analysis appeared only in the late 1980s; see [Hof01a]. The discovery motivated work on the use of exact arithmetic in polyhedral modeling [SI89, For97], as well as for curved surfaces [KKM99a, KKM99b].

Another unavoidable problem arises in systems that use parametric curves and surfaces as shape elements. Here, the representation of two parametric surfaces is usually recorded as a parametric curve. Unfortunately, in general the intersection of two parametric surfaces is not a parametric curve and must be approximated; e.g., [Hos92, PM02]. It is then difficult to interpret such shape constructs correctly.

Binary arithmetic of fixed precision cannot always encode decimal input of fixed precision. An example is 1/10 which requires one decimal but is a periodic binary fraction. For more detail on this subject see Chapter 45.

Work in symbolic algebraic computation (Chapter 37) has foundational importance, for instance in regard to converting between surface representations. Some of the applications of symbolic computation are explored in [BCK88, Cho88, Hof90].

57.2.2 ALGORITHMIC INFRASTRUCTURE

Algorithmic infrastructure is a prominent research subject in solid modeling. Among the many questions addressed is the development of efficient and robust algorithms for carrying out the geometric computations that arise in solid modeling. The problems include point/curve/surface-solid classification [Til80], computing the intersection of two solids, determining the intersection of two surfaces [PM02], interpolating smooth surfaces to eliminate sharp edges on solids, and many more.

Specific algorithms often require specific data structures and representations. Together, algorithms and representations develop by co-evolution that occasionally is disrupted. Past disruptions include the design style of Pro/Engineer that compelled other systems to adopt a similar design vocabulary. It remains to be seen how technologies such as 3D printing and concepts such as query-based system architectures impact the development of the algorithmic infrastructure and its representations. The examples that are described primarily rely on Brep as system representation.

Academic work considers structuring *application programming interfaces* (APIs) that encapsulate the functional capabilities of solid modelers so they can be used in other programs; [ABC+00]. Such APIs play a prominent role in applications because they allow building on existing software functionality and constructing different abstraction hierarchies than the one implemented by a full-service solid modeling system. The work attempts to give a system-independent specification of basic API functionality for solid modeling.

An important consideration when devising infrastructure is that the algorithms are often used by other programs, whether or not there is an API. Therefore, they must be ultra-reliable and in most cases must not require user intervention for exceptional situations. This requirement becomes crucial for more recent work seeking to integrate multiple systems, each addressing different aspects of an overarching application. For instance, one subsystem would be used to design an artifact X, a second analyzes thermal properties of X, a third subsystem analyzes tolerance requirement for given manufacturing processes, and so on. Here, the communication between subsystems raises critical issues, see Section 57.4. Restricting to shape design and manipulations, some of the researched design operations, their implementation, and their presentation in the GUI include the following.

Surface intersection.   Given two bounded areas of two surfaces, determine all intersection curve components. All components of the intersection of the two patches must be correctly identified, including isolated points and singularities. Since this computation is done in $\mathbb{R}^3$, classical algebraic geometry is of limited help. A solution must negotiate well the unreliability of the substratum; see also [PM02].

Offsetting.   Given a surface, its ***offset*** is the set of all points that have fixed minimum distance from the surface. Offsets can have self-intersections that must be culled. There is a technical relationship between offsetting and forming the MAT. Namely, when offsetting a curve or surface by a fixed distance, the self-intersections must lie on the medial axis. Offsetting can be used to determine certain blending surfaces, and is also useful for *shelling* that creates thin-walled solids. Offsetting and shelling are global solid operations, considered in [BW89, For95, PS95, RSB96].

Blending.   Given two intersecting surfaces, a third surface is interpolated between them to smooth the intersection edge. A simple example is shown in Figure 57.2.1. A locally convex blend surface is often called a *round*, and a locally concave one a *fillet*. The blend surface in Figure 57.2.1 is a fillet.

Blending has been considered almost since the beginning of solid modeling, and some intuitive and interesting techniques have been developed over the years. For example, consider blending two primary surfaces f and g. Roll a ball of fixed radius r along the intersection such that it maintains contact with both f and g. Then the surface of the volume swept by the ball can be used as a blending surface, suitably trimmed. Note that the center of the ball lies on the intersection of the offsets, by

FIGURE 57.2.1
Left: two cylinders intersecting in a closed edge. Right: edge blended with a constant-radius, rolling-ball blend; the bounding curves of the blend are shown.

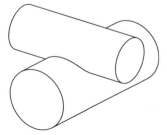 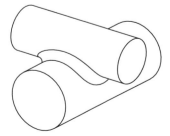

r, of both f and g. In more complicated schemes the radius of the ball is varied along the intersection.

Blends can interfere: Figure 57.2.2 shows the problem of overlapping blends. The fillet and round constructed separately do not meet in the region of overlap. Possible resolutions can be proposed but are difficult to systematize. Multiple surfaces have to be composed and shown to have appropriate global characteristics, primary surfaces have to be cut back appropriately.

When the primary surfaces meet at a vertex tangentially, blending surfaces must "dissipate." Figure 57.2.3 shows several methods to dissipate round and fillet at the end vertices. The examples are from [Bra97] and illustrate the dimensions of the global problem.

Deformations.   Given a solid body, deform it locally or globally. The deformation could be required to obey constraints such as preserving volume or optimizing physical constraints. For example, we could deform the basic shape of a ship hull to minimize drag in fluids of various viscosities.

Shelling.   Given a solid, hollow out the volume so that a thin-wall solid shape remains whose outer surface is part of the boundary of the input solid. The wall thickness is a parameter of the operation. Variations include designating parts of the solid surface as "open." For instance, taking a solid cylinder and designating

FIGURE 57.2.2
Global blend interference [Bra97]: The round of the front edge overlaps with the fillet of the cylinder edge on top (left). Without further action, the two blends do not connect, leaving a gap in the surface. The solution shown in the middle modifies the front round. Other possibilities include modifying the fillet or inserting a separate blend in the overlap region (right).

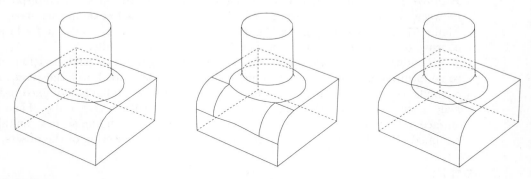

FIGURE 57.2.3
Global blend interference [Bra97]: At ending vertex, the round and the fillet must be merged into a compatible structure. Several solutions are illustrated.

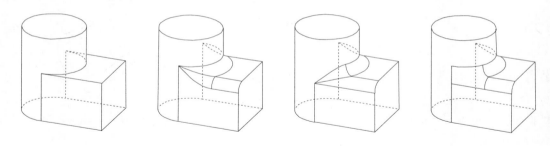

both flat end faces as open the operation creates a hollow tube of the same outside diameter. Conceptually, the operation subtracts an inset of the solid, obtained by shrinking the original solid, an offset operation.

57.2.3 FEATURES AND CONSTRAINTS

GLOSSARY

Form feature: Any stereotypical shape detail that has application significance.

Geometric constraint: Prescribed distance, angle, collinearity, concentricity, etc.

Generic design: Solid design with constraints and parameters without regard to specific values.

Design instance: Resulting solid after substituting specific values for parameters and constraints.

Parametric constraint solving: Solving a system of nonlinear equations, arising in geometric constraint solving, that has a fixed triangular structure.

Variational constraint solving: Solving a system of nonlinear simultaneous equations arising in geometric constraint solving.

In solid modeling, two design paradigms have become standard for manufacturing applications, *feature-based design* and *constraint-based design*. The paradigms expose a need to reconsider solid representations at a different level of abstraction.

The representations reviewed before are for individual, specific solids. However, we need to represent entire *classes* of solids, comprising a generic design. Roughly speaking, solids in a class are built structurally in the same way, from complex shape primitives, and are instantiated subject to constraints that interrelate specific shape elements and parameters. How these families should be defined precisely, how each generic design should be represented, and how designs should be edited are all important research issues of considerable depth.

Neither features nor constraints are new concepts, so there is a sizable literature on both. The confluence of the two issues in solid modeling systems, however, is new and raises a number of questions that have only more recently been articulated and addressed. [SHL92, KRU94] discuss feature work. Constraints are the subject of [BFH+95, HV94, Kra92]. The confluence of the two strands and some of the implications are discussed in [HJ92]. Some of the technical issues that must be addressed are explained in [Hof93a, CH95a], and there is more work emerging on this subject. In particular, Shapiro and Raghothama propose several criteria for defining a family of solids; [RS02a, RS98].

57.2.4 FEATURE-BASED DESIGN

Feature-based design is usually understood to mean designing with shape elements such as slots, holes, pockets, etc., that have significance to manufacturing applications relating to function, manufacturing process, performance, cost, and so on. Focusing on shape primarily, we can conceptualize solid design in terms of three classes of features: generative, modifying, and referencing features; [CCH94, CH95a]. A feature is added to an existing design using attachment attributes and placement conditions. Subsequent editing may change both types of attachment information.

As an example, consider the solid shown to the right in Figure 57.2.4. A hole was added to the design on the left, and this could be specified by giving the diameter of the hole, placing its cross section, a circle, on the side face, and requiring that the hole extend to the next face. Should the slot at which the hole ends be moved or altered by subsequent editing, then the hole would automatically be adjusted to the required extent.

FIGURE 57.2.4

Left: Solid block with a profiled slot. Right: After adding a hole with the attribute "through next face," an edited solid is obtained. If the slot is moved later, the hole will adjust automatically.

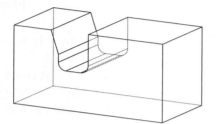 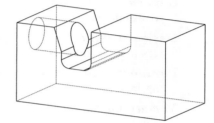

57.2.5 CONSTRAINT-BASED DESIGN

Constraint-based design refers to specifying shape with the help of constraints, when placing features or when defining shape parameters [Owe91, BFH$^+$95]. For instance, assume that we are to design a cross section for use in specifying a solid of revolution. A rough topological sketch is prepared (Figure 57.2.5, left), annotated with constraints, and instantiated to a sketch that satisfies the constraints exactly (Figure 57.2.5, right). Auxiliary geometric structures can be added, such as an axis of rotation. There is an extensive literature on constraint solving, from a variety of perspectives. Surveys with a solid modeling perspective include, e.g., [HJA05, FHJA16].

FIGURE 57.2.5

Geometric constraint solving. Input to the constraint solver shown on the left. Here, the arc should be tangent to the adjacent segments, and the two other segments should be perpendicular. Output of the constraint solver shown on the right.

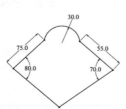 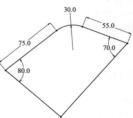

Most solid modeling systems use both features and constraints in the design interface. The constraints on cross sections and other two-dimensional structures are usually unordered, but the constraints on 3D geometry may restrict to a fixed sequence. Solving systems of unordered constraints is also referred to as *variational constraint solving*. Mathematically, it is equivalent to solving a system of nonlinear simultaneous equations. Solving constraints in a fixed sequence is also known as *parametric constraint solving*. The latter is equivalent to solving a system of nonlinear equations that has a triangular structure where each equation introduces a new variable.

A *well-constrained* geometric constraint problem corresponds naturally to a system of nonlinear algebraic equations with a finite set of solutions. In general, there will be several solutions of a single, well-constrained geometric problem. An example is shown in Figure 57.2.6. This raises the question of exactly how a constraint solver should select one of those solutions efficiently, and why.

FIGURE 57.2.6
The well-constrained geometric problem of placing 4 points by 5 distances has two distinct solutions.

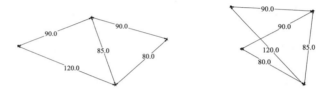

From symbolic computation we know that there are algorithms to convert a system of nonlinear equations that is not triangular, into an equivalent, triangular system. The distinction between parametric and variational constraint solving is therefore artificial in theory. However, full-scale triangularization of systems of nonlinear equations is not tractable in many cases, so the distinction is relevant in practice.

Many constraint solvers proceed in two major phases; see [Owe91, BFH+95, HLS01b, HLS01a].

In the first phase, the constraint problem is abstracted into a graph whose vertices represent the geometric primitives to be placed, and whose edges represent the given constraints. Vertices are labeled by the number of independent coordinates needed to position the represented primitive in a coordinate system. Edges are labeled by the number of independent equations needed to express the represented constraint. The constraint solver analyzes this constraint graph seeking to determine a set of small subproblems that can be solved separately.

In the second phase, the solver processes the subproblems, found by the first phase, solving the associated equations and combining their solutions, thereby determining the solution of the original constraint problem.

In principle, the graph patterns for constraint problems in 2D can be of arbitrary complexity. However, when restricting to problems involving only points, lines and circles, a small set of patterns suffices for the first phase. They can be found with straightforward graph algorithms. Moreover, a small set of simple equations arises in the second phase: when the radius of the circles is given, only univariate quadratic equations must be solved in the second phase. [Owe91].

Spatial constraint solving is very much more demanding than planar constraint solving. In particular, the subsystems of equations that must be solved in the second solver phase ramp up steeply. Spatial constraint problems appear in two different species in the literature.

1. The geometric elements connected by constraints are points, lines and planes in 3-space. The constraints are distance, angle, perpendicularity, parallel, etc. See, e.g., [HV94, GHY04].

2. The geometric elements connected by constraints are as before but they are elements of a solid. The constraints are as before. See, e.g., [JS13]

From an application perspective the second variety appears to be of greater significance. See also Section 57.4.

57.3 ARCHITECTURES AND SYSTEMS

Today, solid modeling technology is at the core of most scientific, engineering, and consumer applications that require unambiguous representation of shape information. The methods for utilizing the rich infrastructure of solid modeling may be divided into three broad categories:

1. application-specific user interfaces that allow maximum leveraging of the constraint-based and feature-based layers;

2. application programming interfaces (API) that support packaging of selective solid modeling capabilities and components, much like any other software library; and

3. plug-compatible functional components to support the current trend towards modularized and distributed system architectures.

Both efficacy and limitations of these methods are often determined by their ability to use more than one representation at the same time and convert between them if necessary.

57.3.1 USER INTERFACES

Ultimately, the functional capabilities of a solid modeling system have to be presented to a user, typically through a *graphical user interface* (GUI). It would be a mistake to dismiss GUI design as a simple exercise. If the GUI merely presents the functionality of the infrastructure literally, an opportunity for operational leveraging has been lost. Instead, the GUI should conceptualize the functionalities an application needs. As in programming language design, this conceptual view can be convenient or inconvenient for a particular application. Research on GUIs therefore is largely done with a particular application area in mind.

   User interfaces are intimately tied to specific solid modeling representations supported at an appropriate level of abstraction. For example, in mechanical engineering product design, an important aspect of the GUI might be to allow the user to specify the shape conveniently and precisely. Early solid modeling systems relied heavily on parametric constructive and procedural representations to express the intended shape which were usually converted to other representations for visualization and downstream applications such finite element meshing and manufacturing process planning [RV83]. Common target representations included boundary representations and various spatial discretizations.

As solid modeling systems gained acceptance in industry, user interfaces began to require direct references to boundary representations to support sketching, reference datums, and direct surface manipulations needed to support advanced automotive and aerospace applications. With time, advanced interfaces emerged to support mechanical design in terms of constraints (on distances, radii, angles, and other dimensions and parameters) and application-specific features (sheet metal, NC machining, structural, molding, etc.) that would combine elements of constructive and boundary representations [CH95b]. To support such user interfaces, the algorithmic infrastructure must be capable of supporting and maintaining consistency of both types of representations within a single system [RS00].

In interfaces for virtual environment definition and navigation, on the other hand, approximate constraints and direct manipulation interfaces would be better. More recently, widespread use of solid modeling in architecture, industrial design, 3D printing, and many consumer applications witnessed renewed popularity of procedural and declarative specifications that are presented to users as functional programming compilers [PPV$^+$03], visual programming languages [M$^+$10], cloud-hosted web browser interfaces, and open source systems [KW14]. Increased popularity of meshes and voxels led to interfaces designed specifically for the creation and editing of such representations and algorithms for many emerging applications, including scanning and shape reconstruction, geometric signal processing, and 3D printing [CCC$^+$08, SS10, DTD$^+$15].

57.3.2 APPLICATION PROGRAMMING INTERFACES (API)

The widespread demand for solid modeling technology in many different applications and industries led to componentization and packaging of different layers into standalone libraries that are accessible via Application Programming Interfaces (APIs). The most popular packages are solid modeling kernels that comprise the substratum and algorithmic layers for boundary representations supporting low-degree polynomial and spline surfaces, notably Parasolid, ACIS, openNURBS, Open Cascade and others. The constraint management layer is usually available as a separate component, for example from D-Cubed [Hof01b]. Modeling kernels based on other representation schemes are also available, but are less popular, e.g., PADL-2 (based on CSG and quadric surface Breps) and Hyperfun (based on declarative implicit functional representations). Solid modeling capabilities are also being exposed in APIs of more general purpose systems that aim at broader computational arenas, notably in computational geometry systems such as CGAL [FP09], visualization toolkits (VTK), and mesh processing systems such as Meshlab, to name a few.

While APIs have been the primary means for integrating solid modeling technology in thousands of applications, it is important to note that APIs are usually fine tuned to the representation and algorithmic infrastructure they encapsulate. As such, they do not cleanly encapsulate the internal implementation details: These APIs and are not interchangeable in that replacing one solid modeling component with another is usually difficult or impossible). Moreover, APIs are not necessarily interoperable when they are based on incompatible mathematical models and representations. [ABC+00] describes a significant effort to design a representation-independent API for solid modeling. The approach is based on the idea that most representations, including CSG, boundary, and spatial discretizations, may be put into a canonical stratified form similar to a cell complex where individual strata are (primitive) sign-invariant k-manifolds in the decomposition of shape [Sha91b]. The results of the effort remained largely academic because such stratifications are nontrivial to compute, lead to excessive fragmentation of geometric information, and are rarely used in practice.

57.3.3 MODEL EXCHANGE AND PLUG-COMPATIBILITY

Given the great diversity of modeling representation schemes, standalone systems, and APIs, the need for interchanging, sharing, and combining solid models from different sources has become greater than ever. To meet this need, there have been, and continue to be, many efforts to find standards for exchanging shape models. These efforts began already in the early days of solid modeling research; [RV82, Fre96]. Abstractly, the problem may be stated as follows:

> Let M be a solid model authored in system C. In order to use M in another (modeling) system C', M has to be converted into an "equivalent" model M' in the native representation of system C'. The model M' would then be used as if it had been authored by C'.

This informal formulation masks the fundamental challenge underlying the task: the notion of equivalence assumes the existence of a rigorous model semantics that is common to both systems and is adhered to when authoring and/or converting the two models. Unfortunately, this is often not the case. Conceptually, there are three plausible approaches to achieve flawless model interoperability and exchange as discussed below: (1) exchange of generic models, (2) exchange of representations of model instances, and (3) query-supported interchangeability of models.

The simplest and the most elegant method to solve the above problem is to treat M and M' as instances of a common generic model corresponding to some specific valuation of parameters and dimensional constraints. The generic model may then be construed as a procedure for generating model M in any system C which is capable of evaluating this procedure. This approach has been advocated for generic Erep models [HJ92], and as a method for interoperability between major commercial CAD systems [SR04]. Because generic models are not fully instantiated, they tend to be compact and contain mostly symbolic information, thereby supporting efficient and robust solutions to the interchange problem. The problem with this approach is the lack of standard semantics and lack of consistent support for generic models in different systems [Sha91a, KPIS08]. Thus, a "slot" feature may well be instantiated differently in systems C and C'. Initial efforts towards standardization of two-dimensional procedural definitions is reported in [KMHP11]. Formal semantics is known for many constructive representations, such as CSG, but they can express only a small subset of widely used generic models.

The second approach attempts to solve the problem directly by exchanging instantiated and fully evaluated representations of M and M'. The approach assumes that the two representations can in fact be converted into each other, either exactly or approximately, but well enough to be considered equivalent in some sense. The representations are imported (exported) in representation-specific file format. One way to avoid a quadratic number of such convertors is to convert M into a neutral format as model M_N that can also be read by C' and that C' can convert to M'. By the far the most popular neutral exchange format is STEP [Kem99] which covers many flavors of boundary representations and some basic constructive representations. Unfortunately, these standardization and data translation efforts continue to have limited success. A percentage of exchanges result in models M' that must be manually repaired subsequently.

The difficulty realizing the first two scenarios is due in part to mathematical facts. It is also in part due to the fact that the systems C and C' may interpret the same data items differently. Consider a system C that models curved surfaces as trimmed nonuniform, rational B-splines (NURBs) and can evaluate them to an accuracy of 10^{-9}, and a system C' that models faceted surfaces only and can evaluate them to an accuracy of 10^{-6}. The evaluation accuracy could be absolute, in system C, but relative in system C'. Moreover, the trim curves of NURB faces in system C can only be approximated in the NURB framework, necessitating algorithms that make sophisticated interpretations when evaluating the model M, based on assumptions which may not be represented in M_N. It is natural then to look for a notion by which a model M, authored in C, is *almost equal* to a model M' authored in C', based on some specific metric. But, again, the metrics concepts fail usually for lack of transitivity.

The third approach to the exchange problem is inspired by the notion of interchangeable parts in mechanical assembly [HSS14] and works as follows. Instead of ascertaining that two part models are *almost the same*, a relation that is not transitive, one ascertains that two part models M and M' are each within allowable tolerance of the reference model M_0. This induces a relation between M and M' that is transitive. Note that the models are never compared to each other, and the common formal semantics is embodied into the measurement procedure that is performed according to the *Geometric Dimensioning and Tolerancing* (GD&T) standard [ASME09]. An analogous approach for solid modeling suggests that different systems C and C' should exchange not models but formally defined and standardized *queries*. The closeness of a given model M to a reference model M_0 is established by queries of M and M_0. Queries are simple geometric predicates. They request the authoring system C to query a model M, thereby avoiding the translation problems characterized above. Examples of standard queries include point-membership classification (PMC) where a query point p is classified as being *in* the interior of the modeled solid, *on* the surface, or *outside* of the solid; distance from a point to the boundary of a solid. A line-solid classification intersects a query line with a solid, and segments the line accordingly. Such classification queries are analogous to testing a physical part with a coordinate measuring machine. In computer science terms, a solid model M is encapsulated as an object model with a query-based interface that, in contrast to APIs, does not expose the internal details of representation or implementations in the authoring system C. The authoring system interprets the solid model, M, thereby actualizing all inferences and particular interpretations of M, whether explicit or implicit in the model.

Queries can be used to make CAD systems interoperate, as well as exchange partial or complete model information [HSS14]. They have also been used to integrate solid modeling systems with engineering analysis systems [FST06, FST11].

57.3.4 REPRESENTATION CONVERSIONS

Representation conversions have influenced the architecture and evolution of solid modeling systems from early days in several important ways [RV83]. They are needed when interchanging solid data between systems that use different native representations. They are also used when certain operations require them, either within the same system or for interoperability between different systems. For instance, the conversion from CSG to Brep supports rendering the solid shapes, using algorithms that are optimized for rendering triangles. Rendering CSG-represented solids directly, by ray casting for example, would be much less efficient. Meshing is another common example where Brep is converted into a finite element mesh used for engineering analysis. The problem may be formulated as follows:

A solid model M is given in a representation X. We seek to represent M in another representation Y, either exactly or approximately.

The statement assumes that M is a well-defined mathematical model, or at least that it is possible to check if two representations X and Y represent the same solid.

Conversions may be broadly divided into two categories: deterministic *evaluations* when the source representation X contains more information than the target information Y, and nondeterministic *comprehensions* when additional information in Y has to be synthesized based on additional assumptions or goals [HSS11].

Computational properties of evaluation procedures are widely researched and are mostly well understood. In abstract terms, suppose that the target representation Y can be described as a finite collection of primitive geometric elements $\{y\}$. Then any evaluation procedure implements a *generate and test* paradigm [Sha97]:

1. Generate a sufficient set of candidates y;

2. test each y against the source representation X;

3. assemble the target representation Y from those elements y that passed the test.

Details vary depending on the mathematical properties of the two representations and the efficiency of the first two steps. We discuss evaluation conversions first.

Boundary evaluation is an evaluation conversion that generates the Brep Y from a constructive representation X. It is used in most commercial systems to evaluate generic and feature-based representations. The procedure is a generalization of the well-known CSG to Brep conversion [RV85], and amounts to generating candidate faces and edges from the primitives in CSG, classifying them with respect to the constructive definition, and merging those passing the test into an optimized boundary data structure. Brep solids are usually converted into polygonal (most commonly, triangle) mesh representations that are required in many applications, including rendering and 3D printing. These polygonizations can also be produced directly from constructive representations, a common approach when dealing with implicit representations [ALJ+15]. Spatial discretizations are commonly computed from constructive, boundary, and mesh representations; generation of a nonconforming regular discretization is straightforward with voxelization or Delaunay tetrahedralization [Si15].

Constructing conforming discretizations is more challenging, particularly when conversion is done for finite element analysis or other numerical treatments of continuum problems. In that case, the problem is not a geometric problem alone: the quality of the discretization must also be judged by nongeometric criteria that derive from the nature of the physical problem and the numerical algorithms used to solve it. This is an active area of research with many approaches based on octree subdivision, on Delaunay triangulation, and on MAT computations.

Algorithms for constructing the MAT from constructive and boundary representations involve generating a sufficient set of bisectors (curves or surfaces) for primitives in the source representation X, trimming them against each other, and assembling the pieces into a complete medial axis [CKM04, MCD11]. Because simple boundary geometry elements can produce very complicated curve segments and surface patches in the MAT, approximation approaches are favored in practice. Some are based on geometric principles, some on a Delaunay triangulation of an approximated boundary, and some on a grid subdivision of ambient space. The conversion from MAT to Brep has been addressed by Vermeer [Ver94] and later by Amenta [ACK01]. Note that a polyhedral MAT produces a solid boundary that can contain spherical, conical, and cylindrical elements.

The *comprehension* conversion procedures are similar to the *evaluation* conversions, except that the step of generating a sufficient set of primitives $\{y\}$, needed to construct the target representation Y, usually involves application-specific heuristics. For example, Breps may be constructed from either spatial discretizations or from meshes by fitting surfaces, but the type of the surfaces to be fitted (splines, algebraic, quadratic, etc.) must be assumed. Constructing generic feature-based representations from Breps is a further generalization of this class of conversions that requires heuristically matching procedural and parametric primitives to sets of boundary surfaces [HPR00, NMLS15].

Constructing a CSG representation from a Brep is a classical problem. For Breps with polygonal faces the problem amounts to constructing a BSP tree. The solution in [SV93] for Brep faces that include natural quadrics (sphere, cone and cylinder) reveals that the core problem for curved boundary faces concerns the first step of generating sufficient candidates $\{y\}$. The set of candidates $\{y\}$, here cells, requires finding additional (nonunique) separating primitives that do not contribute to the boundary and hence are not explicitly present in the boundary representation. This problem is mostly solved for second-degree surfaces, but for higher degree surfaces it is open.

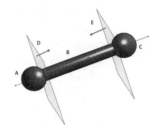

FIGURE 57.3.1

A Brep solid that cannot be converted to a CSG expression when restricting to the three half spaces A, B, and C, and their complements, that contribute nonempty boundary face areas.

Figure 57.3.1 shows an example of conversion for a simple solid: The solid's boundary is a union of three faces: two spherical and one cylindrical. However the solid cannot be represented as a Boolean set combination of the three corresponding halfspaces A, B, and C, and their complements A', B' and C', as shown in the figure. The intersection term $A'BC'$ contains both points inside and points outside the solid. A CSG representation becomes possible when using two additional (nonunique) separating halfspaces D and E. A canonical CSG representation of the solid, as a union of intersection terms, is

$$ABC'D'E + AB'C'D'E + A'BCDE' + A'B'CDE'+ \\ ABC'DE + A'BC'DE + A'BCDE + A'BC'DE,$$

where regularized $\cap$ and $\cup$ operations are represented by multiplication and addition respectively, and X' stands for regularized complement of X. An equivalent optimized CSG expression for the same solid is

$$A + CDE + B.$$

57.4 OPEN PROBLEMS

Most major problems in solid modeling contain a distorting conceptualization aspect. That is, a precise, technical formulation of the problem commits to a specific conceptualization of the larger context that may be contentious. For example, consider the following technical problem:

> Given an implicit algebraic surface S and a distance d, find the "offset" of S by d.

Assuming a precise definition of offset, and a restriction to irreducible algebraic surfaces S, the problem statement ignores the fact that a solid model is not bounded by a single, implicit surface, and that implicit surfaces of high algebraic degree may cause severe computational problems when used in a solid modeler.

CONSTRAINT SOLVING

Geometric constraint solvers trade efficiency for generality. Some very interesting techniques have been developed for planar problems that are fast but not very general. Nevertheless, they are useful in solid modeling applications. They could be extended in various ways without substantially impacting on efficiency. Such extensions, for constraint solving in the plane, include the incorporation of parametric curve segments as geometric elements, more general constraint configurations, as well as relations among distances and angles. The bulk of the work needed is a robust equation solver for the second phase.

Formulating explicit constraints on more general representations of polynomial and rational curves and surfaces is a challenging problem. Recent proposals call to replace explicit constraints with "black box" constraints that are defined implicitly by queries they support [GFM+16].

Recall from Section 57.2.5 the definition of the constraint graph. In a sense, the graph captures structure: by varying the valuation of the dimensional constraints, angle and distance, we obtain a family of constraint problems of a particular structure. Several questions come to mind:

1. Is the family of problems generically well-constrained? That is, does there exist a valuation of the dimensional constraints for which the resulting constraint problem is well-defined. The same question for over- and under-constrained families; see Chapter 61.

2. Constraint valuations for which a problem is well-constrained represent a manifold in a space of high dimension. What is the geometry of that manifold? The same question is of interest for under-constrained families since they include linkages; see Chapter 9.

3. Restricting to points and Euclidean distance, we obtain linkages whose properties have been studied in mathematics; e.g., [Max64, Hen08]. Characterize graphs that are rigid; see Chapter 63.

For points and lines in the plane there is a characterization of generically well-defined graphs [Lam70]. When circles are allowed as well, the result of [Lam70] no longer holds. For spatial constraint problems, with points, lines and planes as primitives, the problem is open. However, when the geometric primitives are drawn from the Brep of a set of solids, then at least a partial characterization is known [JS13].

Spatial geometric constraint solving in particular poses many other open problems, both variational and parametric. The set of parametric/sequential problems for points and planes is straightforward. Minimal variational problems for points and planes are understood; e.g., [Ver94]. One of the sequential construction problems for lines is to find all common tangents of four spheres in space. Equivalently, construct a line at prescribed distances from four fixed points. There are up to 12 common tangents determined by an equation system of degree 24 [HY00].

Constraint problems involving points, planes and lines have been investigated by Gao. Here the number of minimal configurations involving only a few primitives is large: [GHY04] shows that the minimal configurations with four, five and six lines number, respectively, 1, 12, and 494.

FEATURES

Manufacturing applications need cogent definitions of features to accelerate the design process. Such definitions ought to be in terms of generic mechanisms of form and of function, which are largely missing. Most feature definitions rely on specific representations and are not interchangeable. Also needed are mapping algorithms interrelating different feature schemata.

A set of features, say those conceptualizing machining a shape from stock, represents a particular *view* of the shape. In manufacturing applications there are many views, including machining, tolerancing, design view, etc. The problem of altering a design in one view with an automatic update of the other views is challenging. Some approaches have been based on subdividing the shape by superimposing all feature boundaries, and then tracking how the subdivision is affected by changes to one of the features. Abstractly, this problem is a generalization of a representation conversion problem where multiple representations of the same shape must be maintained simultaneously [RS00]. Maintaining multiple views for additively manufactured material structures is much more difficult because each view may not only require a different representation but may also be based on a distinct mathematical model and geometric shape [RRSS16].

SEMANTICS OF CONSTRAINT-BASED DESIGN

A solid shape design in terms of constraints can be changed simply by changing constraint values. To date, all such changes have been specified in terms of the procedures and algorithms that effect the change. What is needed is an abstract definition of shape change under such constraint changes to obtain a semantic definition of generic design and constraint-based editing. Such a definition must be visually intuitive.

Cellular homotopy has been proposed as a basic principle that formalizes the notion of intuitive update for both constructive and boundary representations [RS02b, RS98], but practical applications of this principle require a solution to the persistent naming problem [MP02].

MODEL TOLERANCES AND RECTIFICATION

Because of the substratum problem, Brep data structures can be invalid in the sense that the geometric description does not agree fully with the topological description. For instance, there may be small cracks between adjacent faces, the edge between two adjacent faces may not be where the curve description would place it, and so on. This has motivated work to "heal" the defective surface by closing cracks, eliminating overlaps, and so on. Some approaches sew up cracks with smaller faces, and in the case of polyhedra with triangles. Optimal healing is known to be NP-hard.

An intuitive idea is to assign tolerances to faces, edges and vertices, effectively thickening them so that the surface closes up [Jac95]. The difficulty is to work out what happens when nonadjacent faces merge into adjacent ones. The natural geometric enlargement, moreover, creates mathematically difficult surfaces; for instance, the offset surface of an ellipsoid increases the algebraic degree by a factor of 4. So, an interval-based approach has been proposed in which there is no closed-form description of the enlarged geometric elements [SSP01]. More formally, model rectification requires establishing validity of the enlarged sets[QS06] and its consistency with the intended exact representation [Sha08, SPB04].

INTEROPERABILITY

The broad term of interoperability subsumes a number of problems that have not been fully resolved, including model interchangeability and plug compatibility, system interoperation, and system integration. While progress has been made in all of these areas, interoperability challenges represent a major technological and economic barrier to wider adaption of solid modeling [HSS11]. A key open issue is lack of common formal semantics model that spans the broad spectrum of solid modeling models, representations, and systems.

The departure from a data-centric to a query-based approach as basis of system integration and interoperation is recent and appear promising [HSS14]. More work is needed to better understand the pros and cons. Technical issues that are clear include completeness and soundness of various query systems, how to structure queries appropriately for applications, granularity of the communication substrate, latency, and efficiency, to name a few. There are also nontechnical issues: the total number of queries needed to integrate foreign models may reveal more about the model creation and use than either system would be willing to share. These considerations should become clearer as the approach is tested over time and investigated scenarios are found to be compelling or otherwise.

MODELS OF INTERNAL STRUCTURE

Representation problems for the interior of solid models are closely linked to material modeling applications [PKM13]. Given the diversity of application needs, the absence of unified representations of the solid interior stands out as a key issue that ought to be addressed. To get a feel for the diversity, consider the following.

Laminated, composite objects, such as airplane wings and ship hulls, need to be represented. Embedded sensors, even computers and actuators may have to be added to create active structures. Polycrystalline materials may require different representational constructs, as do porous materials, insulating materials, etc. Functionally graded materials must be represented and analyzed. In medical applications, organ shape and tissue structure must be represented and understood.

These, and other areas of applications all require new ideas on the representation side so that they can fully explore and analyze their research questions computationally [RRSS16]. A unified representation would therefore have significant scientific value.

57.5 SOURCES AND RELATED MATERIAL

RELATED CHAPTERS

REFERENCES

[ABC+00]   C. Armstrong, A. Bowyer, S. Cameron, J. Corney, G. Jared, R. Martin, A. Middle-ditch, M. Sabin, and J. Salmon. *Djinn: A Geometric Interface for Solid Modelling*. Information Geometers, Winchester, 2000.

[ACK01]   N. Amenta, S. Choi, and R.K. Kolluri. The power crust. In *Proc. 6th ACM Sympos. Solid Modeling Appl.*, pages 249–260, 2001.

[AKK+02]   V. Adzhiev, E. Kartasheva, T. Kunii, A. Pasko, and B. Schmitt. Cellular-functional modeling of heterogeneous objects. In *Proc. 7th ACM Sympos. Solid Modeling Appl.*, pages 192–203, 2002.

[ALJ+15]   B.R. de Araújo, D.S. Lopes, P. Jepp, J.A. Jorge, and B. Wyvill. A survey on implicit surface polygonization. *ACM Comp. Surv.*, 47:60, 2015.

[ASME09]   ASME. *Dimensioning and Tolerancing: Y.14.5-2009*. American Society of Mechanical Engineers, New York, 2009.

[AVH+13]   M.K. Abera, P. Verboven, E. Herremans, T. Defraeye, S.W. Fanta, Q.T. Ho, J. Carmeliet, and B.M. Nicolaï. 3D virtual pome fruit tissue generation based on cell growth modeling. *Food Bioprocess Tech.*, 7:542–555, 2013.

[BB97]   J. Bloomenthal and C.L. Bajaj, editors. *Introduction to Implicit Surfaces*. Morgan Kaufmann, San Francisco, 1997.

[BCK88]   B. Buchberger, G.E. Collins, and B. Kutzler. Algebraic methods for geometric reasoning. *Annual Reviews Comp. Sci.*, 3:85–120, 1988.

[BFH+95]   W. Bouma, I. Fudos, C. Hoffmann, J. Cai, and R. Paige. A geometric constraint solver. *Comput.-Aided Des.*, 27:487–501, 1995.

[Bin01]   S. Biner. Thermo-elastic analysis of functionally graded materials using Voronoi elements. *Materials Sci. Engrg: A*, 31:136–146, 2001.

[Blu73]   H. Blum. Biological shape and visual science (part I). *J. Theoret. Biol.*, 38:205–287, 1973.

[BN90]   P. Brunet and I. Navazo. Solid representation and operation using extended octrees. *ACM Trans. Graphics*, 9:170–197, 1990.

[Bra75]   I. Braid. The synthesis of solids bounded by many faces. *Comm. ACM*, 18:209–216, 1975.

[Bra97]   I. Braid. Non-local blending of boundary models. *Comput.-Aided Des.*, 29:89–100, 1997.

[Bri93]   E. Brisson. Representing geometric structures in d dimensions: Topology and order. *Discrete Comput. Geom.*, 9:387–426, 1993.

[Bro82]   C.M. Brown. PADL-2: A technical summary. *IEEE Comput. Graph. Appl.*, 2:69–84, 1982.

[BST04]   A. Biswas, V. Shapiro, and I. Tsukanov. Heterogeneous material modeling with distance fields. *Comput. Aided Geom. Design*, 21:215–242, 2004.

[BW89]   M.I.G. Bloor and M.J. Wilson. Generating blending surfaces with partial differential equations. *Comput.-Aided Des.*, 21:165–171, 1989.

[CCC+08]   P. Cignoni, M. Callieri, M. Corsini, M. Dellepiane, F. Ganovelli, and G. Ranzuglia. Meshlab: An open-source mesh processing tool. In *Proc. Eurographics Italian Chapter Conf.*, pages 129–136, 2008.

[CCH94]   V. Capoyleas, X. Chen, and C.M. Hoffmann. Generic naming in generative, constraint-based design. *Comput.-Aided Des.*, 28:17–26, 1994.

[CH95a]   X. Chen and C.M. Hoffmann. On editability of feature based design. *Comput.-Aided Des.*, 27:905–914, 1995.

[CH95b]   X. Chen and C.M. Hoffmann. Towards feature attachment. *Comput.-Aided Des.*, 27:695–702, 1995.

[Chi88]   H. Chiyokura. *Solid Modeling with Designbase*. Addison-Wesley, Boston, 1988.

[Cho88]   S.-C. Chou. *Mechanical Geometry Theorem Proving*. Vol. 41 of *Mathematics and Its Applications*, Reidel Publishing Co., Dordrecht, 1988.

[CKM04]   T. Culver, J. Keyser, and D. Manocha. Exact computation of the medial axis of a polyhedron. *Computer Aided Geom. Design*, 21:65–98, 2004.

[CMP95]   V. Chandru, S. Manohar, and C.E. Prakash. Voxel-based modeling for layered manufacturing. *IEEE Comput. Graphics Appl.*, 15:42–47, 1995.

[CMS98]   P. Cignoni, C. Montani, and R. Scopigno. A comparison of mesh simplification algorithms. *Computers & Graphics*, 22:37–54, 1998.

[CT00]   M. Chen and J.V. Tucker. Constructive volume geometry. In *Computer Graphics Forum*, 19:281–293, 2000.

[DKP⁺92]   D. Dutta, N. Kikuchi, P. Papalmbros, F. Prinz, and L. Weiss. Project Maxwell: Towards rapid realization of superior products. In *Proc. Solid Freeform Fabrication Sympos.*, pages 54–62, University of Texas at Austin, 1992.

[DPS14]   A. DiCarlo, A. Paoluzzi, and V. Shapiro. Linear algebraic representation for topological structures. *Comput.-Aided Des.*, 46:269–274, 2014.

[DTD⁺15]   E.L. Doubrovski, E.Y. Tsai, D. Dikovsky, J.M.P. Geraedts, H. Herr, and N. Oxman. Voxel-based fabrication through material property mapping: A design method for bitmap printing. *Comput.-Aided Des.*, 60:3–13, 2015.

[Elb15]   G. Elber. Precise construction of micro-structures and porous geometry via functional composition. *Comput.-Aided Des.*, 2015. Manuscript submitted for publication.

[Far88]   G. Farin. *Curves and Surfaces for Computer-Aided Geometric Design*. Academic Press, San Diego, 1988.

[FAR16]   H.J. Fogg, C.G. Armstrong, and T.T. Robinson. Enhanced medial-axis-based block-structured meshing in 2-d. *Comput.-Aided Des.*, 72:87–101, 2016.

[FHJA16]   I. Fudos, C.M. Hoffmann, and R. Joan-Arinyo. Tree-decomposable and underconstrained geometric constraint problems. Preprint, `arXiv:1608.05205`, 2016.

[For95]   M. Forsyth. Shelling and offsetting bodies. In *Proc. 3rd ACM Sympos. Solid Modeling Appl.*, pages 373–381, 1995.

[For97]   S. Fortune. Polyhedral modeling with multi-precision integer arithmetic. *Comput.-Aided Des.*, 29:123–133, 1997.

[FP09]   A. Fabri and S. Pion. CGAL: The computational geometry algorithms library. In *Proc. 17th ACM SIGSPATIAL GIS*, pages 538–539, 2009.

[Fre96]   S. Frechette. Interoperability requirements for CAD data transfer in the AutoSTEP project. Technical Report NISTIR 5844, National Institute of Standards and Technology, Gaithersburg, 1996.

[FST06]   M. Freytag, V. Shapiro, and I. Tsukanov. Field modeling with sampled distances. *Comput.-Aided Des.*, 38:87–100, 2006.

[FST11]   M. Freytag, V. Shapiro, and I. Tsukanov. Finite element analysis in situ. *Finite Elements in Analysis and Design*, 47:957–972, 2011.

[FVP13]   O. Fryazinov, T. Vilbrandt, and A. Pasko. Multi-scale space-variant frep cellular structures. *Comput.-Aided Des.*, 45:26–34, 2013.

[GDC94]   The Geometric Design and Computation Group, University of Utah. *Alpha_1 advanced experimental CAD modeling system*, www.cs.utah.edu/gdc/projects/alpha1/, 1994.

[GFM+16]   G. Gouaty, L. Fang, D. Michelucci, M. Daniel, J.-P. Pernot, R. Raffin, S. Lanquetin, and M. Neveu. Variational geometric modeling with black box constraints and dags. *Comput.-Aided Des.*, 75:1–12, 2016.

[GHY04]   X.-S. Gao, C.M. Hoffmann, and W.-Q. Yang. Solving spatial basic geometric constraint configurations with locus intersection. *Comput.-Aided Des.*, 36:111–122, 2004.

[GLM96]   S. Ghosh, K. Lee, and S. Moorthy. Two scale analysis of heterogeneous elastic-plastic materials with asymptotic homogenization and Voronoi cell finite element model. *Computer Methods Appl. Mechanics Engrg.*, 132:63–116, 1996.

[GQD13]   Q. Ge, H.J. Qi, and M.L. Dunn. Active materials by four-dimension printing. *Applied Physics Letters*, 103:131901, 2013.

[Hen08]   L. Henneberg. *Die graphische Statik der starren Körper*. In F. Klein and C. Müller, editors, *Encyklopädie der Mathematischen Wissenschaften mit Einschluss ihrer Anwendungen*, pages 345–434, Springer, Wiesbaden, 1908.

[HJ92]   C.M. Hoffmann and R. Juan. Erep: An editable, high-level representation for geometric design and analysis. In P. Wilson, M. Wozny, and M. Pratt, editors, *Geometric Modeling for Product Realization*, pages 129–164, North-Holland, Amsterdam, 1992.

[HJA05]   C.M. Hoffmann and R. Joan-Arinyo. A brief on constraint solving. *Comput.-Aided Des.*, 2:655–663, 2005.

[HL93]   J. Hoschek and D. Lasser. *Computer Aided Geometric Design*. A.K. Peters, Natick, 1993.

[HLS01a]   C.M. Hoffmann, A. Lomonosov, and M. Sitharam. Decomposition plans for geometric constraint problems, part II: New algorithms. *J. Symbolic Comput.*, 31:409–427, 2001.

[HLS01b]   C.M. Hoffmann, A. Lomonosov, and M. Sitharam. Decomposition plans for geometric constraint systems, part I: Performance measures for CAD. *J. Symbolic Comput.*, 31:367–408, 2001.

[Hof89]   C.M. Hoffmann. *Geometric and Solid Modeling*. Morgan Kaufmann, San Francisco, 1989.

[Hof90]   C.M. Hoffmann. Algebraic and numerical techniques for offsets and blends. In W. Dahmen, M. Gasca, and C.A. Micchelli, editor, *Computations of Curves and Surfaces*, pages 499–528, Kluwer Academic, Dordrecht, 1990.

[Hof92]   C.M. Hoffmann. Computer vision, descriptive geometry, and classical mechanics. In B. Falcidieno and I. Herman, editors, *Computer Graphics and Mathematics*, pages 229–244, Springer, Berlin, 1992.

[Hof93a]   C.M. Hoffmann. On the semantics of generative geometry representations. In *Proc. 19th ASME Design Automation Conf.*, vol. 2, pages 411–420, 1993.

[Hof93b]   C.M. Hoffmann. On the separability problem of real functions and its significance in solid modeling. In *Computational Algebra*, vol. 151 of *Lecture Notes Pure Appl. Math.*, pages 191–204, Marcel Dekker, New York, 1993.

[Hof95]   C. Hoffmann. Geometric approaches to mesh generation. In I. Babuska, J. Flaherty, W. Henshaw, J. Hopcroft, J. Oliger, and T. Tezduyar, editors, *Modeling, Mesh Generation, and Adaptive Numerical Methods for Partial Differential Equations*, pages 31–51, Springer, Berlin, 1995.

[Hof01a]   C.M. Hoffmann. Robustness in geometric computations. *J. Comput. Inf. Sci. Eng.*, 1:143–155, 2001.

[Hof01b]   C.M. Hoffmann. D-cubed's dimensional constraint manager. *J. Comput. Inf. Sci. Eng.*, 1:100–101, 2001.

[Hol05]   S.J. Hollister. Porous scaffold design for tissue engineering. *Nature Materials*, 4:518–524, 2005.

[Hos92]   M. Hosaka. *Modeling of Curves and Surfaces in CAD/CAM*. Springer, New York, 1992.

[HPR00]   J. Han, M. Pratt, and W.C. Regli. Manufacturing feature recognition from solid models: A status report. *IEEE Trans. Robot. Autom.*, 16:782–796, 2000.

[HSS11]   C.M. Hoffmann, V. Shapiro, and V. Srinivasan. Geometric interoperability for resilient manufacturing. Technical report CSD 11-015, Purdue University, 2011.

[HSS14]   C. Hoffmann, V. Shapiro, and V. Srinivasan. Geometric interoperability via queries. *Comput.-Aided Des.*, 46:148–159, January 2014.

[HV94]   C.M. Hoffmann and P. Vermeer. Geometric constraint solving in $\mathbb{R}^2$ and $\mathbb{R}^3$. In D.Z. Du and F. Hwang, editors, *Computing in Euclidean Geometry*, 2nd edition, World Scientific, Singapore, 1994.

[HY00]   C.M. Hoffmann and B. Yuan. On spatial constraint solving approaches. In *Proc. 3rd Workshop on Automated Deduction in Geometry*, vol. 2061 of *LNCS*, pages 1–15, Springer, Berlin, 2000.

[Jac95]   D.J. Jackson. Boundary representation modelling with local tolerancing. In *Proc. 3rd ACM Sympos. Solid Modeling Appl.*, pages 247–253, 1995.

[JDKK07]   L.J.M. Jacobs, K.C.H. Danen, M.F. Kemmere, and J.T.F. Keurentjes. Quantitative morphology analysis of polymers foamed with supercritical carbon dioxide using Voronoi diagrams. *Comput. Materials Sci.*, 38:751–758, 2007.

[JLP+99]   T.R. Jackson, H. Liu, N.M. Patrikalakis, E.M. Sachs, and M.J. Cima. Modeling and designing functionally graded material components for fabrication with local composition control. *Materials & Design*, 20:63–75, 1999.

[JS13]   A.L.S. John and J. Sidman. Combinatorics and the rigidity of CAD systems. *Comput.-Aided Des.*, 45:473–482, 2013.

[KBDH99]   V. Kumar, D. Burns, D. Dutta, and C. Hoffmann. A framework for object modeling. *Comput.-Aided Des.*, 31:541–556, 1999.

[KD97]   V. Kumar and D. Dutta. An approach to modeling multi-material objects. In *Proc. 4th ACM Sympos. Solid Modeling Appl.*, pages 336–345, 1997.

[Kem99]   S.J. Kemmerer. *STEP: the grand experience*. US Department of Commerce, Technology Administration, National Institute of Standards and Technology, 1999.

[KK13]   A.K.M.B. Khoda and B. Koc. Functionally heterogeneous porous scaffold design for tissue engineering. *Comput.-Aided Des.*, 45:1276–1293, 2013.

[KKM99a]   J. Keyser, S. Krishnan, and D. Manocha. Efficient and accurate B-rep generation of low degree sculptured solids using exact arithmetic: I—representations. *Comput. Aided Geom. Design*, 16:841–859, 1999.

[KKM99b]   J. Keyser, S. Krishnan, and D. Manocha. Efficient and accurate B-rep generation of low degree sculptured solids using exact arithmetic: II—computation. *Comput. Aided Geom. Design*, 16:861–882, 1999.

[KMHP11]   B.C. Kim, D. Mun, S. Han, and M.J. Pratt. A method to exchange procedurally represented 2D CAD model data using ISO 10303 STEP. *Comput.-Aided Des.*, 43:1717–1728, 2011.

[KPIS08]   J. Kim, M.J. Pratt, R.G. Iyer, and R.D. Sriram. Standardized data exchange of CAD models with design intent. *Comput.-Aided Des.*, 40:760–777, 2008.

[Kra92]   G. Kramer. *Solving Geometric Constraint Systems*. MIT Press, Cambridge, 1992.

[KRU94]   F.-L. Krause, E. Rieger, and A. Ulbrich. Feature processing as kernel for integrated CAE systems. In *Proc. IFIP Conf. Feature Modeling and Recognition in Advanced CAD/CAM Systems*, vol. II, pages 693–716, 1994.

[KT07]   X.Y. Kou and S.T. Tan. Heterogeneous object modeling: A review. *Comput.-Aided Des.*, 39:284–301, 2007.

[KT10]   X.Y. Kou and S.T. Tan. A simple and effective geometric representation for irregular porous structure modeling. *Comput.-Aided Des.*, 42:930–941, 2010.

[KW14]   M. Kintel and C. Wolf. Openscad. *GNU General Public License, p GNU General Public License*, 2014.

[Lam70]   G. Laman. On graphs and the rigidity of plane skeletal structures. *J. Engrg. Math.*, 4:331–340, 1970.

[LS15]   X. Liu and V. Shapiro. Random heterogeneous materials via texture synthesis. *Comput. Materials Sci.*, 99:177–189, 2015.

[LS16]   X. Liu and V. Shapiro. Sample-based design of functionally graded material structures. In *ASME Design Engineering Technical Conf. and Computers and Inf. in Engng. Conf.*, pages IDETC2016–60431, 2016.

[M+10]   R. McNeel et al. Grasshopper-generative modeling with rhino. `http://www.grasshopper3d.com`, 2010.

[Män88]   M. Mäntylä. *An Introduction to Solid Modeling*. Computer Science Press, New York, 1988.

[Max64]   J.C. Maxwell. On reciprocal figures and diagrams of forces. *Philos. Mag.*, 27:250–261, 1864.

[MCD11]   S. Musuvathy, E. Cohen, and J. Damon. Computing medial axes of generic 3D regions bounded by B-spline surfaces. *Comput.-Aided Des.*, 43:1485–1495, 2011.

[Mea82]   D. Meagher. Geometric modeling using octree encoding. *Computer Graphics Image Processing*, 19:129–147, 1982.

[MP02]   D. Marcheix and G. Pierra. A survey of the persistent naming problem. In *Proc. 7th ACM Sympos. Solid Modeling Appl.*, pages 13–22, 2002.

[MUSA15]   O. Morgan, K. Upreti, G. Subbarayan, and D.C. Anderson. Higeom: A symbolic framework for a unified function space representation of trivariate solids for isogeometric analysis. *Comput.-Aided Des.*, 65:34–50, 2015.

[MVH+08]   H.K. Mebatsion, P. Verboven, Q.T. Ho, B.E. Verlinden, and B.M. Nicolaï. Modelling fruit (micro)structures, why and how? *Trends Food Sci. Tech.*, 19:59–66, 2008.

[MY14]   M.Y. Ma and H. Ye. An image analysis method to obtain the effective thermal conductivity of metallic foams via a redefined concept of shape factor. *Appl. Thermal Engrg.*, 73:1277–1282, 2014.

[Nay90]   B. Naylor. Binary space partitioning trees as an alternative representation of polytopes. *Comput.-Aided Des.*, 22:250-252, 1990.

[NMLS15]   Z. Niu, R.R. Martin, F.C. Langbein, and M.A. Sabin. Rapidly finding CAD features using database optimization. *Comput.-Aided Des.*, 69:35–50, 2015.

[NR95]   B. Naylor and L. Rogers. Constructing binary space partitioning trees from piecewise Bézier curves. In *Proc. Graphics Interface*, pages 181–191, 1995.

[NSK+97]   M. Näf, G. Székely, R. Kikinis, M.E. Shenton, and O. Kübler. 3D Voronoi skeletons and their usage for the characterization and recognition of 3D organ shape. *Computer Vision and Image Understanding*, 66:147–161, 1997.

[Owe91]   J.C. Owen. Algebraic solution for geometry from dimensional constraints. In *Proc. 1st ACM Sympos. Solid Modeling Found. and CAD/CAM Appl.*, pages 397–407, 1991.

[Oxm11]   N. Oxman. Variable property rapid prototyping. *Virtual Physical Prototyping*, 6:3–31, 2011.

[PASS95]   A. Pasko, V. Adzhiev, A. Sourin, and V. Savchenko. Function representation in geometric modeling: Concepts, implementation and applications. *The Visual Computer*, 11:429–446, 1995.

[PKM13]   J.H. Panchal, S.R. Kalidindi, and D.L. McDowell. Key computational modeling issues in integrated computational materials engineering. *Comput.-Aided Des.*, 45:4–25, 2013.

[PM02]   N.M. Patrikalakis and T. Maekawa. *Shape Interrogation for Computer-Aided Design and Manufacture.* Springer, Berlin, 2002.

[PPV+03]   A. Paoluzzi, V. Pascucci, M. Vicentino, C. Baldazzi, and S. Portuesi. *Geometric Programming for Computer Aided Design.* John Wiley & Sons, New York, 2003.

[PS93]   R.S. Palmer and V. Shapiro. Chain models of physical behavior for engineering analysis and design. *Research Engrg. Design*, 5:161–184, 1993.

[PS95]   A. Pasko and V. Savchenko. Algebraic sums for deformation of constructive solids. In *Proc. 3rd ACM Sympos. Solid Modeling Appl.*, pages 403–408, 1995.

[QS06]   J. Qi and V. Shapiro. ε-topological formulation of tolerant solid modeling. *Comput.-Aided Des.*, 38:367–377, 2006.

[Req77]   A.A.G. Requicha. Mathematical models of rigid solids. Technical Report PAP 28, University of Rochester, 1977.

[RO89]   J.R. Rossignac and M.A. O'Connor. *SGC: A Dimension-Independent Model for Pointsets with Internal Structures and Incomplete Boundaries.* IBM TJ Watson Research Center, Yorktown Heights, 1989.

[Ros07]   D.W. Rosen. Computer-aided design for additive manufacturing of cellular structures. *Computer-Aided Design Appl.*, 4:585–594, 2007.

[Rot82]   S.D. Roth. Ray casting for modeling solids. *Computer Graphics Image Processing*, 18:109–144, 1982.

[RR91]   J.R. Rossignac and A.A.G. Requicha. Constructive non-regularized geometry. *Comput.-Aided Des.*, 23:21–32, 1991.

[RRSS16]   W. Regli, J. Rossignac, V. Shapiro, and V. Srinivasan. The new frontiers in computational modeling of material structures. *Comput.-Aided Des.*, 77:73–85, 2016.

[RS98]   S. Raghothama and V. Shapiro. Boundary representation deformation in parametric solid modeling. *ACM Trans. Graphics*, 17:259–286, 1998.

[RS00]   S. Raghotama and V. Shapiro. Consistent updates in dual representation systems. *Comput.-Aided Des.*, 32:463–477, 2000.

[RS02a]   S. Raghotama and V. Shapiro. Topological framework for part families. In *Proc. 7th ACM Sympos. Solid Modeling Appl.*, pages 1–12, 2002.

[RS02b]   S. Raghothama and V. Shapiro. Topological framework for part families. *J. Comput. Inf. Sci. Eng.*, 2:246–255, 2002.

[RSB96]   A. Rappoport, A. Sheffer, and M. Bercovier. Volume-preserving free-form solids. *IEEE Trans. Visualization Computer Graphics*, 2:19–27, 1996.

[RV77]   A.A.G. Requicha and H.B. Voelcker. Constructive solid geometry. Tech. Report 25, University of Rochester, 1977.

[RV82]   A.A.G. Requicha and H.B. Voelcker. Solid modeling: A historical summary and contemporary assessment. *IEEE Comput. Graphics Appl.*, 2:9–24, 1982.

[RV83]   A.A.G. Requicha and H.B. Voelcker. Solid modeling: Current status and research directions. *IEEE Comput. Graphics Appl.*, 3:25–37, 1983.

[RV85]   A.A.G. Requicha and H.B. Voelcker. Boolean operations in solid modeling: Boundary evaluation and merging algorithms. *Proc. IEEE*, 73:30–44, 1985.

[RW91]   S.J. Rock and M.J. Wozny. A flexible file format for solid freeform fabrication. In *Proc. Solid Freeform Fabrication Sympos.*, pages 1–12, University of Texas at Austin, 1991.

[Sam89a] H.J. Samet. *Applications of Spatial Data Structures: Computer Graphics, Image Processing, and GIS.* Addison–Wesley, Boston, 1989.

[Sam89b] H.J. Samet. *Design and analysis of Spatial Data Structures: Quadtrees, Octrees, and other Hierarchical Methods.* Addison–Wesley, Boston, 1989.

[SAR95]   D.J. Sheehy, C.G. Armstrong, and D.J. Robinson. Computing the medial surface of a solid from a domain Delaunay triangulation. In *Proc. 3rd ACM Sympos. Solid Modeling Appl.*, pages 201–212, 1995.

[Sha91a] J.J. Shah. Assessment of features technology. *Comput.-Aided Des.*, 23:331–343, 1991.

[Sha91b] V. Shapiro. *Representations of Semialgebraic Sets in Finite Algebras Generated by Space Decompositions.* PhD thesis, Sibley School of Mechanical Engineering, Cornell University, 1991.

[Sha97]   V. Shapiro. Errata: Maintenance of geometric representations through space decompositions. *Internat. J. Comput. Geom. Appl.*, 7:383–418, 1997.

[Sha07]   V. Shapiro. Semi-analytic geometry with r-functions. *Acta Numer.*, 16:239, 2007.

[Sha08]   V. Shapiro. Homotopy conditions for tolerant geometric queries. In *Reliable Implementation of Real Number Algorithms: Theory and Practice*, pages 162–180, Springer, Berlin, 2008.

[SHL92]   J. Shah, D. Hsiao, and J. Leonard. A systematic approach for design-manufacturing feature mapping. In P. Wilson, M. Wozny, and M. Pratt, editors, *Geometric Modeling for Product Realization*, pages 205–222, North Holland, Amsterdam, 1992.

[SI89]   K. Sugihara and M. Iri. A solid modeling system free from topological inconsistency. *J. Inform. Process.*, 12:380–393, 1989.

[Si15]   H. Si. TetGen, a Delaunay-based quality tetrahedral mesh generator. *ACM Trans. Math. Software*, 41:11, 2015.

[SPB04]   T. Sakkalis, T.J. Peters, and J. Bisceglio. Isotopic approximations and interval solids. *Comput.-Aided Des.*, 36:1089–1100, 2004.

[SR04]   S. Spitz and A. Rappoport. Integrated feature-based and geometric CAD data exchange. In *Proc. 9th ACM Sympos. Solid Modeling Appl.*, pages 183–190, 2004.

[Sri03]   V. Srinivasan. *Theory of Dimensioning: An Introduction to Parameterizing Geometric Models*. CRC Press, Boca Raton, 2003.

[SS10]   R. Schmidt and K. Singh. Meshmixer: An interface for rapid mesh composition. In *Proc. ACM SIGGRAPH*, article 6, 2010.

[SSND05]   W. Sun, B. Starly, J. Nam, and A. Darling. Bio-CAD modeling and its applications in computer-aided tissue engineering. *Comput.-Aided Des.*, 37:1097–1114, 2005.

[SSP01]   G. Shen, T. Sakkalis, and N. Patrikalakis. Analysis of boundary representation model rectification. In *Proc. 6th ACM Sympos. Solid Modeling Appl.*, pages 149–158, 2001.

[SV93]   V. Shapiro and D. Vossler. Separation for boundary to CSG conversion. *ACM Trans. Graphics*, 12:35–55, 1993.

[Sys01]   S. Systems. *The SGDL Language*. www.sgdl-sys.com, 2001.

[Tak92]   T. Takala. A taxonomy on geometric and topological models. In B. Falcidieno, I. Herman, and C. Pienovi, editors, *Computer Graphics Math.*, pages 147–171, Springer, Berlin, 1992.

[Til80]   R.B. Tilove. Set membership classification: A unified approach to geometric intersection problems. *IEEE Trans. Computers*, 100:874–883, 1980.

[TN87]   W.C. Thibault and B.F. Naylor. Set operations on polyhedra using binary space partitioning trees. In *Proc. 14th ACM Conf. Comp. Graphics*, pages 153–162, 1987.

[TWM85]   J.E. Thompson, Z.U.A. Warsi, and C.W. Mastin. *Numerical Grid Generation*. North Holland, Amsterdam, 1985.

[UMP95]   Unit Manufacturing Process Research Committee. *Unit Manufacturing Processes: Issues and Opportunities in Research*. National Academies Press, Washington, 1995.

[Ver94]   P. Vermeer. *Medial Axis Transform to Boundary Representation Conversion*. PhD thesis, Purdue University, 1994.

[WCR05]   H. Wang, Y. Chen, and D.W. Rosen. A hybrid geometric modeling method for large scale conformal cellular structures. In *Proc. ASME Internat. Design Engineering Technical Conf. and Computers and Inf. in Engng. Conf.*, pages 421–427, 2005.

[Wei86]   K. Weiler. *Topological Structures for Geometric Modeling*. PhD thesis, Rensselaer Polytechnic Inst., 1986.

[WGG99]   B. Wyvill, A. Guy, and E. Galin. Extending the CSG tree. Warping, blending and Boolean operations in an implicit surface modeling system. *Computer Graphics Forum*, 18:149–158, 1999.

[YBSH16]   U. Yaman, N. Butt, E. Sacks, and C. Hoffmann. Slice coherence in a query-based architecture for 3D heterogeneous printing. *Comput.-Aided Des.*, 75:27–38, 2016.

[Zag97]   J. Zagajac. *Engineering Analysis over Subsets*. PhD thesis, The Sibley School of Mechanical and Aerospace Engineering, Cornell University, 1997.

58 COMPUTATION OF ROBUST STATISTICS: DEPTH, MEDIAN, AND RELATED MEASURES

Peter J. Rousseeuw and Mia Hubert

INTRODUCTION

As statistical data sets grow larger and larger, the availability of fast and efficient algorithms becomes ever more important in practice. Classical methods are often easy to compute, even in high dimensions, but they are sensitive to outlying data points. Robust statistics develops methods that are less influenced by abnormal observations, often at the cost of higher computational complexity. Many robust methods, especially those based on ranks, are closely related to geometric or combinatorial problems. An early overview of relations between statistics and geometry was given in [Sha76].

Recently many other (mostly multivariate) statistical methods have been developed that have a combinatorial or geometric character and are computationally intensive. Techniques of computational geometry appear to be very well suited for the development of fast algorithms. Over the last two decades, especially the notion of statistical depth received considerable attention from the computational geometry community. In this chapter we mainly concentrate on depth and multivariate medians, and in Section 58.3 we list other areas of statistics where computational geometry has been of use in constructing efficient algorithms, such as cluster analysis.

58.1 MULTIVARIATE RANKING

A data set consisting of n univariate points is usually ranked in ascending or descending order. Univariate order statistics (i.e., the "kth smallest value out of n") and derived quantities have been studied extensively. The median is defined as the order statistic of rank $(n + 1)/2$ when n is odd, and as the average of the order statistics of ranks $n/2$ and $(n+2)/2$ when n is even. The median and any other order statistic of a univariate data set can be computed in $O(n)$ time. Generalization to higher dimensions is, however, not straightforward.

Alternatively, univariate points may be ranked from the outside inward by assigning the most extreme data points depth 1, the second smallest and second largest data points depth 2, etc. The deepest point then equals the usual median of the sample. The advantage of this type of ranking is that it can be extended to higher dimensions more easily. This section gives an overview of several possible generalizations of depth and the median to multivariate settings. Surveys of statistical applications of multivariate data depth may be found in [LPS99], [ZS00], and [Mos13].

GLOSSARY

Bagdistance: Generalized norm on $\mathbb{R}^d$ based on halfspace depth regions.

Bagplot: Bivariate generalization of the boxplot based on depth regions.

Breakdown value: The smallest fraction of contaminated data points that can move the estimator arbitrarily far away.

Centerpoint: Any point with halfspace depth $\geqslant \lceil n/(d+1) \rceil$.

Deepest fit: Median hyperplane based on regression depth.

Depth: The outside-inward "rank" of a point (not necessarily a data point).

Depth region: The set of all points with depth $\geqslant k$ is called the kth depth region D_k.

Median: The point with maximal depth. When this point is not unique, the median is taken to be the centroid of the depth region with highest depth.

Tukey median: Median based on halfspace depth.

HALFSPACE LOCATION DEPTH

Let $X_n = \{x_1, \ldots, x_n\}$ be a finite set of data points in $\mathbb{R}^d$. The *Tukey depth* or *halfspace depth* (introduced by [Tuk75] and further developed by [DG92]) of any point $\boldsymbol{\theta}$ in $\mathbb{R}^d$ (not necessarily a data point) determines how central the point is inside the data cloud. The halfspace depth of $\boldsymbol{\theta}$ is defined as the minimal number of data points in any closed halfspace determined by a hyperplane through $\boldsymbol{\theta}$:

$$hdepth(\boldsymbol{\theta}; X_n) = \min_{\|\boldsymbol{u}\|=1} \#\{i; \boldsymbol{u}^\tau \boldsymbol{x}_i \geqslant \boldsymbol{u}^\tau \boldsymbol{\theta}\}.$$

Thus, a point lying outside the convex hull of X_n has depth 0, and any data point has depth at least 1. Figure 58.1.1 illustrates this definition for $d = 2$.

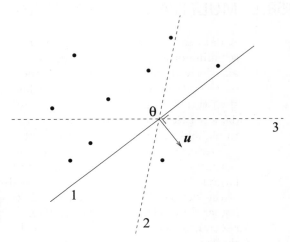

FIGURE 58.1.1
Illustration of the bivariate halfspace depth. Here $\boldsymbol{\theta}$ (which is not a data point itself) has depth 1 because the halfspace determined by $\boldsymbol{u}$ contains only one data point.

The set of all points with depth $\geqslant k$ is called the kth depth region D_k. The halfspace depth regions form a sequence of nested polyhedra. Each D_k is the intersection of all halfspaces containing at least $n - k + 1$ data points. Moreover, every data point must be a vertex of one or more depth regions. When the innermost depth region is a singleton, that point is called the *Tukey median*. When the innermost depth region is larger than a singleton, the Tukey median is defined as its centroid. This makes the Tukey median unique by construction. Note that the Tukey median does not have to be a data point.

Note that the depth regions give an indication of the shape of the data cloud. Based on this idea one can construct the *bagplot* [RRT99], a bivariate version of the univariate boxplot. Figure 58.1.2 shows such a bagplot. The cross in the white disk is the Tukey median. The dark area is an interpolation between two subsequent depth regions, and contains 50% of the data. This area (the "bag") gives an idea of the shape of the majority of the data cloud. Inflating the bag by a factor of 3 relative to the Tukey median yields the "fence" (not shown), and data points outside the fence are called outliers and marked by stars. Finally, the light gray area is the convex hull of the non-outlying data points.

FIGURE 58.1.2
Bagplot of the heart and spleen size of 73 hamsters.

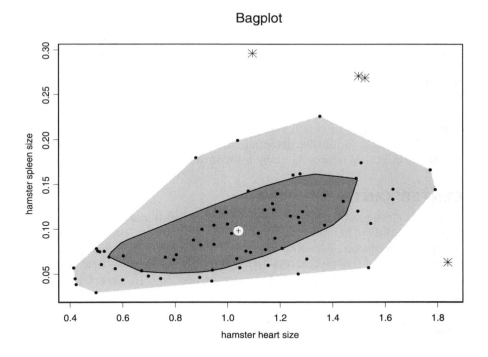

More generally, in the multivariate case one can define the *bagdistance* [HRS15b] of a point x relative to the Tukey median and the bag. Assume that the Tukey median lies in the interior of the bag, not on its boundary (this excludes degenerate cases). Then the bagdistance is the smallest real number λ such that the bag inflated (or deflated) by λ around the Tukey median contains the point x. When the Tukey median equals $\mathbf{0}$, it is shown in [HRS15b] that the bagdistance satisfies all axioms of a norm except that $\|a\boldsymbol{x}\| = |a|\|\boldsymbol{x}\|$ only needs to hold when $a \geqslant 0$. The bagdistance is used for outlier detection [HRS15a] and statistical classification [HRS15b].

An often used criterion to judge the robustness of an estimator is its *breakdown value*. The breakdown value is the smallest fraction of data points that we need to replace in order to move the estimator of the contaminated data set arbitrarily far away. The classical mean of a data set has breakdown value zero since we can move it anywhere by moving one observation. Note that the breakdown value of any translation equivariant estimator can be at most $1/2$. This can be seen as follows: if we replace half of the data points by a far-away translation image of the remaining half, the estimator cannot know which were the original data.

The Tukey depth and the corresponding median have good statistical properties. The Tukey median T^* is a location estimator with breakdown value $\varepsilon_n(T^*; X_n)$ $\geqslant 1/(d+1)$ for any data set in general position. This means that it remains in a predetermined bounded region unless $n/(d+1)$ or more data points are moved. At an elliptically symmetric distribution the breakdown value becomes $1/3$ for large n, irrespective of d. Moreover, the halfspace depth is invariant under all nonsingular affine transformations of the data, making the Tukey median affine equivariant. Since data transformations such as rotation and rescaling are very common in statistics, this is an important property. The statistical asymptotics of the Tukey median have been studied in [BH99].

The need for fast algorithms for the halfspace depth has only grown over the years, since it is currently being applied to a variety of settings such as nonparametric classification [LCL12]. A related development is the fast growing field of functional data analysis, where the data are functions on a univariate interval (e.g., time or wavelength) or on a rectangle (e.g., surfaces, images). Often the function values are themselves multivariate. One can then define the depth of a curve (surface) by integrating the depth over all points as in [CHSV14]. This functional depth can again be used for outlier detection and classification [HRS15a, HRS15b], but it requires computing depths in many multivariate data sets instead of just one.

CENTERPOINTS

There is a close relationship between the Tukey depth and centerpoints, which have been long studied in computational geometry. In fact, Tukey depth extends the notion of centerpoint. A *centerpoint* is any point with halfspace depth $\geqslant \lceil n/(d+1) \rceil$. A consequence of Helly's theorem is that there always exists at least one centerpoint, so the depth of the Tukey median cannot be less than $\lceil n/(d+1) \rceil$.

OTHER LOCATION DEPTH NOTIONS

1. **Simplicial depth** ([Liu90]). The depth of $\boldsymbol{\theta}$ equals the number of simplices formed by $d+1$ data points that contain $\boldsymbol{\theta}$. Formally,

$$sdepth(\boldsymbol{\theta}; X_n) = \#\{(i_1, \ldots, i_{d+1}); \boldsymbol{\theta} \in S[\boldsymbol{x}_{i_1}, \ldots, \boldsymbol{x}_{i_{d+1}}]\}$$

where $S[\boldsymbol{x}_{i_1}, \ldots, \boldsymbol{x}_{i_{d+1}}]$ is the closed simplex with vertices $\boldsymbol{x}_{i_1}, \ldots, \boldsymbol{x}_{i_{d+1}}$. The simplicial median is affine equivariant with a breakdown value bounded above by $1/(d+2)$. Unlike halfspace depth, its depth regions need not be convex, as seen, e.g., in the example in [Mos13].

2. *Oja depth* ([Oja83]). This is also called simplicial volume depth:

$$odepth(\boldsymbol{\theta}; X_n) = \left(1 + \sum_{(i_1,\ldots,i_d)} \{volume\ S[\boldsymbol{\theta}, \boldsymbol{x}_{i_1}, \ldots, \boldsymbol{x}_{i_d}]\}\right)^{-1}.$$

The corresponding median is also affine equivariant, but has zero breakdown value.

3. *Projection depth.* We first define the *outlyingness* ([DG92]) of any point $\boldsymbol{\theta}$ relative to the data set X_n as

$$O(\boldsymbol{\theta}; X_n) = \max_{\|\boldsymbol{u}\|=1} \frac{|\boldsymbol{u}^\tau \boldsymbol{\theta} - \mathrm{med}_i\{\boldsymbol{u}^\tau \boldsymbol{x}_i\}|}{\mathrm{MAD}_i\{\boldsymbol{u}^\tau \boldsymbol{x}_i\}},$$

where the median absolute deviation (MAD) of a univariate data set $\{y_1, \ldots, y_n\}$ is the statistic $\mathrm{MAD}_i\{y_i\} = \mathrm{med}_i|y_i - \mathrm{med}_j\{y_j\}|$. The outlyingness is small for centrally located points and increases if we move toward the boundary of the data cloud. Instead of the median and the MAD, also another pair (T, S) of a location and scatter estimate may be chosen. This leads to different notions of projection depth, all defined as

$$pdepth(\boldsymbol{\theta}; X_n) = (1 + O(\boldsymbol{\theta}; X_n))^{-1}.$$

General projection depth is studied in [Zuo03]. When using the median and the MAD, the projection depth has breakdown value $1/2$ and is affine equivariant. Its depth regions are convex.

4. *Spatial depth* ([Ser02]). Spatial depth is related to multivariate quantiles proposed in [Cha96]:

$$spdepth(\boldsymbol{\theta}; X_n) = 1 - \left\| \frac{1}{n} \sum_{i=1}^{n} \frac{\boldsymbol{x}_i - \boldsymbol{\theta}}{\|\boldsymbol{x}_i - \boldsymbol{\theta}\|} \right\|$$

The spatial median is also called the L^1 median ([Gow74]). It has breakdown value $1/2$, but is not affine equivariant (it is only equivariant with respect to translations, multiplication by a scalar factor, and orthogonal transformations). For a recent survey on the computation of the spatial median see [FFC12].

A comparison of the main properties of the different location depth medians is given in Table 58.1.1.

ARRANGEMENT AND REGRESSION DEPTH

Following [RH99b] we now define the depth of a point relative to an arrangement of hyperplanes (see Chapter 28). A point $\boldsymbol{\theta}$ is said to have zero **arrangement depth** if there exists a ray $\{\boldsymbol{\theta} + \lambda\boldsymbol{u}; \lambda \geqslant 0\}$ that does not cross any of the hyperplanes h_i in the arrangement. (A hyperplane parallel to the ray is counted as intersecting at infinity.) The arrangement depth of any point $\boldsymbol{\theta}$ is defined as the minimum number of hyperplanes intersected by any ray from $\boldsymbol{\theta}$. Figure 58.1.3 shows an arrangement of lines. In this plot, the points $\boldsymbol{\theta}$ and $\boldsymbol{\eta}$ have arrangement depth 0 and the point

TABLE 58.1.1   Comparison of several location depth
medians.

MEDIAN	BREAKDOWN VALUE	AFFINE EQUIVARIANCE
Tukey	worst-case $1/(d+1)$ typically $1/3$	yes
Simplicial	$\leqslant 1/(d+2)$	yes
Oja	$2/n \approx 0$	yes
Projection	$1/2$	yes
Spatial	$1/2$	no

ξ has arrangement depth 2. The arrangement depth is always constant on open cells and on cell edges. It was shown ([RH99b]) that any arrangement of lines in the plane encloses a point with arrangement depth at least $\lceil n/3 \rceil$, giving rise to a new type of "centerpoints."

FIGURE 58.1.3
Example of arrangement depth. In this arrangement of lines, the points $\boldsymbol{\theta}$ and $\boldsymbol{\eta}$ have arrangement depth 0, whereas $\boldsymbol{\xi}$ has arrangement depth 2. (See Figure 58.1.4 for the dual plot.)

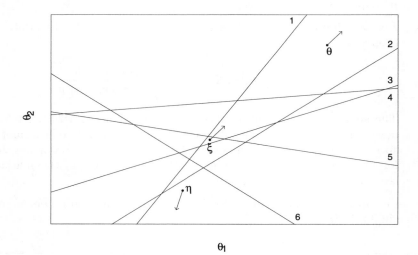

This notion of depth was originally defined ([RH99]) in the dual, as the depth of a regression hyperplane $H_{\boldsymbol{\theta}}$ relative to a point configuration of the form $Z_n = \{(\boldsymbol{x}_1, y_1), \ldots, (\boldsymbol{x}_n, y_n)\}$ in $\mathbb{R}^{d+1}$. **Regression depth** ranks hyperplanes according to how well they fit the data in a regression model, with $\boldsymbol{x}$ containing the predictor variables and y the response. A vertical hyperplane (given by $\boldsymbol{a}^\tau \boldsymbol{x} = \text{constant}$), which cannot be used to predict future response values, is called a "nonfit" and assigned regression depth 0. The regression depth of a general hyperplane $H_{\boldsymbol{\theta}}$ is found by rotating H_θ in a continuous movement until it becomes vertical. The minimum number of data points that is passed in such a rotation is called the regression depth of H_θ. Figure 58.1.4 is the dual representation of Figure 58.1.3. (For instance, the line θ has slope θ_1 and intercept θ_2 and corresponds to the point (θ_1, θ_2) in Figure 58.1.3.) The lines θ and η have regression depth 0, whereas the line ξ has regression depth 2.

FIGURE 58.1.4

Example of the regression depth of a line in a bivariate configuration of points. The lines θ and η have regression depth 0, whereas the line ξ has regression depth 2. (This is the dual of Figure 58.1.3.)

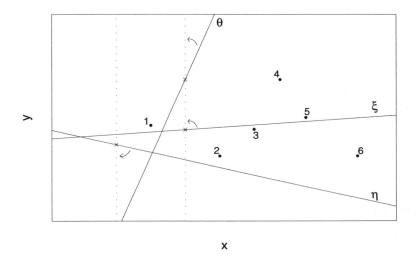

In statistics one is interested in the *deepest fit* or regression depth median, because this is a line (hyperplane) about which the data are well-balanced. The statistical properties of regression depth and the deepest fit are very similar to those of the Tukey depth and median. The bounds on the maximal depth are almost the same. Moreover, for both depth notions the value of the maximal depth can be used to characterize the symmetry of the distribution ([RS04]). The breakdown value of the deepest fit is at least $1/(d+1)$ and under linearity of the conditional median of y given x it converges to $1/3$. In the next section, we will see that the optimal complexities for computing the depth and the median are also comparable to those for halfspace depth. For a detailed comparison of the properties of halfspace and regression depth, see [HRV01].

The arrangement depth region D_k is defined in the primal, as the set of points with arrangement depth at least k. Contrary to the Tukey depth, these depth regions need not be convex. But nevertheless it was proved that there always exists a point with arrangement depth at least $\lceil n/(d+1) \rceil$ ([ABE+00]). An analysis-based proof was given in [Miz02].

ARRANGEMENT LEVELS

Arrangement depth is undirected (isotropic) in the sense that it is defined as a minimum over all possible directions. If we restrict ourselves to vertical directions u (i.e., up or down), we obtain the usual levels of the arrangement (cf. Section 28.2). The absence of preferential directions makes arrangement depth invariant under affine transformations.

58.2 COMPUTING DEPTH

Although the definitions of depth are intuitive, the computational aspects can be quite challenging. Below we will first focus on the bivariate case, and then on higher dimensions.

Algorithms for depth-related measures are often more complex for data sets which are not in general position than for data sets which are. For example, the boundaries of subsequent halfspace depth regions are always disjoint when the data are in general position, but this does not hold for nongeneral position. Preferably, algorithms should be able to handle both the general position case and the nongeneral position case directly. As a quick fix, algorithms which were made for general position can also be applied in the other case if one first adds small random errors to the data points. This is a standard perturbation technique sometimes referred to as dithering. For large data sets, dithering has only a limited effect on the results.

BIVARIATE ALGORITHMS

For the bivariate case several algorithms have been developed early on. Table 58.2.1 gives an overview of algorithms, each of which has been implemented, to compute the depth in a given point θ in $\mathbb{R}^2$. These algorithms are time-optimal, since the problem of computing these bivariate depths has an $\Omega(n \log n)$ lower bound ([ACG$^+$02], [LS03a]).

The algorithms for halfspace and simplicial depth are based on the same technique. First, data points are radially sorted around θ. Then a line through θ is rotated. The depth is calculated by counting the number of points that are passed by the rotating line in a specific manner. The planar arrangement depth algorithm is easiest to visualize in the regression setting. To compute the depth of a hyperplane H_θ with coefficients θ, the data are first sorted along the x-axis. A vertical line L is then moved from left to right and each time a data point is passed, the number of points above and below H_θ on both sides of L is updated.

In general, computing a median is harder than computing the depth in a point, because typically there are many candidate points. For instance, for the bivariate simplicial median the currently best algorithm requires $O(n^4)$ time, whereas its corresponding depth needs only $O(n \log n)$. The simplicial median seems difficult to compute because there are $O(n^4)$ candidate points (namely, all intersections of lines passing through two data points) and the simplicial depth regions have irregular shapes, but of course a faster algorithm may yet be found.

Fortunately, in several important cases the median can be computed without computing the depth of individual points. A linear-time algorithm to compute a

TABLE 58.2.1   Computing the depth of a bivariate point.

DEPTH	TIME COMPLEXITY	SOURCE
Tukey depth	$O(n \log n)$	[RR96a]
Simplicial depth	$O(n \log n)$	[RR96a]
Arrangement/regression depth	$O(n \log n)$	[RH99]

bivariate centerpoint was described in [JM94]. Table 58.2.2 gives an overview of algorithms to compute bivariate depth-based medians. For the bivariate Tukey median the lower bound $\Omega(n \log n)$ was proved in [LS00], and the currently fastest algorithm takes $O(n \log^3 n)$ time ([LS03a]). The lower bound $\Omega(n \log n)$ also holds for the median of arrangement (regression) depth as shown by [LS03b]. Fast algorithms were devised by [LS03b] and [KMR$^+$08].

TABLE 58.2.2   Computing the bivariate median.

MEDIAN	TIME COMPLEXITY	SOURCE
Tukey median	$O(n \log^3 n)$	[LS03a]
Simplicial median	$O(n^4)$	[ALS$^+$03]
Oja median	$O(n \log^3 n)$	[ALS$^+$03]
Regression depth median	$O(n \log n)$	[LS03b]

The computation of bivariate halfspace depth regions has also been studied. The first algorithm [RR96b] required $O(n^2 \log n)$ time per depth region. An algorithm to compute all regions in $O(n^2)$ time is constructed and implemented in [MRR$^+$03]. This algorithm thus also yields the Tukey median. It is based on the dual arrangement of lines where topological sweep is applied. A completely different approach is implemented in [KMV02]. They make direct use of the graphics hardware to approximate the depth regions of a set of points in $O(nW + W^3) + nCW^2/512$ time, where the pixel grid is of dimension $(2W + 1) \times (2W + 1)$. Recently, [BRS11] constructed an algorithm to update halfspace depth and its regions when points are added to the data set.

ALGORITHMS IN HIGHER DIMENSIONS

The calculation of depth regions and medians is more computationally intensive in higher dimensions. In statistical practice such data are quite common, and therefore reliable and efficient algorithms are needed. The computational aspects of depth in higher dimensions are still being explored.

The first algorithms to compute the halfspace and regression depth of a given point in $\mathbb{R}^d$ with $d > 2$ were constructed in [RS98] and require $O(n^{d-1} \log n)$ time. The main idea was to use projections onto a lower-dimensional space. This reduces the problem to computing bivariate depths, for which the existing algorithms have optimal time complexity. In [BCI$^+$08] theoretical output-sensitive algorithms for the halfspace depth are proposed. An interesting computational connection between halfspace depth and multivariate quantiles was provided in [HPS10] and [KM12]. More recently, [DM16] provided a generalized version of the algorithm of [RS98] together with C++ code. For the depth regions of halfspace depth in higher dimensions an algorithm was recently proposed in [LMM14].

For the computation of projection depth see [LZ14]. The simplicial depth of a point in $\mathbb{R}^3$ can be computed in $O(n^2)$ time, and in $\mathbb{R}^4$ the fastest algorithm needs $O(n^4)$ time [CO01]. For higher dimensions, no better algorithm is known than the straightforward $O(n^{d+1})$ method to compute all simplices.

When the number of data points and dimensions are such that the above algorithms become infeasible, one can resort to approximate algorithms. For halfspace depth such approximate algorithms were proposed in [RS98] and [CMW13]. An approximate algorithm in $\mathbb{R}^3$ based on a randomized data structure was proposed in [AC09]. An approximation to the Tukey median using steepest descent can be found in [SR00]. In [VRHS02] an algorithm is described to approximate the deepest regression fit in any dimension. A randomized algorithm for the Tukey median was proposed in [Cha04].

58.3 OTHER STATISTICAL TECHNIQUES

Computational geometry has provided fast and reliable algorithms for many other statistical techniques.

Linear regression is a frequently used statistical technique. The ordinary least squares regression, minimizing the sum of squares of the residuals, is easy to calculate, but produces unreliable results whenever one or more outliers are present in the data. Robust alternatives are often computationally intensive. We here give some examples of regression methods for which geometric or combinatorial algorithms have been constructed.

1. L^1 *regression.* This well-known alternative to least squares regression minimizes the sum of the absolute values of the residuals, and is robust to vertical outliers. Algorithms for L^1 regression may be found in, e.g., [YKII88] and [PK97].

2. *Least median of squares (LMS) regression* ([Rou84]). This method minimizes the median of the squared residuals and has a breakdown value of $1/2$. To compute the bivariate LMS line, an $O(n^2)$ algorithm using topological sweep has been developed [ES90]. An approximation algorithm for the LMS line was constructed in [MNR+97]. The recent algorithm of [BM14] uses mixed integer optimization.

3. *Median slope regression* ([The50], [Sen68]). This bivariate regression technique estimates the slope as the median of the slopes of all lines through two data points. An algorithm with optimal complexity $O(n \log n)$ is given in [BC98], and a more practical randomized algorithm in [DMN92].

4. *Repeated median regression* ([Sie82]). Median slope regression takes the median over all couples (d-tuples in general) of data points. Here, this median is replaced by d nested medians. For the bivariate repeated median regression line, [MMN98] provide an efficient randomized algorithm.

The aim of cluster analysis (Sections 48.5 and 54.1) is to divide a data set into clusters of similar objects. Partitioning methods divide the data into k groups. Hierarchical methods construct a tree (called a dendrogram) such that each cut of the tree gives a partition of the data set. A selection of clustering methods with accompanying algorithms is presented in [SHR97]. The general problem of partitioning a data set into groups such that the partition minimizes a given error function f is NP-hard. However, for some special cases efficient algorithms exist. For a small number of clusters in low dimensions, exact algorithms for partitioning methods can be constructed. Constructing clustering trees is also closely related to geometric problems (see e.g., [Epp97], [Epp98]).

58.4 SOURCES AND RELATED MATERIAL

SURVEYS

All results not given an explicit reference above may be traced in these surveys.

[Mos13]: A survey of multivariate data depth and its statistical applications.

[Sha76]: An overview of the computational complexities of basic statistics problems like ranking, regression, and classification.

[Sma90]: An overview of several multivariate medians and their basic properties.

[ZS00]: A classification of multivariate data depths based on their statistical properties.

RELATED CHAPTERS

Chapter 1: Finite point configurations
Chapter 28: Arrangements
Chapter 54: Pattern recognition

REFERENCES

[ABE+00]   N. Amenta, M. Bern, D. Eppstein, and S.-H. Teng. Regression depth and center points. *Discrete Comput. Geom.*, 23:305–323, 2000.

[AC09]   P. Afshani and T.M. Chan. On approximate range counting and depth. *Discrete Comput. Geom.*, 42:3–21, 2009.

[ACG+02]   G. Aloupis, C. Cortés, F. Gómez, M. Soss, and G.T. Toussaint. Lower bounds for computing statistical depth. *Comput. Stat. Data Anal.*, 40:223–229, 2002.

[ALS+03]   G. Aloupis, S. Langerman, M. Soss, and G.T. Toussaint. Algorithms for bivariate medians and a Fermat-Torricelli problem for lines. *Comput. Geom.*, 26:69–79, 2003.

[BC98]   H. Brönnimann and B. Chazelle. Optimal slope selection via cuttings. *Comput. Geom.*, 10:23–29, 1998.

[BCI⁺08]   D. Bremner, D. Chen, J. Iacono, S. Langerman, and P. Morin. Output-sensitive algorithms for Tukey depth and related problems. *Stat. Comput.*, 18:259–266, 2008.

[BH99]   Z.-D. Bai and X. He. Asymptotic distributions of the maximal depth estimators for regression and multivariate location. *Ann. Statist.*, 27:1616–1637, 1999.

[BM14]   D. Bertsimas and R. Mazumder. Least quantile regression via modern optimization. *Ann. Statist.*, 42:2494–2525, 2014.

[BRS11]   M. Burr, E. Rafalin, and D.L. Souvaine. Dynamic maintenance of halfspace depth for points and contours. Preprint, `arXiv:1109.1517`, 2011.

[Cha96]   P. Chaudhuri. On a geometric notion of quantiles for multivariate data. *J. Amer. Statist. Assoc.*, 91:862–872, 1996.

[Cha04]   T.M. Chan. An optimal randomized algorithm for maximum Tukey depth. In *Proc. ACM-SIAM Sympos. Discrete Algorithms*, 430–436, 2004.

[CHSV14]   G. Claeskens, M. Hubert, L. Slaets, and K. Vakili. Multivariate functional halfspace depth. *J. Amer. Statist. Assoc.*, 109:411–423, 2014.

[CMW13]   D. Chen, P. Morin, and U. Wagner. Absolute approximation of Tukey depth: Theory and experiments. *Comput. Geom.*, 46:566–573, 2013.

[CO01]   A.Y. Cheng and M. Ouyang. On algorithms for simplicial depth. In *Proc. 13th Canadian Conf. on Comp. Geom.*, pages 53–56, Waterloo, 2001.

[DG92]   D.L. Donoho and M. Gasko. Breakdown properties of location estimates based on halfspace depth and projected outlyingness. *Ann. Statist.*, 20:1803–1827, 1992.

[DM16]   R. Dyckerhoff and P. Mozharovskyi. Exact computation of the halfspace depth. *Comput. Statist. Data Anal.*, 98:19–30, 2016.

[DMN92]   M.B. Dillencourt, D.M. Mount, and N.S. Netanyahu. A randomized algorithm for slope selection. *Internat. J. Comput. Geom. Appl.*, 2:1–27, 1992.

[Epp97]   D. Eppstein. Faster construction of planar two-centers. In *Proc. 8th Annu. ACM-SIAM Sympos. Discrete Algorithms*, pages 131–138, 1997.

[Epp98]   D. Eppstein. Fast hierarchical clustering and other applications of dynamic closest pairs. In *Proc. 9th ACM-SIAM Sympos. Discrete Algorithms*, pages 619–628, 1998.

[ES90]   H. Edelsbrunner and D.L. Souvaine. Computing least median of squares regression lines and guided topological sweep. *J. Amer. Statist. Assoc.*, 85:115–119, 1990.

[FFC12]   H. Fritz, P. Filzmoser, and C. Croux. A comparison of algorithms for the multivariate L_1-median. *Comput. Statist.*, 27:393–410, 2012.

[Gow74]   J.C. Gower. The mediancenter. *J. Roy. Statist. Soc. Ser. C*, 32:466–470, 1974.

[HPS10]   M. Hallin, D. Paindaveine, and M. Šiman. Multivariate quantiles and multiple-output regression quantiles: From L_1-optimization to halfspace depth. *Ann. Statist.*, 38:635–669, 2010.

[HRV01]   M. Hubert, P.J. Rousseeuw, and S. Van Aelst. Similarities between location depth and regression depth. In L.T. Fernholz, editor, *Statistics in Genetics and in the Environmental Sciences*, pages 153–162, Birkhäuser, Basel, 2001.

[HRS15a]   M. Hubert, P.J. Rousseeuw, and P. Segaert. Multivariate functional outlier detection. *Stat. Methods Appl.*, 24:177–202, 2015.

[HRS15b]   M. Hubert, P.J. Rousseeuw, and P. Segaert. Multivariate and functional classification using depth and distance. *Adv. Data Anal. Classif.*, online first, 2016.

[JM94]   S. Jadhav and A. Mukhopadhyay. Computing a centerpoint of a finite planar set of points in linear time. *Discrete Comput. Geom.*, 12:291–312, 1994.

[KM12]   L. Kong and I. Mizera. Quantile tomography: Using quantiles with multivariate data. *Statist. Sinica*, 22:1589–1610, 2012.

[KMR⁺08]   M. van Kreveld, J.S.B. Mitchell, P. Rousseeuw, M. Sharir, J. Snoeyink, and B. Speckmann. Efficient algorithms for maximum regression depth. *Discrete Comput. Geom.*, 39:656–677, 2008.

[KMV02]   S. Krishnan, N.H. Mustafa, and S. Venkatasubramanian. Hardware-assisted computation of depth contours. In *Proc. 13th ACM-SIAM Sympos. Discrete Algorithms*, pages 558–567, 2002.

[LS00]   S. Langerman and W. Steiger. Computing a maximal depth point in the plane. In *Proc. Japan Conf. Discrete Comput. Geom.*, pages 46–47, 2000.

[LS03a]   S. Langerman and W. Steiger. Optimization in arrangements. In *Proc. 20th Sympos. Theor. Aspects Comp. Sci.*, vol. 2607 of *LNCS*, pages 50–61, Springer, Berlin, 2003.

[LS03b]   S. Langerman and W. Steiger. The complexity of hyperplane depth in the plane. *Discrete Comput. Geom.*, 30:299–309, 2003.

[LCL12]   J. Li, J.A. Cuesta-Albertos, and R.Y. Liu. DD-classifier: Nonparametric classification procedure based on DD-plot. *J. Amer. Statist. Assoc.*, 107:737–753, 2012.

[Liu90]   R.Y. Liu. On a notion of data depth based on random simplices. *Ann. Statist.*, 18:405–414, 1990.

[LMM14]   X. Liu, K. Mosler, and P. Mozharovskyi. Fast computation of Tukey trimmed regions in dimension $p > 2$. arXiv:1412.5122 [stat.CO], 2014.

[LPS99]   R.Y. Liu, J. Parelius, and K. Singh. Multivariate analysis by data depth: descriptive statistics, graphics and inference. *Ann. Statist.*, 27:783–840, 1999.

[LZ14]   X. Liu and Y. Zuo. Computing projection depth and its associated estimators. *Stat. Comput.*, 24:51–63, 2014.

[Miz02]   I. Mizera. On depth and deep points: a calculus. *Ann. Statist.* 30:1681–1736, 2002.

[MMN98]   J. Matoušek, D.M. Mount, and N.S. Netanyahu. Efficient randomized algorithms for the repeated median line estimator. *Algorithmica*, 20:136–150, 1998.

[MNR⁺97]   D.M. Mount, N.S. Netanyahu, K. Romanik, R. Silverman, and A.Y. Wu. A practical approximation algorithm for the LMS line estimator. In *Proc. 8th Annu. ACM-SIAM Sympos. Discrete Algorithms*, pages 473–482, 1997.

[Mos13]   K. Mosler. Depth statistics. In C. Becker, R. Fried, and S. Kuhnt, editors, *Robustness and Complex Data Structures*, pages 17–34, Springer, Heidelberg, 2013.

[MRR⁺03]   K. Miller, S. Ramaswami, P. Rousseeuw, J.A. Sellarès, D.L. Souvaine, I. Streinu, and A. Struyf. Efficient computation of location depth contours by methods of computational geometry. *Stat. Comput.*, 13:153–162, 2003.

[Oja83]   H. Oja. Descriptive statistics for multivariate distributions. *Statist. Probab. Lett.*, 1:327–332, 1983.

[PK97]   S. Portnoy and R. Koenker. The Gaussian hare and the Laplacian tortoise: computability of squared-error versus absolute-error estimators. *Stat. Sci.*, 12:279–300, 1997.

[RH99]   P.J. Rousseeuw and M. Hubert. Regression depth. *J. Amer. Statist. Assoc.*, 94:388–402, 1999.

[RH99b]   P.J. Rousseeuw and M. Hubert. Depth in an arrangement of hyperplanes. *Discrete Comput. Geom.*, 22:167–176, 1999.

[Rou84]   P.J. Rousseeuw. Least median of squares regression. *J. Amer. Statist. Assoc.*, 79:871–880, 1984.

[RR96a]   P.J. Rousseeuw and I. Ruts. Algorithm AS 307: Bivariate location depth. *J. Roy. Statist. Soc. Ser. C*, 45:516–526, 1996.

[RR96b]   I. Ruts and P.J. Rousseeuw. Computing depth contours of bivariate point clouds. *Comput. Stat. Data Anal.*, 23:153–168, 1996.

[RRT99]   P.J. Rousseeuw, I. Ruts, and J.W. Tukey. The bagplot: A bivariate boxplot. *Amer. Statist.*, 53:382–387, 1999.

[RS98]   P.J. Rousseeuw and A. Struyf. Computing location depth and regression depth in higher dimensions. *Stat. Comput.*, 8:193–203, 1998.

[RS04]   P.J. Rousseeuw and A. Struyf. Characterizing angular symmetry and regression symmetry. *J. Statist. Plann. Inference*, 122:161-173, 2004.

[Sen68]   P.K. Sen. Estimates of the regression coefficient based on Kendall's tau. *J. Amer. Statist. Assoc.*, 63:1379–1389, 1968.

[Ser02]   R. Serfling. A depth function and a scale curve based on spatial quantiles. In Y. Dodge, editor, *Statistical Data Analysis Based on the L_1-Norm and Related Methods*, pages 25–38, Birkhaüser, Basel, 2002.

[Sha76]   M.I. Shamos. Geometry and statistics: Problems at the interface. In J.F. Traub, editor, *Algorithms and Complexity: New Directions and Recent Results*, pages 251–280, Academic Press, Boston, 1976.

[SHR97]   A. Struyf, M. Hubert, and P.J. Rousseeuw. Integrating robust clustering techniques in S-PLUS. *Comput. Stat. Data Anal.*, 26:17–37, 1997.

[Sie82]   A. Siegel. Robust regression using repeated medians. *Biometrika*, 69:242–244, 1982.

[Sma90]   C.G. Small. A survey of multidimensional medians. *Internat. Statistical Review*, 58:263–277, 1990.

[SR00]   A. Struyf and P.J. Rousseeuw. High-dimensional computation of the deepest location. *Comput. Stat. Data Anal.*, 34:415–426, 2000.

[The50]   H. Theil. A rank-invariant method of linear and polynomial regression analysis (parts 1-3). *Nederl. Akad. Wetensch. Ser. A*, 53:386–392, 521–525, 1397–1412, 1950.

[Tuk75]   J.W. Tukey. Mathematics and the picturing of data. In *Proc. Internat. Congr. of Math.*, 2, pages 523–531, Vancouver, 1975.

[VRHS02]   S. Van Aelst, P.J. Rousseeuw, M. Hubert, and A. Struyf. The deepest regression method. *J. Multivar. Anal.*, 81:138–166, 2002.

[YKII88]   P. Yamamoto, K. Kato, K. Imai, and H. Imai. Algorithms for vertical and orthogonal L^1 linear approximation of points. In *Proc. 4th Sympos. Comput. Geom.*, pages 352–361, ACM Press, 1988.

[Zuo03]   Y. Zuo. Projection based depth functions and associated medians. *Ann. Statist.*, 31:1460–1490, 2003.

[ZS00]   Y. Zuo and R. Serfling. General notions of statistical depth function. *Ann. Statist.*, 28:461–482, 2000.

59 GEOGRAPHIC INFORMATION SYSTEMS
Marc van Kreveld

INTRODUCTION

Geographic information systems (GIS) facilitate the input, storage, manipulation, analysis, and visualization of geographic data. Geographic data generally has a location, size, shape, and various attributes, and may have a temporal component as well. Geographical analysis is important for a GIS. It includes combining different spatial themes, relating the dependency of phenomena to distance, interpolating, studying trends and patterns, and more. Without analysis, a GIS could be called a spatial database.

Not all aspects of GIS are relevant to computational geometry. Human-computer interaction, and legal aspects of GIS, are also considered part of GIS research. This chapter focuses primarily on those aspects that are susceptible to algorithms research. Even here, the approach taken within GIS research is different from the approach a computational geometer would take, with much less initial abstraction of the problem, and less emphasis on theoretical efficiency. The GIS research field is multi-disciplinary: it includes researchers from geography, geodesy, cartography, and computer science. The research areas geodesy, surveying, photogrammetry, and remote sensing primarily deal with the data input, storage, and correction aspects of GIS. Cartography mainly concentrates on the visualization aspects.

Section 59.1 deals with spatial data structures important to GIS. Section 59.2 discusses the most common spatial analysis methods. Section 59.3 discusses the visualization of spatial data, in particular automated cartography. Section 59.4 deals with Digital Elevation Models (DEMs) and their algorithms. Section 59.5 discusses algorithmic results on trajectory data. Section 59.6 overviews miscellaneous other occurrences of geometric algorithms in GIS. Section 59.7 discusses the most important contributions that can be made from the computational geometry perspective to research problems in GIS.

59.1 SPATIAL DATA STRUCTURES

GIS store different types of data separately, such as land cover, elevation, and municipality boundaries. Therefore, each such data set is stored in a separate data structure that is tailored to the data, both in terms of representation and searching efficiency.

Geometric data structures for intersection, point location, and windowing are a mainstream topic in computational geometry, and are treated at length in Chapters 38, 40, and 42. This section concentrates on concepts and results that are specific and important to GIS. We overview raster and vector representation of data, problems that appear in data input and correction in a GIS, and a well-known spatial indexing structure.

GLOSSARY

Thematic map layer: Separately stored and manipulatable component of a map that contains the data of only one specific theme or geographic variable.

(Geographic) feature: Any geographically meaningful object.

Raster structure: Representation of geometric data based on a subdivision of the underlying space into a regular grid of square pixels.

Vector structure: Representation of geometric data based on the representation of points with coordinates, and line segments between those points.

Digitizing: Process of transforming cartographic data such as paper maps into digital form by tracing boundaries with a mouse-like device.

Conflation: Process of rectifying a digital data set by comparison with another digital data set that covers the same region (cf. rubber sheeting).

Topological vector structure: Vector structure in which incidence and adjacency of points, line segments, and faces is explicitly represented.

Quadtree: Tree where every internal node has four children, and which corresponds to a recursive subdivision of a square into four subsquares. The standard quadtree is for raster data. Leaves correspond to pixels or larger squares that appear in the recursive subdivision and are uniform in the thematic value.

R-tree: Tree based on a recursive partitioning of a set of objects into subsets, where every internal node stores a number of pairs (BB, S) equal to the number subtrees, where BB is the axis-parallel bounding box of all objects that appear in the subtree S. All leaves have the same depth and store the objects. R-trees have high (but constant) degree and are well-suited for secondary memory.

59.1.1 RASTER AND VECTOR STRUCTURES

Geographic data is composed of geometry, topology, and attributes. The attributes contain the semantics of a geographic feature. There are two essentially different ways to represent the geometric part of geographic data: raster and vector. This distinction is the same as representation in image space and object space in computer graphics (Chapter 52).

Data acquisition and input into a GIS often cause error and imprecision in the data, which must be corrected either manually or in an automated way. Also, the digitizing of paper maps yields unstructured collections of polygonal lines in vector format, to which topological structure is usually added using the GIS.

The topological vector structure obtained could be represented as, for example, a doubly-connected edge list or a quad-edge structure. But for maps with administrative boundaries, where long polygonal boundary lines occur where all vertices have degree 2, such a representation is space-inefficient. It is undesirable to have a separate object for every vertex and edge, with pointers to the incident features. The following variation gives better efficiency. Group maximal chains of degree-2 vertices into single objects, and treat them like an edge in the doubly-connected edge list. More explicitly, a chain stores pointers to the origin vertex (junction or endpoint), the destination vertex (junction or endpoint), the left face, the right face, and the next and previous chains in the two cycles of the faces incident to this

FIGURE 59.1.1

A set of polygons and an example of an R-tree for it.

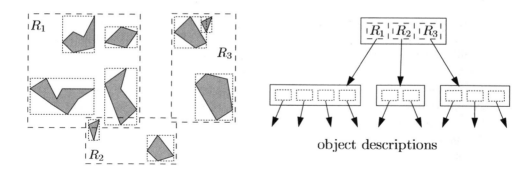

object descriptions

chain. Such a representation allows retrieval of k adjacent faces of a face with m vertices to be reported in $O(k)$ time rather than in $O(m)$ time.

Advances in geographic data modeling and representation include multi-scale models, temporal and spatio-temporal models, fuzzy models, and qualitative representations of location.

59.1.2 R-TREES

The most widely used spatial data structure in GIS is the R-tree of Guttmann [Gut84]. It is a type of box-tree (see [BCG+96]) that has high (but constant) degree internal nodes, with all leaves at the same depth. It permits any type of object to be stored, and supports several types of queries, such as windowing, point location, and intersection. Insertions and deletions are both supported. The definition of R-trees does not specify which objects go in which subtree, and different heuristics for grouping give rise to different versions [BKSS90, KF94, LLE97].

R-trees generally do not have nontrivial worst-case query time bounds, so different versions must be compared experimentally. Only the versions of Agarwal et al. [ABG+02] and Arge et al. [ABHY08] have a query time bound better than linear (close to $O(\sqrt{n})$) when the stored objects are rectangles. These structures are also I/O-efficient.

59.2 SPATIAL ANALYSIS

Spatial analysis is the process of discovering information implicitly present in one or more spatial data sets. It includes common GIS operations such as map overlay, buffer computation, and shortest paths on road networks, but also geostatistical and spatial data analysis functions such as cluster detection, spatial interpolation, and spatial modeling. We discuss the most common forms and results in this section. For cluster analysis and classification, see Chapter 32.

GLOSSARY

Map overlay: The operation of combining two thematic map layers of the same region in order to obtain one new map layer, often with a refinement of the subdivisions used for the input map layers.

Buffer: The region of the plane within a certain specified distance to a geographic feature.

Neighborhood analysis: The study of how relations between geographic features depend on the distance.

Network analysis: The study of distance, reachability, travel time, and similar geographic functions that can be defined for network data (graphs with a geographic meaning).

Cluster analysis: The study of the grouping in sets of geographic features by proximity.

Trend analysis: The study of time-dependent patterns in geographic data.

Spatial interpolation: The derivation of values at locations based on values at other (nearby) locations.

Geostatistics: Statistics for data associated with locations in the plane.

59.2.1 MAP OVERLAY

With map overlay, two or more thematic map layers are combined into one. For example, if one map layer contains elevation contours and another map layer (of the same region) forest types, then their overlay reveals which types of forest occur at which elevations. One layer can also serve as a mask for the other layer. Overlay is essential to locating a region that has various properties that appear in different thematic map layers. In the spatial database literature, map overlay is also called *spatial join.*

Map overlay is commonly solved using a plane sweep like the Bentley-Ottmann algorithm [BO79] for line segment intersection. This leads to an $O((n + k) \log n)$ time algorithm for two planar subdivisions of complexity $O(n)$, and output complexity $O(k)$. However, map overlay of two thematic map layers is essentially an extension of a red-blue line segment intersection problem, and can therefore be solved in optimal $O(n \log n + k)$ time [CEGS94, Cha94, PS94]. In case each subdivision is simply connected, the running time can be improved to $O(n + k)$ [FH95].

Map overlay in GIS must handle imprecise data as well, and therefore overlay methods that include sliver removal have been suggested. Essentially, boundaries that are closer than some pre-specified value are identified in the overlay. This is also called *epsilon filter* or *fuzzy tolerance* [Chr97]. The idea of using an epsilon band around a cartographic line is due to Perkal [Per66].

Since R-trees can also be used as access structures for subdivisions, map overlay can also be performed efficiently using R-trees [BKS93, KBS91, Oos94].

59.2.2 BUFFER COMPUTATION

The buffer of a geographic feature of width ϵ is the same as the Minkowski sum of that feature with a disk of radius ϵ, centered at the origin; see Chapter 50.

Computation can be done with the algorithms mentioned in that chapter.

Buffer computation and map overlay are two main ingredients for urban planning. As an example, three requirements for a new factory may be the proximity of a river, at least some distance to houses, and not in nature preserves. A map with suitable locations is obtained after computing buffers for two of the thematic map layers, and then combining these with each other and the third layer.

59.2.3 SPATIAL INTERPOLATION

Spatial interpolation is one of the main operations in geostatistics. It is the operation of defining values at locations when only values at other locations are known. For example, when ground measurements are taken at various locations, we only know values at a finite set of points, but we would like to know the values everywhere. Several methods exist for this version of the spatial interpolation problem, including triangulation, moving windows, natural neighbors, and Kriging. Triangulation is discussed in a later section. Moving windows is a form of weighted averaging of known (or observed) values within a window around the point with unknown value.

Natural neighbor interpolation is a method based on Voronoi diagrams [Sib81, SBM95]. Suppose the Voronoi diagram of the points with known values is given, and we want to obtain an interpolated value at another location p_0. We determine what Voronoi cell p_0 would "own" if it were inserted in the point set defining the Voronoi diagram. Let A be the area of the Voronoi cell of p_0, and let $A_1, \ldots, A_k$ be the areas removed from the Voronoi cells of the points $p_1, \ldots, p_k$, due to the insertion of p_0. Then, by natural neighbor interpolation, the interpolated value at p_0 is $\sum_{i=1}^{k} (A_i/A) \cdot V(p_i)$, where $V(p_i)$ denotes the known or observed value at p_i. The bivariate function obtained is continuous everywhere, and differentiable except at the points with known values.

Kriging is an interesting method that also applies weighted linear combinations of the known (or observed) values, that is, $V(p_0) = \sum_{i=1}^{n} \lambda_i \cdot V(p_i)$. The λ_i are the weights, and $\sum_{i=1}^{n} \lambda_i = 1$. Furthermore, the weights are chosen so that the estimation variance is less than for any other linear combination of known values. One additional advantage of Kriging is that it provides an estimation error as well [BM98].

Splines, discussed extensively in Chapter 56, can also be used for interpolation. A version used in GIS are the thin-plate splines. They do not necessarily pass through the known values of the points, and can therefore reduce artifacts. The spline function minimizes the sum of two components, one representing the smoothness and the other representing the proximity to the known values of the points [BM98].

59.3 VISUALIZATION OF SPATIAL DATA

Various tasks traditionally performed manually by cartographers can be automated. This not only leads to a more efficient map production process, it may also be necessary during the use of a GIS. A user of a GIS will typically select a several thematic map layers for display, and exactly which layers are selected determines which labels are needed and where they are placed best. Hence, pre-computation of label positions can often not be done off-line or by hand.

GLOSSARY

Choropleth map: Map in which the regions of an administrative subdivision are shown using a particular color scheme to represent a geographic variable.

Isoline map: Map for a continuous spatial phenomenon where curves of equal value for that phenomenon are displayed.

Cartogram: Map in which the area of the regions of an administrative subdivision represent a geographic variable (also called **value-by-area map**).

Rectangular cartogram: Map in which administrative regions are shown as rectangles and the area of the rectangles represent a geographic variable.

Linear cartogram: Map in which travel times between locations are represented by distance on the map.

Schematic map: Map where important locations and connections between them (direct transportation) are shown highly stylized, and where location is preserved only approximately.

Flow map: Map that displays arrows of varying thickness that represent flows between geographic features. Arrows are often bundled at shared departure features or arrival features to enhance the visual representation.

Dot map: Map that shows dots where a dot represents the presence of a phenomenon or of a certain quantity of entities.

Label placement problem: The problem of placing text to annotate features on a map, according to various constraints and optimization criteria.

Line simplification problem: The problem of computing a polygonal line with fewer vertices from another polygonal line, while satisfying given constraints of distance.

Cartographic generalization problem: The problem of computing a map at a coarser (smaller) scale from a data set whose detail would be appropriate for a map at a finer (larger) scale.

59.3.1 LABEL PLACEMENT

Automated label placement has been the topic of considerable research, both within cartography and within the field of algorithms. One can distinguish three types of labels: labels for point objects, labels for line objects, and labels for polygonal objects. Imhof [Imh75] provides many examples of well-placed and poorly-placed labels, demonstrating the many different requirements for practical, high-quality label placement.

The point-label placement problem is the following optimization problem. Given a set of points, each with a specified label (name or other text), place as many labels as possible adjacent to their point, but without overlap between any two labels. One can extend the problem by restricting, or not allowing, overlap with other map features, avoiding ambiguity, and so on. Another version of point labeling is to maximize label size under the condition that all points be labeled. Label

FIGURE 59.3.1
River labeling due to Wolff et al. [WKK+00].

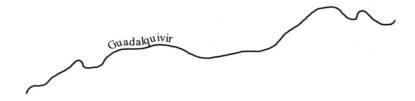

placement for point objects is usually approached as follows. A label is modeled by an axis-parallel rectangle, the bounding box of the text. For each point, define a restricted set of positions considered for its label, the candidates. Typically, the four positions where a corner of the label coincides with the point are chosen. In the label number maximization problem, the problem is abstracted to maximum independent set in a graph where edges represent intersections of two candidate label positions. Tables 59.3.1 and 59.3.2 contain selected results.

TABLE 59.3.1   Point-label placement: size maximization; selected results.

TYPE OF LABEL	POSITIONS	APPROX. FACTOR	TIME	SOURCE
Equal-size square	4	2	$O(n \log n)$	[FW91, WW97]
Equal-size square	2	1	$O(n \log n)$	[FW91]
Arbitrary rectangle	2	1	$O(n \log^2 n)$	[FW91]
Arbitrary rectangle	4	2	$O(n^2 \log n)$	[JC04]
Equal-size disk	touching	$2.98 + \epsilon$	$O(n \log n) + n(1/\epsilon)^{O(1/\epsilon^2)}$	[JBQZ04].

TABLE 59.3.2   Point-label placement: number maximization; selected results.

TYPE OF LABEL	POSITIONS	APPROX. FACTOR	TIME	SOURCE
Rectangle	constant	$O(\log n)$	$O(n \log n)$	[AKS98]
Fixed-height rectangle	constant	2 (or PTAS)	$O(n \log n)$	[AKS98, Cha04]
Fixed-height rectangle	touching	2 (or PTAS)	$O(n \log n)$	[KSW99, PSS+03]
Disk, disk-like	constant	PTAS	$n^{O(1/\epsilon^2)}$	[EJS05]

Combinatorial optimization approaches have also been applied frequently to the point-label placement problem [CMS95, DTB99]. However, experiments show that simple heuristics work well in practice [CMS95, WWKS01].

We next discuss the labeling of line features. Here we distinguish in streets and rivers. The labeling of street patterns yields a combinatorial optimization problem similar to point labeling [NW00, Str01]. River labeling is quite different, because there are several different criteria that constitute a good river label placement. The label should be close to the river, it should follow the shape of the river, it should not have too high curvature, it should be as horizontal as possible, and it should have few inflection points. The algorithm of Wolff et al. [WKK+00] includes all of these criteria; Figure 59.3.1.

The labeling of polygonal features appears for instance when placing the name of a country or lake inside that feature. It is common to either choose horizontal and straight placement, or let the shape follow the main shape of the polygonal feature. In the first case, one can place the label in the middle of the largest scaled copy of the label that fits inside the region. In the second case, one can use the medial axis to retrieve the main shape and place the label along it.

59.3.2 CARTOGRAPHIC GENERALIZATION

Cartographic generalization is the process of transforming and displaying cartographic data with less detail and information (i.e., on a coarser, smaller scale) than the input data contains. Examples include omitting small towns and minor roads, using only one color for nature regions rather than a distinction in forest, heath, moor, etc., aggregating several buildings into one block, and exaggerating the width of a road on a small-scale map. Generalization is a very important research topic in automated cartography [MLW95, MRS07].

The changes to the map data for generalization are done by *generalization operators*. They include selection, aggregation, typification, reclassification, smoothing, displacement, exaggeration, symbolization, collapse, and many others. Detecting a need for generalization is accomplished by computing certain geometric measures on distance, density, and detail. This will trigger the generalization operators to perform transformations. It may happen that one operator causes the need of a change somewhere else on the map, possibly leading to a domino effect. For example, one of the common generalization operators is displacement, to ensure a certain distance between two map features that should not appear adjacent. Moving one feature may cause the need to move another, leading to iterative displacement algorithms [Høj98, LJ01, MP01].

The problem of (polygonal) line simplification (cf. Section 50.3) is often considered a cartographic generalization problem, too. However, if the motivation for line simplification is only data reduction, then line simplification cannot be considered generalization. But since line simplification methods automatically reduce detail in polygonal lines, we will discuss some methods here.

The best-known cartographic line simplification method is due to Douglas and Peucker [DP73]. Starting with a line segment between the endpoints of the polygonal line, it selects the most distant vertex to be added to the simplification, and then continues recursively on the two parts that appear. This process continues until the most distant vertex is closer than some chosen threshold value to the approximating line segment. Theoretically, the method is highly unsatisfactory because it can create self-intersections in the output, requires quadratic time in the worst case, and may need many more segments in the approximation than the optimal approximation. However, it is very simple and usually works well in practice. Hershberger and Snoeyink devised a different algorithm to compute the same approximation which runs in $O(n \log^* n)$ time [HS98].

Weibel [Wei97] and van der Poorten and Jones [PJ99] demonstrate that many aspects are involved in practical line simplification for GIS, and that many different criteria may be used. Buchin et al. [BMS11b] show that local improvements are very effective and can also be used for other generalization operators. The GIS literature contains several more practical approaches.

Guibas et al. [GHMS93] and Estkowski and Michell [EM01] show that minimum vertex simplification is NP-hard when self-intersections are not allowed. Selected algorithmic results are listed in Table 59.3.3.

TABLE 59.3.3   Polygonal line simplification: selected results.

OUTPUT VERTICES	COMPLEXITY	ERROR CRITERION	NOTES	SOURCE
From input	$O(n^2)$	distance	min. link, self-inter.	[CC96]
From input	$O(n^{4/3+\delta})$	vertical distance	min. link, self-inter.	[AV00]
From input	$O(n^2 \log n)$	distance	respects context	[BKS98]
From input	$O(n^3)$	distance	subdivision	[EM01]
From input	$O(n^2)$	distance along path		[GNS07]
From input	$O(n^3), O(n^4)$	area displacement	min. area measures	[BCC$^+$06]
From input	$O(n^2)$	distance	angle constraints	[CDH$^+$05]
Arbitrary	$O(n^2 \log n)$	distance	subdivision	[GHMS93]

59.3.3 THEMATIC MAPS

Topographic maps are general-purpose maps that display a variety of themes of general interest together, like roads, towns, forests, and elevation contours. Thematic maps, on the other hand, concentrate on a particular theme, and may use alternative methods of visualizing the information. A choropleth map could, e.g., show the population densities of the states of the U.S. by coloring each with a color from a well-chosen set of colors, for instance, five saturation values of red. The geographic theme of population density can be seen as a scalar function. Here the points of the plane are aggregated by state.

There are other ways of visualizing scalar functions cartographically, including isoline maps, dot maps, and cartograms. The latter again applies to aggregated regions of the plane. Flow maps visualize a presence and quantity of flow from one (aggregated) region to another. Schematic maps visualize connections between locations, such as subway maps. Dent et al. [DTH09] provide a good overview of several thematic map types.

Cartograms show values for regions by shrinking and expanding those regions, so that the area of each region corresponds to the value represented. The most important usage is the population cartogram, where a region A with a population twice that of region B will be shown twice as large as B. Necessarily, cartograms show a distortion of the geographic space. To keep the regions more or less recognizable, they should keep their shape, location, and adjacency as much as possible.

Several algorithms have been proposed to construct cartograms, given an administrative subdivision and a value for each region. Tobler [Tob86] simply uses scaling on the x- and y-coordinates, which may prevent regions from being shown at the correct size. Dougenick et al. [DCN85] compute a centroid for each region, which is assigned a repelling force if the region should grow and an attracting force if the region should shrink. The forces of all centroids on all boundaries of the map then result in new positions of these boundaries. This is used in an iterative algorithm. Kocmoud and House's approach [KH98] is constraint programming. They

FIGURE 59.3.2

Left, regions with a K_4 as adjacency graph cannot be represented by adjacent rectangles. Right, four rectangles inside a rectangular outline and with specified adjacencies cannot realize the specified areas without changing the adjacencies.

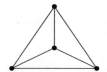 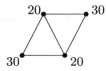

also attempt to preserve the main orientations of the boundaries. Edelsbrunner and Waupotitsch [EW97] give a cartogram construction algorithm based on simplicial complexes in the plane, where paths of triangles are used to define deformations that let one region grow at the expense of the size of another.

The first algorithmic study of rectangular cartograms, where all regions are rectangles, was done by van Kreveld and Speckmann [KS07]; extensions and refinements were presented by Buchin et al. [BSV12]. Broadly speaking, there are two reasons why correct rectangle sizes and correct adjacencies are not always possible, see Figure 59.3.2. To overcome these problems, rectilinear cartograms were introduced, where regions can have more than four corners. De Berg et al. [BMS09] showed that only constantly many corners are needed in rectilinear cartograms. Their bound on the number of corners was improved by Alam et al. [ABF+13] to eight corners per region, which is worst-case optimal.

In linear cartograms, distances are distorted rather than areas. This is useful when a map serves to represent travel time [BK12, KWFP10, SI09]. Alternatively, one can use edge length to represent travel time [BGH+14].

Flow maps show the movement of objects between geographic locations on a map using thick arrows, see Figure 59.3.3. Edge bundling is commonly used to avoid visual clutter. Using a modification of Steiner minimal trees, Buchin et al. [BSV15] modeled this problem and solved it with an approximation algorithm, since the general problem is NP-hard.

Dot maps show values by dots, where one dot represents, e.g., ten thousand people. This allows the distribution of the population to be shown better than in cartograms, but the relative populations for two regions are more difficult to compare. De Berg et al. [BBCM02] show the connection between dot maps and discrepancy, and compare various heuristics to construct dot maps.

Schematic maps are commonly used for public transportation systems. Direct lines, or connections between major stations, are shown with a polygonal line that is highly abstracted: it has only a few segments and often, these segments are restricted to be horizontal, vertical, or have slope +1 or −1. Cabello et al. [CBK05] place the connections incrementally in a pre-computed order, leading to an $O(n \log n)$ time algorithm. Neyer [Ney99] views the problem as a line simplification problem and approximates each connection with the minimum number of segments in the specified orientations. Nöllenburg and Wolff [NW11] give an integral approach to the problem that takes multiple constraints into account and solve it with mixed-integer programming. Brandes and Wagner [BW98] show connections between stations by circular arcs and address the visualization problem as a graph layout problem (Chapter 55).

FIGURE 59.3.3

Flow map showing migration from Colorado (from Buchin et al. [BSV11]).

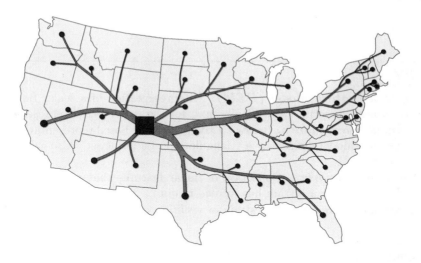

59.3.4 DYNAMIC AND ANIMATED MAPS

Besides computations for traditional paper maps, a more recent trend is to study dynamic maps, animated maps, interactive maps, Web maps, and multimedia maps [KB01, CPG99]. This area leads to a number of new computational issues, where efficiency is very important and quality is less critical. At the same time, the computational problems involved may become harder. For example, where label maximization on static maps is generally NP-hard, a related version on dynamic maps is PSPACE-hard [BG14]. Several other results on dynamic and interactive map labeling exist [BFH01, PPH99].

Zooming out on a map also makes real-time cartographic generalization necessary. The problem is that not only the size of features must be changed, but also the way of visualization. On large-scale maps, cities are shown by polygonal outlines, but on small-scale maps, they are shown by point symbols. The process is called **dynamic** or **on-the-fly generalization** [BNPW10, MG99, GNN13, GNR16, Oos95, SHWZ14]. Ideally, the changes made during zooming should be made in a continuous manner, with no major, sudden changes on the map [Kre01]. In static generalization, the objective is to compute a new representation, but in dynamic generalization, the problem is the computation of the transition.

59.4 DIGITAL ELEVATION MODELS

We have concentrated on types of data based on subdivisions with well-marked boundaries. Another important type of data is the scalar function in two variables. The most common example from geography is elevation above sea level, also called terrain. Three other examples are annual precipitation, nitrate concentration per cubic meter, and average noise level.

There are two common representations for elevation: the regular square grid, or **elevation matrix**, which is a raster representation, and the triangular irregular network (TIN), which is a vector representation. For the latter representation, the Delaunay triangulation is often used.

GLOSSARY

Digital elevation model (DEM):   Representation of a scalar function in two variables. Sometimes specifically used for the raster-based representation.

Triangular irregular network (TIN):   Vector-based representation of a digital elevation model defined by a triangulation of a point set. Also called polyhedral terrain.

Drainage network: Collection of linear features that represent the locations where water on a terrain has formed rivers.

Viewshed analysis: The study of visibility on a terrain.

59.4.1 CONSTRUCTION AND SIMPLIFICATION OF TINS

The problem of simplifying a digital elevation model, or performing raster to vector conversion for a digital elevation model, is a higher-dimensional version of line simplification. The best algorithm known is similar in approach to the Douglas-Peucker algorithm for line simplification given in Section 59.3.2. Assume that the outer boundary of the DEM is rectangular, a set of points with their elevation is given (e.g., based on a regular square grid), and assume that a maximum allowed vertical error $\epsilon > 0$ is specified. An initial coarse simplification of the TIN is a triangulation of the four corners of the rectangle. If that simplification is vertically within a distance ϵ from all points, then it is accepted. Otherwise, the point with largest vertical distance is selected and added to the triangulation, which is restored by flipping to the Delaunay triangulation. The process is then repeated.

The method requires quadratic time in the worst case, but an implementation can be given which, under natural assumptions, takes $O(n \log n)$ time in practice [Hel90, Fjä91, HG95].

Agarwal and Suri [AS98] show that a corresponding optimization problem is NP-hard, and give an approximation algorithm that requires $O(n^8)$ time. If m is the size of the optimal piecewise linear ϵ-approximation of the n given points, then the computed approximation has size $O(m \log m)$. Agarwal and Desikan [AD97] give a cubic time ϵ-approximation algorithm with a worse size bound on the approximation, but with some assumptions the approximation has the same size asymptotically and runs in near-quadratic time.

When a TIN is constructed for modeling terrains, various geometric computation problems arise. When the input is a set of (digitized) contour lines, a triangulation between the contour lines such as the constrained Delaunay triangulation can be used [DP89, Sch98]. Care must be taken that no triangle with all three vertices on the same contour line is created, as this gives undesirable artifacts. Thibault and Gold [TG00] provide a solution that avoids flat triangles by adding vertices on the medial axis or skeleton, which are given intermediate elevations (see also [GD02]). If information on rivers is present too, then these can be included as edges of the TIN using a constrained Delaunay triangulation [MS99]. Other approaches of interest are those by Silva et al. [SMK95], Little and Shi [LS01], and van Kreveld and Silveira [KS11]. These methods concern the construction of a TIN that can integrate or preserve important features like valleys and ridges of the terrain.

Gudmundsson et al. [GHK02] define a class of well-shaped triangulations called higher-order Delaunay triangulations, and use them to create TINs with fewer local minima in the terrain, because such minima generally do not occur. The idea was developed further by de Kok et al. [KKL07].

Multi-resolution terrain modeling has been studied extensively, see Puppo and Scopigno [PS97] for an overview. The approach of de Berg and Dobrindt [BD98] allows multiple levels of detail, and even the combination of multiple levels of detail in one terrain, but still uses a Delaunay triangulation in all cases. See [Ros04] for more information on surface simplification and multi-resolution representations.

59.4.2 VISUALIZATION OF DEMS

Digital elevation models may be visualized in several ways. A traditional way is by contour maps, and the process of deriving a contour map is called ***contouring*** [Wat92].

A perspective view of a digital elevation model can be obtained by back-to-front rendering of the grid elements or triangles. If a vector representation of a perspective view is needed, an algorithm of Katz et al. [KOS92] achieves this for a TIN in $O((n + k) \log n \log \log n)$ time, where n is the number of triangles and k is the complexity of the visibility map.

59.4.3 DERIVED MAPS AND PRODUCTS

In the analysis of terrain—e.g., for land suitability studies—slope and aspect maps are important derived products of a digital elevation model. They are straightforward to compute. Similarly, the plan and profile curvature can be of importance, for example for waterflow and erosion modeling.

The computation of the drainage network, based on the shape of the terrain, has been frequently studied, most often for grid data. Besides the drainage network, watersheds also provide important terrain information. A surprising combinatorial result on TINs of de Berg et al. [BBD+96] is that if water always follows the direction of steepest descent, and the drainage network consists of all points that receive flow from some region with positive area, then triangulations exist with n triangles for which the drainage network has complexity $\Theta(n^3)$, see also de Berg et al. [BHT11].

Viewshed analysis is the study of visibility in the terrain [BCK08, BHT10, HLM+14, KOS92]. Viewshed analysis has applications in urban and touristical planning, and for telecommunication, for example, to place a small number of antennas so that every point on the terrain has direct visibility to at least one antenna [BMM+04, Fra02]. Other visibility results for terrains may be found in Section 33.8.3.

The computation of shortest paths between two points on a terrain is a problem of both theoretical and practical interest. The approach of Alexandrov et al. [AMS05, LMS97] is significant both theoretically and practically, because it also deals with the weighted version. The main idea is to place Steiner points on edges to convert the problem into a graph problem, and then apply Dijkstra's algorithm. This gives a simple approximation algorithm for least-cost paths.

59.5   TRAJECTORY DATA

Trajectory data is a type of geographic data that has become of increasing importance. Due to high-quality GPS and RFID technology, large data sets of trajectories have been collected and stored, with applications in pedestrian and vehicle movement analysis, weather forecasting, behavioral ecology, security, and more. The processing and analysis of such data gives rise to algorithmic problems that have been studied in the areas of GIS and computational geometry [ZZ11]. Geometrically, a trajectory can be seen as a polygonal line, but the time component implies that different methods are needed from those for polygonal line processing. Most algorithmic research models a trajectory by a sequence of time-stamped points (the vertices) and assumes linear interpolation in both space and time in between these vertices. This makes location change over time in a piecewise-linear manner and velocity in a piecewise-constant manner.

Since processing and analyzing trajectories may be time-consuming, trajectory simplification can be important as a preprocessing step. One could consider any algorithm for polygonal line simplification to simplify trajectories, but it is better to use a dedicated trajectory simplification method [CWT06, GKM+09, LK11].

59.5.1 TRAJECTORY SEGMENTATION

Just like images can be segmented into semantically meaningful parts, so can trajectories. A segmentation of a trajectory partitions it into parts where each part corresponds to a type of motion that could indicate a type of behavior, a mode of transport, an environment in which movement takes place, or a reason for movement. The parts of a segmentation are subtrajectories, but confusingly, they are also referred to as segments.

The formal, algorithmic study of trajectory segmentation was initiated by Buchin et al. [BDKS11], who distinguish attribute values associated to each time instance of the trajectory (location at time t, speed at time t, etc.) and criteria relating to these attributes that specify when two time instances may not be in the same segment of the segmentation. The objective is a minimum-segment segmentation. For a class of criteria, an optimal segmentation can be computed in $O(n \log n)$ time. There are, however, criteria that are important in practice which are not in this class. Examples are some relaxed criteria that allow some small deviations to their strict versions. Follow-up research has identified wider classes of criteria that also allow polynomial-time optimal segmentation [ADK+13, ABB+14].

59.5.2 TRAJECTORY SIMILARITY

Determining the similarity two trajectories is important, also for clustering of trajectories. A similarity measure can be based on the geometry of the trajectories only, in which case the Hausdorff distance or Fréchet distance may be used. If the time aspect should influence similarity then other similarity measures are needed. One of the most widely used measures is called *dynamic time warping* [BC94]. Unfortunately, dynamic time warping does not satisfy the triangle inequality, it requires quadratic time to compute, and it uses only the distances between vertices. A large body of follow-up research has tried to overcome these problems.

When time cannot be conceptually stretched to get a better match, simpler similarity measures can be used [CW99, NP06]. These can be used for subtrajectory similarity computation: find the subtrajectories of two given trajectories of at least a given length whose distance is minimum [BBKL11]. Other research allows rotation and scaling for computing the similarity of two complete trajectories [TDS⁺07]. Yet another direction is to incorporate the context when computing similarity [BDS14].

59.5.3 PATTERNS IN TRAJECTORY DATA

When a collection of trajectories is given, analysis involves finding patterns in these trajectories. One of the most common data analysis tasks is clustering, and indeed, trajectory clustering has been studied extensively [HLT10, NP06]. However, most patterns in trajectory data concern parts of trajectories, for instance trajectories that move together for a certain length of time (and not necessarily the full duration of the trajectory). This pattern is referred to as a flock. For a flocking model that limits the involved trajectories to the ones that can fit together in a fixed-size disk, computing a maximum-duration flock is NP-hard [GK06], but approximation algorithms exist that do not fix the size of the disk exactly. Various other models for flocks and accompanying algorithms have been presented [BGHW08, GKS07, KMB05]. Besides finding flocks, it is also possible the study the changes in the grouping structure. The Reeb graph has been identified as a useful structure for this [BBK⁺13].

Patterns may involve identifying individuals that take on a specific role. The most common pattern of this type is the leadership pattern [AGLW08, GKS07]. A model for leadership describes a leader as an entity such that there are other entities that move in the direction of the leader and are close by. A related pattern is the single-file movement pattern [BBG08].

Patterns may involve the discovery of frequently visited places or hotspots [BDGW10, DGPW11, GKS13]. In these patterns, the region of interest may be given, or to be determined from the trajectories while only the size and shape is given. In some versions, we are interested in the longest consecutive visit, in other versions we are interested in the longest total visit.

59.6 OTHER GEOMETRIC ALGORITHMS IN GIS

The previous sections gave a brief overview of the main topics in which geometric algorithms play a role. There are a number of other topics, the most important of which are mentioned in this section.

GPS data from cars have been used as a source to construct road networks. This requires large-scale averaging, because many traces represent the same road [AW12, AKPW15, CGHS10, GSBW11]. To analyze the similarity of two road networks, an appropriate measure is needed. Here the geometric aspects of the network should be taken into account, and not just the graph structure [AFHW15]. By studying the average number of times a network is intersected by a random line, point location and ray-shooting navigation queries can be done efficiently [EGT09]. Generally, road networks are not planar graphs, and therefore many algorithmic results on planar graphs do not automatically transfer to road networks. Using other natural properties of road networks, several algorithmic results apply nevertheless [EG08] Clusters of points in networks are stretches of the network whose point density is relatively high. One can consider paths, trees, or other connected subgraphs of the network [BCG$^+$10].

The search for I/O-efficient algorithms for GIS-related problems has led to a substantial amount of research [AVV07]. Besides the R-tree and its variants mentioned earlier, terrain analysis has been the main application area for such algorithms [AAMS08].

Motivated by geographic information retrieval, the Internet can be used as a data source for vernacular regions. The delineation of such regions involves finding a reasonable boundary for a given set of inliers and outliers [BMS11a, RBK$^+$08].

Privacy considerations play a major role in location-based services. With this in mind, data-oblivious geometric algorithms can be devised which allow secure multi-party computation protocols [EGT10].

59.7 ALGORITHMIC CHALLENGES IN GIS

The application area GIS is the source of a number of interesting research problems. Many of these are simply-stated algorithmic problems, such as finding the most efficient algorithm for a well-defined problem, or finding the best approximation factor for some computationally hard problem. But from the application perspective, the study of relatively simple solutions for problems in which a number of different requirements must be satisfied or optimized simultaneously is more important. For example, label placement with high cartographic quality has to be achieved with no overlap between different labels, no or little overlap of a label with other map features, clear association between a feature and its label, and avoidance of areas that are too dense with text. There is no simple problem statement with optimization that can capture this. As a second example, in realistic terrain reconstruction, seven constraints have been listed [Sch98]. Such constraints cannot be formulated straightforwardly as algorithmic optimization. It is usually more important which requirements can be handled simultaneously and effectively, than how efficient a solution is.

Challenges for algorithms research on GIS problems include developing methods that deal with multiple criteria simultaneously, either as a whole solution or as part of an optimization approach such as genetic or evolutionary algorithms. The appropriate formulation of the GIS problem itself, and comparison of results based on different formulations, are also issues of major importance.

A more specific research direction is the study of movement patterns where entities are not modeled as single points, like a model for a moving glacier [SBL$^+$12]. In extension to this, the computations involved in the modeling of geographic processes is a good source of problems for computational geometers. Another research direction is the handling of data imprecision in a computational model that is valid, relevant, and simple enough to be accepted in GIS.

59.8 SOURCES AND RELATED MATERIAL

BOOKS

[Sam06]: Book on data structures for GIS and other applications.

[HCC11, LGMR11]: Two general GIS books.

[WD04]: A GIS book with a computing focus.

[DTH09, RMM+95]: Two books that focus on cartography aspects of GIS.

[Lau14, ZZ11]: Two books on movement and trajectories.

OTHER

Other surveys: computational geometry and GIS [DMP00], terrain modeling [HT14], patterns in trajectories [GLW08].

Journals: Journal of Spatial Information Science, ACM Transactions on Spatial Algorithms and Systems, GeoInformatica, International Journal of Geographical Information Science (IJGIS), Transactions in GIS, Cartography & GIS.

Conference proceedings: SIGSPATIAL Conference, GIScience, Symposium on Spatial and Temporal Databases (SSTD), International Symposium on Spatial Data Handling (SDH), Auto-Carto, International Cartographic Conference (ICC), Conference on Spatial Information Theory (COSIT).

RELATED CHAPTERS

Chapter 23: Computational topology of graphs on surfaces
Chapter 27: Voronoi diagrams and Delaunay triangulations
Chapter 32: Proximity algorithms
Chapter 51: Robotics
Chapter 52: Computer graphics
Chapter 54: Pattern recognition

REFERENCES

[AAMS08]   P.K. Agarwal, L. Arge, T. Mølhave, and B. Sadri. I/O-efficient efficient algorithms for computing contours on a terrain. In *Proc. 24th Sympos. Comput. Geom.*, pages 129–138, ACM Press, 2008.

[ABB+14]   S.P.A. Alewijnse, K. Buchin, M. Buchin, A. Kölzsch, H. Kruckenberg, and M.A. Westenberg. A framework for trajectory segmentation by stable criteria. In *Proc. 22nd ACM SIGSPATIAL Int. Conf. Advances Geographic Inform. Syst.*, pages 351–360, 2014.

[ABF+13]   M.J. Alam, T. Biedl, S. Felsner, M. Kaufmann, S.G. Kobourov, and T. Ueckerdt.

Computing cartograms with optimal complexity. *Discrete Comput. Geom.*, 50:784–810, 2013.

[ABG⁺02]   P. Agarwal, M. de Berg, J. Gudmundsson, M. Hammar, and H.J. Haverkort. Box-trees and R-trees with near-optimal query time. *Discrete Comput. Geom.*, 28:291–312, 2002.

[ABHY08]   L. Arge, M. de Berg, H.J. Haverkort, and K. Yi. The priority R-tree: A practically efficient and worst-case optimal R-tree. *ACM Trans. Algorithms*, 4:9, 2008.

[AD97]   P.K. Agarwal and P.K. Desikan. An efficient algorithm for terrain simplification. In *Proc. 8th ACM-SIAM Sympos. Discrete Algorithms*, pages 139–147, 1997.

[ADK⁺13]   B. Aronov, A. Driemel, M. van Kreveld, M. Löffler, and F. Staals. Segmentation of trajectories for non-monotone criteria. In *Proc. 24th ACM-SIAM Sympos. Discrete Algorithms*, pages 1897–1911, 2013.

[AFHW15]   M. Ahmed, B.T. Fasy, K.S. Hickmann, and C. Wenk. A path-based distance for street map comparison. *ACM Trans. Spatial Algorithms Systems*, 1:3, 2015.

[AGLW08]   M. Andersson, J. Gudmundsson, P. Laube, and T. Wolle. Reporting leaders and followers among trajectories of moving point objects. *GeoInformatica*, 12:497–528, 2008.

[AKPW15]   M. Ahmed, S. Karagiorgou, D. Pfoser, and C. Wenk. A comparison and evaluation of map construction algorithms using vehicle tracking data. *GeoInformatica*, 19:601–632, 2015.

[AKS98]   P.K. Agarwal, M. van Kreveld, and S. Suri. Label placement by maximum independent set in rectangles. *Comput. Geom.*, 11:209–218, 1998.

[AMS05]   L. Aleksandrov, A. Maheshwari, and J.-R. Sack. Determining approximate shortest paths on weighted polyhedral surfaces. *J. ACM*, 52:25–53, 2005.

[AS98]   P.K. Agarwal and S. Suri. Surface approximation and geometric partitions. *SIAM J. Comput.*, 27:1016–1035, 1998.

[AV00]   P.K. Agarwal and K.R. Varadarajan. Efficient algorithms for approximating polygonal chains. *Discrete Comput. Geom.*, 23:273–291, 2000.

[AVV07]   L. Arge, D.E. Vengroff, and J.S. Vitter. External-memory algorithms for processing line segments in geographic information systems. *Algorithmica*, 47:1–25, 2007.

[AW12]   M. Ahmed and C. Wenk. Constructing street networks from GPS trajectories. In *Proc. 20th Eur. Sympos. Algorithms*, vol. 7501 of *LNCS*, pages 60–71, Springer, Berlin, 2012.

[BBCM02]   M. de Berg, J. Bose, O. Cheong, and P. Morin. On simplifying dot maps. In *Proc. 18th Eur. Workshop Comput. Geom.*, pages 96–100, 2002.

[BBD⁺96]   M. de Berg, P. Bose, K. Dobrint, M. van Kreveld, M.H. Overmars, M. de Groot, T. Roos, J. Snoeyink, and S. Yu. The complexity of rivers in triangulated terrains. In *Proc. 8th Canad. Conf. Comput. Geom.*, pages 325–330, 1996.

[BBG08]   K. Buchin, M. Buchin, and J. Gudmundsson. Detecting single file movement. In *Proc. 16th ACM SIGSPATIAL Int. Sympos. Advances Geographic Inform. Syst.*, page 33, 2008.

[BBK⁺13]   K. Buchin, M. Buchin, M. van Kreveld, B. Speckmann, and F. Staals. Trajectory grouping structure. In *Proc. 13th Algorithms Data Structures Sympos.*, vol. 8037 of *LNCS*, pages 219–230, Springer, Berlin, 2013.

[BBKL11]   K. Buchin, M. Buchin, M. van Kreveld, and J. Luo. Finding long and similar parts of trajectories. *Comput. Geom.*, 44:465–476, 2011.

[BC94]   D.J. Berndt and J. Clifford. Using dynamic time warping to find patterns in time series. In *Knowledge Discovery in Databases: Papers from the 1994 AAAI Workshop. Technical Report WS-94-03*, pages 359–370, 1994.

[BCC+06]   P. Bose, S. Cabello, O. Cheong, J. Gudmundsson, M. van Kreveld, and B. Speckmann. Area-preserving approximations of polygonal paths. *J. Discrete Algorithms*, 4:554–566, 2006.

[BCG+96]   G. Barequet, B. Chazelle, L. Guibas, J. Mitchell, and A. Tal. BOXTREE: A hierarchical representation for surfaces in 3D. *Computer Graphics Forum*, 15:387396, 1996.

[BCG+10]   K. Buchin, S. Cabello, J. Gudmundsson, M. Löffler, J. Luo, G. Rote, R.I. Silveira, B. Speckmann, and T. Wolle. Finding the most relevant fragments in networks. *J. Graph Algorithms Appl.*, 14:307–336, 2010.

[BCK08]   B. Ben-Moshe, P. Carmi, and M.J. Katz. Approximating the visible region of a point on a terrain. *GeoInformatica*, 12:21–36, 2008.

[BD98]   M. de Berg and K. Dobrindt. On levels of detail in terrains. *Graphical Models Image Processing*, 60:1–12, 1998.

[BDGW10]   M. Benkert, B. Djordjevic, J. Gudmundsson, and T. Wolle. Finding popular places. *Internat. J. Comput. Geom. Appl.*, 20:19–42, 2010.

[BDKS11]   M. Buchin, A. Driemel, M. van Kreveld, and V. Sacristán. Segmenting trajectories: A framework and algorithms using spatiotemporal criteria. *J. Spatial Inform. Sci.*, 3:33–63, 2011.

[BDS14]   M. Buchin, S. Dodge, and B. Speckmann. Similarity of trajectories taking into account geographic context. *J. Spatial Inform. Sci.*, 9:101–124, 2014.

[BFH01]   B. Bell, S. Feiner, and T. Höllerer. View management for virtual and augmented reality. In *Proc. ACM Sympos. User Interface Software Tech.*, pages 101–110, 2001.

[BG14]   K. Buchin and D.H.P. Gerrits. Dynamic point labeling is strongly PSPACE-complete. *Internat. J. Comput. Geom. Appl.*, 24:373, 2014.

[BGH+14]   K. Buchin, A. van Goethem, M. Hoffmann, M. van Kreveld, and B. Speckmann. Travel-time maps: Linear cartograms with fixed vertex locations. In *Proc. 8th Conf. Geographic Inform. Sci.*, vol. 8728 of *LNCS*, pages 18–33, Springer, Berlin, 2014.

[BGHW08]   M. Benkert, J. Gudmundsson, F. Hübner, and T. Wolle. Reporting flock patterns. *Comput. Geom.*, 41:111–125, 2008.

[BHT10]   M. de Berg, H.J. Haverkort, and C.P. Tsirogiannis. Visibility maps of realistic terrains have linear smoothed complexity. *J. Comput. Geom.*, 1:57–71, 2010.

[BHT11]   M. de Berg, H.J. Haverkort, and C.P. Tsirogiannis. Implicit flow routing on terrains with applications to surface networks and drainage structures. In *Proc. 22nd ACM-SIAM Sympos. Discrete Algorithms*, pages 285–296, 2011.

[BK12]   S. Bies and M. van Kreveld. Time-space maps from triangulations. In *Proc. 20th Sympos. Graph Drawing*, vol. 7704 of *LNCS*, pages 511–516, Springer, Berlin, 2012.

[BKS93]   T. Brinkhoff, H.-P. Kriegel, and B. Seeger. Efficient processing of spatial joins using R-trees. In *Proc. ACM SIGMOD Conf. Management Data*, pages 237–246, 1993.

[BKS98]   M. de Berg, M. van Kreveld, and S. Schirra. Topologically correct subdivision simplification using the bandwidth criterion. *Cartography and GIS*, 25:243–257, 1998.

[BKSS90]   N. Beckmann, H.-P. Kriegel, R. Schneider, and B. Seeger. The R*-tree: An efficient and robust access method for points and rectangles. In *Proc. ACM SIGMOD Conf. Management Data*, pages 322–331, 1990.

[BM98]   P.A. Burrouh and R.A. McDonnell. *Principles of Geographical Information Systems.* Oxford University Press, 1998.

[BMM⁺04]   B. Ben-Moshe, J.S.B. Mitchell, M.J. Katz, and Y. Nir. Visibility preserving terrain simplification—an experimental study. *Comput. Geom.*, 28:175–190, 2004.

[BMS09]   M. de Berg, E. Mumford, and B. Speckmann. On rectilinear duals for vertex-weighted plane graphs. *Discrete Math.*, 309:1794–1812, 2009.

[BMS11a]   M. de Berg, W. Meulemans, and B. Speckmann. Delineating imprecise regions via shortest-path graphs. In *Proc. 19th ACM SIGSPATIAL Int. Sympos. Advances Geographic Inform. Syst.*, pages 271–280, 2011.

[BMS11b]   K. Buchin, W. Meulemans, and B. Speckmann. A new method for subdivision simplification with applications to urban-area generalization. In *Proc. 19th ACM SIGSPATIAL Int. Sympos. Advances Geographic Inform. Syst.*, pages 261–270, 2011.

[BNPW10]   K. Been, M. Nöllenburg, S.-H. Poon, and A. Wolff. Optimizing active ranges for consistent dynamic map labeling. *Comput. Geom.*, 43:312–328, 2010.

[BO79]   J.L. Bentley and T.A. Ottmann. Algorithms for reporting and counting geometric intersections. *IEEE Trans. Comput.*, C-28:643–647, 1979.

[BSV11]   K. Buchin, B. Speckmann, and K. Verbeek. Flow map layout via spiral trees. *IEEE Trans. Vis. Comp. Graph.*, 17:2536–2544, 2011.

[BSV12]   K. Buchin, B. Speckmann, and S. Verdonschot. Evolution strategies for optimizing rectangular cartograms. In *Proc. 7th Conf. Geographic Inform. Sci.*, pages 29–42, 2012.

[BSV15]   K. Buchin, B. Speckmann, and K. Verbeek. Angle-restricted steiner arborescences for flow map layout. *Algorithmica*, 72:656–685, 2015.

[BW98]   U. Brandes and D. Wagner. Using graph layout to visualize train interconnection data. In *Proc. 6th Sympos. Graph Drawing*, vol. 1547 in *LNCS*, pages 44–56, Springer, Berlin, 1998.

[CBK05]   S. Cabello, M. de Berg, and M. van Kreveld. Schematization of networks. *Comput. Geom.*, 30:223–228, 2005.

[CC96]   W.S. Chan and F. Chin. Approximation of polygonal curves with minimum number of line segments or minimum error. *Internat. J. Comput. Geom. Appl.*, 6:59–77, 1996.

[CDH⁺05]   D.Z. Chen, O. Daescu, J. Hershberger, P.M. Kogge, N. Mi, and J. Snoeyink. Polygonal path simplification with angle constraints. *Comput. Geom.*, 32:173–187, 2005.

[CEGS94]   B. Chazelle, H. Edelsbrunner, L.J. Guibas, and M. Sharir. Algorithms for bichromatic line segment problems and polyhedral terrains. *Algorithmica*, 11:116–132, 1994.

[CGHS10]   D. Chen, L.J. Guibas, J. Hershberger, and J. Sun. Road network reconstruction for organizing paths. In *Proc. 21st ACM-SIAM Sympos. Discrete Algorithms*, pages 1309–1320, 2010.

[Cha94]   T.M. Chan. A simple trapezoid sweep algorithm for reporting red/blue segment intersections. In *Proc. 6th Canad. Conf. Comput. Geom.*, pages 263–268, 1994.

[Cha04]   T.M. Chan. A note on maximum independent sets in rectangle intersection graphs. *Inform. Process. Lett.*, 89:19–23, 2004.

[Chr97]   N. Chrisman. *Exploring Geographic Information Systems.* Wiley, New York, 1997; second edition, 2002.

[CMS95]   J. Christensen, J. Marks, and S. Shieber. An empirical study of algorithms for point-feature label placement. *ACM Trans. Graph.*, 14:203–232, 1995.

[CPG99]   W. Cartwright, M. Peterson, and G. Gartner, editors. *Multimedia Cartography.* Springer, Berlin, 1999.

[CW99]   K.K.W. Chu and M.H. Wong. Fast time-series searching with scaling and shifting. In *Proc. 18th ACM Sympos. Principles Database Syst.*, pages 237–248, 1999.

[CWT06]   H. Cao, O. Wolfson, and G. Trajcevski. Spatio-temporal data reduction with deterministic error bounds. *VLDB J.*, 15:211–228, 2006.

[DCN85]   J.A. Dougenik, N.R. Chrisman, and D.R. Niemeyer. An algorithm to construct continuous area cartograms. *The Professional Geographer*, 37:75–81, 1985.

[DGPW11]   B. Djordjevic, J. Gudmundsson, A. Pham, and T. Wolle. Detecting regular visit patterns. *Algorithmica*, 60:829–852, 2011.

[DMP00]   L. De Floriani, P. Magillo, and E. Puppo. Applications of computational geometry to geographic information systems. In J.-R. Sack and J. Urrutia, editors, *Handbook of Computational Geometry*, pages 333–388, Elsevier, Amsterdam, 2000.

[DP73]   D.H. Douglas and T.K. Peucker. Algorithms for the reduction of the number of points required to represent a digitized line or its caricature. *Canadian Cartographer*, 10(2):112–122, 1973.

[DP89]   L. De Floriani and E. Puppo. A survey of constrained Delaunay triangulation algorithms for surface representation. In G.G. Pieroni, editor, *Issues on Machine Vision*, pages 95–104, Springer, Vienna, 1989.

[DTB99]   S. van Dijk, D. Thierens, and M. de Berg. On the design of genetic algorithms for geographical applications. In *Proc. Genetic Evol. Comput. Conf.*, pages 188–195, Morgan Kaufmann, San Francisco, 1999.

[DTH09]   B.D. Dent, J. Torguson, and T.W. Hodler. *Cartography: Thematic Map Design*, 6th edition. McGraw Hill, New York, 2009.

[EG08]   D. Eppstein and M.T. Goodrich. Studying (non-planar) road networks through an algorithmic lens. In *Proc. 16th ACM SIGSPATIAL Int. Sympos. Advances Geographic Inform. Syst.*, page 16, 2008.

[EGT09]   D. Eppstein, M.T. Goodrich, and L. Trott. Going off-road: Transversal complexity in road networks. In *Proc. 17th ACM SIGSPATIAL Int. Sympos. Advances Geographic Inform. Syst.*, pages 23–32, 2009.

[EGT10]   D. Eppstein, M.T. Goodrich, and R. Tamassia. Privacy-preserving data-oblivious geometric algorithms for geographic data. In *Proc. 18th ACM SIGSPATIAL Int. Sympos. Advances Geographic Inform. Syst.*, pages 13–22, 2010.

[EJS05]   T. Erlebach, K. Jansen, and E. Seidel. Polynomial-time approximation schemes for geometric intersection graphs. *SIAM J. Comput.*, 34:1302–1323, 2005.

[EM01]   R. Estkowski and J.S.B. Mitchell. Symplifying a polygonal subdivision while keeping it simple. In *Proc. 17th Sympos. Comput. Geom.*, pages 40–49, ACM press, 2001.

[EW97]   H. Edelsbrunner and R. Waupotitsch. A combinatorial approach to cartograms. *Comput. Geom.*, 7:343–360, 1997.

[FH95]   U. Finke and K.H. Hinrichs. Overlaying simply connected planar subdivisions in linear time. In *Proc. 11th Sympos. Comput. Geom.*, pages 119–126, ACM Press, 1995.

[Fjä91]   P.-O. Fjällström. Polyhedral approximation of bivariate functions. In *Proc. 3rd Canad. Conf. Comput. Geom.*, pages 187–190, 1991.

[Fra02]   W.R. Franklin. Siting observers on terrain. In *Proc. 10th Sympos. Spatial Data Handling*, pages 109–120, Springer, Berlin, 2002.

[FW91]   M. Formann and F. Wagner. A packing problem with applications to lettering of maps. In *Proc. 7th Sympos. Comput. Geom.*, pages 281–288, ACM Press, 1991.

[GD02]   C. Gold and M. Dakowicz. Terrain modelling based on contours and slopes. In *Proc. 10th Sympos. Advances Spatial Data Handling*, pages 95–107, Springer, Berlin, 2002.

[GHK02]   J. Gudmundsson, M. Hammar, and M. van Kreveld. Higher order Delaunay triangulations. *Comput. Geom.*, 23:85–98, 2002.

[GHMS93]   L.J. Guibas, J.E. Hershberger, J.S.B. Mitchell, and J.S. Snoeyink. Approximating polygons and subdivisions with minimum link paths. *Internat. J. Comput. Geom. Appl.*, 3:383–415, 1993.

[GK06]   J. Gudmundsson and M. van Kreveld. Computing longest duration flocks in trajectory data. In *14th ACM Int. Sympos. Geographic Inform. Syst.*, pages 35–42, 2006.

[GKM+09]   J. Gudmundsson, J. Katajainen, D. Merrick, C. Ong, and T. Wolle. Compressing spatio-temporal trajectories. *Comput. Geom.*, 42:825–841, 2009.

[GKS07]   J. Gudmundsson, M. van Kreveld, and B. Speckmann. Efficient detection of patterns in 2D trajectories of moving points. *GeoInformatica*, 11:195–215, 2007.

[GKS13]   J. Gudmundsson, M. van Kreveld, and F. Staals. Algorithms for hotspot computation on trajectory data. In *SIGSPATIAL 2013 Int. Conf. Advances Geographic Inform. Syst.*, pages 134–143, 2013.

[GLW08]   J. Gudmundsson, P. Laube, and T. Wolle. Movement patterns in spatio-temporal data. In S. Shekhar and H. Xiong, editors, *Encyclopedia of GIS*, pages 726–732, Springer, Berlin, 2008.

[GNN13]   A. Gemsa, B. Niedermann, and M. Nöllenburg. Trajectory-based dynamic map labeling. In *Proc. 24th Int. Sympos. Algorithms Comput.*, vol. 8283 of *LNCS*, pages 413–423, Springer, Berlin, 2013.

[GNR16]   A. Gemsa, M. Nöllenburg, and I. Rutter. Consistent labeling of rotating maps. *J. Comput. Geom.*, 7:308—331, 2016.

[GNS07]   J. Gudmundsson, G. Narasimhan, and M. Smid. Distance-preserving approximations of polygonal paths. *Comput. Geom.*, 36:183–196, 2007.

[GSBW11]   X. Ge, I. Safa, M. Belkin, and Y. Wang. Data skeletonization via Reeb graphs. In *Proc. 24th Int. Conf. Neural Inform. Process. Syst.*, pages 837–845, Curran Associates, Red Hook, 2011.

[Gut84]   A. Guttmann. R-Trees: a dynamic indexing structure for spatial searching. In *Proc. ACM-SIGMOD Int. Conf. Management of Data*, pages 47–57, 1984.

[HCC11]   I. Heywood, S. Cornelius, and S. Carver. *An Introduction to Geographical Information Systems*. Prentice Hall, 4th edition, 2011.

[Hel90]   M. Heller. Triangulation algorithms for adaptive terrain modeling. In *Proc. 4th Sympos. Spatial Data Handling*, pages 163–174, 1990.

[HG95]   P.S. Heckbert and M. Garland. Fast polygonal approximation of terrains and height fields. Report CMU-CS-95-181, Carnegie Mellon University, 1995.

[HLM+14]   F. Hurtado, M. Löffler, I. Matos, V. Sacristán, M. Saumell, R.I. Silveira, and F. Staals. Terrain visibility with multiple viewpoints. *Internat. J. Comput. Geom. Appl.*, 24:275–306, 2014.

[HLT10]   J. Han, Z. Li, and L.A. Tang. Mining moving object, trajectory and traffic data. In *Proc. 15th Int. Conf. Database Syst. Advanced Appl.*, pages 485–486, 2010.

[Høj98]   P. Højholt. Solving local and global space conflicts in map generalization using a finite element method adapted from structural mechanics. In *Proc. 8th Int. Sympos. Spatial Data Handling*, pages 679–689, 1998.

[HS98]   J. Hershberger and J. Snoeyink. Cartographic line simplification and polygon CSG formulae in $O(n \log^* n)$ time. *Comput. Geom.*, 11:175–185, 1998.

[HT14]   H.J. Haverkort and L. Toma. Terrain modeling for the geosciences. In *Computing Handbook: Computer Science and Software Engineering*, 3rd edition, chap. 31, pages 1–21, Taylor & Francis, 2014.

[Imh75]   E. Imhof. Positioning names on maps. *The American Cartographer*, 2:128–144, 1975.

[JBQZ04]   M. Jiang, S. Bereg, Z. Qin, and B. Zhu. New bounds on map labeling with circular labels. In *Proc. 15th Int. Sympos. Algorithms Comput.*, vol. 3341 of *LNCS*, pages 606–617, Springer, Berlin, 2004.

[JC04]   J.-W. Jung and K.-Y. Chwa. Labeling points with given rectangles. *Inform. Process. Lett.*, 89:115–121, 2004.

[KB01]   M.-J. Kraak and A. Brown, editors. *Web Cartography: Developments and Prospects*. Taylor & Francis, 2001.

[KBS91]   H.-P. Kriegel, T. Brinkhoff, and R. Schneider. The combination of spatial access methods and computational geometry in geographic database systems. In *Proc. 2nd Sympos. Advances in Spatial Databases*, vol. 525 of *LNCS*, pages 5–21, Springer, Berlin. 1991.

[KF94]   I. Kamel and C. Faloutsos. Hilbert R-tree: An improved R-tree using fractals. In *Proc. 20th VLDB Conf.*, pages 500–510, 1994.

[KH98]   C.J. Kocmoud and D.H. House. A constrained-based approach to constructing continuous cartograms. In *Proc. 8th Int. Sympos. Spatial Data Handling*, pages 236–246, 1998.

[KKL07]   T. de Kok, M. van Kreveld, and M. Löffler. Generating realistic terrains with higher-order Delaunay triangulations. *Comput. Geom.*, 36:52–65, 2007.

[KMB05]   P. Kalnis, N. Mamoulis, and S. Bakiras. On discovering moving clusters in spatio-temporal data. In *Proc. 9th Sympos. Advances Spatial Temporal Databases*, pages 364–381, 2005.

[KOS92]   M.J. Katz, M.H. Overmars, and M. Sharir. Efficient hidden surface removal for objects with small union size. *Comput. Geom.*, 2:223–234, 1992.

[Kre01]   M. van Kreveld. Smooth generalization for continuous zooming. In *Proc. 20th Int. Cartographic Conf.*, vol. 3, pages 2180–2185, 2001.

[KS07]   M. van Kreveld and B. Speckmann. On rectangular cartograms. *Comput. Geom.*, 37:175–187, 2007.

[KS11]   M. van Kreveld and R.I. Silveira. Embedding rivers in triangulated irregular networks with linear programming. *Int. J. Geographical Inform. Sci.*, 25:615–631, 2011.

[KSW99]   M. van Kreveld, T. Strijk, and A. Wolff. Point labeling with sliding labels. *Comput. Geom.*, 13:21–47, 1999.

[KWFP10]   C. Kaiser, F. Walsh, C.J.Q. Farmer, and A. Pozdnoukhov. User-centric time-distance representation of road networks. In *Proc. 6th Int. Conf. Geographic Inform. Sci.*, pages 85–99, 2010.

[Lau14]   P. Laube. *Computational Movement Analysis*. Springer Briefs in Computer Science, Springer, Berlin, 2014.

[LGMR11]   P.A. Longley, M.F. Goodchild, D.J. Maguire, and D.W. Rhind. *Geographic Information Systems and Science*, 3rd edition. Wiley, New York, 2011.

[LJ01]   M. Lonergan and C.B. Jones. An iterative displacement method for conflict resolution in map generalization. *Algorithmica*, 21:287–301, 2001.

[LK11]   W.-C. Lee and J. Krumm. Trajectory preprocessing. In *Computing with Spatial Trajectories*, pages 3–33. 2011.

[LLE97]   S.T. Leutenegger, M.A. Lopez, and J. Edington. STR: A simple and efficient algorithm for R-tree packing. In *Proc. 13th IEEE Int. Conf. Data Eng.*, pages 497–506, 1997.

[LMS97]   M. Lanthier, A. Maheshwari, and J.-R. Sack. Approximating weighted shortest paths on polyhedral surfaces. In *Proc. 13th Sympos. Comput. Geom.*, pages 274–283, ACM Press, 1997.

[LS01]   J.J. Little and P. Shi. Structural lines, TINs and DEMs. *Algorithmica*, 21:243–263, 2001.

[MG99]   W.A. Mackaness and E. Glover. The application of dynamic generalization to virtual map design. In *Proc. 19th Int. Cartographic Conf.*, 1999. CD-ROM.

[MLW95]   J.-C. Müller, J.-P. Lagrange, and R. Weibel, editors. *GIS and Generalization—Methodology and Practice*, volume 1 of *GISDATA*. Taylor & Francis, 1995.

[MP01]   W.A. Mackaness and R. Purves. Automated displacement for large numbers of discrete map objects. *Algorithmica*, 21:302–311, 2001.

[MRS07]   W.A. Mackaness, A. Ruas, and L.T. Sarjakoski, editors. *Generalisation of Geographic Information: Cartographic Modelling and Applications*. Elsevier, Amsterdam, 2007.

[MS99]   M. McAllister and J. Snoeyink. Extracting consistent watersheds from digital river and elevation data. In *Proc. ASPRS/ACSM Annual Conference*, 1999.

[Ney99]   G. Neyer. Line simplification with restricted orientations. In *Proc. 6th Workshop Algorithms Data Structures*, vol. 1663 of *LNCS*, pages 13–24, Springer, 1999.

[NP06]   M. Nanni and D. Pedreschi. Time-focused clustering of trajectories of moving objects. *J. Intell. Inf. Syst.*, 27:267–289, 2006.

[NW00]   G. Neyer and F. Wagner. Labeling downtown. In *Proc. Italian Conf. Algorithms and Complexity*, vol. 1767 of *LNCS*, pages 113–125, Springer, Berlin, 2000.

[NW11]   M. Nöllenburg and A. Wolff. Drawing and labeling high-quality metro maps by mixed-integer programming. *IEEE Trans. Vis. Comp. Graph.*, 17:626–641, 2011.

[Oos94]   P. van Oosterom. An R-tree based map-overlay algorithm. In *Proc. 5th Eur. Conf. Geogr. Inform. Syst.*, pages 318–327, EGIS, 1994.

[Oos95]   P. van Oosterom. The GAP-tree, an approach to "on-the-fly" map generalization of an area partitioning. In J.-C. Müller, J.-P. Lagrange, and R. Weibel, editors, *GIS and Generalization—Methodology and Practice*, number 1 in GISDATA, Taylor & Francis, 1995.

[Per66]   J. Perkal. On the length of empirical curves. Discussion paper 10, Ann Arbor Michigan Inter-University Community of Mathematical Geographers, 1966.

[PJ99]   P. van der Poorten and C.B. Jones. Customisable line generalisation using Delaunay triangulation. In *Proc. 19th Int. Cartographic Conf.*, 1999. CD-ROM.

[PPH99]   I. Petzold, L. Plümer, and M. Heber. Label placement for dynamically generated screen maps. In *Proc. 19th Int. Cartographic Conf.*, pages 893–903, 1999.

[PS94]   L. Palazzi and J. Snoeyink. Counting and reporting red/blue segment intersections. *CVGIP: Graph. Models Image Process.*, 56:304–311, 1994.

[PS97]   E. Puppo and R. Scopigno. Simplification, LOD and multiresolution principles and applications. In *Proc. EUROGRAPHICS*, vol. 16. Blackwell, 1997.

[PSS$^+$03]   S.-H. Poon, C.-S. Shin, T. Strijk, T. Uno, and A. Wolff. Labeling points with weights. *Algorithmica*, 38:341–362, 2003.

[RBK$^+$08]   I. Reinbacher, M. Benkert, M. van Kreveld, J.S.B. Mitchell, J. Snoeyink, and A. Wolff. Delineating boundaries for imprecise regions. *Algorithmica*, 50:386–414, 2008.

[RMM$^+$95]   A.H. Robinson, J. Morrison, P.C. Muehrcke, A.J. Kimerling, and S.C. Guptill. *Elements of Cartography*, 6th edition. John Wiley & Sons, New York, 1995.

[Ros04]   J. Rossignac. Surface simplification and 3D geometry compression. Chap. 54 in J.E. Goodman and J. O'Rourke, *Handbook of Discrete and Computational Geometry*, 2nd edition, CRC Press, Boca Raton, 2004.

[Sam06]   H. Samet. *Foundations of Multidimensional and Metric Data Structures*. Morgan Kaufmann, San Francisco, 2006.

[SBM95]   M. Sambridge, J. Braun, and H. McQueen. Geophysical parameterization and interpolation of irregular data using natural neighbours. *Geophys. J. Int.*, 122:837–857, 1995.

[SBL$^+$12]   J. Shamoun-Baranes, E. van Loon, R.S. Purves, B. Speckmann, D. Weiskopf, and C.J. Camphuysen. Analysis and visualization of animal movement. *Biology Letters*, 2012:6–9, 2012.

[Sch98]   B. Schneider. Geomorphologically sound reconstruction of digital terrain surfaces from contours. In *Proc. 8th Int. Sympos. Spatial Data Handling*, pages 657–667, 1998.

[SHWZ14]   N. Schwartges, J.-H. Haunert, A. Wolff, and D. Zwiebler. Point labeling with sliding labels in interactive maps. In *Connecting a Digital Europe through Location and Place*, pages 295–310, Springer, Berlin, 2014.

[SI09]   E. Shimizu and R. Inoue. A new algorithm for distance cartogram construction. *Int. J. Geographical Inform. Sci.*, 23:1453–1470, 2009.

[Sib81]   R. Sibson. A brief description of natural neighbour interpolation. In V. Barnet, editor, *Interpreting Multivariate Data*, pages 21–36, Wiley, New York, 1981.

[SMK95]   C. Silva, J.S.B. Mitchell, and A.E. Kaufman. Automatic generation of triangular irregular networks using greedy cuts. In *Proc. IEEE Conf. Vis.*, pages 201–208, 1995.

[Str01]   T. Strijk. *Geometric Algorithms for Cartographic Label Placement*. PhD thesis, Utrecht University, Department of Computer Science, 2001.

[TDS$^+$07]   G. Trajcevski, H. Ding, P. Scheuermann, R. Tamassia, and D. Vaccaro. Dynamics-aware similarity of moving objects trajectories. In *15th ACM Sympos. Geographic Inform. Syst.*, page 11, 2007.

[TG00]   D. Thibault and C.M. Gold. Terrain reconstruction from contours by skeleton construction. *GeoInformatica*, 4:349–373, 2000.

[Tob86]   W.R. Tobler. Pseudo-cartograms. *The American Cartographer*, 13:43–50, 1986.

[Wat92]   D.F. Watson. *Contouring: A Guide to the Analysis and Display of Spatial Data.* Pergamon, Oxford, 1992.

[WD04]   M. Worboys and M. Duckham. *GIS: A Computing Perspective.* Taylor & Francis, 2nd edition, 2004.

[Wei97]   R. Weibel. A typology of constraints to line simplification. In M.J. Kraak and M. Molenaar, editors, *Proc. 7th Int. Symposium on Spatial Data Handling*, pages 533–546, Taylor & Francis, 1997.

[WKK+00]   A. Wolff, L. Knipping, M. van Kreveld, T. Strijk, and P.K. Agarwal. A simple and efficient algorithm for high-quality line labeling. In P.M. Atkinson and D.J. Martin, editors, *Innovations in GIS VII: GeoComputation*, chap. 11, pages 147–159, Taylor & Francis, 2000.

[WW97]   F. Wagner and A. Wolff. A practical map labeling algorithm. *Comput. Geom.*, 7:387–404, 1997.

[WWKS01]   F. Wagner, A. Wolff, V. Kapoor, and T. Strijk. Three rules suffice for good label placement. *Algorithmica*, 30:334–349, 2001.

[ZZ11]   Y. Zheng and X. Zhou, editors. *Computing with Spatial Trajectories.* Springer, New York, 2011.

60 GEOMETRIC APPLICATIONS OF THE GRASSMANN-CAYLEY ALGEBRA

Neil L. White

INTRODUCTION

Grassmann-Cayley algebra is first and foremost a means of translating synthetic projective geometric statements into invariant algebraic statements in the bracket ring, which is the ring of projective invariants. A general philosophical principle of invariant theory, sometimes referred to as **Gram's theorem**, says that any projectively invariant geometric statement has an equivalent expression in the bracket ring; thus we are providing here the practical means to carry this out. We give an introduction to the basic concepts, and illustrate the method with several examples from projective geometry, rigidity theory, and robotics.

60.1 BASIC CONCEPTS

Let P be a $(d-1)$-dimensional projective space over the field F, and V the canonically associated d-dimensional vector space over F. Let S be a finite set of n points in P and, for each point, fix a homogeneous coordinate vector in V. We assume that S spans V, hence also that $n \geq d$. Initially, we choose all of the coordinates to be distinct, algebraically independent indeterminates in F, although we can always specialize to the actual coordinates we want in applications. For $p_i \in S$, let the coordinate vector be $(x_{1,i}, \ldots, x_{d,i})$.

GLOSSARY

Bracket: A $d \times d$ determinant of the homogeneous coordinate vectors of d points in S. Brackets are relative projective invariants, meaning that under projective transformations their value changes only in a very predictable way (in fact, under a basis change of determinant 1, they are literally invariant). Hence brackets may also be thought of as coordinate-free symbolic expressions. The bracket of $u_1, \ldots, u_d$ is denoted by $[u_1, \ldots, u_d]$.

Bracket ring: The ring B generated by the set of all brackets of d-tuples of points in S, where $n = |S| \geq d$. It is a subring of the ring $F[x_{1,1}, x_{1,2}, \ldots, x_{d,n}]$ of polynomials in the coordinates of points in S.

Straightening algorithm: A normal form algorithm in the bracket ring.

Join of points: An exterior product of k points, $k \leq d$, computed in the exterior algebra of V. We denote such a product by $a_1 \vee a_2 \vee \cdots \vee a_k$, or simply $a_1 a_2 \cdots a_k$, rather than $a_1 \wedge a_2 \wedge \cdots \wedge a_k$, which is commonly used in exterior algebra. A concrete version of this operation is to compute the Plücker coordinate vector of (the subspace spanned by) the k points, that is, the vector whose components are all $k \times k$ minors (in some prespecified order) of the $d \times k$ matrix whose columns are the homogeneous coordinates of the k points.

Extensor of step k, or **decomposable k-tensor:**   A join of k points. Extensors of step k span a vector space $V^{(k)}$ of dimension $\binom{d}{k}$. (Note that not every element of $V^{(k)}$ is an extensor.)

Antisymmetric tensor: Any element of the direct sum $\Lambda V = \oplus_k V^{(k)}$.

Copoint: Any antisymmetric tensor of step $d-1$. A copoint is always an extensor.

Join: The exterior product operation on ΛV. The join of two tensors can always be reduced by distributivity to a linear combination of joins of points.

Integral: $E = u_1 u_2 \cdots u_d$, for any vectors $u_1, u_2, \ldots, u_d$ such that $[u_1, u_2, \ldots, u_d] = 1$. Every extensor of step d is a scalar multiple of the integral E.

Meet: If $A = a_1 a_2 \cdots a_j$ and $B = b_1 b_2 \cdots b_k$, with $j + k \geq d$, then

$$A \wedge B = \sum_\sigma \operatorname{sgn}(\sigma)[a_{\sigma(1)}, \ldots, a_{\sigma(d-k)}, b_1, \ldots, b_k] a_{\sigma(d-k+1)} \cdots a_{\sigma(j)}$$
$$\equiv [\overset{\bullet}{a}_1, \ldots, \overset{\bullet}{a}_{d-k}, b_1, \ldots, b_k] \, \overset{\bullet}{a}_{d-k+1} \cdots \overset{\bullet}{a}_j .$$

The sum is taken over all permutations σ of $\{1, 2, \ldots, j\}$ such that $\sigma(1) < \sigma(2) < \cdots < \sigma(d-k)$ and $\sigma(d-k+1) < \sigma(d-k+2) < \cdots < \sigma(j)$. Each such permutation is called a **shuffle** of the $(d - k, j - (d - k))$ split of A, and the dots represent such a signed sum over all the shuffles of the dotted symbols.

Grassmann-Cayley algebra: The vector space $\Lambda(V)$ together with the operations $\vee$ and $\wedge$.

PROPERTIES OF GRASSMANN-CAYLEY ALGEBRA

(i) $A \vee B = (-1)^{jk} B \vee A$ and $A \wedge B = (-1)^{(d-k)(d-j)} B \wedge A$, if A and B are extensors of steps j and k.

(ii) $\vee$ and $\wedge$ are associative and distributive over addition and scalar multiplication.

(iii) $A \vee B = (A \wedge B) \vee E$ if $\operatorname{step}(A) + \operatorname{step}(B) = d$.

(iv)   A meet of two extensors is again an extensor.

(v)   The meet is dual to the join, where duality exchanges points and copoints.

(vi)   **Alternative Law:**   Let $a_1, a_2, \ldots, a_k$ be points and $\gamma_1, \gamma_2, \ldots, \gamma_s$ copoints. Then if $k \geq s$,

$$(a_1 a_2 \cdots a_k) \wedge (\gamma_1 \wedge \gamma_2 \wedge \cdots \wedge \gamma_s) = [\overset{\bullet}{a}_1, \gamma_1][\overset{\bullet}{a}_2, \gamma_2] \cdots [\overset{\bullet}{a}_s, \gamma_s] \overset{\bullet}{a}_{s+1} \vee \cdots \vee \overset{\bullet}{a}_k .$$

Here the dots refer to all shuffles over the $(1, 1, \ldots, 1, k - s)$ split of $a_1 \cdots a_k$, that is, a signed sum over all permutations of the a's such that the last $k - s$ of them are in increasing order.

60.2 GEOMETRY ↔ G.-C. ALGEBRA → BRACKET ALGEBRA

If X is a projective subspace of dimension $k - 1$, pick a basis $a_1, a_2, \ldots, a_k$ and let $A = a_1 a_2 \cdots a_k$ be an extensor. We call $X = \overline{A}$ the **support** of A.

(i)   If $A \neq 0$ is an extensor, then A determines $\overline{A}$ uniquely.

(ii) If $\overline{A} \cap \overline{B} \neq \emptyset$, then $\overline{A \vee B} = \overline{A} + \overline{B}$.

(iii) If $\overline{A} \cup \overline{B}$ spans V, then $\overline{A \wedge B} = \overline{A} \cap \overline{B}$.

TABLE 60.2.1   Examples of geometric conditions and corresponding Grassmann-Cayley algebra statements.

GEOMETRIC CONDITION	DIM	G.-C. ALGEBRA STATEMENT	BRACKET STATEMENT
Point $\overline{a}$ is on the line $\overline{bc}$ (or $\overline{b}$ is on $\overline{ac}$, etc.)	2	$a \wedge bc = 0$	$[abc] = 0$
Lines $\overline{ab}$ and $\overline{cd}$ intersect	3	$ab \wedge cd = 0$	$[abcd] = 0$
Lines $\overline{ab}$, $\overline{cd}$, $\overline{ef}$ concur	2	$ab \wedge cd \wedge ef = 0$	$[\overset{\bullet}{a}cd][\overset{\bullet}{b}ef] = 0$
Planes $\overline{abc}$, $\overline{def}$, and $\overline{ghi}$ have a line in common	3	$abc \wedge def \wedge ghi = 0$	$[\overset{\bullet}{a}def][\overset{\bullet}{b}ghi][\overset{\bullet}{c}xyz] = 0 \ \forall x, y, z$
The intersections of $\overline{ab}$ with $\overline{cd}$ and of $\overline{ef}$ with $\overline{gh}$ are collinear with $\overline{i}$	2	$(ab \wedge cd) \vee (ef \wedge gh) \vee i = 0$	$[\overset{\bullet}{a}cd][\overset{\diamond}{e}gh][\overset{\bullet}{b}\overset{\diamond}{f}i] = 0$

The geometric conditions in Table 60.2.1 should be interpreted projectively. For example, the concurrency of three lines includes as a special case that the three lines are mutually parallel, if one prefers to interpret the conditions in affine space. Degenerate cases are always included, so that the concurrency of three lines includes as a special case the equality of two or even all three of the lines, for example.

Most of the interesting geometric conditions translate into Grassmann-Cayley conditions of step 0 (or, equivalently, step d), and therefore expand into bracket conditions directly. When the Grassmann-Cayley condition is not of step 0, as in the example in Table 60.2.1 of three planes in three-space containing a common line, then the Grassmann-Cayley condition may be joined with an appropriate number of universally quantified points to get a conjunction of bracket conditions. The joined points may also be required to come from a specified basis to make this a conjunction of a finite number of bracket conditions.

In this fashion, any incidence relation in projective geometry may be translated into a conjunction of Grassmann-Cayley statements, and, conversely, Grassmann-Cayley statements may be translated back to projective geometry just as easily, provided they involve only join and meet, not addition.

Many identities in the Grassmann-Cayley algebra yield algebraic, coordinate-free proofs of important geometric theorems. These proofs typically take the form "the left-hand side of the identity is 0 if and only if the right-hand side of the identity is 0," and the resulting equivalent Grassmann-Cayley conditions translate to interesting geometric conditions as above.

TABLE 60.2.2   Examples of Grassmann-Cayley identities and corresponding geometric theorems, in dimension 2.

GEOMETRIC THEOREM	G.-C. ALGEBRA IDENTITY
Desargues's theorem: Derived points $ab \wedge a'b'$, $ac \wedge a'c'$, and $bc \wedge b'c'$ are collinear if and only if abc or $a'b'c'$ are collinear or aa', bb', and cc' concur.	$(ab \wedge a'b') \vee (ac \wedge a'c') \vee (bc \wedge b'c') =$ $[abc][a'b'c'](aa' \wedge bb' \wedge cc')$
Pappus's theorem and Pascal's theorem: If abc and $a'b'c'$ are both collinear sets, then $(bc' \wedge b'c)$, $(ca' \wedge c'a)$, and $(ab' \wedge a'b)$ are collinear.	$[ab'c'][a'bc'][a'b'c][abc]$ $-[abc'][ab'c][a'bc][a'b'c']$ $= (bc' \wedge b'c) \vee (ca' \wedge c'a) \vee (ab' \wedge a'b)$
Pappus's theorem (alternate version): If $aa'x$, $bb'x$, $cc'x$, $ab'y$, $bc'y$, and $ca'y$ are collinear, then ac', ba', cb' concur.	$aa' \wedge bb' \wedge cc' + ab' \wedge bc' \wedge ca'$ $+ac' \wedge ba' \wedge cb' = 0$
Fano's theorem: If no three of a, b, c, d are collinear, then $(ab \wedge cd)$, $(bc \wedge ad)$, and $(ca \wedge bd)$ are collinear if and only if char $F = 2$.	$(ab \wedge cd) \vee (bc \wedge ad) \vee (ca \wedge bd)$ $= 2\,[abc][abd][acd][bcd]$

The identities in Table 60.2.2 are proved by expanding both sides, using the rules for join and meet, and then verifying the equality of the resulting expressions by using the straightening algorithm of bracket algebra (see [Stu93]).

The right-hand side of the identity for the first version of Pappus's theorem is also the Grassmann-Cayley form of the geometric construction used in Pascal's theorem, and hence is 0 if and only if the six points lie on a common conic (Pappus's theorem being the degenerate case of Pascal's theorem in which the conic consists of two lines). Hence the left-hand side of the same identity is the bracket expression that is 0 if and only if the six points lie on a common conic. In particular, if abc and $a'b'c'$ are both collinear, we see immediately from the underlined brackets that the left-hand side is 0.

Numerous other projective geometry incidence theorems may be proved using the Grassmann-Cayley algebra. We illustrate this with an example modified from [RS76]. Other examples may be found in the same reference.

THEOREM 60.2.1

In 3-space, if triangles abc and $a'b'c'$ are in perspective from the point d, then the lines $a'bc \wedge ab'c'$, $b'ca \wedge bc'a'$, $c'ab \wedge ca'b'$, and $a'b'c' \wedge abc$ are all coplanar.

Proof. We prove the general case, where a, b, c, d, a', b', c' are all distinct, triangles abc and $a'b'c'$ are nondegenerate, and d is in neither the plane abc nor the plane $a'b'c'$. Then, since a, a', d are collinear, we may write $\alpha a' = \beta a + d$ for nonzero scalars α and β. Since we are using homogeneous coordinates for points, a, and similarly a', may be replaced by nonzero scalar multiples of themselves without changing the geometry. Thus, without loss of generality, we may write $a' = a + d$. Similarly, $b' = b + d$ and $c' = c + d$. Now

$$L_1 := a'bc \wedge ab'c' = [a'ab'c']bc - [bab'c']a'c + [cab'c']a'b$$

$$= [dabc]bc + [badc]ca + [cabd]ab + [badc]cd + [cabd]db$$

$$= [abcd](-bc - ac + ab + cd - bd).$$

Similarly,

$$L_2 := b'ca \wedge bc'a' = [abcd](ac + ab + bc + ad - cd),$$

$$L_3 := c'ab \wedge ca'b' = [abcd](-ab + bc - ac + bd - ad),$$

$$L_4 := a'b'c' \wedge abc = [abcd](bc - ac + ab).$$

Now we check that any two of these lines intersect. For example,

$$L_1 \wedge L_2 = [abcd]^2(-bc - ac + ab + cd - bd) \wedge (ac + ab + bc + ad - cd) = 0.$$

However, this shows only that either all four lines are coplanar or all four lines concur. To prove the former, it suffices to check that the intersection of $\overline{L_1}$ and $\overline{L_4}$ is distinct from that of $\overline{L_2}$ and $\overline{L_4}$. Notice that $L_1 \wedge L_4$ does not tell us the point of intersection, because $\overline{L_1}$ and $\overline{L_4}$ do not jointly span V, by our previous computation. But if we choose a generic vector x representing a point in general position, it follows from $\overline{L_1} \neq \overline{L_4}$, which must hold in our general case, that $(L_1 \vee x) \wedge L_4$ is nonzero and does represent the desired point of intersection. Then we compute

$$(L_1 \vee x) \wedge L_4 = [abcd]^2(-bcx - acx + abx + cdx - bdx) \wedge (bc - ac + ab)$$

$$= [abcd]^2(2[abcx] - [bcdx] - [acdx])(c - b)$$

$$= \alpha(c - b)$$

for some nonzero scalar α. Similarly, $(L_2 \vee x) \wedge L_4 = \beta(c - a)$ for some nonzero scalar β. By the nondegeneracy of the triangle abc, these two points of intersection are distinct. $\qquad\square$

60.3 CAYLEY FACTORIZATION: BRACKET ALGEBRA → GEOMETRY

(1) Projective geometry	
$\updownarrow$	
(2) Grassmann-Cayley algebra	↑ Cayley factorization
$\downarrow$	
(3) Bracket algebra	
$\downarrow$	
(4) Coordinate algebra	

$(1)\leftrightarrow(2)\to(3)$ in the chart above is explained in Section 60.2 above, with $(2)\to(1)$ being straightforward only in the case of a Grassmann-Cayley expression involving only joins and meets. $(3)\to(4)$ is the trivial expansion of a determinant into a polynomial in its d^2 entries. $(4)\to(3)$ is possible only for invariant polynomials (under the special linear group); see [Stu93] for an algorithm.

PHILOSOPHY OF INVARIANT THEORY: It is best for many purposes to avoid level (4), and to work instead with the symbolic coordinate-free expressions on levels (2) and (3).

Cayley factorization, (3)→(2), refers to the translation of a bracket polynomial into an equivalent Grassmann-Cayley expression involving only joins and meets. The input polynomial must be homogeneous (i.e., each point must occur the same number of times in the brackets of each bracket monomial of the polynomial), and Cayley factorization is not always possible. No practical algorithm is known in general, but an algorithm [Whi91] is known that finds such a factorization—or else announces its impossibility—in the multilinear case (each point occurs exactly once in each monomial). This algorithm is practical up to about 20 points.

MULTILINEAR CAYLEY FACTORIZATION

The multilinear Cayley factorization (MCF) algorithm is too complex to present here in detail; instead, we give an example and indicate roughly how the algorithm proceeds on the example.

Let

$$P = -[acj][deh][bfg] - [cdj][aeh][bfg] - [cdj][abe][fgh]$$
$$+ [acj][bdf][egh] - [acj][bdg][efh] + [acj][bdh][efg].$$

Note that P is multilinear in the 9 points. The MCF algorithm now looks for sets of points $x, y, \ldots, z$ such that the extensor $xy \cdots z$ could be part of a Cayley factorization of P. For this choice of P, it turns out that no such set larger than a pair of elements occurs. An example of such a pair is a, d; in fact, if d is replaced by a in P, leaving two a's in each term of P, although in different brackets, the resulting bracket polynomial is equal to 0, as can be verified using the straightening algorithm. The MCF algorithm, using the straightening algorithm as a subroutine, finds that $(a, d), (b, h), (c, j), (f, g)$ are all the pairs with this property.

The algorithm now looks for combinations of these extensors that could appear as a meet in a Cayley factorization of P. (For details, see [Whi91].) It finds in our example that $ad \wedge cj$ is such a combination. As soon as a single such combination is found, an algebraic substitution involving a new variable, $z = ad \wedge cj$, is performed, and a new bracket polynomial of smaller degree involving this new variable is derived; the algorithm then begins anew on this polynomial. If no such combination is found, the input bracket polynomial is then known to have no Cayley factorization. In our example, this derived polynomial turns out to be $P = [zef][gbh] - [zeg][fbh]$, which of necessity is still multilinear. The MCF algorithm proceeds to find (and we can directly see by consulting Table 60.2.1) that $P = ze \wedge fg \wedge bh$. Thus, our final Cayley factorization is output as

$$P = ((ad \wedge cj) \vee e) \wedge fg \wedge bh.$$

It is significant that this algorithm requires no backtracking. For example, once $ad \wedge cj$ is found as a possible meet in a Cayley factorization of P, it is known that if P has a Cayley factorization at all, then it must also have one using the factor $ad \wedge cj$; hence we are justified in factoring it out, i.e., substituting a new variable for it. Other Cayley factorizations may be possible, for example,

$$P = ((fg \wedge bh) \vee (ad \wedge cj)) \wedge e.$$

Note that these two factorizations have the same geometric meaning.

60.4 APPLICATIONS

60.4.1 ROBOTICS

GLOSSARY

Robot arm: A set of rigid bodies, or links, connected in series by joints that allow relative movement of the successive links, as described below. The first link is regarded as fixed in position, or tied to the ground, while the last link, called the **end-effector**, is the one that grasps objects or performs tasks.

Revolute joint: A joint between two successive links of a robot arm that allows only a rotation between them. In simpler terms, a revolute joint is a hinge connecting two links.

Prismatic joint: A joint between two successive links of a robot arm that allows only a translational movement between the two links.

Screw joint: A joint between two successive links of a robot arm that allows only a screw movement between the two links.

TABLE 60.4.1   Modeling instantaneous robotics.

ROBOTICS CONCEPT	GRASSMANN-CAYLEY EQUIVALENT
Revolute joint on axis $\overline{ab}$	$\alpha(a \vee b)$, a 2-extensor
Rotation about line $\overline{ab}$	$\beta(a \vee b)$
Motion of point p in rotation about line $\overline{ab}$	$\beta(a \vee b) \vee p$
Screw joint	indecomposable 2-tensor
Prismatic joint	2-extensor at infinity
Motion space of the end-effector,	span of the extensors
where $j_1, j_2, \ldots, j_k$ are joints in series	$< j_1, j_2, \ldots, j_k >$

We are considering here only the instantaneous kinematics or statics of robot arms, that is, positions and motions at a given instant in time. A robot arm has a **critical configuration** if the joint extensors become linearly dependent. If the arm has six joints in three-space, a critical configuration means a loss of full mobility. If the arm has a larger number of joints, criticality is defined as any six of the joint extensors becoming linearly dependent. This can mean severe problems with the driving program in real-life robots, even when the motion space retains full dimensionality.

In one sense, criticality is trivial to determine, since we need only compute a determinant function, called the **superbracket**, on the six-dimensional space $\Lambda^2(V)$. However, if we want to know all the critical configurations of a given robot arm, this becomes a nontrivial question, that of determining all of the zeroes of the superbracket. To answer it, we need to express the superbracket in terms of ordinary brackets. This has been done in [MW91], where the superbracket of the six 2-extensors $a_1a_2, b_1b_2, \ldots, f_1f_2$ is given by

$$[[a_1a_2, b_1b_2, c_1c_2, d_1d_2, e_1e_2, f_1f_2]] =$$

$$- [a_1a_2b_1b_2][c_1c_2\overset{\bullet}{d_1}\overset{\circ}{e_1}][\overset{\bullet}{d_2}\overset{\circ}{e_2}f_1f_2]$$

$$+ [a_1a_2\overset{\bullet}{b_1}\overset{\circ}{c_1}][\overset{\bullet}{b_2}\overset{\circ}{c_2}d_1d_2][e_1e_2f_1f_2]$$

$$- [a_1a_2\overset{\bullet}{b_1}\overset{\circ}{c_1}][\overset{\bullet}{b_2}d_1d_2\overset{\triangleleft}{e_1}][\overset{\circ}{c_2}\overset{\triangleleft}{e_2}f_1f_2]$$

$$+ [a_1a_2\overset{\bullet}{b_1}\overset{\circ}{d_1}][\overset{\bullet}{b_2}c_1c_2\overset{\triangleleft}{e_1}][\overset{\circ}{d_2}\overset{\triangleleft}{e_2}f_1f_2].$$

(Here the dots, diamonds, and triangles have the same meaning as the dots in Section 60.1.)

Consider the particular example of the six-revolute-joint robot arm illustrated in Figure 60.4.1, whose first two joints lie on intersecting lines, whose third and fourth joints are parallel, and whose last two joints also lie on intersecting lines. The larger cylinders in the figure represent the revolute joints. To express the superbracket, we must choose two points on each joint axis. We may choose $b_1 = a_2$, $d_1 = c_2$ (where this point is at infinity), and $f_1 = e_2$, as shown by the black dots. The thin cylinders represent the links; for example, the first link, between a_2 and b_2, is connected to the ground (not shown) by joint a_1a_2, and can therefore only rotate around the axis $\overline{a_1a_2}$.

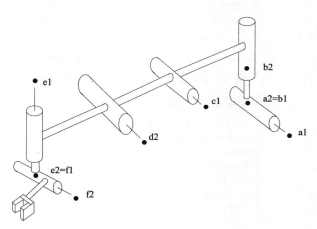

FIGURE 60.4.1
Six-revolute-joint robot arm.

Plugging in and deleting terms with a repeated point inside a bracket, we get

$$- [a_1a_2b\overset{\bullet}{c_1}][a_2c_2d_2e_2][\overset{\bullet}{c}e_1e_2f_2] \tag{60.4.1}$$

$$+ [a_1a_2b\overset{\bullet}{c_2}][a_2c_1c_2e_2][\overset{\bullet}{d}e_1e_2f_2] \tag{60.4.2}$$

$$= [\overset{\bullet}{c}a_1a_2b_2][\overset{\bullet}{d}a_2c_2e_2][\overset{\bullet}{c}e_1e_2f_2], \tag{60.4.3}$$

where each of (60.4.1) and (60.4.2) has two terms because of the dotting, and the same four terms constitute (60.4.3), since two of the six terms generated by the dotting are zero because of the repetition of c_2 in the second bracket.

Finally, we recognize (60.4.3) as the bracket expansion of

$$(c_1d_2c_2) \wedge (a_1a_2b_2) \wedge (a_2c_2e_2) \wedge (e_1e_2f_2).$$

We then recognize that the geometric conditions for criticality are any positions that make this Grassmann-Cayley expression 0, namely

(i) one or more of the planes $\overline{c_1 c_2 d_2}, \overline{a_1 a_2 b_2}, \overline{a_2 c_2 e_2}, \overline{e_1 e_2 f_2}$ is degenerate, or

(ii) the four planes have nonempty intersection.

Notice that in an actual robot arm of the type we are considering, none of the degeneracies in (i) can actually occur.

See Section 51.1 for more information.

60.4.2 BAR FRAMEWORKS

Consider a generically $(d-1)$-isostatic graph G (see Section 61.1 of this Handbook), that is, a graph for which almost all realizations in $(d-1)$-space as a bar framework are minimally first-order rigid. Since first-order rigidity is a projective invariant (see Theorem 61.1.25), we would like to know the projective geometric conditions that characterize all of its nonrigid (first-order flexible) realizations. By Gram's theorem, these conditions must be expressible in terms of bracket conditions, and [WW83] shows that the first-order flexible realizations are characterized by the zeroes of a single bracket polynomial C_G, called the *pure condition* (see Theorem 61.1.27). Furthermore, [WW83] gives an algorithm to construct the pure condition C_G directly from the graph G. Then we require Cayley factorization to recover the geometric incidence condition, if it is not already known. Consider the following examples, illustrated in Figure 60.4.2.

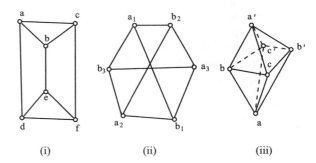

FIGURE 60.4.2
Three examples of bar frameworks.   (i)   (ii)   (iii)

(i) The graph G is the edge skeleton of a triangular prism, realized in the plane. We have $C_G = [abc][def]([abe][dfc] - [dbe][afc])$, and we may recognize the factor in parentheses as the third example in Table 60.2.1. Thus $C_G = 0$, and the framework is first-order flexible, if and only if one of the triangles $\overline{abc}$ or $\overline{def}$ is degenerate, or the three lines $\overline{ad}, \overline{be}, \overline{cf}$ are concurrent, or one or more of these lines is degenerate.

(ii) The graph G is $K_{3,3}$, a complete bipartite graph, realized in the plane. Then $C_G = [a_1 a_2 a_3][a_1 b_2 b_3][b_1 a_2 b_3][b_1 b_2 a_3] - [b_1 b_2 b_3][b_1 a_2 a_3][a_1 b_2 a_3][a_1 a_2 b_3]$, and this is the second example in Table 60.2.2. Thus $C_G = 0$, and the framework is first-order flexible, if and only if the six points lie on a common conic or, equivalently by Pascal's theorem, the three points $\overline{a_1 b_2} \wedge a_2 b_1$, $\overline{a_1 b_3} \wedge a_3 b_1$, $\overline{a_2 b_3} \wedge a_3 b_2$ are collinear.

(iii) The graph G is the edge skeleton of an octahedron, realized in Euclidean 3-space. Then $C_G = [abc'a'][bca'b'][cab'c'] + [abc'b'][bca'c'][cab'a']$, and this can be recognized directly as the expansion of the Grassmann-Cayley expression $abc \wedge a'bc' \wedge a'b'c \wedge ab'c'$. Thus $C_G = 0$, and the framework is first-order flexible, if and only if the four alternating octahedral face planes $\overline{abc}$, $\overline{a'bc'}$, $\overline{a'b'c}$, and $\overline{ab'c'}$ concur, or any one or more of these planes is degenerate. This, in turn, is equivalent to the same condition on the other four face planes, $\overline{abc'}$, $\overline{ab'c}$, $\overline{a'bc}$, $\overline{a'b'c'}$.

60.4.3 BAR-AND-BODY FRAMEWORKS

A *bar-and-body framework* consists of a finite number of $(d-1)$-dimensional rigid bodies, free to move in Euclidean $(d-1)$-space, and connected by rigid bars, with the connections at the ends of each bar allowing free rotation of the bar relative to the rigid body; i.e., the connections are "universal joints." Each rigid body may be replaced by a first-order rigid bar framework in such a way that the result is one large bar framework, thus in one sense reducing the study of bar-and-body frameworks to that of bar frameworks. Nevertheless, the combinatorics of bar-and-body frameworks is quite different from that of bar frameworks, since the original rigid bodies are not allowed to become first-order flexible in any realization, contrary to the case with bar frameworks.

A generically isostatic bar-and-body framework has a pure condition, just as a bar framework has, whose zeroes are precisely the special positions in which the framework has a first-order flex. However, this pure condition is a bracket polynomial in the *bars* of the framework, as opposed to a bracket polynomial in the vertices, as was the case with bar frameworks. An algorithm to directly compute the pure condition for a bar-and-body framework, somewhat similar to that for bar frameworks, is given in [WW87]. We illustrate with the example in Figure 60.4.3, consisting of three rigid bodies and six bars in the plane. We may interpret the word "plane" here as "real projective plane."

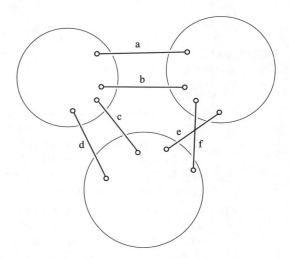

FIGURE 60.4.3
A bar-and-body framework.

Hence $V = \mathbb{R}^3$, and we let $W = \Lambda^2(V) \cong V^* \cong \mathbb{R}^3$. We think of the endpoints of the bars as elements of V, and hence the lines determined by the bars are two-extensors of these points, or elements of W. The algorithm produces the pure condition $[abc][def] - [abd][cef]$. This bracket polynomial may be Cayley factored as $ab \wedge cd \wedge ef$, as seen in Table 60.2.1. Now we switch to thinking of $a, b, \ldots, f$ as 2-extensors in V rather than elements of W, and recall that there is a duality between V and W, hence between $\Lambda(V)$ and $\Lambda(W)$. Thus, the framework has a first-order flex if and only if $(a \wedge b) \vee (c \wedge d) \vee (e \wedge f) = 0$ in $\Lambda(V)$. Hence the desired geometric condition for the existence of a first-order flex is that the three points $\overline{a \wedge b}$, $\overline{c \wedge d}$, and $\overline{e \wedge f}$ are collinear. Now $\overline{a \wedge b}$ is just the center of relative (instantaneous) motion for the two bodies connected by those two bars: think of fixing one of the bodies and then rotating the other body about this center; the lengths of the two bars are instantaneously preserved. The geometric result we have obtained is just a restatement of the classical theorem of Arnhold-Kempe that in any flex of three rigid bodies, the centers of relative motion of the three pairs of bodies must be collinear.

60.4.4 AUTOMATED GEOMETRIC THEOREM-PROVING

J. Richter-Gebert [RG95] uses Grassmann-Cayley algebra to derive bracket conditions for projective geometric incidences in order to produce coordinate-free automatic proofs of theorems in projective geometry. By introducing two circular points at infinity, the same can be done for theorems in Euclidean geometry [CRG95].

Richter-Gebert's technique is to reduce each hypothesis to a *binomial* equation, that is, an equation with a single product of brackets on each side. For example, as we have seen, the concurrence of three lines $\overline{ab}, \overline{cd}, \overline{ef}$ may be rewritten as $[acd][bef] = [bcd][aef]$. Similarly, the collinearity of three points $\overline{a}, \overline{b}, \overline{c}$ may be expressed as $[abd][bce] = [abe][bcd]$, avoiding the much more obvious expression $[abc] = 0$ since it is not of the required form. If all binomial equations are now multiplied together, and provided they were appropriately chosen in the first place, common factors may be canceled (which involves nondegeneracy assumptions, so that the common factors are nonzero), resulting in the desired conclusion. A surprising array of theorems may be cast in this format, and this approach has been successfully implemented.

More recent work along similar lines, extending it especially to conic geometry, is by H. Li and Y. Wu [LW03a, LW03b].

60.4.5 COMPUTER VISION

Much of computer vision study involves projective geometry, and hence is very amenable to the techniques of the Grassmann-Cayley algebra. One reference that explicitly applies these techniques to a system of up to three pinhole cameras is Faugeras and Papadopoulo [FP98].

60.5 SOURCES AND RELATED MATERIAL

SURVEYS

[DRS74] and [BBR85]: These two papers survey the properties of the Grassmann-Cayley algebra (called the "double algebra" in [BBR85]).

[Whi95]: A more elementary survey than the two above.

[Whi94]: Emphasizes the concrete approach via Plücker coordinates, and gives more detail on the connections to robotics.

RELATED CHAPTERS

Chapter 9: Geometry and topology of polygonal linkages
Chapter 51: Robotics
Chapter 61: Rigidity and scene analysis
Chapter 62: Rigidity of symmetric frameworks

REFERENCES

[BBR85]   M. Barnabei, A. Brini, and G.-C. Rota. On the exterior calculus of invariant theory. *J. Algebra*, 96:120–160, 1985.

[CRG95]   H. Crapo and J. Richter-Gebert. Automatic proving of geometric theorems. In N. White, editor, *Invariant Methods in Discrete and Computational Geometry*, pages 167–196. Kluwer, Dordrecht, 1995.

[DRS74]   P. Doubilet, G.-C. Rota, and J. Stein. On the foundations of combinatorial theory: IX, combinatorial methods in invariant theory. *Stud. Appl. Math.*, 53:185–216, 1974.

[FP98]   O. Faugeras and T. Papadopoulo. Grassmann-Cayley algebra for modelling systems of cameras and the algebraic equations of the manifold of trifocal tensors. *Philos. Trans. Roy. Soc. London*, Ser. A, 356:1123–1152, 1998.

[LW03a]   H. Li and Y. Wu Automated short proof generation for projective geometric theorems with Cayley and bracket algebras, I. Incidence geometry. *J. Symbolic Comput.*, 36:717–762, 2003.

[LW03b]   H. Li and Y. Wu Automated short proof generation for projective geometric theorems with Cayley and bracket algebras, II. Conic geometry. *J. Symbolic Comput.*, 36:763–809, 2003.

[MW91]   T. McMillan and N.L. White. The dotted straightening algorithm. *J. Symbolic Comput.*, 11:471–482, 1991.

[RG95]   J. Richter-Gebert. Mechanical theorem proving in projective geometry. *Ann. Math. Artif. Intell.*, 13:139–172, 1995.

[RS76]   G.-C. Rota and J. Stein. Applications of Cayley algebras. In *Colloquio Internazionale sulle Teorie Combinatorie*, pages 71–97, Accademia Nazionale dei Lincei, 1976.

[Stu93]   B. Sturmfels. *Algorithms in Invariant Theory*. Springer-Verlag, New York, 1993.

[Whi91]   N. White. Multilinear Cayley factorization. *J. Symbolic Comput.*, 11:421–438, 1991.

[Whi94]   N. White. Grassmann-Cayley algebra and robotics. *J. Intell. Robot. Syst.*, 11:91–107, 1994.

[Whi95]   N. White. A tutorial on Grassmann-Cayley algebra. In N. White, editor, *Invariant Methods in Discrete and Computational Geometry*, pages 93–106. Kluwer, Dordrecht, 1995.

[WW83]   N. White and W. Whiteley. The algebraic geometry of stresses in frameworks. *SIAM J. Algebraic Discrete Methods*, 4:481–511, 1983.

[WW87]   N. White and W. Whiteley. The algebraic geometry of motions in bar-and-body frameworks. *SIAM J. Algebraic Discrete Methods*, 8:1–32, 1987.

61 RIGIDITY AND SCENE ANALYSIS
Bernd Schulze and Walter Whiteley

INTRODUCTION

Rigidity and flexibility of frameworks (motions preserving lengths of bars) and scene analysis (liftings from plane polyhedral pictures to spatial polyhedra) are two core examples of a general class of geometric problems:

(a) Given a discrete configuration of points, lines, planes, ... in Euclidean space, and a set of geometric constraints (fixed lengths for rigidity, fixed incidences, and fixed projections of points for scene analysis), what is the set of solutions and what is its local form: discrete? k-dimensional?

(b) Given a structure satisfying the constraints, is it unique, or at least locally unique, up to trivial changes, such as congruences for rigidity, or vertical scale for liftings?

(c) How does this answer depend on the combinatorics of the structure and how does it depend on the specific geometry of the initial data or object?

The rigidity of frameworks examines points constrained by fixed distances between pairs of points, using vocabulary and linear techniques drawn from structural engineering: bars and joints, first-order rigidity and first-order flexes, and static rigidity and static self-stresses (Section 61.1). Section 61.2 describes some extended structures and their applications. Scene analysis and the dual concept of parallel drawings are described in Section 61.3. Finally, Section 61.4 describes reciprocal diagrams which form a fundamental geometric connection between liftings of polyhedral pictures and self-stresses in frameworks which continue to be used in structural engineering.

These core problems have a wide range of applications across many areas of applied geometry. The methods used and the results obtained for these problems serve as a model for what might be hoped for other sets of constraints (plane first-order results) and as a warning of the complexity that does arise (higher dimensions and broader forms of rigidity). The subject has a rich history, stretching back into at least the middle of the 19th century, in structural and mechanical engineering. Other independent rediscoveries and connections have arisen in crystallography and scene analysis. The range of constraints and applications has continued to expand over the last two decades. For more general geometric reconstruction problems, see Chapter 34.

61.1 RIGIDITY OF BAR FRAMEWORKS

Given a set of points in space, with certain distances to be preserved, what other configurations have the same distances? If we make small changes in the distances, will there be a small (linear scale) change in the position? What is the structure, locally and globally, of the algebraic variety of these "realizations"?

We begin with the simplest linear theory: first-order rigidity, and the equivalent dual static rigidity, which are the linearized (and therefore linear algebra) version of rigidity. Generic rigidity refers to first-order rigidity of "almost all" geometric positions of the underlying combinatorial structure. After the initial results presenting first-order rigidity (Section 61.1.1), the study divides into the combinatorics of generic rigidity, using graphs (Section 61.1.2); a few basic results on rigid and flexible frameworks (Section 61.1.3); and the geometry of special positions in first-order rigidity, using projective geometry (Section 61.1.4).

61.1.1 FIRST-ORDER RIGIDITY

GLOSSARY

Configuration of points in d-space: An assignment $p = (p_1, \ldots, p_v)$ of points $p_i \in \mathbb{R}^d$ to an index set V, where $v = |V|$.

Congruent configurations: Two configurations p and q in d-space, on the same set V, related by an isometry T of $\mathbb{R}^d$ (with $T(p_i) = q_i$ for all $i \in V$).

Bar framework in d-space $G(p)$ (or ***(bar-joint) framework***): A graph $G = (V, E)$ (no loops or multiple edges) and a configuration p in d-space for the vertices V, such that $p_i \neq p_j$ for each edge $\{i, j\} \in E$ (Figure 61.1.1(a)). An edge $\{i, j\} \in E$ is called a ***bar*** in $G(p)$,

First-order flex (or ***infinitesimal motion***): For a bar framework $G(p)$, an assignment of velocities $p' : V \to \mathbb{R}^d$, such that for each edge $\{i, j\} \in E$: $(p_i - p_j) \cdot (p'_i - p'_j) = 0$ (Figure 61.1.1(c,d), where the arrows represent nonzero velocities).

Trivial first-order flex: A first-order flex p' that is the derivative of a flex of congruent frameworks (Figure 61.1.1(c)). (There is a fixed skew-symmetric matrix S (a rotation) and a fixed vector t (a translation) such that, for all vertices $i \in V$, $p'_i = p_i S + t$.)

First-order flexible (or ***infinitesimally flexible) framework***: A framework $G(p)$ with a nontrivial first-order flex (Figure 61.1.1(d)).

First-order rigid (or ***infinitesimally rigid) framework***: A bar framework $G(p)$ for which every first-order flex is trivial (Figures 61.1.1(a), 61.1.2(a)).

Rigidity matrix: For a framework $G(p)$ in d-space, $R_G(p)$ is the $|E| \times d|V|$ matrix for the system of equations: $(p_i - p_j) \cdot (p'_i - p'_j) = 0$ in the unknown velocities p'_i. The first-order flex equations are expressed as

$$R_G(p){p'}^T = \begin{bmatrix} \vdots & \ddots & \vdots & \cdots & \vdots & \ddots & \vdots \\ 0 & \cdots & (p_i - p_j) & \cdots & (p_j - p_i) & \cdots & 0 \\ \vdots & \ddots & \vdots & \cdots & \vdots & \ddots & \vdots \end{bmatrix} \times {p'}^T = 0^T.$$

Self-stress: For a framework $G(p)$, a row dependence ω for the rigidity matrix: $\omega R_G(p) = 0$. Equivalently, an assignment of scalars ω_{ij} to the edges such that at each vertex i, $\sum_{\{j| \ \{i,j\} \in E \ \}} \omega_{ij}(p_i - p_j) = 0$ (placing these "internal forces" $\omega_{ij}(p_i - p_j)$ in equilibrium at vertex i). $\omega_{ij} < 0$ is tension, $\omega_{ij} < 0$ is compression.

Independent framework: A bar framework $G(p)$ for which the rigidity matrix
has independent rows. Equivalently, there is only the zero self-stress.

Isostatic framework: A framework that is first-order rigid and independent.

Generically rigid graph in d-space: A graph G for which the frameworks $G(p)$
are first-order rigid on an open dense subset of configurations p in d-space (Figures 61.1.1(a) and 61.1.2(a)).

Special position of a graph G in d-space: Any configuration $p \in \mathbb{R}^{dv}$ such that
the rigidity matrix $R_G(p)$ has rank smaller than the maximum rank (the rank at a
configuration with algebraically independent coordinates); See Figures 61.1.1(d)
and 61.1.2(b) for examples. The special positions form an algebraic variety. A
configuration p is ***regular*** if it is not special (that is $R_G(p)$ is maximum rank
over all configurations).

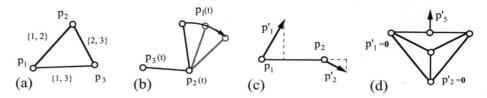

FIGURE 61.1.1
*A rigid framework (a); a flexible framework (b); a first-order flex (c) and a first-order
flexible, but rigid framework (d).*

BASIC CONNECTIONS

Because the constraints $|p_i - p_j| = |q_i - q_j|$ are algebraic in the coordinates of the
points (after squaring), we can work with the Jacobian matrix formed by the partial
derivatives of these equations — the rigidity matrix of the framework.

The dimension of the space of trivial first-order motions of a framework in d-
space is $\binom{d+1}{2}$ provided $|V| \geq d$ (the velocities generated by d translations and by
$\binom{d}{2}$ rotations form a basis).

THEOREM 61.1.1   First-order Rank

*A framework $G(p)$ with $|V| \geq d$ is first-order rigid if and only if the rigidity matrix
$R_G(p)$ has rank $d|V| - \binom{d+1}{2}$.*

*A framework $G(p)$ with few vertices, $|V| \leq d$, is isostatic if and only if the
rigidity matrix $R_G(p)$ has rank $\binom{v}{2}$ (if and only if G is the complete graph on V
and the points p_i do not lie in an affine space of dimension $|V| - 2$).*

First-order rigidity is linear algebra, with first-order rigid frameworks, self-
stresses, and isostatic frameworks playing the roles of spanning sets, linear dependence, and bases of the row space for the rigidity matrix of the complete graph on
the configuration p.

There is a dual theory of static rigidity for bar frameworks. Where first-order
rigidity focuses on the kernel of the rigidity matrix (first-order flexes) and on the
column space and column rank, static rigidity focuses on the cokernel of the rigidity
matrix (the self-stresses) and on the row space of the rigidity matrix (the resolvable
static loads). Methods from both approaches are widely used [CW82, Whi84a,

Whi96], although in this chapter we present the results primarily in the vocabulary of first-order rigidity.

THEOREM 61.1.2   Isostatic Frameworks

For a framework $G(p)$ in d-space, with $|V| \geq d$, the following are equivalent:

(a) *$G(p)$ is isostatic (first-order rigid and independent);*

(b) *$G(p)$ is first-order rigid with $|E| = d|V| - \binom{d+1}{2}$;*

(c) *$G(p)$ is independent with $|E| = d|V| - \binom{d+1}{2}$;*

(d) *$G(p)$ is first-order rigid, and removing any one bar (but no vertices) leaves a first-order flexible framework.*

First-order rigidity of a framework $G(p)$ is a robust property: a small change in the configuration p preserves this rigidity. Independence implies that the lengths of bars are robust: any small change in these distances can be realized by a nearby configuration. On the other hand, self-stresses mean that one of the distances is algebraically dependent on the others: many small changes in the distances will have no realizations, or no nearby realizations.

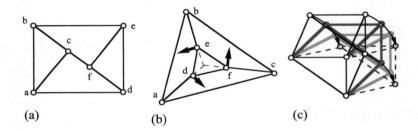

(a)   (b)   (c)

FIGURE 61.1.2
The graph G is realized as (a) a first-order rigid framework; as (b) a first-order flexible but rigid framework; as (c) a finitely flexible framework.

We note that the regular configurations of a graph G are an open dense subset.

THEOREM 61.1.3   Generic Rigidity Theorem

For a graph G and a fixed dimension d the following are equivalent:

(a) *G is generically rigid in d-space;*

(b) *for each configuration $p \in \mathbb{R}^{dv}$ using algebraically independent numbers over the rationals as coordinates, the framework $G(p)$ is first-order rigid;*

(c) *$G(p)$ is first-order rigid for all regular configurations;*

(d) *$G(p)$ is first-order rigid for some configuration.*

61.1.2 COMBINATORICS FOR GENERIC RIGIDITY

The major goal in generic rigidity is a combinatorial characterization of graphs that are generically rigid in d-space. The companion problem is to find efficient combinatorial algorithms to test graphs for generic rigidity. For the plane (and the line), this is solved. Beyond the plane the results are essentially incomplete, but some significant partial results are available.

GLOSSARY

Generically d-independent: A graph G for which some (equivalently, almost all) configurations p produce independent frameworks in d-space.

Generically d-isostatic graph: A graph G for which some (equivalently, almost all) configurations p produce isostatic frameworks in d-space.

Generic d-circuit: A graph G that is dependent for all configurations p in d-space but for all edges $\{i,j\} \in E$, $G - \{i,j\}$ is generically independent in d-space.

Complete bipartite graph: A graph $K_{m,n} = (A \cup B, A \times B)$, where A and B are disjoint sets of cardinality $|A| = m$ and $|B| = n$.

Triangulated d-pseudomanifold: A finite set of d-simplices (sets of $d+1$) with the property that each d subset of a simplex (facet) occurs in exactly two simplices, any two simplices are connected by a path of simplices and shared facets, and any two simplices sharing a vertex are connected through other simplices at this vertex. (For example, the triangles, edges, and vertices of a closed triangulated 2-surface without boundary form a 2-pseudomanifold.) See Section 16.3.

Henneberg d-construction for a graph G: A sequence $(V_d, E_d), \ldots, (V_n, E_n)$ of graphs, such that (Figure 61.1.6(a)):

(i) For each index $d < j \le n$, (V_j, E_j) is obtained from (V_{j-1}, E_{j-1}) by

vertex addition: attaching a new vertex by d edges (Figure 61.1.3(a) for $d = 2$), or

edge splitting: replacing an edge from (V_{j-1}, E_{j-1}) with a new vertex joined to its ends and to $d-1$ other vertices (Figure 61.1.3(b) for $d = 2$); and

(ii) (V_d, E_d) is the complete graph on d vertices, and $(V_n, E_n) = G$.

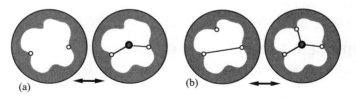

FIGURE 61.1.3
Both vertex addition (a) and edge split (b) preserve generic rigidity and generic independence.

Proper 3Tree2 partition: A partition of the edges of a graph into three trees, such that each vertex is attached to exactly two of these trees and no nontrivial subtrees of distinct trees T_i have the same support (i.e., the same vertices) (Figure 61.1.6(b)).

Proper 2Tree partition: A partition of the edges of a graph into two spanning trees, such that no nontrivial subtrees of distinct trees T_i have the same support (i.e., the same vertices) (Figure 61.1.6(c)).

d-connected graph: A graph G such that removing any $d - 1$ vertices (and all incident edges) leaves a connected graph. (Equivalently, a graph such that any two vertices can be connected by at least d paths that are vertex-disjoint except for their endpoints.)

BASIC PROPERTIES IN ALL DIMENSIONS

THEOREM 61.1.4 Necessary Counts and Connectivity Theorem

If a graph G is generically d-isostatic, then, if $|V| \geq d$, $|E| = d|V| - \binom{d+1}{2}$, and for every $V' \subseteq V$ with $|V'| \geq d$, the number of edges in the subgraph of G induced by V' is at most $d|V'| - \binom{d+1}{2}$.

If G is a generically d-isostatic graph with $|V| > d$, then G is d-connected.

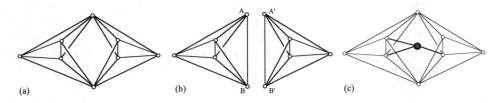

(a)   (b)   (c)

FIGURE 61.1.4
The double banana framework (a) is generically flexible, though it satisfies the basic counts of Theorem 61.1.4. It is generically dependent (b) and remains both flexible and dependent even when extended to be 3-connected (c).

For dimensions 1 and 2, the first count alone is sufficient for generic rigidity (see below). For dimensions $d > 2$, these two conditions are not enough to characterize the generically d-isostatic graphs (Table 61.2.1). Figure 61.1.4(a) shows a generically flexible counterexample for the sufficiency of the counts in dimension 3.

THEOREM 61.1.5 Bipartite Graphs [Whi84b]

A complete bipartite graph $K_{m,n}$, with $m > 1$, is generically rigid in dimension d if and only if $m + n \geq \binom{d+2}{2}$ and $m, n > d$.

INDUCTIVE CONSTRUCTIONS FOR ISOSTATIC GRAPHS

Inductive constructions for graphs that preserve generic rigidity are used both to prove theorems for general classes of frameworks and to analyze particular graphs [TW85, NR14].

THEOREM 61.1.6 Vertex Addition Theorem

The set of generically d-isostatic graphs is closed under vertex addition and under the deletion of vertices of degree d (Figure 61.1.3(a) for $d = 2$).

THEOREM 61.1.7 Edge Split Theorem

The set of generically d-isostatic graphs is closed under edge splitting. Conversely, given a vertex v_0 of degree $d + 1$ there is some pair j, k not in E and adjacent to v_0 such that adding that edge gives a d-isostatic graph. (Figure 61.1.3(b) for $d = 2$).

THEOREM 61.1.8 Construction Theorem

If a graph G is obtained by a Henneberg d-construction, then G is generically d-isostatic (Figure 61.1.6(a) for $d = 2$).

THEOREM 61.1.9 Gluing Theorem

If $G_1 = (V_1, E_1)$ and $G_2 = (V_2, E_2)$ are generically d-rigid graphs sharing at least d vertices, then $G = (V_1 \cup V_2, E_1 \cup E_2)$ is generically d-rigid.

Vertex splitting is defined in Figure 61.1.5 [Whi91].

THEOREM 61.1.10 Vertex Splitting Theorem

If the graph G' is a vertex split of a generically d-isostatic graph G on d edges (Figure 61.1.5(a) for $d = 3$) or a vertex split on $d - 1$ edges (Figure 61.1.5(b) for $d = 3$), then G' is generically d-isostatic.

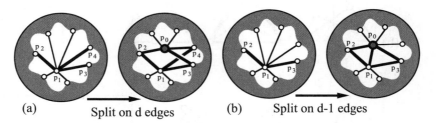

(a) Split on d edges (b) Split on d-1 edges

FIGURE 61.1.5
Two forms of vertex split (a) on d edges and (b) $d - 1$ edges preserve generic rigidity and generic independence.

PLANE ISOSTATIC GRAPHS

Many plane results are expressed in terms of trees in the graph, building on a simpler correspondence between rigidity on the line and the connectivity of the graph. The earliest currently known complete proof of what is now known as Laman's Theorem (Theorem 61.1.11(b)) is [Poll27], though earlier statements of the result appear in the literature without complete proofs. The following is folklore.

THEOREM 61.1.11 Line Rigidity

For graph G and configuration p on the line with $p_i \neq p_j$ for all $\{i, j\} \in E$, the following are equivalent:

(a) *$G(p)$ is minimal among rigid frameworks on the line with these vertices;*
(b) *$G(p)$ is isostatic on the line;*
(c) *G is a spanning tree on the vertices;*
(d) *$|E| = |V| - 1$ and for every nonempty subset E' with vertices V', $|E'| \leq |V'| - 1$.*

THEOREM 61.1.12   Plane Isostatic Graphs Theorem

For a graph G with $|V| \geq 2$, the following are equivalent:

(a) *G is generically isostatic in the plane;*

(b) *$|E| = 2|V| - 3$, and for every subgraph (V', E') with $|V'| \geq 2$ vertices, $|E'| \leq 2|V'| - 3$ (Laman's theorem; see also Table 61.2.1);*

(c) *there is a Henneberg 2-construction for G (Henneberg's theorem);*

(d) *E has a proper 3Tree2 partition (Crapo's theorem);*

(e) *for each $\{i, j\} \in E$, the multigraph obtained by doubling the edge $\{i, j\}$ is the union of two spanning trees (Recski's theorem).*

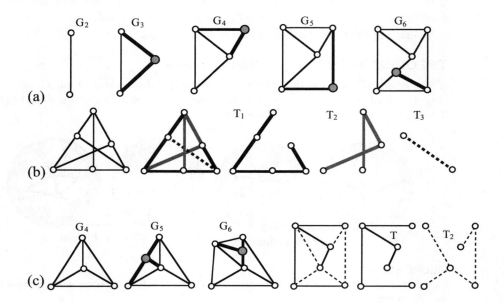

FIGURE 61.1.6

Figure (a) gives a Henneberg 2-construction for the graph in Figure 61.1.2. Figure (b) gives a decomposition into 3 trees, two at each vertex. Figure (c) gives inductive steps for a generic 2-circuit which decomposes into two spanning trees.

THEOREM 61.1.13   Plane 2-Circuits Theorem

For a graph G with $|V| \geq 2$, the following are equivalent:

(a) *G is a generic 2-circuit;*

(b) *$|E| = 2|V| - 2$, and for every proper subset E' on vertices V', $|E'| \leq 2|V'| - 3$;*

(c) *there is a construction for G from K_4, using only edge splitting and gluing (Berg and Jordán's theorem);*

(d) *E has a proper 2Tree partition.*

Figure 61.1.6(c) shows the construction for a 2-circuit, and an associated 2Tree partition. For 2-circuits with planar graphs, the planar dual is also a 2-circuit. The inductive techniques given above, and others, form dual pairs of constructions for these planar 2-circuits, pairing edge splits with vertex splits [BCW02].

THEOREM 61.1.14   Sufficient Connectivity [LY82]

If a graph G is 6-connected, then G is generically rigid in the plane.

There are 5-connected graphs that are not generically rigid in the plane. However, it was shown in [JJ09] that the 6-connectivity condition in Theorem 61.1.14 can be replaced by the weaker condition of "6-mixed connectivity."

ALGORITHMS FOR GENERIC 2-RIGIDITY

Each of the combinatorial characterizations has an associated polynomial time algorithm for verifying whether a graph is generically 2-isostatic:

(i) Counts: This can be checked by an $O(|V|^2)$ algorithm based on bipartite matchings or network flows on an associated graph [Sug86]. An efficient and widely used implementation is the "pebble game" [LS08].

(ii) 2-construction: Existence of a 2-construction can be checked by an $O(2^{|V|})$ algorithm, but a proposed 2-construction can be verified in $O(|V|)$ time.

(iii) 3Tree2 covering: Existence can be checked by an $O(|V|^2)$ matroid partition algorithm [Cra90].

(iv) 2Tree partition: All required 2Tree partitions can be found by a matroidal algorithm of order $O(|V|^3)$.

GENERICALLY RIGID GRAPHS IN HIGHER DIMENSIONS

Most of the results are covered by the initial summary for all dimensions d. Special results apply to the graphs of triangulated polytopes, as well as more general triangulated surfaces (see Section 16.3 for definitions).

THEOREM 61.1.15   Triangulated Pseudomanifolds Theorem

For $d \geq 2$, the graph of a triangulated d-pseudomanifold is generically $(d+1)$-rigid.

In particular, the graph of any closed triangulated 2-surface without boundary is generically rigid in 3-space (Fogelsanger's theorem), and the graph of any triangulated sphere is generically 3-isostatic (Gluck's theorem) [Fog88, Whi96]. Beyond the triangulated spheres in 3-space, most of these graphs are not isostatic, but are dependent. These results can all be proven by inductions using vertex splitting (and reduction techniques based on edge contractions).

These results have also been modified to take a triangulated sphere (and manifolds) and replace some discs by rigid blocks, and make other discs into holes. The resulting structures are *block and hole polyhedra* [FW13, CKP15a, CKP15b]). Here is a sample theorem, among a range of recent results.

THEOREM 61.1.16   Isostatic Towers [FW13]

A block and hole sphere, with one k-gon hole and one k-gon block, is generically isostatic if and only if there are k vertex disjoint paths connecting the hole and the block.

As a consequence of the recently proved Molecular Theorem 61.2.4, we describe a new class of generically 3-rigid graphs in Theorem 61.2.5. Further notable results regarding the generic independence of graphs in Euclidean d-space are the following.

THEOREM 61.1.17   Generic Independence in d-Space

Let $G = (V, E)$ be a graph with $|V| \geq d$ and let $G(p)$ be a generic framework in $\mathbb{R}^d$.

(a) *If $1 \leq d \leq 4$, and G is K_{d+2}-minor free, then $G(p)$ is independent [Nev07];*

(b) *if d is even, and for every $V' \subseteq V$ with $|V'| \geq 2$, the number of edges in the subgraph of G induced by V', $i(V')$, is at most $(\frac{d}{2}+1)|V'| - (d+1)$, then $G(p)$ is independent [JJ05a].*

   If $d = 3$ and $i(V') \leq \frac{1}{2}(5|V'| - 7)$ for all $V' \subseteq V$ with $|V'| \geq 2$, then $G(p)$ is independent [JJ05a].

It is unknown whether the statement in (a) also holds for $d = 5$, but it is known to be false for $d \geq 6$. It is conjectured that the statement in (b) for even d also holds for all $d \geq 2$.

OPEN PROBLEMS

There is no combinatorial characterization of generically 3-isostatic graphs. There are a variety of conjectures towards such characterizations. Some older and recent work offers conjectures [JJ05b, JJ06]. We present one of the conjectures [TW85].

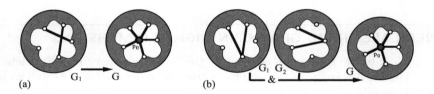

FIGURE 61.1.7

Two possible ways (a, b) to generate 5-valent vertices which are conjectured to preserve first-order rigidity.

CONJECTURE 61.1.18   *3-D Replacement Conjecture*

The X-replacement in Figure 61.1.7(a) takes a graph G_1 that is generically rigid in 3-space to a graph G that is generically rigid in 3-space.

   The double V-replacement in Figure 61.1.7(b) takes two graphs G_1, G_2 that are generically rigid in 3-space to a graph G that is generically rigid in 3-space.

Every 3-isostatic graph is generated by an "extended Henneberg 3-construction," which uses these two moves along with edge splitting and vertex addition. What is unproven is that *only* 3-isostatic graphs are generated in this way.

The plane analogue of X-replacement is true for plane generic rigidity (without adding the fifth bar) [TW85, BCW02], and the 4-space analogue is false for some graphs (with two extra bars added in this analogue). If these conjectured steps prove correct in 3-space, then we would have inductive techniques to generate the graphs of all isostatic frameworks in 3-space, but the algorithm would be exponential.

For 4-space, there is no conjecture that has held up against the known counter-examples based on generically 4-flexible complete bipartite graphs such as $K_{7,7}$. We note that a related constraint matroid from multi-variate splines shares many of these inductive constructions, as well as gaps [Whi96]. However, X-replacement is known to work in the analogue of all dimensions there.

It is known that $d(d + 1)$ connectivity is the minimum connectivity which is sufficient for body-bar frameworks. There is no confirmed polynomial bound on the minimum connectivity sufficient for generic rigidity in d-space for $d > 2$. We do have conjectures.

CONJECTURE 61.1.19   *Sufficient Connectivity Conjecture*

If a graph G is $d(d + 1)$-connected, then G is generically rigid in d-space.

A graph can be checked for generic 3-rigidity by a "brute force" $O(2^{2^{|V|}})$ time algorithm. Assign the points independent variables as coordinates, form the rigidity matrix, then check the rank by symbolic computation. On the other hand, if numerical coordinates are chosen for the points "at random," then the rank of this numerical matrix ($O(|E|^3)$) will be the generic value, with probability 1. This gives a randomized polynomial-time algorithm, but there is no known deterministic algorithm that runs in polynomial, or even exponential, time.

61.1.3 RIGID AND FLEXIBLE FRAMEWORKS

GLOSSARY

Bar equivalence: Two frameworks $G(p)$ and $G(q)$ such that all bars have the same length in both configurations: $|p_i - p_j| = |q_i - q_j|$ for all bars $\{i, j\} \in E$.

Analytic flex: An analytic function $p(t) : [0, 1) \to \mathbb{R}^{vd}$ such that $G(p(0))$ is bar-equivalent to $G(p(t))$ for all t (i.e., all bars have constant length).

Flexible framework: A bar framework $G(p)$ in $\mathbb{R}^d$ with an analytic flex $p(t)$ such that $p(0) = p$ but p is not congruent to $p(t)$ for all $t > 0$ (Figure 61.1.1(b)).

Rigid framework: A bar framework $G(p)$ in d-space that is not flexible (Figure 61.1.1(a,d)).

BASIC CONNECTIONS

Because the constraints $|p_i - p_j| = |q_i - q_j|$ are algebraic in the coordinates of the points (after squaring), many alternate definitions of a "rigid framework" are equivalent. These connections depend on results such as the curve selection theorem of algebraic geometry or the inverse function theorem. A key early paper is [AR78].

THEOREM 61.1.20   Alternate Rigidity Definitions

For a bar framework $G(p)$ the following conditions are equivalent:

(a) *the framework is rigid;*

(b) *for every continuous path, or* **continuous flex** *of $G(p)$, $p(t) \in \mathbb{R}^{vd}$, $0 \leq t < 1$ and $p(0) = p$, such that $G(p(t))$ is bar-equivalent to $G(p)$ for all t, $p(t)$ is congruent to p for all t;*

(c) *there is an $\epsilon > 0$ such that if $G(p)$ and $G(q)$ are bar-equivalent and $|p - q| < \epsilon$, then p is congruent to q.*

Essentially, the first derivative of a nontrivial analytic flex is a nontrivial first-order flex: $D_t\big((p_i(t) - p_j(t))^2 = c_{ij}\big)\big|_{t=0} \Rightarrow 2(p_i - p_j) \cdot (p_i' - p_j') = 0$. (If this first derivative is trivial, then the earliest nontrivial derivative is a first-order motion.) This result is related to general forms of the inverse function theorem.

THEOREM 61.1.21   First-order Rigid to Rigid

If a bar framework $G(p)$ is first-order rigid, then $G(p)$ is rigid.

Rigidity and first-order rigidity are equivalent in most situations [AR78].

THEOREM 61.1.22   Rigid to First-order Rigid

If a bar framework $G(p)$ is regular, then $G(p)$ is first-order rigid if and only if $G(p)$ is rigid.

The other way to express this equivalence is that if a framework $G(p)$ has a first-order flex at a regular configuration p, then $G(p)$ is flexible.

Some first-order flexes are not the derivatives of analytic flexes (Figures 61.1.1(d) and 61.1.2(b)). However, a nontrivial first-order flex for a framework does guarantee a pair of nearby noncongruent, bar-equivalent frameworks. The averaging technique can also be used for alternative proofs that first-order rigidity implies rigidity.

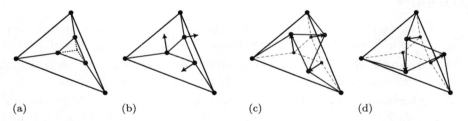

(a)   (b)   (c)   (d)

FIGURE 61.1.8
Given an infinitesimal motion of a framework (a), the velocity vectors (b) push forward (c) and pull backwards (d) to create two equivalent, non-congruent frameworks (de-averaging).

THEOREM 61.1.23   Averaging Theorem

If the points of a configuration p affinely span d-space, then the assignment p' is a nontrivial first-order flex of $G(p)$ if and only if the frameworks $G(p + p')$ and $G(p - p')$ are bar-equivalent and not congruent (Figure 61.1.8).

Whereas first-order rigidity is projectively invariant, rigidity itself is not projectively invariant—or even affinely invariant. It is a purely Euclidean property.

61.1.4 GEOMETRY OF FIRST-ORDER RIGIDITY

GLOSSARY

Projective transform of a d-configuration p: A d-configuration q on the same vertices, such that there is an invertible matrix T of size $(d+1) \times (d+1)$ making $T(p_i, 1) = \lambda_i(q_i, 1)$ (where $(p_i, 1)$ is the vector p_i extended with an additional 1 — the affine coordinates of p_i).

Affine spanning set for d-space: A configuration p of points such that every point $q_0 \in \mathbb{R}^d$ can be expressed as an affine combination of the p_i: $q_0 = \sum_i \lambda_i p_i$, with $\sum_i \lambda_i = 1$. (Equivalently, the affine coordinates $(p_i, 1)$ span the vector space $\mathbb{R}^{d+1}$.)

Cone graph: The graph $G * u$ obtained from $G = (V, E)$ by adding a new vertex u and the $|V|$ edges (u, i) for all vertices $i \in V$.

Cone projection from p_0: For a $(d+1)$-configuration p on V, a configuration $q = \Pi_0(p)$ in d-space (placed as a hyperplane in $(d+1)$-space) on the vertices $V \setminus \{0\}$, such that $p_i \neq p_0$ is on the line $q_i p_0$ for all $i \neq 0$ (see also Chapter 63.2).

BASIC RESULTS

THEOREM 61.1.24   First-Order Flex Test

If the points of a configuration p on the vertices V affinely span d-space, then a first-order motion p' is nontrivial if and only if there is some pair h, k (not a bar) such that: $(p_h - p_k) \cdot (p'_h - p'_k) \neq 0$.

Projective invariance of static and infinitesimal rigidity has been known for more than 150 years, but is often forgotten [CW82].

THEOREM 61.1.25   Projective Invariance

If a framework $G(p)$ is first-order rigid (isostatic, independent) and $q = T(p)$ is a projective transform of p, then $G(q)$ is first-order rigid (isostatic, independent, respectively).

The following result provides an alternate proof of projective invariance as well as a corresponding generic result for cones.

THEOREM 61.1.26   Coning Theorem

A framework $G(\Pi_0 p)$ is first-order rigid (isostatic, independent) in d-space if and only if the cone $(G * u)(p)$ is first-order rigid (isostatic, independent, respectively) in $(d+1)$-space.

The special positions of a graph in d-space are rare, since they form a proper algebraic variety (essentially generated by minors of the rigidity matrix with variables for the coordinates of points). For a generically isostatic graph, this set of special positions can be described by the zeros of a single polynomial [WW83].

THEOREM 61.1.27   Pure Condition

For any graph G that is generically isostatic in d-space, there is a homogeneous polynomial $C_G(x_{1,1}, \ldots, x_{1,d}, \ldots, x_{|V|,1}, \ldots, x_{|V|,d})$ such that $G(p)$ is first-order flexible if and only if $C_G(p_1, \ldots, p_{|V|}) = 0$. C_G is of degree $(val_i + 1 - d)$ in the variables $(x_{i,1}, \ldots, x_{i,d})$ for each vertex i of degree val_i in the graph.

Since Grassmann algebra (Chapter 60) is the appropriate language for these projective properties, these pure conditions C_G are polynomials in the Grassmann algebra. Section 60.4 contains several examples of these polynomial conditions.

THEOREM 61.1.28   Quadrics for Bipartite Graphs [Whi84b]

For a complete bipartite graph $K_{m,n}$ and $d > 1$, the framework $K_{m,n}(p)$, with $p(A)$ and $p(B)$ each affinely spanning d-space, is first-order flexible if and only if all the points $p(A \cup B)$ lie on a quadric surface of d-space (Figure 61.1.9).

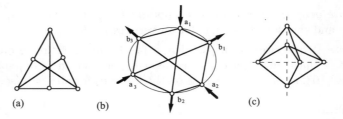

FIGURE 61.1.9
Three realizations of the complete bipartite graph $K_{3,3}$ which are (a) first-order rigid; (b) first-order flexible on a conic; and (c) flexible.

The following classical result describes an important open set of configurations that are regular for triangulated spheres [Whi84a].

THEOREM 61.1.29   Extended Cauchy Theorem

If $G(p)$ consists of the vertices and edges of a convex simplicial d-polytope, then $G(p)$ is first-order rigid in d-space.

If $G(p)$ consists of the vertices and edges of a strictly convex polyhedron in 3-space, then $G(p)$ is independent.

We recall that Steinitz's theorem guarantees that every 3-connected planar graph has a realization as the edges of a strictly convex polyhedron in 3-space, which gives Gluck's theorem. Connelly [Con78] gives a nonconvex (but not self-intersecting) triangulated sphere (with nine vertices) that is flexible. For many graphs, such as a triangulated torus (Theorem 61.1.15), we do not have even one specific configuration that gives a first-order rigid framework, only the guarantee that "almost all" configurations will work.

The Carpenter's Rule problem on straightening plane embedded polygonal paths and convexifying plane embedded polygons uses independence of appropriate bar frameworks (see Chapter 9).

There are further significant consequences of the underlying projective geometry of first-order rigidity [CW82]. The concepts of first-order rigidity and first-order flexibility, as well as the dual statics, can be expressed in any of the Cayley-Klein metrics that are extracted from the shared underlying projective space. This family includes the spherical metric, the hyperbolic metric, and others [SW12]. It is possible to express first-order rigidity in entirely projective terms that are essentially independent of the metric. In this way, the points "at infinity" in the Euclidean space can be fully integrated into first-order rigidity. For example, throughout Section 61.2.1 on Body-Bar and Body-Hinge structures, points at infinity can be included in the Cayley Algebra for bars, hinges, and centers of motion. However, in some metrics such as the hyperbolic metric, there is a singular set (the sphere at infinity, also known as the *absolute*) on which rigidity equations have distinct properties. This transfer goes back to Pogorelov and has been reworked in [SW07].

THEOREM 61.1.30   Transfer of Metrics

For a given graph G and a fixed point p in projective space of dimension d, the framework G(p) is first-order rigid in Euclidean space if and only if G(p) is first-order rigid in any alternate Cayley-Klein metric, with p not containing points on the absolute.

The most extreme projective transformation is polarity. The infinitesimal rigidity of specific body-bar and body-hinge frameworks are invariant under polarity (see Section 61.2.1).

In the plane, polarity switches points and lines of edges of a framework to lines and points of intersection [Whi89]. Surprisingly, this transformation turns plane infinitesimal motions into lifting motions of the lines into 3-space, which preserve the intersections — a result related to parallel drawings and liftings (Section 61.3). This transformation also takes tensegrity frameworks into weavings of lines in the plane. In both settings, the plane results correspond to interesting spatial results about when line configurations with specified intersections, or specific over-under patterns, can separate in 3-space or can only occur in flat (plane) formations.

61.2 RIGIDITY OF OTHER RELATED STRUCTURES

A number of related structures have also been investigated for first-order rigidity. In particular, body-bar, body-hinge, and molecular structures have important practical applications in engineering, robotics, and chemistry. For these structures, there exist basically complete combinatorial theories for all dimensions. Moreover, in mechanical and structural engineering it is common to analyze the rigidity and flexibility of pinned frameworks rather than "free-floating" frameworks. Finally, a new area of application of rigidity theory is the control of formations of autonomous agents. A notion of "directed rigidity" has recently been developed for this purpose.

61.2.1 BODY-BAR AND BODY-HINGE STRUCTURES

GLOSSARY

Body-bar framework in d-space $G(b)$:   An undirected multigraph G (with no loops) and a bar-configuration $b : E \to \Lambda^2(\mathbb{R}^{d+1})$ that maps each edge $e = \{i, j\}$

to the 2-extensor $\hat{p}_{e,i} \vee \hat{p}_{e,j}$ which indicates the Plücker coordinates of the bar connecting the point $p_{e,i}$ in the body i and the point $p_{e,j}$ in the body j. (For any $p \in \mathbb{R}^d$, $\hat{p}$ denotes the homogeneous coordinates of p, i.e., $\hat{p} = \begin{pmatrix} p \\ 1 \end{pmatrix} \in \mathbb{R}^{d+1}$.) See also Chapter 60. Note that while an edge $\{i, j\}$ is an unordered pair, $\hat{p}_{e,i} \vee \hat{p}_{e,j}$ is ordered. For deciding whether $G(b)$ is first-order rigid, however, we just need the linear space spanned by $\hat{p}_{e,i} \vee \hat{p}_{e,j}$ and hence this ordering is irrelevant.

First-order flex of a body-bar framework $G(b)$: A map $m : V \to \mathbb{R}^{\binom{d+1}{2}}$ satisfying $(m(i) - m(j)) \cdot b(e) = 0$ for all $e = \{i, j\} \in E$. A first-order flex m of $G(b)$ is ***trivial*** if $m(i) = m(j)$ for all $i, j \in V$ (i.e., if each vertex has the same "screw center" $m(i)$), and $G(b)$ is ***first-order rigid*** if every first-order flex of $G(b)$ is trivial.

Body-hinge framework in d-space $G(h)$: An undirected graph G and a hinge-configuration $b : E \to \Lambda^{d-1}(\mathbb{R}^{d+1})$ that maps each edge $e = \{i, j\}$ to the $(d-1)$-extensor $\hat{p}_{e,1} \vee \hat{p}_{e,2} \vee \cdots \vee \hat{p}_{e,d-1}$, which indicates the Plücker coordinates of a hinge, i.e., a $(d-1)$-dimensional simplex determined by the points $p_{e,1}, \ldots, p_{e,d-1}$ in the bodies of i and j.

First-order flex of a body-hinge framework $G(h)$: A map $m : V \to \mathbb{R}^{\binom{d+1}{2}}$ satisfying $(m(i) - m(j)) \in \text{span}\{h(e)\}$ for all $e = \{i, j\} \in E$. A first-order flex m of $G(h)$ is ***trivial*** if $m(i) = m(j)$ for all $i, j \in V$, and $G(h)$ is ***first-order rigid*** if every first-order flex of $G(b)$ is trivial.

Panel-hinge framework: A body-hinge framework $G(h)$ with the property that for each $v \in V$, all of the $(d-2)$-dimensional affine subspaces $h(e)$ for the edges e incident to v are contained in a common $(d-1)$-dimensional affine subspace (i.e., a hyperplane).

Molecular framework: A body-hinge framework $G(h)$ with the property that for each $v \in V$, all of the hinge spaces $h(e)$ for the edges e incident to v contain a common point.

BODY-BAR FRAMEWORKS

Three-dimensional body-bar frameworks provide a useful model for many physical structures, linkages and robotic mechanisms. Tay established a combinatorial characterization of the multigraphs that are first-order rigid for almost all realizations as a body-bar framework in $\mathbb{R}^d$, for any dimension d [Tay84, WW87].

THEOREM 61.2.1   Tay's Theorem

For a multigraph G the following are equivalent:

  (a) *for some line assignment of bars $b_{i,j}$ in d-space to the edges $\{i, j\}$ of G, the body-bar framework $G(b)$ is first-order rigid;*

  (b) *for almost all bar configurations b, the body-bar framework $G(b)$ is first-order rigid;*

  (c) *G contains $\binom{d+1}{2}$ edge-disjoint spanning trees.*

Condition (c) in Theorem 61.2.1 can be checked efficiently via a pebble game algorithm in $O(|V||E|)$ time. Note that a d-dimensional body-bar framework in which the attachment points of the bars affinely span d-space can be modeled as a bar-joint framework, with each body replaced by a first-order rigid bar-joint framework on these points. Therefore, Theorem 61.2.1 offers a fast combinatorial algorithm for the generically first-order rigid graphs in dimension $d \geq 3$ for the special class of body-bar frameworks. [WW87] offers some initial geometric criteria for the infinitesimal flexibility of generically isostatic multigraphs.

BODY-HINGE FRAMEWORKS

A body-hinge framework is a collection of rigid bodies, indexed by V, which are connected in pairs along hinges (lines in 3-space), indexed by edges of a graph. The bodies each move, preserving the contacts at the hinges. Of particular interest in applications are body-hinge frameworks in 3-space. Such structures could be modeled as bar-joint frameworks, with each hinge replaced by a pair of joints and each body replaced by a first-order rigid framework on the joints of its hinges (and other joints if desired). Since a hinge in 3-space removes 5 of the 6 relative degrees of freedom between a pair of rigid bodies, a 3-dimensional body-hinge framework can also be modeled as a body-bar framework by replacing each hinge with 5 independent bars, each intersecting the hinge line. It was shown by Tay and Whiteley in [TW84] that this special geometry of the bar configuration does not induce any additional first-order flexibility. The following result says that Tay's theorem regarding "generic" body-bar frameworks also applies to "generic" body-hinge frameworks [TW84].

THEOREM 61.2.2   Tay and Whiteley's Theorem

For a graph G the following are equivalent:

(a) *for some hinge assignment $h_{i,j}$ in d-space to the edges $\{i, j\}$ of G, the body-hinge framework $G(h)$ is first-order rigid;*

(b) *for almost all hinge assignments h, the body-hinge framework $G(h)$ is first-order rigid;*

(c) *if each edge of the graph is replaced by $\binom{d+1}{2} - 1$ copies, the resulting multigraph contains $\binom{d+1}{2}$ edge-disjoint spanning trees.*

The following geometric result for body-hinge frameworks in 3-space was shown in [CW82] and extended to block and hole polyhedra in [FRW12].

THEOREM 61.2.3   Spherical Flexes and Stresses

Given an abstract spherical structure (see Section 61.3) $S = (V, F; \underline{E})$, and an assignment of distinct points $p_i \in \mathbb{R}^3$ to the vertices, the following two conditions are equivalent:

(a) *the bar framework $G(p)$ on $G = (V, E)$ has a nontrivial self-stress;*

(b) *the body-hinge framework on the dual graph $G^* = (F, E^*)$ with hinge lines $p_i p_j$ for each edge $\{i, j\}$ of G is first-order flexible.*

PANEL-HINGE AND MOLECULAR FRAMEWORKS

Panel-hinge and molecular frameworks are body-hinge frameworks with special types of hinge configurations. While 3-dimensional panel-hinge frameworks are common structures in engineering, the polar molecular frameworks provide a useful model for biomolecules, with the bodies and hinges representing the atoms and lines of the bond lines, respectively [Whi05]. Crapo and Whiteley showed in [CW82] that first-order rigidity is invariant under projective polarity.

Tay and Whiteley conjectured that the special geometry of the hinge configurations in panel-hinge frameworks (or in their projective duals) do not give rise to any added first-order flexibility [TW84]. This conjecture was confirmed by Katoh and Tanigawa [KT11]; the 2-dimensional case of this theorem was proven by Jackson and Jordán in [JJ08b].

THEOREM 61.2.4   Panel-Hinge and Molecular Theorem

For a graph G the following are equivalent:

(a) *G can be realized as a first-order rigid body-hinge framework in $\mathbb{R}^d$;*
(b) *G can be realized as a first-order rigid panel-hinge framework in $\mathbb{R}^d$;*
(c) *G can be realized as a first-order rigid molecular framework in $\mathbb{R}^d$;*
(d) *if each edge of G is replaced by $\binom{d+1}{2} - 1$ copies, the resulting multigraph contains $\binom{d+1}{2}$ edge-disjoint spanning trees.*

In the special case of dimension $d = 3$, this result says that if each body is realized with all hinges of each body concurrent in a single point (molecular structures), the "generic" rigidity is still measured by the existence of six spanning trees in the corresponding multigraph. The pebble game algorithm for checking this condition is implemented in several software packages (such as FIRST [JRKT01]) for analyzing protein flexibility [Whi05].

For $d = 3$ there is an equivalent result expressed in terms of bar-joint frameworks. The *square* of a graph G is obtained from G by joining each pair of vertices of distance two in G in the graph.

COROLLARY 61.2.5   Square Graphs

For a graph G, $5G$ contains 6 edge-disjoint spanning trees if and only if G^2 is first-order rigid in 3-space.

Recent work of [JJ07, JJ08a] derives some further corollaries. (1) G^2 is independent if and only if G^2 satisfies $|E'| \leq 3|V'| - 6$ for all vertex-induced subgraphs (V', E') of G^2 with $|V'| \geq 3$. This was conjectured by Jacobs, and one direction was previously known. (2) The Molecular Theorem implies that there exists an efficient algorithm to identify the maximal rigid subgraphs of a molecular bar and joint graph. Moreover, it was shown in [Jor12] that a 7-connected molecular graph (G^2) is generically rigid in 3-space.

61.2.2 FRAMEWORKS SUPPORTED ON SURFACES

Recent work has analyzed the rigidity of 3-dimensional bar frameworks whose joints are constrained to lie on a 2-dimensional irreducible algebraic variety embedded in $\mathbb{R}^3$, with a particular focus on classical surfaces such as spheres, cylinders and cones. Any first-order motion of such a framework must have the property that all of its velocity vectors are tangential to the surface. The following theorem summarizes the key combinatorial results for such structures. The proofs are based on Henneberg-type inductive constructions. We refer the reader to [NOP12, NOP14] for details.

THEOREM 61.2.6 Frameworks on Surfaces

Let $G = (V, E)$ be a graph, and let $\mathcal{M}$ be an irreducible algebraic surface in $\mathbb{R}^3$. Further, let $k \in \{1, 2\}$ be the dimension of the group of Euclidean isometries supported by $\mathcal{M}$. Then almost all realizations $G(p)$ of G with $p : V \to \mathcal{M}$ are isostatic on $\mathcal{M}$ if and only if G is a complete graph on $1, 2, 3$ or 4 vertices, or satisfies $|E| = 2|V| - k$ and for every non-empty subgraph (V', E'), $|E'| \leq 2|V'| - k$.

Note that the small complete graphs are special cases, as they are considered isostatic in this context, even though their realizations may have tangential first-order flexes which are not tangential isometries of the surface. Theorem 61.2.6 takes care of the cases where $\mathcal{M}$ is a circular cylinder ($k = 2$) or a circular cone, torus, or surface of revolution ($k = 1$), for example. For $k = 3$, this is the transfer of Laman's Theorem to the sphere (or concentric spheres) [SW12]. Compare to Table 61.2.1. It is not yet known how to extend these results to surfaces with $k = 0$.

61.2.3 FRAMEWORKS IN NON-EUCLIDEAN NORMED SPACES

Another recent research strand is the development of a rigidity theory for a general non-Euclidean normed space, such as $\mathbb{R}^d$ equipped with an ℓ^q norm $\| \cdot \|_q$, or a polyhedral norm $\| \cdot \|_{\mathcal{P}}$, where the unit ball is a polyhedron $\mathcal{P}$. The mathematical foundation for this theory, as well as basic tools and methods, were established in [Kit15, KP14, KS15]. For the ℓ^q norms, where $1 \leq q \leq \infty$, $q \neq 2$ the following analogue of Laman's theorem was proved in [KP14].

THEOREM 61.2.7 Rigid graphs in $(\mathbb{R}^2, \| \cdot \|_q)$, where $1 < q < \infty$, $q \neq 2$

Let G be a graph and q be an integer with $1 < q < \infty$, $q \neq 2$. Almost all realizations $G(p)$ of G in $(\mathbb{R}^2, \| \cdot \|_q)$ are isostatic if and only if G satisfies $|E| = 2|V| - 2$ and for every non-empty subgraph (V', E'), $|E'| \leq 2|V'| - 2$.

Since the polyhedral norms (such as the well-known ℓ^1 or ℓ^∞ norms) are not smooth, some new features emerge in the rigidity theory for these norms. For example, the set of regular realizations of a graph G is no longer dense. Moreover, there exist configurations p, where the 'rigidity map' which records the edge lengths of the framework $G(p)$ is not differentiable. For a space $(\mathbb{R}^d, \| \cdot \|_{\mathcal{P}})$, where d is an arbitrary dimension, the first-order rigidity of frameworks may be characterized in terms of edge-colourings of G which are induced by the placement of each bar relative to the facets of the unit ball (see [Kit15, KP14]).

For $d = 2$, there also exists an analogue of Laman's theorem for polyhedral norms [Kit15, KP14]. A framework $G(p)$ in $(\mathbb{R}^d, \| \cdot \|_{\mathcal{P}})$ is *well-positioned* if the rigidity map is differentiable at p. This is the case if and only if $p_i - p_j$ is contained in the interior of the conical hull of some facet of the unit ball, for each edge $\{i, j\}$.

THEOREM 61.2.8 Rigid graphs in $(\mathbb{R}^2, \| \cdot \|_{\mathcal{P}})$

Let G be a graph and $\| \cdot \|_{\mathcal{P}}$ be a polyhedral norm in $\mathbb{R}^2$. There exists a well-positioned isostatic framework $G(p)$ in $(\mathbb{R}^2, \| \cdot \|_{\mathcal{P}})$ if and only if G satisfies $|E| = 2|V| - 2$ and for every non-empty subgraph (V', E'), $|E'| \leq 2|V'| - 2$.

Theorems 61.2.7 and 61.2.8 have not yet been extended to higher dimensions.

61.2.4 PINNED GRAPHS AND DIRECTED GRAPHS

GLOSSARY

Pinned graph: A simple graph $G = (V_I \cup V_P, E)$ whose vertex set is partitioned into the sets V_I and V_P, and whose edge set E has the property that each edge in E is incident to at least one vertex in V_I. The vertices in V_I are called **inner** and the vertices in V_P are called **pinned**.

Pinned framework in d-space: A triple (G, p, P), where $G = (V_I \cup V_P, E)$ is a pinned graph, and $p : V_I \to \mathbb{R}^d$ and $P : V_P \to \mathbb{R}^d$ are configurations of points in d-space.

First-order rigid pinned framework: A pinned framework (G, p, P) whose pinned rigidity matrix $R_G(p, P)$ has rank $d|V_I|$, where $R_G(p, P)$ is obtained from the standard rigidity matrix of $G(p, P)$ by removing the columns corresponding to the vertices in V_P. Further, (G, p, P) is **independent** if the rows of $R_G(p, P)$ are independent, and **isostatic** if it is first-order rigid and independent.

Pinned d-isostatic graph: A pinned graph $G = (V_I \cup V_P, E)$ for which some (equivalently, almost all) realizations (G, p, P) in $\mathbb{R}^d$ are isostatic.

Pinned graph composition: For two pinned graphs $G = (V_I \cup V_P, E)$ and $H = (W_I \cup W_P, F)$, and an injective map $c : V_P \to W_I \cup W_P$, the pinned graph $C(G, H)$ obtained from G and H by identifying the pinned vertices V_P of G with their images $c(V_P)$.

d-Assur graph: A d-isostatic pinned graph $G = (V_I \cup V_P, E)$ which is minimal in the sense that for all proper subsets of vertices $V_{I'} \cup V_{P'}$, where $V_{I'} \neq \emptyset$, $V_{I'} \cup V_{P'}$ induces a pinned subgraph $G' = (V_{I'} \cup V_{P'}, E')$ with $|E'| \leq d|V_{I'}| - 1$.

Orientation: For a simple graph $G = (V, E)$, an assignment of a direction (or-dering) to each edge in E. The resulting structure is called a **directed graph**.

d-directed orientation: For a pinned graph G, an orientation of G such that every inner vertex has out-degree exactly d and every pinned vertex has out-degree exactly 0.

Strongly connected: A directed graph G is called strongly connected if and only if for any two vertices i and j in G, there is a directed path from i to j and from j to i. The **strongly connected components** of G are the strongly connected subgraphs which are maximal in the sense that a subgraph cannot be enlarged to another strongly connected subgraph by including additional vertices and its incident edges.

Formation in d-space $G(p)$: A directed graph $G = (V, E)$ (no loops or multiple edges) and a configuration p in d-space for the vertices V.

Persistent formation: A formation $G(p)$ with the property that there exists an open neighborhood $N_\epsilon(p)$ of p so that every configuration q in $N_\epsilon(p)$ with the distance set defined by $d_{ij} = |p_i - p_j|$ is congruent to p.

Generically persistent graph in d-space: A directed graph G for which the formations $G(p)$ are persistent for almost all configurations p in d-space.

Minimally persistent graph: A directed graph which is generically persistent and has the property that the removal of any edge yields a directed graph which is not generically persistent.

Leader-follower graph: A directed graph $G = (V, E)$ which has a vertex v_0 of out-degree 0 (called the *leader*), a vertex v_1 of out-degree 1 (called the *follower*), and $(v_1, v_0) \in E$.

FIRST ORDER RIGIDITY OF PINNED GRAPHS

We have necessary conditions for a pinned graph to be d-isostatic, as well as necessary and sufficient conditions in the plane. These are presented in Table 61.2.1. For pinned frameworks in the plane with generic inner joints, the pins only need to be in general position for isostaticity.

TABLE 61.2.1

STRUCTURE	NECESSARY COUNTS IN d-SPACE	CHARACTERIZATION IN THE PLANE
bar-joint	$|E| = d|V| - \binom{d+1}{2}, \ |V| \geq d$	$|E| = 2|V| - 3, \ |V| \geq 2$
(61.1.2)	$|E'| \leq d|V'| - \binom{d+1}{2}, \ |V'| \geq d$	$|E'| \leq 2|V'| - 3, \ |V'| \geq 2$
pinned	$|E| = d|V_I|$	$|E| = 2|V_I|$
bar-joint	$|E'| \leq d|V_{I'}|, \ |V_{P'}| \geq d$	$|E'| \leq 2|V_{I'}|, \ |V_{P'}| \geq 2$
(61.2.4)	$|E'| \leq d|V_{I'}| - \binom{d+1-k}{2}, \ |V_{P'}| = k,$	$|E'| \leq 2|V_{I'}| - \binom{3-k}{2}, \ |V_{P'}| = k,$
	$0 \leq k < d$	$0 \leq k < 2$

For dimension $d \geq 3$, the conditions in Table 61.2.1 are not sufficient for G to be pinned d-isostatic. The bar-joint frameworks in 3-space (also referred to as "floating frameworks" in the engineering community) provide the core counterexamples (see Figure 61.1.4).

Where the counts give necessary and sufficient conditions ($d = 2$ in Table 61.2.1, and parallel drawing for $d \geq 2$ in Table 61.3.1), the counts alone define a matroid ([Whi96] Appendix A, [Whi88b]). Similarly, the counts for body-bar frameworks in Section 61.2.1 define a matroid. For $d \geq 3$, the counts are no longer enough to define the full independence structure of a matroid [Whi96].

ASSUR DECOMPOSITION OF PINNED GRAPHS

An important and widely used method in the analysis and synthesis of mechanisms is the decomposition of mechanical linkages into fundamental minimal components, each of which permits a simplified analysis. This layered approach for analysing mechanisms was originally developed by Assur and has now been reworked using rigidity theory. A mechanical linkage is converted into a pinned isostatic framework (G, p, P) in $\mathbb{R}^d$ by (i) transforming it into a flexible pinned framework with a designated driver (e.g., a piston changing the length between a pair of vertices) and (ii) replacing the driver by a rigid bar. The resulting d-isostatic pinned graph G decomposes uniquely into a partially ordered set of minimal pinned isostatic components (Assur graphs). G is reconstructed from these components by repeated pinned graph compositions (following the partial order) [SSW10a, SSW13].

This d-Assur decomposition of G may be algorithmically obtained as follows [SSW13]. Find a d-directed orientation of the d-isostatic pinned graph by applying the $d|V|$ pebble-game algorithm to G [LS08]. In this directed graph, we condense all pinned vertices into a single pinned vertex (making it into a sink), and then determine the strongly connected components of this graph (via the $O(|E|)$ time Tarjan's algorithm). When we extend the strongly connected components by the outgoing edges from the component, we obtain the d-Assur decomposition. The partial order is the acyclic directed graph obtained from contracting each strongly connected component to a single vertex, and discarding any multiple edges that may arise (Figure 61.2.1(c,d)).

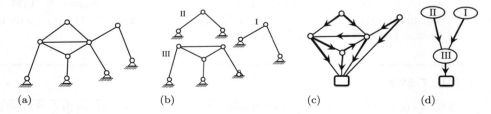

(a)   (b)   (c)   (d)

FIGURE 61.2.1
A pinned 2-isostatic graph G (a) and its 2-Assur decomposition (b). The directed graph in (c) is obtained from a 2-directed orientation on G by contracting all pinned vertices to a single vertex. (d) presents the partial order of the strongly connected components.

THEOREM 61.2.9 d-Assur Decomposition [SSW13]

Given a pinned d-isostatic graph G, there is a d-directed orientation of G. With any such d-directed orientation, the following decompositions are equivalent:

(a) *the d-Assur decomposition of G;*

(b) *the strongly connected decomposition into extended components associated with the d-directed orientation (with all pinned vertices identified);*

(c) *the block-triangular decomposition of the pinned rigidity matrix into a maximal number of components for some linear order extending the partial order of (i) or equivalently (ii).*

THEOREM 61.2.10   Plane Assur Graphs [SSW10a]

Let $G = (V_I \cup V_P, E)$ be a pinned isostatic graph. Then the following are equivalent:

(a) *G is 2-Assur;*

(b) *if the set V_P is contracted to a single vertex, inducing the unpinned graph (or multigraph) G^* with edge set E, then G^* is a generic 2-circuit;*

(c) *either G has a single inner vertex of degree 2 or each time we delete a vertex, the resulting pinned graph has a non-zero motion of all inner vertices (in generic position);*

(d) *deletion of any edge from G results in a pinned graph that has a non-zero motion of all inner vertices (in generic position).*

The purely combinatorial condition (b) can be checked by fast pebble-game algorithms. It also follows from (b) that all 2-Assur graphs on at least 5 vertices can be obtained from basic Assur graphs by a sequence of simple graph moves (such as edge splitting) and pin-rearrangements. This is a corollary of the inductive construction of plane circuits in [BJ02]. Such inductive constructions allow engineers to easily generate basic building blocks for synthesizing new linkages. Condition (c) allows a quick check of the 2-Assur property for smaller graphs, and (d) guarantees that a driver inserted for an arbitrary edge will (generically) put all inner vertices into motion, as will putting one pin into motion.

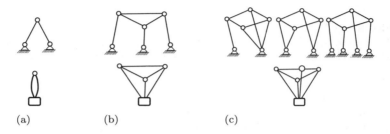

(a)   (b)   (c)

FIGURE 61.2.2
2-Assur graphs and their corresponding generic 2-circuits. One circuit can correspond to a family of Assur graphs (c).

Alternatively, 2-Assur graphs can be characterized geometrically, based on special singular realizations. See [SSW10b] for details.

For dimension $d \geq 3$, a good combinatorial characterization of d-Assur graphs is not known. The fundamental property of 2-Assur graphs given in Theorem 61.2.10 does not, in general, hold for d-Assur graphs where $d \geq 3$ (see [SSW13]). The decomposition of a structure into "Assur components" has been extended to bar frameworks with symmetries [NSSW14] and to tensegrity frameworks [STB+09].

CONTROL OF FORMATIONS

Another application of rigidity theory is the distributed control of groups of mobile robots that enable them to "hold formation" in the absence of a centralised control or global coordinate system (e.g., systems moving indoors, underwater or in other GPS-denied environments). See [AYFH08, OPA15] for recent surveys on this topic. Rigidity-based methods are just made for this task, as they allow the robust control using only distance (or angle) measurements. The task of sensing and maintaining a prescribed distance between a pair of agents may either be shared by both agents or assigned to a single designated agent. The first case is modelled via an undirected graph, and techniques from rigidity theory can be applied directly. For the second case, we assign an orientation to each edge, where the edge (i, j) directed from i to j indicates that agent i is responsible for maintaining its distance to agent j. An adapted "directed rigidity" (or "persistence") theory is required to study the control of formations of this type (see [HADB07]).

Intuitively, a formation is persistent if, provided that each agent is trying to satisfy all the distance constraints for which it is responsible, all the agents can in fact accomplish this task, and, consequently, the shape of the entire formation is preserved as the formation is moving continuously. In the following, we restrict attention to formations in $\mathbb{R}^2$, and we say that a formation $G(p)$ (or directed graph G) is rigid if the bar framework (undirected graph) obtained from $G(p)$ (G) by

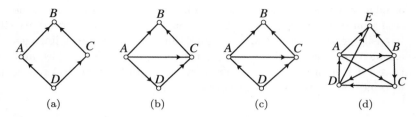

FIGURE 61.2.3

Formations in $\mathbb{R}^2$: (a) is not rigid and hence not persistent; (b) is rigid, but not persistent, since A may be unable to satisfy its distance constraints; (c) is minimally persistent; (d) is rigid but not persistent by [BJ08] due to the K_4 subgraph induced by A, B, C, D.

removing the orientation of the edges is (generically) 2-rigid. Persistence beyond minimal persistence has some advantages in situations where some directed links may be broken by obstacles or lost data.

THEOREM 61.2.11   Combin. Characterization of Persistence in 2D [HADB07]

A directed graph is generically persistent if and only if all those subgraphs, obtained by removing outgoing edges from vertices with out-degree larger than 2 until all the vertices have out-degree smaller than or equal to 2, are rigid.

Theorem 61.2.11 readily provides a (non-polynomial) algorithm to check the persistence of a directed graph. It remains open whether persistence can also be checked in polynomial time. For special classes, efficient algorithms are known [HADB07, BJ08].

Clearly, rigidity is a necessary, but not sufficient condition for persistence. It is an open problem to decide whether all generically rigid graphs have persistent orientations. However, affirmative answers exist for various classes of graphs, including wheels, power graphs of cycles, complete graphs, generically minimally rigid graphs, and generically rigid graphs with $|E| = 2|V| - 2$ [HADB07, BJ08].

The directed graphs that are both generically minimally rigid and persistent are precisely the minimally persistent graphs [HADB07]. Further, a directed graph is minimally persistent if and only if it is generically minimally rigid and no vertex has more than two outgoing edges. For leader follower formations, this orientation can be efficiently found by applying the pebble-game algorithm and drawing pebbles to the leader-follower edge, as the ground. However, an additional cycle-reversal operation is needed to sequentially build *all* minimally persistent directed graphs.

Real-life scenarios often require formations to be reconfigured due to changes in the environment, the presence of obstacles along the paths of agents, or unexpected losses of agents, for example. For this, various types of splitting, merging and closing ranks operations have been developed for formations. For formations modeled by *undirected* graphs, we refer the reader to [EAM+04, OM02]. For *directed* graphs, initial results can be found in [HYFA08]. With leader-follower formations, Assur decompositions also provide an additional tool.

Recent work confirms that many of the results for generically persistent graphs extend to 3- and higher-dimensional spaces [YHF+07]. While the concept of persistence is applicable in dimension $d \geq 3$, it is not always sufficient to guarantee the desired stability of the formation [YHF+07]. A natural question is the extension of a persistence theory to formations based on other types of geometric constraint systems, such as 2- or higher-dimensional formations consisting of points, lines or other rigid building blocks, linked by (possibly mixed) angle, direction (bearings) and distance constraints. See the references in [OPA15]) and Section 61.3.

61.2.5 TENSEGRITY FRAMEWORKS

In a tensegrity framework, we replace some (or all) of the equalities for bars with inequalities for the distances—corresponding to *cables* (the distance can shrink but not expand) and *struts* (the distance can expand but not shrink). The study of these inequalities introduces techniques from linear programming. Further results for tensegrity frameworks appear in Chapter 63.

GLOSSARY

Signed graph: A graph with a partition of the edges into three classes, written $G_\pm = (V; E_-, E_0, E_+)$.

Tensegrity framework $G_\pm(p)$ **in** $\mathbb{R}^d$: A signed graph $G_\pm = (V; E_-, E_0, E_+)$ and a configuration p on V.

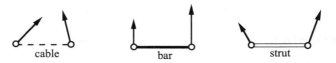

FIGURE 61.2.4
A tensegrity framework has (a) cables that can get shorter; (b) bars that cannot change length, and (c) struts that can get longer.

Cables, bars, struts: For a tensegrity framework $G_\pm(p)$, the members of E_-, of E_0, and of E_+, respectively. In figures, cables are indicated by dashed lines, struts by double thin lines, and bars by single thick lines (see Figure 61.2.4).

$G_\pm(p)$ **dominates** $G_\pm(q)$**:** For each edge, the appropriate condition holds:

$$|p_i - p_j| \geq |q_i - q_j| \quad \text{when} \quad \{i, j\} \in E_-$$
$$|p_i - p_j| = |q_i - q_j| \quad \text{when} \quad \{i, j\} \in E_0$$
$$|p_i - p_j| \leq |q_i - q_j| \quad \text{when} \quad \{i, j\} \in E_+.$$

Rigid tensegrity framework $G_\pm(p)$**:** For every analytic path $p(t)$ in $\mathbb{R}^{vd}$, $0 \leq t < 1$, if $p(0) = p$ and $G(p)$ dominates $G(p(t))$ for all t, then p is congruent to $p(t)$ for all t.

First-order flex of a tensegrity framework $G_\pm$: An assignment $p' : V \to \mathbb{R}^d$ of velocities to the vertices such that, for each edge $\{i, j\} \in E$ (Figure 61.2.4),

$$(p_j - p_i) \cdot (p'_j - p'_i) \leq 0 \quad \text{for cables} \quad \{i, j\} \in E_-$$
$$(p_j - p_i) \cdot (p'_j - p'_i) = 0 \quad \text{for bars} \quad \{i, j\} \in E_0$$
$$(p_j - p_i) \cdot (p'_j - p'_i) \geq 0 \quad \text{for struts} \quad \{i, j\} \in E_+.$$

Trivial first-order flex: A first-order flex p' of a tensegrity framework $G_\pm(p)$ such that $p'_i = Sp_i + t$ for all vertices i, with a fixed skew-symmetric matrix S and vector t.

First-order rigid: A tensegrity framework $G_\pm(p)$ is first-order rigid if every first-order flex is trivial, and **first-order flexible** otherwise.

Proper self-stress on a tensegrity framework $G_\pm(p)$:   An assignment ω of scalars to the edges of G such that:

    (a) $\omega_{ij} \geq 0$ for cables $\{i,j\} \in E_-$;

    (b) $\omega_{ij} \leq 0$ for struts $\{i,j\} \in E_+$; and

    (c)   for each vertex i, $\sum_{\{j \mid \{i,j\}\in E\}} \omega_{ij}(p_j - p_i) = 0$.

Strict self-stress: A proper self-stress ω with strict inequalities in (a) and (b).

Underlying bar framework: For a tensegrity framework $G_\pm(p)$, the bar framework $G(p)$ on the unsigned graph $G = (V, E)$, where $E = E_- \cup E_0 \cup E_-$ (Figure 61.2.5(a,b)).

BASIC RESULTS

The equivalent definitions of "rigidity" and the basic connections between rigidity and first-order rigidity all transfer directly to tensegrity frameworks [RW81].

THEOREM 61.2.12   First-Order Stress Test

A tensegrity framework $G_\pm(p)$ is first-order rigid if and only if the underlying bar framework $G(p)$ is first-order rigid and there is a strict self-stress on $G_\pm(p)$ (Figure 61.2.5 (a,b).

This connection to self-stresses means that any first-order rigid tensegrity framework with at least one cable or strut has $|E| > d|V| - \binom{d+1}{2}$ edges.

THEOREM 61.2.13   Reversal Theorem

A tensegrity framework $G_\pm(p)$ is first-order rigid if and only if the reversed framework $G^r_\pm(p)$ is first-order rigid, where the graph $G^r_\pm$ interchanges cables and struts (Figure 61.2.5 (a,c)).

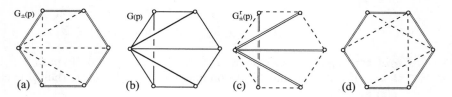

FIGURE 61.2.5
The tensegrity framework (a) is built with a proper stress on the first-order rigid bar framework (b). Reversing the stress (c) is also first-order rigid. (d) is another tensegrity polygon which is also first-order rigid.

There is no single "generic" behavior for a signed graph $G_\pm$. If some configuration produces a first-order rigid framework for a graph $G_\pm$, then the set of all such configurations is open but not dense. The algebraic variety of "special positions" of the underlying unsigned graph divides the configuration space into open subsets, in some of which all configurations are rigid, and in others, none are. The required sign pattern for a self-stress can change as you cross such a boundary [WW83].

The first-order rigidity of a tensegrity framework is projectively invariant, with the proviso that a cable (strut) $\{i, j\}$ is switched to a strut (cable) whenever $\lambda_i \lambda_j < 0$ for the projective transformation.

THEOREM 61.2.14   Stress Existence

If a tensegrity framework $G_\pm(p)$ with at least one cable or strut is rigid, then there is a nonzero proper self-stress.

A number of results relate minima of quadratic energy functions to the rigidity of tensegrity frameworks. These energy results are not invariant under projective transformations, but such rigidity is preserved under "small" affine transformations. This is one result, drawn from results on second-order rigidity [CW96].

THEOREM 61.2.15   Rigidity Stress Test

A tensegrity framework $G_\pm(p)$ is rigid if, for each nontrivial first-order motion p' of $G_\pm(p)$, there is a proper self-stress $\omega^{p'}$ making $\sum_{ij} \omega_{ij}^{p'} (p'_i - p'_j) \cdot (p'_i - p'_j) > 0$.

61.3   SCENE ANALYSIS

The problem of reconstructing spatial objects (polyhedra or polyhedral surfaces) from a single plane picture is basic to several applications. This section summarizes the combinatorial results for "generic pictures" (Section 61.3.1). Section 61.3.2 presents a polar "parallel configurations" interpretation of the same abstract mathematics and Section 61.3.3 presents connections to other fields of discrete geometry. In addition, liftings and scene analysis have strong connections to the recent work on 'affine rigidity' [GGLT13].

Within approximation theory, the study of multivariate splines considers surfaces which are piecewise polynomial of bounded degree, extending the polyhedral scenes which are piecewise linear (see Chapter 56). There are many analogies between the theory of rigidity and the theory of multivariate splines at the level of matroidal combinatorics and geometry which are described in [Whi96, Whi98].

61.3.1 COMBINATORICS OF PLANE POLYHEDRAL PICTURES

GLOSSARY

Polyhedral incidence structure S: An abstract set of *vertices* V, an abstract set of *faces* F and a set of *incidences* $I \subset V \times F$.

d-scene for an incidence structure $S = (V, F; I)$:   A pair of location maps, $p : V \rightarrow \mathbb{R}^d$, $p_i = (x_i, \ldots, z_i, w_i)$ and $P : F \rightarrow \mathbb{R}^d$, $P^j = (A^j, \ldots, C^j, D^j)$, such that, for each $(i, j) \in I$: $A^j x_i + \ldots + C^j z_i + w_i + D^j = 0$. (We assume that no hyperplane is vertical, i.e., is parallel to the vector $(0, 0, \ldots, 0, 1)$.)

(d−1)-*picture* of an incidence structure S: A location map $r : V \to \mathbb{R}^{d-1}$, $r_i = (x_i, \dots, z_i)$ (Figure 61.3.1(a)).

Lifting of a $(d-1)$-picture $S(r)$:   A d-scene $S(p, P)$ with vertical projection $\Pi(p) = r$ (Figure 61.3.1(b)). (I.e., if $p_i = (x_i, \dots, z_i, w_i)$, then $r_i = (x_i, \dots, z_i) = \Pi(p_i)$).

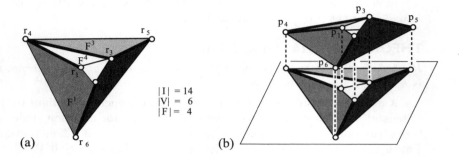

$|I| = 14$
$|V| = 6$
$|F| = 4$

FIGURE 61.3.1
A generic polyhedral picture and an associated sharp scene.

Lifting matrix for a picture $S(r)$:   The $|I| \times (|V| + d|F|)$ coefficient matrix $M_S(r)$ of the system of equations for liftings of a picture $S(r)$: for each $(i, j) \in I$, $A^j x_i + \dots + C^j z_i + w_i + D^j = 0$, where the variables are ordered:

$$\dots, w_i, \dots \ ; \ \dots, A^j, \dots, C^j, D^j, \dots.$$

Sharp picture: A $(d-1)$-picture $S(r)$ that has a lifting $S(p, P)$ with a distinct hyperplane for each face (Figure 61.3.1(a,b)).

BASIC RESULTS

Since the incidence equations are linear, there is no distinction between "continuous liftings" and "first-order liftings." Since the rank of the lifting matrix is determined by a polynomial process on the entries, "generic properties" of pictures have several characterizations. These were conjectured by Sugihara and proven in [Whi88a]. The larger overview of the problems can be found in [Sug86].

We find the usual correspondence of one generic sharp picture and all generic realizations being sharp.

The generic properties of a structure are robust: all small changes in such a sharp picture are also sharp pictures and small changes in the points of a sharp picture require only small changes in the sharp lifting. Even special positions of such structures will always have nontrivial liftings, although these may not be sharp. However, up to numerical round-off, all pictures "are generic." Other structures that are not generically sharp (Figure 61.3.2(a)) may have sharp pictures in special positions (Figure 61.3.2(b)), but a small change in the position of even one point can destroy this sharpness.

The incidence equations allow certain "trivial" changes to a lifted scene that will preserve the picture—generated by adding a single plane H^0 to all existing planes: $P^j_* = H^0 + P^j$; and by changes in vertical scale in the scene: $w^*_i = \lambda w_i$. This space of ***lifting equivalences*** has dimension $d+1$, provided the points of the

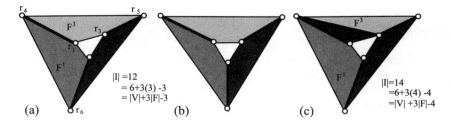

FIGURE 61.3.2
The generic picture (a) has only trivial liftings, while the special position (b) has sharp liftings. The modified picture (c) has sharp liftings (see Figure 61.3.1(b)).

scene do not lie in a single hyperplane. These results are connected to entries in Table 61.2.1.

THEOREM 61.3.1   Picture Theorem

A generic picture of an incidence structure $S = (V, F; I)$ with at least two faces has a sharp lifting, unique up to lifting equivalence, if and only if $|I| = |V| + d|F| - (d+1)$ and, for all subsets I' of incidences on at least two faces, $|I'| \leq |V'| + d|F'| - (d+1)$ (Figure 61.3.2(a,c)).

A generic picture of an incidence structure $S = (V, F; I)$ has independent rows in the lifting matrix if and only if for all nonempty subsets I' of incidences, $|I'| \leq |V'| + d|F'| - d$ (Figure 61.3.2(a)).

ALGORITHMS

Any part of a structure with $|I'| = |V'| + d|F'| - d$ independent incidences will be forced to be coplanar over a picture with algebraically independent coordinates for the points. If the structure is not generically sharp, then an effective, robust lifting algorithm consists of selecting a subset of vertices for which the incidences are sharp, then "correcting" the position of the other vertices based on calculations in the resulting scene. This requires effective algorithms for selecting such a set of incidences. Sugihara and Imai have implemented $O(|I|^2)$ time algorithms for finding maximal generically sharp (independent) structures using modified bipartite matching on the incidence structure [Sug86].

61.3.2 PARALLEL DRAWINGS

The mathematical structure defined for polyhedral pictures has another, dual interpretation: the polar of a "point constrained by one projection" is a "hyperplane constrained by an assigned normal." Two configurations sharing the prescribed normals are "parallel drawings" of one another [Whi88a]. These geometric patterns, used by engineering draftsmen in the nineteenth century, have reappeared in a number of branches of discrete geometry. This dual interpretation also establishes a basic connection between the geometry and combinatorics of scene analysis and the geometry and combinatorics of first-order rigidity of frameworks.

GLOSSARY

Parallel d-scenes for an incidence structure: Two d-scenes $S(p, P)$, $S(q, Q)$ such that for each face j, $P^j||Q^j$ (that is, the first $d - 1$ coordinates are equal) (Figure 61.3.3).   (For convenience, not necessity, we stick with the "nonvertical" scenes of the previous section.)

Nontrivially parallel d-scene for a d-scene $S(p, P)$:   A parallel d-scene $S(q, Q)$, such that the configuration q is not a translation or dilatation of the configuration p (Figure 61.3.3 for $d = 2$).

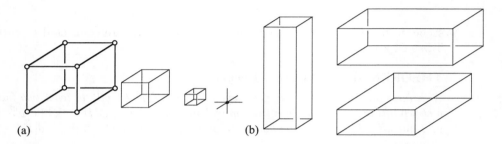

FIGURE 61.3.3
The cube graph has trivial parallel drawings (a), and a space of non-trivial drawings (b).

Directions for the faces: An assignment of d-vectors $D^j = (A^j, \dots, C^j)$ to $j \in F$.

d-scene realizing directions D: A d-scene $S(p, P)$ such that for each face $j \in F$, the first $d - 1$ coordinates of P^j and D^j coincide.

Parallel drawing matrix for directions D in d-space: The $|I| \times (|V| + d|F|)$ matrix $M_S(D)$ for the system of equations for each incidence $(i, j) \in I$: $A^j x_i + B^j y_i + \dots + C^j z_i + w_i + D^j = 0$, where the variables are ordered:

$$\dots, D^j, \dots \ ; \ \dots, x_i, y_i, \dots, z_i, w_i, \dots.$$

DL-framework: A *mixed graph* $G_{DL} = (V; D, L)$ *with two classes of edges (not necessarily disjoint) D (for directions) and L (for lengths) and a configuration $p : V \to \mathbb{R}^2$. A first-order flex is a map $p' : V \to \mathbb{R}^2$ such that (i) for an edge in D: $(p_i - p_j)^{\perp} \cdot (p'_i - p'_j) = 0$ and (ii) for an edge in L: $(p_i - p_j) \cdot (p'_i - p'_j) = 0$. The trivial first-order flexes are translations, and a framework $G_{DL}(p)$ is* **tight** *if all first-order flexes are trivial (Figure 61.3.5). In $\mathbb{R}^d$ we have a multi-graph where D can contain up to $d - 1$ copies of a pair and $(p_i - p_j)^{\perp}$ represents a vector selected from the $(d - 1)$-dimensional space of normals to the vector $(p_i - p_j)$.*

BASIC RESULTS

All results for polyhedral pictures dualize to parallel drawings. Again, for parallel drawings there is no distinction between continuous changes and first-order changes. The trivially parallel drawings, generated by d translations and one dilation towards a point, form a vector space of dimension $d + 1$, provided there are at least two distinct points (Figure 61.3.3(a)). (A trivially parallel drawing may even have all points coincident, though the faces will still have assigned directions (Figure 61.3.3(a)).)

THEOREM 61.3.2   Parallel Drawing Theorem for Scenes

For generic selections of the directions D in d-space for the faces, a structure $S = (V, F; I)$ has a realization $S(p, P)$ with all points p distinct if and only if, for every nonempty set I' of incidences involving at least two points $V(I')$ and faces $F(I')$, $|I'| \leq d|V(I')| + |F(I')| - (d + 1)$ (Figure 61.3.3(a)).

In particular, a configuration p, P with distinct points realizing generic directions for the incidence structure is unique, up to translation and dilatation, if and only if $|I| = d|V| + |F| - (d + 1)$ and $|I'| \leq d|V'| + |F'| - (d + 1)$.

Of course other nontrivially parallel drawings will also occur if the rank is smaller than $d|V'| + |F'| - (d + 1)$ (Figure 61.3.3(b), with a generic rank 1 less than required for $d = 2$, and a geometric rank, as drawn, 2 less than required).

Figure 61.3.3 may also be interpreted as the parallel drawings of a "cube in 3-space." For spherical polyhedra, there is an isomorphism between the nontrivially parallel drawings in 3-space (the parallel drawings modulo the trivial drawings) and the nontrivially parallel drawings in a plane projection [CW94]. Only the dimension (4 vs. 3) of the trivially parallel drawings will change with the projection.

61.3.3 CONNECTIONS TO OTHER FIELDS

FIRST-ORDER MOTIONS AND PARALLEL DRAWINGS

For any plane framework, if we turn the vectors of a first-order motion 90° (say clockwise), they become the vectors joining p to a parallel drawing q of the framework (Figure 61.3.4(a,b)). The converse is also true: a result that is folklore in the structural engineering community.

THEOREM 61.3.3   Parallel Drawing Test for Plane First-order Flexes

A plane framework $G(p)$ has a nontrivial first-order flex if and only if the configuration $G(p)$ has a nontrivially parallel drawing $G(q)$ (Figure 61.3.4(b,c)).

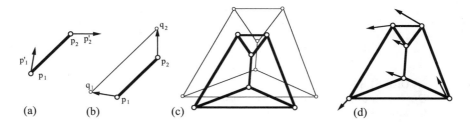

FIGURE 61.3.4
In the plane, first-order flexes (a,d) correspond to parallel drawings (c,d), by taking the vectors of one and turning them all 90°.

In this transfer, the translations go to translations, and the rotations become dilations, or scalings.

Because of this connection, combinatorial and geometric results for plane first-order rigidity and for plane parallel drawings have numerous deep connections. For example, Laman's theorem (Theorem 61.1.12(b)) is a corollary of the parallel drawing theorem, for $d = 2$. In higher dimensions, the connection is one-way: a nontrivially parallel drawing of a "framework" (the "direction of an edge" is represented by $d-1$ facets through the two points) induces one (or more) nontrivial first-order motions of the corresponding bar framework. The theory of parallel drawing in higher dimensions is more complete and has simpler algorithms than the theory of first-order rigidity in higher dimensions, generalizing almost all results for plane first-order rigidity and plane parallel drawings [Whi96]. In Table 61.3.1, we give the counts that are necessary and sufficient for independence of parallel drawings of graphs and multi-graphs over all dimensions.

TABLE 61.3.1

STRUCTURE	NECESSARY COUNTS IN d-SPACE	CHARACTERIZATION IN THE PLANE
parallel	$|D| = d|V| - (d+1),\ |V| \geq 2$	$|D| = 2|V| - 3,\ |V| \geq 2$
drawing	$|D'| \leq d|V'| - (d+1),\ |V'| \geq 2$	$|D'| \leq 2|V'| - 3,\ |V'| \geq 2$
(61.3.3)	(Also sufficient in d-space!)	
mixed (DL)	$|D| + |L| = d|V| - d,\ |V| \geq 1$	$|D| + |L| = 2|V| - 2,\ |V| \geq 1$
(61.3.3)	$|D'| + |L'| \leq d|V'| - d,\ |V'| \geq 1$	$|D'| + |L'| \leq 2|V'| - 2,\ |V'| \geq 1$
	$|D'| \leq d|V'| - (d+1),\ |V'| \geq 2$	$|D'| \leq 2|V'| - 3,\ |V'| \geq 2$
	$|L'| \leq d|V'| - \binom{d+1}{2},\ |V'| \geq d$	$|L'| \leq 2|V'| - 3,\ |V'| \geq 2$

DIRECTION-LENGTH FRAMEWORKS IN THE PLANE

We can combine the information from parallel drawing and first-order rigidity through the direction-length or DL-frameworks (Figure 61.3.5). Notice that the same pair of vertices can be both a direction and a length.

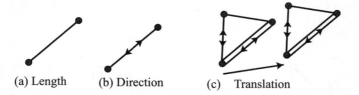

(a) Length   (b) Direction   (c) Translation

FIGURE 61.3.5
DL-frameworks have both length constraints (a) and direction constraints (b). The only trivial motions are translations (c).

Since the only trivial first-order flexes are translations, a *tight* framework in $d = 2$ will have the minimal count of $|D| + |L| = 2|V| - 2$. Note, however, that the maximum size of a pure graph with only lengths (or with only directions) is still $|D| \leq 2|V| - 3$ and $|L| \leq 2|V| - 3$ for $|V| \geq 2$. These are both necessary and sufficient (Table 61.3.1). The extended necessary counts for $d \geq 2$ also appear in Table 61.3.1.

An additional observation is that swapping directions and lengths for $d = 2$ preserves the counts and generic tightness. Working with the corresponding rigidity matrix, this swapping actually holds for any specific geometric realization [SW99]. These conditions can be checked by running the pebble game three times (once for D, once for L, and once overall).

ANGLES IN CAD

In plane computer-aided design, many different patterns of constraints (lengths, angles, incidences of points and lines, etc.) are used to design or describe configurations of points and lines, up to congruence or local congruence. With distances between points, the geometry becomes that of first-order rigidity. If angles and incidences are added, even the problems of "generic rigidity" of constraints are unsolved (and perhaps not solvable in polynomial time). However, special designs, mixing lengths, distances of points to lines, and trees of angles have been solved, using direct extensions of the techniques and results for plane frameworks, and plane parallel drawings, and the DL-frameworks ([SW99]).

For the specific constraints of distances between points, angles between lines, and distances between points and lines, some new combinatorial characterizations have recently been published [JO16]. Recent work has developed combinatorial characterizations for point-line frameworks, via projections of frameworks from the sphere, with points at infinity becoming lines which have angle constraints, and point-line distances as added constraints [EJN+17].

MINKOWSKI DECOMPOSABILITY

By a theorem of Shephard, a polytope is decomposable as the Minkowski sum of two simpler polyhedra if and only if the faces and vertices of the polytope (or the edges and vertices of the polytope) have a nontrivially parallel drawing. Many characterizations of Minkowski indecomposable polytopes can be deduced directly from results for parallel d-scenes (or equivalently, for polyhedral pictures of the polar polytope). This includes the projective invariance of Minkowski decomposability.

61.4 RECIPROCAL DIAGRAMS

The reciprocal diagram is a geometric construction that has appeared, independently over a 140-year span [Max64, Cre72], in areas such as "graphical statics" (drafting techniques for resolving forces), scene analysis, and computational geometry.

Continuing work in structural engineering is developing reciprocal diagrams as a tool for design and as a tool for analysis [BMMM16, MBMK16, BO07, Tac12]. Related papers extend reciprocal diagrams to 3-space and higher dimensions [Mic08, Ryb99].

GLOSSARY

Abstract spherical polyhedron $S = (V, F; \underline{E})$: For a 2-connected planar graph $G_S = (V, E_S)$, embedded on a sphere (or in the plane), we record the vertices as V and the regions as faces F, and rewrite the directed edges $\underline{E}$ as ordered 4-tuples $\underline{e} = \langle h, i; j, k \rangle$, where the edge from vertex h to vertex i has face j on the right and face k on the left. (The reversed edge $-\underline{e} = \langle i, h; k, j \rangle$ runs from i to h, with k on the right.)

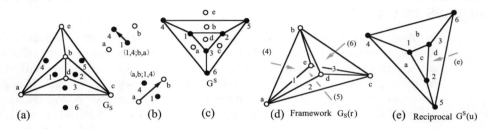

FIGURE 61.4.1

A planar graph, with its dual (a,c), (b,d) will have reciprocal drawings, with dual edges mutually perpendicular, if and only if there is a self-stress.

Dual abstract spherical polyhedron: The abstract spherical polyhedron S^* formed by switching the roles of V and F, and switching the pairs of indices in each ordered edge $\underline{e} = \langle h, i; j, k \rangle$ into $\underline{e}^* = \langle j, k; i, h \rangle$. (Also the abstract spherical polyhedron formed by the dual planar graph $G^S = (F, E^S)$ of the original planar graph (Figure 61.4.1(a,c)).)

Proper spatial spherical polyhedron: An assignment of points $p_i = (x_i, y_i, z_i)$ to the vertices and planes $P^j = (A^j, B^j, D^j)$ to the faces of an abstract spherical polyhedron $(V, F; \underline{E})$, such that if vertex i and face j share an edge, then the point lies on the plane: $A^j x_i + B^j y_i + z_i + D^j = 0$; and at each edge the two vertices are distinct points and the two faces have distinct planes.

Projection of a proper spatial polyhedron $S(p, P)$: The plane framework $G_S(r)$, where r is the vertical projection of the points p (i.e., $r_i = \Pi p_i = (x_i, y_i)$) (Figure 61.4.2).

Gradient diagram of a proper spatial polyhedron $S(p, P)$: The plane framework $G^S(s)$, where $s_j = (A^j, B^j)$ is (minus) the gradient of the plane P^j (Figure 61.4.2).

Reciprocal diagrams: For an abstract spherical polyhedron S, two frameworks $G_S(r)$ and $G^S(s)$ on the graph and the dual graph of the polyhedron, such that for each directed edge $\langle h, i; j, k \rangle \in \underline{E}$, $(r_h - r_i) \cdot (s_j - s_k) = 0$ (Figure 61.4.1(d,e)).

BASIC RESULTS

Reciprocal diagrams have deep connections to both of our previous topics:

(a) Given a spatial scene on a spherical structure, with no faces vertical, the vertical projection and the gradient diagram are reciprocal diagrams. (This follows because the difference of the gradients at an edge is a vector perpendicular to the vertical plane through the edge.)

(b) Given a pair of reciprocal diagrams on $S = (V, F; \underline{E})$, then for each edge $\underline{e} = \langle h, i; j, k \rangle$ the scalars ω_{ij} defined by $\omega(r_h - r_i) = (s_j - s_k)^\perp$ (where $^\perp$ means rotate by 90° clockwise) form a self-stress on the framework $G_S(r)$. (This follows because the closed polygon of a face in $G^S(s)$ is, after $^\perp$, the vector sum for the "vertex equilibrium" in the self-stress condition.)

These facts can be extended to other oriented polyhedra and their projections. The real surprise is that, for spherical polyhedra, the converses hold and all these concepts are equivalent (an observation dating back to Clerk Maxwell and the drafting techniques of graphical statics). The first complete proof appears to be in [KW04].

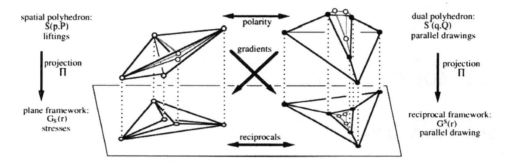

FIGURE 61.4.2
For planar graphs, the reciprocal pairs are entwined with a pair of polar polyhedra by projection and lifting.

THEOREM 61.4.1   Maxwell's Theorem

For an an abstract spherical polyhedron $(V, F; \underline{E})$, the following are equivalent:

(a) *The framework $G_S(r)$, with the vertices of each edge distinct, has a self-stress nonzero on all edges;*

(b) *$G_S(r)$ has a reciprocal framework $G^S(s)$ with the vertices of each edge distinct;*

(c) *$G_S(r)$ is the vertical projection of a proper spatial polyhedron $S(p, P)$;*

(d) *$G_S(r)$ is the gradient diagram of a proper spatial polyhedron $S^*(q, Q)$.*

There are other refinements of this theorem, that connect the space of self-stresses of $G_S(r)$ with the space of parallel drawings (and first-order flexes) of the reciprocal $G^S(s)$, the space of polyhedra $S(p, P)$ with the same projection$^\perp$, and the space of parallel drawings of $S^*(q, Q)$ [CW94] (Figure 61.4.2).

A second refinement connects the local convexity of the edge of the polyhedron with the sign of the self-stress.

THEOREM 61.4.2   Convex Self-stress

The vertical projection of a strictly convex polyhedron, with no faces vertical, produces a plane framework with a self-stress that is < 0 on the boundary edges (the edges bounding the infinite region of the plane) and > 0 on all edges interior to this boundary polygon.

A plane Delaunay triangulation also has a basic "reciprocal" relationship to the plane Voronoi diagram: the edges joining vertices at the centers of the regions are perpendicular to edges of the polygon of the Voronoi regions surrounding the vertex. This pair of reciprocals is directly related to the projection of a spatial convex polyhedral cap, as are generalized Voronoi diagrams [WABC13]. See Section 27.1.

This pattern of "reciprocal constructions" and the connection to liftings to polytopes in the next dimension generalizes to higher dimensions [CW94]. For example, for Voronoi diagrams and Delaunay simplicial complexes, the edges of one are perpendicular to facets of the other, in all dimensions. Moreover, for appropriate sphere-like homology, the existence of a reciprocal corresponds to the existence of nontrivial liftings [CW94, Ryb99]. Such geometric structures are also related to k-rigidity and to combinatorial proofs of the g-theorem in polyhedral combinatorics [TW00].

61.5 SOURCES AND RELATED MATERIALS

SURVEYS AND BASIC SOURCES

All results not given an explicit reference can be traced through these surveys:

[CW96]: A presentation of basic results for concepts of rigidity between first-order rigidity and rigidity for tensegrity frameworks.

[CWW14]: A recent book with a range of contributions on rigidity.

[Wiki]: A range of older preprints, including chapters from a draft book on rigidity.

[DT99]: Conference papers related to applications of rigidity theory.

[GSS93]: An older monograph devoted to combinatorial results for the graphs of generically rigid frameworks, with an extensive bibliography on many aspects of rigidity.

[Ros00]: An older thesis that explores in depth multiple topics of this chapter and their connections.

[SSS17]: A new handbook on constraint theory — to appear shortly.

[Sug86]: A monograph on the reconstruction of spatial polyhedral objects from plane pictures.

[Whi96]: An expository article presenting matroidal aspects of first-order rigidity, scene analysis, and multivariate splines.

RELATED CHAPTERS

Chapter 9: Geometry and topology of polygonal linkages
Chapter 27: Voronoi diagrams and Delaunay triangulations
Chapter 34: Geometric reconstruction problems
Chapter 51: Robotics
Chapter 55: Graph drawing
Chapter 60: Geometric applications of the Grassmann-Cayley algebra

Chapter 62: Rigidity of symmetric frameworks
Chapter 63: Global rigidity

REFERENCES

[AR78]   L. Asimow and B. Roth. The rigidity of graphs. *Trans. AMS*, 245:279–289, 1978.

[AYFH08] B.D.O. Anderson, C. Yu, B. Fidan, and J.M. Hendrickx. Rigid graph control architectures for autonomous formations. *IEEE Control Syst. Mag.*, 28:48–63, 2008.

[BCW02]   L. Berenchtein, L. Chávez, and W. Whiteley. Inductive constructions for 2-rigidity: Bases and circuits via tree partitions. Manuscript, York University, Toronto, 2002.

[BJ02]   A.R. Berg and T. Jordán. A proof of Connelly's conjecture on 3-connected circuits of the rigidity matroid. *J. Combin. Theory Ser. B.*, 88:77–79, 2003.

[BJ08]   J. Bang-Jensen and T. Jordán. On persistent directed graphs. *Networks*, 52:271–276, 2008.

[BMMM16] A. Baker, A. McRobie, T. Mitchell, and M. Mazurek. Mechanisms and states of self-stress of planar trusses using graphic statics, part I: the fundamental theorem of linear algebra and the Airy stress function. *Int. J. Space Structures*, 31:85–101, 2016.

[BO07]   P. Block and J. Ochsendorf. Thrust network analysis: a new methodology for three-dimensional equilibrium. *J. Int. Assoc. Shell & Spatial Structures*, 48:167–173, 2007.

[CKP15a] J. Cruickshank, D. Kitson, and S. Power. The generic rigidity of triangulated blocks and holes. *J. Combin. Theory Ser. B*, to appear, 2016.

[CKP15b] J. Cruickshank, D. Kitson, and S. Power. The rigidity of a partially triangulated torus. Preprint, arXiv:1509.00711, 2015.

[Con78]   R. Connelly. A flexible sphere. *Math. Intelligencer*, 1:130–131, 1978.

[Cra90]   H. Crapo. On the generic rigidity of structures in the plane. Technical report, RR-1278, INRIA, Rocquencourt, 1990.

[Cre72]   L. Cremona. *Graphical Statics*; English translation of the 1872 edition. Oxford University Press, London, 1890.

[CW82]   H. Crapo and W. Whiteley. Statics of frameworks and motions of panel structures: a projective geometric introduction. *Structural Topology*, 6:43–82, 1982.

[CW94]   H. Crapo and W. Whiteley. Spaces of stresses, projections and parallel drawings for spherical polyhedra. *Beitr. Algebra Geom.*, 35:259–281, 1994.

[CW96]   R. Connelly and W. Whiteley. Second-order rigidity and pre-stress stability for tensegrity frameworks. *SIAM J. Discrete Math.*, 9:453–492, 1996.

[CWW14]   R. Connelly, A. Ivić Weiss, and W. Whiteley, editors. *Rigidity and Symmetry*. Vol. 70 of *Fields Inst. Comm.*, Springer, New York, 2014.

[DT99]   P.M. Duxbury and M.F. Thorpe, editors. *Rigidity Theory and Applications*, Kluwer/Plenum, New York, 1999.

[EAM+04] T. Eren, B.D.O. Anderson, A.S. Morse, W. Whiteley, and P.N. Belhumeur. Operations on rigid formations of autonomous agents. *Comm. Inform. Syst.*, 3:223–258, 2004.

[EJN+17] Y. Eftekhari, B. Jackson, A. Nixon, B. Schulze, S. Tanigawa, and W. Whiteley. Point-hyperplane frameworks, slider joints, and rigidity preserving transformations. Preprint, `arXiv:1703.06844`, 2017.

[Fog88]   A. Fogelsanger. *The Generic Rigidity of Minimal Cycles*. Ph.D. Thesis, Cornell University, Ithaca, 1988.

[FRW12]   W. Finbow, E. Ross, and W. Whiteley. The rigidity of spherical frameworks: swapping blocks and holes. *SIAM J. Discrete Math.*, 26:280–304, 2012.

[FW13]   W. Finbow and W. Whiteley. Isostatic block and hole frameworks. *SIAM J. Discrete Math.*, 27:991–1020, 2013.

[GGLT13] S.J. Gortler, C. Gotsman, L. Liu, and D.P. Thurston. On affine rigidity. *J. Comput. Geom.*, 4:160–181, 2013.

[GSS93]   J. Graver, B. Servatius, and H. Servatius. *Combinatorial Rigidity*. Vol. 2 of *AMS Monogr.*, AMS, Providence, 1993.

[HADB07] J.M. Hendrickx, D.O. Anderson, J.-C. Delvenne, and V.D. Blondel. Directed graphs for the analysis of rigidity and persistence in autonomous agent systems. *Int. J. Robust Nonlinear Control*, 17:960–981, 2007.

[HYFA08] J.M. Hendrickx, C. Yu, B. Fidan, and B.D.O. Anderson. Rigidity and persistence for ensuring shape maintenance of multiagent meta-formations. *Asian J. Contr.*, 53:968–979, 2008.

[JJ05a]   B. Jackson and T. Jordán. The d-dimensional rigidity matroid of sparse graphs. *J. Combin. Theory Ser. B*, 95:118–133, 2005.

[JJ05b]   B. Jackson and T. Jordán. The Dress conjectures on the rank in the 3-dimensional rigidity matroid. *Adv. Appl. Math.*, 35:355–367, 2005.

[JJ06]   B. Jackson and T. Jordán. On the rank function of the 3-dimensional rigidity matroid. *Internat. J. Comput. Geom. Appl.*, 16:415–429, 2006.

[JJ07]   B. Jackson and T. Jordán. Rigid components in molecular graphs, *Algorithmica*, 48:399–412, 2007.

[JJ08a]   B. Jackson and T. Jordán. On the rigidity of molecular graphs. *Combinatorica*, 28:645–658, 2008.

[JJ08b]   B Jackson and T. Jordán. Pin-collinear body-and-pin frameworks and the Molecular Conjecture. *Discrete Comput. Geom.*, 40:258–278, 2008.

[JJ09]   B. Jackson and T. Jordán. A sufficient connectivity condition for generic rigidity in the plane. *Discrete Appl. Math.*, 157:1965–1968, 2009.

[JO16]   B. Jackson and J.C. Owen. A characterisation of the generic rigidity of 2-dimensional point-line frameworks. *J. Combin. Theory Ser. B*, 119:96–121, 2016.

[Jor12]   T. Jordán. Highly connected molecular graphs are rigid in three dimensions. *Inform. Process. Lett.*, 112:356–359, 2012.

[JRKT01] D.J. Jacobs, A.J. Rader, L.A. Kuhn, and M.F. Thorpe. Protein flexibility predictions using graph theory. *Proteins*, 44:150–165, 2001.

[Kit15]   D. Kitson. Finite and infinitesimal rigidity with polyhedral norms. *Discrete Comput. Geom.*, 54:390–411, 2015.

[KP14]   D Kitson and S. Power. Infinitesimal rigidity for non-Euclidean bar-joint frameworks. *Bull. London Math. Soc.*, 46:685–697, 2014.

[KS15]   D. Kitson and B. Schulze. Maxwell-Laman counts for bar-joint frameworks in normed spaces. *Linear Algebra Appl.*, 481:313–329, 2015.

[KT11]   N. Katoh and S. Tanigawa. A proof of the Molecular Conjecture. *Discrete Comput. Geom.*, 45:647–700, 2011.

[KW04]   F. Klein and K. Weighardt. Über Spannungsflächen und reziproke Diagramme, mit besonderer Berücksichtigung der Maxwellschen Arbeiten. *Archiv der Mathematik and Physik III*, 1904. Reprint in *Felix Klein Gesammelte Mathematische Abhandlungen*, pages 660–691, Springer, Berlin, 1922.

[LS08]   A. Lee and I. Streinu. Pebble game algorithms and sparse graphs. *Discrete Math.*, 308:1425–1437, 2008.

[LY82]   L. Lovász and Y. Yemini. On generic rigidity in the plane. *SIAM J. Algebraic Discrete Methods*, 3:91–98, 1982.

[Max64]   J.C. Maxwell. On reciprocal diagrams and diagrams of forces. *Phil. Mag. Series*, 26:250–261, 1864.

[MBMK16]   A. McRobie, A. Baker, T. Mitchell, and M. Konstantatou. Mechanisms and states of self-stress of planar trusses using graphic statics, part II: applications and extensions. *Int. J. Space Structures*, to appear, 2016.

[Mic08]   A. Micheletti. On generalized reciprocal diagrams for self-stressed frameworks. *Int. J. Space Structures*, 23:153–166, 2008.

[Nev07]   E. Nevo. On embeddability and stresses of graphs. *Combinatorica*, 27:465–472, 2007.

[NOP12]   A. Nixon, J.C. Owen, and S.C. Power. Rigidity of frameworks supported on surfaces. *SIAM J. Discrete Math.*, 26:1733–1757, 2012.

[NOP14]   A. Nixon, J.C. Owen, and S.C. Power. A characterisation of generically rigid frameworks on surfaces of revolution. *SIAM J. Discrete Math.*, 28:2008–2028, 2014.

[NR14]   A. Nixon and E. Ross. One brick at a time: a survey of inductive constructions in rigidity theory. In: *Rigidity and Symmetry*, vol. 70 of *Fields Inst. Comm.*, pages 303–324, Springer, New York, 2014.

[NSSW14]   A. Nixon, B. Schulze, A. Sljoka, and W. Whiteley. Symmetry adapted Assur decompositions. *Symmetry*, 6:516–550, 2014.

[OM02]   R. Olfati-Saber and R.M. Murray. Graph rigidity and distributed formation stabilization of multi-vehicle systems. In *Proc. IEEE Conf. Decision Control*, pages 2965–2971, 2002.

[OPA15]   K.-K. Oh, M.-C. Park and H.-S. Ahn. A survey of multi-agent formation control. *Automatica*, 53:424–440, 2015.

[Poll27]   H. Pollaczek-Geiringer. Über die Gliederung ebener Fachwerke. *Zeitschrift für Angewandte Mathematik und Mechanik*7:58-72, 1927.

[Ros00]   L. Ros. *A Kinematic-Geometric Approach to Spatial Interpretation of Line Drawings.* PhD thesis, Technical University of Catalonia, 2000.

[RW81]   B. Roth and W. Whiteley. Tensegrity frameworks. *Trans. Amer. Math. Soc.*, 265:419–446, 1981.

[Ryb99]   K. Rybnikov. Lifting and stresses of cell complexes. *Discrete Comput. Geom.*, 21:481–517, 1999.

[SSS17]   M. Sitharam, A. St. John, and J. Sidman, editors. *Handbook of Geometric Constraint Systems Principles.* CRC Press, Boca Raton, to appear.

[STB+09]   O. Shai, I. Tehori, A. Bronfeld, M. Slavutin, and U. Ben-Hanan. Adjustable tensegrity robot based on assur graph principle. *ASME 2009*, 10:257–261, 2009.

[SSW10a]   B. Servatius, O. Shai, and W. Whiteley. Combinatorial characterization of the Assur graphs from engineering. *European J. Combin.*, 31:1091–1104, 2010.

[SSW10b]   B. Servatius, O. Shai, and W. Whiteley. Geometric properties of Assur graphs. *European J. Combin.*, 31:1105–1120, 2010.

[SSW13]   O. Shai, A. Sljoka, and W. Whiteley. Directed graphs, decompositions, and spatial linkages. *Discrete Appl. Math.*, 161:3028–3047, 2013.

[Sug86]   K. Sugihara. *Machine Interpretation of Line Drawings.* MIT Press, Cambridge, 1986.

[SW99]   B. Servatius and W. Whiteley. Constraining plane configurations in CAD: combinatorics of directions and lengths. *SIAM J. Discrete Math.*, 12:136–153, 1999.

[SW07]   F. Saliola and W. Whiteley. Some notes on the equivalence of first-order rigidity in various geometries. Preprint, arXiv:0709.3354, 2007.

[SW12]   B. Schulze and W. Whiteley. Coning, symmetry and spherical frameworks. *Discrete Comput. Geom.*, 48:622–657, 2012.

[Tac12]   T. Tachi. Design of infinitesimally and finitely flexible origami based on reciprocal figures. *J. Geom. Graphics*, 26:223–234, 2012.

[Tay84]   T.-S. Tay. Rigidity of multigraphs I: linking rigid bodies in n-space. *J. Combin. Theory Ser. B*, 26:95–112, 1984.

[TW84]   T.-S. Tay and W. Whiteley. Recent advances in the generic rigidity of structures. *Structural Topology*, 9:31–38, 1984.

[TW85]   T.-S. Tay and W. Whiteley. Generating isostatic frameworks. *Structural Topology*, 11:20–69, 1985.

[TW00]   T.-S. Tay and W. Whiteley. A homological approach to skeletal rigidity. *Adv. Appl. Math.*, 25:102–151, 2000.

[WABC13]   W. Whiteley, P.F. Ash, E. Bolker, and H. Crapo. Convex polyhedra, Dirichlet tessellations, and spider webs. In *Shaping Space: Exploring Polyhedra in Nature, Art, and the Geometrical Imagination*, Part III, pages 231–251, Springer New York, 2013.

[Whi84a]   W. Whiteley. Infinitesimally rigid polyhedra I: statics of frameworks. *Trans. Amer. Math. Soc.*, 285:431–465, 1984.

[Whi84b]   W. Whiteley. Infinitesimal motions of a bipartite framework. *Pacific J. Math.*, 110:233–255, 1984.

[Whi88a]   W. Whiteley. A matroid on hypergraphs, with applications in scene analysis and geometry. *Discrete Comput. Geom.*, 4:75-95, 1988.

[Whi88b]   W. Whiteley. The union of matroids and the rigidity of frameworks. *SIAM J. Discrete Math.*, 1:237–255, 1988.

[Whi89]   W. Whiteley. Rigidity and polarity II. Weaving lines and plane tensegrity frameworks. *Geom. Dedicata*, 30:255–279, 1989.

[Whi91]   W. Whiteley. Vertex splitting in isostatic frameworks. *Struct. Topol.*, 16:23–30, 1991.

[Whi96]   W. Whiteley. Some matroids from discrete applied geometry. In J. Bonin, J. Oxley, and B. Servatius, editors, *Matroid Theory*, vol. 197 of *Contemp. Math.*, pages 171–311, AMS, Providence, 1996.

[Whi98]   W. Whiteley. An analogy in geometric homology: rigidity and cofactors on geometric graphs. In B. Sagan and R. Stanley, editors, *Mathematical Essays in Honor of Gian-Carlo-Rota*, pages 413–437, Birkhauser, Boston, 1998.

[Whi05]   W. Whiteley. Counting out to the flexibility of molecules. *J. Phys. Biol.*, 2:1–11, 2005.

[Wiki]   W. Whiteley. *Wiki of Preprint Resources in Rigidity Theory*. http://wiki.math.yorku.ca/index.php/Resources_in_Rigidity_Theory.

[WW83]   N. White and W. Whiteley. Algebraic geometry of stresses in frameworks. *SIAM J. Alg. Disc. Meth.*, 4:53–70, 1983.

[WW87]   N. White and W. Whiteley. The algebraic geometry of motions of bar and body frameworks. *SIAM J. Alg. Disc. Meth.*, 8:1–32, 1987.

[YHF+07]   C. Yu, J.M. Hendrickx, B. Fidan, B.D.O. Anderson, and V.D. Blondel. Three and higher dimensional autonomous formations: Rigidity, persistence and structural persistence. *Automatica*, 43:387–402, 2007.

62 RIGIDITY OF SYMMETRIC FRAMEWORKS
Bernd Schulze and Walter Whiteley

INTRODUCTION

Since symmetry is ubiquitous in both man-made structures (e.g., buildings or mechanical linkages) and in structures found in nature (e.g., proteins or crystals), it is natural to consider the impact of symmetry on the rigidity and flexibility properties of frameworks. The special geometry induced by various symmetry groups often leads to added first-order (and sometimes even continuous) flexibility in the structure. These phenomena have been studied in the following two settings:

(1) **Forced Symmetry:** The framework is symmetric (with respect to a certain group) and must maintain this symmetry throughout its motions.

(2) **Incidental Symmetry:** The framework is symmetric (with respect to a certain group), but is allowed to move in unrestricted ways.

The key tool for analyzing the forced-symmetric rigidity properties of a symmetric framework is its corresponding group-labeled quotient graph (or "gain graph"). In particular, using very simple counts on the number of vertex and edge orbits under the group action (i.e., vertices and edges of the gain graph), we can often detect symmetry-preserving first-order flexibility in symmetric frameworks that are generically rigid without symmetry. For configurations which are regular modulo the given symmetry, these first-order flexes even extend to continuous flexes. By introducing (gain-)sparsity counts for all subgraphs of a gain graph, Laman-type combinatorial characterizations of all (symmetry-)regular forced-symmetric rigid frameworks have also been obtained for various symmetry groups. Moreover, these combinatorial results have been extended to "body-bar frameworks" with an arbitrary symmetry group in d-dimensional space.

Analyzing the rigidity of incidentally symmetric frameworks is more challenging and relies on tools from group representation theory. However, very simple necessary conditions for an incidentally symmetric framework to be first-order rigid can still be derived. These can be formulated in terms of counts on the number of vertices and edges that remain unshifted under the various symmetry operations of the framework. Similar techniques have also successfully been applied to the theory of scene analysis, and are expected to be of wider use in other areas of discrete geometry. A number of combinatorial characterizations of incidentally symmetric first-order rigid bar-joint and body-bar frameworks have also recently been obtained via an extension of some key tools from the forced-symmetric theory (such as the orbit rigidity matrix).

We discuss forced-symmetric frameworks in Section 62.1, and incidentally symmetric frameworks in Section 62.2. Finally, in Section 62.3, we discuss the rigidity of infinite periodic frameworks.

62.1 FORCED-SYMMETRIC FRAMEWORKS

GLOSSARY

Automorphism of a graph: For a simple graph $G = (V, E)$, a permutation $\pi : V \to V$ such that $\{i, j\} \in E$ if and only if $\{\pi(i), \pi(j)\} \in E$. The group of all automorphisms of G is denoted by $\mathrm{Aut}(G)$.

Γ-*symmetric graph:* For a group Γ, a simple graph $G = (V, E)$ for which there exists a group action $\theta : \Gamma \to \mathrm{Aut}(G)$. The action θ is ***free*** if $\theta(\gamma)(i) \neq i$ for all $i \in V$ and all non-trivial $\gamma \in \Gamma$.

Vertex orbit: For a Γ-symmetric graph $G = (V, E)$ and $i \in V$, the set $\Gamma i = \{\theta(\gamma)(i) \mid \gamma \in \Gamma\}$. Analogously, the ***edge orbit*** of G for $e = \{i, j\} \in E$, is the set $\Gamma e = \{\{\theta(\gamma)(i), \theta(\gamma)(j)\} \mid \gamma \in \Gamma\}$.

Quotient graph: For a Γ-symmetric graph $G = (V, E)$, the multigraph G/Γ with vertex set $V/\Gamma = \{\Gamma i \mid i \in V\}$ and edge set $E/\Gamma = \{\Gamma e \mid e \in E\}$.

Quotient Γ-gain graph: Let $G = (V, E)$ be a Γ-symmetric graph, where the group action $\theta : \Gamma \to \mathrm{Aut}(G)$ is free. Each edge orbit Γe connecting Γi and Γj in the quotient graph G/Γ can be written as $\{\{\theta(\gamma)(i), \theta(\gamma) \circ \theta(\alpha)(j)\} \mid \gamma \in \Gamma\}$ for a unique $\alpha \in \Gamma$. For each Γe, orient Γe from Γi to Γj in G/Γ and assign to it the gain α. The resulting oriented quotient graph $G_0 = (V_0, E_0)$, together with the gain labeling $\psi : E_0 \to \Gamma$ described above, is the quotient Γ-gain graph (G_0, ψ) of G. (See also Figure 62.1.1.)

(Note that (G_0, ψ) is unique up to choices of representative vertices, and that the orientation is only used as a reference orientation and may be changed, provided that we also modify ψ so that if an edge has gain α in one orientation, then it has gain α^{-1} in the other direction.)

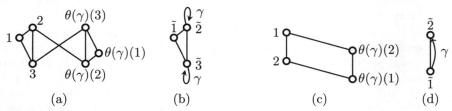

(a)   (b)   (c)   (d)

FIGURE 62.1.1
$\mathbb{Z}_2$-*symmetric graphs (a,c), and their quotient $\mathbb{Z}_2$-gain graphs (b,d), where $\mathbb{Z}_2 = \{id, \gamma\}$. (The orientation and gain labeling is omitted for all edges with gain id.) The triangle in (b) is balanced, whereas any edge set containing a loop in (b) is unbalanced. The edge set in (d) is also unbalanced.*

Gain of a closed walk: For a quotient Γ-gain graph (G_0, ψ) and a closed walk
$$W = \tilde{v}_1, \tilde{e}_1, \tilde{v}_2, \ldots, \tilde{v}_k, \tilde{e}_k, \tilde{v}_1$$
of (G_0, ψ), the group element $\psi(W) = \Pi_{i=1}^{k} \psi(\tilde{e}_i)^{\mathrm{sign}(\tilde{e}_i)}$, where $\mathrm{sign}(\tilde{e}_i) = 1$ if $\tilde{e}_i$ is directed from $\tilde{v}_i$ to $\tilde{v}_{i+1}$, and $\mathrm{sign}(\tilde{e}_i) = -1$ otherwise.

Subgroup induced by an edge subset: For a quotient Γ-gain graph (G_0, ψ), a subset $F \subseteq E_0$, and a vertex $\tilde{i}$ of the vertex set $V(F) \subseteq V_0$ induced by F, the subgroup $\langle F \rangle_{\psi, \tilde{i}} = \{\psi(W) \mid W \in \mathcal{W}(F, \tilde{i})\}$ of Γ, where $\mathcal{W}(F, \tilde{i})$ is the set of closed walks starting at $\tilde{i}$ using only edges of F.

Balanced edge set: A (possibly disconnected) subset of the edge set of a quotient Γ-gain graph (G_0, ψ) with the property that all of its connected components are balanced, where a connected edge subset F of E_0 is balanced if $\langle F \rangle_{\psi, \tilde{i}} = \{\mathrm{id}\}$ for some $\tilde{i} \in V(F)$ (or equivalently, $\langle F \rangle_{\psi, \tilde{i}} = \{\mathrm{id}\}$ for all $\tilde{i} \in V(F)$). A subset of E_0 is called ***unbalanced*** if it is not balanced (that is, if it contains an unbalanced cycle).

Cyclic edge set: A (possibly disconnected) subset of the edge set of a quotient Γ-gain graph (G_0, ψ) with the property that all of its connected components are cyclic, where a connected edge subset F of E_0 is cyclic if $\langle F \rangle_{\psi, \tilde{i}}$ is a cyclic subgroup of Γ for some $\tilde{i} \in V(F)$ (or equivalently, for all $\tilde{i} \in V(F)$).

(k, ℓ, m)-gain-sparse: For non-negative integers k, ℓ, m with $m \leq \ell$, a quotient Γ-gain graph (G_0, ψ) satisfying

$$|F| \leq \begin{cases} k|V(F)| - \ell, & \text{for all non-empty balanced } F \subseteq E_0, \\ k|V(F)| - m, & \text{for all non-empty } F \subseteq E_0. \end{cases}$$

If (G_0, ψ) also satisfies $|E_0| = k|V_0| - m$, then it is called ***(k, ℓ, m)-gain-tight***. For example, the $\mathbb{Z}_2$-gain graphs in Figure 62.1.1 (b) and (d) are $(2, 3, 1)$-gain-tight and $(2, 3, 2)$-gain-tight, respectively.

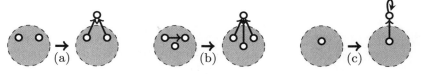

FIGURE 62.1.2
Examples of the three types of Γ-symmetric Henneberg construction moves. The gain labeling is omitted for all edges.

Γ-symmetric Henneberg construction: For a quotient Γ-gain graph (G_0, ψ), a sequence $(H_1, \psi_1), \ldots, (H_n, \psi_n)$ of Γ-gain graphs such that:

(i) For each index $1 < j \leq n$, (H_j, ψ_j) is obtained from (H_{j-1}, ψ_{j-1}) by

vertex addition: attaching a new vertex $\tilde{v}$ by two new non-loop edges $\tilde{e}_1$ and $\tilde{e}_2$. If $\tilde{e}_1$ and $\tilde{e}_2$ are parallel, then $\psi_j(\tilde{e}_1) \neq \psi_j(\tilde{e}_2)$ (assuming that $\tilde{e}_1$ and $\tilde{e}_2$ are directed to $\tilde{v}$ (see Figure 62.1.2(a))).

edge splitting: replacing an edge (possibly a loop) $\tilde{e}$ of (H_{j-1}, ψ_{j-1}) with a new vertex $\tilde{v}$ joined to its end(s) by two new edges $\tilde{e}_1$ and $\tilde{e}_2$, such that the tail of $\tilde{e}_1$ is the tail of $\tilde{e}$ and the tail of $\tilde{e}_2$ is the head of $\tilde{e}$, and $\psi(\tilde{e}_1) \cdot \psi(\tilde{e}_2)^{-1} = \psi(\tilde{e})$, and finally adding a third edge $\tilde{e}_3$ oriented from a vertex $\tilde{z}$ of H_{j-1} to $\tilde{v}$ so that every two-cycle $\tilde{e}_i \tilde{e}_k$, if it exists, is unbalanced in (H_j, ψ_j) (see Figure 62.1.2(b)).

loop extension: attaching a new vertex $\tilde{v}$ to a vertex of H_{j-1} by a new edge with any gain, and adding a new loop $\tilde{l}$ incident to $\tilde{v}$ with $\psi(\tilde{l}) \neq id$ (see Figure 62.1.2(c)).

(ii) (H_1, ψ_1) is one vertex with one unbalanced loop, and $(H_n, \psi_n) = (G_0, \psi)$.

Γ-symmetric framework: For a graph $G = (V, E)$, a group action $\theta : \Gamma \to \mathrm{Aut}(G)$, and a homomorphism $\tau : \Gamma \to O(\mathbb{R}^d)$, a framework $G(p)$ (as defined in Chapter 61, with $p : V \to \mathbb{R}^d$ a configuration of points in $\mathbb{R}^d$) satisfying

$$\tau(\gamma)(p_i) = p_{\theta(\gamma)(i)} \qquad \text{for all } \gamma \in \Gamma \text{ and all } i \in V.$$

Symmetry group of a framework: For a Γ-symmetric framework, the group $\tau(\Gamma) = \{\tau(\gamma) \mid \gamma \in \Gamma\}$ of isometries of $\mathbb{R}^d$.

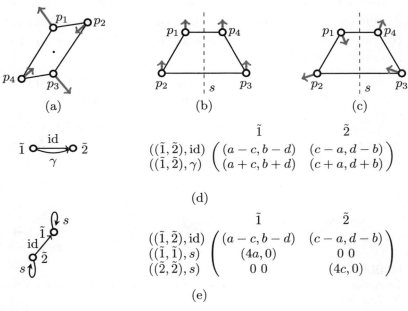

FIGURE 62.1.3

First-order flexes of frameworks in $\mathbb{R}^2$ with $\mathcal{C}_2 = \{id, \gamma\}$ (half-turn) and $\mathcal{C}_s = \{id, s\}$ (reflection) symmetry: (a) a fully $\mathcal{C}_2$-symmetric non-trivial first-order flex; (b) a fully $\mathcal{C}_s$-symmetric trivial first-order flex; (c) a non-trivial first-order flex which is not fully $\mathcal{C}_s$-symmetric (but "anti-symmetric"); (d) The quotient $\mathbb{Z}_2$-gain graph corresponding to the framework in (a) and its orbit rigidity matrix (with $p_1 = (a, b), p_2 = (c, d), p_3 = (-a, -b)$, and $p_4 = (-c, -d)$); (e) the quotient $\mathbb{Z}_2$-gain graph corresponding to the framework in (b,c) and its orbit rigidity matrix (with $p_1 = (a, b), p_2 = (c, d), p_3 = (-c, d),$ and $p_4 = (-a, b)$).

Orbit rigidity matrix: For a Γ-symmetric framework $G(p)$ (with respect to the free action $\theta : \Gamma \to \mathrm{Aut}(G)$ and $\tau : \Gamma \to O(\mathbb{R}^d)$), where (G_0, ψ) is the quotient Γ-gain graph of G, the $|E_0| \times d|V_0|$ matrix $O(G_0, \psi, p)$ defined as follows. Choose a representative vertex $\tilde{i}$ for each vertex Γi in V_0. The row corresponding to the edge $\tilde{e} = (\tilde{i}, \tilde{j})$, $\tilde{i} \neq \tilde{j}$, with gain $\psi(\tilde{e})$ in E_0 is then given by

$$\left(0 \ldots 0 \quad \overbrace{p(\tilde{i}) - \tau(\psi(\tilde{e}))p(\tilde{j})}^{\tilde{i}} \quad 0 \ldots 0 \quad \overbrace{p(\tilde{j}) - \tau(\psi(\tilde{e}))^{-1}p(\tilde{i})}^{\tilde{j}} \quad 0 \ldots 0\right).$$

If $\tilde{e} = (\tilde{i}, \tilde{i})$ is a loop at $\tilde{i}$, then the row corresponding to $\tilde{e}$ is given by

$$\left(0 \ldots 0 \quad \overbrace{2p(\tilde{i}) - \tau(\psi(\tilde{e}))p(\tilde{i}) - \tau(\psi(\tilde{e}))^{-1}p(\tilde{i})}^{\tilde{i}} \quad 0 \ldots 0 \quad 0 \quad 0 \ldots 0\right).$$

Fully Γ-symmetric first-order flex: For a Γ-symmetric framework $G(p)$, a first-order flex $p' : V \to \mathbb{R}^d$ of $G(p)$ satisfying

$$\tau(\gamma)p_i' = p_{\theta(\gamma)(i)}' \qquad \text{for all } \gamma \in \Gamma \text{ and all } i \in V.$$

Fully Γ-symmetric self-stress: For a Γ-symmetric framework $G(p)$, a self-stress ω_{ij} satisfying $\omega_e = \omega_f$ for all edges e, f in the same edge orbit Γe.

Forced Γ-symmetric first-order rigid framework: A Γ-symmetric framework for which every fully Γ-symmetric first-order flex is trivial.

Forced Γ-symmetric isostatic framework: A forced Γ-symmetric first-order rigid framework $G(p)$ whose orbit rigidity matrix $O(G_0, \psi, p)$ has independent rows (i.e., $G(p)$ has no fully Γ-symmetric self-stress).

Γ-regular framework: A Γ-symmetric framework $G(p)$ (with respect to $\theta : \Gamma \to \text{Aut}(G)$ and $\tau : \Gamma \to O(\mathbb{R}^d)$) whose orbit rigidity matrix has maximal rank among all Γ-symmetric frameworks $G(q)$ (with respect to θ and τ).

BASIC RESULTS

A key reason for the interest in forced Γ-symmetric first-order rigidity is that for almost all Γ-symmetric realizations of a given graph as a bar-joint framework, a fully Γ-symmetric first-order flex extends to a *continuous* flex which preserves the symmetry of the framework throughout the path [GF07, Sch10d]. Results on forced Γ-symmetric first-order rigidity therefore provide important tools for detecting hidden continuous flexibility in symmetric frameworks. (See, e.g., Figure 62.1.4.)

THEOREM 62.1.1 Γ-Regular Rigidity Theorem

A Γ-regular framework $G(p)$ has a non-trivial fully Γ-symmetric first-order flex if and only if $G(p)$ has a non-trivial continuous flex which preserves the symmetry of $G(p)$ throughout the path.

A fundamental tool for studying the forced Γ-symmetric first-order rigidity properties of a framework $G(p)$ is the orbit rigidity matrix [SW11].

THEOREM 62.1.2 The Orbit Rigidity Matrix

Let $G(p)$ be a Γ-symmetric framework (with respect to $\theta : \Gamma \to \text{Aut}(G)$ and $\tau : \Gamma \to O(\mathbb{R}^d)$). The kernel of the orbit rigidity matrix $O(G_0, \psi, p)$ is isomorphic to the space of fully Γ-symmetric first-order flexes of $G(p)$, and the kernel of $O(G_0, \psi, p)^T$ is isomorphic to the space of fully Γ-symmetric self-stresses of $G(p)$.

As an immediate consequence we obtain the following basic result.

THEOREM 62.1.3   Rank of the Orbit Rigidity Matrix

A Γ-symmetric framework (with respect to the free action $\theta : \Gamma \to \mathrm{Aut}(G)$ and $\tau : \Gamma \to O(\mathbb{R}^d)$) with $|V| \geq d$ is forced Γ-symmetric first-order rigid if and only if rank $O(G_0, \psi, p) = d|V_0| - \mathrm{triv}_{\tau(\Gamma)}$, where $\mathrm{triv}_{\tau(\Gamma)}$ is the dimension of the space of fully Γ-symmetric trivial first-order flexes of $G(p)$.

Note that $\mathrm{triv}_{\tau(\Gamma)}$ can easily be computed for any symmetry group in any dimension. For $d = 2, 3$, $\mathrm{triv}_{\tau(\Gamma)}$ can also be read off directly from the character table of the symmetry group $\tau(\Gamma)$ [AH94, ACP70]. For example, for $d = 3$ and $\Gamma = \mathbb{Z}_2$, we have $\mathrm{triv}_{\tau(\Gamma)} = 2$ if $\tau(\Gamma) = \mathcal{C}_2$ (the velocities generated by a translation along the half-turn axis and a rotation about the axis form a basis), and $\mathrm{triv}_{\tau(\Gamma)} = 3$ if $\tau(\Gamma) = \mathcal{C}_s$ (the velocities generated by two independent translations along the mirror and a rotation about the axis perpendicular to the mirror form a basis).

The following result provides simple necessary counting conditions for a framework to be forced Γ-symmetric isostatic for all symmetry groups in all dimensions [JKT16].

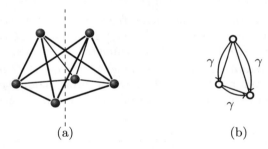

(a)   (b)

FIGURE 62.1.4
A flexible "Bricard octahedron" with half-turn symmetry (a) and its quotient $\mathbb{Z}_2$-gain graph, where $\mathbb{Z}_2 = \{id, \gamma\}$ (b). (The orientation and gain labeling is omitted for all edges with gain id). While generic realizations of the octahedral graph (without symmetry) are isostatic in 3-space, the symmetry-preserving continuous flexibility of the $\mathbb{Z}_2$-regular realization of this graph shown in (a) is easily detected using Theorems 62.1.1 and 62.1.4, since $|E_0| = 6$, $|V_0| = 3$, and hence $|E_0| = 6 < 7 = 3|V_0| - \mathrm{triv}_{\mathcal{C}_2}$.

THEOREM 62.1.4   Necessary Counting Conditions

Let $G(p)$ be a forced Γ-symmetric isostatic framework with respect to the free action $\theta : \Gamma \to \mathrm{Aut}(G)$ and $\tau : \Gamma \to O(\mathbb{R}^d)$, where $|V| \geq d$. Then the quotient Γ-gain graph (G_0, ψ) of G satisfies

(a) $|E_0| = d|V_0| - \mathrm{triv}_{\tau(\Gamma)}$

(b) $|F| \leq d|V(F)| - \mathrm{triv}_{\tau((\langle F \rangle_{\psi, \tilde{i}})}(p(F))$ *for all $F \subseteq E_0$ and all $\tilde{i} \in V(F)$,*

where $\tilde{i} \in V_0$ is identified with its representative vertex, and $\mathrm{triv}_{\tau((\langle F \rangle_{\psi, \tilde{i}})}(p(F))$ is the dimension of the space of fully $(\langle F \rangle_{\psi, \tilde{i}})$-symmetric trivial first-order flexes of the configuration $p(F) = \{\tau(\gamma)(p(\tilde{i})) \mid \tilde{i} \in V(F), \gamma \in \Gamma\}$.

For a number of symmetry groups $\tau(\Gamma)$ in the plane, these counts have also been shown to be sufficient for a Γ-regular framework to be forced Γ-symmetric isostatic (see Theorems 62.1.5 and 62.1.6).

Finally, we note that while the orbit rigidity matrix $O(G_0, \psi, p)$ has a particularly simple form if the action $\theta : \Gamma \to \text{Aut}(G)$ is free, it can also be constructed for frameworks $G(p)$, where θ is not free [SW11]. In this case, the counts in Theorem 62.1.4 need to be adjusted accordingly. For example, for half-turn symmetry $\mathcal{C}_2$ in 3-space, the count in Theorem 62.1.4(a) becomes $|E_0| = 3|V_0 \setminus V_0'| + |V_0'| - \text{triv}_{\mathcal{C}_2}$, where V_0' is the set of vertices that are fixed by the half-turn. This is because each vertex in V_0' is in an orbit on its own and has only one degree of freedom, as it must remain on the half-turn axis. These adjustments are straightforward, but they lead to significantly messier gain-sparsity counts in Theorem 62.1.4. While all of the results in this section are expected to extend to frameworks where the action θ is not free, these problems have not yet been fully investigated.

COMBINATORIAL RESULTS

All Γ-regular realizations of G (i.e., *almost all* Γ-symmetric realizations of G) share the same fully Γ-symmetric rigidity properties. Therefore, for Γ-regular frameworks, forced Γ-symmetric first-order rigidity is a purely combinatorial concept, and hence a property of the underlying quotient Γ-gain graph. For forced-symmetric rigidity in the plane, Laman-type theorems have been established for all cyclic groups and all dihedral groups of order $2n$, where n is odd [JKT16, MT11, MT12, MT15].

THEOREM 62.1.5   Reflectional or Rotational Symmetry in the Plane

Let $n \geq 2$, and let $G(p)$ be a $\mathbb{Z}_n$-regular framework with respect to the free action $\theta : \mathbb{Z}_n \to \text{Aut}(G)$ and $\tau : \mathbb{Z}_n \to O(\mathbb{R}^2)$. Then the following are equivalent:

(a) *$G(p)$ is forced $\mathbb{Z}_n$-symmetric isostatic;*

(b) *the quotient $\mathbb{Z}_n$-gain graph (G_0, ψ) of G is $(2, 3, 1)$-gain-tight;*

(c) *(G_0, ψ) has a Γ-symmetric Henneberg construction.*

Note that the count $|E_0| = 2|V_0| - 1$ reflects the fact that $\text{triv}_{\tau(\mathbb{Z}_n)} = 1$ for all cyclic groups $\mathbb{Z}_n$, $n \geq 2$. If $\tau(\mathbb{Z}_n)$ describes rotational symmetry, then a first-order rotation about the origin forms a basis, and if $\tau(\mathbb{Z}_2)$ describes mirror symmetry, then a first-order translation along the mirror line forms a basis (see also Figure 62.1.3 (b)).

Examples of $(2, 3, 1)$-gain-tight $\mathbb{Z}_2$-gain graphs are shown in Figures 62.1.1 (b) and 62.1.3 (e). By Theorem 62.1.5, $\mathbb{Z}_2$-regular realizations of the corresponding "covering graphs" are forced $\mathbb{Z}_2$-symmetric isostatic (but still flexible). See also Figures 62.1.1 (a) (and 62.2.2 (b) with one edge removed) and 62.1.3 (c), respectively.

For the dihedral groups of order $2n$, where n is odd, we have the following result [JKT16].

THEOREM 62.1.6   Dihedral Symmetry in the Plane

Let $G(p)$ be a D_{2n}-regular framework with respect to the free action $\theta : D_{2n} \to \text{Aut}(G)$ and $\tau : D_{2n} \to O(\mathbb{R}^2)$, where $n \geq 3$ is an odd integer, and $\mathcal{C}_{nv} = \tau(D_{2n})$ describes the dihedral symmetry group of order $2n$ in $\mathbb{R}^2$. Then $G(p)$ is forced

D_{2n}-symmetric isostatic if and only if the quotient D_{2n}-gain graph (G_0, ψ) of G satisfies

(a) $|E_0| = 2|V_0|$

(b) $|F| \leq \begin{cases} 2|V(F)| - 3 & \text{for all non-empty balanced } F \subseteq E_0, \\ 2|V(F)| - 1 & \text{for all non-empty unbalanced and cyclic } F \subseteq E_0, \\ 2|V(F)| & \text{for all } F \subseteq E_0. \end{cases}$

Analogous to Theorem 62.1.5, the D_{2n}-gain graphs satisfying the counts in Theorem 62.1.6 can also be characterized via an inductive Henneberg-type construction sequence. However, this construction sequence requires some additional base graphs and some additional gain-graph operations [JKT16].

For the dihedral groups D_{2n}, where n is an even integer, combinatorial characterizations for forced D_{2n}-symmetric rigidity have not yet been obtained. A famous example which shows that the counts in the above theorem are not sufficient for a D_{2n}-regular framework to be forced D_{2n}-symmetric isostatic is the realization of the complete bipartite graph $K_{4,4}$ shown in Figure 62.1.5 [SW11]. (The motion of this framework is also known as Bottema's mechanism in the engineering literature.) See [JKT16] for further examples.

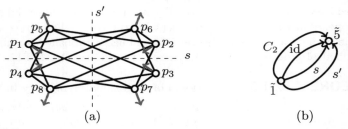

(a)   (b)

FIGURE 62.1.5
A D_{2n}-regular realization of $K_{4,4}$ and its quotient D_{2n}-gain graph, where $n = 2$. The fully D_{2n}-symmetric first-order flex extends to a symmetry-preserving continuous flex.

Since a combinatorial characterization of isostatic generic bar-joint frameworks (without symmetry) in dimension 3 and higher has not yet been established, there are also no known characterizations of forced Γ-regular isostatic frameworks for any symmetry group in dimension 3 or higher. However, such combinatorial characterizations have been obtained for any symmetry group in any dimension for the special class of "body-bar frameworks" (rigid full-dimensional bodies, connected in pairs by stiff bars) [Tan15]. The underlying combinatorial structure of a body-bar framework is a multigraph whose vertices and edges represent the rigid bodies and stiff bars, respectively. The notions of a Γ-symmetric graph and a Γ-symmetric bar-joint framework can naturally be extended to multigraphs and body-bar frameworks.

THEOREM 62.1.7 Body-Bar Frameworks in d-Space

Let G be a multigraph which is Γ-symmetric with respect to the free action $\theta : \Gamma \to \text{Aut}(G)$. Further, let $\tau : \Gamma \to O(\mathbb{R}^d)$ be a homomorphism. Then all Γ-regular body-bar realizations $G(q)$ of G are forced Γ-symmetric isostatic if and only if the quotient Γ-gain graph (G_0, ψ) of G satisfies

(a) $|E_0| = \binom{d+1}{2}|V_0| - \mathrm{triv}_{\tau(\Gamma)}$

(b) $|F| \leq \binom{d+1}{2}|V(F)| - \mathrm{triv}_{\tau(\langle F\rangle_{\psi, \tilde{i}})}(q(F))$ for all $F \subseteq E_0$.

Analogous to the non-symmetric situation (recall Section 61.1.3), Γ-regular body-bar, body-hinge, and even molecular realizations of G are conjectured to share the same forced Γ-symmetric rigidity properties [PRS⁺14]. This has so far only been verified for body-bar and body-hinge frameworks with $\mathbb{Z}_2 \times \cdots \times \mathbb{Z}_2$ symmetry, where the group acts freely on the vertices and edges [ST14] (see also Theorem 62.2.8).

CONJECTURE 62.1.8 Symmetric Molecular Conjecture (Version I)

The orbit rigidity matrix of a Γ-regular body-bar realization of a multigraph G has the same rank as the orbit rigidity matrix of a Γ-regular molecular realization of G.

This conjecture has important practical applications, as many proteins exhibit non-trivial (rotational) symmetries. The most common ones are half-turn symmetry ($\mathcal{C}_2$) and dihedral symmetry of order 4 and 6 generated by two half-turns ($\mathcal{D}_2$) and a half-turn and a three-fold rotation ($\mathcal{D}_3$), respectively. It turns out these three groups are also the symmetries which give rise to (symmetry-preserving) flexibility in body-bar frameworks which have the minimal number of edges to satisfy the necessary Maxwell count $|E| \geq 6|V| - 6$ for rigidity (under the assumption that the group acts freely on the vertices and edges) [SSW14]. See Table 62.1.1 for details. It is therefore intended to refine the ProFlex/FIRST algorithms for testing protein flexibility based on this conjecture.

Note that there also exists a second version of the Symmetric Molecular Conjecture, which concerns the first-order rigidity of incidentally symmetric frameworks (see Section 62.2).

TABLE 62.1.1 Symmetry-induced flexibility in body-bar frameworks with $\mathcal{C}_2$, $\mathcal{D}_2$ and $\mathcal{D}_3$ symmetry. (We use the standard Schoenflies notation for symmetry groups.)

$\tau(\Gamma)$	$\mathrm{triv}_{\tau(\Gamma)}$	$|E|$	$|E_0|$	$6|V_0| - \mathrm{triv}_{\tau(\Gamma)}$	$6|V_0| - \mathrm{triv}_{\tau(\Gamma)} - |E_0|$
$\mathcal{C}_1$	6	$6|V| - 6$	$6|V_0| - 6$	$6|V_0| - 6$	0
$\mathcal{C}_2$	2	$6|V| - 6$	$6|V_0| - 3$	$6|V_0| - 2$	1
$\mathcal{C}_3$	2	$6|V| - 6$	$6|V_0| - 2$	$6|V_0| - 2$	0
$\mathcal{C}_6$	2	$6|V| - 6$	$6|V_0| - 1$	$6|V_0| - 2$	-1
$\mathcal{D}_2$	0	$6|V| - 4$	$6|V_0| - 1$	$6|V_0|$	1
$\mathcal{D}_3$	0	$6|V| - 6$	$6|V_0| - 1$	$6|V_0|$	1

GEOMETRIC RESULTS

The forced Γ-symmetric rigidity properties of frameworks can also be transferred to other metric spaces, such as the spherical or hyperbolic space, via the technique of coning [SW12]. Recall from Section 61.1.3 that the cone graph of a graph G is the graph $G * u$ obtained from G by adding the new vertex u and the edges $\{u, v_i\}$ for all vertices $v_i \in V$.

THEOREM 62.1.9   Symmetric Coning

Let $G(p)$ be a framework in $\mathbb{R}^d$, and embed $G(p)$ into the hyperplane $x_{d+1} = 1$ of $\mathbb{R}^{d+1}$ via $\overline{p}_i = (p_i, 1) \in \mathbb{R}^{d+1}$. Further, let $(G * u)(\overline{p}^*)$ be the framework obtained from $G(p)$ by coning G and placing the new cone vertex at the origin of $\mathbb{R}^{d+1}$. Then

(a) *$G(p)$ is Γ-symmetric with respect to $\theta : \Gamma \to \mathrm{Aut}(G)$ and $\tau : \Gamma \to O(\mathbb{R}^d)$ if and only if $(G * u)(\overline{p}^*)$ is Γ-symmetric with respect to $\theta^* : \Gamma \to \mathrm{Aut}(G * u)$ defined by $\theta^*(\gamma)|_V = \theta(\gamma)$ and $\theta^*(\gamma)(u) = u$ for all $\gamma \in \Gamma$, and $\tau^* : \Gamma \to O(\mathbb{R}^{d+1})$ defined by $\tau^*(\gamma) = \begin{pmatrix} \tau(\gamma) & 0 \\ 0 & 1 \end{pmatrix}$ for all $\gamma \in \Gamma$.*

(b) *$G(p)$ has a non-trivial fully Γ-symmetric first-order flex (self-stress) in $\mathbb{R}^d$ if and only if $(G * u)(\overline{p}^*)$ has a non-trivial fully Γ-symmetric first-order flex (self-stress) in $\mathbb{R}^{d+1}$.*

(c) *If $(G * u)(q)$ is a Γ-symmetric framework (with respect to θ^* and τ^*) obtained from $(G * u)(\overline{p}^*)$ by moving the vertices of a vertex orbit of G along their corresponding cone rays to $p(u)$ (the origin), then $G(p)$ has a non-trivial fully Γ-symmetric first-order flex (self-stress) if and only if $(G * u)(q)$ has a non-trivial fully Γ-symmetric first-order flex (self-stress).*

As a simple corollary of Theorem 62.1.9 we obtain the following result.

THEOREM 62.1.10   Transfer between Euclidean and Spherical Space

Let q be a configuration of points on the unit sphere $\mathbb{S}^d$ (with no points on the equator) such that the projection $\pi(q)$ of the points from the origin (the center of the sphere) onto the hyperplane $x_{d+1} = 1$ of $\mathbb{R}^{d+1}$ (and then projected back to $\mathbb{R}^d$), is equal to the configuration p. Then $G(p)$ has a non-trivial fully Γ-symmetric first-order flex (self-stress) in $\mathbb{R}^d$ if and only if $G(q)$ has a non-trivial fully Γ-symmetric first-order flex (self-stress) in $\mathbb{S}^d$.

It turns out that we may even transfer *continuous* flexibility between metrics via the technique of symmetric coning.

THEOREM 62.1.11   Transfer of Continuous Symmetry-Preserving Flexes

If $G(p)$ is a Γ-regular framework in $\mathbb{R}^d$, and $G(q)$ is a Γ-symmetric framework in $\mathbb{R}^{d+1}$ such that the projection $\pi(q)$ of q (as defined above) is equal to p, then $G(p)$ has a non-trivial symmetry-preserving continuous flex if and only if $G(q)$ does.

In particular, Theorem 62.1.11 allows us to transfer continuous flexibility between the d-sphere (with no points on the equator) and Euclidean d-space.

The transfer of fully Γ-symmetric first-order (and continuous) rigidity and flexibility properties from Euclidean space to other Cayley-Klein metrics, such as hyperbolic space, is carried out analogously [SW12].

62.2   INCIDENTALLY SYMMETRIC FRAMEWORKS

GLOSSARY

Group representation: For a group Γ and a linear space X, a homomorphism

$\rho : \Gamma \to \mathrm{GL}(X)$. The space X is called the ***representation space*** of ρ. (Note that two representations are considered equivalent if they are similar.)

ρ-invariant subspace: For a representation $\rho : \Gamma \to \mathrm{GL}(X)$, a subspace $U \subseteq X$ satisfying $\rho(\gamma)(U) \subseteq U$ for all $\gamma \in \Gamma$.

Irreducible representation: A group representation $\rho : \Gamma \to \mathrm{GL}(X)$ with the property that X and $\{0\}$ are the only ρ-invariant subspaces of X.

Intertwining map: For two representations ρ_1 and ρ_2 of a group Γ (with respective representation spaces X and Y), a linear map $T : X \to Y$ such that $T\rho_1(\gamma) = \rho_2(\gamma)T$ for all $\gamma \in \Gamma$. The set of all intertwining maps of ρ_1 and ρ_2 forms a linear space which is denoted by $\mathrm{Hom}_\Gamma(\rho_1, \rho_2)$.

Tensor product: For two representations ρ_1 and ρ_2 of a group Γ, the representation $\rho_1 \otimes \rho_2$ defined by $\rho_1 \otimes \rho_2(\gamma) = \rho_1(\gamma) \otimes \rho_2(\gamma)$ for all $\gamma \in \Gamma$.

External representation: For a Γ-symmetric framework $G(p)$ (with respect to $\theta : \Gamma \to \mathrm{Aut}(G)$ and $\tau : \Gamma \to O(\mathbb{R}^d)$) the representation $\tau \otimes P_V : \Gamma \to \mathbb{R}^{d|V|}$, where $P_V : \Gamma \to \mathrm{GL}(\mathbb{R}^{|V|})$ assigns to $\gamma \in \Gamma$ the permutation matrix of the permutation $\theta(\gamma)$ of V; that is, $P_V(\gamma) = [\delta_{i,\theta(\gamma)(j)}]_{i,j}$, where δ denotes the Kronecker delta.

Internal representation: For a Γ-symmetric graph G (with respect to $\theta : \Gamma \to \mathrm{Aut}(G)$), the representation $P_E : \Gamma \to \mathrm{GL}(\mathbb{R}^{|E|})$ which assigns to $\gamma \in \Gamma$ the permutation matrix of the permutation $\theta(\gamma)$ of E.

Fixed vertex: For a Γ-symmetric graph G (with respect to $\theta : \Gamma \to \mathrm{Aut}(G)$) and an element $\gamma \in \Gamma$, a vertex i with $\theta(\gamma)(i) = i$. Similarly, an edge $e = \{i, j\}$ of G is fixed by γ if $\theta(\gamma)(e) = e$, i.e., if either $\theta(\gamma)(i) = i$ and $\theta(\gamma)(j) = j$ or $\theta(\gamma)(i) = j$ and $\theta(\gamma)(j) = i$.

Inc-Γ-regular framework: A Γ-symmetric framework $G(p)$ (with respect to $\theta : \Gamma \to \mathrm{Aut}(G)$ and $\tau : \Gamma \to O(\mathbb{R}^d)$) whose rigidity matrix has maximal rank among all Γ-symmetric frameworks $G(q)$ (with respect to θ and τ).

Character: For a representation ρ of a group Γ, the row vector $\chi(\rho)$ whose ith component is the trace of $\rho(\gamma_i)$ for some fixed ordering $\gamma_1, \ldots, \gamma_{|\Gamma|}$ of the elements of Γ.

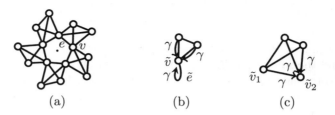

(a) (b) (c)

FIGURE 62.2.1
A $\mathbb{Z}_5$-symmetric graph (a) and its corresponding quotient $\mathbb{Z}_5$-gain graph (b) whose edge set is near-balanced. A balanced split is shown in (c). The orientation and gain labeling is omitted for all edges with gain id, and γ denotes rotation by $2\pi/5$.

Near-balanced edge set: For a quotient Γ-gain graph (G_0, ψ), a vertex $\tilde{v}$ of (G_0, ψ), and a partition $\{E_1, E_2, E_{12}\}$ of the edges of (G_0, ψ) incident with $\tilde{v}$, where E_{12} is the set of loops at $\tilde{v}$, a ***split*** of (G_0, ψ) is a quotient Γ-gain graph (G_0', ψ) obtained from (G_0, ψ) by splitting $\tilde{v}$ into two vertices $\tilde{v}_1$ and $\tilde{v}_2$ so that

$\tilde{v}_i$ is incident to the edges in E_i for $i = 1, 2$, and the loops in E_{12} are replaced by directed edges from $\tilde{v}_1$ to $\tilde{v}_2$, without changing any gains. (By the definition of a quotient Γ-gain graph, the gain of a loop is freely invertible, so we may choose the original gain or its inverse for any edge replacing a loop.) A connected subset F of (G_0, ψ) is near-balanced if it is unbalanced and there is a split of (G_0, ψ) in which F becomes a balanced set. See also Figure 62.2.1.

SYMMETRY-ADAPTED COUNTING RULES

Using methods from group representation theory, the rigidity matrix of a Γ-symmetric framework $G(p)$ can be transformed into a block-diagonalized form [KG00, Sch09, Sch10a]. This is a fundamental result, as it can be used to break up the rigidity analysis of a symmetric framework into a number of independent subproblems, one for each block of the rigidity matrix. The block-diagonalization of the rigidity matrix is obtained by showing that it intertwines two representations of the group Γ associated with the edges and vertices of the graph G (also known as the external and internal representation in the engineering community).

THEOREM 62.2.1   Intertwining Property of the Rigidity Matrix

Let $G(p)$ be a Γ-symmetric framework with respect to $\theta : \Gamma \to \mathrm{Aut}(G)$ and $\tau : \Gamma \to O(\mathbb{R}^d)$. Then the rigidity matrix of $G(p)$, $R_G(p)$, lies in $\mathrm{Hom}_\Gamma(\tau \otimes P_V, P_E)$.

By Theorem 62.2.1 and Schur's lemma, there exist invertible matrices S and T such that $T^\top R_G(p)S$ is block-diagonalized. More precisely, if $\rho_0, \ldots, \rho_r$ are the irreducible representations of Γ, then for an appropriate choice of symmetry-adapted bases, the rigidity matrix takes on the following block form

$$T^\top R_G(p)S := \widetilde{R}_G(p) = \begin{pmatrix} \widetilde{R}_0(G(p)) & & \mathbf{0} \\ & \ddots & \\ \mathbf{0} & & \widetilde{R}_r(G(p)) \end{pmatrix},$$

where the submatrix block $\widetilde{R}_i(G(p))$ corresponds to the irreducible representation ρ_i of Γ. This block-diagonalization of the rigidity matrix corresponds to a decomposition $\mathbb{R}^{d|V|} = X_0 \oplus \cdots \oplus X_r$ of the space $\mathbb{R}^{d|V|}$ into a direct sum of $(\tau \otimes P_V)$-invariant subspaces X_i, and a decomposition $\mathbb{R}^{|E|} = Y_0 \oplus \cdots \oplus Y_r$ of the space $\mathbb{R}^{|E|}$ into a direct sum of P_E-invariant subspaces Y_i. The spaces X_i and Y_i are associated with ρ_i, and the submatrix $\widetilde{R}_i(G(p))$ is of size $\dim (Y_i) \times \dim (X_i)$.

Note that the submatrix block $\widetilde{R}_0(G(p))$ which corresponds to the trivial irreducible representation ρ_0 (with $\rho_0(\gamma) = 1$ for all $\gamma \in \Gamma$) is equivalent to the orbit rigidity matrix discussed in the previous section. The entries of the orbit rigidity matrix can be written down explicitly (see Section 62.1) without using any methods from group representation theory.

THEOREM 62.2.2 $(\tau \otimes P_V)$-Invariance of the Trivial Flex Space

Let $G(p)$ be a Γ-symmetric framework with respect to $\theta : \Gamma \to \mathrm{Aut}(G)$ and $\tau : \Gamma \to O(\mathbb{R}^d)$. Then the space of trivial first-order flexes $\mathcal{T}(G, p)$ of $G(p)$ is a $(\tau \otimes P_V)$-invariant subspace of $\mathbb{R}^{d|V|}$.

We denote by $(\tau \otimes P_V)^{(\mathcal{T})}$ the subrepresentation of $\tau \otimes P_V$ with representation space $\mathcal{T}(G, p)$. The space $\mathcal{T}(G, p)$ may now also be written as a direct sum $\mathcal{T} = T_0 \oplus \cdots \oplus T_r$ of $(\tau \otimes P_V)$-invariant subspaces, and for each $i = 1, \ldots, r$, we obtain the necessary condition dim $(Y_i) = $ dim $(X_i) - $ dim (T_i) for a Γ-symmetric framework to be isostatic. Using basic results from character theory, these conditions can be written in a more succinct form as follows [FG00, OP10, Sch09, Sch10a].

THEOREM 62.2.3 Symmetry-Extended Necessary Counting Conditions

Let $G(p)$ be an isostatic framework which is Γ-symmetric with respect to θ and τ. Then the following character equation holds.

$$\chi(P_E) = \chi(\tau \otimes P_V) - \chi((\tau \otimes P_V)^{(\mathcal{T})}).$$

It is well known from group representation theory that the character of any representation of Γ can be written uniquely as the linear combination of the characters of the irreducible representations of Γ. So suppose that $\chi(P_E) = \alpha_0 \chi(\rho_0) + \cdots + \alpha_r \chi(\rho_r)$, and $\chi(\tau \otimes P_V) - \chi((\tau \otimes P_V)^{(\mathcal{T})}) = \beta_0 \chi(\rho_0) + \cdots + \beta_r \chi(\rho_r)$, where $\alpha_i \beta_i \in \mathbb{N} \cup \{0\}$ for all $i = 0, \ldots, r$. If a Γ-symmetric framework $G(p)$ is not isostatic, then it follows from Theorem 62.2.3 that $\alpha_i \neq \beta_i$ for some i. If $\alpha_i < \beta_i$, then $G(p)$ has a non-trivial "ρ_i-symmetric" first-order flex belonging to the space X_i, and if $\alpha_i > \beta_i$, then $G(p)$ has a non-zero "ρ_i-symmetric" self-stress belonging to the space Y_i.

Consider, for example, the framework $G(p)$ with half-turn symmetry $\mathcal{C}_2$ in Figure 62.2.2 (b). We have $\chi(P_E) = (9, 3)$, $\chi(P_V) = (6, 0)$, $\chi(\tau) = (2, -2)$, and $\chi(\tau \otimes P_V) = (12, 0)$. Moreover, $\chi((\tau \otimes P_V)^{(\mathcal{T})}) = (3, -1)$ (see Table 62.2.1). Thus,

$$\chi(P_E) = (9, 3) \neq (9, 1) = (12, 0) - (3, -1) = \chi(\tau \otimes P_V) - \chi((\tau \otimes P_V)^{(\mathcal{T})}),$$

and hence, by Theorem 62.2.3, $G(p)$ is not isostatic. Let ρ_0 be the trivial ("fully-symmetric") irreducible representation of $\mathcal{C}_2$ (which assigns 1 to both the identity and the half-turn), and let ρ_1 be the non-trivial ("anti-symmetric") irreducible representation of $\mathcal{C}_2$ (which assigns 1 to the identity and -1 to the half-turn). Then we have $(9, 3) = 6\rho_0 + 3\rho_1$ and $(9, 1) = 5\rho_0 + 4\rho_1$, and hence we may conclude that $G(p)$ has an anti-symmetric first-order flex and a fully-symmetric self-stress.

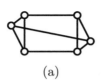 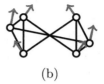

(a)   (b)

FIGURE 62.2.2
Realizations of the triangular prism graph with half-turn symmetry in the plane. The framework in (a) is isostatic, whereas the framework in (b) is first-order flexible, as detected by Theorem 62.2.3.

The equation in Theorem 62.2.3 comprises of one equation for each $\gamma \in \Gamma$. If we consider each of these equations independently, then we may obtain very simple necessary conditions for a Γ-symmetric framework $G(p)$ to be isostatic in terms of the number of vertices and edges of G that are fixed by the elements of Γ [CFG$^+$09].

THEOREM 62.2.4   Conditions for Individual Group Elements

Let $G(p)$ be an isostatic framework which is Γ-symmetric with respect to θ and τ, and let $|V_\gamma|$ and $|E_\gamma|$ denote the number of vertices and edges of G that are fixed by γ, respectively. Then, for every $\gamma \in \Gamma$, we have

$$|E_\gamma| = \text{trace}(\tau(\gamma)) \cdot |V_\gamma| - \text{trace}((\tau \otimes P_V)^{(\mathcal{T})}(\gamma)).$$

By considering standard bases for the spaces of first-order translations and rotations, the numbers $\text{trace}((\tau \otimes P_V)^{(\mathcal{T})}(\gamma))$, $\gamma \in \Gamma$, can easily be computed for any symmetry group $\tau(\Gamma)$ [Sch09]. The calculations of characters for the symmetry-extended counting rule for isostatic frameworks in the plane, for example, are shown in Table 62.2.1. In this table we again use the Schoenflies notation for symmetric structures; in particular, the symbols s and C_n denote a reflection and a rotation by $2\pi/n$, respectively.

TABLE 62.2.1   Calculations of characters for the symmetry-extended counting rule for isostatic frameworks in the plane.

	Id	$C_n, n > 2$	C_2	s
$\chi(P_E)$	$|E|$	$|E_{C_n}|$	$|E_{C_2}|$	$|E_s|$
$\chi(\tau \otimes P_V)$	$2|V|$	$(2\cos\frac{2\pi}{n})|V_{C_n}|$	$-2|V_{C_2}|$	0
$\chi((\tau \otimes P_V)^{(\mathcal{T})})$	3	$2\cos\frac{2\pi}{n} + 1$	-1	-1

For the identity element of Γ, Theorem 62.2.4 simply recovers the standard non-symmetric count. For the non-trivial symmetry operations, however, we obtain additional necessary conditions. In particular, for isostatic frameworks in the plane, we have the following result.

THEOREM 62.2.5   Restrictions on Fixed Structural Elements in the Plane

Let $G(p)$ be an isostatic framework which is Γ-symmetric with respect to θ and τ. Then the following hold.

(a) *If $C_2 \in \tau(\Gamma)$, then $|V_{C_2}| = 0$ and $|E_{C_2}| = 1$;*

(b) *if $C_3 \in \tau(\Gamma)$, then $|V_{C_3}| = 0$;*

(c) *if $s \in \tau(\Gamma)$, then $|E_s| = 1$;*

(d) *there does not exist a rotation $C_n \in \tau(\Gamma)$ with $n > 3$.*

By Theorem 62.2.5 there are only 5 non-trivial symmetry groups which allow an isostatic framework in the plane, namely the rotational groups $\mathcal{C}_2$ and $\mathcal{C}_3$, the reflectional group $\mathcal{C}_s$, and the dihedral groups $\mathcal{C}_{2v}$ and $\mathcal{C}_{3v}$ of order 4 and 6.

For 3-dimensional frameworks, all symmetry groups are possible. In fact, there exist infinite families of triangulated convex polyhedra for every symmetry group in 3-space, and the 1-skeleta of these structures are isostatic by Cauchy-Dehn's rigidity theorem (recall Chapter 61). However, restrictions to the placement of structural components still apply.

Analogous symmetry-extended counting rules have also been established for various other types of geometric constraint systems, such as body-bar frameworks

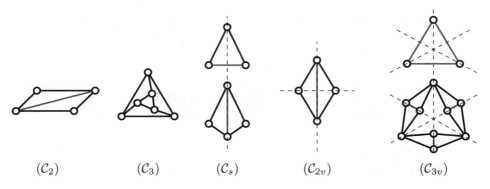

$(\mathcal{C}_2)$ $\qquad\qquad$ $(\mathcal{C}_3)$ $\qquad\qquad$ $(\mathcal{C}_s)$ $\qquad\qquad$ $(\mathcal{C}_{2v})$ $\qquad\qquad$ $(\mathcal{C}_{3v})$

FIGURE 62.2.3
Examples of symmetric isostatic frameworks in the plane, where fixed edges are shown in gray colour.

[GSW10], body-hinge frameworks [GF05, SFG14], and infinite periodic frameworks [GF14], as well as for frameworks in non-Euclidean normed spaces [KiS15]. Moreover, similar methods have recently also been applied to analyze liftings of symmetric pictures to polyhedral scenes [KaS17].

COMBINATORIAL RESULTS

It is easy to see that the set of all inc-Γ-regular realizations of a graph G (as a barjoint framework) forms a dense open subset of all Γ-symmetric realizations of G, and that all inc-Γ-regular realizations share the same first-order rigidity properties. It was conjectured in [CFG$^+$09] that for the five non-trivial symmetry groups which allow isostatic frameworks in the plane, the standard Laman counts, together with the additional conditions in Theorem 62.2.4, are also sufficient for inc-Γ-regular realizations of G to be isostatic. This conjecture has been proved for the groups $\mathcal{C}_2$, $\mathcal{C}_3$, and $\mathcal{C}_s$ [Sch10b, Sch10c], but it remains open for the dihedral groups.

THEOREM 62.2.6   Symmetric Laman's Theorem

Let $G(p)$ be an inc-Γ-regular framework with respect to $\theta : \Gamma \to \mathrm{Aut}(G)$ and $\tau : \Gamma \to O(\mathbb{R}^2)$. Then $G(p)$ is isostatic if and only if $|E| = 2|V| - 3$ and for every subgraph (V', E') with $|V'| \geq 2$ vertices, $|E'| \leq 2|V'| - 3$ (Laman's conditions), and

(a) *for $\tau(\Gamma) = \mathcal{C}_2$, we have $|V_{C_2}| = 0$ and $|E_{C_2}| = 1$;*

(b) *for $\tau(\Gamma) = \mathcal{C}_3$, we have $|V_{C_3}| = 0$;*

(c) *for $\tau(\Gamma) = \mathcal{C}_s$, we have $|E_s| = 1$.*

There also exist alternative characterizations for inc-Γ-regular isostaticity for these three groups. These are given in terms of symmetric Henneberg-type inductive construction sequences and in terms of symmetric 3Tree2 partitions (recall Section 61.1.2). See [Sch10b, Sch10c] for details.

Since an isostatic Γ-symmetric framework must obey certain restrictions on the number of vertices and edges that are fixed by the various elements in Γ, a first-order rigid Γ-symmetric framework usually does not contain an isostatic Γ-symmetric subframework on the same vertex set (see the frameworks in Figure 62.2.4, for example). Consequently, Theorem 62.2.6 can in general not be used to decide whether a given inc-Γ-regular framework $G(p)$ is first-order rigid.

 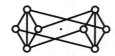

FIGURE 62.2.4
First-order rigid Γ-symmetric frameworks in $\mathbb{R}^2$ with respective symmetry groups $\mathcal{C}_3$, $\mathcal{C}_2$, and $\mathcal{C}_s$, which do not contain a Γ-symmetric isostatic subframework on the same vertex set.

However, this problem may be solved by analyzing each of the block matrices of the block-decomposed rigidity matrix $\widetilde{R}_G(p)$. Clearly, if for every $i = 0, \ldots, r$, $G(p)$ does not have any ρ_i-symmetric non-trivial first-order flex, then the block-diagonalization of $\widetilde{R}_G(p)$ and the corresponding decomposition of the infinitesimal flex space guarantees that $G(p)$ is first-order rigid. To make each of the block-matrices $\widetilde{R}_i(G(p))$ combinatorially accessible, an equivalent ρ_i-symmetric orbit rigidity matrix has recently been described in explicit form for each i [ST15]. This approach has provided complete combinatorial characterizations of first-order rigid inc-Γ-regular frameworks for a variety of groups, both in the plane and in higher dimensions. These results are important, because symmetric first-order rigid (rather than isostatic) frameworks are ubiquitous in human designs, and in some natural settings, such as proteins.

THEOREM 62.2.7   Inc-$\mathbb{Z}_2$-Regular First-Order Rigidity in the Plane

Let $G(p)$ be an inc-$\mathbb{Z}_2$-regular framework with respect to the free action $\theta : \mathbb{Z}_2 \to \mathrm{Aut}(G)$ and $\tau : \mathbb{Z}_2 \to O(\mathbb{R}^2)$, where $\tau(\mathbb{Z}_2) = \mathcal{C}_2$ or $\mathcal{C}_s$. Then $G(p)$ is first-order rigid if and only if the quotient Γ-gain graph of G contains a spanning $(2, 3, i)$-gain-tight subgraph (H_i, ψ_i) for each $i = 1, 2$.

For the group $\mathbb{Z}_3$, first-order rigid inc-$\mathbb{Z}_3$-regular frameworks in the plane (where the action $\theta : \mathbb{Z}_3 \to \mathrm{Aut}(G)$ is free) can be characterized in terms of spanning isostatic $\mathbb{Z}_3$-symmetric subframeworks using Theorem 62.2.6(c).

For any cyclic group $\mathbb{Z}_k$ of order $k \geq 4$, there always exists a non-trivial irreducible representation ρ_i for which there does not exist any ρ_i-symmetric trivial first-order flex. Moreover, the ρ_0-symmetric trivial first-order flex space is of dimension 1 for all of these groups (recall Theorem 62.1.5). Therefore, for $k \geq 4$, the quotient $\mathbb{Z}_k$-gain graph (G_0, ψ) of a $\mathbb{Z}_k$-symmetric first-order rigid framework $G(p)$ in the plane (where $\theta : \mathbb{Z}_k \to \mathrm{Aut}(G)$ is free) must contain a spanning $(2, 3, 0)$-gain-tight subgraph and a spanning $(2, 3, 1)$-gain-tight subgraph. Further necessary conditions were established in [Ike15, IT15, IT13, ST15]. In particular, the edge set of the $(2, 3, 0)$-gain-tight subgraph cannot be near-balanced (see Figure 62.2.1).

It was also shown in [Ike15, IT13] that for an inc-$\mathbb{Z}_k$-regular framework $G(p)$, where $k \geq 6$, the existence of a spanning generically 2-isostatic subgraph is not sufficient for $G(p)$ to be first-order rigid (see Figure 62.2.5). However, if for the graph G, we have $|E| = 2|V|$, and $k \geq 5$ is a prime number less than 1000, then $G(p)$ is in fact first-order rigid if and only if G contains a spanning generically 2-

isostatic subgraph. For complete combinatorial characterizations of inc-$\mathbb{Z}_k$-regular first-order rigid frameworks in the plane, where k is odd, see [Ike15, IT15, IT13].

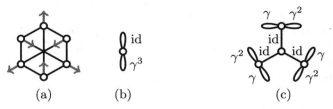

(a)   (b)   (c)

FIGURE 62.2.5
$\mathbb{Z}_6$-symmetric realizations of the generically 2-isostatic graph $K_{3,3}$ are first-order flexible (a); the quotient $\mathbb{Z}_6$-gain graph of $K_{3,3}$ is shown in (b), where γ denotes rotation by $2\pi/6$. (c) A quotient $\mathbb{Z}_k$-gain graph with $\mathbb{Z}_k = \langle\gamma\rangle$ whose covering graph is generically rigid without symmetry, but becomes first-order flexible for $\mathbb{Z}_k$-symmetric realizations, where $k \geq 7$. The directions of edges are omitted in the gain graphs.

For the remaining symmetry groups $\tau(\Gamma)$ in the plane, analogous combinatorial descriptions of inc-Γ-regular first-order rigid bar-joint frameworks have not yet been established. Due to the well-known difficulties for the non-symmetric case, there are also no extensions of these results to bar-joint frameworks in dimension $d \geq 3$. However, combinatorial characterizations for inc-Γ-regular first-order rigidity have been obtained for *body-bar* and *body-hinge* frameworks in arbitrary dimension for the groups $\mathbb{Z}_2 \times \cdots \times \mathbb{Z}_2$ [ST14]. These characterizations are given in terms of gain-sparsity counts as well as in terms of subgraph packing conditions for the corresponding quotient gain graphs. Since the statements of these theorems are fairly complex, we only provide a sample result for body-hinge frameworks in 3-space with half-turn symmetry, the most common symmetry found in proteins.

THEOREM 62.2.8 Inc-$\mathbb{Z}_2$-Regular Body-Hinge Frameworks in 3-Space

Let $G(h)$ be an inc-$\mathbb{Z}_2$-regular body-hinge framework in $\mathbb{R}^3$ with respect to the action $\theta : \mathbb{Z}_2 \to \mathrm{Aut}(G)$ which is free on the vertex set and the edge set of the multigraph G, and $\tau : \mathbb{Z}_2 \to O(\mathbb{R}^3)$, where $\tau(\mathbb{Z}_2) = \mathcal{C}_2$. Then $G(h)$ is first-order rigid if and only if the quotient $\mathbb{Z}_2$-gain graph of G contains six edge-disjoint subgraphs, two of which are spanning trees, and each of the other four has the property that each connected component contains exactly one cycle, which is unbalanced.

The results in [ST14] suggest that inc-Γ-regular body-bar and body-hinge realizations of the same multigraph share the same first-order rigidity properties. Moreover, it is conjectured that the following stronger version of Conjecture 62.1.8 for incidentally symmetric structures also holds [PRS$^+$14].

CONJECTURE 62.2.9 Symmetric Molecular Conjecture (Version II)

The rigidity matrix of an inc-Γ-regular body-bar realization of a multigraph G has the same rank as the rigidity matrix of an inc-Γ-regular molecular realization of G.

GEOMETRIC RESULTS

It follows immediately from Theorem 61.1.20 (Coning Theorem) that a Γ-symmetric framework $G(p)$ with respect to $\theta : \Gamma \to \mathrm{Aut}(G)$ and $\tau : \Gamma \to O(\mathbb{R}^d)$ is first-order rigid if and only if the Γ-symmetric cone framework $(G * u)(\overline{p}^*)$ with respect to $\theta^* : \Gamma \to \mathrm{Aut}(G * u)$ and $\tau^* : \Gamma \to O(\mathbb{R}^{d+1})$ (as defined in Section 62.1) is first-order rigid.

In particular, this allows us to transfer first-order rigidity and flexibility of incidentally symmetric frameworks between Euclidean, spherical, and hyperbolic space. Note, however, that for a nontrivial group Γ, a nontrivial first-order flex of an inc-Γ-regular framework usually does not extend to a continuous flex.

ALGORITHMS

Using simple modifications of the algorithms presented in Chapter 61, we can check whether inc-Γ-regular realizations of a graph in $\mathbb{R}^2$ are isostatic in polynomial time. For example, we may check Laman's conditions in $O(|V|^2)$ time and then check the additional symmetry conditions on the number of fixed structural components (see Theorem 62.2.6) in constant time. Alternatively, we may use an $O(|V|^2)$ matroid partition algorithm to check whether there exists a proper symmetric 3Tree2 partition.

To verify whether a Γ-symmetric graph is incidentally symmetric first-order rigid (or just forced Γ-symmetric rigid), we need to check the corresponding (k, ℓ, m)-gain-sparsity counts for its quotient Γ-gain graph (recall Theorems 62.1.5, 62.1.6 and 62.2.7, for example). This can be done in polynomial time as follows, provided that $0 \le \ell \le 2k - 1$ (see also [BHM+11, JKT16]).

In the first step, we may verify that the gain graph is $(k, 0)$-sparse (or (k, m)-sparse if $m > 0$) using a standard "pebble game algorithm" [BJ03, LS08]. In the second step, we then need to test whether every edge set violating the (k, ℓ)-sparsity count induces a certain subgroup of Γ (or whether it is near-balanced, if necessary). It suffices to test this for every circuit in the matroid induced by the (k, ℓ)-sparsity count, and these circuits can be enumerated in polynomial time (see [Sey94] e.g.).

As we will see in the next section (Section 62.3), this algorithm may also be used to decide the rigidity of infinite periodic frameworks.

Recall that for regular (finite) bar-joint frameworks in dimension $d \ge 3$, we do not have a polynomial time algorithm to test for rigidity. However, polynomial time algorithms do exist for deciding the rigidity of regular *body-bar* or *body-hinge* frameworks in d-space. The same is true in the symmetric situation, where we can check the conditions in Theorem 62.2.8, for example, in $O(|V_0|^{5/2}|E_0|)$ time via a matroid partition algorithm [ST14].

Finally, we note that for a given Γ-symmetric framework $G(p)$, there exists a randomized polynomial time algorithm to check whether $G(p)$ is Γ-regular or inc-Γ-regular.

62.3 PERIODIC FRAMEWORKS

GLOSSARY

Periodic graph: For a free abelian group Γ of rank d, a simple infinite graph $\tilde{G} = (\tilde{V}, \tilde{E})$ with finite degree at every vertex for which there exists a group action $\theta : \Gamma \to \mathrm{Aut}(\tilde{G})$ which is free on the vertex set of $\tilde{G}$ and such that the quotient graph $\tilde{G}/\Gamma$ is finite. (Note that Γ is isomorphic to $\mathbb{Z}^d$.)

L-periodic framework: For a periodic graph $\tilde{G}$ (with respect to $\theta : \Gamma \to \mathrm{Aut}(\tilde{G})$), the pair $(\tilde{G}, \tilde{p})$, also denoted by $\tilde{G}(\tilde{p})$, where $\tilde{p} : \tilde{V} \to \mathbb{R}^d$ is a map, $\mathcal{T}(\mathbb{R}^d)$ is the group of translations of $\mathbb{R}^d$ (which is identified with the space $\mathbb{R}^d$ of translation vectors), and $L : \Gamma \to \mathcal{T}(\mathbb{R}^d)$ is a faithful representation with the property that

$$\tilde{p}_i + L(\gamma) = \tilde{p}_{\theta(\gamma)(i)} \qquad \text{for all } \gamma \in \Gamma \text{ and all } i \in \tilde{V}.$$

Periodic first-order flex: Fix an isomorphism $\Gamma \simeq \mathbb{Z}^d$, and let $\tilde{G}(\tilde{p})$ be an L-periodic framework (with respect to $\theta : \Gamma \to \mathrm{Aut}(\tilde{G})$). Choose a set of representatives, $v_1, \ldots, v_a$, for the vertex orbits of $\tilde{G}$, and a set of representatives, $(v_i, \theta(\gamma_\beta)v_j)$, for the b edge orbits of $\tilde{G}$. Further, let $x_i = \tilde{p}(v_i)$ for each i, and let $\mu_k = L(\gamma_k)$, $k = 1, \ldots, d$, be the translation vectors in $\mathbb{R}^d$ which correspond to the standard basis $\gamma_1, \ldots, \gamma_d$ of Γ. Let $\gamma_\beta = \sum_{k=1}^d c_\beta^k \gamma_k$ for $c_\beta^k \in \mathbb{Z}$, and $\mu(\beta) = \sum_{k=1}^d c_\beta^k \mu_k$. A vector $(y_1, \ldots, y_a, \nu_1, \ldots, \nu_d) \in \mathbb{R}^{da+d^2}$ is a periodic first-order flex of $\tilde{G}(\tilde{p})$ if

$$\langle (x_j + \mu(\beta)) - x_i, (y_j + \nu(\beta)) - y_i \rangle = 0 \quad \text{for } \beta = 1, \ldots, b,$$

where $\langle \cdot, \cdot \rangle$ represents the inner product, and $\nu(\beta) = \sum_{k=1}^d c_\beta^k \nu_k$. We refer to the matrix corresponding to the linear system above as the **periodic rigidity matrix** of $\tilde{G}(\tilde{p})$.

First-order rigid periodic framework: An L-periodic framework $\tilde{G}(\tilde{p})$ for which every periodic first-order flex is trivial. A first-order rigid periodic framework with no redundant constraints is called **isostatic**.

$\mathbb{Z}^d$-regular: An L-periodic framework $\tilde{G}(\tilde{p})$ in $\mathbb{R}^d$ whose periodic rigidity matrix has maximal rank among all L'-periodic realizations $\tilde{G}(\tilde{p}')$ of $\tilde{G}$ in $\mathbb{R}^d$ (with any choice of L').

BASIC RESULTS

Rigidity analyses of infinite periodic frameworks have found numerous applications in both mathematics (in the theory of periodic packings, e.g.) and in other scientific areas such as crystallography, materials science, and engineering. Most of the theoretical work has focused on *forced-periodic* frameworks, i.e., periodic frameworks which must maintain the periodicity throughout their motions. In the following, we will therefore restrict our attention to this setting. However, the rigidity and flexibility of infinite structures is currently a highly active research area and new tools and methods for more general rigidity analyses of such structures are developing quickly. In particular, new insights into the rigidity and flexibility properties of periodic or crystallographic frameworks have recently been gained via novel applications of methods from real and functional analysis [OP11, Pow14a, Pow14b].

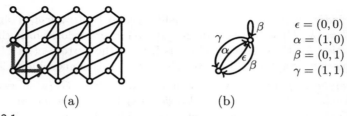

$$\epsilon = (0,0)$$
$$\alpha = (1,0)$$
$$\beta = (0,1)$$
$$\gamma = (1,1)$$

(a)   (b)

FIGURE 62.3.1
Part of a first-order rigid periodic framework in the plane (a) and its quotient $\mathbb{Z}^2$-gain graph (b).

The description of the first-order rigidity of periodic frameworks provided in the previous section is based on the approach taken in [BS10] (see also [GH03, Pow14a]). An alternative mathematical formulation is given in [Ros11, Ros14a]. The main difference between the two models is that in [BS10] a periodic framework is considered as a realization of an infinite graph with a periodic group action, whereas in [Ros11, Ros14a] it is considered as a realization of a finite graph within a fundamental domain of $\mathbb{R}^d$. Since in the latter model, the orientation of the periodic lattice is fixed, rotations have been eliminated from the space of trivial motions. In other words, realizations of finite graphs in a fundamental domain constitute equivalence classes of L-periodic frameworks $\tilde{G}(\tilde{p})$ under rotation.

When studying the first-order rigidity of periodic frameworks, we may allow different types of flexibility in the lattice representation of the periodicity group. The definitions provided in the previous section assume that the lattice representation is fully flexible. While this seems to be the most natural set-up, motions of periodic structures with a fixed lattice representation or with various types of partially flexible lattice representations are also of interest in applications. In particular, these results are useful to analyze motions of structures at shorter time scales.

The following basic result is the analog of Theorem 62.1.1. While it is stated for periodic frameworks with a fully flexible lattice representation, it also holds for other types of lattice flexibility, including the fixed lattice [MT13, Ros11].

THEOREM 62.3.1 $\mathbb{Z}^d$-Regular Rigidity Theorem

A $\mathbb{Z}^d$-regular L-periodic framework $\tilde{G}(\tilde{p})$ has a non-trivial periodic first-order flex if and only if $\tilde{G}(\tilde{p})$ has a non-trivial continuous flex which preserves the periodicity throughout the motion.

Using the periodic rigidity matrix (and its counterparts for the fixed or the partially flexible lattice representations), we may immediately derive some basic necessary counting conditions for a periodic framework in d-space to be first-order rigid. The following theorem states these conditions for the fully flexible and the fixed lattice representation.

THEOREM 62.3.2   Necessary Counting Conditions for Periodic Rigidity

Let $\tilde{G}(\tilde{p})$ be an L-periodic isostatic framework in d-space. Then the quotient graph $\tilde{G}_0$ of $\tilde{G}$ satisfies

(a) *for $L(\mathbb{Z}^d)$ fully flexible:* $|\tilde{E}_0| = d|\tilde{V}_0| + \binom{d}{2}$;

(b) *for $L(\mathbb{Z}^d)$ fixed:* $|\tilde{E}_0| = d|\tilde{V}_0| - d$.

TABLE 62.3.1   Types of lattice deformation in 2- and 3-space.

Lattice Deformation	l (for d=2)	l (for d=3)
flexible	3	6
distortional	2	5
scaling	2	3
hydrostatic	1	1
fixed	0	0

More generally, for an L-periodic framework $\tilde{G}(\tilde{p})$ in d-space to be isostatic, its quotient graph must satisfy the count $|\tilde{E}_0| = d|\tilde{V}_0| + l - d$, where l is the dimension of the space of permissible lattice deformations. Table 62.3.1 summarizes these counts for some fundamental types of lattice deformations in dimensions 2 and 3. A "distortional change" in the lattice keeps the volume of the fundamental domain fixed but allows the shape of the lattice to change, a "scaling change" keeps the angles of the fundamental domain fixed but allows the scale of the translations to change independently, and finally, a "hydrostatic change" keeps the shape of the lattice unchanged but allows scalings to change the volume.

Analogous counts for various types of "crystallographic frameworks" (i.e., periodic frameworks with additional symmetry) can be found in [RSW11]. Note that in addition to these overall counts we may also derive further necessary counting conditions for first-order rigidity by considering all edge-induced subgraphs of the given quotient graph. However, as in the case of finite symmetric frameworks, these gain-sparsity counts are more complex, as we will see in the next section.

COMBINATORIAL RESULTS

For a given periodicity group Γ, combinatorial characterizations of first-order rigid periodic frameworks have been studied at multiple different levels. At the simplest level, we may ask whether a given quotient graph $\tilde{G}/\Gamma$ is the quotient graph of a first-order rigid periodic framework for *some* gain assignment of the edges of $\tilde{G}/\Gamma$. In other words, we seek a characterization for first-order rigidity up to generic liftings of edges from $\tilde{G}/\Gamma$ to a covering periodic graph. The following theorem summarizes the key results for this problem in the case where the lattice representation of Γ is either fully flexible or fixed [BS11, Whi88].

THEOREM 62.3.3   Periodic Rigidity for Generic Liftings

A quotient graph $\tilde{G}/\Gamma$ is the quotient graph of an isostatic L-periodic framework $\tilde{G}(\tilde{p})$ in d-space for some gain assignment of the edges of $\tilde{G}/\Gamma$ if and only if

(a) *for $L(\mathbb{Z}^d)$ fully flexible: $\tilde{G}/\Gamma$ satisfies $|\tilde{E}_0| = d|\tilde{V}_0| + \binom{d}{2}$ and contains a spanning subgraph with $d|\tilde{V}_0| - d$ edges which has the property that every subgraph with m edges and n vertices satisfies $m \leq dn - d$.*

(b) *for $L(\mathbb{Z}^d)$ fixed: $\tilde{G}/\Gamma$ satisfies $|\tilde{E}_0| = d|\tilde{V}_0| - d$ and every subgraph of $\tilde{G}/\Gamma$ with m edges and n vertices satisfies $m \leq dn - d$.*

Note that the condition in Theorem 62.3.3(b) is equivalent to the condition that $\tilde{G}/\Gamma$ is the union of d edge-disjoint spanning trees.

For the fully flexible lattice, Theorem 62.3.3(a) has also been extended to body-bar frameworks [BST15].

THEOREM 62.3.4   Generic Liftings to Periodic Body-Bar Frameworks

A quotient graph $\tilde{G}/\Gamma$ with $|\tilde{E}_0| = \binom{d+1}{2}(|\tilde{V}_0| - 1) + d^2$ is the quotient graph of an isostatic periodic body-bar framework in d-space for some gain assignment of the edges of $\tilde{G}/\Gamma$ if and only if it satisfies one and hence both of the following equivalent conditions.

(a) *$\tilde{G}/\Gamma$ decomposes into the disjoint union of two spanning subgraphs, one with $d|\tilde{V}_0| - d$ edges and the property that every subgraph with m edges and n vertices satisfies $m \leq dn - d$, and the other one with $\binom{d}{2}|\tilde{V}_0| + \binom{d+1}{2}$ edges and the property that every subgraph with m edges and n vertices satisfies $m \leq \binom{d}{2}n + \binom{d}{2}$.*

(b) *$\tilde{G}/\Gamma$ contains the disjoint union of two spanning subgraphs, one with $d|\tilde{V}_0| - d$ edges and the property that every subgraph with m edges and n vertices satisfies $m \leq dn - d$, and the other one with $\binom{d}{2}|\tilde{V}_0|$ edges and the property that every subgraph with m edges and n vertices satisfies $m \leq \binom{d}{2}n$.*

Next we will consider combinatorial characterizations of first-order rigid periodic frameworks at a more discerning level. Specifically, for a given periodicity group Γ, we will study Γ-regular first-order rigidity by analyzing *quotient Γ-gain graphs*, i.e., quotient graphs that are equipped with an orientation and a gain labeling of the edges, rather than just quotient graphs. In other words, the gain assignment of the edges, and hence the lifting of the edges from the quotient graph to the covering periodic graph is now part of the initial data. Note that the formal definition of a quotient Γ-gain graph is completely analogous to the one given in Section 62.1 for finite symmetric frameworks (see also Figure 62.3.1). We begin by summarizing the main results for bar-joint frameworks in the plane. For the fully flexible lattice representation, the following result was obtained using periodic direction networks [MT13].

THEOREM 62.3.5   Periodic Rigidity for the Fully Flexible Lattice in the Plane

Let $\tilde{G}(\tilde{p})$ be a $\mathbb{Z}^2$-regular L-periodic framework in $\mathbb{R}^2$. Then $\tilde{G}(\tilde{p})$ is isostatic if and only if the quotient $\mathbb{Z}^2$-gain graph of $\tilde{G}$ satisfies

(a) *$|\tilde{E}_0| = 2|\tilde{V}_0| + 1$;*

(b) *$|F| \leq 2|V(F)| - 3 + 2k(F) - 2(c(F) - 1)$ for all non-empty $F \subseteq \tilde{E}_0$,*

where $k(F)$ is the $\mathbb{Z}^2$-rank of F, i.e., the rank of the subgroup of $\mathbb{Z}^2$ induced by F (which is defined analogously as for finite groups (see Section 62.1), and $c(F)$ is the number of connected components of the subgraph induced by F.

Using a Henneberg-type recursive construction sequence for quotient $\mathbb{Z}^2$-gain graphs, the following analogous result for the fixed lattice was established in [Ros15].

THEOREM 62.3.6   Periodic Rigidity for the Fixed Lattice in the Plane

Let $\tilde{G}(\tilde{p})$ be a $\mathbb{Z}^2$-regular L-periodic framework in $\mathbb{R}^2$, where $L(\mathbb{Z}^2)$ is non-singular and has to remain fixed. Then $\tilde{G}(\tilde{p})$ is isostatic if and only if the quotient $\mathbb{Z}^2$-gain graph of $\tilde{G}$ satisfies

(a) $|\tilde{E}_0| = 2|\tilde{V}_0| - 2$;

(b) $|F| \leq 2|V(F)| - 3$ *for all non-empty $F \subseteq \tilde{E}_0$ with $\mathbb{Z}^2$-rank equal to 0;*

(c) $|F| \leq 2|V(F)| - 2$ *for all non-empty $F \subseteq \tilde{E}_0$.*

Similar results also exist for periodic frameworks with partially flexible lattice representations (such as lattice representations with one degree of freedom [NR15] or with a fixed angle or fixed area fundamental domain [MT14a]).

There also exist various extensions of these results to crystallographic frameworks which must maintain the full crystallographic symmetry throughout any motion. In particular, for crystallographic groups Γ generated by translations and rotations in the plane, combinatorial characterizations for forced Γ-symmetric first-order rigidity (with a fully flexible lattice representation) are presented in [MT14b]. To obtain these results, one needs to carefully keep track of the types of lattice flexibility that are compatible with the subgroups of Γ induced by the various edge sets.

For crystallographic *body-bar frameworks* with a *fixed* lattice representation in an arbitrary-dimensional Euclidean space, combinatorial characterizations for forced Γ-symmetric first-order rigidity were established in [Ros14b, Tan15]. Analogous to Theorem 62.1.7, these characterizations are given in terms of gain-sparsity counts for the underlying multigraphs. In particular, for *periodic* body-bar frameworks, we have the following result.

THEOREM 62.3.7   Periodic Fixed-Lattice Body-Bar Frameworks in d-Space

Let $\tilde{G}(\tilde{b})$ be a $\mathbb{Z}^d$-regular L-periodic body-bar framework $\mathbb{R}^d$, where $L(\mathbb{Z}^d)$ is non-singular and has to remain fixed. Then $\tilde{G}(\tilde{b})$ is isostatic if and only if the quotient $\mathbb{Z}^d$-gain graph of $\tilde{G}$ satisfies

(a) $|\tilde{E}_0| = \binom{d+1}{2}|\tilde{V}_0| - d$;

(b) $|F| \leq \binom{d+1}{2}|V(F)| - \binom{d+1}{2} + \sum_{i=1}^{k(F)}(d-i)$ *for all non-empty $F \subseteq \tilde{E}_0$,*

where $k(F)$ is the $\mathbb{Z}^d$-rank of F, i.e., the rank of the subgroup of $\mathbb{Z}^d$ induced by F.

Extensions of this result to periodic body-hinge or molecular structures have not yet been investigated.

In order to gain an understanding of "incidentally periodic" first-order rigid frameworks, some recent work has also investigated the problem of characterizing periodic frameworks which are first-order rigid *for any choice of the periodicity lattice.* Such frameworks are called "ultra-rigid." An algebraic characterization for ultra-rigidity in arbitrary-dimensional Euclidean space has been obtained in [MT14c]. For the special case when the number of edge orbits is as small as possible for ultra-rigidity in dimension 2, a combinatorial characterization is also given in [MT14c]. All of these results apply to both the fully flexible and the fixed lattice representation.

Finally, when we handle and analyze a real crystal, it is finite. It is a *fragment* that might embed into a theoretical, infinite periodic structure. A recent paper [Whi14] examines when the rigidity, or flexibility, of a sufficiently large fragment will extend to some, or all, of the infinite periodic structures into which the fragment might be embedded.

62.4 SOURCES AND RELATED MATERIALS

BASIC SOURCES

The following are some key references for the various aspects of the study of rigid and flexible structures with symmetry:

[SW11]: A basic introduction to forced-symmetric rigidity and the orbit rigidity matrix.

[JKT16]: A key source for basic properties of gain graphs and combinatorial characterizations for forced-symmetric rigidity in the plane.

[SW12]: A description of the transfer of forced-symmetric rigidity properties between Euclidean and spherical space (as well as other Cayley-Klein metrics).

[CG16]: A new textbook on rigidity theory which contains a thorough discussion of the first-order rigidity analysis of incidentally symmetric frameworks.

[FG00]: A (somewhat informal) description of the symmetry-extended necessary counting conditions for a symmetric bar-joint framework to be isostatic. A rigorous mathematical treatment of this theory can be found in [Sch10a], for example.

[Sch10b]: An article on combinatorial characterizations of (incidentally) symmetry-regular isostatic bar-joint frameworks. In particular, it establishes a symmetry-extended Laman's theorem for the simplest group in the plane (three-fold rotational symmetry).

[ST14]: An introduction to "phase-symmetric orbit rigidity matrices" and the theory of incidentally symmetric first-order rigid frameworks.

[BS10]: A basic introduction to the rigidity and flexibility of periodic frameworks. See [Ros11] for an alternative approach.

[MT13]: A detailed description of combinatorial characterizations of $\mathbb{Z}^2$-regular rigid periodic frameworks.

RELATED CHAPTERS

REFERENCES

[AH94]   S.L. Altmann and P. Herzig. *Point-Group Theory Tables*. Clarendon Press, Oxford, 1994.

[ACP70]   P.W. Atkins, M.S. Child and C.S.G. Phillips. *Tables for Group Theory*. Oxford University Press, 1970.

[BHM⁺11]   M. Berardi, B. Heeringa, J. Malestein and L. Theran. Rigid components in fixed-lattice and cone frameworks. In *Proc. 23rd Canad. Conf. Comput. Geom.*, pages 1–6, 2011.

[BJ03]   A.R. Berg and T. Jordán. Algorithms for graph rigidity and scene analysis. In *Proc. 11th Eur. Sympos. Algorithms*, vol. 2832 of *LNCS*, pages 78–89, Springer, Berlin, 2003.

[BS10]   C. Borcea and I. Streinu. Periodic frameworks and flexibility. *Proc. Royal Soc. A*, 466:2633–2649, 2010.

[BS11]   C. Borcea and I. Streinu. Minimally rigid periodic graphs. *Bull. Lond. Math. Soc.*, 43:1093–1103, 2011.

[BST15]   C. Borcea, I. Streinu, and S. Tanigawa. Periodic body-and-bar frameworks. *SIAM J. Discrete Math.*, 29:93–112, 2015.

[CFG⁺09]   R. Connelly, P.W. Fowler, S.D. Guest, B. Schulze and W. Whiteley. When is a symmetric pin-jointed framework isostatic? *Internat. J. Solids Structures*, 46:762–773, 2009.

[CG16]   R. Connelly and S.D. Guest. *Frameworks, Tensegrities and Symmetry: Understanding Stable Structures*. Cambridge University Press, to appear, 2016.

[FG00]   P.W. Fowler and S.D. Guest. A symmetry extension of Maxwell's rule for rigidity of frames. *Internat. J. Solids Structures*, 37:1793–1804, 2000.

[GF05]   S.D. Guest and P.W. Fowler. A symmetry-extended mobility rule. *Mechanism and Machine Theory*, 40:1002–1014, 2005.

[GF07]   S.D. Guest and P.W. Fowler. Symmetry conditions and finite mechanisms. *Mechanics of Materials and Structures*, 2(2), 2007.

[GF14]   S.D. Guest and P.W. Fowler. Symmetry-extended counting rules for periodic frameworks. *Phil. Trans. Royal Soc. A*, 372:20120029, 2014.

[GH03]   S.D. Guest and J.W. Hutchinson. On the determinacy of repetitive structures. *J. Mechanics and Physics of Solids*, 51:383–391, 2003.

[GSW10]   S.D. Guest, B. Schulze, and W. Whiteley. When is a symmetric body-bar structure isostatic? *Internat. J. Solids Structures*, 47:2745–2754, 2010.

[Ike15]   R. Ikeshita. *Infinitesimal Rigidity of Symmetric Frameworks*. Master's thesis, Department of Mathematical Informatics, University of Tokyo, 2015.

[IT13]   R. Ikeshita and S. Tanigawa. Inductive constructions of sparse group-labeled graphs with applications to rigidity. In *25th Workshop on Topological Graph Theory*, Yokohama, 2013.

[IT15]   R. Ikeshita and S. Tanigawa. Count matroids on group-labeled graphs. Preprint, arXiv:1507.01259, 2015.

[JKT16]   T. Jordán, V.E. Kaszanitzky, and S. Tanigawa. Gain-sparsity and symmetry-forced rigidity in the plane. *Discrete Comput. Geom.*, 55:314–372, 2016.

[KG00]   R.D. Kangwai and S.D. Guest. Symmetry-adapted equilibrium matrices. *Internat. J. Solids Structures*, 37:1525–1548, 2000.

[KaS17]   V. Kaszanitzky and B. Schulze. Lifting symmetric pictures to polyhedral scenes. *Ars. Math. Contemp.*, 13:31–47, 2017.

[KiS15]   D. Kitson and B. Schulze. Maxwell-Laman counts for bar-joint frameworks in normed spaces. *Linear Algebra Appl.*, 481:313–329, 2015.

[LS08]   A. Lee and I. Streinu. Pebble game algorithms and sparse graphs. *Discrete Math.*, 308:1425–1437, 2008.

[MT11]   J. Malestein and L. Theran. Generic rigidity of frameworks with orientation-preserving crystallographic symmetry. Preprint, arXiv:1108.2518, 2011.

[MT12]   J. Malestein and L. Theran. Generic rigidity of reflection frameworks. Preprint, arXiv:1203.2276, 2012.

[MT13]   J. Malestein and L. Theran. Generic combinatorial rigidity of periodic frameworks. *Adv. Math.*, 233:291–331, 2013.

[MT14a]   J. Malestein and L. Theran. Generic rigidity with forced symmetry and sparse colored graphs. In R. Connelly, A.I. Weiss, and W. Whiteley, *Rigidity and Symmetry*, vol. 70 of *Fields Inst. Comm.*, Springer, New York, 2014.

[MT14b]   J. Malestein and L. Theran. Frameworks with forced symmetry II: orientation-preserving crystallographic groups. *Geometriae Dedicata*, 170:219–262, 2014.

[MT14c]   J. Malestein and L. Theran. Ultrarigid periodic frameworks. Preprint, arXiv:1404.2319, 2014.

[MT15]   J. Malestein and L. Theran. Frameworks with forced symmetry I: reflections and rotations. *Discrete Comput. Geom.*, 54:339–367, 2015.

[NR15]   A. Nixon and E. Ross Periodic rigidity on a variable torus using inductive constructions. *Electron. J. Combin.*, 22:P1, 2015.

[OP10]   J.C. Owen and S.C. Power. Frameworks, symmetry and rigidity. *Internat. J. Comput. Geom. Appl.*, 20:723–750, 2010.

[OP11]   J.C. Owen and S.C. Power. Infinite bar-joint frameworks, crystals and operator theory. *New York J. Math.*, 17:445–490, 2011.

[PRS⁺14]   J. Porta, L. Ros, B. Schulze, A. Sljoka, and W. Whiteley. On the Symmetric Molecular Conjectures. *Computational Kinematics, Mechanisms and Machine Science*, 15:175–184, 2014

[Pow14a]   S.C. Power. Crystal frameworks, symmetry and affinely periodic flexes. *New York J. Math.*, 20:665–693, 2014.

[Pow14b]   S.C. Power. Polynomials for crystal frameworks and the rigid unit mode spectrum. *Phil. Trans. Royal Soc. A*, 372:20120030, 2014.

[Ros11]   E. Ross. *The Rigidity of Periodic Frameworks as Graphs on a Torus*. Ph.D. thesis, Department of Mathematics and Statistics, York University, 2011.

[Ros14a]   E. Ross. The rigidity of periodic frameworks as graphs on a fixed torus. *Contrib. Discrete Math.*, 9:11-45, 2014.

[Ros14b]   E. Ross. The rigidity of periodic body-bar frameworks on the three-dimensional fixed torus. *Phil. Trans. Royal Soc. A*, 372:20120112, 2014.

[Ros15]   E. Ross. Inductive constructions for frameworks on a two-dimensional fixed torus. *Discrete Comput. Geom.*, 54:78–109, 2015

[RSW11]   E. Ross, B. Schulze, and W. Whiteley. Finite motions from periodic frameworks with added symmetry. *Internat. J. Solids Structures*, 48:1711–1729, 2011

[Sch09]   B. Schulze. *Combinatorial and Geometric Rigidity with Symmetry Constraints*. Ph.D. thesis, Department of Mathematics and Statistics, York University, 2009.

[Sch10a]   B. Schulze. Block-diagonalized rigidity matrices of symmetric frameworks and applications. *Beiträge Algebra Geometrie*, 51:427–466, 2010.

[Sch10b] B. Schulze. Symmetric versions of Laman's Theorem. *Discrete Comput. Geom.*, 44:946–972, 2010.

[Sch10c] B. Schulze. Symmetric Laman theorems for the groups C_2 and C_s. *Electron. J. Combin.*, 17:R154, 2010.

[Sch10d] B. Schulze. Symmetry as a sufficient condition for a finite flex. *SIAM J. Discrete Math.*, 24:1291–1312, 2010.

[SFG14] B. Schulze, P.W. Fowler, and S.D. Guest. When is a symmetric body-hinge structure isostatic? *Internat. J. Solids Structures*, 51:2157–2166, 2014.

[SSW14] B. Schulze, A. Sljoka, and W. Whiteley. How does symmetry impact the flexibility of proteins? *Phil. Trans. Royal Soc. A*, 372:2012004, 2014.

[ST14] B. Schulze and S. Tanigawa. Linking rigid bodies symmetrically. *European J. Combin.*, 42:145–166, 2014.

[ST15] B. Schulze and S. Tanigawa. Infinitesimal rigidity of symmetric frameworks. *SIAM J. Discrete Math.*, 29:1259–1286, 2015.

[SW11] B. Schulze and W. Whiteley. The orbit rigidity matrix of a symmetric framework. *Discrete Comput. Geom.*, 46:561–598, 2011.

[SW12] B. Schulze and W. Whiteley. Coning, symmetry, and spherical frameworks. *Discrete Comput. Geom.*, 48:622–657, 2012.

[Sey94] P.D. Seymour. A note on hyperplane generation. *J. Combin. Theory Ser. B*, 61:88–91, 1994.

[Tan15] S. Tanigawa. Matroids of gain graphs in applied discrete geometry. *Trans. AMS*, 367:8597–8641, 2015.

[Whi88] W. Whiteley. The union of matroids and the rigidity of frameworks. *SIAM J. Discrete Math.*, 1:237–255, 1988.

[Whi14] W. Whiteley. Fragmentary and incidental behavior of columns, slabs and crystals, *Phil. Trans. Royal Soc. A*, 372:20120032, 2014.

63 GLOBAL RIGIDITY

Tibor Jordán and Walter Whiteley

INTRODUCTION

Chapter 61 described the basic theory of infinitesimal rigidity of bar and joint structures and a number of related structures. In this chapter, we consider the stronger properties of:

(a) Global Rigidity: given a discrete configuration of points in Euclidean d-space, and a set of fixed pairwise distances, is the set of solutions unique, up to congruence in d-space?

(b) Universal Rigidity: given a discrete configuration of points in Euclidean d-space, and a set of fixed pairwise distances, is the set of solutions unique, up to congruence in all dimensions $d' \geq d$?

(c) How do global rigidity and universal rigidity depend on the combinatorial properties of the associated graph, in which the vertices and edges correspond to points and fixed pairwise distances, respectively, and how do they depend on the specific geometry of the initial configuration?

To study global rigidity, we use vocabulary and techniques drawn from (i) structural engineering: bars and joints, redundant first-order rigidity, static self-stresses (linear techniques); as well as from (ii) minima for energy functions with their companion stress matrices. There are both global rigidity theorems which hold for almost all realizations of a graph G based on combinatorial properties of the graph; and global rigidity theorems that hold for some specific realizations, and depend on the particular details of the geometry of (G, p).

Specifically, there are many globally rigid frameworks where the underlying graph is not generically globally rigid. It is computationally hard to test the global rigidity of such a particular framework.

For universal rigidity, we adapt the stress techniques from semidefinite programming. Even when universal rigidity occurs for some generic realizations of a graph, it may not occur for all such realizations, so it is not a generic property in the broad sense. However, it is weakly generic in the sense that a graph can have a full dimensional open subset of universally rigid realizations. Recent results show a strong connection: there exists a generic globally rigid realization of a graph if and only if there exists a generic universally rigid realization of that graph. So there is little difference in the combinatorics for generic realizations. However, a recent algorithm tests the universal rigidity of a given geometric framework (G, p), whereas no algorithm exists for testing global rigidity of a specific framework. Thus universal rigidity can be a valuable tool for confirming the global rigidity of a specific geometric framework (G, p).

Some results and techniques for universal rigidity were developed for tensegrity frameworks, where the bars with fixed lengths are replaced by cables (which can only become shorter) and struts (which can only become longer). These have a narrower set of stresses and relevant stress matrices, and provide some additional insights into the behaviour of real structures and the ways the techniques are applied.

Global rigidity has a significant range of applications, such as localization in sensor networks and molecular conformations. Some applications and extensions also involve variations of the structure. Work on global rigidity of symmetric structures (Chapter 62) is in its initial stages.

63.1 BASICS FOR GLOBAL AND UNIVERSAL RIGIDITY OF GRAPHS

Global rigidity results have both a combinatorial form, belonging to graphs, and a geometric form, depending on the special geometry of the realizations. The same two forms are found for universal rigidity.

63.1.1 BASICS FOR GLOBALLY RIGID GRAPHS

We begin with some basic results that apply to generic frameworks and therefore almost all frameworks on a given graph. As such, they can be presented in terms of the graphs.

GLOSSARY FOR GLOBAL AND UNIVERSAL RIGIDITY

Configuration of points in d-space: A map $p : V \to \mathbb{R}^d$ that assigns points $p_i \in \mathbb{R}^d$, $1 \le i \le n$, to an index set $V = \{1, 2, ..., n\}$.

Generic configuration: A configuration for which the set of the $d|V|$ coordinates of the points is algebraically independent over the rationals.

Congruent configurations: Two configurations p and q in d-space, on the same set V, related by an isometry T of $\mathbb{R}^d$ (with $T(p_i) = q_i$ for all $i \in V$).

Bar-and-joint framework in d-space (or **framework:**) A pair (G, p) of a graph $G = (V, E)$ (no loops or multiple edges) and a configuration p in d-space for the vertex set V. We shall assume that there are no 0-length edges, that is, $p(u) \neq p(v)$ for all $uv \in E$.

Realization of graph G in d-space: A d-dimensional framework (G, p).

Globally rigid framework: A framework (G, p) in d-space for which every d-dimensional realization (G, q) of G with the same edge lengths as in (G, p) is congruent to (G, p).

Rigidity matrix: For a framework (G, p) in d-space, $R(G, p)$ is the $|E| \times d|V|$ matrix for the system of equations: $(p_i - p_j) \cdot (p'_i - p'_j) = 0$ in the unknown velocities p'_i. The first-order flex equations are expressed as

$$R(G,p){p'}^T = \begin{bmatrix} \vdots & \ddots & \vdots & \cdots & \vdots & \ddots & \vdots \\ 0 & \cdots & (p_i - p_j) & \cdots & (p_j - p_i) & \cdots & 0 \\ \vdots & \ddots & \vdots & \cdots & \vdots & \ddots & \vdots \end{bmatrix} \times {p'}^T = 0^T.$$

Equilibrium stress: For a framework (G, p), an assignment of scalars ω_{ij} to the edges such that at each vertex i,

$$\sum_{\{j|\ \{i,j\}\in E\ \}} \omega_{ij}(p_i - p_j) = 0.$$

Equivalently, a row dependence for the rigidity matrix: $\omega R(G, p) = 0$.

Stress matrix: For a framework (G, p) in d-space and equilibrium stress ω the stress matrix Ω is the $|V| \times |V|$ symmetric matrix in which the entries are defined so that $\Omega[i, j] = -\omega_{ij}$ for the edges, $\Omega[i, j] = 0$ for nonadjacent vertex pairs, and $\Omega[i, i]$ is calculated so that each row and column sum is equal to zero.

Independent framework: A framework (G, p) for which the rigidity matrix has independent rows. Equivalently, there is only the zero equilibrium stress for (G, p).

Universally rigid framework: A framework (G, p) in d-space for which every other d'-dimensional realization (G, q) of G with the same edge lengths as in (G, p) is congruent to (G, p).

BASICS FOR GLOBAL RIGIDITY OF GENERIC FRAMEWORKS

It is a hard problem to decide if a given framework is globally rigid. Saxe [Sax79] showed that it is NP-hard to decide if even a 1-dimensional framework is globally rigid. See Example 63.1.5 below for some subtleties that may arise. The problem becomes tractable, however, if we consider generic frameworks, that is, frameworks (G, p) for which p is generic. For these frameworks we have the following fundamental necessary and sufficient conditions in terms of stresses and stress matrices due to Connelly [Con05] (sufficiency) and Gortler, Healy and Thurston [GHT10] (necessity), respectively.

THEOREM 63.1.1 Stress Matrix Condition for Global Rigidity

Let (G, p) be a generic framework in $\mathbb{R}^d$ on at least $d + 2$ vertices. (G, p) is globally rigid in $\mathbb{R}^d$ if and only if (G, p) has an equilibrium stress ω for which the rank of the associated stress matrix Ω is $|V| - d - 1$.

Theorem 63.1.1 implies that global rigidity is a generic property in the following sense.

THEOREM 63.1.2 Generic Global Rigidity Theorem

For a graph G and a fixed dimension d the following are equivalent:

(a) *(G, p) is globally rigid for some generic configuration $p \in \mathbb{R}^d$;*

(b) *(G, p) is globally rigid for all generic configurations $p \in \mathbb{R}^d$.*

(c) *(G, p) is globally rigid for an open dense subset of configurations $p \in \mathbb{R}^d$.*

63.1.2 BASICS FOR UNIVERSAL RIGIDITY OF FRAMEWORKS

There are several key basic theorems for universally rigid generic frameworks. These depend on a stress matrix Ω being positive semi-definite (PSD) with maximal rank

$n - d - 1$, where $n = |V|$. These properties can be broken down into two stages: dimensional rigidity, where the stress matrix has full rank $n - d - 1$, leaving possible affine motions within the space, and super stability where the stress matrix is PSD and the directions of the framework do not lie on a conic (eliminating affine motions within the space).

BASIC RESULTS FOR UNIVERSAL RIGIDITY

For generic frameworks we have a more refined stress matrix condition, similar to that of Theorem 63.1.1, due to Gortler and Thurston [GT14a].

THEOREM 63.1.3   Stress Matrix Condition for Universal Rigidity

Let (G, p) be a generic framework in $\mathbb{R}^d$ on at least $d + 2$ vertices. Then (G, p) is universally rigid if and only if there exists an equilibrium stress ω for (G, p) for which the associated stress matrix Ω is positive semi-definite and has rank $|V| - d - 1$.

Universal rigidity of frameworks is not a generic property. Example 63.1.5 below illustrates that the four-cycle $K_{2,2}$ has generic realizations which are universally rigid as well as generic realizations which are not universally rigid on the line. This leads to two different problems concerning the combinatorial aspects of universal rigidity. Given an integer $d \geq 1$, characterize the graphs G (i) for which *every* generic realization of G in $\mathbb{R}^d$ is universally rigid, (ii) for which *some* generic realization of G in $\mathbb{R}^d$ is universally rigid.

The graphs satisfying (i) will be called *generically universally rigid* in $\mathbb{R}^d$. The characterization of these graphs is an unsolved problem, even in $\mathbb{R}^1$.

For each universally rigid generic framework (G, q) in $\mathbb{R}^d$, there is an open neighbourhood $U(q)$ of q in $\mathbb{R}^{d|V|}$, such that for all $p \in U(q)$, the framework (G, p) is universally rigid [CGT16a]. For this reason, the graphs satisfying (ii) are called *openly universally rigid* or *weakly generically universally rigid* (*WGUR*, for short).

BASIC RESULTS CONNECTING GLOBAL AND UNIVERSAL RIGIDITY

It is clear that a framework that is universally rigid is also globally rigid and hence every weakly generically universally rigid graph is generically globally rigid. What may be surprising is that recent results [CGT16a] confirm that these two families of graphs are the same, for every fixed dimension. The following theorem summarizes this equivalence.

THEOREM 63.1.4   Equivalence of Generic Global Rigidity and Weak Generic Universal Rigidity

For a graph G on at least $d + 2$ vertices and a fixed dimension d, the following are equivalent:

(i) *a graph G is generically globally rigid in $\mathbb{R}^d$;*

(ii) *a graph is weakly generically universally rigid in $\mathbb{R}^d$;*

(iii) *there exists a generic framework (G, q) in $\mathbb{R}^d$ which is universally rigid;*

(iv) *there exists a generic framework (G, p) in $\mathbb{R}^d$ which is globally rigid;*

(v) *there exists a generic framework (G, p) in $\mathbb{R}^d$ with a stress matrix $\Omega(G, p)$ which has rank $n - d - 1$;*

(vi) *there exists a generic framework (G, q) in $\mathbb{R}^d$ with a PSD stress matrix Ω which has rank $n - d - 1$.*

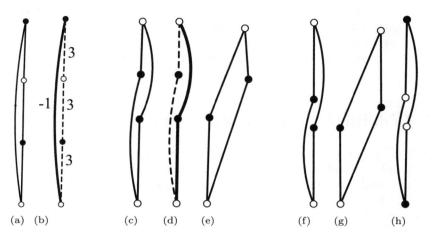

FIGURE 63.1.1

A range of frameworks on the graph $K_{2,2}$. Figures (a,b) present a universally rigid framework, where (b) uses the convention of dashed lines for tension, and thick lines for compression. Figures (c,d,e) present a framework which is globally rigid but not universally rigid as (e) illustrates. Figures (f,g,h) present a framework which is not globally rigid as (f) and (h) have the same edge lengths on the line.

We illustrate these definitions and connections with a collection of related simple frameworks on the line.

EXAMPLE 63.1.5 $K_{2,2}$ on the line

Consider the simple 'bow' framework of Figure 63.1.1(a). Intuitively, this is universally rigid (and therefore globally rigid). We can confirm this with a positive semi-definite (PSD) stress matrix, for the simple stress of Figure 63.1.1(b):

$$\Omega(G, p) = \begin{bmatrix} 2 & -3 & 0 & 1 \\ -3 & 6 & -3 & 0 \\ 0 & -3 & 6 & -3 \\ 1 & 0 & -3 & 2 \end{bmatrix}$$

One can check that this is PSD by taking the $k \times k$ determinants from the top corner, down the diagonal $1 \le k \le 4$. Moreover, Ω has rank $2 = 4 - 1 - 1$, as required for universal (and global) rigidity in dimension 1. Although this specific framework is not generic, a small perturbation of the vertices will make it generic, and will preserve the PSD property. So all the theorems above apply to $K_{2,2}$. It is generically globally rigid and is weakly generically universally rigid.

We know that all generic frameworks on $K_{2,2}$ on the line will be globally rigid. Consider the framework in Figure 63.1.1(c) which is globally rigid. However, Figure 63.1.1(e) shows another realization with the same lengths, in the plane, so it is not universally rigid. It has a stress matrix Ω for the stress with signs illustrated in Figure 63.1.1(d) which has rank 2, but this cannot be PSD (by Figure 63.1.1(e)).

While the framework in Figure 63.1.1(c) is globally rigid, the framework in Figure 63.1.1(f) is in a special position with the same pattern and is not globally rigid. It can move through the plane as a parallelogram (Figure 63.1.1 (g) to Figure 63.1.1 (h)) to another framework, which is also on the line but not congruent. Figures 63.1.1 (f,h) have different stresses, each with a stress matrix of rank 2, but these frameworks are not globally rigid, reminding us that the theorems above are only guaranteed to be sufficient for generic frameworks.

63.2   COMBINATORICS FOR GENERIC GLOBAL RIGIDITY

The major goal in generic global rigidity is a combinatorial characterization of graphs with globally rigid generic realizations in d-space. The companion problem is to find efficient combinatorial algorithms to test graphs for generic global rigidity. For the plane (and the line), this is solved. Beyond the plane the results are incomplete, but some significant partial results are available.

Because of Theorem 63.1.4, all results for generic global rigidity are also results for weak generic universal rigidity. We will usually not mention this extension. For geometric—not generic—frameworks, there will be some key differences in the techniques and results.

GLOSSARY

Globally rigid graph in $\mathbb{R}^d$:   A graph G for which some (or equivalently, all) generic configurations p produce globally rigid frameworks (G, p) in d-space.

Edge splitting operation (d-dimensional):   Replaces an edge of graph G with a new vertex joined to the end vertices of the edge and to $d - 1$ other vertices.

d-connected graph: A graph G such that removing any $d - 1$ vertices (and all incident edges) leaves a connected graph. (Equivalently, a graph such that any two vertices can be connected by at least d paths that are vertex-disjoint except for their endpoints.)

k-tree-connected graph: A graph G which contains k edge-disjoint spanning trees.

Highly k-tree-connected graph: A graph G for which the removal of any edge leaves a k-tree-connected graph.

Rigid graph in $\mathbb{R}^d$:   A graph G for which some (or equivalently, all) generic configurations p produce rigid frameworks (G, p) in d-space.

Redundantly rigid graph in $\mathbb{R}^d$:   A graph $G = (V, E)$ for which $G - e$ is rigid in $\mathbb{R}^d$ for all $e \in E$.

M-circuit (or generic circuit) in $\mathbb{R}^d$:   A graph $G = (V, E)$ for which a generic realization (G, p) in $\mathbb{R}^d$ is dependent, but $(G - e, p)$ is independent in d-space for all $e \in E$.

M-connected graph in $\mathbb{R}^d$:   A graph $G = (V, E)$ for which every edge pair $e, f \in E$ belongs to a subgraph H of G which is an M-circuit.

Cone graph: The graph $G * u$ obtained from $G = (V, E)$ by adding a new vertex u and the $|V|$ edges (u, i) for all vertices $i \in V$ (Figure 63.2.4.)

BASIC PROPERTIES FOR GLOBAL RIGIDITY OF GRAPHS IN ALL DIMENSIONS

The following necessary conditions, due to Hendrickson [Hen82], provide a basic link between local and global rigidity.

THEOREM 63.2.1 Hendrickson's Necessary Conditions

Let G be a globally rigid graph in $\mathbb{R}^d$. Then either G is a complete graph on at most $d + 1$ vertices, or G is

(a) *$(d + 1)$-connected; and*
(b) *redundantly rigid in $\mathbb{R}^d$.*

It is clear that the $d + 1$ connectivity condition is necessary for the broader class of general position frameworks which are globally rigid (or universally rigid). For a weaker converse see Theorem 63.5.4.

The necessity of redundant rigidity can also be observed from the Stress Matrix Condition Theorem 63.1.1. If an edge is not redundant the pair has a zero entry in the stress matrix. Removing the edge makes a smaller graph which is also globally rigid. On the other hand, removing the edge at a generic configuration, makes the graph flexible—a contradiction.

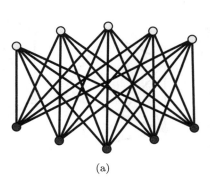
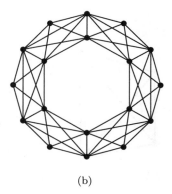

(a)   (b)

FIGURE 63.2.1
The two smallest known Hendrickson graphs in 3-space.

These necessary conditions together are also sufficient to imply the global rigidity of the graph in $\mathbb{R}^d$ for $d = 1, 2$, as we shall see below. This is not the case, however, for dimensions $d \geq 3$. We say that a graph G is a *Hendrickson graph* in $\mathbb{R}^d$ if it satisfies the necessary conditions (a) and (b) of Theorem 63.2.1 in $\mathbb{R}^d$ but it is not globally rigid in $\mathbb{R}^d$. For $d = 3$, Connelly [Con91] showed that the complete bipartite graph $K_{5,5}$ is a Hendrickson graph. He also constructed similar examples (specific complete bipartite graphs on $\binom{d+2}{2}$ vertices) for all $d \geq 3$. Frank and Jiang [FJ11] found two more (bipartite) Hendrickson graphs in $\mathbb{R}^4$ as well as infinite families in $\mathbb{R}^d$ for $d \geq 5$. Jordán, Király, and Tanigawa [JKT16] constructed infinite families of Hendrickson graphs for all $d \geq 3$ (see Figure 63.2.1 (b)). Further examples can be obtained by using the observation that the cone graph of a d-dimensional Hendrickson graph is a $d + 1$-dimensional Hendrickson graph.

There is a generic global rigidity result for complete bipartite frameworks which combines results in [CG17, Con91].

THEOREM 63.2.2

A complete bipartite graph $K_{m,n}$ is generically globally rigid in $\mathbb{R}^d$ if and only if $m, n \geq d + 1$ and $m + n \geq \binom{d+2}{2} + 1$.

The examples with $m, n \geq d + 1$ and $m + n = \binom{d+2}{2}$ include a number of Hendrickson graphs which are not globally rigid, such as $K_{5,5}$ in 3-space, and $K_{6,9}$, $K_{7,8}$ in 4-space.

INDUCTIVE CONSTRUCTIONS FOR GLOBAL RIGIDITY

Inductive constructions for graphs that preserve generic global rigidity are used both to prove theorems for general classes of frameworks and to analyze particular graphs.

Adding a new vertex of degree $d+1$ preserves global rigidity in $\mathbb{R}^d$. By applying a sequence of this operation to the base graph K_{d+1} we obtain the $(d+1)$-lateration graphs, which are thus sparse globally rigid graphs in $\mathbb{R}^d$. A finer and more useful operation that preserves global rigidity is edge splitting.

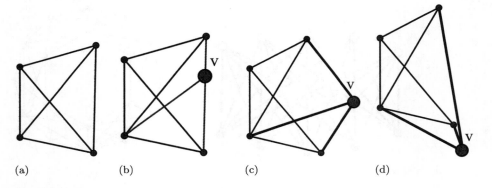

(a)　　　　(b)　　　　(c)　　　　(d)

FIGURE 63.2.2

From these planar frameworks: (a) is universally rigid (it has a complete graph). The edge split (b) is globally rigid and (b,c) are universally rigid. Figure (d) is globally rigid, but not universally rigid (it lifts in $\mathbb{R}^3$).

THEOREM 63.2.3　Edge Split Theorem [Con05, JJS06]

Let $G = (V, E)$ be a graph obtained from a globally rigid graph H by a d-dimensional edge splitting operation. Then G is globally rigid in $\mathbb{R}^d$.

Another operation merges two graphs. For example, it is easy to see that the union of two graphs G_1, G_2 that are globally rigid in $\mathbb{R}^d$ is also globally rigid in

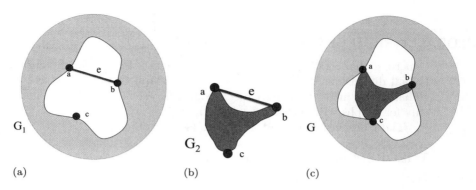

FIGURE 63.2.3
If G_1 (a) and G_2 (b) are globally rigid graphs sharing the edge e, then gluing on $d+1$ vertices creates a globally rigid graph without e.

$\mathbb{R}^d$, provided they share at least $d+1$ vertices. In fact it remains globally rigid even if we delete the edges of G_1 spanned by their common vertices before taking the union. An even stronger statement holds for graphs which have exactly $d+1$ vertices in common.

THEOREM 63.2.4   Gluing Theorem [Con11]

If $G_1 = (V_1, E_1)$ and $G_2 = (V_2, E_2)$ are globally rigid graphs in $\mathbb{R}^d$ sharing at least $d+1$ vertices, then $G = (V_1 \cup V_2, E_1 \cup E_2 - G_1[V_1 \cap V_2])$ is globally rigid in $\mathbb{R}^d$.

If $G_1 = (V_1, E_1)$ and $G_2 = (V_2, E_2)$ are globally rigid graphs in $\mathbb{R}^d$ sharing exactly $d+1$ vertices and some edge e, then $G = (V_1 \cup V_2, E_1 \cup E_2 - e)$ is globally rigid in $\mathbb{R}^d$ (Figure 63.2.3).

The first statement of Theorem 63.2.4 extends to pairs of geometric globally rigid frameworks and to pairs of universally rigid frameworks, provided the overlapping vertices affinely span the space.

The following operation can also be used to construct globally rigid graphs from smaller graphs. Let G be a graph, X be a subset of $V(G)$, and let H be a graph on X (whose edges may not be in G). The pair (H, X) is called a *rooted minor* of (G, X) if H can be obtained from G by deleting and contracting edges of G, that is, if there is a partition $\{U_v | v \in X\}$ of $V(G)$ into $|X|$ subsets such that $v \in U_v$ and $G[U_v]$ are connected for all $v \in X$, and G has an edge between U_u and U_v for each $uv \in E(H)$. Let $K(X)$ denote the complete graph on the vertex set X.

THEOREM 63.2.5   Rooted Minor Theorem [Tan15]

Let G_1 and G_2 be graphs with $X = V(G_1) \cap V(G_2)$ and H be a graph on X. Suppose that $|X| \geq d+1$, G_1 is rigid in $\mathbb{R}^d$, (H, X) is a rooted minor of (G_2, X), $G_1 \cup H$ and $G_2 \cup K(X)$ are globally rigid or $G_1 \cup K(X)$ and $G_2 \cup H$ are globally rigid in $\mathbb{R}^d$. Then $G_1 \cup G_2$ is globally rigid in $\mathbb{R}^d$.

Note that the Rooted Minor Theorem implies the Gluing Theorem as well as the Edge Split Theorem.

GLOBALLY RIGID GRAPHS IN THE PLANE

Generic global rigidity has been completely characterized for dimensions up to 2. For all dimensions d it is easy to see that a graph G on at most $d + 1$ vertices is globally rigid in $\mathbb{R}^d$ if and only if G is complete. Thus we shall formulate the results only for graphs with at least $d + 2$ vertices.

The 1-dimensional result is folklore (see also Figure 63.1.1).

THEOREM 63.2.6   Global Rigidity on the Line

For a graph G with $|V| \geq 3$ the following are equivalent:

(a) *G is globally rigid in $\mathbb{R}^1$;*

(b) *G is 2-connected;*

(c) *there is a construction for G from K_3, using only edge splitting and edge addition.*

Note that 2-connected graphs are redundantly rigid in $\mathbb{R}^1$. The equivalence of (b) and (c) is clear by using the well-known ear-decompositions of 2-connected graphs.

The 2-dimensional result is based on the Edge Split Theorem (i.e., (c) implies (a) below), Hendrickson's Necessary Conditions (i.e., (a) implies (b)) and an inductive construction of 3-connected redundantly rigid graphs due to Jackson and Jordán [JJ05], which shows that (b) implies (c).

THEOREM 63.2.7   Global Rigidity in the Plane

For a graph G with $|V| \geq 4$ the following are equivalent:

(a) *G is globally rigid in $\mathbb{R}^2$;*

(b) *G is 3-connected and redundantly rigid in $\mathbb{R}^2$;*

(c) *there is a construction for G from K_4, using only edge splitting and edge addition.*

Note that if G has $|E| = 2|V| - 2$, that is, if G is a 3-connected M-circuit, then the edge splitting operation alone suffices in (c). This fact and an inductive construction for M-circuits can be found in Berg and Jordán [BJ03].

Since 6-connected graphs are redundantly rigid in $\mathbb{R}^2$, we obtain the following sufficient condition.

THEOREM 63.2.8   Sufficient Connectivity in the Plane

If a graph G is 6-connected, then G is globally rigid in $\mathbb{R}^2$.

The bound 6 above on vertex-connectivity is the best possible (see [LY82] for 5-connected graphs which are not even rigid). Jackson and Jordán [JJ09b] strengthened Theorem 63.2.8 by showing that 6-connectivity can be replaced by a weaker connectivity property, involving a mixture of vertex- and edge-connectivity parameters.

We remark that the characterization of globally rigid graphs in the plane has been used to deduce a number of variations (e.g., for zeolites [Jor10a], squares of graphs) and to solve related optimization problems (e.g., minimum cost anchor placement [Jor10b]).

GENERICALLY GLOBALLY RIGID GRAPHS IN HIGHER DIMENSIONS

There are some additional generic results for higher dimensions. A useful observation, proved by Connelly and Whiteley [CW10], is that coning preserves global rigidity.

THEOREM 63.2.9   Generic Global Rigidity Coning

A graph G is globally rigid in $\mathbb{R}^d$ if and only if the cone $G * u$ is globally rigid in $\mathbb{R}^{d+1}$.

Tanigawa [Tan15] established another interesting connection between rigidity and global rigidity. We say that G is *vertex-redundantly rigid* in $\mathbb{R}^d$ if $G - v$ is rigid in $\mathbb{R}^d$ for all $v \in V(G)$.

THEOREM 63.2.10   Vertex-redundant Rigidity Implies Global Rigidity

If G is vertex-redundantly rigid in $\mathbb{R}^d$ then it is globally rigid in $\mathbb{R}^d$.

The concept of (vertex-)redundant rigidity can be generalized to d-dimensional k-*rigid* graphs, for all $k \geq 2$: these graphs remain rigid in $\mathbb{R}^d$ after removing any set of at most $k - 1$ vertices. Kaszanitzky and Király [KK16] solve a number of extremal questions related to these families. One can also consider higher degrees of redundant rigidity with respect to edge removal as well as similar notions for global rigidity. An interesting open problem is whether these graph properties can be tested in polynomial time for $d \geq 2$. Kobayashi et al. [KHKS16] show that a version of this problem for body-hinge graphs, in which sets of hinges are removed without losing global rigidity, is tractable.

Theorem 63.2.10 can be used to deduce a new sufficient condition for global rigidity as well as to solve an augmentation problem, see [Jor17].

THEOREM 63.2.11

If G is rigid in $\mathbb{R}^{d+1}$ then it is globally rigid in $\mathbb{R}^d$.

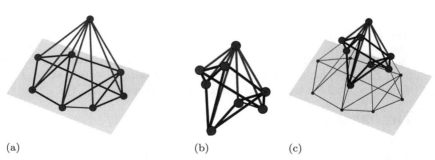

(a)   (b)   (c)

FIGURE 63.2.4

A redundantly rigid plane framework cones to a redundantly rigid framework in three-space (a). A redundantly rigid cone (b) slices back to a redundantly rigid framework in the plane (c).

THEOREM 63.2.12

Every rigid graph in $\mathbb{R}^d$ on $|V|$ vertices can be made globally rigid in $\mathbb{R}^d$ by adding at most $|V| - d - 1$ edges.

The bound is tight for all $d \geq 1$: consider the complete bipartite graphs $K_{n-d,d}$.

A number of other related structures have also been investigated for generic global rigidity. These structures, which appear in engineering, robotics, and chemistry, can be defined in $\mathbb{R}^d$ for all $d \geq 2$. They consist of rigid bodies which are connected by disjoint bars (a body-bar framework) or by hinges (a body-hinge structure) that constrain the relative motion of the corresponding pairs of bodies to a rotation about an affine subspace of dimension $d - 2$. For example in 3-space a hinge is a line (segment) which restricts the relative motion of the two bodies connected by the hinge to a rotation about the shared hinge line.

The *underlying multigraph H* of such a framework has one vertex for each body and one edge for each bar (resp. hinge), connecting the vertices of the bodies it connects. (See Section 61.2 for definitions and infinitesimal rigidity results.)

These frameworks can be modeled as bar-and-joint frameworks, with each body replaced by a large enough rigid framework on a complete graph. In the case of a body-bar framework these complete graphs are disjoint and the connecting bars correspond to vertex-disjoint bars between the complete subgraphs. For body-hinge frameworks the complete graphs corresponding to the bodies are constructed so that if two bodies are connected by a hinge then the complete graphs share $d - 1$ vertices. The underlying graphs of these special bar-and-joint frameworks are called (d-dimensional) *body-bar graphs* and *body-hinge graphs*, respectively.

We shall formulate the results concerning the global rigidity of generic body-bar and body-hinge frameworks in terms of body-bar and body-hinge graphs, respectively. For these graphs, global rigidity has been fully characterized in $\mathbb{R}^d$ for all $d \geq 1$ in terms of their underlying multigraphs. The body-bar result is due to Connelly, Jordán, and Whiteley [CJW13].

THEOREM 63.2.13   Body-Bar Global Rigidity

Let G be a d-dimensional body-bar graph on at least $d + 2$ vertices with underlying multigraph H and let $d \geq 1$ be an integer. Then the following are equivalent:

(a) *G is globally rigid in $\mathbb{R}^d$;*

(b) *G is redundantly rigid in $\mathbb{R}^d$;*

(c) *H is highly $\binom{d+1}{2}$-tree-connected.*

For a multigraph H and integer k we use kH to denote the multigraph obtained from H by replacing each edge e of H by k parallel copies of e. The body-hinge version was solved by Jordán, Király, and Tanigawa [JKT16].

THEOREM 63.2.14   Body-Hinge Global Rigidity

Let G be a d-dimensional body-hinge graph on at least $d + 2$ vertices with underlying multigraph H and let $d \geq 3$ be an integer. Then the following are equivalent:

(a) *G is globally rigid in $\mathbb{R}^d$;*

(b) *$(\binom{d+1}{2} - 1)H$ is highly $\binom{d+1}{2}$-tree-connected.*

The two-dimensional characterization is slightly different.

THEOREM 63.2.15   Body-Hinge Global Rigidity in the Plane

A 2-dimensional body-hinge graph G on at least 4 vertices with underlying graph H is globally rigid in $\mathbb{R}^2$ if and only if H is 3-edge-connected.

A recent result, due to Jordán and Tanigawa [JT17], characterizes globally rigid braced triangulations in $\mathbb{R}^3$.

GLOBALLY LINKED PAIRS AND THE NUMBER OF REALIZATIONS

Even if a graph is not globally rigid, some parts of it may have a unique realization with the given edge lengths. One notion that can be used to analyse these parts is as follows. We say that a pair $\{u, v\}$ of vertices of G is *globally linked* in G in $\mathbb{R}^d$ if for all generic d-dimensional realizations (G, p) we have that the distance between $q(u)$ and $q(v)$ is the same in all realizations (G, q) equivalent with (G, p). Thus a graph G is globally rigid in $\mathbb{R}^d$ if and only if all pairs of vertices in G are globally linked in $\mathbb{R}^d$. It is not hard to see that a pair $\{u, v\}$ is globally linked in G in $\mathbb{R}^1$ if and only if there exist two openly vertex-disjoint paths from u to v in G, which is equivalent to $\{u, v\}$ sharing a globally rigid subgraph. It follows that global linkedness is a generic property on the line.

In higher dimensions global linkedness is not a generic property and it remains an open problem to find a combinatorial characterization of globally linked pairs even in $\mathbb{R}^2$ (see Figure 63.2.5). This section lists some of the partial results that have been proven in the planar case by Jackson, Jordán, and Szabadka [JJS06, JJS14].

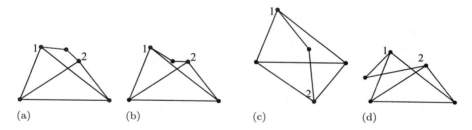

FIGURE 63.2.5
The pair $\{1, 2\}$ is globally linked in (a,b) although the framework is not globally rigid. In (c,d), with the same graph, $\{1, 2\}$ are not globally linked, so being globally linked is not a generic property of the graph.

THEOREM 63.2.16

Let G and H be graphs such that G is obtained from H by a two-dimensional edge splitting operation on edge xy and vertex w. If $H - xy$ is rigid in $\mathbb{R}^2$ and that $\{u, v\}$ is globally linked in H in $\mathbb{R}^2$, then $\{u, v\}$ is globally linked in G in $\mathbb{R}^2$.

For the family of M-connected graphs, global linkedness has been characterized as follows, see [JJS06]. Note that globally rigid graphs are M-connected and M-connected graphs are redundantly rigid in $\mathbb{R}^2$, see [JJ05].

THEOREM 63.2.17

Let $G = (V, E)$ be an M-connected graph in $\mathbb{R}^2$ and let $u, v \in V$. Then $\{u, v\}$ is globally linked in G if and only if there exist three openly vertex-disjoint paths from u to v in G (Figure 63.2.6).

In minimally rigid graphs, or more broadly, in all independent graphs, there are no globally linked pairs, other than the adjacent pairs of vertices [JJS14].

THEOREM 63.2.18

Let $G = (V, E)$ be an independent graph in $\mathbb{R}^2$ and $u, v \in V$. Then $\{u, v\}$ is globally linked in G in $\mathbb{R}^2$ if and only if $uv \in E$.

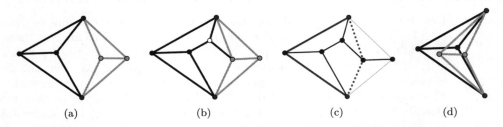

(a)   (b)   (c)   (d)

FIGURE 63.2.6

Edge splitting preserves globally linked pairs, as well as M-circuits (a,b). (c) Three openly vertex disjoin paths. (d) A framework with $h = 2$ and its two noncongruent realizations.

A related observation is that if a pair $\{u, v\}$ is globally linked in G in $\mathbb{R}^d$, then either the edge uv is present in G, or it is in an M-circuit in $G + uv$ in $\mathbb{R}^d$.

One can also determine the maximum number of pairwise equivalent but noncongruent generic realizations of an M-connected graph. Given a rigid generic framework (G, p), let $h(G, p)$ denote the number of distinct congruence classes of frameworks which are equivalent to (G, p). Given a rigid graph G, let $h(G) = \max\{h(G, p)\}$, where the maximum is taken over all generic frameworks (G, p).

Borcea and Streinu [BS04] investigated the number of realizations of minimally rigid generic frameworks (G, p) in the plane. Their results imply that $h(G) \leq 4^n$ for all rigid graphs G. They also construct an infinite family of generic minimally rigid frameworks (G, p) for which $h(G, p)$ has order $12^{\frac{n}{3}}$ which is approximately $(2.28)^n$. One can determine the exact value of $h(G, p)$ for all generic realizations (G, p) of an M-connected graph $G = (V, E)$. For $u, v \in V$, let $b(u, v)$ denote the number of connected components of $G - \{u, v\}$ and put $c(G) = \sum_{u,v \in V}(b(u, v) - 1)$. The next equality was verified by Jackson, Jordán, and Szabadka [JJS06].

THEOREM 63.2.19

Let G be an M-connected graph in $\mathbb{R}^2$. Then $h(G, p) = 2^{c(G)}$ for all generic realizations (G, p) of G.

It follows that $h(G) \leq 2^{\frac{n-2}{2}-1}$ for all M-connected graphs G. A family of graphs attaining this bound is a collection of K_4's joined along a common edge.

OPEN PROBLEMS ON GENERIC GLOBAL RIGIDITY

As noted earlier, the major open problem in this area is to find a combinatorial characterization of (and an efficient deterministic algorithm for testing) globally rigid graphs in $\mathbb{R}^d$ for $d \geq 3$. There are several related conjectures.

The next conjecture is open for $d \geq 3$.

CONJECTURE 63.2.20 Sufficient Connectivity Conjecture

For every $d \geq 1$ there is a (smallest) integer $f(d)$ such that every $f(d)$-connected graph G is globally rigid in $\mathbb{R}^d$.

Some of the results mentioned earlier imply that $f(1) = 2$ and $f(2) = 6$. By replacing globally rigid by rigid in Conjecture 63.2.20 we obtain a conjecture of Lovász and Yemini [LY82] from 1982. By Theorem 63.2.10 the two conjectures are equivalent.

Note that no degree of vertex-connectivity suffices to imply generic universal rigidity (c.f. Theorem 63.5.2).

A molecular model in three-space treats each atom and its set of neighbours as pairwise adjacent joints (due to dihedral constraints). Thus bonds may be interpreted as hinges between such complete subframeworks or simply as rigid bars between the corresponding atoms. In the former case we obtain molecular-hinge frameworks (which are geometrically singular body-hinge frameworks since the lines of all bonds of an atom are concurrent in the center of the atom) while in the latter case (where all pairs of neighbours of the atom are also connected by rigid bars) we obtain bar-and-joint frameworks whose underlying graph is a square of another graph, that is, a *molecular graph*. (See Section 61.2 for related results.) These models are central to applications of (global) rigidity to protein structures with thousands of atoms [Whi99].

CONJECTURE 63.2.21 Molecular Graph Global Rigidity Conjecture

Let G be a graph with no cycles of length at most four. Then G^2 is generically globally rigid in $\mathbb{R}^3$ if and only if G^2 is 4-connected and the multigraph $5G$ is highly 6-tree connected.

A graph G is *minimally globally rigid* in $\mathbb{R}^d$ if it is globally rigid but removing any edge from G leaves a graph which is not globally rigid.

CONJECTURE 63.2.22 Minimally Globally Rigid Conjecture

Let G be minimally globally rigid in $\mathbb{R}^d$. Then

(a) *$|E| \leq (d+1)|V| - \binom{d+2}{2}$ and*

(b) *the minimum degree of G is at most $2d + 1$.*

This conjecture has been verified for $d = 1, 2$ (with slightly better bounds); see [Jor17]. The bounds on the edge number and the minimum degree would be close to being tight for all d, see the complete bipartite graph $K_{d+1,n-d-1}$.

Let v be a vertex of degree at least $d - 1$ in G with neighbour set $N(v)$. The *d-dimensional vertex splitting* operation at vertex v in G partitions $N(v)$ into three (possibly empty) sets A_1, C, A_2 with $|C| = d - 1$, deletes vertex v and its incident edges, and adds two new vertices v_1, v_2 and new edges from v_i to all vertices in $A_i \cup C$, for $i = 1, 2$, as well as the edge $v_1 v_2$ (the so-called *bridging edge*). The operation is *nontrivial* if A_1, A_2 are both nonempty. The vertex splitting operation preserves rigidity. It is conjectured to preserve global rigidity, provided that it is nontrivial. The 2-dimensional version of this conjecture was verified in [JS09]. For higher dimensions this conjecture is open even in the following weaker form.

CONJECTURE 63.2.23 Vertex Split Conjecture

Let H be globally rigid in $\mathbb{R}^d$ and let G be obtained from H by a nontrivial vertex *splitting operation with bridging edge e. If $G - e$ is rigid in $\mathbb{R}^d$ then G is globally rigid in $\mathbb{R}^d$.*

A recent result of [JT17] shows that if G is obtained from a globally rigid graph H with maximum degree at most $d + 2$ by a sequence of nontrivial vertex splitting operations, then G is globally rigid in $\mathbb{R}^d$.

There is a conjectured characterization of globally linked pairs in $\mathbb{R}^2$. Sufficiency follows from Theorem 63.2.17.

CONJECTURE 63.2.24   Globally Linked Pairs Conjecture

Let $G = (V, E)$ be a graph and let $u, v \in V$. Then $\{u, v\}$ is globally linked in G in $\mathbb{R}^2$ if and only if $uv \in E$ or there is an M-connected subgraph H of G with $\{u, v\} \subseteq V(H)$ for which there exist three openly vertex-disjoint paths from u to v in H.

ALGORITHMS FOR GENERIC GLOBAL RIGIDITY

The Stress Matrix Condition for Generic Global Rigidity (Theorem 63.1.1) implies that there is a randomized polynomial time algorithm for testing whether a given graph G is globally rigid in $\mathbb{R}^d$, for any fixed d. See [CW10, GHT10] for the details.

We have deterministic polynomial-time algorithms for the cases in which global rigidity is well characterized: low dimensions ($d = 1, 2$) and special classes of graphs (body-bar, body-hinge). These algorithms are based on testing low dimensional (redundant) rigidity (see Chapter 61), checking whether a graph is k-vertex-connected (for which there exist well-known efficient network flow-based algorithms in general, and linear time algorithms up to $k = 3$) and testing (high) k-tree-connectivity.

For the latter problem, the tree partitions of Theorems 63.2.13 and 63.2.14 can be computed by matroidal algorithms of order $O(|V|^3)$ time.

We also have algorithms for identifying finer substructures (rigid, redundantly rigid or M-connected components), see e.g., [Jor16].

63.3 VARIATIONS AND CONNECTIONS TO OTHER FIELDS

GLOSSARY

Direction-length framework: A pair (G, p) of a loopless edge-labeled graph $G = (V; D, L)$ and a configuration p in d-space for the vertex set V. The edge set of G consists of direction edges D and length edges L that represent direction and length constraints, respectively. A direction edge uv fixes the gradient of the line through $p(u)$ and $p(v)$, whereas a length edge uv specifies the distance between the points $p(u)$ and $p(v)$.

Mixed graph: A graph $G = (V; D, L)$ whose edge set consists of direction edges D and length edges L.

Pure graph: A (mixed) graph $G = (V; D, L)$, in which $L = \emptyset$ (a direction-pure graph) or $D = \emptyset$ (a length-pure graph), respectively.

Congruent realizations: Two realizations (G, p) and (G, q) of G for which (G, q) can be obtained from (G, p) by a translation and, possibly, a dilation by -1.

Globally rigid direction-length framework: A direction-length framework (G, p) in d-space for which every other d-dimensional realization (G, p) with the same edge lengths (for $e \in L$) and edge directions (for $e \in D$) as in (G, p) is congruent to (G, p).

Direction balanced graph: A 2-connected mixed graph in which both sides of any 2-vertex-separation contain a direction edge.

63.3.1 OTHER CONSTRAINTS

In computer-aided design, many different patterns of constraints (lengths, angles, incidences of points and lines, etc.) are used to design or describe configurations of points and lines, up to congruence. With distances between points, the geometry becomes that of global rigidity, with several positive results on the generic behaviour. For some other constraints the corresponding combinatorial problems (concerning "generic" configurations) are unsolved, and perhaps not solvable in polynomial time.

However, for special designs, e.g., for direction constraints in $\mathbb{R}^d$ and even for the combination of distance and direction constraints in $\mathbb{R}^2$, these problems have been solved, using extensions of the techniques and results for bar-and-joint frameworks.

DIRECTION-LENGTH FRAMEWORKS

By the Generic Global Rigidity Theorem (Theorem 63.1.2) global rigidity is a generic property for length-pure frameworks (with respect to the global rigidity definition for bar-and-joint frameworks). Whiteley [Whi96] showed that global rigidity is equivalent to first-order rigidity for direction-pure frameworks in $\mathbb{R}^d$. This implies that global rigidity is a generic property for direction-pure frameworks in $\mathbb{R}^d$ for all $d \geq 1$ (with respect to the adapted global rigidity definition allowing translations and dilations).

It is not known whether global rigidity is a generic property for direction-length frameworks, even in $\mathbb{R}^2$. In the plane we do have a number of partial results. In particular, we have a list of necessary conditions [JJ10a, JK11]. Let (G, p) be a generic direction-length framework with at least three vertices and $G = (V; D, L)$. If (G, p) is globally rigid in $\mathbb{R}^d$ then (i) G is not pure, (ii) G is rigid, (iii) G is 2-connected, (iv) G is direction balanced, (v) the only 2-edge-cuts in G consist of incident direction edges, and (vi) if $|L| \geq 2$ then $G - e$ is rigid for all $e \in L$.

The characterization of generically rigid mixed graphs gives rise to a rigidity matroid defined on the edge set of a mixed graph $G = (V; D, L)$ in which M-circuits and M-connected graphs are well-characterized.

Jackson and Jordán [JJ10a] characterized global rigidity for mixed graphs whose edge set is a circuit in the two-dimensional (mixed) rigidity matroid.

THEOREM 63.3.1 Mixed Global Rigidity for Plane Circuits

Let (G, p) be a generic realization of a mixed graph whose rigidity matroid is a circuit. Then (G, p) is globally rigid in $\mathbb{R}^2$ if and only if G is direction balanced.

Recently Clinch [Cli16] extended Theorem 63.3.1 to M-connected mixed graphs.

THEOREM 63.3.2   Mixed Global Rigidity for Plane Connected Matroids

Let (G, p) be a generic realization of a mixed graph whose rigidity matroid is connected. Then (G, p) is globally rigid in $\mathbb{R}^2$ if and only if G is direction balanced.

The proofs of the previous theorems rely on inductive constructions of the corresponding families of mixed graphs, using the mixed versions of the two-dimensional extension operations [JJ10b]. The d-dimensional versions of these operations are treated in Nguyen [Ngu12].

A recent result by Clinch, Jackson, and Keevash [CJK16] characterizes mixed graphs with the property that *all* generic realizations in $\mathbb{R}^2$ are globally rigid. There is also a special case of d-dimensional direction-length frameworks where global rigidity is fully understood.

THEOREM 63.3.3

Let G be a mixed graph in which all pairs of adjacent vertices are connected by both a length and a direction edge, and let (G, p) be a d-dimensional generic realization of G. Then (G, p) is globally rigid if and only if G is 2-connected.

GLOBAL RIGIDITY OF FRAMEWORKS ON SURFACES

Another direction of research investigates frameworks in $\mathbb{R}^3$ whose vertices are constrained to lie on 2-dimensional surfaces. (See Section 61.2 for related results.) Generic rigidity has been characterized for various surfaces and there are some partial results concerning generic global rigidity on the sphere, cylinder, cone, and ellipsoid. In fact the spherical case has been settled by a result of Connelly and Whiteley, who proved that a generic framework on the sphere is globally rigid if and only if the corresponding generic framework is globally rigid in the plane [CW10].

Hendrickson's necessary conditions have been extended to the cylinder and the cone: a generic globally rigid framework with at least five vertices must be redundantly rigid and 2-connected. A sufficient condition, in terms of the corresponding stress matrix has also been found by Jackson and Nixon [JN15b]. Jackson, McCourt and Nixon [JMN14] conjecture that the necessary conditions above are also sufficient to guarantee global rigidity on the cylinder and the cone. The missing piece is an inductive construction for the graph family in question using operations which are known to preserve global rigidity.

63.3.2 CONNECTIONS TO OTHER FIELDS

There are a number of situations where global rigidity (or universal rigidity) is a key part of the computational process. These problems are often expressed in terms of "distance geometry." Dependence among the distances being used is a synonym for having a self-stress in the corresponding framework.

LOCALIZATION

An important problem is computing the location of sensors, scattered randomly in the plane, using pairwise distances. This problem is a setting for applying global rigidity [JJ09a, AEG+06, SY07]. The computation needs access to a set of distance measurements which form a globally rigid graph, in order to have a unique solution. Typically, a few locations are known; these are called anchors and they function like pinned vertices. We are seeking a globally rigid pinned framework [Jor10b]. Given the expected errors in the measurements, the calculations reasonably assume the vertices are generic.

There are several simple approaches which have been used to make the calculations manageable. One is to find a trilateral subgraph: a graph which is formed inductively from the anchors by adding 3-valent vertices. Each step of the calculation has modest overhead and the locations are built up, one at a time. If the sensors have many neighbours with measured distances, then we can select such a trilateration graph with high probability [AEG+06].

MOLECULAR CONFORMATIONS

Another significant computation problem is finding the shape (conformation) of a molecule from NMR (nuclear magnetic resonance) data. NMR measures (approximately) a set of pairwise distances, typically between pairs of hydrogen atoms. Again one needs to calculate the (relative) locations with data that is redundant at least in some parts of most molecules. Some parts will be underdetermined (e.g., tails of a protein may not be fixed). Some of the programs for optimizing these calculations use "smoothing" through small universally rigid subgraphs to refine the errors in the measurements (e.g., [CH88]).

63.4 GEOMETRY OF GLOBAL RIGIDITY

Unlike infinitesimal rigidity, graphs which are not generically globally rigid can have special position realizations which are globally rigid and even universally rigid, provided that they have a nontrivial stress with an appropriate stress matrix and some additional features. There are layers of analysis here that we shall return to in the following sections, including the study of super stability and tensegrity frameworks.

GLOSSARY

Special position of a graph G in d-space: Any configuration $p \in \mathbb{R}^{d|V|}$ such that the rigidity matrix $R(G, p)$, or any submatrix, has rank smaller than the maximum rank (the rank at a configuration with algebraically independent coordinates). These form an algebraic variety (see Chapter 61).

Affine spanning set for d-space: A configuration p of points such that every point $q \in \mathbb{R}^d$ can be expressed as an affine combination of the p_i: $q = \sum_i \lambda_i p_i$, with $\sum_i \lambda_i = 1$. (Equivalently, the affine coordinates $(p_i, 1)$ span the vector space $\mathbb{R}^{d+1}$.)

BASIC GEOMETRIC RESULTS

There are some basic results for specific geometric frameworks. The stronger geometric results will appear for Universal Rigidity. The following three theorems are due to Connelly and Whiteley [CW10].

THEOREM 63.4.1 Stress Matrix and Infinitesimal Rigidity Certificate

If there is a realization (G, p) of G in $\mathbb{R}^d$ which is infinitesimally rigid with an equilibrium stress ω for which the associated stress matrix Ω has rank $|V| - d - 1$ then G is generically globally rigid in $\mathbb{R}^d$.

This theorem does not guarantee that (G, p) is itself globally rigid.

THEOREM 63.4.2 Stability Lemma

Given a framework (G, p) which is globally rigid and infinitesimally rigid in $\mathbb{R}^d$, there is an open neighborhood U of p in $\mathbb{R}^{d|V|}$ such that for all $q \in U$ the framework (G, q) is globally rigid and infinitesimally rigid.

The 2-dimensional case of the Stability Lemma has been generalized in [JJS14]: if (G, p) is infinitesimally rigid in $\mathbb{R}^2$, then there is an open neighbourhood U of p in $\mathbb{R}^{d|V|}$ for which $h(G, q) \leq h(G, p)$ for all $q \in U$.

An immediate corollary of the Stability Lemma is a different kind of global rigidity certificate.

THEOREM 63.4.3 Global Rigidity and Infinitesimal Rigidity Certificate

If there is a realization (G, p) of G which is globally rigid and infinitesimally rigid in $\mathbb{R}^d$ then G is generically globally rigid in $\mathbb{R}^d$.

63.5 UNIVERSALLY RIGID FRAMEWORKS

We return to universal rigidity of a framework (G, p), working in several layers, which turn out to be connected by important theorems and classes of PSD stress matrices. Here are several basic theorems for generic universally rigid frameworks. These depend on a stress matrix Ω being positive semi-definite (PSD) with maximal rank $n - d - 1$, where there are n vertices.

We begin with a few results for particular classes of graphs. We then follow two related but narrower concepts: dimensional rigidity and super stability. While not all universally rigid frameworks are super stable, a recent result of Connelly and Gortler [CG15] shows that every universally rigid framework can be iteratively built by a nested sequence of super stable structures.

We first present the results for bar frameworks. In the next section, we attend to the signs of the coefficients in the stress, and present the results for tensegrity frameworks, which is the form in which many of them appear in engineering and tensegrity installations.

The complexity of testing the universal rigidity of frameworks is open.

GLOSSARY

Dimensionally rigid framework: A framework (G, p) in d-space with an affine span of dimension d, is dimensionally rigid in $\mathbb{R}^d$ if every framework (G, q) equivalent to (G, p) in any dimension has an affine span of dimension at most d.

Edge directions lie on a conic at infinity: The edge directions of a framework (G, p) lie on a conic at infinity if there is a symmetric matrix Q such that $m_{i,j} Q m_{i,j}^T = 0$ for all $m_{i,j} = (p_i - p_j)$ with $\{i, j\} \in E$.

Super stable framework: A framework (G, p) in d-space for which there is a stress matrix which is positive semi-definite with rank $|V| - d - 1$ and the edge directions do not lie on a conic at infinity.

Strictly separated by a quadric: Two sets of points P, Q in $\mathbb{R}^d$, with affine coordinates, for which there exists a quadric represented by a symmetric $(d + 1) \times (d + 1)$ matrix Q such that $p_i^T Q p_i > 0$ for all $p_i \in P$ and $q_j^T Q q_j < 0$ for all $q_j \in Q$.

BASIC RESULTS FOR UNIVERSAL RIGIDITY

We have described some basic results in Section 63.1.1. Several specific classes of graphs have more complete answers.

Connelly and Gortler characterized universally rigid complete bipartite frameworks for all $d \geq 1$ [CG17], extending the 1-dimensional result from [JN15a]. These geometric results lie behind some results for global rigidity cited earlier in Theorem 63.2.2.

THEOREM 63.5.1   Complete Bipartite Frameworks

Let G be a complete bipartite graph and let (G, p) be a d-dimensional realization of G in general position, with $m + n > d + 1$. Then (G, p) is not universally rigid if and only if the two vertex classes are strictly separated by a quadric.

THEOREM 63.5.2   Bipartite Universal Rigidity [JN15a]

The only generically universally rigid bipartite graph in $\mathbb{R}^d$ is K_2, for all $d \geq 1$.

There is also a sufficient condition which works for squares of graphs.

THEOREM 63.5.3   Universally Rigid Squares [GGLT13]

Let G be a $(d + 1)$-connected graph. Then every generic d-dimensional realization of G^2 is universally rigid.

THEOREM 63.5.4   General Position Realizations [Alf17]

If a graph G is $(d + 1)$-connected, then there exists a general position configuration p in $\mathbb{R}^d$ such that the framework (G, p) is universally rigid (therefore globally rigid).

DIMENSIONAL RIGIDITY

Dimensional rigidity, introduced by Alfakih [Alf07], is a concept that says a framework, with the given edge lengths, always lives in a restricted dimension. Again, beyond the small complete graphs, this is linked to properties of stress matrices. This is weaker than universal rigidity, as there may be noncongruent affinely equivalent frameworks. However, the concepts are clearly linked, and dimensional rigidity can be a key step towards proving universal rigidity, see e.g., [CG15].

THEOREM 63.5.5   Stress matrices for dimensional rigidity [Alf11]

1. *A framework (G, p) with n vertices whose affine span is d-dimensional with $n \geq d + 2$, is dimensionally rigid if and only if (G, p) has a nonzero PSD stress matrix.*

2. *If a framework (G, p) with n vertices in $\mathbb{R}^d$ has an equilibrium stress with a PSD stress matrix of rank $n - d - 1$, then (G, p) is dimensionally rigid in $\mathbb{R}^d$.*

THEOREM 63.5.6   [AY13, CGT16b]

A framework (G, p) in $\mathbb{R}^d$, with d-dimensional affine span, has a nontrivial affine flex if and only if it has an equivalent noncongruent affine image in $\mathbb{R}^d$ if and only if the edge directions lie on a conic at infinity.

THEOREM 63.5.7   Dimensional Rigidity

If a framework (G, p) with n vertices in $\mathbb{R}^d$ is dimensionally rigid in $\mathbb{R}^d$, and (G, q) is equivalent to (G, p), then q is an affine image of p.

As a result, we have the following strong connection to universal rigidity.

THEOREM 63.5.8   Dimensional Rigidity and Universal Rigidity

A framework (G, p) with n vertices in $\mathbb{R}^d$ is universally rigid if and only if it is dimensionally rigid and the edge directions do not lie on a conic at infinity.

SUPER STABILITY

The focus of the stress matrix approach remains finding a positive semi-definite stress matrix of appropriate rank. This is captured by the notion of super stability, defined by the following result of Connelly.

THEOREM 63.5.9   Super Stability

Let (G, p) be a framework whose affine span is all of $\mathbb{R}^d$, with an equilibrium stress ω and stress matrix Ω. Suppose further that

1. *Ω is positive semi-definite (PSD);*
2. *the rank of Ω is $n - d - 1$;*
3. *the member directions of (G, p) do not lie on a conic at infinity.*

Then (G, p) is universally rigid.

A framework (G, p) satisfying conditions (1-3) is called *super stable*.

THEOREM 63.5.10   Generic Global Rigidity Implies Super Stability [CGT16a]

If G is generically globally rigid in $\mathbb{R}^d$, then there exists a framework (G, p) in $\mathbb{R}^d$ that is infinitesimally rigid in $\mathbb{R}^d$ and super stable. Moreover, every framework within a small enough open neighborhood of (G, p) will be infinitesimally rigid in $\mathbb{R}^d$ and super stable, including some generic framework.

Not all universally rigid frameworks are super stable. However [CG15] shows that every universally rigid framework can be constructed through a nested sequence of affine spaces, each with an associated super stable framework. Such a nested sequence with associated super stable frameworks is a certificate for universal (global) rigidity of the framework and it is algorithmically efficient to confirm that the certificate is correct. What is not algorithmically efficient, is finding the sequence. Failure to find this sequence also generates a certificate that the framework is not universally rigid—by finding a higher dimensional set which includes equivalent, but not congruent, frameworks. Again, this is not efficient.

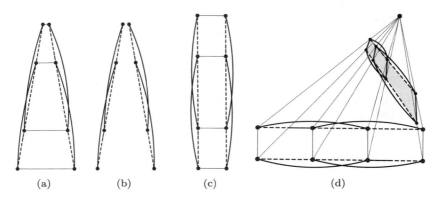

(a)   (b)   (c)   (d)

FIGURE 63.5.1
A universally rigid framework (a) will have super stable components (b), and members that have no self-stress at the first level of iteration. (c) shows a projectively equivalent framework which is not even rigid, though it is dimensionally rigid. (d) shows a cone of both (a) and (c) which is globally rigid, with a nonglobally rigid base.

Universal rigidity of a framework (G, p) is not preserved by projective transformations, because the property that the edge directions lie on a conic at infinity may change when a new "infinity" appears under the projection. That this is the only failure was confirmed by [Alf14, AN13]. Moreover, when the edge vectors at d vertices linearly span the space, the edge directions automatically avoid conics at infinity [Alf14].

63.5.1 PROJECTIVE TRANSFORMATIONS, CONING AND CHANGE OF METRIC

Given rigidity properties of a framework (G, p) in dimension d, we can often extend these properties to a coned framework in $(d+1)$-space, and then project to a different hyperplane to obtain another d-dimensional realization (G, q) as a slice of the cone, preserving the properties. This is a concrete form of a projective transformation from (G, p) to (G, q). Alternatively, there can be a direct analysis of the impact of a projective transformation from p to $q = f(p)$ on the various properties [SW07]. As Chapter 61 describes, these projective transformations preserve the first-order rigidity of (G, p), and have a clearly described impact on the coefficients of any equilibrium stress of (G, p). These processes also provide the tools to confirm the transfer of these basic properties among Euclidean, spherical, Minkowskian, and hyperbolic metrics, all of which live in a common projective space. The geometric transfer, for a specific projective configuration $\tilde{p}$ in any of the metrics, then gives corresponding combinatorial transfers of generic properties.

Here we look at the extensions of these techniques and results to global rigidity, dimensional rigidity, super stability, and universal rigidity for bar-and-joint frameworks.

GLOSSARY

Projective transform of a configuration p of points in d-space: A configuration q on the same vertices in d-space such that there is an invertible $(d+1) \times (d+1)$ matrix T for which $T(p_i, 1) = \lambda_i(q_i, 1)$ for all $1 \leq i \leq n$ (where λ_i is a scalar and $(p_i, 1)$ is the vector p_i extended with an additional 1—the affine coordinates of p_i).

Cone slice from v_0: For a $(d+1)$-dimensional configuration p of the vertices $\{v_0, v_1, \ldots, v_n\}$, a configuration q in d-space (placed as a hyperplane in $(d+1)$-space) of the vertices $\{v_1, .., v_n\}$, such that q_i is on the line through p_0, p_i for all $1 \leq i \leq n$. We use $q = \Pi_0(p)$ to denote that q is obtained from p by a cone slice from v_0 (Figure 63.2.4(b,c)).

CONING

The Generic Global Rigidity Coning Theorem (Theorem 63.2.9) fails as a statement for specific geometric frameworks, as there are both examples where the plane framework is globally rigid and the cone is not [CW10], and examples where the cone framework is globally rigid and the projection back to the plane is not globally rigid (or even rigid) [CG15]. See Figure 63.5.1 for illustrative examples.

On the other hand there are strong geometric results for dimensional rigidity and super stability. In the following statements on coning (due to Connelly and Gortler [CG17]) we shall use (G, p) and $(G * u, q)$ to denote a framework in d-space and a corresponding cone framework in $(d + 1)$-space. Note this includes all cone frameworks with the same cone slice.

THEOREM 63.5.11   Dimensional Rigidity Coning

The framework (G, p) is dimensionally rigid in $\mathbb{R}^d$ if and only if the cone framework $(G * u, q)$ is dimensionally rigid in $\mathbb{R}^{d+1}$.

THEOREM 63.5.12   Super Stability Coning

The framework (G, p) is super stable in $\mathbb{R}^d$ if and only if the cone framework $(G * u, q)$ is super stable in $\mathbb{R}^{d+1}$.

There is a standard *sliding* operation on cone frameworks, where vertices p_i slide along the line through p_0 and p_i, perhaps even through p_0 to the other side along the same line, so that they remain distinct from the cone vertex p_0. The following result from [CG17] follows from the fact that sliding does not change the projected framework of the cone to $\mathbb{R}^d$.

COROLLARY 63.5.13 Sliding in Cones

Let $(G * u, p)$ be a cone framework in $\mathbb{R}^{d+1}$ and let $(G * u, q)$ be obtained from it by sliding. Then $(G * u, p)$ is super stable (resp. dimensionally rigid) if and only if $(G * u, q)$ is super stable (resp. dimensionally rigid).

THEOREM 63.5.14 Universal Rigidity Coning

If the framework (G, p) is super stable in $\mathbb{R}^d$ then the cone framework $(G * u, q)$ is universally rigid in $\mathbb{R}^{d+1}$.

If the cone framework $(G * u, q)$ is universally rigid in $\mathbb{R}^{d+1}$ then the framework (G, p) is dimensionally rigid in $\mathbb{R}^d$.

The only failure for projecting a universally rigid cone framework to a universally rigid framework can come from the appearance of affine motions due to the projection having edge directions on a conic at infinity. If, for example, the framework (G, p) is in general position, this cannot happen, as was observed by Alfakih [AY13, Alf17].

PROJECTIVE TRANSFORMATIONS

A general projection in $\mathbb{R}^d$ can be formed by coning to $\mathbb{R}^{d+1}$, rotating, and reprojecting, perhaps several times. As a result, the coning results guarantee the projective invariance of key properties.

THEOREM 63.5.15 Projective Invariance of Dimensional Rigidity and Super Stability

A framework (G, p) is super stable (resp. dimensionally rigid) in $\mathbb{R}^d$ if and only if every invertible projective transformation which keeps vertices finite is super stable (resp. dimensionally rigid) in $\mathbb{R}^d$.

This projective invariance is almost always true for universal rigidity. The key is whether the directions of the edges meet the set X which is going to infinity, in a conic.

THEOREM 63.5.16 Projection and Universal Rigidity

If a framework (G, p) is universally rigid in $\mathbb{R}^d$, and X is a hyperplane avoiding all vertices, such that the edge directions do not meet it in a conic, then any invertible projective transformation T which takes X to infinity makes $(G, T(p))$ universally rigid.

When a nested sequence of affine spaces, with a sequence of PSD matrices is used to iteratively demonstrate the dimensional rigidity (universal rigidity) of a given framework (G, p), the projected framework $(G, T(p))$ has the projected affine sequence to demonstrate its dimensional rigidity (universal rigidity).

It is well known that global rigidity is not projectively invariant for specific frameworks (G, p). However, the rank of any stress matrix Ω on the framework is invariant under projective transformations [CW10]. We might move to a configuration where the rank of the stress matrix is not sufficient to guarantee global rigidity.

THEOREM 63.5.17

If a framework (G, p) is globally rigid in $\mathbb{R}^d$ then there is an open neighborhood $O(p)$ among projective images of p such that (G, q) is globally rigid in $\mathbb{R}^d$ for all $q \in O(p)$.

CHANGE OF METRIC

It is a classical result that infinitesimal rigidity, and associated properties (even with symmetry) transfer from Euclidean metric to all the other metric spaces built by Cayley-Klein metrics on the shared projective space [SW07, CW10]. See also Chapters 61 and 62. It is natural to ask how the properties of global rigidity and universal rigidity transfer. Coning, along with sliding, takes frameworks from the Euclidean metric in $\mathbb{R}^d$ to the Spherical metric in $\mathbb{S}^d$.

Given a projective configuration p in d-dimensional projective space, a corresponding configuration in $\mathbb{R}^d$ (all vertices finite) is called $R(p)$ and a corresponding configuration in $\mathbb{S}^d$ is called $S(p)$.

THEOREM 63.5.18   Super Stability Transfer to the Spherical Metric

For a given graph G and a fixed configuration p in projective space of dimension d, the framework $(G, R(p))$ is super stable in Euclidean metric space $\mathbb{R}^d$ if only if $(G, S(p))$ is super stable in the spherical metric $\mathbb{S}^d$.

THEOREM 63.5.19   Global and Universal Rigidity Transfer to the Spherical Metric

A given graph G is generically globally rigid in Euclidean metric space if and only if G is generically globally rigid in the spherical metric.

For a given graph G and a fixed configuration p in projective space of dimension d, (G, p) is universally rigid in the spherical metric if and only if the framework (G, p) is universally rigid in Euclidean metric space, for almost all projections.

Basically, all results and methods transfer from Euclidean metric space to the spherical metric.

Gortler and Thurston [GT14b] considered generic global rigidity in complex space and pseudo-Euclidean metrics (metrics with more general signatures). Here are a few results for generic configurations.

THEOREM 63.5.20   Transfer of Metrics: Complex Space

A graph G is globally rigid at some (all) generic configurations in Euclidean metric space if and only if G is globally rigid at some (all) generic configurations in complex metric space of the same dimension.

THEOREM 63.5.21   Transfer of Metrics: Pseudo-Euclidean

If a graph G is globally rigid at generic configurations in Euclidean metric space, then G is globally rigid at all generic configurations in every alternate Cayley-Klein metric.

If a graph G contains a d-simplex, then G is globally rigid at some generic configurations in Euclidean d-space if and only if G is globally rigid at every generic configuration in every d-dimensional alternate Cayley-Klein metric.

The gap in Theorem 63.5.21 is that there may be additional graphs that are globally rigid for *some* generic configurations in a pseudo-Euclidean space but not globally rigid at other generic configurations in the same space, and are therefore *never* globally rigid at a generic configuration in Euclidean metric space. Alternatively, it is an *open problem* in the pseudo-Euclidean metric whether global rigidity is "generic": *does one generic globally rigid framework imply that all generic frameworks are globally rigid?*

One technique for these transfers among metrics is the *Pogorelov map*, which takes any pair of equivalent frameworks (G, p), (G, q) in one metric M to another pair (G, p'), (G, q') in another metric M' through the process of averaging to an infinitesimally flexible framework $(G, p + q)$ in M. By first-order principles, this transfers to confirm that $(G, p+q)$ is infinitesimally flexible in M', and de-averages to an equivalent pair (G, p'), (G, q') in M' [GT14b, CW10].

63.6 TENSEGRITY FRAMEWORKS

In a tensegrity framework, we replace some (or all) of the equalities for bars with inequalities for the distances, corresponding to *cables* (the distance can shrink but not expand) and *struts* (the distance can expand but not shrink). See Chapter 61 for basic results on the infinitesimal rigidity of tensegrity frameworks. Many results on universal rigidity transfer directly from results for bar frameworks, as soon as we align the cables and struts with the sign pattern of the self-stress with a PSD stress matrix.

GLOSSARY

Tensegrity graph: A graph $T = (V; B, C, S)$ with a partition (or labelling) of the edges into three classes, called *bars*, *cables*, and *struts*. The edges of T may be called *members*. In figures, cables are indicated by dashed lines, struts by thick lines, and bars by single thin lines.

Tensegrity framework in $\mathbb{R}^d$: A pair (T, p), where T is a tensegrity graph and p is a configuration p of the vertex set of T in $\mathbb{R}^d$.

(T, p) **dominates** (T, q)**:** For each member of T, the appropriate condition holds:

$$|p_i - p_j| = |q_i - q_j| \quad \text{when} \quad \{i, j\} \in B$$
$$|p_i - p_j| \geq |q_i - q_j| \quad \text{when} \quad \{i, j\} \in C$$
$$|p_i - p_j| \leq |q_i - q_j| \quad \text{when} \quad \{i, j\} \in S.$$

Globally rigid tensegrity framework: A d-dimensional tensegrity framework (T, p) for which any other realization (T, q) in $\mathbb{R}^d$, dominated by (T, p) is congruent to p.

Universally globally rigid tensegrity framework: A d-dimensional tensegrity framework (T, p) for which any other realization (T, q) in any dimension, dominated by (T, p) is congruent to p.

Proper equilibrium stress on a tensegrity framework (T, p): An assignment ω of scalars to the members of T such that:

(a) $\omega_{ij} \geq 0$ for cables $\{i, j\} \in C$;

(b) $\omega_{ij} \leq 0$ for struts $\{i, j\} \in S$; and

(c)   for each vertex i, $\sum_{\{j \mid \{i,j\} \in E\}} \omega_{ij}(p_j - p_i) = 0$.

Strict proper equilibrium stress: A proper equilibrium stress ω with the inequalities in (a) and (b) strict.

Underlying bar framework: For a tensegrity framework (T, p), where $T = (V; B, C, S)$, the bar framework $(\bar{G}, p)$ on the graph $\bar{G} = (V, E)$ with $E = B \cup C \cup S$.

Spiderweb: A labelled graph $G_- = (V_0, V_1; C)$, with pinned vertices V_0 and cable members only, with $C \subset V_1 \times [V_0 \cup V_1]$, and a configuration p for $V_0 \cup V_1$.

Spiderweb self-stress on (G_-, p): An assignment ω of nonnegative scalars to C such that for each unpinned vertex $i \in V_1$, $\sum_{\{j \mid \{i,j\} \in C\}} \omega_{ij}(p_j - p_i) = 0$.

Spiderweb flex: A flex $p(t)$ for (G_-, p) with all pinned vertices fixed.

63.6.1 BASIC RESULTS FOR TENSEGRITIES

All the definitions and results for dimensional rigidity and superstability of frameworks extend to proper equilibrium stresses of tensegrity frameworks. The following result confirms that super stability transfers from bar frameworks to tensegrity frameworks. These transfers flow from [CG17, CGT16a, CGT16b].

THEOREM 63.6.1 Dimensional Rigidity of Tensegrity Frameworks

Let (T, p) be a d-dimensional tensegrity framework, where the affine span of the joint positions is all of $\mathbb{R}^d$, with a proper equilibrium stress ω and stress matrix Ω. Suppose further that

(i) *Ω is positive semi-definite,*

(ii) *the rank of Ω is $n - d - 1$.*

Then (T, p) is dimensionally rigid.

THEOREM 63.6.2 The Fundamental Theorem of Tensegrity Frameworks

Let (T, p) be a d-dimensional tensegrity framework, where the affine span of the joint positions is all of $\mathbb{R}^d$, with a proper equilibrium stress ω and stress matrix Ω. Suppose further that

(i) *Ω is positive semi-definite,*

(ii) *the rank of Ω is $n - d - 1$,*

(iii) *the stressed member directions of (G, p) do not lie on a conic at infinity.*

Then (T, p) is universally rigid.

For example, Connelly [Con82] proved that a tensegrity framework obtained from a planar polygon by putting a joint at each vertex, a cable along each edge, and struts connecting other vertices (a *tensegrity polygon*) such that the resulting tensegrity has some nonzero proper equilibrium stress satisfies conditions (i)-(iii) and hence it is universally globally rigid. Geleji and Jordán [GJ13] characterized the tensegrity polygons for which all convex realizations in $\mathbb{R}^2$ possess a nonzero proper equilibrium stress.

The following result of Bezdek and Connelly [Con06] is an initial analogous result in 3-space.

THEOREM 63.6.3

If a tensegrity framework in $\mathbb{R}^3$ has cables along the edges of a convex centrally symmetric polyhedron, and struts connecting antipodal vertices, then it is super stable.

Note that within these theorems are many examples of universally rigid tensegrity frameworks which are *not infinitesimally rigid*. There is a version of the stability theorem that extends even to these special position frameworks.

COROLLARY 63.6.4   Tensegrity Stability Theorem

If we take a globally rigid framework where Ω is positive semi-definite of the required rank, then within the variety of projectively equivalent frameworks there is an open neighborhood $O(p)$ within which all frameworks are universally rigid.

A special result for the modified spiderwebs further illustrates the role of tensegrity frameworks.

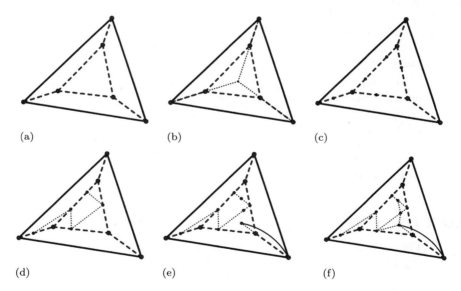

(a)   (b)   (c)

(d)   (e)   (f)

FIGURE 63.6.1

The development of super stable tensegrity frameworks. Figure (a) has a spiderweb self-stress, due to the geometry of the three concurrent edges (b). It is super stable but not infinitesimally rigid. Figure (c,d) shows ways of extending the framework that preserve universal rigidity, although the added edges in (d) will not have a nonzero self-stress, following the inductions below. Moreover, we can add vertices outside a cable edge (Figure (e) and continue to add edges and new vertices, preserving the universal rigidity.

THEOREM 63.6.5   Spiderweb Rigidity

Any spiderweb (G_-, p) in d-space with a spiderweb self-stress, positive on all cables, is super stable in d-space.

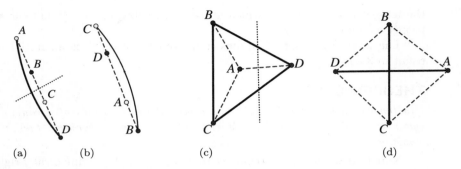

FIGURE 63.6.2
If a universally rigid tensegrity framework is projected, the edges cut by the line going to infinity change class and their stress changes sign (a,b), (c,d)

THEOREM 63.6.6   Projective Invariance of Super Stability of Tensegrity Frameworks

Let $f : X \to \mathbb{R}^d$ be an invertible projective transformation, where X is a $(d-1)$-dimensional affine subspace of $\mathbb{R}^d$, and suppose that for each i, $p_i \notin X$. Further assume that the member directions of (G,p) do not lie on a conic at infinity, and that the member directions of $(G, f(p))$ do not lie on a conic on X for a projective transformation f. Then the tensegrity framework (G,p) is super stable if and only if $(G, f(p))$ is super stable, where the strut/cable designation for $\{i,j\}$ changes only when the line segment $[p_i, p_j]$ intersects X and bars go to bars.

Coning also extends to tensegrity frameworks. Make all the edges incident to the cone vertex bars, and initially preserve the designation of cables and struts, with configurations near the original $\mathbb{R}^d$ (relative to the cone vertex p_0). Then switch the cable and strut designation under sliding only when a vertex slides across the cone vertex (one vertex at a time). This process will transfer all the relevant tensegrity results from the Euclidean metric to tensegrities in the spherical metric, with the same caveats as for bar frameworks.

INDUCTIVE CONSTRUCTIONS FOR TENSEGRITIES

The *tensegrity gluing* of two tensegrity graphs $T_1 = (V_1; B_1, C_1, S_1)$ and $T_2 = (V_2; B_2, C_2, S_2)$ is the tensegrity graph T with vertices $V = V_1 \cup V_2$, $B = B_1 \cup B_2 \cup (C_1 \cap S_2) \cup (S_1 \cap C_2)$, $C = C_1 \cup C_2 - B$, and $S = S_1 \cup S_2 - B$.

The following is a corollary of the previous gluing results.

COROLLARY 63.6.7   Tensegrity Gluing

If (T_1, p_1) and (T_2, p_2) are universally rigid (resp. dimensionally rigid) tensegrity frameworks in $\mathbb{R}^d$ sharing at least $d+1$ points affinely spanning $\mathbb{R}^d$ then the tensegrity glued framework $(T, p_1 \cup p_2)$ is universally rigid (resp. dimensionally rigid) in $\mathbb{R}^d$.

Appendix A of [CW96] includes a process of: (i) adding a new vertex p_c anywhere along the interior of a cable $[p_i p_j]$, splitting the cable into two cables $\{i, c\}$, $\{c, j\}$ and (ii) inserting a new vertex p_s along the exterior of a strut $[p_i p_j]$ (say along the p_j side) splitting the strut into a strut $\{i, s\}$ and a cable $\{s, j\}$. We note that the strut insertion is the same as a cable insertion on a projective image of the framework. See Figure 63.6.1(c,d,e).

These insertions provide inductive steps that preserve dimensional rigidity, super stability, and universal rigidity of specific geometric tensegrity frameworks. This provides a geometric inductive process for building up universally rigid frameworks, onto which one can then hang further—even nongeneric—frameworks by geometric gluing.

Applied just to cables, the insertions take a spiderweb to a spiderweb. Again, one can insert vertices and then glue in another spiderweb, iterating to create larger spiderwebs.

63.7 SOURCES AND RELATED MATERIALS

SURVEYS AND BASIC SOURCES

We refer the reader to the following books, book chapters, survey articles and recent substantial articles for a more detailed overview of this field.

[CG16]: A forthcoming book on frameworks and tensegrities with a number of connections to global rigidity.

[Alf14]: Basic results for universal rigidity and dimensional rigidity.

[CG15]: Key recent article for super stability and the iterative constructions, with extensions to tensegrity frameworks.

[CG17]: Basic results for complete bipartite frameworks and current results in universal rigidity, including coning and projection.

[JJ09a]: A survey chapter on graph theoretic techniques in sensor network localization.

[AEG+06]: An introduction to the links between sensor network localization and combinatorial rigidity.

[Jor10b]: A survey chapter on rigid and globally rigid pinned frameworks.

[Jor16]: A short monograph on generic rigidity and global rigidity in the plane.

[Whi96]: An expository article presenting matroidal aspects of first-order rigidity, redundant rigidity, scene analysis, and multivariate splines.

[Wiki]: A wiki site with a number of preprints, including original papers on global rigidity.

RELATED CHAPTERS

Chapter 9: Geometry and topology of polygonal linkages
Chapter 57: Solid modeling
Chapter 61: Rigidity and scene analysis
Chapter 62: Symmetry and rigidity

REFERENCES

[AEG+06]   J. Aspnes, T. Eren, D.K. Goldenberg, A.S. Morse, W. Whiteley, Y.R. Yang, B.D.O. Anderson, and P.N. Belhumeur. A theory of network localization. *IEEE Trans. Mobile Comput.*, 5:1663–1678, 2006.

[Alf07]   A.Y. Alfakih. On dimensional rigidity of bar-and-joint frameworks. *Discrete Appl. Math.*, 155:1244–1253, 2007.

[Alf11]   A.Y. Alfakih. On bar frameworks, stress matrices and semidefinite programming. *Math. Progr. Ser. B*, 129:113–128, 2011.

[Alf14]   A.Y. Alfakih. Local, dimensional and universal rigidities: A unified Gram matrix approach. In R. Connelly, A.I. Weiss, and W. Whiteley, editors, *Rigidity and Symmetry*, vol. 70 of *Fields Inst. Comm.*, pages, 41-60, Springer, New York, 2014.

[Alf17]   A.Y. Alfakih. Graph connectivity and universal rigidity of bar frameworks. *Discrete Appl. Math.*, 217:707–710, 2017.

[AN13]   A.Y. Alfakih and V.-H. Nguyen. On affine motions and universal rigidity of tensegrity frameworks. *Linear Algebra Appl.*, 439:3134–3147, 2013.

[AY13]   A.Y. Alfakih and Y. Ye. On affine motions and bar frameworks in general positions. *Linear Algebra Appl.*, 438:31–36, 2013.

[BJ03]   A.R. Berg and T. Jordán. A proof of Connelly's conjecture on 3-connected circuits of the rigidity matroid. *J. Combin. Theory Ser. B*, 88:77–97, 2003.

[BS04]   C. Borcea and I. Streinu. The number of embeddings of minimally rigid graphs. *Discrete Comput. Geom.*, 31:287–303, 2004.

[CH88]   G.M. Crippen and T.F. Havel. *Distance Geometry and Molecular Conformation*. Research Studies Press, Taunton, and Wiley, New York, 1988.

[CJK16]   K. Clinch, B. Jackson, and P. Keevash. Global rigidity of 2-dimensional direction-length frameworks. Preprint, `arXiv:1607.00508`, 2016.

[Cli16]   K. Clinch. Global rigidity of 2-dimensional direction-length frameworks with connected rigidity matroids. Preprint, `arXiv:1608.08559`, 2016.

[Con82]   R. Connelly. Rigidity and energy. *Invent. Math.*, 66:11–33, 1982.

[Con91]   R. Connelly. On generic global rigidity. In: *Applied Geometry and Discrete Mathematics*, vol. 4 of *DIMACS Ser. Discrete Math. Theoret. Comp. Sci.*, pages 147–155, AMS, Providence, 1991.

[Con05]   R. Connelly. Generic global rigidity. *Discrete Comput. Geom.* 33:549–563, 2005.

[Con06]   R. Connelly (with K. Bezdek). Stress matrices and M matrices. *Oberwolfach Reports*, 3:678–680, 2006. Extended version available at `http://www.math.cornell.edu/~connelly/oberwolfach.connelly.pdf`.

[Con11]   R. Connelly. Combining globally rigid frameworks. *Proc. Steklov Inst. Math.*, 275:191–198, 2011.

[CG15]   R. Connelly and S.J. Gortler. Iterative universal rigidity. *Discrete Comput. Geom.*, 53:847–877, 2015.

[CG16]   R. Connelly and S.D. Guest. *Frameworks, Tensegrities and Symmetry: Understanding Stable Structures*. Cambridge University Press, 2016.

[CG17]   R. Connelly and S.J. Gortler. Universal rigidity of complete bipartite graphs. *Discrete Comput. Geom.*, 57:281–304, 2017.

[CGT16a]   R. Connelly, S.J. Gortler, and L. Theran. Generic global and universal rigidity, Preprint, `arXiv:1604.07475`, 2016.

[CGT16b]   R. Connelly, S.J. Gortler, and L. Theran. Affine rigidity and conics at infinity. Preprint, `arXiv:1605.07911`, 2016.

[CJW13]   R. Connelly, T. Jordán, and W. Whiteley. Generic global rigidity of body-bar frame-works. *J. Combin. Theory Ser. B.*, 103:689–705, 2013.

[CW96]   R. Connelly and W. Whiteley. Second-order rigidity and pre-stress stability for tensegrity frameworks. *SIAM J. Discrete Math.*, 9:453–492, 1996.

[CW10]   R. Connelly and W. Whiteley. Global rigidity: the effect of coning, *Discrete Comput. Geom.*, 43:717–735, 2010.

[FJ11]   S. Frank and J. Jiang. New classes of counterexamples to Hendrickson's global rigidity conjecture. *Discrete Comput. Geom.*, 45:574–591, 2011.

[GGLT13]   S.J. Gortler, C. Gotsman, L. Liu, and D.P. Thurston. On affine rigidity. *J. Comput. Geom.*, 4:160–181, 2013.

[GHT10]   S.J. Gortler, A. Healy, and D.P. Thurston. Characterizing generic global rigidity. *Amer. J. Math.*, 132:897–939, 2010.

[GJ13]   J. Geleji and T. Jordán. Robust tensegrity polygons. *Discrete Comput. Geom.*, 50:537–551, 2013.

[GT14a]   S.J. Gortler and D.P. Thurston. Characterizing the universal rigidity of generic frameworks, *Discrete Comput. Geom.*, 51:1017–1036, 2014.

[GT14b]   S.J. Gortler and D.P. Thurston. Generic global rigidity in complex and pseudo-Euclidean spaces. In R. Connelly, A. Ivic Weiss, and W. Whiteley, editors, *Rigidity and Symmetry*, vol. 70 of *Fields Inst. Comm.*, pages 131–154, Springer, New York, 2014.

[Hen82]   B. Hendrickson. Conditions for unique graph realizations. *SIAM J. Comput.* 21:65–84, 1992.

[JJ05]   B. Jackson and T. Jordán. Connected rigidity matroids and unique realizations of graphs. *J. Combin. Theory Ser. B*, 94:1–29, 2005.

[JJ09a]   B. Jackson and T. Jordán. Graph theoretic techniques in the analysis of uniquely localizable sensor networks. In G. Mao and B. Fidan, editors, *Localization Algorithms and Strategies for Wireless Sensor Networks*, pages 146–173, IGI Global, Hershey, 2009.

[JJ09b]   B. Jackson and T. Jordán. A sufficient connectivity condition for generic rigidity in the plane. *Discrete Appl. Math.*, 157:1965–1968, 2009.

[JJ10a]   B. Jackson, T. Jordán. Globally rigid circuits of the direction-length rigidity matroid. *J. Combin. Theory Ser. B*, 100:1–22, 2010.

[JJ10b]   B. Jackson, T. Jordán. Operations preserving global rigidity of generic direction-length frameworks. *Internat. J. Comput. Geom. Appl.*, 20:685–706, 2010.

[JJS06]   B. Jackson, T. Jordán, and Z. Szabadka. Globally linked pairs of vertices in equivalent realizations of graphs. *Discrete Comput. Geom.*, 35:493–512, 2006.

[JJS14]   B. Jackson, T. Jordán, and Z. Szabadka. Globally linked pairs of vertices in rigid frameworks. In R. Connelly, A. Ivic Weiss, W. Whiteley, editors, *Rigidity and Symmetry*, vol. 70 of *Fields Inst. Comm.*, pages 177–203, Springer, New York, 2014.

[JK11]   B. Jackson and P. Keevash. Necessary conditions for the global rigidity of direction-length frameworks. *Discrete Comput. Geom.*, 46:72–85, 2011.

[JKT16]   T. Jordán, C. Király, and S. Tanigawa. Generic global rigidity of body-hinge frameworks. *J. Combin. Theory Ser. B*, 117:59–76, 2016.

[JMN14]   B. Jackson, T.A. McCourt, and A. Nixon. Necessary conditions for the generic global rigidity of frameworks on surfaces. *Discrete Comput. Geom.*, 52:344–360, 2014.

[JN15a]   T. Jordán and V.-H. Nguyen. On universally rigid frameworks on the line. *Contrib. Discrete Math.*, 10:10–21, 2015.

[JN15b]   B. Jackson and A. Nixon. Stress matrices and global rigidity of frameworks on surfaces. *Discrete Comput. Geom.*, 54:586–609, 2015.

[Jor10a]   T. Jordán. Generically globally rigid zeolites in the plane. *Inform. Process. Lett.*, 110:841–844, 2010.

[Jor10b]   T. Jordán. Rigid and globally rigid graphs with pinned vertices. In G.O.H. Katona, A. Schrijver, T. Szőnyi, editors, *Fete of Combinatorics and Computer Science*, vol. 20 of *Bolyai Soc. Math. Studies*, pages 151–172, Springer, Berlin, 2010.

[Jor16]   T. Jordán. Combinatorial rigidity: graphs and matroids in the theory of rigid frameworks. In *Discrete Geometric Analysis*, vol. 34 of *MSJ Memoirs*, pages 33–112, 2016.

[Jor17]   T. Jordán. Extremal problems and results in combinatorial rigidity. In *Proc. 10th Japanese-Hungarian Sympos. on Discrete Math. Appl.*, pages 297–304, 2017.

[JS09]   T. Jordán and Z. Szabadka. Operations preserving the global rigidity of graphs and frameworks in the plane. *Comput. Geom.*, 42:511–521, 2009.

[JT17]   T. Jordán and S. Tanigawa. Global rigidity of triangulations with braces. Technical report TR-2017-06, Egerváry Research Group, Budapest, 2017.

[KHKS16]   Y. Kobayashi, Y. Higashikawa, N. Katoh, and A. Sljoka. Characterizing redundant rigidity and redundant global rigidity of body-hinge graphs. *Inform. Process. Lett.*, 116:175–178, 2016.

[KK16]   V.E. Kaszanitzky and C. Király. On minimally highly vertex-redundantly rigid graphs. *Graphs Combin.*, 32:225–240, 2016.

[LY82]   L. Lovász and Y. Yemini. On generic rigidity in the plane. *SIAM J. Algebr. Discrete Methods*, 3:91–98, 1982.

[Ngu12]   V.-H. Nguyen. 1-extensions and global rigidity of generic direction-length frameworks, *Internat. J. Comput. Geom. Appl.*, 22:577–591, 2012.

[Sax79]   J.B. Saxe. Embeddability of weighted graphs in k-space is strongly NP-hard. Tech. Report CMU-CS-80-102, Carnegie Mellon University, Pittsburgh, 1979.

[SW07]   F. Saliola and W. Whiteley. Some notes on the equivalence of first-order rigidity in various geometries. Preprint, `arXiv:0709.3354`, 2007.

[SY07]   A.M. So and Y. Ye. Theory of semidefinite programming for sensor network localization. *Math. Program.*, 109:367–384, 2007.

[Tan15]   S. Tanigawa. Sufficient conditions for the global rigidity of graphs. *J. Combin. Theory Ser. B*, 113:123–140, 2015.

[Whi96]   W. Whiteley. Some matroids from discrete applied geometry. In J. Bonin, J. Oxley, and B. Servatius, editors, *Matroid Theory*, vol. 197 of *Contemp. Math.*, pages 171–311, AMS, Providence, 1996.

[Whi99]   W. Whiteley. Rigidity of molecular structures: generic and geometric analysis. In P.M. Duxbury and M.F. Thorpe, editors, *Rigidity Theory and Applications*, Kluwer/Plenum, New York, 1999.

[Wiki]   W. Whiteley. Wiki of Preprint Resources in Rigidity Theory. `http://wiki.math.yorku.ca/index.php/Resources_in_Rigidity_Theory`.

64 CRYSTALS, PERIODIC AND APERIODIC
Marjorie Senechal

INTRODUCTION

Are you looking for the chapter on "Crystals and Quasicrystals"? Look no further: you have found it. Today the word "crystal" spans the periodic and the aperiodic alike. Just as in the nineteenth century "pseudogeometry"[1] soon became "non-Euclidean geometry," we speak of aperiodic crystals now.

64.1 CLASSICAL CRYSTALLOGRAPHY

At the turn of the 19th century a French mineralogist, R.J. Haüy, accidentally dropped a friend's fine specimen of calcite. It shattered into shards but, fortunately, chance favors the prepared mind. The hapless Haüy did not wail "tout est perdu!," he shouted "tout est trouvé!" instead. For, sweeping up the shards, he noticed that they were rhombohedra of different sizes but with the same interfacial angles. He'd found the answer to the problem he'd been pondering: why do some crystals of the same species have different external forms? Why, for example, are some pyrite crystals cubes, and others irregular pentagonal dodecahedra? Every crystal, Haüy quickly surmised, is a stack of countless identical, subvisible, building blocks, laid face to face, row after row, layer upon layer. The polyhedral forms we see are the stacks, which can be finished off in different ways.

Haüy's theory took hold, and with it the periodicity paradigm, which held (until the late 1970s) that the atoms in crystals are arranged in three-dimensional periodic patterns. Representing his blocks by the points at their centers, Haüy's building blocks became lattices. In this section we review the achievements of 19th century mathematical crystallography from that point forward.

GLOSSARY

Lattice: A group of translations of $\mathbb{R}^n$ generated by n linearly independent vectors.

Point lattice: The orbit of a point $x \in \mathbb{R}^n$ under the action of a lattice.

Basis for a lattice L: A set of n linearly independent vectors that generate L.

Dual lattice L^* of a lattice L: $L^* = \{\vec{y} \in \mathbb{R}^n : \vec{y} \cdot \vec{x} \in \mathbb{Z}, \vec{x} \in L\}$, where $\cdot$ denotes the usual scalar product.

Crystallographic group: A group of isometries that acts transitively on an infinite, discrete, point set.

---
[1] A term used by Poincaré and others

Unit cell (of a point lattice):   A parallelepiped whose edges are a (vector) basis for the lattice.

Point group (of a lattice L):   A group of isometries that fix a point of L.

Voronoi cell (of a point $x \in L$):   The polytope $V_L(x) \subset \mathbb{R}^n$ whose points are at least as close to x as to any other point of L. (See also Chapters 3 and 22.) By construction, $V_L(x)$ is invariant under the point group of x, whereas the unit cell of L may not be.

Voronoi tiling (of a lattice L):   The tiling whose tiles are the Voronoi cells of the points of L.

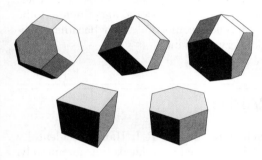

FIGURE 64.1.1

The five combinatorial types of Voronoi cells for lattices in $\mathbb{R}^3$ are, from the upper left clockwise, the truncated octahedron, the rhombic dodecahedron, the elongated rhombic dodecahedron, the cube, and the hexagonal prism.

MAIN RESULTS

1. $L^{**} = L$.

2. Lattices are classified by their symmetry groups and the combinatorial structure of their Voronoi cells. There are five lattices in $\mathbb{R}^2$; the fourteen lattices in $\mathbb{R}^3$ are called Bravais lattices after Auguste Bravais (1811–1863) who first enumerated them [Bra49].

3. By construction, the Voronoi cells of a lattice are congruent convex polytopes that fit together face-to-face, and the lattice acts transitively on the tiling. Every convex polytope that tiles in this fashion is centrally symmetric, its facets—its $(n-1)$-dimensional faces—are centrally symmetric, and each belt—set of parallel $(n-2)$-dimensional faces—has size four or six. The converse is also true [McM80].

4. Corollaries: (a) Easy: the Voronoi cell of a point lattice in $\mathbb{R}^2$ is a centrosymmetric quadrilateral or hexagon. (b) Not at all easy: in $\mathbb{R}^3$ there are five combinatorial types of lattice Voronoi cells (see Figure 64.1.1).

5. There are 17 crystallographic groups in $\mathbb{R}^2$, 230 in $\mathbb{R}^3$, and 4894 in $\mathbb{R}^4$ (see [BBN+78]).

6. **Bieberbach's Theorem.** A crystallographic group G in any dimension is a product of a translation group T and a finite group of isometries, where T is the maximal abelian subgroup of G. Thus an orbit of G is a union of a finite number of congruent lattices (Figure 64.1.2). This theorem solved part of Hilbert's 18th problem. See [Yan01] and [Sen96] for further discussion and references.

7. The order of the rotation subgroup of a point group of a lattice in $\mathbb{R}^2$ and $\mathbb{R}^3$ is 2, 3, 4, or 6. This theorem, which concerns lattices, not (material) crystals, was nevertheless called *The Crystallographic Restriction* before aperiodic crystals were discovered. (See [Sen96].)

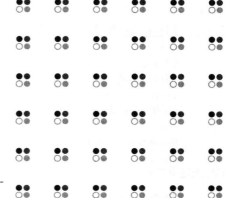

FIGURE 64.1.2
An orbit of a crystallographic group is a union of congruent lattices.

Table 64.1.1 shows the possible orders m, $2 \le m \le 13$, of rotational symmetries for point groups of lattices, and the lowest dimension $d(m)$ in which they can occur. We see that five-fold rotations, as well as n-fold rotations with $n > 6$, are "forbidden" in $\mathbb{R}^2$ and $\mathbb{R}^3$. (This table is easily computed from the formula in [Sen96, p. 51].)

TABLE 64.1.1 m-fold rotational symmetries.

m	$d(m)$	m	$d(m)$	m	$d(m)$	m	$d(m)$
2	1	5	4	8	4	11	12
3	2	6	2	9	6	12	4
4	2	7	6	10	4	13	12

REMARK: LATTICES AND CRYSTAL FORMS

Like Haüy, Bravais tried to link crystal form to crystal growth. "Bravais' Law" (see [Aut13]) states:

> The faces that appear on a crystal are parallel to the lattice planes of greatest density.

For a brief discussion of this "law" and the physical assumptions behind it, see [Sen90]. It follows from those assumptions that the visible, polyhedral shape of the grown crystal is the Voronoi cell of its dual lattice. For periodic crystals with relatively simple structures, agreement with reality is reasonably good.

OPEN PROBLEM

Voronoi's conjecture that every polytope that tiles $\mathbb{R}^n$ by translation is an affine image of the Voronoi cell of a lattice in $\mathbb{R}^n$ has been proved for zonotopes ([Erd99] and certain other special cases) but the general case remains open for $n > 4$ ([Mag15].

64.2 DELAUNAY SETS

Classical mathematical crystallography, outlined above, developed symbiotically with group theory and focused on atomic patterns as wholes. In the 1930s B.N. Delaunay and A.D. Aleksandrov, together with the crystallographer N.N. Padurov, took a local approach, beginning with very general point sets [DAP34]. Delauany sets, as these sets are called today, can be (and are being) used to model gases, liquids and liquid crystals, as well as solid materials.

GLOSSARY

Delaunay set: A Delaunay set is a point set $\Lambda \subset \mathbb{R}^n$ that is uniformly discrete and relatively dense. That is, Λ satisfies two conditions:

1. There is a real number $r > 0$ such that every open ball of radius r contains at most one point of Λ;
2. There is a real number $R > 0$ such that every closed ball of radius greater than R contains at least one point of Λ.

q-Star $St(x,q)$ of $x \in \Lambda$: $St(x,q) := \Lambda \cap B(x,q)$, where $B(x,q)$ is the ball of radius q and center x.

q-Atlas of Λ: A set of representatives of the translation classes of the q-stars of Λ.

Star of $x \in \Lambda$: $\lim_{q \to \infty} St(x,q)$.

Patch-counting function $N_\Lambda(q)$:   The size of the q-atlas of Λ.

Regular system of points: A Delaunay set whose stars are congruent; equivalently, an orbit of an infinite group of isometries. The union of a finite number of regular systems is said to be *multiregular*.

EXAMPLES

- Any bi-infinite set of points on a line with a finite set of distinct interpoint spacings $\ell_1, \ldots, \ell_k$ is a Delaunay set, with $r = \min(\ell_i)$ and $R = \max(\ell_i)$.

- Figure 64.2.1 shows a portion of a Delaunay set in $\mathbb{R}^2$ and, for an arbitrarily chosen value of q, the q-stars of several of its points.

MAIN RESULTS

1. A Delaunay set $\Lambda \subset \mathbb{R}^n$ is countably infinite.

2. For any $x, y \in \Lambda$, the distance $|x - y|$ is at least r.

3. The distance from $x \in \Lambda$ to any vertex of $V(x)$ is at most R.

4. For every $x \in \Lambda$, $B(x, r) \subset V(x) \subset B(x, 2R)$. Thus the Voronoi tiling is normal (see Chapter 3).

5. The Local Theorem: There is a real number k such that if $N_\Lambda(2Rk) = 1$, then Λ is a regular system of points [DDSG76].

6. If an orbit of a group of isometries of $\mathbb{R}^n$ is a Delaunay set, then the group is crystallographic [DLS98].

The first four results above are easy exercises.

FIGURE 64.2.1
Left: a portion of a Delaunay set in $\mathbb{R}^2$. Right: q-stars of seven of its points. Note that, in this example, $q > R$.

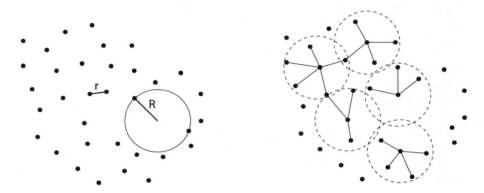

A CLASSIFICATION OF DELAUNAY SETS

The discovery of aperiodic crystals ([Wol74],[JJ77], [SBG$^+$84] brought Delaunay sets renewed attention. We present here a classification proposed by J. Lagarias in [Lag99], [Lag00], and [LP02]. As above, Λ is a Delaunay set in $\mathbb{R}^n$.

GLOSSARY

Difference set of Λ: The vector set $\Lambda - \Lambda = \{x - y; x, y \in \Lambda\}$.

Finite type: The patch-counting function $N_\Lambda(q)$ is finite for every positive real number q.

Repetitive: For every $q > 0$, the stars of the q-atlas are relatively dense in $\mathbb{R}^n$. That is, for each star $St(x, q)$ there is an $R_s > 0$ such that every ball of radius R_s contains a copy of the star.

Linearly repetitive: $N_\Lambda(q) = O(q)$.

Complexity function $M_\Lambda(q)$ of Λ: $M_\Lambda(q)$ is the minimal radius of a ball in $\mathbb{R}^n$ such that every ball of that radius contains the center of a copy of every q-star ([LP02]).

Meyer set: A point set Λ for which $\Lambda - \Lambda$ is a Delaunay set [Mey95].

MAIN RESULTS

Like the classification of Delaunay sets, these results are due to J. Lagarias [Lag99].

1. If Λ has $m = n + k$ generators, each of its points can be associated to an integral m-tuple. This defines an injection Φ from Λ to a lattice $L \subset \mathbb{R}^m$ (crystallographers refer to L as "superspace").
2. Λ is of finite type if and only if $\Lambda - \Lambda$ is finitely generated, closed and discrete.
3. If Λ is of finite type, then $|\Phi(x) - \Phi(x')| < C|x - x'|$; that is, the distance between points in Λ is proportional to the distance between their addresses in $\mathbb{R}^m$.
4. If Λ is linearly repetitive, then there is a linear map $\tilde{L}(x)$ such that $|\Phi(x) - \tilde{L}(x)| = o(|x|)$.
5. If Λ is a Meyer set, then $|\Phi(x) - \tilde{L}(x)| \leq C$; that is, the addresses of the points of Λ lie in a bounded strip in $\mathbb{R}^m$.

The last result above suggests that, for Meyer sets, we can reverse the process: instead of lifting a Delaunay set from $\mathbb{R}^n$ to a lattice in a higher-dimensional space $\mathbb{R}^m$, we can construct Delaunay sets in $\mathbb{R}^n$ by projecting bounded strips in R^m onto an n-dimensional subspace.

More precisely, let L be a lattice of rank $m = k + n$ in $\mathbb{R}^m$; let $p_{\|}$ and $p_{\perp}$ be the orthogonal projections into a n-dimensional subspace $\mathcal{E} = \mathbb{R}^n$ and its orthogonal complement $\mathcal{E}^{\perp} = \mathbb{R}^k$, respectively. Assume that $p_{\|}$, restricted to L, is one-to-one and $p_{\perp}(L)$ is everywhere dense in $\mathbb{R}^k$. Let Ω be a bounded subset of $\mathbb{R}^k$ with nonempty interior. The points of L for which $p_{\perp}(x) \in \Omega$ lie in such a strip, and their projection onto $\mathcal{E}$ is a Meyer set. Equivalently, we can place a copy of the window Ω at every point of L; the Meyer set is the projection onto $\mathcal{E}$ of the lattice points whose windows it cuts. (In this latter construction the window is often called a "density.")

GLOSSARY

Cut and project set: Let L, $p_{\|}$, $p_{\perp}$, and Ω be as above. The set

$$\Lambda(\Omega) = \{p_{\|}(x) \mid x \in L, \; p_{\perp}(x) \in \Omega\} \tag{64.2.1}$$

is called a cut-and-project set.

Window: The bounded set $\Omega \subset \mathcal{E}^{\perp}$ is the window of the projection. When Ω is a translate of $p_{\perp}(V_{\Lambda})$, the window is said to be **canonical**.

The window of a cut-and-project set contains detailed information about the q-stars of the set.

The ingredients for a one-dimensional cut-and-project set are shown in Figure 64.2.2. Here $m = 2$, $n = k = 1$, and L is a square lattice. The subspace $\mathcal{E}$ is a solid line of positive slope which, we assume, is irrational (to guarantee that the projections meet our requirements). The window Ω is the thick line segment in $\mathcal{E}^{\perp}$. To construct the model set we project onto $\mathcal{E}$ those lattice points x for which $p_{\perp}(x) \in \Omega$ (equivalently, x lies in the cylinder bounded by the dotted lines). Note that the window in Figure 64.2.2 is *not* canonical.

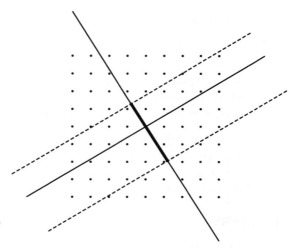

FIGURE 64.2.2
Ingredients for a one-dimensional model set.
The subspace $\mathcal{E}$ is the solid line; the window
Ω is the thick line segment in $\mathcal{E}^{\perp}$.

OPEN PROBLEMS

1. For $\mathbb{R}^2$, the value of k in the local theorem is 2. For $\mathbb{R}^3$, it is at most 10 [Dol17]. Find k for each $n > 2$.

2. State and prove the local theorem for Delaunay sets in other spaces (e.g., spherical, hyperbolic). (A local theorem for multiregular system of points is discussed in [DLS98].)

3. Formulate and prove appropriate "local theorems" for repetitive and model sets.

64.3 WHAT IS A CRYSTAL?

The relation between a point set representing the atomic pattern of a crystal and the crystal's diffraction pattern can be summarized in a commutative ("Wiener") diagram [Sen96]. Here $\updownarrow$ denotes Fourier transformation and $\rho(x)$ is the tempered distribution $\rho(x) = \sum_{x \in X} \delta_x$, with δ_x the Dirac delta at x.

$$\rho(x) \overset{\text{autocorrelation}}{\longrightarrow} \rho(x) * \overline{\rho(-x)}$$

$$\updownarrow \qquad\qquad\qquad \updownarrow$$

$$\hat{\rho}(s) \overset{\text{squaring}}{\longrightarrow} |\hat{\rho}(s)|^2$$

In crystallography, the x-ray intensities $|\hat{\rho}(s)|^2$ are observed (photographically or, today, detected digitally) and the task is to deduce $\rho(s)$ from it. This is not a straightforward exercise: $\rho(s)$ cannot be determined directly, since it is complex and $|\hat{\rho}(s)|^2$ is not, nor is the mapping $|\hat{\rho}(s)|^2 \to \rho(s)$ unique. Nevertheless crystallographers have developed techniques to "solve" periodic crystal structures and their constituent molecules, even such complex molecules as proteins and DNA.

This work has revolutionized biology and materials science, and earned many Nobel prizes.[2]

Aperiodic crystals "satisfy" the Wiener diagram too, but their solution demands new definitions, technologies, and mathematical tools. Soon after their discovery, the International Union of Crystallography (IUCr) appointed a Commission of Aperiodic Crystals to define "crystal" more broadly. The Commission proposed a working definition to stimulate research [IUCr92]:

A crystal is a solid with an essentially discrete diffraction pattern.

But which atomic arrangements (or point sets) produce such patterns?

Note: In this section we denote point sets by X because the working definition does not require that X is Delaunay (or that $X - X$ be finitely generated).

GLOSSARY

Diffractive point set: A point set for which the autocorrelation measure $\gamma_X = \rho(x) * \overline{\rho(-x)}$ is uniquely defined (see [Hof95], [Lag00]).

Diffraction measure of a diffractive point set: The Fourier transform $\hat{\gamma}_X$ of γ_X. By Lebesgue's decomposition theorem, $\hat{\gamma}_X$ can be uniquely written as a sum of discrete, singular continuous, and absolutely continuous measures:

$$\hat{\gamma} = \hat{\gamma}_d + \hat{\gamma}_{sc} + \hat{\gamma}_{ac}.$$

Bragg peaks: The crystallographers' term for the discrete component $\hat{\gamma}_d$, which is a countable sum of weighted Dirac deltas.

Crystal (IUCr working definition):   Any discrete point set $X \subset \mathbb{R}^3$ such that $\hat{\gamma}_d$ is relatively dense in $\mathbb{R}^3$.

Poisson comb: A crystal for which $\hat{\gamma} = \hat{\gamma}_d$.

Periodic crystal: A crystal whose symmetry group includes a maximal abelian subgroup of translations.

Aperiodic crystal: A crystal whose symmetry group does not include translations. Aperiodic crystals include modulated crystals ([Wol74], [JJ77]) and so-called quasicrystals (the crystals described below).

Icosahedral crystal: A crystal whose diffraction patterns exhibit 5-fold, 3-fold, and 2-fold rotational symmetries (the rotations of the icosahedron). Icosahedral crystals are aperiodic in three linearly independent directions.

Octagonal crystal, decagonal crystal, dodecagonal crystal: A crystal with octagonal, decagonal, dodecagonal diffraction symmetry. Such crystals are aperiodic in two directions, periodic in the third.

MAIN RESULTS

1. Every lattice L is a Poisson comb and $\hat{\gamma}_d = L^*$.

2. Every regular system of points is a Poisson comb.

3. Every Meyer set (and thus every cut-and-project set) is a Poisson comb [Str05].

---

[2]See http://www.iucr.org/people/nobel-prize.

4. There are Poisson combs which are not Delaunay sets ([BMP99]).

5. There are Poisson combs for which $\hat{\gamma}_d$ is not finitely generated [Gri15].

The IUCr's working definition of "crystal" has stimulated much research and much has been learned, but we still do not have both necessary and sufficient conditions for a point set to be a crystal.

OPEN PROBLEM

Find necessary and sufficient conditions on a Delaunay set Λ for $\hat{\gamma}_d$ to be relatively dense in $\mathbb{R}^n$ (see also [Sen06]).

64.4 MODELING REAL CRYSTALS

Since Haüy, crystallographers have used tilings to model both the growth and form of crystals. The growth problem is: Why and how do crystals self-assemble from fundamental units? The form problem is: How do these fundamental units link together to form atomic patterns, and which patterns do they form? For aperiodic crystals, there is no simple rule (like Bravais' Law) that suggests the answer to either question. To oversimplify, there are two approaches: tiling models and cluster models, corresponding to the debate among physicists over the relative roles of energy and entropy in crystal growth.

In tiling models, energy is assumed to be encoded in the tiles' implicit or explicit matching rules, such as "fit the tiles together face to face," or the much more complex rules for Penrose and other aperiodic tilings. But the latter are too complex to model realistic interactions among atoms.

The alternate approach, which foregrounds entropy, begins with spontaneously-formed nanoclusters and studies how they grow and link together. For physical reasons, these initial nanoclusters are often icosahedral. Thus in entropy-driven models, the icosahedron, famously banned from the crystal kingdom by the "crystallographic restriction," takes center stage. This requires new geometrical tools ([Man07], [Sen15]). Here we discuss two examples that seem to point the way.

EXAMPLE 1: The Yb-Cd "quasicrystal"

The first aperiodic crystal structure to be "solved" (in the sense of pinpointing the positions of its atoms) was the "Yb-Cd quasicrystal" (ytterbium and cadmium). The atomic pattern is not a tiling. Instead, its fundamental building unit, which the authors call RTH complexes, overlap and pack together leaving gaps ([TGY⁺07]).

An RTH complex is a set of nested atomic clusters, where a "cluster" is "a set of close atoms distributed on fully occupied high symmetry orbits" (see [GPQK00]). The innermost cluster is a set of four cadmium atoms at the vertices of a regular tetrahedron. This is surrounded by 12 ytterbium atoms at the vertices of a regular icosahedron. Continuing outward, the next three clusters are comprised of cadmium atoms at the vertices of a regular dodecahedron, a semi-regular icosidodecahedron, and, outermost, a rhombic triacontahedron. Cadmium atoms are also situated at or near the midpoints of the edges of the triacontahedron.

RTH complexes overlap in well-defined ways: the convex hulls of the overlap

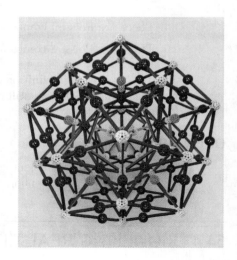

FIGURE 64.4.1
Jean Taylor's ZomeTool model of an RTH complex.
Photograph by Jean Taylor.

regions are identical oblate golden rhombohedra. Their packing is also well-defined: the gaps between complexes are golden rhombohedra too (of both kinds). All "gap" rhombohedra have "midpoint" atoms in their edges, and the acute rhombohedra have two Yb atoms inside them. A full description of the intricate, interlinked, pattern formed by this "soft packing" is beyond the scope of this chapter; for more details of its geometry, see [ST13].

We note that overlapping nested clusters (of several types) have been proposed as models for aperiodic crystal structure from the beginning; in addition to the references cited above see, for example, [Els89] and [Bur92].

EXAMPLE 2: A simulation

Like the YbCd crystal of Example 1, the real aperiodic crystals found to date are binary or ternary. This might suggest that the phenomenon depends upon a mix of atoms. But at least theoretically this is not the case: the Glotzer group of chemical engineers at the University of Michigan has simulated the self-assembly of a single-component icosahedral crystal for a suitably chosen interatomic potential [EDP15].

In this simulation, nearest-neighbor bonds point (approximately and statistically) in five-fold directions, indicating the formation of icosahedral clusters; diffraction shows the icosahedral symmetry of the structure as a whole. Thus the simulated crystal closely matches a cut-and-project model. Writing $b_1, \ldots, b_6$ for vectors from the center to the five-fold vertices of an icosahedron, the investigators selected bond vectors v closely aligned to them, i.e., those satisfying, for a suitable value of ϵ,

$$|v \cdot b| > (1 - \epsilon)||v|| \, ||b_i||$$

and found six-dimensional addresses for their particles by an iterative process of determining nearest-neighbor paths.

OPEN PROBLEMS

These examples suggest new problems in discrete geometry (the first task is to formulate them rigorously).

- Develop a theory of crystal growth by self-assembling nanoscale particles. As noted in [KG07], this will not be the classical model of layer-upon-layer outward from a "seed": modeling the growth of aperiodic crystals requires a new paradigm.
- Create a catalog of nested clusters (for examples see, e.g., [GPQK00] and [SD12]).
- Develop a theory of "soft packings"—a suitable mix of tiling, packing, and covering—that illuminates the linking of nested clusters in aperiodic crystals [BL15]. This will entail new definitions of density, kissing number, and so on.
- Are we missing something? The cut-and-project construction is very general: such sets can have rotational symmetry of any order. Yet the only rotational symmetries found in aperiodic crystals (so far) are octagonal, decagonal, dodecahedral, and icosahedral. Is there a real crystallographic restriction? What is it, and why?
- The RTH complex is not rigid: The tetrahedron flips among its possible inscriptions in the dodecahedron, and the dodecahedron is distorted by the flipping. (The three outer clusters have nearly undistorted icosahedral symmetry and their axes are aligned.) The flipping and the consequent distortions of the surrounding dodecahedra apparently drive the formation of this crystal in ways still not fully understood. Also in the simulation described above, the particles are in constant motion. These examples suggest we study Delaunay sets whose points vibrate and drift.

64.5 APERIODIC ORDER BEYOND CRYSTALS

In this chapter we have discussed concepts and techniques of discrete geometry that seem useful today for understanding the growth and form of real, and in particular aperiodic, crystals. Thus the fast-growing field of tiling dynamical systems has been left aside, as have self-similar tilings and point sets. Nor have we attempted to sketch the field of "aperiodic order" beyond crystals, which has burgeoned since the last edition of this Handbook.

Aperiodic order includes, for example, point sets for which $\hat{\gamma}_d = 0$ or for which $\hat{\gamma}$ does not exist (i.e., point sets which are not diffractive). Indeed, the diffraction spectrum is inadequate for a deeper study of long-range order. The measure γ is a function of Λ's two-point correlations (hence the importance of $\Lambda - \Lambda$ in crystallography.) But two-point correlation masks subtle differences:

- The Rudin-Shapiro sequence, which is generated by recursion, and the Bernoulli coin-flipping sequence have the same diffraction spectrum. This is perplexing: surely a deterministic pattern is more orderly than a random set! In fact we do find differences if we look more deeply. Although their diffraction spectra are identical, their dynamical spectra are not [HB00].
- The family of generalized Thue-Morse sequences, which are generated by substitution rules, have self-similarities that do not appear in the pure-point component in their diffraction measures. They are, however, revealed by looking at the two-point correlations of pairs [Gri15].

These examples are just the beginning. In the 21st century, crystallography is merging with materials science; the question *What is a crystal?* may become as academic as *What is a planet?* In the next edition of this Handbook, applications of "aperiodic order beyond crystals" will have a chapter of their own.

64.6 SOURCES AND RESOURCES

SURVEYS AND COLLECTIONS

[BG13]: A mathematically sophisticated survey of the rapidly developing theory of aperiodic order.

[BM00]: A multi-authored survey of major problems in the field, as seen at the turn of the 21st century.

[FI08]: A multi-authored survey of "quasicrystals" as a subfield of metal physics.

[HH15]: A selection of influential articles on crystallography, classical and modern. This book is especially useful for mathematicians seeking an overview of the field through the eyes of its practitioners.

[Moo97]: The proceedings of a NATO conference held Waterloo, Canada, in 1995.

[Sen96]: An overview and gentle introduction to the relations between these subjects, as of 1995.

RELATED CHAPTERS

Chapter 2: Packing and covering
Chapter 3: Tilings
Chapter 7: Lattice points and lattice polytopes
Chapter 20: Polyhedral maps
Chapter 27: Voronoi diagrams and Delaunay triangulations

REFERENCES

[Aut13]   A. Authier. *Early Days of X-ray Crystallography.* Oxford University Press, 2013.

[BBN+78]   H. Brown, R. Bülow, J. Neubüser, H. Wondratschek, and H. Zassenhaus. *Crystallographic Groups of Four-dimensional Space.* Wiley, New York, 1978.

[BG13]   M. Baake and U. Grimm. *Theory of Aperiodic Order: A Mathematical Invitation.* Cambridge University Press, 2013.

[BL15]   K. Bezdek and Z. Lángi. Density bounds for outer parallel domains of unit ball packings. *Proc. Steklov Inst. Math.,* 288:209–225, 2015.

[BM00]   M. Baake and R.V. Moody, editors. *Directions in Mathematical Quasicrystals.* Vol. 13 of CRM Monograph Series, AMS, Providence, 2000.

[BMP99]   M. Baake, R.V. Moody, and P.A.B. Pleasants. Diffraction from visible lattice points and k^{th} power free integers. *Discrete Math.* 221:3–45, 1999.

[Bra49]   A. Bravais. *On the Systems Formed by Points Regularly Distributed on a Plane or in Space.* Crystallographic Society of America, 1849.

[Bur92]   S. Burkov. Modeling decagonal quasicrystal: random assembly of interpenetrating decagonal clusters. *J. Phys. (Paris)*, 2:695-706, 1991.

[DAP34]   B.N. Delone, A.D. Aleksandrov, and N.N. Padurov. *Mathematical Foundations of the Structural Analysis of Crystals* (in Russian). ONTI, Leningrad, 1934.

[DDSG76]   B.N. Delone, N.P. Dolbilin, M.I. Shtogrin, and R.V. Galiulin. A local test for the regularity of a system of points. *Dokl. Akad. Nauk. SSSR*, 227:19–21, 1976. English translation: *Soviet Math. Dokl.*, 17:319–322, 1976.

[DLS98]   N.P. Dolbilin, J.C. Lagarias, and M. Senechal. Multiregular point systems. *Discrete Comput. Geom.*, 20:477–498, 1998.

[Dol17]   N. Dolbilin. Delone sets in $\mathbb{R}^3$: Regularity conditions (in Russian). *Fundam. Prikl. Mat.*, to appear, 2017.

[EDP15]   M. Engel, P.F. Damasceno, C.L. Phillips, and S.C. Glotzer. Computational self-assembly of a one-component icosahedral quasicrystal. *Nature Mater.*, 14:109–116, 2015.

[Els89]   V. Elser. The growth of the icosahedral phase. In *Fractals, Quasicrystals, Chaos, Knots and Algebraic Quantum Mechanics*. Volume 235 of the series NATO ASI Series, pages 121–138. Kluwer, Dordrecht, 1988.

[Erd99]   R. Erdahl, Zonotopes, dicings and Voronoi's conjecture on parallelohedra. *European J. Comb.*, 20:527–549, 1999.

[FI08]   T. Fujiwara and Y. Ishii, editors. *Quasicrystals*. Elsevier, Amsterdam, 2008.

[Gri15]   U. Grimm, Aperiodic crystals and beyond. *Acta Cryst.* B71:258–274, 2015.

[GPQK00]   D. Gratias, F. Puyraimond, M. Quiquandon, and A. Katz. Atomic clusters in icosahedral *F*-type quasicrystals. *Phys. Rev. B*, 63, 2000.

[HB00]   M. Höffe and M. Baake. Surprises in diffuse scattering. *Z. Krist.*, 215:441–444, 2000.

[HH15]   I. Hargittai and B. Hargittai. *Science of Crystal Structures: Highlights in Crystallography*. Springer, Heidelberg, 2015.

[Hof95]   A. Hof. On diffraction by aperiodic structures. *Comm. Math. Phys.*, 169:25–43, 1995.

[IUCr92]   Terms of reference of the IUCr Commission on Aperiodic Crystals. *Acta Cryst.*, A48:928, 1992.

[JJ77]   A. Janner and T. Janssen. Symmetry of periodically distorted crystals. *Phys. Rev. B*, 15:643–658, 1977.

[KG07]   A. Keys and S.C. Glotzer. How do quasicrystals grow? *Phys. Rev. Lett.*, 99:235503, 2007.

[Lag00]   J.C. Lagarias. Mathematical quasicrystals and the problem of diffraction. In M. Baake and R.V. Moody, editors, *Directions in Mathematical Quasicrystals*. Volume 13 of *CRM Monograph Series*, pages 61–93, AMS, Providence, 2000.

[Lag99]   J.C. Lagarias. Geometric models for quasicrystals I. Delone sets of finite type. *Discrete Comput. Geom.*, 21:161–191, 1999.

[LP02]   J.C. Lagarias and P.A.B. Pleasants. Local complexity of Delone sets and crystallinity. *Bull. Canadian Math. Soc.*, 45:634–652, 2002.

[Mag15]   A. Magazinov. Voronoi's conjecture for extensions of Voronoi parallelohedra. *Russian Mathematical Surveys*, 69:763–764, 2014.

[Man07]   B. Mann. 23 mathematical challenges. In presentation *Defense Sciences Offices: The Heart and Soul of the Far Side* given at DARPA, Washington, 2007.

[McM80]   P. McMullen. Convex bodies which tile space by translation. *Mathematica*, 27:113–121: 1980.

[Mey95]   Y. Meyer. Quasicrystals, Diophantine approximation and algebraic numbers. In F. Axel and D. Gratias, editors, *Beyond Quasicrystals.*, pages 3–16. Collection du Centre de Physique des Houches, Les Éditions de Physique, Springer-Verlag, Berlin, 1995.

[Moo97]   R.V. Moody. *The Mathematics of Long-Range Aperiodic Order.* Volume 489 of NATO Advanced Science Institutes Series C., Kluwer, Dordrecht, 1997.

[SBG+84]   D. Shechtman, I. Blech, D. Gratias, and J.W. Cahn. Metallic phase with long-range orientational order and no translational symmetry. *Phys. Rev. Lett.*, 53:1951–1953, 1984.

[SD12]   W. Steurer and S. Deloudi. Cluster packing from a higher dimensional perspective. *Structural Chemistry*, 23:1115–1120, 2012.

[Sen90]   M. Senechal. *Crystalline Symmetries: An Informal Mathematical Introduction*, Adam Hilger, IOP Publishing, 1990.

[Sen96]   M. Senechal. *Quasicrystals and Geometry.* Cambridge Univ. Press, 1996.

[Sen06]   M. Senechal. What is ... a quasicrystal? *Notices AMS*, 53:886-887, 2006.

[Sen15]   M. Senechal. Delaunay sets and condensed matter: the dialogue continues. *Proc. Steklov Inst. Math.*, 288:259–264, 2015.

[ST13]   M. Senechal and J.E. Taylor. Quasicrystals: The view from Stockholm. *Math. Intelligencer*, 35:1–9, 2013.

[Str05]   N. Strungaru. Almost periodic measures and long-range order in Meyer sets. *Discrete Comput. Geom.*, 33:483–505, 2005.

[TGY+07]   H. Takakura, C.P. Gómez, A. Yamamoto, M. de Boissieu, and A.P. Tsai. Atomic structure of the binary icosahedral Yb–Cd quasicrystal. *Nature Materials*, 6:58–63, 2007.

[Wol74]   P.M. de Wolff. The pseudo-symmetry of modulated crystal structures. *Acta Cryst.*, A30:777–785, 1974.

[Yan01]   B. Yandell. *The Honors Class.* A.K. Peters, Natick, 2001.

65 COMPUTATIONAL TOPOLOGY FOR STRUCTURAL MOLECULAR BIOLOGY

Herbert Edelsbrunner and Patrice Koehl

INTRODUCTION

The advent of high-throughput technologies and the concurrent advances in information sciences have led to a data revolution in biology. This revolution is most significant in molecular biology, with an increase in the number and scale of the "omics" projects over the last decade. *Genomics* projects, for example, have produced impressive advances in our knowledge of the information concealed into genomes, from the many genes that encode for the proteins that are responsible for most if not all cellular functions, to the noncoding regions that are now known to provide regulatory functions. *Proteomics* initiatives help to decipher the role of post-translation modifications on the protein structures and provide maps of protein-protein interactions, while *functional genomics* is the field that attempts to make use of the data produced by these projects to understand protein functions. The biggest challenge today is to assimilate the wealth of information provided by these initiatives into a conceptual framework that will help us decipher life. For example, the current views of the relationship between protein structure and function remain fragmented. We know of their sequences, more and more about their structures, we have information on their biological activities, but we have difficulties connecting this dotted line into an informed whole. We lack the experimental and computational tools for directly studying protein structure, function, and dynamics at the molecular and supra-molecular levels. In this chapter, we review some of the current developments in building the computational tools that are needed, focusing on the role that geometry and topology play in these efforts. One of our goals is to raise the general awareness about the importance of geometric methods in elucidating the mysterious foundations of our very existence. Another goal is the broadening of what we consider a geometric algorithm. There is plenty of valuable no-man's-land between combinatorial and numerical algorithms, and it seems opportune to explore this land with a computational-geometric frame of mind.

65.1 BIOMOLECULES

GLOSSARY

DNA: Deoxyribo Nucleic Acid. A double-stranded molecule found in all cells that is the support of genetic information. Each strand is a long polymer built from four different building blocks, the nucleotides. The sequence in which these nucleotides are arranged contains the entire information required to describe cells and their functions. The two strands are complementary to each other, allowing for repair should one strand be damaged.

RNA: Ribo Nucleic Acid. A long polymer much akin to DNA, being also formed as sequences of four types of nucleotides. RNAs can serve as either carrier of information (in their linear sequences), or as active functional molecules whose activities are related to their 3-dimensional shapes.

Protein: A long polymer, also called a *polypeptide chain*, built from twenty different building blocks, the amino acids. Proteins are active molecules that perform most activities required for cells to function.

Genome: Genetic material of a living organism. It consists of DNA, and in some cases, of RNA (RNA viruses). For humans, it is physically divided into 23 *chromosomes*, each forming a long double-strand of DNA.

Gene: A gene is a segment of the genome that encodes a functional RNA or a protein product. The transmission of genes from an organism to its offsprings is the basis of the heredity.

Central dogma: The Central Dogma is a framework for understanding the transfer of information between the genes in the genome and the proteins they encode for. Schematically, it states that "DNA makes RNA and RNA makes protein."

FIGURE 65.1.1
The DNA gets replicated as a whole. Pieces of DNA referred to as genes are transcribed into pieces of RNA, which are then translated into proteins.

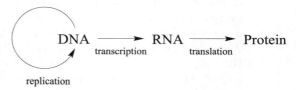

Replication: Process of producing two identical replicas of DNA from one original DNA molecule.

Transcription: First step of gene expression, in which a particular segment of DNA (gene) is copied into an RNA molecule.

Translation: Process in which the messenger RNA produced by transcription from DNA is decoded by a ribosome to produce a specific amino acid chain, or protein.

Protein folding: Process in which a polypeptide chain of amino acid folds into a usually unique globular shape. This 3D shape encodes for the function of the protein.

Intrinsically disordered protein (IDP):   A protein that lacks a fixed or ordered three-dimensional structure or shape. Despite their lack of stable structure, they form a very large and functionally important class of proteins.

INFORMATION TRANSFER: FROM DNA TO PROTEIN

One of the key features of biological life is its ability to self-replicate. Self-replication is the behavior of a system that yields manufacturing of an identical copy of itself. Biological cells, given suitable environments, reproduce by cell division. During cell division, the information defining the cell, namely its genome, is replicated and then transmitted to the daughter cells: this is the essence of heredity. Interestingly, the entire machinery that performs the replication as well as the compendium that defines the process of replication are both encoded into the genome itself. Understanding the latter has been at the core of molecular biology. Research in this domain has led to a fundamental hypothesis in biology: the Central Dogma. We briefly describe it in the context of information transfer.

The genome is the genetic material of an organism. It consists of DNA, and in a few rare cases (mostly some viruses), of RNA. The DNA is a long polymer whose building blocks are nucleotides. Each nucleotide contains two parts, a backbone consisting of a deoxyribose and a phosphate, and an aromatic base, of which there are four types: adenine (A), thymine (T), guanine (G) and cytosine (C). The nucleotides are linked together to form a long chain, called a *strand*. Cells contain strands of DNA in pairs that are mirrors of each other. When correctly aligned, A pairs with T, G pairs with C, and the two strands form a double helix [WC53]. The geometry of this helix is surprisingly uniform, with only small, albeit important structural differences between regions of different sequences. The order in which the nucleotides appear in one DNA strand defines its sequence. Some stretches of the sequence contain information that can be transcribed first into an RNA molecule and then translated into a protein (Central Dogma). These stretches are called *genes*. It is estimated, for example, that the human genome contains around 20,000 genes [PS10], which represent 1-3% of the whole genome. For a long time, the remainder was considered to be nonfunctional, and therefore dubbed to be "junk" DNA. This view has changed, however, with the advent of the genomic projects. For example, the international Encyclopedia of DNA Elements (ENCODE) project has used biochemical approaches to uncover that at least 80% of human genomic DNA has biochemical activity [ENC12]. While this number has been recently questioned as being too high [Doo13, PG14], as biochemical activities may not imply function, it remains that a large fraction of the noncoding DNA plays a role in regulation of gene expression.

DNA replication is the biological process of generating two identical copies of DNA from one original DNA molecule. This process occurs in all living organisms; it is the basis for heredity. As DNA is made up of two complementary strands wound into a double helix, each strand serves as a template for the production of the complementary strand. This mechanism was first suggested by Watson and Crick based on their model of the structure for DNA [WC53]. As replication is the mechanism that ensures transfer of information from one generation to the other, most species have developed control systems to ensure its fidelity. Replication is performed by DNA polymerases. The function of these molecular machines is not quite perfect, making about one mistake for every ten million base pairs copied [MK08]. Error correction is a property of most of the DNA polymerases. When an incorrect base pair is recognized, DNA polymerase moves backwards by one base pair of DNA, excises the incorrect nucleotide and replaces it with the correct one. This process is known as *proofreading*. It is noteworthy that geometry plays an important role here. Incorporation of the wrong nucleotide leads to changes in the shape of the DNA, and it is this change in geometry that the polymerase detects. In addition to the proofreading process, most cells rely on post-replication mismatch repair mechanisms to monitor the DNA for errors and correct them. The combination of the intrinsic error rates of polymerases, proofreading, and post-replication mismatch repair usually enables replication fidelity of less than one mistake for every billion nucleotides added [MK08]. We do note that this level of fidelity may vary between species. Unicellular organisms that rely on fast adaptation to survive, such as bacteria, usually have polymerases with much lower levels of fidelity [Kun04].

Transcription is the first step in the transfer of information from DNA to its end product, the protein. During this step, a particular segment of DNA is copied into RNA by the enzyme RNA polymerase. RNA molecules are very similar to DNA, being formed as sequences of four types of nucleotides, namely A, G, C, and uracil (U), which is a derivative of thymine. In contrast to the double-stranded DNA, RNA is mostly found to be singled-stranded. This way, it can adopt a large variety of conformations, which remain difficult to predict based on the RNA sequence [SM12]. Interestingly, RNA is considered an essential molecule in the early steps of the origin of life [Gil86, Cec93].

Translation is the last step in gene expression. In translation, the messenger RNA produced by transcription from DNA is decoded by a ribosome to produce a specific amino acid chain, or polypeptide. There are 20 types of amino acids, which share a common *backbone* and are distinguished by their chemically diverse *side-chains*, which range in size from a single hydrogen atom to large aromatic rings and can be charged or include only nonpolar saturated hydrocarbons. The order in which amino acids appear defines the *primary sequence*, also referred to as the *primary structure*, of the polypeptide. In its native environment, the polypeptide chain adopts a unique 3-dimensional shape, in which case it is referred to as a *protein*. The shape defines the *tertiary* or *native structure* of the protein. In this structure, nonpolar amino acids have a tendency to re-group and form the core, while polar amino acids remain accessible to the solvent.

We note that the scenario "DNA makes RNA and RNA makes protein" captured by the Central Dogma is reminiscent of the Turing machine model of computing, in which information is read from an input tape and the results of the computations are printed on an output tape.

FROM SEQUENCE TO FUNCTION

Proteins, the end products of the information encoded in the genome of any organism, play a central role in defining the life of this organism as they catalyze most biochemical reactions within cells and are responsible, among other functions, for the transport of nutrients and for signal transmission within and between cells. Proteins become functional only when they adopt a 3-dimensional shape, the so-called tertiary, or native structure of the protein. This is by no means different from the macroscopic world: most proteins serve as tools in the cell and as such either have a defined or adaptive shape to function, much like the shapes of the tools we use are defined according to the functions they need to perform. Understanding the shape (the geometry) of a protein is therefore at the core of understanding how cells function. From the seminal work of Anfinsen [Anf73], we know that the sequence fully determines the 3-dimensional structure of the protein, which itself defines its function. While the key to the decoding of the information contained in genes was found more than fifty years ago (the genetic code), we have not yet found the rules that relate a protein sequence to its structure [KL99, BS01]. Our knowledge of protein structure therefore comes from years of experimental studies, either using X-ray crystallography or NMR spectroscopy. The first protein structures to be solved were those of hemoglobin and myoglobin [KDS+60, PRC+60].

As of June 2016, there are more than 110,000 protein structures in the database of biomolecular structures [BWF+00]; see `http://www.rcsb.org`. This number remains small compared to the number of existing proteins. There is therefore a lot of effort put into predicting the structure of a protein from the knowledge of its sequence: one of the "holy grails" in molecular biology, namely the protein structure prediction problem [Dil07, EH07, DB07, Zha08, Zha09]. Efforts to solve this problem currently focus on protein sequence analysis, as a consequence of the wealth of sequence data resulting from various genome-sequencing projects, either completed or ongoing. As of May 2016, there were more than 550,000 protein sequences deposited in SwissProt-Uniprot version 2016-05, the fully annotated repository of protein sequences. Data produced by these projects have already led to significant improvements in predictions of both protein 3D structures and functions; see for example [MHS11]. However, we still stand at the dawn of understanding the information encoded in the sequence of a protein.

It is worth noting that if the paradigm shape-defines-function is the rule in biology, intrinsically disordered proteins form a significant class of exceptions, as they lack stable structures [DW05, DSUS08]. Shape, however, remains important for those proteins, although it is its flexibility and plasticity that is of essence, as shown for example in the case of P53 [OMY+09].

65.2 GEOMETRIC MODELS

The shape of a protein and its chemical reactivity are highly correlated as the latter depends on the positions of the nuclei and electrons within the protein: this correlation is the rationale for high-resolution experimental and computational studies of the structures and shapes of proteins. Early crystallographers who studied proteins could not rely (as it is common nowadays on computers and computer graphics programs for representation and analysis of their structures. They had developed a large array of finely crafted physical models that allowed them to have a feeling for the shapes of these molecules. These models, usually made out of painted wood, plastic, rubber, and/or metal, were designed to highlight different properties of the protein under study. The current models in computer graphics programs mimic those early models. The *cartoon diagrams*, also called *ribbon diagrams* or *Richardson diagrams* [Ric85] show the overall path and organization of the protein backbone in 3D. Cartoon diagrams are generated by interpolating a smooth curve through the polypeptide backbone. In the *stick models*, atoms are represented as points (sometimes as small balls) attached together by sticks that represent the chemical bonds. These models capture the stereochemistry of the protein. In the *space-filling models*, such as those of Corey-Pauling-Koltun (CPK) [CP53, Kol65], atoms are represented as balls, whose sizes are set to capture the volumes occupied by the atoms. The radii of those balls are set to the van der Waals radii of the atoms. The *CPK model* has now become standard in the field of macromolecular modeling: a protein is represented as the union of a set of balls, whose centers match with the atomic centers and radii defined by van der Waals radii. The structure of a protein is then fully defined by the coordinates of these centers, and the radii values. The *macromolecular surface* is the geometric surface or boundary of these unions of balls. Note that other definitions are possible; this will be discussed in more detail below.

GLOSSARY

FIGURE 65.2.1

Three representations (diagrams) of the same protein, the HIV-1 protease (Protein Data Bank [BWF+00], identifier: 3MXE). The cartoon diagram on the left characterizes the geometry of the backbone of the protein, the stick diagram in the middle shows the chemical bonds, and the space-filling diagram highlights the space occupied by the protein. The three diagrams complement each other in their representation of relevant information.

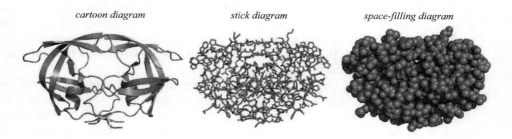

cartoon diagram   stick diagram   space-filling diagram

Cartoon diagram: Model that represents the overall path and organization of the protein backbone in 3D. The cartoon diagram is generated by interpolating a smooth curve through the protein backbone.

Stick diagram: Model that represents the chemical connectivity in a protein by displaying chemical bonds as sticks (edges). Atoms are usually just vertices where the edges meet. Some stick diagrams use balls to represent those vertices, the ball-and-stick models.

Space-filling diagram: Model that represents a protein by the space it occupies. Most commonly, each atom is represented by a ball (a solid sphere), and the protein is the union of these balls.

FIGURE 65.2.2

Three most common molecular surface models for representing proteins (2D examples). Dashed, red circles represent the probe solvent spheres.

vdW surface   accessible surface   molecular surface

Van der Waals surface: Boundary of space-filling diagram defined as the union of balls with van der Waals radii. The sizes of these balls are chosen to reflect the transition from an attractive to a repulsive van der Waals force.

Solvent-accessible surface: Boundary of space-filling diagram in which each van der Waals ball is enlarged by the radius of the solvent sphere. Alternatively, it is the set of centers of solvent spheres that touch but do not otherwise intersect the van der Waals surface.

Molecular surface: Boundary of the portion of space inaccessible to the solvent. It is obtained by rolling the solvent sphere about the van der Waals surface.

Power distance: Square length of tangent line segment from a point x to a sphere with center z and radius r. It is also referred to as the *weighted square distance* and formally defined as $\|x - z\|^2 - r^2$.

Voronoi diagram: Decomposition of space into convex polyhedra. Each polyhedron corresponds to a sphere in a given collection and consists of all points for which this sphere minimizes the power distance. This decomposition is also known as the *power diagram* and the *weighted Voronoi diagram.*

Delaunay triangulation: Dual to the Voronoi diagram. For generic collections of spheres, it is a simplicial complex consisting of tetrahedra, triangles, edges, and vertices. This complex is also known as the *regular triangulation*, the *coherent triangulation*, and the *weighted Delaunay triangulation.*

Dual complex: Dual to the Voronoi decomposition of a union of balls. It is a subcomplex of the Delaunay triangulation.

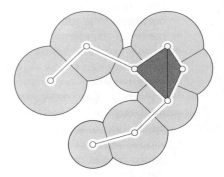

FIGURE 65.2.3
Each Voronoi polygon intersects the union of disks in a convex set, which is the intersection with its defining disk. The drawing shows the Voronoi decomposition of the union and the dual complex superimposed.

Growth model: Rule for growing all spheres in a collection continuously and simultaneously. The rule that increases the square radius r^2 to $r^2 + t$ at time t keeps the Voronoi diagram invariant at all times.

Alpha complex: The dual complex at time $t = \alpha$ for a collection of spheres that grow while keeping the Voronoi diagram invariant. The *alpha shape* is the underlying space of the alpha complex.

Filtration: Nested sequence of complexes. The prime example here is the sequence of alpha complexes.

ALTERNATIVE SURFACE REPRESENTATIONS

While geometric models for the molecular surface provide a deterministic description of the boundary for the shape of a biomolecule, models using implicit or parametric surfaces may be favorable for certain applications [Bli82, ZBX11]. The *implicit molecular surface models* use a level set of a scalar function $f \colon \mathbb{R}^3 \to \mathbb{R}$ that maps each point from the 3-dimensional space to a real value [OF03, CCW06, CP13]. The most common scalar function used for macromolecular surfaces is a summation of Gaussian functions [GP95]. Other scalar functions, such as polynomial and Fermi-Dirac switching functions, have been used as well [LFSB03]. Bates *et al.* [BWZ08] proposed the *minimal molecular surface* as a level set of a scalar function that is the output from a numerical minimization procedure. *Parametric surface models* specify each point on the macromolecular surface by a pair of real value variables. Piecewise polynomials such as Non-Uniform Rational B-spline (NURBS) and Bernstein-Bézier have been proposed to generate parametric representations for molecular surfaces [BLMP97, ZBX11]. Spherical harmonics and their extensions parametrize the macromolecular surface using spherical coordinates and provide a compact analytical representation of macromolecular shapes [MG88, DO93a, DO93b].

We note that neither implicit nor parametric macromolecular surface models are independent from the geometric models based on the union of balls, as they usually have a set of parameters that are tuned such that they provide a reasonable approximation of the surface of the latter.

SPACE-FILLING DIAGRAMS

Our starting point is the *van der Waals force*. These forces capture interactions between atoms and molecules and mostly include attraction and repulsion. At short range up to a few Angstrom, this force is attractive but significantly weaker than covalent or ionic bonds. At very short range, the force is strongly repulsive. We can assign *van der Waals radii* to the atoms so that the force changes from attractive to repulsive when the corresponding spheres touch [GR01]. The *van der Waals surface* is the boundary of the space-filling diagram made up of the balls with van der Waals radii. In the 1970s, Richards and collaborators extended this idea to capture the interaction of a protein with the surrounding solvent [LR71, Ric77]. The *solvent-accessible surface* is the boundary of the space-filling diagram in which the balls are grown by the radius of the sphere that models a single solvent molecule. Usually the solvent is water, represented by a sphere of radius 1.4 Angstrom. The *molecular surface* is obtained by rolling the solvent sphere over the van der Waals surface and filling in the inaccessible crevices and cusps. This surface is sometimes referred to as the *Connolly surface*, after the creator of the first software representing this surface by a collection of dots [Con83]. We mention that this surface may have sharp edges, namely when the solvent sphere cannot quite squeeze through an opening of the protein and thus forms a circular or similar curve feature on the surface.

DUAL STRUCTURES

We complement the space-filling representations of proteins with geometrically dual structures. A major advantage of these dual structures is their computational convenience. We begin by introducing the *Voronoi diagram* of a collection of balls or spheres, which decomposes the space into convex polyhedra [Vor07]. Next we intersect the union of balls with the Voronoi diagram and obtain a decomposition of the space-filling diagram into convex *cells*. Indeed, these cells are the intersections of the balls with their corresponding Voronoi polyhedra. The *dual complex* is the collection of simplices that express the intersection pattern between the cells: we have a vertex for every cell, an edge for every pair of cells that share a common facet, a triangle for every triplet of cells that share a common edge, and a tetrahedron for every quadruplet of cells that share a common point [EKS83, EM94]. This exhausts all possible intersection patterns in the assumed generic case. We get a natural embedding if we use the sphere centers as the vertices of the dual complex.

GROWTH MODEL

One and the same Voronoi diagram corresponds to more than just one collection of spheres. For example, if we grow the square radius r_i^2 of the ith sphere to $r_i^2 + t$, for every i, we get the same Voronoi diagram. Think of t as time parametrizing this particular growth model of the spheres. While the Voronoi diagram remains fixed, the dual complex changes. The cells in which the balls intersect the Voronoi polyhedra grow monotonically with time, which implies that the dual complex can acquire but not lose simplices. We thus get a nested sequence of dual complexes,

$$\emptyset = K_0 \subseteq K_1 \subseteq \ldots \subseteq K_m = D,$$

which begins with the empty complex at time $t = -\infty$ and ends with the Delaunay triangulation [Del34] at time $t = \infty$. We refer to this sequence as a *filtration* of the Delaunay triangulation and think of it as the dual representation of the protein at all scale levels.

ALPHA SHAPE THEORY

The dual structures and the growth model introduced above form the basis of the alpha shape theory and its applications to molecular shapes. Alpha shapes have a technical definition that was originally introduced to formalize the notion of 'shape' for a set of points [EKS83]. It can be seen as a generalization of the convex hull of this set. One *alpha complex* is a subcomplex of the Delaunay triangulation, and the corresponding *alpha shape* is the union of the simplices in the alpha complex. Such an alpha shape is characterized by a parameter, α, that corresponds to the parameter t defined above. This parameter controls the level of detail that is desired: the set of all alpha values leads to a family of shapes capturing the intuitive notion of 'crude' versus 'fine' shape of the set.

In its applications to structural biology, the set of points corresponds to the collection of atoms of the molecule of interest, with each atom assigned a weight corresponding to its van der Waals radius. The Delaunay triangulation of this set of weighted points is computed. Most applications require the alpha complex corresponding to $\alpha = 0$, as the corresponding alpha shape best represents the space-filling diagram (either delimited by the vdW surface or by the solvent accessible surface). The alpha complex, K_0, can then be used to measure the molecular shape. The complete filtration can also be used to characterize the topology of the bio-molecule, as captured by the simplices of the dual complexes and of the Delaunay triangulation. This will be discussed below.

65.3 MOLECULAR SKIN OF A PROTEIN

We introduce yet another surface bounding a space-filling diagram of sorts. The *molecular skin* is the boundary of the union of infinitely many balls. Besides the balls with van der Waals radii representing the atoms, we have balls interpolating between them that give rise to blending patches and, all together, to a tangent-continuous surface. The molecular skin is rather similar in appearance to the molecular surface but uses hyperboloids instead of tori to blend between the spheres [Ede99]. The smoothness of the surface permits a mesh whose triangles are all approximately equiangular [CDES01]. Applications of this mesh include the representation of proteins for visualization purposes and the solution of differential equations defined over the surface by finite-element and other numerical methods.

GLOSSARY

Molecular skin: Surface of a molecule that is geometrically similar to the molecular surface but uses hyperboloid instead of torus patches for blending. In mathematical terms, it is the boundary of the union of interpolated spheres, which we construct from the set of given spheres as follows. Supposing we have two spheres with centers a_1, a_2 and radii r_1, r_2 such that the distance between the two centers is smaller than $\sqrt{2}(r_1 + r_2)$. For each real number $0 \leq \lambda \leq 1$, the corresponding *interpolated sphere* is obtained by first increasing the radii to $\sqrt{2}r_1$ and $\sqrt{2}r_2$, second fixing the new center to $a_3 = (1 - \lambda)a_1 + \lambda a_2$, third choosing the new radius such that the sphere passes through the circle in which the given two spheres intersect, and fourth shrinking to radius $r_3/\sqrt{2}$. If the distance between the centers is larger than $\sqrt{2}(r_1 + r_2)$, then we extend the construction to include spheres with imaginary radii, which correspond to empty balls and therefore do not contribute to the surface we construct.

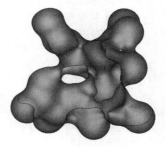

FIGURE 65.3.1
Cutaway view of the skin of a small molecule. We see a blend of sphere and hyperboloid patches. The surface is inside-outside symmetric: it can be defined by a collection of spheres on either of its two sides.

Mixed complex: Decomposition of space into shrunken Voronoi polyhedra, shrunken Delaunay tetrahedra, and shrunken products of corresponding Voronoi polygons and Delaunay edges as well as Voronoi edges and Delaunay triangles. It decomposes the skin surface into sphere and hyperboloid patches.

Maximum normal curvature: The larger absolute value $\kappa(x)$ of the two principal curvatures at a point x of the surface.

ε-sample: A collection S of points on the molecular skin $\mathbb{M}$ such that every point $x \in \mathbb{M}$ has a point $u \in S$ at distance $\|x - u\| \leq \varepsilon/\kappa(x)$.

Restricted Delaunay triangulation: Dual to the restriction of the (3-dimensional) Voronoi diagram of S to the molecular skin $\mathbb{M}$.

Shape space: Locally parametrized space of shapes. The prime example here is the $(k-1)$-dimensional space generated by k shapes, each specified by a collection of spheres in $\mathbb{R}^3$.

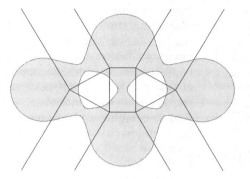

FIGURE 65.3.2
The skin curve defined by four circles in the plane. The mixed complex decomposes the curve into pieces of circles and hyperbolas.

TRIANGULATED MOLECULAR SKIN

The molecular skin has geometric properties that can be exploited to construct a numerically high-quality mesh and to maintain that mesh during deformation. The most important of these is the continuity of the *maximum normal curvature* function $\kappa \colon \mathbb{M} \to \mathbb{R}$. To define it, consider the 1-parameter family of geodesics passing through x, and let $\kappa(x)$ be the maximum of their curvatures at x. We use this function to guide the local density of the points distributed over $\mathbb{M}$ that are used as vertices of the mesh. Given such a collection S of points, we construct a mesh using its Voronoi diagram restricted to $\mathbb{M}$. The polyhedra decompose the surface into patches, and the mesh is constructed as the dual of this decomposition [Che93]. As proved in [ES97], the mesh is homeomorphic to the surface if the pieces of the restricted Voronoi diagram are topologically simple sets of the appropriate dimensions. In other words, the intersection of each Voronoi polyhedron, polygon, or edge with $\mathbb{M}$ is either empty or a topological disk, interval, or single point. Because of the smoothness of $\mathbb{M}$, this topological property is implied if the points form an ε-sampling, with $\varepsilon = 0.279$ or smaller [CDES01]. An alternative approach to triangulating the molecular skin can be found in [KV07].

DEFORMATION AND SHAPE SPACE

The variation of the maximum normal curvature function can be bounded by the one-sided Lipschitz condition $|1/\kappa(x) - 1/\kappa(y)| \leq \|x - y\|$, in which the distance is measured in $\mathbb{R}^3$. The continuity over $\mathbb{R}^3$ and not just over $\mathbb{M}$ is crucial when it comes to maintaining the mesh while changing the surface. This leads us to the topic of deformations and shape space. The latter is constructed as a parametrization of the deformation process. The deformation from a shape A_0 to another shape A_1 can be written as $\lambda_0 A_0 + \lambda_1 A_1$, with $\lambda_1 = 1 - \lambda_0$. Accordingly, we may think of the unit interval as a 1-dimensional shape space. We can generalize this to a k-dimensional shape space as long as the different ways of arriving at $(\lambda_0, \lambda_1, \ldots, \lambda_k)$, with $\sum \lambda_i = 1$ and $\lambda_i \geq 0$ for all i, all give the same shape $A = \sum \lambda_i A_i$. How to define deformations so that this is indeed the case is explained in [CEF01].

65.4 CONNECTIVITY AND SHAPE FEATURES

Protein connectivity is often understood in terms of its covalent bonds, in particular along the backbone. In this section, we discuss a different notion, namely the topological connectivity of the space assigned to a protein by its space-filling diagram. We mention *homeomorphisms, homotopies, homology groups* and *Euler characteristics*, which are common topological concepts used to define and talk about connectivity. Of particular importance are the homology groups and their ranks, the *Betti numbers*, as they lend themselves to efficient algorithms. In addition to computing the connectivity of a single space-filling diagram, we study how the connectivity changes when the balls grow. The sequence of space-filling diagrams obtained this way corresponds to the filtration of dual complexes introduced earlier. We use this filtration to define basic shape features, such as pockets in proteins and interfaces between complexed proteins and molecules.

GLOSSARY

Topological equivalence: Equivalence relation between topological spaces defined by *homeomorphisms*, which are continuous bijections with continuous inverses.

Homotopy equivalence: Weaker equivalence relation between topological spaces $\mathbb{X}$ and $\mathbb{Y}$ defined by maps $f \colon \mathbb{X} \to \mathbb{Y}$ and $g \colon \mathbb{Y} \to \mathbb{X}$ whose compositions $g \circ f$ and $f \circ g$ are homotopic to the identities on $\mathbb{X}$ and on $\mathbb{Y}$.

Deformation retraction: A homotopy between the identity on $\mathbb{X}$ and a retraction of $\mathbb{X}$ to $\mathbb{Y} \subseteq \mathbb{X}$ that leaves $\mathbb{Y}$ fixed. The existence of the deformation implies that $\mathbb{X}$ and $\mathbb{Y}$ are homotopy equivalent.

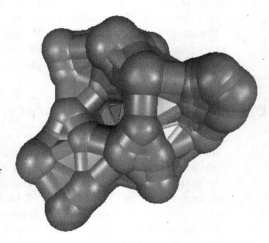

FIGURE 65.4.1
Snapshot during the deformation retraction of the space-filling representation of gramicidin to its dual complex. The spheres shrink to vertices while the intersection circles become cylinders that eventually turn into edges.

Homology groups: Quotients of cycle groups and their boundary subgroups. There is one group per dimension. The kth *Betti number*, β_k, is the rank of the kth homology group.

Euler characteristic: The alternating sum of Betti numbers: $\chi = \sum_{k \geq 0} (-1)^k \beta_k$.

Voids: Bounded connected components of the complement. Here, we are primarily interested in voids of space-filling diagrams embedded in $\mathbb{R}^3$.

Pockets: Maximal regions in the complement of a space-filling diagram that become voids before they disappear. Here, we assume the growth model that preserves the Voronoi diagram of the spheres.

Persistent homology groups: Quotients of the cycle groups at some time t and their boundary subgroups a later time $t + p$. The ranks of these groups are the *persistent Betti numbers*.

Protein complex: Two or more docked proteins. A complex can be represented by a single space-filling diagram of colored balls.

Molecular interface: Surface consisting of bichromatic Voronoi polygons that separate the proteins in the complex. The surface is retracted to the region in which the proteins are in close contact.

FIGURE 65.4.2
Molecular interface of the neurotoxic vipoxin complex. The surface has nonzero genus, which is unusual. In this case, we have genus equal to three, which implies the existence of three loops from each protein that are linked with each other. The linking might explain the unusually high stability of the complex, which remains for years in solution. The piecewise linear surface has been smoothed to improve visibility.

CLASSIFICATION

The connectivity of topological spaces is commonly discussed by forming equivalence classes of spaces that are connected the same way. Sameness may be defined as being homeomorphic, being homotopy equivalent, having isomorphic homology groups, or having the same Euler characteristic. In this sequence, the classification gets progressively coarser but also easier to compute. Homology groups seem to be a good compromise as they capture a great deal of connectivity information and have fast algorithms. The classic approach to computing homology groups is algebraic and considers the incidence matrices of adjacent dimensions. Each matrix is reduced to *Smith normal form* using a Gaussian-elimination-like reduction algorithm. The ranks and torsion coefficients of the homology groups can be read off directly from the reduced matrices [Mun84]. Depending on which coefficients we use and exactly how we reduce, the running time can be anywhere between cubic in the number of simplices and exponential or worse.

INCREMENTAL ALGORITHM

Space-filling diagrams are embedded in $\mathbb{R}^3$ and enjoy properties that permit much faster algorithms. To get started, we use the existence of a deformation retraction from the space-filling diagram to the dual complex, which implies that the two have isomorphic homology groups [Ede95]. The embedding in $\mathbb{R}^3$ prohibits nonzero torsion coefficients [AH35]. We therefore limit ourselves to Betti numbers, which we compute incrementally, by adding one simplex at a time in an order that agrees with the filtration of the dual complexes. When we add a k-dimensional simplex, σ, the kth Betti number goes up by one if σ belongs to a k-cycle, and the $(k-1)$st Betti number goes down by one if σ does not belong to a k-cycle. The two cases can be distinguished in a time that, for all practical purposes, is constant per operation, leading to an essentially linear time algorithm for computing the Betti numbers of all complexes in the filtration [DE95].

PERSISTENCE

To get a handle on the stability of a homology class, we observe that the simplices that create cycles can be paired with the simplices that destroy cycles. The *persistence* is the time lag between the creation and the destruction [ELZ02]. The idea of pairing lies also at the heart of two types of shape features relevant in the study of protein interactions. A *pocket* in a space-filling diagram is a portion of the outside space that becomes a void before it disappears [EFL98, Kun92]. It is represented by a triangle-tetrahedron pair: the triangle creates a void and the tetrahedron is the last piece that eventually fills that same void. The *molecular interface* consists of all bichromatic Voronoi polygons of a protein complex. To identify the essential portions of this surface, we again observe how voids are formed and retain the bichromatic polygons inside pockets while removing all others [BER06]. A different geometric formalization of the same biochemical concept can be found in [VBR⁺95].

Preliminary experiments in the 1990s suggested that the combination of molecular interfaces and the idea of persistence can be used to predict the hot-spot residues in protein-protein interactions [Wel96]. In the 2000s, persistent homology was used to characterize structural changes in membrane fusion over the course of a simulation [KZP⁺07]. More recently, persistent homology has been used for extracting molecular topological fingerprints (MTFs) of proteins, based on the persistence of molecular topological invariants. These fingerprints have been used for protein characterization, identification, and classification [XW14, XW15a], as well as for cryo-EM data analysis [XW15b].

65.5 DENSITY MAPS IN STRUCTURAL BIOLOGY

Continuous maps over manifolds arise in a variety of settings within structural molecular biology. One is *X-ray crystallography*, which is the most common method for determining the 3-dimensional structure of proteins [BJ76, Rho00]. The key to X-ray crystallography is to obtain first pure crystals of the protein of interest. The crystalline atoms cause a beam of incident X-rays to diffract into many specific directions. By measuring the angles and intensities of these diffracted beams, a crystallographer can produce a 3-dimensional picture of the density of electrons within the crystal. From this electron density, the mean positions of the atoms can be determined, the chemical bonds that connect them, as well as their disorder. The first two protein structures to be solved using this technique were those of hemoglobin and myoglobin [KDS+60, PRC+60]. Of the 110,000 protein structures present in the database of biomolecular structures (PDB) as of June 2016, more than 99,000 were determined using X-ray crystallography.

A second setting is *molecular mechanics*, whose central focus is to develop insight on the forces that stabilize biomolecular structures. Describing the state of a biomolecule in terms of its energy landscape, the native state corresponds to a large basin, and it is mostly the structure of this basin that is of interest. Theoretically, the laws of quantum mechanics completely determine the energy landscape of any given molecule by solving Schrödinger's equation. In practice, however, only the simplest systems such as the hydrogen atom have an exact, explicit solution to this equation and modelers of large molecular systems must rely on approximations. Simulations are based on a space-filling representation of the molecule, in which the atoms interact through empirical forces. There is increased interest in the field of structure biology to map the results of those simulations onto the structures of the molecules under study, to help visualize their properties. We may, for example, be interested in the electrostatic potential induced by a protein and visualize it as a density map over 3-dimensional space or over a surface embedded in that space.

As a third setting, we mention the *protein docking* problem. Given two proteins, or a protein and a ligand, we try to fit protrusions of one into the cavities of the other [Con86]. We make up continuous functions related to the shapes of the surfaces and identify protrusions and cavities as local extremes of these functions. Morse theory is the natural mathematical framework for studying these maps [Mil63, Mat02, LLY+15].

GLOSSARY

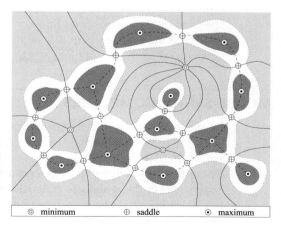

FIGURE 65.5.1

Portion of the Morse-Smale complex of a Morse-Smale function over a 2-manifold. The solid stable 1-manifolds and the dashed unstable 1-manifolds are shown together with two dotted level sets. Observe that all 2-dimensional regions of the complex are quadrangular.

⊚ minimum   ⊕ saddle   ⊙ maximum

Morse function: Generic smooth map on a Riemannian manifold, $f : \mathbb{M} \to \mathbb{R}$. In particular, the genericity assumption includes the fact that all critical points are nondegenerate and have different function values.

Gradient, Hessian: The vector of first derivatives and the matrix of second derivatives.

Critical point: Point at which the gradient of f vanishes. It is **nondegenerate** if the Hessian is invertible. The **index** of a nondegenerate critical point is the number of negative eigenvalues of the Hessian.

Integral line: Maximal curve whose velocity vectors agree with the gradient of the Morse function. Two integral lines are either disjoint or the same.

Stable manifold: Union of integral lines that converge to the same critical point. We get **unstable manifolds** if we negate f and thus effectively reverse the gradient.

Morse-Smale complex: Collection of cells obtained by intersecting stable with unstable manifolds. We require f to be a *Morse-Smale function* satisfying the additional genericity assumption that these intersections are transversal.

Cancellation: Local change of the Morse function that removes a pair of critical points. Their indices are necessarily contiguous.

CRITICAL POINTS

Classic Morse theory applies only to generic smooth maps on manifolds, $f: \mathbb{M} \to \mathbb{R}$. Maps that arise in practice are rarely smooth and generic or, more precisely, the information we are able to collect about maps is rarely enough to go beyond a piecewise linear representation. This is however no reason to give up on applying the underlying ideas of Morse theory. To illustrate this point, we discuss critical points, which for smooth functions are characterized by a vanishing gradient: $\nabla f = 0$. If we draw a small circle around a noncritical point u on a 2-manifold, we get one arc along which the function takes on values less than $f(u)$ and a complementary arc along which the function is greater than or equal to $f(u)$. Call the former arc the *lower link* of u. We get different lower links for critical points: the entire circle for a *minimum*, two arcs for a *saddle*, and the empty set for a *maximum*. A typical representation of a piecewise linear map is a triangulation with function values specified at the vertices and linearly interpolated over the edges and triangles. The lower link of a vertex can still be defined and the criticality of the vertex can be determined from the topology of the lower link [Ban67].

MORSE-SMALE COMPLEXES

In the smooth case, each critical point defines a *stable manifold* of points that converge to it by following the gradient flow. Symmetrically, it defines an *unstable manifold* of points that converge to it by following the reversed gradient flow. These manifolds define decompositions of the manifold into simple cells [Tho49]. Extensions of these ideas to construct similar cell decompositions of manifolds with piecewise linear continuous functions can be found in [EHZ03]. In practice, it is essential to be able to simplify these decompositions, which can be done by canceling critical points in pairs in the order of increasing persistence [ELZ02].

65.6 MEASURING BIOMOLECULES

Protein dynamics is multi-scale: from the jiggling of atoms (pico-seconds), to the domain reorganizations in proteins (micro to milliseconds), to protein folding and diffusion (milli-second to seconds), and finally to binding and translocation (seconds to minutes). Connecting these different scales is a central problem in polymer physics that remains unsolved despite numerous theoretical and computational developments; see [Gue07, DP11]. Computer simulations play an essential role in all corresponding multi-scale methods, as they provide information at the different scales. Usually, computer simulations of protein dynamics start with a large system containing the protein and many water molecules to mimic physiological conditions. Given a model for the physical interactions between these molecules, their space-time trajectories are computed by numerically solving the equations of motion. These trajectories however are limited in scope. Current computing technologies limit the range of time-scales that can be simulated to the microsecond level, for systems that contain up to hundred thousands of atoms [VD11]. There are many directions that are currently explored to extend these limits, from hardware solutions including the development of specialized computers [SDD+07], or by harnessing the power of graphics processor units [SPF+07], to the development of simplified models that are computationally tractable and remain physically accurate. Among such models are those that treat the solvent implicitly, reducing the solute-solvent interactions to their mean-field characteristics. These so-called *implicit solvent models* are often applied to estimate free energy of solute-solvent interactions in structural and chemical processes, folding or conformational transitions of proteins and nucleic acids, association of biological macromolecules with ligands, or transport of drugs across biological membranes. The main advantage of these models is that they express solute-solvent interactions as a function of the solute degrees of freedom alone, more specifically of its volume and surface area. Methods that compute the surface area and volume of a molecule, as well as their derivatives with respect to the position of its atoms are therefore of great interest for computational structural biology.

GLOSSARY

Indicator function: It maps a point x to 1 if $x \in P$ and to 0 if $x \notin P$, in which P is some fixed set. Here, we are interested in convex polyhedra P and can therefore use the alternating sum of the number of faces of various dimensions visible from x as an indicator; see [Ede95] for details.

Inclusion-exclusion: Principle used to compute the volume of a union of bodies as the alternating sum of volumes of k-fold intersections, for $k \geq 1$.

Stereographic projection: Mapping of a sphere minus a point to Euclidean space. The map preserves spheres and angles. We are primarily interested in the case in which both the sphere and the Euclidean space are 3-dimensional.

Atomic solvation parameters: Experimentally determined numbers that assess the hydrophobicity of different atoms [EM86].

Weighted area: Area of the boundary of a space-filling diagram in which the contribution of each individual ball is weighted by its atomic solvation parameter.

FIGURE 65.6.1

Stereographic projection from the north pole. The preimage of a circle in the plane is a circle on the sphere, which is the intersection of the sphere with a plane. By extension, the preimage of a union of disks is the intersection of the sphere with the complement of a convex polyhedron.

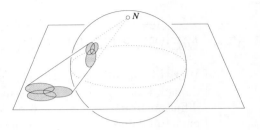

Also a function $A\colon \mathbb{R}^{3n} \to \mathbb{R}$ obtained by parametrizing a space-filling diagram by the coordinates of its n ball centers.

Weighted-area derivative: The linear map $DA_{\mathbf{z}}\colon \mathbb{R}^{3n} \to \mathbb{R}$ defined by $DA_{\mathbf{z}}(\mathbf{t}) = \langle \mathbf{a}, \mathbf{t} \rangle$, in which $\mathbf{z} \in \mathbb{R}^{3n}$ specifies the space-filling diagram, $\mathbf{t} \in \mathbb{R}^{3n}$ lists the coordinates of the motion vectors, and $\mathbf{a} = \nabla A(\mathbf{z})$ is the gradient of A at $\mathbf{z}$. It is also the map $DA\colon \mathbb{R}^{3n} \to \mathbb{R}^{3n}$ that maps $\mathbf{z}$ to $\mathbf{a}$.

Weighted volume: Volume of a space-filling diagram in which the contribution of each individual ball is weighted by its atomic solvation parameter. Also a function $V\colon \mathbb{R}^{3n} \to \mathbb{R}$ obtained by parametrizing a space-filling diagram by the coordinates of its n ball centers.

Weighted-volume derivative: The linear map $DV_{\mathbf{z}}\colon \mathbb{R}^{3n} \to \mathbb{R}$ defined by $DV_{\mathbf{z}}(\mathbf{t}) = \langle \mathbf{v}, \mathbf{t} \rangle$, in which $\mathbf{z} \in \mathbb{R}^{3n}$ specifies the space-filling diagram, $\mathbf{t} \in \mathbb{R}^{3n}$ lists the coordinates of the motion vectors, and $\mathbf{v} = \nabla V(\mathbf{z})$ is the gradient of V at $\mathbf{z}$. It is also the map $DV\colon \mathbb{R}^{3n} \to \mathbb{R}^{3n}$ that maps $\mathbf{z}$ to $\mathbf{v}$.

GEOMETRIC INCLUSION-EXCLUSION

Work on computing the volume and the area of a space-filling diagram $F = \bigcup_i B_i$ can be divided into approximate [Row63] and exact methods [Ric74]. According to the principle of inclusion-exclusion, the volume of F can be expressed as an alternating sum of volumes of intersections:

$$\operatorname{vol} F = \sum_{\Lambda} (-1)^{\operatorname{card} \Lambda + 1} \operatorname{vol} \bigcap_{i \in \Lambda} B_i,$$

in which Λ ranges over all nonempty subsets of the index set. The size of this formula is exponential in the number of balls, and the individual terms can be quite complicated. Most of the terms are redundant, however, and a much smaller formula based on the dual complex K of the space-filling diagram F has been given [Ede95]:

$$\operatorname{vol} F = \sum_{\sigma \in K} (-1)^{\dim \sigma} \operatorname{vol} \bigcap \sigma,$$

in which $\bigcap \sigma$ denotes the intersection of the $\dim \sigma + 1$ balls whose centers are the vertices of σ. The proof is based on the Euler formula for convex polyhedra and uses stereographic projection to relate the space-filling diagram in $\mathbb{R}^3$ with a convex polyhedron in $\mathbb{R}^4$. Precursors of this result include the existence proof of a polynomial size inclusion-exclusion formula [Kra78] and the presentation of such a formula using the simplices in the Delaunay triangulation [NW92]. We note that it is not difficult to modify the formula to get the weighted volume: decompose the terms $\text{vol} \cap \sigma$ into the portions within the Voronoi cells of the participating balls and weight each portion accordingly.

DERIVATIVES

The relationship between the weighted- and unweighted-volume derivatives is less direct than that between the weighted and unweighted volumes. Just to state the formula for the weighted-volume derivative requires more notation than we are willing to introduce here. Instead, we describe the two geometric ingredients, both of which can be computed by geometric inclusion-exclusion [EK03]. The first ingredient is the area of the portion of the disk spanned by the circle of two intersecting spheres that belongs to the Voronoi diagram. This facet is the intersection of the disk with the corresponding Voronoi polygon. The second ingredient is the weighted average vector from the center of the disk to the boundary of said facet. The weight is the infinitesimal contribution to the area as we rotate the vector to sweep out the facet. A similar approach allows for the computation of the weighted area derivative [BEKL04].

VOIDS AND POCKETS

A *void* V is a maximal connected subset of space that is disjoint from and completely surrounded by the union of balls. Its surface area is easily computed by identifying the sphere patches on the boundary of the union that also bound the void. It helps to know that there is a deformation retraction from the union of balls $\bigcup_i B_i$ to the dual complex K [Ede95]. Similarly, there is a corresponding void in K represented by a connected set of simplices in the Delaunay triangulation, that do not belong to K. This set U is open and its boundary (the simplices added by closure) forms what one may call the dual complex of the boundary of V. The volume and surface area of the void V are then computed based on U [Ede95].

It would be interesting to generalize these ideas to pockets as defined in [EFL98]. In contrast to a void, a *pocket* is not completely surrounded by the balls but connected to the outside through narrow channels. Again we have a corresponding set of simplices in the Delaunay triangulation that do not belong to the dual complex, but this set is partially closed at the places the pocket connects to the outside. The inclusion-exclusion formulas still apply, but there are cases in which the cancellation of terms near the connecting channel is not complete and leads to slightly incorrect measurements.

65.7 SHAPE IN STRUCTURAL BIOLOGY

As bio-molecules are usually represented as unions of balls, it is not surprising to see geometric algorithms being adapted to characterize the shapes of molecules. We have discussed above the *alpha shape theory* [EM94], whose first applications in biology focused on computing the volume and surface area of molecular shapes [LEF$^+$98] as well as on characterizing the cavities and pockets formed by a molecule

[LEW98, EFL98]. While these applications of the alpha shape theory remain popular in structural biology — with new and improved software implementations being released regularly, such as CASTp [DOT+06], Volume [CKL11], and UnionBall [MK11] — many applications in new domains have been proposed. Here we review a few of these applications.

GLOSSARY

Atom packing: A measure of how tight atoms are packed within a protein or nucleic acid structure.

Binding site: Region of a protein in which a ligand can bind. These regions often correspond to the cavities and pockets of the protein, though there are examples of binding sites that sit at the surface of the protein.

Structure alignment: Collection of monotonically increasing maps to the integers, one per chain of points modeling a protein backbone.

Protein docking: Process in which a protein forms a complex with another molecule. The complex usually exists only temporarily and facilitates an interaction between the molecules.

STATISTICS OF PROTEIN STRUCTURE GEOMETRY

The experimental determination of a protein structure at the atomic level remains a difficult problem. There is hope however that theoretical and computational techniques will supplement experimental methods and enable protein structure prediction at the near atomic level [KL99]. Many of these techniques rely on the knowledge derived from the analysis of the geometry of known protein structures. Such an analysis requires an objective definition of atomic packing within a molecular structure. The alpha shape theory has proved a useful approach for deriving such a definition. For example, Singh *et al.* [STV96] used the Delaunay complex to define nearest neighbors in protein structures and to derive a four-body statistical potential. This potential has been used successfully for fold recognition, decoy structure determination, mutant analysis, and other studies; see [Vai12] for a full review. The potentials considered in these studies rely on the tetrahedra defined by the Delaunay triangulation of the points representing the atoms. In parallel, Zomorodian and colleagues have shown that it is possible to use the alpha shape theory to filter the list of pairwise interactions to generate a much smaller subset of pairs that retains most of the structural information contained in a proteins [ZGK06]. The alpha shape theory has also been used to compute descriptors for the shapes [WBKC09] and surfaces [TDCL09, TL12] of proteins.

SIMILARITY AND COMPLEMENTARITY

The alpha shape theory allows for the detection of simplices characterizing the geometry of a protein structure. Those simplices include points, edges, triangles, and tetrahedra connecting atoms of this protein structure. It is worth mentioning that it is possible to use those simplices to compare two protein structures and even to derive a structural alignment between them [RSKC05].

As the function of a protein is related to its geometry and as function usually involves binding to a partner protein, significant efforts have been put into characterizing the geometry of protein-ligand interactions, where *ligands* include small molecules, nucleic acids, and other proteins. Among these efforts, a few relate to the applications of the alpha shape theory. They have recently been extended to characterize binding sites at the surface of proteins [TDCL09, TL12]. The alpha shape theory has also been used to characterize the interfaces in protein-protein complexes [BER06] as well as protein-DNA interactions [ZY10].

We mention a geometric parallel between finding a structural alignment between two proteins and predicting the structure of their interactions. While the former is based on the identification of similar geometric patterns between the two structures, the latter is based on the identification of complementary patterns between the surfaces of the two structures. As mentioned above, geometric patterns based on the Delaunay triangulation have been used for structural alignment. In parallel, similar patterns have recently been used to predict protein-protein interactions [EZ12].

CHARACTERIZING MOLECULAR DYNAMICS

All the applications described above relate to the static geometry of molecules. Bio-molecules however are dynamics. A *molecular dynamics simulation* is designed to capture this dynamics: it follows the Newtonian dynamics of the molecule as a function of time, generating millions of snapshots over the course of its trajectory. The alpha shape theory has proved useful to characterize the geometric changes that occur during such a trajectory. For example, using the concept of topological persistence [ELZ02], Kasson et al. characterized structural changes in membrane fusion over the course of a simulation [KZP$^+$07].

65.8 SOURCES AND RELATED MATERIAL

FURTHER READING

For background reading in **algorithms** we recommend: [CLR90], which is a comprehensive introduction to combinatorial algorithms; [Gus97], which is an algorithms text specializing in bioinformatics; [Str93], which is an introduction to linear algebra; and [Sch02], which is a numerical algorithms text in molecular modeling.

For background reading in **geometry** we recommend: [Ped88], which is a geometry text focusing on spheres; [Nee97], which is a lucid introduction to geometric transformations; [Fej72], which studies packing and covering in two and three dimensions; and [Ede01], which is an introduction to computational geometry and topology, focusing on Delaunay triangulations and mesh generation.

For background reading in **topology** we recommend: [Ale61], which is a compilation of three classical texts in combinatorial topology; [Gib77], which is a very readable introduction to homology groups; [Mun84], which is a comprehensive text in algebraic topology; and [Mat02], which is a recent introduction to Morse theory.

For background reading in **biology** we recommend: [ABL$^+$94], which is a basic introduction to molecular biology on the cell level; [Str88], which is a fundamental text in biochemistry; and [Cre93], which is an introduction to protein sequences, structures, and shapes.

RELATED CHAPTERS

REFERENCES

[ABL⁺94]   B. Alberts, D. Bray, J. Lewis, M. Raff, K. Roberts, and J.D. Watson. *Molecular Biology of the Cell*. Garland, New York, 1994.

[AH35]   P. Alexandroff and H. Hopf. *Topologie I*. Julius Springer, Berlin, 1935.

[Ale61]   P. Alexandroff. *Elementary Concepts of Topology*. Dover, New York, 1961.

[Anf73]   C.B. Anfinsen. Principles that govern protein folding. *Science*, 181:223–230, 1973.

[Ban67]   T.F. Banchoff. Critical points and curvature for embedded polyhedra. *J. Differential Geom.*, 1:245–256, 1967.

[BEKL04]   R. Bryant, H. Edelsbrunner, P. Koehl, and M. Levitt. The weighted area derivative of a space filling diagram. *Discrete Comput. Geom.*, 32:293–308, 2004.

[BER06]   Y.E.A. Ban, H. Edelsbrunner, and J. Rudolph. Interface surfaces for protein-protein complexes. *J. ACM*, 53:361–378, 2006.

[BJ76]   T. Blundell and L. Johnson. *Protein Crystallography*. Academic Press, New York, 1976.

[Bli82]   J.F. Blinn. A generalization of algebraic surface drawing. *ACM Trans. Graph.*, 1:235–256, 1982.

[BLMP97]   C. Bajaj, H.Y. Lee, R. Merkert, and V. Pascucci. NURBS based B-rep models for macromolecules and their properties. In *Proc. 4th ACM Sympos. Solid Modeling Appl., 1997*, pages 217–228.

[BS01]   D. Baker and A. Sali. Protein structure prediction and structural genomics. *Science*, 294:93–96, 2001.

[BWF⁺00]   H.M. Berman, J. Westbrook, Z. Feng, G. Gilliland, T.N. Bhat, H. Weissig, *et al.* The Protein Data Bank. *Nucl. Acids. Res.*, 28:235–242, 2000.

[BWZ08]   P.W. Bates, G.-W. Wei, and S. Zhao. Minimal molecular surfaces and their applications. *J. Comp. Chem.*, 29:380–391, 2008.

[CCW06]   T. Can, C.-I. Chen, and Y.-F. Wang. Efficient molecular surface generation using level set methods. *J. Molec. Graph. Modeling*, 25:442–454, 2006.

[CDES01]   H.-L. Cheng, T.K. Dey, H. Edelsbrunner, and J. Sullivan. Dynamic skin triangulation. *Discrete Comput. Geom.*, 25:525–568, 2001.

[Cec93]   T.R. Cech. The efficiency and versatility of catalytic RNA: implications for an RNA world. *Gene*, 135:33–36, 1993.

[CEF01]   H.-L. Cheng, H. Edelsbrunner, and P. Fu. Shape space from deformation. *Comput. Geom.*, 19:191–204, 2001.

[Che93]   L.P. Chew. Guaranteed-quality mesh generation for curved surfaces. In *Proc. 9th Sympos. Comput. Geom.*, pages 274–280, ACM Press, 1993.

[CKL11]   F. Cazals, H. Kanhere, and S. Loriot. Computing the volume of union of balls: a certified algorithm. *ACM Trans. Math. Software*, 38:3, 2011.

[CLR90]   T.H. Cormen, C.E. Leiserson, and R.L. Rivest. *Introduction to Algorithms*. MIT Press, Cambridge, 1990.

[Con83]   M.L. Connolly. Analytic molecular surface calculation. *J. Appl. Crystallogr.*, 6:548–558, 1983.

[Con86]   M.L. Connolly. Measurement of protein surface shape by solid angles. *J. Molecular Graphics*, 4:3–6, 1986.

[CP13]   S.W. Chen and J.-L. Pellequer. Adepth: new representation and its implications for atomic depths of macromolecules. *Nucl. Acids Res.*, 41:W412–W416, 2013.

[CP53]   R.B. Corey and L. Pauling. Molecular models of amino acids, peptides and proteins. *Rev. Sci. Instr.*, 24:621–627, 1953.

[Cre93]   T.E. Creighton. *Proteins*. Freeman, New York, 1993.

[DB07]   R. Das and D. Baker. Macromolecular modeling with Rosetta. *Annu. Rev. Biochem.*, 77:363–382, 2008.

[DE95]   C.J.A. Delfinado and H. Edelsbrunner. An incremental algorithm for Betti numbers of simplicial complexes on the 3-sphere. *Comput. Aided Geom. Design*, 12:771–784, 1995.

[Del34]   B. Delaunay. Sur la sphère vide. *Izv. Akad. Nauk SSSR, Otdelenie Matematicheskikh i Estestvennykh Nauk*, 7:793–800, 1934.

[Dil07]   K.A. Dill, S.B. Ozkan, T.R. Weikl, J.D. Chodera, and V.A. Voelz. The protein folding problem: when will it be solved? *Curr. Opin. Struct. Biol.*, 17:342–346, 2007.

[DO93a]   B.S. Duncan and A.J. Olson. Approximation and characterization of molecular surfaces. *Biopolymers*, 33:219–229, 1993.

[DO93b]   B.S. Duncan and A.J. Olson. Shape analysis of molecular surfaces. *Biopolymers*, 33:231–238, 1993.

[Doo13]   W.F. Doolittle. Is junk DNA bunk? A critique of ENCODE. *Proc. Natl. Acad. Sci. (USA)*, 110:5294–5300, 2013.

[DOT⁺06]   J. Dundas, Z. Ouyang, J. Tseng, A. Binkowski, Y. Turpaz, and J. Liang. CASTp: computed atlas of surface topography of proteins with structural and topographical mapping of functionally annotated residues. *Nuc. Acids Res.*, 34:W116–W118, 2006.

[DP11]   J.J. de Pablo. Coarse-grained simulations of macromolecules: from DNA to nanocomposites. *Ann. Rev. Phys. Chem.*, 62:555–574, 2011.

[DSUS08]   A.K. Dunker, I. Silman, V.N. Uversky, and J.L. Sussman. Function and structure of inherently disordered proteins. *Curr. Opin. Struct. Biol.*, 18:756–764, 2008.

[DW05]   H.J. Dyson and P.E. Wright. Intrinsically unstructured proteins and their functions. *Nat. Rev. Mol. Cell Biol.*, 6:197–208, 2005.

[Ede95]   H. Edelsbrunner. The union of balls and its dual shape. *Discrete Comput. Geom.*, 13:415–167, 1995.

[Ede99]   H. Edelsbrunner. Deformable smooth surface design. *Discrete Comput. Geom.*, 21:87–115, 1999.

[Ede01]   H. Edelsbrunner. *Geometry and Topology for Mesh Generation*. Cambridge University Press, 2001.

[EFL98]   H. Edelsbrunner, M.A. Facello, and J. Liang. On the definition and the construction of pockets in macromolecules. *Discrete Appl. Math.*, 88:83–102, 1998.

[EH07]   A. Elofsson and G. von Heijine. Membrane protein structure: prediction versus reality. *Annu. Rev. Biochem.*, 76:125–140, 2007.

[EHZ03]   H. Edelsbrunner, J. Harer, and A. Zomorodian. Hierarchy of Morse-Smale complexes for piecewise linear 2-manifolds. *Discrete Comput. Geom.*, 30:87–107, 2003.

[EK03]   H. Edelsbrunner and P. Koehl. The weighted volume derivative of a space-filling diagram. *Proc. Natl. Acad. Sci. (USA)*, 100:2203–2208, 2003.

[EKS83]   H. Edelsbrunner, D.G. Kirkpatrick, and R. Seidel. On the shape of a set of points in the plane. *IEEE Trans. Inform. Theory*, 29:551–559, 1983.

[ELZ02]   H. Edelsbrunner, D. Letscher, and A. Zomorodian. Topological persistence and simplification. *Discrete Comput. Geom.*, 28:511–533, 2002.

[EM86]   D. Eisenberg and A. McLachlan. Solvation energy in protein folding and binding. *Nature*, 319:199–203, 1986.

[EM94]   H. Edelsbrunner and E.P. Mücke. Three-dimensional alpha shapes. *ACM Trans. Graphics*, 13:43–72, 1994.

[ENC12]   The ENCODE Project Consortium. An integrated encyclopedia of DNA elements in the human genome. *Nature*, 489:57–74, 2012.

[ES97]   H. Edelsbrunner and N.R. Shah. Triangulating topological spaces. *Internat. J. Comput. Geom. Appl.*, 7:365–378, 1997.

[EZ12]   L. Ellingson and J. Zhang. Protein surface matching by combining local and global geometric information. *PLoS One*, 7:e40540, 2012.

[Fej72]   L. Fejes Tóth. *Lagerungen in der Ebene auf der Kugel und im Raum*. Second edition, Springer-Verlag, Berlin, 1972.

[Gib77]   P.J. Giblin. *Graphs, Surfaces, and Homology. An Introduction to Algebraic Topology*. Chapman and Hall, London, 1977.

[Gil86]   W. Gilbert. The RNA world. *Nature*, 319:618, 1986.

[GP95]   J.A. Grant and B.T. Pickup. A Gaussian description of molecular shape. *J. Phys. Chem.*, 99:3503–3510, 1995.

[GR01]   M. Gerstein and F.M. Richards. Protein geometry: distances, areas, and volumes. In M.G. Rossman and E. Arnold, editors, *The International Tables for Crystallography*, Vol. F, Chapter 22, pages 531–539. Kluwer, Dordrecht, 2001.

[Gue07]   M.G. Guenza. Theoretical models for bridging timescales in polymer physics. *J. Phys. Condens. Matter*, 20:033101, 2007.

[Gus97]   D. Gusfield. *Algorithms on Strings, Trees, and Sequences*. Cambridge University Press, 1997.

[KDS$^+$60] J.C. Kendrew, R.E. Dickerson, B.E. Strandberg, R.G. Hart, D.R. Davies and D.C. Philips. Structure of myoglobin: a three dimensional Fourier synthesis at 2 angstrom resolution. *Nature*, 185:422–427, 1960.

[KL99]   P. Koehl and M. Levitt. A brighter future for protein structure prediction. *Nature Struct. Biol.*, 6:108–111, 1999.

[Kol65]   W.L. Koltun. Precision space-filling atomic models. *Biopolymers*, 3:665–679, 1965.

[Kra78]   K.W. Kratky. The area of intersection of n equal circular disks. *J. Phys. A*, 11:1017–1024, 1978.

[Kun04]   T.A. Kunkel. DNA replication fidelity. *J. Biol. Chem.*, 279:16895–16898, 2004.

[Kun92]   I.D. Kuntz. Structure-based strategies for drug design and discovery. *Science*, 257:1078–1082, 1992.

[KV07]   N. Kruithof and G. Vegter. Meshing skin surfaces with certified topology. *Comput. Geom.*, 36:166–182, 2007.

[KZP$^+$07] P.M. Kasson, A. Zomorodian, S. Park, N. Singhal, L. Guibas and V.S. Pande. Persistent voids: a new structural metric for membrane fusion. *Bioinformatics*, 23:1753–1759, 2007.

[LEF$^+$98] J. Liang, H. Edelsbrunner, P. Fu, P.V. Sudhakar, and S. Subramaniam. Analytical shape computation of macromolecules. I. Molecular area and volume through alpha shape. *Proteins: Struct. Func. Genet.*, 33:1–17, 1998.

[LEW98]   J. Liang, H. Edelsbrunner, and C. Woodward. Anatomy of protein pockets and cavities: measurement of binding site geometry and implications for ligand design. *Prot. Sci.*, 7:1884–1897, 1998.

[LFSB03]   M.S. Lee, M. Feig, F.R. Salsbury and C.L. Brooks. New analytic approximation to the standard molecular volume definition and its application to generalized Born calculations. *J. Comp. Chem.*, 24:1348–1356, 2003.

[LLY+15]   H. Liu, F. Lin, J.-L. Yang, H.R. Wang, and X.-L. Liu. Applying side-chain flexibility in motifs for protein docking. *Genomics Insights*, 15:1–10, 2015.

[LR71]   B. Lee and F.M. Richards. The interpretation of protein structures: estimation of static accessibility. *J. Molecular Biol.*, 55:379–400, 1971.

[Mat02]   Y. Matsumoto. *An Introduction to Morse Theory*. Amer. Math. Soc., Providence, 2002.

[MG88]   N.L. Max and E.D. Getzoff. Spherical harmonic molecular surfaces. *IEEE Comput. Graph. Appl.*, 8:42–50, 1988.

[MHS11]   D.S. Marks, T.A. Hopf, and C. Sander. Protein structure prediction from sequence variation. *Nature Biotechnology*, 30:1072–1080, 2011.

[Mil63]   J. Milnor. *Morse Theory*. Princeton University Press, 1963.

[MK08]   S.D. McCulloch and T.A. Kunkel. The fidelity of DNA synthesis by eukaryotic replicative and translesion synthesis polymerases. *Cell Research*, 18:148–161, 2008.

[MK11]   P. Mach and P. Koehl. Geometric measures of large biomolecules: surface, volume, and pockets. *J. Comp. Chem.*, 32:3023–3038, 2011.

[Mun84]   J.R. Munkres. *Elements of Algebraic Topology*. Addison-Wesley, Redwood City, 1984.

[Nee97]   T. Needham. *Visual Complex Analysis*. Clarendon Press, Oxford, 1997.

[NW92]   D.Q. Naiman and H.P. Wynn. Inclusion-exclusion-Bonferroni identities and inequalities for discrete tube-like problems via Euler characteristics. *Ann. Statist.*, 20:43–76, 1992.

[OF03]   S. Osher and R. Fedkiw. *Level Sets Methods and Dynamic Implicit Surfaces*. Spinger-Verlag, New York, 2003.

[OMY+09]   C.J. Oldfield, J. Meng, J.Y. Yang, M.Q. Yang, V.N. Uversky, *et al.* Flexible nets: disorder and induced fit in the associations of p53 and 14-3-3 with their partners. *BMC Genomics*, 9:S1, 2009.

[Ped88]   D. Pedoe. *Geometry: A Comprehensive Course*. Dover, New York, 1988.

[PG14]   A.F. Palazzo and T.R. Gregory. The case for junk DNA. *PLoS Genetics*, 10:e1004351, 2014.

[PRC+60]   M. Perutz, M. Rossmann, A. Cullis, G. Muirhead, G. Will, and A. North. Structure of hemoglobin: a three-dimensional Fourier synthesis at 5.5 angstrom resolution, obtained by X-ray analysis. *Nature*, 185:416–422, 1960.

[PS10]   M. Pertea and S.L. Salzberg. Between a chicken and a grape: estimating the number of human genes. *Genome Biology*, 11:206–212, 2010.

[Rho00]   G. Rhodes. *Crystallography Made Crystal Clear*. Second edition, Academic Press, San Diego, 2000.

[Ric74]   F.M. Richards. The interpretation of protein structures: total volume, group volume distributions and packing density. *J. Molecular Biol.*, 82:1–14, 1974.

[Ric77]   F.M. Richards. Areas, volumes, packing, and protein structures. *Ann. Rev. Biophys. Bioeng.*, 6:151–176, 1977.

[Ric85]   J.S. Richardson. Schematic drawings of protein structures. *Methods in Enzymology*, 115:359–380, 1985.

[Row63]   J.S. Rowlinson. The triplet distribution function in a fluid of hard spheres. *Molecular Phys.*, 6:517–524, 1963.

[RSKC05]   J. Roach, S. Sharma, K. Kapustina, and C.W. Carter. Structure alignment via De-launay tetrahedralization. *Proteins: Struct. Func. Bioinfo.*, 60:66–81, 2006.

[Sch02]   T. Schlick. *Molecular Modeling and Simulation.* Springer-Verlag, New York, 2002.

[SDD⁺07]   D.E. Shaw, M.M. Deneroff, R. Dror, *et al.* Anton, a special-purpose machine for molecular dynamics simulation. *ACM SIGARCH Computer Architecture News*, 35:1–12, 2007.

[SM12]   M.G. Seetin and M.H. Mathews. RNA structure prediction: an overview of methods. *Methods Mol. Biol.*, 905:99–122, 2012.

[SPF⁺07]   J.E. Stone, J.C. Philipps, P.L. Freddolino, D.J. Hardy, L.G. Trabuco and K. Schul-ten. Accelerating molecular modeling applications with graphics processors. *J. Comp. Chem.*, 28:2618–2640, 2007.

[Str93]   G. Strang. *Introduction to Linear Algebra.* Wellesley-Cambridge Press, Wellesley, 1993.

[Str88]   L. Stryer. *Biochemistry.* Freeman, New York, 1988.

[STV96]   R.K. Singh, A. Tropsha, and I.I. Vaisman. Delaunay tessellation of proteins: Four body nearest-neighbor propensities of amino acid residues. *J. Comput. Biol.*, 3:213–221, 1996.

[TDCL09]   Y.Y. Tseng, C. Dupree, Z.J. Chen and W.H. Li. SplitPocket: identification of protein functional surfaces and characterization of their spatial patterns. *Nucl. Acids Res.*, 37:W384–W389. 2009.

[Tho49]   R. Thom. Sur une partition en cellules associée à une fonction sur une variété. *C. R. Acad. Sci. Paris*, 228:973–975, 1949.

[TL12]   Y.Y. Tseng and W.H. Li Classification of protein functional surfaces using structural characteristics. *Proc. Natl. Acad. Sci. (USA)* 109:1170–1175, 2012.

[Vai12]   I.I. Vaisman. Statistical and computational geometry of biomolecular structure. In J.E. Gentle, W.K. Hardle, and Y. Mori, editors, *Handbook of Computational Statistics*, pages 1095–1112, Springer-Verlag, New York, 2012.

[VBR⁺95]   A. Varshney, F.P. Brooks, Jr., D.C. Richardson, W.V. Wright and D. Minocha. Defin-ing, computing, and visualizing molecular interfaces. In *Proc. IEEE Visualization, 1995*, pages 36–43.

[VD11]   M. Vendruscolo and C.M. Dobson. Protein dynamics: Moore's law in molecular biol-ogy. *Current Biology*, 21:R68–R70, 2011.

[Vor07]   G.F. Voronoi. Nouvelles applications des paramètres continus à la théorie des formes quadratiques. *J. Reine Angew. Math.*, 133:97–178, 1907, and 134:198–287, 1908.

[WBKC09]   J.A. Wilson, A. Bender, T. Kaya and P.A. Clemons. Alpha shapes applied to molec-ular shape characterization exhibit novel properties compared to established shape descriptors. *J. Chem. Inf. Model.*, 49:2231–2241, 2009.

[WC53]   J.D. Watson and F.H.C. Crick. Molecular structure of nucleic acid: a structure for deoxyribose nucleic acid. Genetic implications of the structure of deoxyribonucleic acid. *Nature*, 171:737–738 and 964–967, 1953.

[Wel96]   J.A. Wells. Binding in the growth hormone receptor complex. *Proc. Nat. Acad. Sci. (USA)*, 93:1–6, 1996.

[XW14]   K. Xia and G.-W. Wei. Persistent homology analysis of protein structure, flexibility, and folding. *Int. J. Numer. Meth. Biomed. Engng.*, 30:814–844, 2014.

[XW15a]   K. Xia and G.-W. Wei. Multidimensional persistence in biomolecular data. *J. Comput. Chem.*, 20:1502–1520, 2015.

[XW15b]   K. Xia and G.-W. Wei. Persistent homology for cryo-EM data analysis. *Int. J. Numer. Meth. Biomed. Engng.*, 31, 2015.

[ZBX11]   W. Zhao, C. Bajaj and G. Xu. An Algebraic spline model of molecular surfaces for energetic computations. *IEEE/ACM Trans. Comput. Biol. Bioinform.*, 8:1458–1467, 2011.

[ZGK06]   A. Zomorodian, L. Guibas, and P. Koehl. Geometric filtering of pairwise atomic interactions applied to the design of efficient statistical potentials. *Comput. Aided Graph. Design*, 23:531–544, 2006.

[Zha08]   Y. Zhang. Progress and challenges in protein structure prediction. *Curr. Opin. Struct. Biol.*, 18:342–348, 2008.

[Zha09]   Y. Zhang. Protein structure prediction: when is it useful? *Curr. Opin. Struct. Biol.*, 19:145–155, 2009.

[ZY10]   W. Zhou and H. Yan. A discriminatory function for prediction of protein-DNA interactions based on the alpha shape modeling. *Bioinformatics*, 26:2541–2548, 2010.

66 GEOMETRY AND TOPOLOGY OF GENOMIC DATA

Andrew J. Blumberg and Raúl Rabadán

INTRODUCTION

Since the turn of the century there have been remarkable and revolutionary developments in the acquisition of biological data. The publication of the first draft of the human genome [VAM⁺01] in 2001 was a milestone in this revolution, heralding rapid transformation in almost every realm of biology. A sampling of the rapid progress on fundamental problems includes the enumeration of genomic variations in thousands of individuals [C⁺12a], detailed molecular characterization of thousands of cancers [MFB⁺08], single cell characterization of tumors [PTT⁺14], study of developmental processes [TCG⁺14], and the elucidation of the three-dimensional structure of DNA in the nucleus of cells [DMRM13, LABW⁺09]. Biology has become an extremely data-rich discipline.

In stark contrast with physics, where comparatively simple models of fundamental physical processes have been extremely successful, many theoretical and qualitative aspects of biological processes remain obscure. As a consequence, progress in biology is particularly dependent on quantitative methods to extract biologically relevant information from the enormous mass of biological data. The sheer amount of data poses difficulties, and in addition to its abundance, high-throughput genomic and transcriptomic data typically resides in very high-dimensional spaces (e.g., on the order of the number of genes in the organism, which is typically in the tens of thousands), is often extremely noisy, and is plagued with systematic errors.

Standard approaches to handling genomic data include conventional clustering algorithms and linear statistical inference procedures; e.g., PCA, matrix factorization, spectral clustering, and so forth. The purpose of this chapter is to provide an introduction to novel approaches to organizing and describing the structure of large-scale biological data using tools from geometry and topology. We focus on the area of genetics and genomics, and we will illuminate these techniques using concrete applications to relevant biological problems, including the evolution of viruses and cancer.

OVERVIEW

Most of the problems in genetics/genomics can be formulated in terms of two spaces: the space of genotypes (the genetic makeup of an organism) and the space of phenotypes (the set of observable characteristics of the organism). In a particular situation, one then seeks to characterize the structure of the genotype space, the phenotype space and the function that connects them. For example, many questions can be cast as the identification of the mutations that cause a particular effect (a disease, for instance).

Roughly speaking, the points of the genotype spaces can be thought of as strings of DNA (i.e., long words on the alphabet $\{A, C, G, T\}$). Genomes are not stable objects; they change due to evolutionary processes. For example, as the genome of an organism is copied to produce a progeny, mistakes are made. Comparing the genomes of related organisms can be viewed as looking at a collection of nearby points in this space of genomes. Phylogenetic trees often provide a common encoding of this data, and the secondary genomic space of trees is a basic object of study. In this context, in Section 66.3 we will study how cancers develop with the accumulation of mutations in somatic cells, and how statistics on trees can be used to understand how seasonal influenza evolves.

However, more drastic changes can occur. For instance, sections of the genome can be lost or duplicated. Sometimes even the genomic information can be scrambled. These phenomena happen in cancer cells, for instance, leading to dramatic changes. An even more interesting phenomenon, pervasive across the different domains of life, is that distinct organisms can exchange genomic information. Bacteria can exchange genomic material through diverse mechanisms, viruses borrow material from hosts, and eukaryotes from the same and sometimes from distant species can combine genomes. Trees are not able to capture this kind of evolutionary dynamics; instead, we will represent these processes using tools from topological data analysis. Specifically, in Section 66.4 we illustrate the use of persistent homology to study the evolution of viruses and humans.

The phenotypic spaces are more varied. The phenotype is the set of observable characteristics of an organism; examples of these characteristics are expression of different mRNAs, the expression of proteins, the shape of these proteins, the shape of the cell, the ability to grow and replicate, the susceptibility to different stimuli, the ability to respond to an infection, the size and weight of a multicellular organism, and the capacity to cause a disease, among many, many others. Each of these characteristics constitutes a potentially interesting quantity to measure and study. In this context, in Section 66.4.2 we explain the use of multiscale clustering and the Mapper algorithm to study expression data from cancer tumors.

66.1   TREE SPACES

In 1859, Charles Darwin proposed the tree as a metaphor for the process of species generation through branching of ancestral lineages [Dar59]. Since then, tree structures have been used pervasively in biology, where terminal branches can represent a diverse set of biological entities and taxonomic units including individual or families of genes, organisms, populations of related organisms, species, genera, or higher taxa.

A large set of evolutionary processes are captured well by tree-like structures: in particular, clonal evolution events that start from asexual reproduction of a single organism (the primordial clone), which mutates and differentiates into a large progeny [KCK+14, ZKBR14]. Examples of these processes include single gene phylogeny in non-recombinant viruses, evolution of bacteria that are not involved in horizontal gene transfer events, and metazoan development from a single germ cell. Recent developments in genomics allow the study of such events in exquisite detail, particularly at single cell/clone levels [NKT+11, SSA+13, ESK+15].

Given raw genomic data (e.g., sequences of base pairs), there are now many

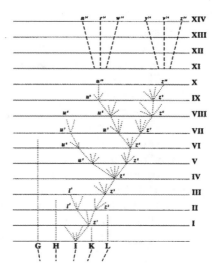

FIGURE 66.1.1

The generation of species according to Darwin. *The tree was chosen as a way of capturing the divergence of species, with time running upwards. The root of tree on the right represents the original species that diversify according to a branching process, through progeny diversification and selection. Some of the branches constitute the modern species (the top two branches), while some other branches do not persist (extinct species).*

ways of producing phylogenetic trees representing evolutionary relationships in the data. Popular methods include nearest-neighbor joining, maximum likelihood estimators, and Bayesian methods [Fel03]. Although the subject of constructing optimal trees from genomic data is an active and interesting area, herein we will assume that a method for generating trees has been fixed.

Our primary focus is on understanding the geometry of the space of phylogenetic trees, which is induced by a metric (distance function). There are various possible ways to understand the distance between two trees, but we will use a particularly natural metric on the set of phylogenetic trees constructed by Billera, Holmes, and Vogtmann [BHV01].

66.1.1 THE BILLERA-HOLMES-VOGTMANN SPACES OF PHYLOGENETIC TREES

GLOSSARY

Phylogenetic tree: A phylogenetic tree with m leaves is a connected graph with no cycles and weighted edges, having m distinguished vertices of degree 1 and labeled $\{1, \ldots, m\}$. All the other vertices are of degree ≥ 3.

External edge: An edge in a phylogenetic tree that terminates in a leaf.

Internal edge: An edge in a phylogenetic tree that does not terminate in a leaf; i.e., that is not an external edge.

Geodesic metric space: A metric space (X, d) is a geodesic metric space if any two points x and y can be joined by a path (i.e., a continuous map $\gamma: [0, 1] \to X$)

with length precisely $d(x, y)$. Here the length of the path γ is defined to be $\sup \sum_{i=1}^{n} d(\gamma(0), \gamma(x_1)) + d(\gamma(x_1), \gamma(x_2)) + \ldots + d(\gamma(x_{k-1}), \gamma(1))$, where the sup is taken over all sequences $0 < x_1 < x_2 < \ldots < x_k < 1$ as k varies.

Comparison triangle: For a fixed Riemannian manifold M, given a triangle $T = [p, q, r]$ in (X, d), a comparison triangle is a triangle $\tilde{T}$ in M with the same edge lengths.

Comparison point: For a triangle $T = [p, q, r]$ in (X, d), given a point z on the edge $[p, q]$, a comparison point in the comparison triangle $\tilde{T}$ is a point $\tilde{z}$ on the corresponding edge $[\tilde{p}, \tilde{q}]$ such that $d(\tilde{z}, \tilde{p}) = d(p, z)$ and $d(\tilde{z}, \tilde{q}) = d(q, z)$.

CAT(0) *space:* A triangle T in M satisfies the CAT(0) inequality if for every pair of points x and y in T and comparison points $\tilde{x}$ and $\tilde{y}$ on a comparison triangle $\tilde{T}$ in Euclidean space, we have $d(x, y) \leq d(\tilde{x}, \tilde{y})$. If every triangle in M satisfies the CAT(0) inequality then we say that M is a CAT(0) space.

CAT(κ) *space:* More generally, let M_κ denote the unique two-dimensional Riemannian manifold with curvature κ. The diameter of M_κ will be denoted D_κ. Then we say that a D_κ-geodesic metric space M is CAT(κ) if every triangle in M with perimeter $\leq 2D_\kappa$ satisfies the inequality above for the corresponding comparison triangle in M_κ. A n-dimensional Riemannian manifold M that is sufficiently smooth has sectional curvature $\leq \kappa$ if and only if M is CAT(κ). For example, Euclidean spaces are CAT(0), spheres are CAT(1), and hyperbolic spaces are CAT(-1).

The space BHV_m of isometry classes of rooted phylogenetic trees with m labelled leaves where the nonzero weights are on the internal branches was introduced and studied by Billera, Holmes, and Vogtmann [BHV01]. The space BHV_m is constructed by gluing together $(2m - 3)!!$ positive orthants $\mathbb{R}^m_{\geq 0}$; each orthant corresponds to a particular binary tree with labelled leaves, with the coordinates specifying the lengths of the edges. If any of the coordinates are 0, the tree is obtained from a binary tree by collapsing some of the edges. We glue orthants together such that a (non-binary) tree is on the boundary between two orthants when it can be obtained by collapsing edges from either tree geometry.

The metric on BHV_m is induced from the standard Euclidean distance on each of the orthants. For two trees t_1 and t_2 which are both in a given orthant, the distance $d_{\text{BHV}_m}(t_1, t_2)$ is defined to be the Euclidean distance between the points specified by the weights on the edges. For two trees which are in different orthants, there exist (many) paths connecting them which consist of a finite number of straight line segments in each quadrant. The length of such a path is the sum of the lengths of these line segments, and the distance $d_{\text{BHV}_m}(t_1, t_2)$ is then the minimum length over all such paths. For many points, the shortest path goes through the "cone point," the star tree in which all internal edges are zero.

Allowing potentially nonzero weights for the m external leaves corresponds to taking the Cartesian product with an m-dimensional orthant. We will focus on the space

$$\Sigma_m = \text{BHV}_{m-1} \times (\mathbb{R}^{\geq 0})^m,$$

which we refer as the evolutionary moduli space. (The $m - 1$ index arises from the fact that we consider unrooted trees.) There is a metric on Σ_m induced from the metric on BHV_{m-1}. Specifically, for a tree t, let $t(i)$ denote the length of the external edge associated to the vertex i. Then

$$d_{\Sigma_m}(t_1, t_2) = \sqrt{d_{\text{BHV}_{m-1}}(\bar{t}_1, \bar{t}_2) + \sum_{i=1}^{m}(t_1(i) - t_2(i))^2},$$

where $\bar{t}_i$ denotes the tree in BHV_{m-1} obtained by forgetting the lengths of the external edges (e.g., see [OP11]). As explained in [BHV01, §4.2], efficiently computing the pairwise distance between any two points on Σ_m is a nontrivial problem, although there exists a polynomial-time algorithm [OP11].

Although metric spaces often arise in contexts in which there is not an evident notion of geometry, Alexandrov observed that a good notion of *curvature* makes sense in any geodesic metric space [Ale57]. The idea is that the curvature of a space can be detected by considering the behavior of the area of triangles, and triangles can be defined in any geodesic metric spaces. The connection between curvature and area of triangles comes from the observation that given side lengths $(\ell_1, \ell_2, \ell_3) \subset \mathbb{R}^3$, a triangle with these side lengths on the surface of the Earth is "fatter" than the corresponding triangle on a Euclidean plane. To be precise, we consider the distance from a vertex of the triangle to a point p on the opposite side—a triangle is fat or thin if this distance is larger or smaller respectively than in the corresponding Euclidean triangle.

A remarkably productive observation of Gromov is that many geometric properties of Riemannian manifolds are shared by $CAT(\kappa)$ spaces. In particular, $\text{CAT}(\kappa)$ spaces with $\kappa \leq 0$ (referred to as *non-positively curved metric spaces*) admit unique geodesics joining each pair of points x and y, balls $B_\epsilon(x)$ are convex and contractible for all x and $\epsilon \geq 0$, and centroids exist and are unique. As a consequence, there exist well-defined notions of mean and variance of a set of points, and more generally one can attempt to perform statistical inference, as we explain below.

Gromov also gave criteria for determining when a metric space produced by gluing together Euclidean cubes (i.e., a *cubical complex*) is a CAT(0) space. The spaces BHV_n and Σ_m are evidently cubical complexes (essentially by construction). The main result of Billera, Holmes, and Vogtmann uses these criteria to show that the metric d_{Σ_n} on BHV_n endows this space with a (global) CAT(0) structure. An easy corollary of this result is that Σ_m is also a CAT(0) space.

66.1.2 THE PROJECTIVE EVOLUTIONARY MODULI SPACE

GLOSSARY

Join of two topological spaces: For topological spaces X and Y, the join $X \star Y$ is the quotient of the space $X \times Y \times [0,1]$ subject to the equivalence relation $(x, y, 0) \sim (x, y', 0)$ and $(x, y, 1) \sim (x', y, 1)$.

In evolutionary applications, we are often interested in classifying and comparing distinct behaviors by understanding the relative lengths of edges: rescaling edge lengths should not change the relationship between the branches [ZKBR14]. Motivated by this consideration, we define $\mathbb{P}\Sigma_m$ to be the subspace of Σ_m consisting of the points $\{t_i\}$ in each orthant for which the constraint $\sum_i t_i = 1$ holds. We denote the space of trees with a fixed sum of internal edges by τ_{m-1}.

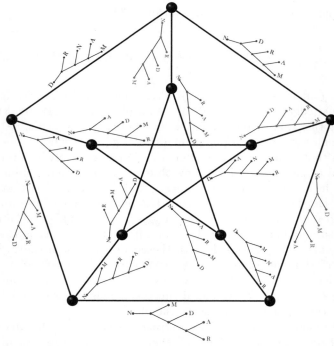

FIGURE 66.1.2
The 15 possible tree topologies arranged on the edges of τ_4. *The Petersen graph represents captures the 15 different topologies (edges) together with transitions between topologies (nodes) corresponding to collapsing internal branches.*

The space of m external branches whose lengths sum to 1 is the standard $m - 1$ dimensional simplex Δ_{m-1} in $\mathbb{R}^m$. The constraint that the length of internal branches plus the external branches sum to 1 implies that

$$\mathbb{P}\Sigma_m = \tau_{m-1} \star \Delta_m.$$

There are various possible natural metrics to consider on $\mathbb{P}\Sigma_m$. The simplest way to endow $\mathbb{P}\Sigma_m$ with a metric is to use the induced intrinsic metric specified by paths in Σ_m constrained to lie entirely within $\mathbb{P}\Sigma_m$. The theory of polyhedral complexes implies that $\mathbb{P}\Sigma_m$ is a complete geodesic metric space [BH99, I.7.19], and it is evidently separable.

Moreover, with the induced intrinsic metric, $\mathbb{P}\Sigma_m$ is in fact a CAT(0) space; although τ_{m-1} has points which are not connected by unique geodesics, the join with Δ_m introduces a new "cone direction" that changes the geometry.

THEOREM 66.1.1 [ZKBR16]

The projective moduli space $\mathbb{P}\Sigma_m$ endowed with the intrinsic metric is a CAT(0) space.

Computing the intrinsic metric on $\mathbb{P}\Sigma_m$ is an interesting open problem; for small m, using ϵ-nets provides a tractable approximation.

66.1.3 STATISTICS AND INFERENCE IN Σ_m AND $\mathbb{P}\Sigma_m$

GLOSSARY

Fréchet mean and variance: Given a fixed set of n trees $\{T_0, \ldots T_{n-1}\} \subseteq \Sigma_m$, the Fréchet mean T is the unique tree that minimizes the quantity

$$E = \sum_{i=0}^{n-1} d_{\Sigma_m}(T_i, T)^2.$$

The variance of T is the ratio $\frac{E}{n}$.

Statistical inference and summarization: Given a collection of trees generated by data, statistical inference refers to procedures for reliably estimating properties of the underlying distribution that generated the data. Summarization methods produce succinct descriptions of the underlying distribution (e.g., moments of distributions); these are often useful for inference. For example, inference methods based on the sample mean can be used to try to distinguish between two underlying distributions.

Genomic data produces a collection of points in the space of phylogenetic trees. Therefore, two central questions are how to understand the structure of such data and how to perform statistical inference and summarization in this context. Discussion of statistical inference in Σ_m was initiated in [BHV01], and subsequently Holmes has written extensively on this topic [Hol03, Hol03, Hol05] (and see also [FOP+13]). More generally, Sturm explains how to study probability measures on general CAT(0) spaces [Stu03]. He shows that there are reasonable notions of moments of distribution, expectation of random variables, and analogues of the law of large numbers on CAT(0) spaces.

There are efficient algorithms for computing the mean and variance [MOP15]. The situation for the central limit theorem is less satisfactory. Barden, Le, and Owen study central limit theorems for Fréchet means in Σ_m [BHO13]; as they explain, the situation exhibits non-classical behavior and the limiting distributions depend on the codimension of the simplex in which the mean lies. Finally, there has been some work on principal components analysis (PCA) in Σ_m [Nye11].

However, in contrast to classical statistics on $\mathbb{R}^n$, we do not know many sensible analytically defined distributions on the evolutionary moduli spaces. In general, the behavior of distributions on Σ_m is somewhat perverse due to the pathological behavior near the origin due to the exponential growth in the mass of an ϵ ball. A much more tractable source of distributions on Σ_m and $\mathbb{P}\Sigma_m$ arise from resampling from a given set of empirical data points.

Furthermore, in both Σ_m and $\mathbb{P}\Sigma_m$ we can perform exploratory data analysis by clustering, and use nearest neighbor classification and similar algorithms for supervised learning. In addition, topological data analysis methods in tree space hold a lot of promise. Another approach to inference in these metric spaces is to study test quantities that take values in $\mathbb{R}^n$; for example, the distance from the centroid in Σ_m. Distributions for such statistics must be estimated by resampling.

66.2 TREE DIMENSIONALITY REDUCTION

When analyzing large numbers of genomes, phylogenetic trees are often too complex to visualize and analyze as they can contain thousands of branches. We now explain a method of dimensionality reduction that projects a single tree in Σ_m or $\mathbb{P}\Sigma_m$ to a "forest" of trees in Σ_n, for $n < m$ [ZKBR16]. Subsampling leaves of a large tree yields a distribution of smaller trees that can still capture properties of the more complex structure. This procedure makes it easy to visualize and analyze high-dimensional data, and avoids scalability issues with algorithms for working with the spaces of phylogenetic trees. We will let $\mathcal{E}_m$ denote either Σ_m or $\mathbb{P}\Sigma_m$.

66.2.1 STRUCTURED DIMENSIONALITY REDUCTION

GLOSSARY

Tree projection: For $S \subseteq \{1, \ldots, m\}$, we define the tree projection function

$$\Psi_S \colon \mathcal{E}_m \to \Sigma_{|S|}$$

by specifying $\Psi_S(t)$ to be the unique tree obtained by taking the induced subgraph of t on the leaves that have labels in S and then deleting vertices of degree 2. An edge e created by vertex deletion is assigned weight $w_1 + w_2$, where the w_i are the weights of the incident edges for the deleted vertex. (It is easy to check that the order of vertex deletion does not change the resulting tree.)

Tree dimensionality reduction: Let $\mathcal{D}(\Sigma_m)$ denote the set of distributions on Σ_m. For $1 \leq k < m$, we define the tree dimensionality reduction map

$$\Psi_k \colon \mathcal{E}_m \to \mathcal{D}(\Sigma_k)$$

as the assignment that takes $T \in \mathcal{E}_m$ to the empirical distribution induced by Ψ_S as S varies over all subsets of $\{1, \ldots, m\}$ of cardinality k. Alternatively, we might prefer to consider the map

$$\Psi'_k \colon \mathcal{E}_m \to \prod_{S \subseteq \{1,\ldots,m\}, |S|=k} \Sigma_k.$$

Sequential tree dimensionality reduction: Let $\mathcal{C}_k(\Sigma_m)$ denote the set of piecewise linear curves in Σ_m; equivalently, $\mathcal{C}_k(\Sigma_m)$ can be thought as the set of ordered sequences in Σ_m of cardinality k. For $1 \leq k < m$, define the map

$$\Psi_C \mathcal{E} \to \mathcal{C}_{m-k}(\Sigma_k)$$

as the assignment that takes $t \in \mathcal{E}_m$ to the curve induced by Ψ_S as S varies over the subsets $\{1, \ldots, k\}$, $\{2, \ldots, k+1\}$, etc.

Using Ψ_S as a building block, we can describe a number of dimensionality reduction procedures on trees. (See figure 66.2.1 for an example of Ψ_S.) The most basic process is simply to exhaustively subsample the labels; this is the tree dimensionality

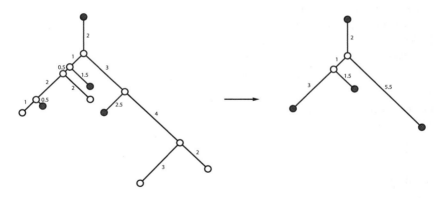

FIGURE 66.2.1
Tree dimensionality reduction on four leaves. *Selecting leaves in a large tree generates smaller trees preserving the metric structure. Repeating this operation one can generate a collection of small trees that capture information from the larger tree.*

reduction map Ψ_k. Of course, in practice we approximate Ψ_k using Monte Carlo approximations.

However, there is often additional structure in the data that can be used. For instance, in many natural examples the leaves have a chronological ordering. Sliding windows over ordered labels then induces an ordering on subtrees generated by Ψ_S. The ordering makes it sensible to consider the associated trees as forming a piecewise-linear curve in Σ_k; we refer to this as sequential tree dimensionality reduction. (Note that given a set of points in Σ_k it is always reasonable to form the associated piecewise-linear curve because each pair of points is connected by a unique geodesic.) Of course, there are many variants of Ψ_C depending on the precise strategy for windowing that is employed.

Remark.   The map Ψ is an operation on tree spaces. However, there is an alternative form of tree dimensionality reduction that instead subsamples the raw data to produce a forest of smaller phylogenetic trees directly. A natural question to ask is whether these two procedures produce the same output. We can give an affirmative answer in the case of neighbor-joining. The situation is more complicated for other methods of producing trees.

Tree dimensionality reduction transforms questions about comparison of trees or analysis of finite sets of trees to questions about comparisons and analysis of sets of clouds of trees.

66.2.2 STABILITY OF TREE DIMENSIONALITY REDUCTION

In order to apply tree dimensionality reduction to real data, we would like to know that small random perturbation of the original sample results in a distribution of subsamples that is "close" in some sense.

The basic result in this direction is that for $S \subseteq \{1, \ldots, m\}$, the projection Ψ_S preserves paths and the increase in length is bounded. To understand the nature of the bound, consider the impact of the addition of edge lengths that occurs when a degree 2 vertex is removed during the reduction process. In the simplest case, we are considering the map $\mathbb{R}^2 \to \mathbb{R}$ specified by $(x, y) \mapsto x + y$, and in general, we are looking at $\mathbb{R}^n \mapsto \mathbb{R}$ specified by $(x_1, x_2, \ldots, x_n) \mapsto \sum_{i=1}^{n} x_i$. Squaring both sides, it is clear that

$$\partial_{\mathbb{R}^n}((x_i), (y_i))^2 \leq \partial_{\mathbb{R}}(\sum_{i=1}^{n} x_i, \sum_{i=1}^{n} y_i)^2.$$

On the other hand, since

$$\partial_{\mathbb{R}}(\sum_{i=1}^{n} x_i, \sum_{i=1}^{n} y_i) \leq n(\max_i |x_i - y_i|) \leq n\partial_{\mathbb{R}^n}((x_i), (y_i)),$$

the addition of edge lengths can result in an expansion bounded by the size of the sum.

For a rooted tree T, let depth(T) denote the length of the longest path from a leaf to the root.

PROPOSITION 66.2.1

Let $S \subseteq \{1, \ldots, m\}$ such that $|S| > 1$ and let $\gamma \colon [0, 1] \to \mathcal{E}_m$ be a path from T to T'. Then $\gamma \circ \Psi_S \colon [0, 1] \to \Sigma_{|S|}$ is a path from $\Psi_S(T)$ to $\Psi_S(T')$ and $|\gamma'| \leq \max(\text{depth}(T), \text{depth}(T'))|\gamma|$.

Figure 66.2.2 indicates synthetic experimental data about the spread produced by Ψ_S as S varies. Using Proposition 66.2.1, we can now easily deduce theorems that provide the theoretical support for the use of tree dimensionality reduction. The following theorem is an immediate consequence of Proposition 66.2.1, choosing the path realizing the distance between T and T'.

THEOREM 66.2.2

For $T, T' \in \mathcal{E}_m$ such that $d_{\mathcal{E}_m}(T, T') \leq \epsilon$, then for any $S \subseteq \{1, \ldots, m\}$ such that $|S| > 1$,

$$d_{\Sigma_{|S|}}(\Psi_S(T), \Psi_S(T')) \leq \max(\text{depth}(T), \text{depth}(T'))\epsilon.$$

Moreover, this bound is tight.

66.3 THE TREE SPACES OF GLIOBLASTOMAS AND THE IN-FLUENZA VIRUS

In this section, we explain two substantial applications of the tree space methodology described in the preceding section [WCL+16, ZKBR16].

66.3.1 THE TREE SPACE OF GLIOBLASTOMAS: EVOLUTION UNDER THERAPY

GLOSSARY

Glioblastoma (GBM) or Grade IV astrocytoma: The most common brain tumor in adults.

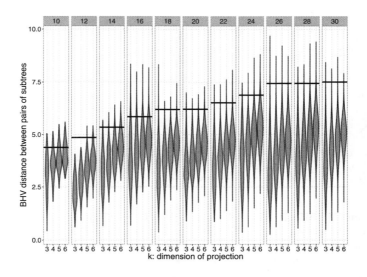

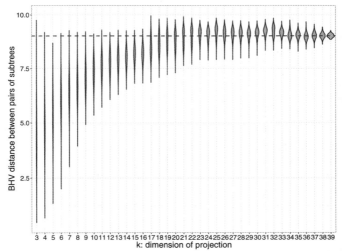

FIGURE 66.2.2
Distributions of subsample distances under the tree projection operation.
*(Top) Pairs of m-dimensional phylogenies with known distance (horizontal black lines)
are projected into distributions of low dimensional trees. The distances between elements
of the projections can exceed the original inter-phylogeny distance (ϵ), but rarely approach
the upper bound. (Bottom) As the dimension of the projection operator approaches that
of the initial phylogenies, there is a decrease in the variance of the distribution of subtree
distances and its median approaches the 40-dimensional $d_{BHV}(T, T')$.*

Temozolomide (TMZ): An alkylating agent used in treatment of brain tumors.
It introduces damage in DNA, triggering death of tumor cells.

Hypermutation: Increase of several orders of magnitude in mutation rates. In
GBM, it is associated to inactivation of repair mechanisms to DNA damage.

Tumors evolve through the accumulation of mutations. Sequencing tumors provides a way to understand the molecular mechanisms driving this process, to under-

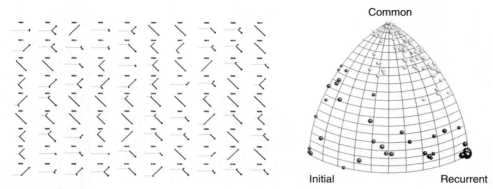

FIGURE 66.3.1

Genomic information of tumors from a patient generates a tree representing the different phases of tumor evolution. *Cohorts of patients can be represented by a forest (left). In yellow are mutations that are common in diagnosis and relapse, in red the ones that are specific to diagnosis and in black specific to relapse. The forest specifies points in the evolutionary moduli space (right).*

stand mechanisms of resistance to therapies, and to identify potential therapeutic alternatives tailored to the genetic background of specific tumors. Glioblastoma (or GBM) is one of the most common and most aggressive types of brain tumors. Median survival after initial diagnosis is slightly more than a year. The standard of care consists of surgery followed by radiotherapy and an alkylating agent, temozolomide (TMZ). Tumors invariably recur, resulting in the death of the patient. Understanding how these tumors evolve, what the effects of therapy are, and the mechanisms of relapse are central research questions.

To study how GBM evolves, longitudinal tumor/matched normal samples in 114 GBM patients were sequenced, both at relapse and diagnosis [WCL+16]. Comparison of the mutational profile in the three samples provides mutations that are in common (founder mutations), mutations that are specific to diagnosis, and mutations that are specific to recurrence. Particular mutations that are specific to diagnosis could be associated to sensitivity to the therapy, and mutations associated to recurrence could yield clues about the mechanisms of resistance. From each of the 114 samples, there are three numbers corresponding to the three different types of mutations. Each triple defines a simple tree with three branches. That is, each patient is represented by a tree and the genomic information of the 114 patients forms a forest in $\mathbb{P}\Sigma_3$ (figure 66.3.1). Here, the upper corner represents the fraction of mutations that are common to both samples; the left corner represents the fraction exclusive to the untreated sample, and the right corner the fraction exclusive to recurrence.

Clustering can provide a guide to different patterns of how tumors evolve in different patients. Using a variety of clustering techniques (including k-means, spectral clustering, and density-based spatial clustering) and resampling-based cross-validation, one can identify three stable clusters. The yellow group represents the limiting case where few mutations are lost from diagnosis. This is similar to the classical model of linear tumor evolution, where mutations accumulate in clones that drive recurrence. The abundance of points far from the right edge of the diagram suggests that in most patients, the dominant clones prior to treatment appear to be replaced by new clones that do not share many of the same mutations. If many mutations in the initial sample are lost at recurrence, this suggests that the clone dominant at recurrence originated (i.e., diverged from the clone dominant at diagnosis) a relatively long time before the initial sample was taken. This is an interesting finding, as it suggests that a different clone to the one that caused the initial tumor is responsible for the recurrent tumor.

Patient histories identified in the black cluster, correspond to particular trees with very long branches associated to relapse tumors. These long branches, a phenomenon called hypermutation, was present only in patients treated with TMZ, and patients in this cluster are associated with significantly longer survival (more than two years). All these patients harbor mutations in the mismatch repair pathway, mostly inactivating mutations in MSH6. The MSH6 protein plays an essential role in repairing damaged DNA, and fixing potential mistakes in the replication of DNA. These tumors cannot effectively repair the damage caused by the therapy (TMZ), accumulating many more mutations in branches of the tree associated to the relapse. Interestingly, the mutations are also different from mutations in non-hypermutated branches: hypermutated recurrent tumors are highly enriched with C to T (and G to A) transitions, occur in a CC/GG motif, and are mostly found in highly expressed genes.

66.3.2 TREE DIMENSIONALITY REDUCTION: INFLUENZA VIRUS EVOLUTION

GLOSSARY

Influenza A virus: A member of the Orthomyxoviridae family of viruses, segmented antisense RNA viruses.

hemagglutinin (HA): A glycoprotein found on the surface of influenza viruses. Responsible for recognition of cell receptors and fusion of viral and endosomal membranes. It is the primary target of neutralizing antibodies.

Influenza A is a single-stranded RNA virus that every year infects roughly 10% of the world population. The virus genome evolves rapidly, changing the antigenic presentation of the virus. This continuous evolution requires yearly updates of the strains used to formulate the influenza vaccine, one of the most effective ways of preventing morbidity. As a consequence, there is a significant effort from national and international organizations to collect genomic and antigenic data from this virus. Large collections of viral genomic data are publicly available (currently more than 100,000 genomes), including annotation of geographic location and time when the virus was isolated. Taking a large phylogenetic tree constructed from viral genomic data and applied tree dimensionality reduction to produce forests of smaller trees; variability in these distributions appears to predict vaccine efficacy [ZKBR16].

Using thousands of sequences of the hemagglutinin gene (HA), the gene coding for the protein associated to the main adapted immune response to the virus, collected in the United States between 1993 and 2015, we relate HA sequences from one season to those from the succeeding ones. One can pick random strains from five consecutive seasons, and using neighbor-joining based on hamming distances, generated un-rooted trees that were projected into $\mathbb{P}\Sigma_5$ (Figure 66.3.3). The majority of spaces for HA show linear evolution of influenza from one season to another, indicating genetic drift as the virus's dominant evolutionary process.

FIGURE 66.3.2
Evolution of influenza A virus presenting clear seasonal variation. *Identifying statistical patterns in large trees is often difficult. This phylogenetic tree of the hemagglutinin (HA) segment from selected H3N2 influenza viruses across 15 seasons can be sub-sampled for statistical analysis in lower dimensional projections.*

One can detect distinct clusters of trees in HA evolutionary spaces that indicate clonal replacement and reemergence of strains in the 2002-2003 season genetically similar to those circulating in the 1999-2000 season.

A natural hypothesis is that elevated HA genetic diversity in circulating influenza predicts poor vaccine performance in the subsequent season. One can apply tree dimensionality reduction based on windowing to generate a predictor for this question. In Figure 66.3.4 one can observe the prediction of vaccine efficacy using the variance of the distribution of trees generated by a lagging window of length 3. In our notation, a window labeled year y would include the flu season of $(y - 1, y)$ and preceding years. The vaccine effectiveness figures represent season $(y, y + 1)$. It is clear, both from the left and right panels, that lower variance in a temporal window predicts increased future vaccine effectiveness, with a Spearman correlation of -0.52 and p-value of 0.02. The lone outlier season came in 1997-1998 [GED06], when the vaccine efficacy was lower than expected. In this season the dominant circulating strain was A/Sydney/5/97 while the vaccine strain was A/Wuhan/359/95. The analysis can be carried out with length-4 or length-5 windows to yield a similar result. Noteworthy is the fact that this association rests only on aligned nucleotide sequence, making no direct use of HA epitope or HI assay data. The correlation between variance of tree distributions and vaccine effectiveness allows us to esti-

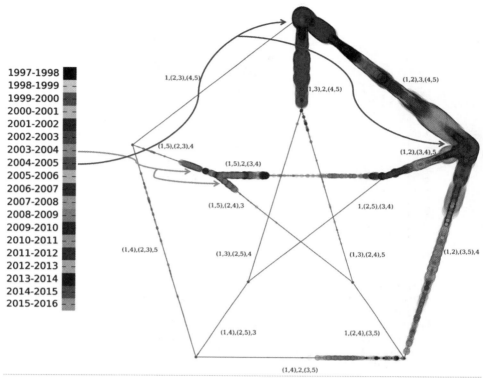

FIGURE 66.3.3
Temporally windowed subtrees in $\mathbb{P}\Sigma_5$. *Using a common set of axes for projective tree space, one can superimpose the distributions of trees derived from windows five seasons long. 1,089 full-length HA segments (H3N2) were collected in New York state from 1993 to 2016. Two consecutive seasons of poor vaccine effectiveness are 2003-2004 and 2004-2005, highlighted with green and gray arrows respectively. The green distribution strongly pairs the 1999-2000 and 2003-2004 strains, hinting at a reemergence.*

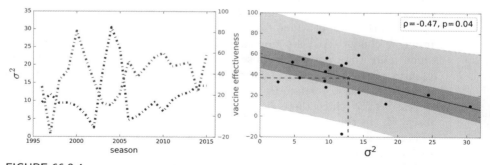

FIGURE 66.3.4
Diversity in recent circulating HA predicts vaccine failure. *Negative correlation observed between vaccine efficacy in season $(t, t+1)$ and the variance in trees generated from seasons $(t-1,t), (t-2,t-1), (t-3,t-2)$.*

mate the influenza vaccine effectiveness for future seasons based on genomic data. In particular, for the 2016-2017 season, the variance in tree space was 12.77, corresponding to an approximate effectiveness of 36%.

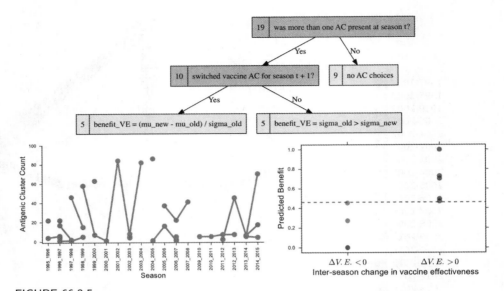

FIGURE 66.3.5

Stratification of trees on predicted antigenic cluster. *Trees are categorically labeled using the predicted [DDL$^+$12] antigenic cluster (AC) of their most recent isolate. (Top) A simple decision rule is fit [SL09] to the 10 seasons in which more than one AC is observed, explaining the change in vaccine effectiveness (VE) in the following season. (Bottom) Colors encode different AC labels associated to the HA sequences collected over time. Plotting the predicted VE change given by the decision rule against the historical data, for the 10 relevant seasons.*

We also used our approach to retrospectively examine the annual W.H.O. decisions to either keep or change the H3N2 component of the Northern hemisphere vaccine, and whether the choice resulted in superior or inferior vaccine effectiveness in the following season. A coarse-grained view of the antigenic features of our HA isolates can be obtained using the work of [DDL$^+$12], who defined a clustering of antigenic phenotype and trained a naive Bayes model, that maps HA protein sequence to these labels. One can begin by labeling our phylogenetic trees using the antigenic cluster (AC) assignments of the classifier, and selecting those seasons in which more than one AC is observed. The goal is to define a mapping between features of the different AC distributions and the change in vaccine effectiveness of the next season relative to the present. Figure 66.3.5 indicates that in 9 of the 19 seasons for which there is a significant amount of data, only a single AC was observed. In this case it does not make sense to ask whether a change in H3N2 vaccine strain should have considered for the subsequent season, since no antigenic diversity was detected. The other 10 seasons were represented as vectors comprised of 4 features: distance between the centroids of the older / newer AC distributions, standard deviation of the older AC distribution, standard deviation of the newer AC distribution, and whether the vaccine strain for the subsequent season was changed from that of the current season. One can associate a binary label to each of the 10 seasons: whether the change in vaccine effectiveness was positive or negative.

One can exhaustively search for a decision tree mapping the feature vectors to the binary label, based on a fitness function maximizing area under the receiver operating characteristic [SL09]. A decision rule that achieves perfect classification on this data set is depicted in Figure 66.3.5. Due to the very small size of the data set, caution is warranted when interpreting the results. Nonetheless, the results do suggest that if a vaccine strain is unchanged, a higher variance in the old AC distribution predicts improvement in vaccine effectiveness ($\Delta V.E. > 0$), while if a vaccine strain is changed, then the new AC distribution being well-resolved from the old AC predicts $\Delta V.E. > 0$.

66.4 TOPOLOGICAL DATA ANALYSIS

Modern algebraic topology arose in order to provide quantitative tools for studying the "shape" of geometric objects, in the sense that it assigns algebraic invariants (e.g., numbers) to geometric objects in a way that depends only on the relative positions of points and lines, not their absolute relationships. The goal of topological data analysis is to recover information about the shape of data by employing the methods of algebraic topology. To do this, two major problems need to be tackled. First, we need a scheme to transform a discrete set of points into a richer topological space in order to compute topological invariants. Second, the *feature scale* of the data must be accounted for; namely, what is the relationship between the size of meaningful features of the shape and the distances between the points. Topological data analysis and notably the idea of *persistence* provide answers to these questions. Persistent homology is reviewed in Chapter 24 of this Handbook so we give only a very brief and rapid review herein.

66.4.1 FROM FINITE METRIC SPACES TO FILTERED SYSTEMS OF COMPLEXES TO PERSISTENT HOMOLOGY

GLOSSARY

Finite metric space: A metric space with finitely many points. A natural geometric example of a finite metric space is a finite set of points $\{x_0, x_1, \ldots, x_k\} \subset \mathbb{R}^n$ equipped with the induced Euclidean metric. A natural biological example of a finite metric space is a finite set of genome sequences and the Hamming metric.

Abstract simplicial complex: A set X of finite non-empty sets such that if A is an element of X then so is every subset of A. We say that such a set is a k-simplex if it has cardinality k, and we think of this data as a collection of triangles glued together along faces; a face is just a subset.

Vietoris-Rips complex: Fix $\epsilon > 0$. The Vietoris-Rips complex $\mathrm{VR}_\epsilon((X, \partial X))$ is the abstract simplicial complex with:

- vertices the points of X, and
- a simplex $[v_0, v_1, \ldots, v_k]$ when $\partial_X(v_i, v_j) < \epsilon$ for all $0 \leq i, j \leq k$.

To apply the invariants of algebraic topology, we need to associate a topological space to a finite metric space. One way to do this involves the use of simplicial complexes, which are combinatorial models of spaces. To any finite metric space (X, ∂_X), there are many strategies for associating simplicial complexes. The simplest and most flexible is the Vietoris-Rips complex.

The Vietoris-Rips complex is entirely determined by its 1-skeleton; a simplex is in $\mathrm{VR}_\epsilon((X, \partial_X))$ if all of its 1-dimensional faces are. The advantage of the Vietoris-Rips complex is that its construction makes sense for any finite metric space. A key property of the Rips complex is that it is functorial in ϵ. For $\epsilon < \epsilon'$ and any metric space (X, ∂_X), there is an induced simplicial map

$$\mathrm{VR}_\epsilon((X, \partial_X)) \to \mathrm{VR}_{\epsilon'}((X, \partial_X)).$$

The idea of *persistence* is to systematically encode the way that topological invariants of a finite metric space change as the feature scale ϵ varies [ELZ02, EH08, Car14]. Somewhat unexpectedly, this idea turns out to reveal an interesting and novel algebraic structure.

We start by observing that as ϵ varies, we obtain a filtered system of simplicial complexes; there are only a finite number of values of ϵ for which the Vietoris-Rips complex changes. More generally, assume that we have a filtered system of simplicial complexes. To this filtered complex, we now wish to assign algebraic invariants. One obvious thing to do is clustering: if we use single-linkage clustering and keep track of how the clusters merge as ϵ changes, we recover the standard dendrogram of hierarchical clustering. Thinking of this another way, we are keeping track of the connected components of the simplicial complexes as ϵ varies. A natural generalization of connected components to higher dimensions is given by homology groups. Since homology H_k is also functorial, for any coefficient ring R we have induced maps of R-modules

$$H_k(\mathrm{VR}_\epsilon((X, \partial_X))) \to H_k(\mathrm{VR}_{\epsilon'}((X, \partial_X))).$$

Definition 1 *Let X be a filtered system of simplicial complexes. For $i, p \in \mathbb{R}$, the (p, i)th persistent k-th homology group of X is the image*

$$PH_k^{i,p} = im\left(H_k(X_i) \to H_k(X_{i+p})\right).$$

An element $\gamma \in H_k(X_i)$ is said to *be born* at i if it is not in any persistent homology group $PH_k^{q,i-q}$ for any $q > 0$. The element γ is said to *die* at ℓ if it becomes zero in $H_k(X_\ell)$. This suggests that we can think of the information contained in the persistent homology groups as a series of elements with intervals representing their lifetime. A potential difficulty with making this precise is that one has to choose generators (i.e., bases) for each homology group consistently in order to keep track of the lifetimes of elements.

When R is a field, this can be handled precisely and efficiently: the structure theorem of Zomorodian-Carlsson tells us that the persistent homology is precisely represented by a "barcode," i.e., a multiset of intervals in $\mathbb{R}$ [ZC05]. (See Figure 66.5.7 for examples of barcode output.)

66.4.2 MULTISCALE CLUSTERING AND MAPPER

For very large data sets, the general techniques of topological data analysis described above can be difficult to apply. The number of simplices grows too rapidly for computation of higher (persistent) homology to be practical. However, clustering algorithms are still useful and tractable in this context. In this section, we describe a method for multiscale clustering that combines ideas of persistence with clustering; this is the Mapper algorithm of Singh, Memoli, and Carlsson [SMC07].

Roughly speaking, the idea of Mapper is to perform clustering at different scales and keep track of how the clusters change as the scale varies. The basic setup is analogous to that of Morse theory and the Reeb graph; we assume the data is presented as a finite metric space (X, ∂_X) along with a filter function $f \colon X \to \mathbb{R}$. In addition, we assume that we have chosen a cover $\mathcal{C} = \{U_\alpha\}$ of the range of f in $\mathbb{R}$; although typically this cover is taken to be a collection of overlapping closed intervals, more exotic choices are possible.

We now proceed as follows:

1. Cluster each inverse image $f^{-1}(U_\alpha)$ for all $U_\alpha \in \mathcal{C}$; denote by $C_{\alpha,i}$ the ith cluster.

2. Form a graph where the vertices are given by the clusters $C_{\alpha,i}$ as α and i vary and there is an edge between $C_{\alpha,i}$ and $C_{\alpha',j}$ when

$$U_\alpha \cap U_{\alpha'} \neq \emptyset \quad \text{and} \quad C_{\alpha,i} \cap C_{\alpha',j} \neq \emptyset.$$

3. Finally, we assign a color to each vertex in the graph according to the average value of f on $C_{\alpha,i}$.

The choice of filter function and the cover are both essential aspects of tuning the results. Common choices of filter function include density measures and eccentricity measures.

66.4.3 STATISTICS AND INFERENCE IN TOPOLOGICAL DATA ANALYSIS

A key question that arises is how to interpret the results of topological data analysis. In some low-dimensional situations (for example, the horizontal gene transfer examples below), it is in fact possible to ascribe direct geometric significance to the results of the computation of persistent homology. In other cases, the output of topological data analysis is very hard to interpret directly and we regard the output as a feature to be used as input to machine learning or other inference procedures (see Section 66.5.2 below for an example).

Another issue is the question of statistical inference and noise in the context of topological data analysis. The space of barcodes that are the output of persistent homology calculations is itself a metric space, and in this context there are stability theorems (akin to Theorem 66.2.2) that relate a metric on the set of compact metric spaces up to isometry with the barcode metric [CEH07]. Roughly speaking, these stability theorems tell us that small perturbations in the input lead to small perturbations in the output, justifying some confidence in the reliability of these procedures. However, persistent homology groups are not robust in the sense that small numbers of arbitrarily bad outliers can cause difficulties.

In light of this, statistical inference seems essential. Distributions of barcodes produced by finite sampling from the data provide a more robust inference procedure [BGMP14, FLR+14, CFL+14, MBT+15, TMMH14, CFL+15], and there has been some work on the bootstrap in this context [CFL+13]. A fair amount is now known about the convergence of such homological statistics to "ground truth" as the sample size increases. However, the statistical properties of the space of barcodes make it difficult to study directly, and as in the situation of the phylogenetic tree spaces, it often makes more sense to consider associated invariants in $\mathbb{R}^n$ or at least a real vector space. There are a variety of approaches to this, ranging from ad hoc projections from barcode space to the more structured approach given by persistence landscapes [Bub15]. (See Chapter 25 of this Handbook for a comprehensive treatment of this subject.)

66.5 BEYOND TREES

GLOSSARY

Vertical evolution: The direct transmission of genetic material from parent to offspring. The genomic material is inherited from a single parent.

Horizontal evolution: Any non-vertical type of acquisition of genomic material, leading to merging of distinct clades to form a new hybrid lineage.

Reticulate network: A graph representation of evolutionary relationships that can accommodate horizontal events.

As described above, molecular phylogenetics—tree building—has become the standard tool for inferring evolutionary relationships. Yet a tree structure is a faithful representation only if there has been no exchange of genetic material among the ancestors of sampled organisms. Specifically, phylogenetic trees are not able to capture horizontal or reticulate evolutionary events, which occur when distinct clades merge together to form a new hybrid lineage. Evidence from genomic data has increasingly suggested that such events are ubiquitous—notable examples including species hybridization, bacterial gene transfer, and meiotic recombination. Influenza undergoes horizontal evolution through reassortment, and HIV through recombination.

The resulting evolutionary representation is no longer a phylogenetic tree but a different structure called a *reticulate network*. This structure is not a tree in the mathematical sense, which precludes the existence of cycles. A cycle in the reticulate network corresponds to a single horizontal evolutionary event (Figure 66.5.1, bottom).

To place these notions into more concrete terms, we can consider the example of a triple reassortment in Figure 66.5.2. Here, we consider three parental viruses with genomes comprised of three different genes and a unique phylogenetic history. All three can undergo a reassortment where each parental virus donates a different gene. The resulting reticulate network is a merging of the three parental phylogenies. Here, we can see that no single phylogenetic tree can faithfully represent the evolutionary history of the reassortant.

The phylogenetic standard for detecting reassortment is to construct a phylogenetic tree for each gene within the genome and comparing each pair of trees for any conflicts in tree structure. Researchers followed this method to derive the triple reassortment history of H7N9 avian influenza [GCH⁺13]. Identifying discrepancies in tree structure by eye is an arduous and error-prone process. Moreover, the tree may not necessarily be robust to noise in the data or the method of tree

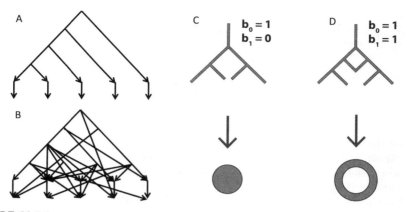

FIGURE 66.5.1

Reticulate networks contain loops. *Idealized, simple phylogenetic trees (A) contrast with more realistic, complex reticulate structures (B). Intuition linking algebraic topology to evolution: a phylogenetic tree can be compressed into a point (C). The same cannot be done for a reticulate structure without destroying the hole at the center (B).*

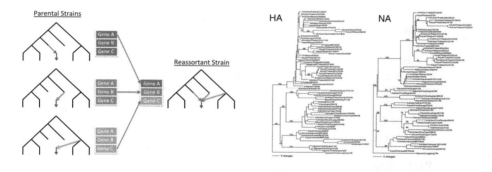

FIGURE 66.5.2

Reticulate network representing the reassortment of three parental strains. *The reticulate network results from merging the three parental phylogenetic trees (left). Indeed, incompatibility between tree topologies inferred from different genes (right) is a criterion used for the identification of evolutionary events involving genomic material exchange.*

construction. The process can be further complicated in settings of complex reticulate patterns, as depicted in Figure 66.5.1 (figure from Doolittle [Doo99]). For example, the simplified universal tree of life stands in stark contrast to the much more complicated but more accurate reticulate "tree" of life [Doo99].

Given a finite metric space obtained from genomic data, i.e., genomic sequences separated from each other by some genetic distance, applying the persistent homology approach to this finite metric space one can uncover the topology of the underlying space and in particular determine if it comes from a tree [CCR13].

In particular, if a metric space is derived from a tree there are strong constraints on the distances between the leaves. These constrains, called the four point conditions or Buneman conditions [Bun74], state that:

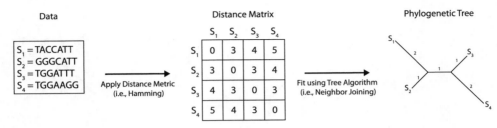

FIGURE 66.5.3

Pipeline for analyzing genomic data using persistent homology. *Staring from some sequences, one can compute distances reflecting the similarity between organisms. These distances provide a finite metric space, that, in some cases, could be summarized by a phylogenetic tree. Distances between branches can be estimated by the addition of weights along the shortest path.*

Definition 2 *there is a unique tree if and only if for every four leaves, i, j, k and l the distance between them satisfies one of the following conditions:*

- $d_{ij} + d_{kl} = d_{ik} + d_{jl} \geq d_{il} + d_{jk}$,

- $d_{il} + d_{jk} = d_{ij} + d_{kl} \geq d_{ik} + d_{jl}$,

- $d_{il} + d_{jk} = d_{ik} + d_{jl} \geq d_{ij} + d_{kl}$.

In settings of vertical evolution, a phylogenetic tree can be continuously deformed into a single point or connected component. The same action cannot be performed for a reticulate network without destroying the loops or cycles in the structure. The active hypothesis then is that the presence of these holes results directly from horizontal evolutionary events (Figure 66.5.1).

This idea can be formalized into the following easy observation [CCR13]:

THEOREM 66.5.1

Let M be any tree-like finite metric space, and let $\epsilon \geq 0$. Then the complex $V_\epsilon(M)$ is a disjoint union of acyclic complexes. In particular, $H_i(V_\epsilon(M)) = \{0\}$ for $i \geq 1$.

In other words, the presence of homology above dimension zero indicates that the metric space does not satisfy the tree-like metric properties. The generators of such homology classes correspond to subsets of genomes whose derived distances do not satisfy the tree condition, indicating that a non-treelike evolutionary process has occurred.

66.5.1 VIRAL EVOLUTION: INFLUENZA A

GLOSSARY

Reassortment: A process of generating novel gene combinations in segmented viruses, through co-infections in the same cell.

Pandemic: A global outbreak of an infectious disease.

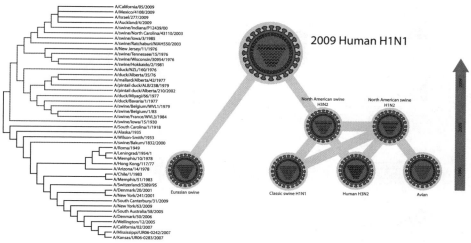

FIGURE 66.5.4

An example of distinct evolutionary modes, the influenza A virus. *(Left) By the rapid accumulation of mutations, influenza A changes its genome and protein. In a particular region of the genome, the evolution can be well described by an evolutionary tree. (Right) However, the genome of the influenza A virus is composed of different segments. When different viruses co-infect the same host they can generate a progeny with genomic material from both parents. That happens often, generating new viruses, such as the virus responsible for the H1N1 pandemic in 2009.*

Segmental co-segregation: The preferential association between different viral segments in a reassortment.

Influenza A is a virus that infects many different kinds of hosts, including pigs, seals, birds, humans, bats, and other mammals. Indeed, the highest genetic diversity of these viruses is found in birds, mostly waterfowl, of the order of Anseriformes (ducks, swans and geese), Passeriformes and Charadriiformes (including gulls). In fact, such waterfowl represent the virus's natural reservoir perpetuating the vast biodiversity of influenza, including all different subtypes (H1 to H14 and N1 to N9). Except for a few high pathogenic H5 and H7 viruses, and in contrast to mammals, the wild duck shows no severe clinical signs upon infection of the virus, which replicates in the waterfowl gut and sheds through fecal matter in the water supply [WBG$^+$92].

As we discussed in Section 66.3, influenza evolves by accumulating mutations at a high pace. Estimates of evolutionary rates, or changes per unit time, indicate that influenza evolves at a rate of $\sim 10^{-3}$ substitutions per nucleotide per year. The genome of this virus is 13,000 bases long, accumulating around a dozen mutations per year. These mutations change the antigenic presentation of the virus, making stable vaccine designs an extremely challenging task.

Influenza viruses also evolve via more complicated processes. The genomic information of the virus is contained in eight different segments, the viral analog of chromosomes, containing one of two genes per segment. When two viruses co-infect the same host cell, they can generate a progeny containing segments from both parental strains [RR07, RLK08]. This phenomenon, called reassortment, shuffles the genomic material of different viruses. Reassortment is the underlying mecha-

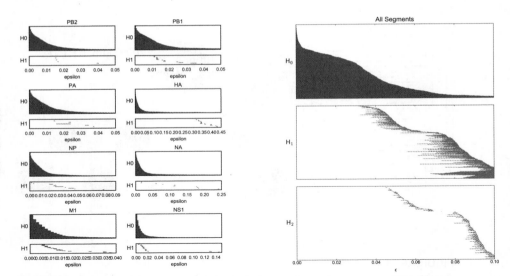

FIGURE 66.5.5

Influenza evolves through mutations and reassortment. *When the persistent homology is computed for finite metric spaces derived from only one segment, the homology is zero-dimensional, suggesting a treelike process (left). However, when different segments are put together, the structure is more complex, revealing nontrivial homology in different dimensions (right). (3,105 influenza whole genomes were analyzed. Data from 1956 to 2012; all influenza A subtypes.)*

nism behind influenza pandemics, the arrival of novel viral strains to the human population, usually containing segments from multiple hosts. Pandemic influenza refers to strains endemic to animal hosts like avian and swine that obtain the requisite host-determinant mutations to infect and adapt to human hosts, thereby spreading on a global scale. A calamitous example of a confirmed influenza pandemic was the Spanish flu epidemic of 1918 that claimed the lives of over 60 million people worldwide. In 2009, a swine-origin H1N1 virus marked the first pandemic of the twenty-first century [TKR09, SPB+09]. Such pandemics are often the result of reassortments that rapidly acquire the necessary functional alleles conferring heightened host infectivity or specificity.

When studying the finite metric spaces obtained from a single segment of the flu virus (e.g., hemagglutinin) using persistent homology [CCR13], one observes that most of the information is contained in the zero-dimensional homology (Figure 66.5.5), indicating that treelike metrics are good approximations to these spaces. However when studying the persistent homology for finite metric spaces combining several segments, a large number of homology classes appear at dimension one and higher, indicating pervasive reassortments.

But the persistent homology information provides information about more than the obstructions for treelike metrics. Indeed, if a few sequences generate a nontrivial class, one could infer that a reassortment event took place among the ancestors of these isolates [CCR13]. The other relevant information is the birth and death times of the class, which provides information about how genetically distant parental viruses were. Numbers of higher-dimensional classes and their sizes (birth, deaths and size of bars in barcode diagram) provide a useful way to generate test statistics. For instance, one can estimate how often different patterns of the eight

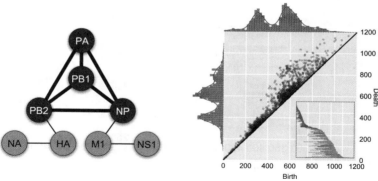

FIGURE 66.5.6

Persistent homology view of reassortment in influenza reveals segment coseg-
regation and different scales of vertical exchange of genomic material. *Left:*
Nonrandom association of flu segments was measured by testing against a null model of
equal reassortment. We identify significant cosegregation within PA, PB1, PB2, NP, con-
sistent with the cooperative function of the polymerase complex. Right: The persistence
diagram for concatenated avian flu sequences reveals bimodal topological structure. Anno-
tating each interval as intra- or inter-subtype identifies a genetic barrier to reassortment
at intermediate scales.

segments of influenza cosegregates, in other words, whether there is any preference
(due to structural or selection reasons) among the potential combinations. It is
easy to see significant co-segregation within several segments that interact forming
the polymerase complex PA, PB1, PB2, NP [CCR13]. This finding is consistent
with the cooperative function of the different proteins' polymerase complex (Figure
66.5.6, left): in order for the virus to efficiently replicate, all units of the polymerase
should work together.

The size of the bars associated to nonzero homology classes is also indicative
of the type of reassortment events that could occur. The persistence diagram for
whole genomes of avian flu sequences reveals a bimodal topological structure (Figure 66.5.6, right). In other words, there are small bars and larger bars. Inspection
of generators of different bars immediately reveals two types of reassortment processes. There could be mixing of viruses that are closely related, i.e., belong to the
same subtype (two H5N1 viruses, for instance), that are associated to small bars.
But there could also be mixing between the genomic material of distant viruses
belonging to two different subtypes (H5N1 and H7N2, for instance), that generate
large bars.

66.5.2 PERSISTENT HOMOLOGY ESTIMATORS IN POPULATION GENET-ICS

GLOSSARY

> **Population genetics:** The study of distributions of frequencies of different alleles
> in a population.
>
> **Coalescence:** A stochastic model that reconstructs the history of a set of indi-
> viduals within a larger population.

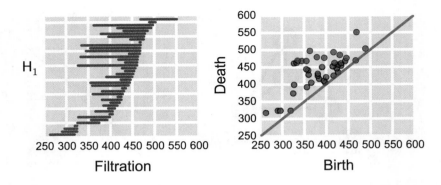

FIGURE 66.5.7
Two representations of the same topological invariants, computed using persistent homology. *Left: Barcode diagram. Right: Persistence diagram. Data was generated from a coalescent simulation with* $n = 100$, $\rho = 72$, *and* $\theta = 500$.

Many simple stochastic models generate complex data that cannot be readily visualized as a manifold or summarized by a small number of topological features. Finite metric spaces extracted from these models will generate persistence diagrams whose complexity increases with the number of sampled points. Nevertheless, the collection of measured topological features may exhibit additional structure, providing useful information about the underlying data generating process. While the persistence diagram is itself a summary of the topological information contained in a sampled point cloud, to perform inference, further summarization may be appropriate, e.g., by considering distributions of properties defined on the diagram. In other words, we are less interested in learning the topology of a particular sample, but rather in understanding the expected topological signal of different model parameters. We explain how this can be applied to models from population genetics, via the coalescent with recombination [ERCR14].

The coalescent process is a stochastic model that generates the genealogy of individuals sampled from an evolving population [Wak09]. The genealogy is then used to simulate the genetic sequences of the sample. This model is essential to many methods commonly used in population genetics. Starting with a present-day sample of n individuals, each individual's lineage is traced backward in time, towards a mutual common ancestor. Two separate lineages collapse via a coalescence event, representing the sharing of an ancestor by the two lineages. The stochastic process ends when all lineages of all sampled individuals collapse into a single common ancestor. In this process, if the total (diploid) population size N is sufficiently large, then the expected time before a coalescence event, in units of $2N$ generations, is approximately exponentially distributed:

$$P(T_k = t) \approx \binom{k}{2} e^{-\binom{k}{2}t}, \tag{66.5.1}$$

where T_k is the time that it takes for k individual lineages to collapse into $k - 1$ lineages.

After generating a genealogy, the genetic sequences of the sample can be simulated by placing mutations on the individual branches of the lineage. The number of mutations on each branch is Poisson-distributed with mean $\theta t/2$, where t is the

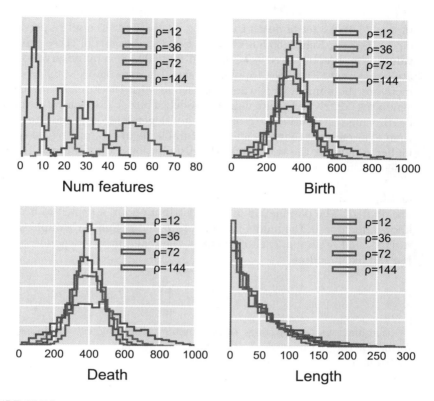

FIGURE 66.5.8

Distributions of statistics defined on the H_1 persistence diagram for different model parameters. *Top left: Number of features. Top right: Birth time distribution. Bottom left: Death time distribution. Bottom right: Feature length distribution. Data generated from 1000 coalescent simulations with $n = 100$, $\theta = 500$, and variable ρ.*

branch length and θ is the population-scaled mutation rate. In this model, the average *genetic distance* between any two sampled individuals, defined by the number of mutations separating them, is θ.

The coalescent with recombination is an extension of this model that allows different genetic loci to have different genealogies. Looking backward in time, recombination is modeled as a splitting event, occurring at a rate determined by population-scaled recombination rate ρ, such that an individual has a different ancestor at different loci. Evolutionary histories are no longer represented by a tree, but rather by an *ancestral recombination graph*. Recombination is the component of the model generating nontrivial topology by introducing deviations from a contractibile tree structure, and is the component which we would like to quantify. Coalescent simulations were performed using ms [Hud02].

The persistence diagram from a typical coalescent simulation is shown in Figure 66.5.7. Examining the diagram, it would be difficult to classify the observed features into signal and noise. Instead, we use the information in the diagram to construct a statistical model in order to infer the parameters, θ and ρ, which generated the data. Note that we consider inference using only H_1 invariants, but the ideas easily generalize to higher dimensions. We consider the following properties of the persistence diagram: the total number of features, K; the set of birth times, $(b_1, \ldots, b_K)$; the set of death times, $(d_1, \ldots, d_K)$; and the set of persistence lengths, $(l_1, \ldots, l_K)$. In Figure 66.5.8 we show the distributions of these properties for four values of ρ, keeping fixed $n = 100$ and $\theta = 500$. Several observations are immediately apparent. First, the topological signal is remarkably stable. Second, higher ρ increases the number of features, consistent with the intuition that recombination generates nontrivial topology in the model. Third, the mean values of the birth and death time distributions are only weakly dependent on ρ and are slightly smaller than θ, suggesting that θ defines a natural scale in the topological space. However, higher ρ tightens the variance of the distributions. Finally, persistence lengths are independent of ρ.

Examining Figure 66.5.8, we can postulate: $K \sim \text{Pois}(\zeta)$, $b_k \sim \text{Gamma}(\alpha, \xi)$, and $l_k \sim \text{Exp}(\eta)$. Death time is given by $d_k = b_k + l_k$, which is incomplete Gamma distributed. The parameters of each distribution are assumed to be an *a priori* unknown function of the model parameters, θ and ρ, and the sample size, n. Keeping n fixed, and assuming each element in the diagram is independent, we can define the full likelihood as

$$p(D|\theta, \rho) = p(K|\theta, \rho) \prod_{k=1}^{K} p(b_k|\theta, \rho) p(l_k|\theta, \rho). \qquad (66.5.2)$$

Simulations over a range of parameter values suggest the following functional forms for the parameters of each distribution. The number of features is Poisson distributed with expected value

$$\zeta = a_0 \log\left(1 + \frac{\rho}{a_1 + a_2\rho}\right). \qquad (66.5.3)$$

Birth times are Gamma distributed with shape parameter

$$\alpha = b_0\rho + b_1 \qquad (66.5.4)$$

and scale parameter

$$\xi = \frac{1}{\alpha}(c_0 \exp(-c_1\rho) + c_2). \qquad (66.5.5)$$

These expressions appear to hold well in the regime $\rho < \theta$, but break down for large ρ. The length distribution is exponentially distributed with shape parameter proportional to mutation rate, $\eta = \alpha\theta$. The coefficients in each of these functions are calibrated using simulations, and could be improved with further analysis. This model has a simple structure and standard maximum likelihood approaches can be used to find optimal values of θ and ρ.

66.5.3 HUMAN RECOMBINATION

GLOSSARY

Meiosis: The fundamental cellular process of gamete formation in sexually reproducing organisms involving their reduction of diploid to haploid germ cells.

Meiotic recombination: The production of new genomic material during meiosis from combinations of parental chromosomes.

Non-tree like events are not only a pervasive phenomena among rapidly evolving organisms, such as virus and bacteria, but also a major mechanism employed by sexually reproducing organisms to ensure genetic diversity of offspring as well as genome integrity. Cells in sexually reproducing organisms contain two copies of most chromosomes, or autosomes. Each copy differs slightly in sequence, but has the same overall structure. In humans we have 22 pairs of chromosomes, as well as sex chromosomes—a pair of X chromosomes for females and an X and Y chromosome for males. Each of these 23 pairs of chromosomes is inherited from a different parent. In the process of meiosis, the cells become haploid, containing only one chromosome of each pair, with different regions randomly selected from the paternal or maternal copy.

Since the work of Morgan using the fruit fly, *Drosophila melanogaster*, as a model, we have a quantitative understanding of how often the chromosomal crossover occurs in meiosis. Morgan was able to establish a link between the probability of crossover and how far away in chromosomal position two different loci were. In humans, recombinations occur at an average rate of one crossover per chromosome per generation. A more quantitative measure of these rates estimates the probability for two different loci in a chromosome to undergo a crossover event. One defines a centi-Morgan (cM) as having a chance of 1% per generation. The average rate in humans is about 1cM per megabase.

However, genetic versus chromosomal distance approaches do not allow a high-resolution mapping below millions of bases, as they require many generations to track many meiotic events. Pedigree and linkage-disequilibrium analysis provide a much more refined view of where recombination occurs [KJK04]. Pedigree analysis studies families of related individuals along several generations. Linkage disequilibrium (LD) is a measure of how the variability of two genomic loci is associated. If there is no recombination between loci, two mutations in the same chromosome will always travel together. If recombination occurs very frequently, the presence of a particular allele will provide little information about nearby mutations. The simplest measure of LD is $D_{ij} = f_{ij} - f_i f_j$, where f_{ij} is the frequency of observing two alleles i and j together, f_i is the frequency of observing the i allele.

It has been found that recombination occurs preferentially at narrow genomic regions known as recombination hotspots [ACTB07, PP10, BIM13]. In mammals, recombination hotspots are specified by binding sites of the meiosis-specific H3K4 tri-methyltransferase PRDM9 [BBG+10, MBT+10, PPP10]. Nevertheless, due to the great genomic complexity of eukaryotes, their recombination landscape is actually the result of a hierarchical combination of factors that operate at different genomic scales. High-resolution mapping of meiotic double-strand breaks (DSBs) in yeast and mice [SGB+11, KSB+12, PSK+11, FSM+14] reveal fine-scale variation in recombination rates within hotspots as well as frequent recombination events occurring outside hotspots [PSK+11].

In humans, population-based recombination maps are an invaluable tool in the study of recombination [MMH+04][MBF+05]. Fine-scale mapping and annotation of human recombination is now possible due to the large number of genomes published by consortia such as the 1,000 Genomes Project [C+12a] and ENCODE [C+12b]. High-resolution datasets, such as those obtained by chromatin immuno-precipitation (ChIP-seq), bisulfite, or RNA sequencing methods, reveal a wide array

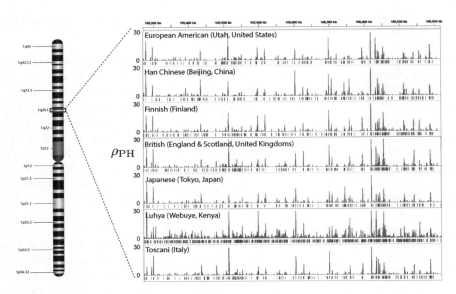

FIGURE 66.5.9

Recombination rate estimates across distant human populations. *Position-wise recombination rates for each of the 7 populations, for the cytogenic band 1q24.1. Blue bars represent recombination rates estimated with a variable-size sliding window containing 14 segregating sides. Below, red segments represent genomic regions where a 500-bp window detects recombination* $(b_1 > 0)$.

of biological features associated to small genomic regions and these can be used to associate locations where recombination occurs to other molecular biological phenomena.

Establishing statistical associations with such narrow (and often clustered) biological features and analyzing the large number of sequences usually present in these datasets is becoming an important challenge for traditional methods of recombination rate estimation, such as linkage disequilibrium-based methods. Robust and scalable methods for detecting and quantifying recombination rates at multiple scales, such as those described in the previous section, are thus particularly useful in this context.

The persistent homology estimators of recombination that were introduced in the previous section can be easily implemented on a sliding window, and therefore adapted to the very long eukaryotic genomes [CLR16, CRE$^+$16]. To avoid local variation of the estimator sensitivity resulting from local variations of the mutation rate, the window can be taken to contain a fixed number of polymorphic sites. Using windows with a smaller or larger number of polymorphic sites allows us to study recombination at smaller or larger scales, respectively. In Figure 66.5.9 a snapshot produced using such an approach is shown for the cytogenic band 1q24.1 of the human genome [CRE$^+$16]. A window containing 14 polymorphic sites was slid, in steps of 7 polymorphic sites, across the entire genome of 647 individuals from 7 different populations (2 Asian, 1 African and 4 of European ancestry) genotyped for the 1,000 Genomes Project [C$^+$12a]. The resulting b_1 at each position was converted into a population-scaled recombination rate per base, using the methods that we have described in the previous section.

Studies of human recombination like this one, involving a large number of individuals from multiple populations, provide insights into the population genetics of human recombination, for instance, revealing differences between populations in the selective forces that drive genetic diversity. Specifically, the above specific recombination maps show differences in the size and location of recombination hotspots of distinct populations, despite a large overall degree of consistency. Hierarchical clustering of these recombination maps reproduces the standard evolutionary relationships between these populations, providing a good consistency check of persistent homology estimators of recombination. More broadly, topological methods can be used to describe gene flow across populations, where migration and admixture now also appear as high-dimensional loops in the evolutionary space.

From a more fundamental point of view, constructing high-resolution recombination maps, such as those described here, allows for the identification of potentially novel genomic associations with recombination, based on annotations catalogued in resources like the ENCODE database [C+12b]. For instance, looking at the density of 500-bp windows on which b_1 is non-zero reveals details on the local fine-scale structure of the recombination rate. Comparison with the binding sites of 118 transcription factors, obtained by ChIP-seq in 91 human cell lines [C+12b], reveals substantial differences in the recombination rate at the binding sites of various specific transcription factors with respect to neighbouring regions [CRE+16]. These differences are also supported by correlated variations of predicted PRDM9 binding motifs as well as epigenetic markers in sperm. Similarly, comparison with the location of endogenous transposable elements reveals enhancements at the insertion loci of some of the most recent families of transposable elements in humans [CRE+16]. High-resolution recombination maps can thus provide a window into the molecular processes that take place in germ cells during meiotic recombination.

66.6 CANCER GENOMICS

GLOSSARY

Cancer: The abnormal uncontrolled cell growth of somatic cells.

Cross-sectional studies: Studies of a cohort of patients with similar diagnosis.

Gene expression: A readout of the number of RNA molecules in a sample.

The model of clonal evolution through somatic mutations is a first approximation to how tumors evolve, but it is far from comprehensive. Many other factors play a crucial role, including surrounding cells (the microenvironment), epigenetic alterations, and immune responses. Nevertheless it is clear that each tumor is the result of a different mutational history. Generally, strategies to identify the most relevant players are based on the idea that if a gene is found mutated more than it would be by random chance, then the mutations in that gene have probably been selected for along the history of the tumor because they allow the clone to grow faster than the surrounding cells. Figure 66.6.1 on the right, shows a circos plot (a common representation of somatic alterations) in glioblastomas, the most common brain tumor in adults. The plot shows recurrent mutations in chromosome 7 containing an oncogene (EGFR) and chromosome 17 containing a tumor suppressor

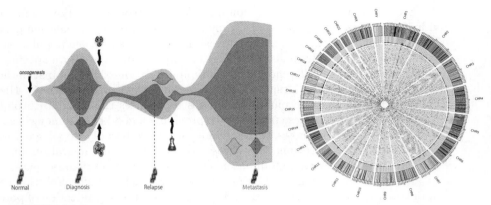

FIGURE 66.6.1

Cancer understood as a clonal process associated to the accumulation of mutations. *Left: A single clone starts acquiring mutations that lead to uncontrolled growth. These mutations activate and de-activate processes necessary for oncogenesis and tumor progression. The cartoon on the left shows a population of clones expanding (time runs from left to right). Treatment can reduce the initial clone but resistant clones could be generated leading to resistance to therapy. Right: 150 glioblastoma genomes from The Cancer Genome Atlas are visualized in a circos plot [FTC+ 13]. Each patient tumor contains different mutations, and protein-changing mutations are represented as red dots along a circle. The location along the circle indicates the genomic location of the mutation. Cancer genomes from multiple patients can be visualized as multiple concentric circles located at different radii. Commonly mutated genes appear as radial rays emanating from the center. Between the external depiction of chromosomes and the internal depiction of individual mutations, there is a histogram showing the number of times that a particular gene has been mutated in the cohort. The spikes at chromosome 7 and chromosome 17 correspond to mutations in EGFR and TP53, respectively.*

(TP53). In the following application, we will show how expression data can be used to stratify patients and find genes relevant to distinct tumor types.

66.6.1 STRATIFYING PATIENTS USING CROSS-SECTIONAL DATA

One important question from both the basic biology and clinical points of views is: how can the molecular data (mutations, fusions, expression, etc.) be used to classify patients into different subtypes? Is there a single tumor type for each cell of origin (like lung adenocarcinoma, or uroepithelial cancer) or are there more refined distinctions? From the basic biology point of view this is important because common molecular features might indicate specific pathways that are activated or deactivated, and what alterations are related to these pathways. From the clinical point of view, patient stratification determines what patients could be sensitive or resistant to a particular therapy, and what their risk is of progression or metastasis. Thus clinical questions can be translated into a problem of understanding the shape or structure of molecular data where each data point is a patient. Information on many patients is usually referred as cross-sectional data. The "dual" or "transpose" problem, looking at different genes in the space of patients, is probably more interesting biologically, as genes differentially expressed in a group of patients can reveal common deregulated pathways (Figure 66.6.2).

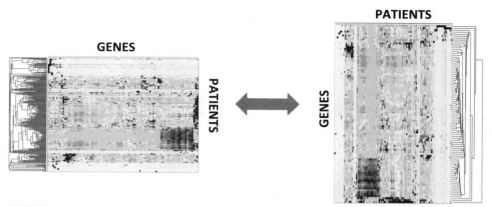

FIGURE 66.6.2

Two dual views of the same data. *Hierarchical clustering of expression point cloud data corresponding to 200 peripheral T-cell lymphoma patients. The data can be seen as two "dual" point cloud data. In the space of genes each point is a patient and clustering of points correspond to clustering of patients. On the dual space, the space of patients, each point is a gene and clustering points correspond to clustering genes.*

Expression is a very rich source of data as it associates to each transcript a real value representing the number of copies of mRNA present in the tumor. Most of the data corresponds to the ensemble of cells present in the sample and as such represents a complex mixture of expression levels from different tumor cells and even different types of non-tumor cells. The typical structure of expression data is a point in a very high-dimensional real vector space, typically of the order of 22,000 potential transcripts. Each patient represents a point in this space, and the cross-sectional data corresponds to point cloud data. The question of stratification is usually posed as a clustering problem: how many groups of patients are there presenting similar expression profiles?

Expression-based classification of tumors have been a dominant theme since the first microarray experiments and there is an extensive literature in the topic [HDC+01, VHV+02] that we do not have the scope and time to discuss. All these earlier approaches are in some way or another based on the idea of clustering patients and genes. It could be, however, that the point cloud data does not cluster neatly. Indeed, that is generally the case due to many biological and technical factors. For example, not every tumor activates or suppresses different pathways with the same strength, resulting in a more continuous structure from suppression to activation at different levels. There is also the infiltration of non-tumor cells into the tumor sample. The degree of infiltration varies, as does the type of infiltrating cell. These and other factors contribute to generating large continuous structures that are not represented by discrete clusters.

A possible approach to this problem is to apply multiscale clustering techniques. Nicolau, Levine, and Carlsson [NLC11] studied microarray gene expression data from 295 breast cancers as well as normal breast tissue. Expression data was normalized and represented by Mapper [SMC07]. In this study the filter function used was a measure of the deviation of the expression of the tumor samples relative to normal controls. Clusters of points of overlapping intervals in the image of this function were represented as nodes, and edges corresponded to shared points between different clusters, as shown in the top left of Figure 66.6.3. In this figure,

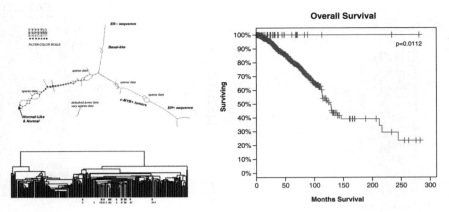

FIGURE 66.6.3
Stratifying patients using cross-sectional expression data. *Top left: Mapper representation of the gene expression data from 295 breast tumors. Lens function was generated as a deviation from expression in normal breast tissue controls. Blue colors correspond to samples with close similarity to normal tissues (left). Tumors with expression profiles significantly different from normal tissue are represented by the two arms on the right-hand side. The upper arm is characterized by low expression of estrogen receptor (ER-). The lower branch contains samples with high expression of c-MYB+. These c-MYB+ tumors cannot be identified using standard clustering (in lower left figure hierarchical clustering split c-MYB+ tumors, represented in red). Independent validation using 960 breast invasive carcinomas from The Cancer Genome Atlas of two of the highest expressed genes in c-MYB+ tumors, DNALI1 and C9ORF116, show very good prognosis for these tumors.*

blue colors correspond to samples with close similarity to normal tissues, as on the left part of the figure. On the right-hand side, the samples divide into two branches. The lower branch, named in their manuscript as c-MYB+ tumors, constitutes 7.5% of the cohort (22 tumors). These tumors are some of the most distinct from normal and are characterized by high expression of genes including c-MYB, ER, DNALI1 and C9ORF116. Hierarchical clustering fails to identify this particular subset of tumors (see bottom left of Figure 66.6.3), instead segregating these tumors into separate clusters with low confidence. These tumors did not correspond to any known breast cancer subtype. Interestingly, all patients with c-MYB+ tumors had very good survival and no metastasis.

To validate these observations in an independent cohort, we looked at samples with high expression of DNALI1 and C9ORF116 (more than a 2-fold overexpression) in 960 breast invasive carcinomas from The Cancer Genome Atlas. Of these, 32 showed high expression in these genes and had excellent survival (Figure 66.6.3, right), confirming the observation of [NLC11]. These tumors did not have TP53 mutations or deletions but were associated with GATA3 mutations.

66.7 SOURCES AND RELATED MATERIAL

FURTHER READING

[BHV01]: The original reference on the metric space of phylogenetic trees; very clearly written.

[Gro81]: Extremely influential introduction of the ideas of metric geometry.

[BBI01, BH99]: Textbooks on metric geometry and non-positively curved spaces; both provide clear and comprehensive treatments.

RELATED CHAPTERS

Chapter 21: Topological methods in discrete geometry
Chapter 23: Computational topology of graphs on surfaces
Chapter 25: High-dimensional topological data analysis

REFERENCES

[ACTB07]   N. Arnheim, P. Calabrese, and I. Tiemann-Boege. Mammalian meiotic recombination hot spots. *Annu. Rev. Genet.*, 41:369–399, 2007.

[Ale57]   A.D. Alexandrov. Über eine Verallgemeinerung der Riemannschen Geometrie. *Schr. Forschungsinst. Math. Berlin*, 1:33–84, 1957.

[BBG$^+$10]   F. Baudat, J. Buard, C. Grey, A. Fledel-Alon, C. Ober, M. Przeworski, G. Coop, and B. de Massy. Prdm9 is a major determinant of meiotic recombination hotspots in humans and mice. *Science*, 327:836–840, 2010.

[BBI01]   D. Burago, Y. Burago, and S. Ivanov. *A Course in Metric Geometry*. AMS, Providence, 2001.

[BGMP14]   A.J. Blumberg, I. Gal, M.A. Mandell, and M. Pancia. Robust statistics, hypothesis testing, and confidence intervals for persistent homology on metric measure spaces. *Found. Comput. Math.*, 14:745–789, 2014.

[BH99]   M.R. Bridson and A. Häfliger. *Metric Spaces of Non-positive Curvature*. Vol. 319 of *Grundlehren der mathematischen Wissenschaften*, Springer, Berlin, 1999.

[BHO13]   D. Barden, L. Huiling, and M. Owen. Central limit theorems for Fréchet means in the space of phylogenetic trees. *Electron. J. Probab.*, 18:1–25, 2013.

[BHV01]   L.J. Billera, S.P. Holmes, and K. Vogtmann. Geometry of the space of phylogenetic trees. *Adv. Appl. Math.*, 27:733–767, 2001.

[BIM13]   F. Baudat, Y. Imai, and B. de Massy. Meiotic recombination in mammals: localization and regulation. *Nature Reviews Genetics*, 14:794–806, 2013.

[Bub15]   P. Bubenik. Statistical topological data analysis using persistence landscapes. *J. Mach. Learn. Res.*, 16:77–102, 2015.

[Bun74]   P. Buneman. A note on the metric properties of trees. *J. Combin. Theory Ser. B*, 17:48–50, 1974.

[C$^+$12a]   G.P. Consortium et al. An integrated map of genetic variation from 1,092 human genomes. *Nature*, 491:56–65, 2012.

[C$^+$12b]   E.P. Consortium et al. An integrated encyclopedia of DNA elements in the human genome. *Nature*, 489:57–74, 2012.

[Car14]   G. Carlsson. Topological pattern recognition for point cloud data. *Acta Numer.*, 23:289–368, 2014.

[CCR13]   J.M. Chan, G. Carlsson, and R. Rabadan. Topology of viral evolution. *Proc. Natl. Acad. Sci. USA*, 110:18566–18571, 2013.

[CEH07]   D. Cohen-Steiner, H. Edelsbrunner, and J. Harer. Stability of persistence diagrams. *Discrete Comput. Geom.*, 37:103–120, 2007.

[CFL$^+$13]   F. Chazal, B. Fasy, F. Lecci, A. Rinaldo, A. Singh, and L. Wasserman. On the bootstrap for persistence diagrams and landscapes. *Model. Anal. Inf. Sist.*, 20:96–105, 2013.

[CFL$^+$14]   F. Chazal, B.T. Fasy, F. Lecci, B. Michel, A. Rinaldo, and L.A. Wasserman. Robust topological inference: Distance to a measure and kernel distance. Preprint, `arXiv/1412.7197`, 2014.

[CFL$^+$15]   F. Chazal, B. Fasy, F. Lecci, B. Michel, A. Rinaldo, and L.A. Wasserman. Subsampling methods for persistent homology. In *Proc. 32nd Int. Conf. Machine Learning*, pages 2143–2151, 2015.

[CLR16]   P.G. Camara, A.J. Levine, and R. Rabadan. Inference of ancestral recombination graphs through topological data analysis. *PLOS Comput. Biol.*, 12:e1005071, 2016.

[CRE$^+$16]   P.G. Camara, D.I. Rosenbloom, K.J. Emmett, A.J. Levine, and R. Rabadan. Topological data analysis generates high-resolution, genome-wide maps of human recombination. *Cell Systems*, 3:83–94, 2016.

[Dar59]   C. Darwin. On the origin of species. *Murray, London*, 360, 1859.

[DDL$^+$12]   X. Du, L. Dong, Y. Lan, et al. Mapping of h3n2 influenza antigenic evolution in china reveals a strategy for vaccine strain recommendation. *Nature Comm.*, 3:709, 2012.

[DMRM13]   J. Dekker, M.A. Marti-Renom, and L.A. Mirny. Exploring the three-dimensional organization of genomes: interpreting chromatin interaction data. *Nature Reviews Genetics*, 14:390–403, 2013.

[Doo99]   W.F. Doolittle. Phylogenetic classification and the universal tree. *Science*, 284:2124–2128, 1999.

[EH08]   H. Edelsbrunner and J. Harer. Persistent homology—a survey. In J. E. Goodman, J. Pach, and R. Pollack, editors, *Surveys on Discrete and Computational Geometry*, vol. 453 of *Contemp. Math.*, pages 257–282, AMS, Providence, 2008.

[ELZ02]   H. Edelsbrunner, D. Letscher, and A. Zomorodian. Topological persistence and simplification. *Discrete Comput. Geom.*, 28:511–533, 2002.

[ERCR14]   K. Emmett, D. Rosenbloom, P. Camara, and R. Rabadan. Parametric inference using persistence diagrams: A case study in population genetics. Preprint, `arXiv:1406.4582`, 2014.

[ESK$^+$15]   P. Eirew, A. Steif, J. Khattra, et al. Dynamics of genomic clones in breast cancer patient xenografts at single-cell resolution. *Nature*, 518:422–426, 2015.

[Fel03]   J. Felsenstein. *Inferring Phylogenies*. Sinauer Associates, Sunderland, 2003.

[FLR$^+$14]   B.T. Fasy, F. Lecci, A. Rinaldo, L. Wasserman, S. Balakrishnan, and A. Singh. Confidence sets for persistence diagrams. *Ann. Statist.*, 42:2301–2339, 2014.

[FOP$^+$13]   A. Feragen, M. Owen, J. Petersen, M.M. Wille, L.H. Thomsen, A. Dirksen, and M. de Bruijne. Tree-space statistics and approximations for large-scale analysis of anatomical trees. In *Proc. 23rd Int. Conf. Inform. Process. Medical Imaging*, vol. 7917 of *LNCS*, pages 74–85, Springer, Berlin, 2013.

[FSM$^+$14]   K.R. Fowler, M. Sasaki, N. Milman, S. Keeney, and G.R. Smith. Evolutionarily diverse determinants of meiotic DNA break and recombination landscapes across the genome. *Genome Research*, gr–172122, 2014.

[FTC+13]   V. Frattini, V. Trifonov, J.M. Chan, et al. The integrated landscape of driver genomic alterations in glioblastoma. *Nature Genetics*, 45:1141–1149, 2013.

[GCH+13]   R. Gao, B. Cao, Y. Hu, et al. Human infection with a novel avian-origin influenza a (h7n9) virus. *New England J. Medicine*, 368:1888–1897, 2013.

[GED06]   V. Gupta, D.J. Earl, and M.W. Deem. Quantifying influenza vaccine efficacy and antigenic distance. *Vaccine*, 24:3881–3888, 2006.

[Gro81]   M. Gromov. *Metric Structures for Riemannian and Non-Riemannian Spaces*. Birkhäuser, Basel, 1981.

[HDC+01]   I. Hedenfalk, D. Duggan, Y. Chen, et al. Gene-expression profiles in hereditary breast cancer. *New England J. Medicine*, 344(8):539–548, 2001.

[Hol03]   S. Holmes. Bootstrapping phylogenetic trees: theory and methods. *Statist. Sci*, 18:241–255, 2003.

[Hol05]   S. Holmes. Statistical approach to tests involving phylogenies. In O. Gascuel, editor, *Mathematics of Evolution and Phylogeny*, pages 91–120, Oxford University Press, 2005.

[Hud02]   R.R. Hudson. Generating samples under a wright–fisher neutral model of genetic variation. *Bioinformatics*, 18:337–338, 2002.

[KCK+14]   H. Khiabanian, Z. Carpenter, J. Kugelman et al. Viral diversity and clonal evolution from unphased genomic data. *BMC Genomics*, 15:S17, 2014.

[KJK04]   L. Kauppi, A.J. Jeffreys, and S. Keeney. Where the crossovers are: recombination distributions in mammals. *Nature Reviews Genetics*, 5:413–424, 2004.

[KSB+12]   P.P. Khil, F. Smagulova, K.M. Brick, R.D. Camerini-Otero, and G.V. Petukhova. Sensitive mapping of recombination hotspots using sequencing-based detection of ssDNA. *Genome Research*, 22:957–965, 2012.

[LABW+09]   E. Lieberman-Aiden, N.L. van Berkum, L. Williams, et al. Comprehensive mapping of long-range interactions reveals folding principles of the human genome. *Science*, 326:289–293, 2009.

[MBF+05]   S. Myers, L. Bottolo, C. Freeman, G. McVean, and P. Donnelly. A fine-scale map of recombination rates and hotspots across the human genome. *Science*, 310:321–324, 2005.

[MBT+10]   S. Myers, R. Bowden, A. Tumian, R.E. Bontrop, C. Freeman, T.S. MacFie, G. McVean, and P. Donnelly. Drive against hotspot motifs in primates implicates the prdm9 gene in meiotic recombination. *Science*, 327:876–879, 2010.

[MBT+15]   E. Munch, P. Bendich, K. Turner, S. Mukherjee, J. Mattingly, and J. Harer. Probabilistic Fréchet means and statistics on vineyards. *Electron. J. Statist.*, 9:1173–1204.

[MFB+08]   R. McLendon, A. Friedman, D. Bigner, et al. Comprehensive genomic characterization defines human glioblastoma genes and core pathways. *Nature*, 455:1061–1068, 2008.

[MMH+04]   G.A. McVean, S.R. Myers, S. Hunt, P. Deloukas, D.R. Bentley, and P. Donnelly. The fine-scale structure of recombination rate variation in the human genome. *Science*, 304:581–584, 2004.

[MOP15]   E. Miller, M. Owen, and J.S. Provan. Polyhedral computational geometry for averaging metric phylogenetic trees. *J. Adv. Appl. Math.*, 68:51–91, 2015.

[NKT+11]   N. Navin, J. Kendall, J. Troge, et al. Tumour evolution inferred by single-cell sequencing. *Nature*, 472:90–94, 2011.

[NLC11]   M. Nicolau, A.J. Levine, and G. Carlsson. Topology based data analysis identifies a subgroup of breast cancers with a unique mutational profile and excellent survival. *Proc. Natl. Acad. Sci. USA*, 108:7265–7270, 2011.

[Nye11]   T.M.W. Nye. Principal components analysis in the space of phylogenetic trees. *Ann. Statis.*, 39:2716–2739, 2011.

[OP11]   M. Owen and J.S. Provan. A fast algorithm for computing geodesic distances in tree space. *IEEE/ACM Trans. Comput. Biol. Bioinf.*, 8:2–13, 2011.

[PP10]   K. Paigen and P. Petkov. Mammalian recombination hot spots: properties, control and evolution. *Nature Reviews Genetics*, 11:221–233, 2010.

[PPP10]   E.D. Parvanov, P.M. Petkov, and K. Paigen. Prdm9 controls activation of mammalian recombination hotspots. *Science*, 327:835–835, 2010.

[PSK⁺11]   J. Pan, M. Sasaki, R. Kniewel, et al. A hierarchical combination of factors shapes the genome-wide topography of yeast meiotic recombination initiation. *Cell*, 144:719–731, 2011.

[PTT⁺14]   A.P. Patel, I. Tirosh, J.J. Trombetta, et al. Single-cell rna-seq highlights intratumoral heterogeneity in primary glioblastoma. *Science*, 344:1396–1401, 2014.

[RLK08]   R. Rabadan, A.J. Levine, and M. Krasnitz. Non-random reassortment in human influenza a viruses. *Influenza and Other Respiratory Viruses*, 2:9–22, 2008.

[RR07]   R. Rabadan and H. Robins. Evolution of the influenza a virus: Some new advances. *Evol. Bioinform. Online*, 3:299–307, 2007.

[SGB⁺11]   F. Smagulova, I.V. Gregoretti, K. Brick, P. Khil, R.D. Camerini-Otero, and G.V. Petukhova. Genome-wide analysis reveals novel molecular features of mouse recombination hotspots. *Nature*, 472:375–378, 2011.

[SL09]   M. Schmidt and H. Lipson. Distilling free-form natural laws from experimental data. *Science*, 324:81–85, 2009.

[SMC07]   G. Singh, F. Mémoli, and G. Carlsson. Topological methods for the analysis of high dimensional data sets and 3d object recognition. In *Eurographics Sympos. Point-Based Graphics*, pages 91–100, 2007.

[SPB⁺09]   A. Solovyov, G. Palacios, T. Briese, W.I. Lipkin, and R. Rabadan. Cluster analysis of the origins of the new influenza A (H1N1) virus. *Euro. Surveill.*, 14:19224, 2009.

[SSA⁺13]   A.K. Shalek, R. Satija, X. Adiconis, et al. Single-cell transcriptomics reveals bimodality in expression and splicing in immune cells. *Nature*, 498:236–240, 2013.

[Stu03]   K.-T. Sturm. Probability measures on metric spaces of nonpositive. In P. Auscher, T. Coulhon, and A, Grigoryan, editors, *Heat Kernels and Analysis on Manifolds, Graphs, and Metric Spaces*, vol. 338 of *Contemp. Math.*, pages 357-390, AMS, Providence, 2003.

[TCG⁺14]   C. Trapnell, D. Cacchiarelli, J. Grimsby, P. Pokharel, S. Li, M. Morse, N.J. Lennon, K.J. Livak, T.S. Mikkelsen, and J.L. Rinn. The dynamics and regulators of cell fate decisions are revealed by pseudotemporal ordering of single cells. *Nature Biotechnology*, 32:381–386, 2014.

[TKR09]   V. Trifonov, H. Khiabanian, and R. Rabadan. Geographic dependence, surveillance, and origins of the 2009 influenza A (H1N1) virus. *New England J. Medicine*, 361:115–119, 2009.

[TMMH14]   K. Turner, Y. Mileyko, S. Mukherjee, and J. Harer. Fréchet means for distributions of persistence diagrams. *Discrete Comput. Geom.*, 52:44–70, 2014.

[VAM+01]   J.C. Venter, M.D. Adams, E.W. Myers, et al. The sequence of the human genome. *Science*, 291:1304–1351, 2001.

[VHV+02]   M.J. van de Vijver, Y.D. He, L.J. van't Veer, et al. A gene-expression signature as a predictor of survival in breast cancer. *New England J. Medicine*, 347:1999–2009, 2002.

[Wak09]   J. Wakeley. *Coalescent Theory*. Roberts & Company, Greenwood Village, 2009.

[WBG+92]   R.G. Webster, W.J. Bean, O.T. Gorman, T.M. Chambers, and Y. Kawaoka. Evolution and ecology of influenza a viruses. *Microbiological Reviews*, 56:152–179, 1992.

[WCL+16]   J. Wang, E. Cazzato, E. Ladewig, et al. Clonal evolution of glioblastoma under therapy. *Nature Genetics*, 48:768–776, 2016.

[ZC05]   A. Zomorodian and G. Carlsson. Computing persistent homology. *Discrete Comput. Geom.*, 33:249–274, 2005.

[ZKBR14]   S. Zairis, H. Khiabanian, A.J. Blumberg, and R. Rabadan. Moduli spaces of phylogenetic trees describing tumor evolutionary patterns. In *Proc. Int. Conf. Brain Informatics Health*, vol. 8609 of *LNCS*, pages 528–539, Springer, Berlin, 2014.

[ZKBR16]   S. Zairis, H. Khiabanian, A.J. Blumberg, and R. Rabadan. Genomic data analysis in tree spaces. Preprint, **arXiv:1607.07503**, 2016.

Part VIII

GEOMETRIC SOFTWARE

67 SOFTWARE

Michael Joswig and Benjamin Lorenz

INTRODUCTION

This chapter is intended as a guide through the ever-growing jungle of geometry software. Software comes in many guises. There are the fully fledged systems consisting of a hundred thousand lines of code that meet professional standards in software design and user support. But there are also the tiny code fragments scattered over the Internet that can nonetheless be valuable for research purposes. And, of course, the very many individual programs and packages in between.

Likewise today we find a wide group of users of geometry software. On the one hand, there are researchers in geometry, teachers, and their students. On the other hand, geometry software has found its way into numerous applications in the sciences as well as industry. Because it seems impossible to cover every possible aspect, we focus on software packages that are interesting from the researcher's point of view, and, to a lesser extent, from the student's.

This bias has a few implications. Most of the packages listed are designed to run on UNIX/Linux[1] machines. Moreover, the researcher's genuine desire to understand produces a natural inclination toward open source software. This is clearly reflected in the selections below. Major exceptions to these rules of thumb will be mentioned explicitly.

In order to keep the (already long) list of references as short as possible, in most cases only the Web address of each software package is listed rather than manuals and printed descriptions, which can be found elsewhere. At least for the freely available packages, this allows one to access the software directly. On the other hand, this may seem careless, since some Web addresses are short-lived. This disadvantage usually can be compensated by relying on modern search engines.

The chapter is organized as follows. We start with a discussion of some technicalities (independent of particular systems). Since, after all, a computer is a technical object, the successful use of geometry software may depend on such things. The main body of the text consists of two halves. First, we browse through the topics of this Handbook. Each major topic is linked to related software systems. Remarks on the algorithms are added mostly in areas where many implementations exist. Second, some of the software systems mentioned in the first part are listed in alphabetical order. We give a brief overview of some of their features. The libraries CGAL [F+16] and LEDA [led] are discussed in depth in Chapter 68. In particular, the CGAL implementations for the fundamental geometric algorithms in dimensions

---

[1]No attempt is made to comment on differences between various UNIX platforms. According to distrowatch.com, today's default UNIX-like platform is probably close to Debian/Ubuntu Linux. Most other Linux distributions are not very different for our purposes. FreeBSD and its derivative MacOS X come quite close. Restricting to one kind of operating system is less of a problem nowadays. This is due to a multitude of virtualization concepts that provide the missing links among several platforms. We would like to mention docker [doc] as one project that is becoming increasingly popular.

two and three mark the state of the art and should be considered as the standard reference.

This chapter is a snapshot as of early 2016. It cannot be complete in any sense. Even worse, the situation is changing so rapidly that the information given will be outdated soon. All this makes it almost impossible for the nonexpert to get any impression of what software is available. Therefore, this is an attempt to provide an overview in spite of the obvious complications. For historical reasons we kept pointers to some codes or web pages, which can be interesting, even if they are a bit outdated.

GLOSSARY

Software can have various forms from the technical point of view. In particular, the amount of technical knowledge required by the user varies considerably. The notions explained below are intended as guidelines.

Stand-alone software: This is a program that usually comes "as-is" and can be used immediately if properly installed. No programming skills are required.

Libraries: A collection of software components that can be accessed by writing a main program that calls functions implemented in the library. Good libraries come with example code that illustrates how (at least some of) the functions can be used. However, in order to exploit all the features, the user is expected to do some programming work. On the other hand, libraries have the advantage that they can be integrated into existing code. Some stand-alone programs can also be used as libraries; if they appear in this category, too, then there are substantial differences between the two versions, or the library has additional functionality.

General-purpose systems and their modules: Computer algebra systems like `Maple` [map15], `Mathematica` [mat15] and `SageMath` [S$^+$16] form integrated software environments with elaborate user interfaces that incorporate numerous algorithms from essentially all areas of mathematics. This chapter lists only functionality or extensions that the authors find particularly noteworthy.

Additional Web services: There are very many software overviews on the Web. A few of them that are focused on a specific topic are mentioned in the main text. Some webpages offer additional pieces of source code. Even better, nowadays websites may offer access to some compute server running in the background. This is one of the two most common ways for mathematical software to arrive at mobile devices (the other being `JavaScript`). A short list of particularly useful websites is given further below.

GENERAL SOURCES

For each of the major general-purpose computer algebra systems there exists a website with many additional packages and individual solutions. See the Web addresses of the respective products. In particular, `Sage` [S$^+$16] offers a free cloud service that is suitable for collaborative work. A particularly powerful general source is `Wolfram Alpha` [Wol16], which employs `Mathematica` [mat15] and calls itself a "computational knowledge engine." It is not even restricted to mathematics.

There are several major websites that are of general interest to the discrete and computational geometry community. Some of them also collect references to software, which are updated more or less frequently. One popular example is Eppstein's "Geometry Junkyard" [Epp12].

For those who are beginning to learn how to develop geometry software it will probably be too hard to do so by reading the source code of mature systems only. O'Rourke's book [O'R98] can help fill this gap. Its numerous example programs in C and Java are also electronically available [O'R00].

"The Stony Brook Algorithm Repository" maintained by Skiena [Ski01] is still useful to some extent. Section 1.6 is dedicated to computational geometry, and it contains links to implementations.

In addition to software for doing geometry it is sometimes equally important to get at interesting research data. A particularly comprehensive, and fast growing, site is `Imaginary` [ima]. The `DGD Gallery`, where "DGD" is short for "Discretization in Geometry and Dynamics," is a new site with geometric models of various kinds [dgd].

ARITHMETIC

Depending on the application, issues concerning the arithmetic used for implementing a geometric algorithm can be essential. Using any kind of exact arithmetic is expensive, but the overhead induced also strongly depends on the application. A principal choice for exact arithmetic is *unlimited precision integer* or *rational* arithmetic as implemented in the `GNU Multiprecision Library` (GMP) [gmp15]. Several projects also use the `Multiple Precision Integers and Rationals` (MPIR) [mpi15] that was forked from `GMP`. They are implemented as C-libraries, but they can be used from many common programming languages via corresponding bindings. However, some problems require nonrational constructions. To cover such instances, libraries like `Core` [YD10] and LEDA ([led], Chapter 68) offer special data types that allow for exact computation with certain radical expressions.

Geometric algorithms often rely on a few primitives like: Decide whether a point is on a hyperplane or, if not, tell which side it lies on. Thus exact coordinates for geometric objects are sometimes less important than their true relative position. It is therefore natural to use techniques like interval arithmetic. This may be based on packages for multiple precision floating-point numbers such as `GNU Multiple Precision Floating-Point Reliably` (MPFR) [mpf16]. *Floating-point filters* can be understood as an improved kind of interval arithmetic that employs higher precision or exact methods if needed. For more detailed information see Chapter 45.

Yet another arithmetic concept is the following: Compute with machine size integers but halt (or trigger an exception) if an overflow occurs. Typically such an implementation depends on the hardware and thus requires at least a few lines of assembler code. Useful applications for such an approach are situations where the overflow signals that the computation is expected to become too large to finish in any reasonable amount of time. For instance, `hull` [Cla] uses exact integer arithmetic for convex hull computation and signals an overflow. Going one step further, `Normaliz` [BSSR16] first computes with machine size integers but when an overflow occurs only the corresponding computation step, e.g., a vector or matrix multiplication, will be restarted with arbitrary precision. When the result still does not fit the machine size, the whole computation will be restarted with arbitrary precision integers.

Instead of using a form of exact arithmetic, some implementations perform combinatorial post-processing in order to repair flawed results coming from rounding errors. An example is the convex hull code `qhull` by Barber, Dobkin, and Huhdanpaa [BDH16]. Usually, this is only partially successful; see the discussion on the corresponding Web page [BDH01] in `qhull`'s documentation.

FURTHER TECHNICAL REMARKS

While the programming language in which a software package is implemented often does not affect the user, this can obviously become an issue for the administrator who does the installation. Many of the software systems listed below are distributed as source code written in C or C++. Additionally, some of the larger packages are offered as precompiled binaries for common platforms.

C is usually easy. If the source code complies with the ANSI standard, it should be possible to compile it with any C compiler. Some time ago the situation was different for C++. However, current compilers (e.g., `gcc` and `clang`) quickly converge to the C++14 and C++1z standards.

While it is by no means a new programming language, in recent years Python became increasingly popular. For basic mathematical computations there are some interesting libraries, including `numpy` and `scipy`. Further, `SageMath` uses (the old version 2 of) Python as its scripting language. Currently, there is a bit of a hype around the Julia language, which is interesting for doing mathematics and became popular for scientific computing. For computational geometry there is a GitHub site [jul16] with code for basic constructions.

JavaScript deserves to be mentioned, too. Currently, this is the only programming language supported by most web browsers without the need of plugins or extensions. This makes it particularly interesting for software to be deployed on mobile devices.

67.1 SOFTWARE SORTED BY TOPIC

This section should give a first indication of what software to use for solving a given problem. The subsections reflect the overall structure of the whole Handbook. References to `CGAL` [F+16] and `LEDA` [led] usually are omitted, since these large projects are covered in detail in Chapter 68.

67.1.1 COMBINATORIAL AND DISCRETE GEOMETRY

This section deals with software handling the combinatorial aspects of finitely many objects, such as points, lines, or circles, in Euclidean space. Polytopes are described in Section 67.1.2.

STAND-ALONE SOFTWARE

The simplest geometric objects are clearly points. Therefore, essentially all geometry software can deal with them in one way or another. A key concept to many nontrivial properties of finite point sets in $\mathbb{R}^d$ is the notion of an *oriented matroid*.

For oriented matroid software and the computation of the set of all triangulations of a given point set, see TOPCOM by Rambau [Ram14]. In order to have correct combinatorial results, arbitrary precision arithmetic is essential.

Stephenson's CirclePack [Ste09] can create, manipulate, store, and display circle packings.

Lattice points in convex polytopes are related to combinatorial optimization and volume computations; see Beck and Robins [BR07]. Further there is a connection to commutative algebra, e.g., via monomial ideals and Gröbner bases; see Bruns and Gubeladze [BG09]. Software packages that specialize in lattice point computations include barvinok [Ver16], LattE [BBDL+15], and Normaliz [BSSR16]. Moreover, polymake offers interfaces to several of the above and adds further functionality. Various volume computation algorithms for polytopes, using exact and floating-point arithmetic, are implemented in vinci by Büeler, Enge, and Fukuda [BEF03].

Dynamic geometry software allows the creation of geometrical constructions from points, lines, circles, and so on, which later can be rearranged interactively. Objects depending, e.g., on intersections, change accordingly. Among other features, such systems can be used for visualization purposes and, in particular, also for working with polygonal linkages. An open source system in this area is GeoGebra [Geo15]. Commercial products include Cabrilog's Cabri 3D [Cab07], the Geometer's Sketchpad [Geo13] as well as Cinderella [RGK16] by Kortenkamp and Richter-Gebert.

Graph theory certainly is a core topic in discrete mathematics and therefore naturally plays a role in discrete and computational geometry. There is an abundance of algorithms and software packages, but they are not especially well suited to geometry, and so they are skipped here. Often symmetry properties of geometric objects can be reduced to automorphisms of certain graphs. While the complexity status of the graph isomorphism problem remains open, bliss [JK15] by Junttila and Kaski or McKay's nauty [MP16] work quite well for many practical purposes.

LIBRARIES

Ehrhart polynomials and integer points in polytopes are also accessible via Loechner's PolyLib [Loe10] and the Integer Set Library [isl16].

ADDITIONAL WEB SERVICES

The new VaryLab web site [RSW] offers interactive functions for surface optimization. This includes, e.g., discrete conformal mappings of euclidean, spherical, and hyperbolic surfaces based on circle patterns and circle packings. For polyominoes, see Eppstein's Geometry Junkyard [Epp03] and Chapter 14.

67.1.2 POLYTOPES AND POLYHEDRA

In this section we discuss software related to the computational study of convex polytopes. The distinction between polytopes and unbounded polyhedra is not essential since, up to a projective transformation, each polyhedron is the Minkowski sum of an affine subspace and a polytope.

A key problem in the algorithmic treatment of polytopes is the convex hull problem, which is addressed in Section 67.1.4.

STAND-ALONE SOFTWARE

polymake [GJ+16] is a comprehensive framework for dealing with polytopes in terms of vertex or facet coordinates as well as on the combinatorial level. The system offers a wide functionality, which is not restricted to polytopes and which is further augmented by interfacing to many other programs operating on polytopes. Among the combinatorial algorithms implemented is a fast method for enumerating all the faces of a polytope given in terms of the vertex-facet incidences by Kaibel and Pfetsch [KP02].

Normaliz by Bruns et al. [BSSR16] was initially designed for combinatorial computations in commutative algebra. However, it offers many functions for vector configurations, lattice polytopes, and rational cones.

Triangulations of polytopes can be rather large and intricate. Rambau's TOP-COM [Ram14] is primarily designed to examine the set of all triangulations of a given polytope (or arbitrary point configurations). Pfeifle and Rambau [PR03] combined TOPCOM with polymake to compute secondary fans and secondary polytopes; see also Section 16.6. Gfan by Jensen is software for dealing with Gröbner fans and derived objects, but it can also compute secondary fans [Jen].

The combinatorial equivalence of polytopes can be reduced to a graph isomorphism problem. As mentioned above, graph isomorphism can be checked by nauty [MP16] or bliss [JK15].

The Geometry Center's Geomview [Geo14] or jReality [WGB+09] can both be used for, among others, the visualization of 3-polytopes and (Schlegel diagrams of) 4-polytopes.

LIBRARIES

Normaliz [BSSR16] is, in fact, primarily a C++ library, but it comes with executables ready for use. The Parma Polyhedral Library (PPL) offers a fast implementation of the double description method for convex hull computations [BHZ16]. PolyLib [Loe10] is a library for working with rational polytopes; it is primarily designed for computing Ehrhart polynomials. polymake [GJ+16] comes with a C++ template library for container types that extend the Standard Template Library (STL). This allows one to access all the functionality, including the interfaced programs, from the programmer's own code. Further, the library offers a variety of container classes suitable for the manipulation of polytopes.

GENERAL PURPOSE SYSTEMS AND THEIR MODULES

convex by Franz [Fra09] is a package for the investigation of rational polytopes and polyhedral fans in Maple.

ADDITIONAL WEB SERVICES

The website Polyhedral.info [pol] contains a long list of codes that are related to polyhedral geometry and applications.

67.1.3 COMBINATORIAL AND COMPUTATIONAL TOPOLOGY

Recent years saw an increasing use of methods from computational topology in discrete and computational geometry. A basic operation is to compute the homology of a finite simplicial complex (or similar). Although polynomial time methods (in the size of the boundary matrices) are known for most problems, the (worst case exponential) elimination methods seem to be superior in practice; see Dumas et al. [DHSW03]. Implementations with a focus on the combinatorics include the GAP package simpcomp by Effenberger and Spreer [ES16] as well as polymake [GJ$^+$16].

STAND-ALONE SOFTWARE

The Computational Homology Project CHomP [cho] is specifically designed for the global analysis of nonlinear spaces and nonlinear dynamics.

SnapPea by Weeks [Wee00] is a program for creating and examining hyperbolic 3-manifolds. On top of this, Culler and Dunfield built SnapPy (which is pronounced ['snæpaɪ]) [CDW]. A comprehensive system for normal surface theory of 3-manifolds is Regina by Burton and his co-authors [BBP$^+$14].

Geomview's [Geo14] extension package Maniview can be used to visualize 3-manifolds from within.

LIBRARIES

Persistent homology is a particularly powerful computational topology technique applicable to a wide range of applications; see Chapter 24. PHAT is a C++ library with methods for computing the persistence pairs of a filtered cell complex represented by an ordered boundary matrix with mod 2 coefficients [BKR]. Another C++ library with a similar purpose is Dionysus [Mor]. A somewhat broader software project for topological data analysis is GUDHI [BGJV], and it also does persistence. For those who prefer Java over C or C++ there is JavaPlex [TVJA14].

ADDITIONAL WEB SERVICES

The CompuTop.org Software Archive of Dunfield [Dun] collects software for low-dimensional topology.

67.1.4 ALGORITHMS FOR FUNDAMENTAL GEOMETRIC OBJECTS

The computation of convex hulls and Delaunay triangulations/Voronoi diagrams is of key importance. For correct combinatorial output it is crucial to rely on arbitrary-precision arithmetic. On the other hand, some applications, e.g., volume computation, are content with floating-point arithmetic for approximate results. Some algorithmically more advanced but theoretically yet basic topics in this section are related to topology and real algebraic geometry.

In our terminology the *convex hull problem* asks for enumerating the facets of the convex hull of finitely many points in $\mathbb{R}^d$. The dual problem of enumerating the vertices and extremal rays of the intersection of finitely many halfspaces is equivalent by means of cone polarity. There is the related problem of deciding which points among a given set are extremal, that is, vertices of the convex hull. This can be solved by means of linear optimization.

STAND-ALONE SOFTWARE

Many convex hull algorithms are known, and there are several implementations. However, there is currently no algorithm for computing the convex hull in time polynomial in the combined input and output size, unless the dimension is considered constant. The behavior of each known algorithm depends greatly on the specific combinatorial properties of the polytope on which it is working. One way of summarizing the computational results from Avis, Bremner, and Seidel [ABS97] and Assarf et al. [AGH+15] is: Essentially for each known algorithm there is a family of polytopes for which the given algorithm is superior to any other, and there is a second family for which the same algorithm is inferior to any other. For these families of polytopes we do have a theoretical, asymptotic analysis that explains the empirical results; see Chapter 26. Moreover, there are families of polytopes for which none of the existing algorithms performs well. Which algorithm or implementation works best for certain purposes will thus depend on the class of polytopes that is typical in those applications. For an overview of general convex hull codes see Table 67.1.1.[2] Note that `cddlib` and `lrslib` are both listed under exact arithmetic but can also be used with floating-point arithmetic.

Additionally, there are specialized codes: `azove` by Behle [Beh07] is designed to compute the vertices of a polytope with 0/1-coordinates from an inequality description by iteratively solving linear programs.

TABLE 67.1.1   Overview of convex hull codes.

Exact arithmetic		
PROGRAM	ALGORITHM	REMARKS
`beneath_beyond`	beneath-beyond [Ede87, 8.3.1]	Part of `polymake` [GJ+16]
`cddlib` [Fuk07]	double description [JT13, 5.2]	
`lrslib` [Avi15]	reverse search [AF92]	
`normaliz` [BSSR16]	pyramid decomposition [BIS16]	
`porta` [CL09]	double description	
`ppl` [BHZ16]	double description	
Non-exact arithmetic		
PROGRAM	ALGORITHM	REMARKS
`cddf+` [Fuk07]	double description	
`hull` [Cla]	randomized incremental [CMS93]	Assumes input in gen. pos.; Exact computation unless Overflow signaled
`qhull` [BDH16]	quickhull [BDH96]	

The computation of Delaunay triangulations in d dimensions can be reduced to a $(d+1)$-dimensional convex hull problem; see Section 26.1. Thus, in principle, each of the convex hull implementations can be used to generate Voronoi diagrams. Additionally, however, some codes directly support Voronoi diagrams, notably Clarkson's `hull` [Cla], `qhull` by Barber, Dobkin, and Huhdanpaa [BDH16], and, among the programs with exact rational arithmetic, `lrs` by Avis [Avi15].

The following codes are specialized for 2-dimensional Voronoi diagrams: Shewchuk's `Triangle` [She05] and Fortune's `voronoi` [For01]. See also `cdt` by Lischinski [Lis98] for incremental constrained 2-dimensional Delaunay triangulation. For 3-dimensional problems there is `Detri` by Mücke [Müc95]. Delaunay triangulations

---

[2]We call an implementation *exact* if it, intentionally (but there may be programming errors, of course), gives correct results for *all* possible inputs. The nonexact convex hull codes use floating-point arithmetic or more advanced methods, but for each of them some input is known that makes them fail. The quality of the output of the nonexact convex hull codes varies considerably.

and, in particular, constrained Delaunay triangulations, play a significant role in meshing. Therefore, several of the Voronoi/Delaunay packages also have features for meshing and vice versa.

For the special case of triangulating a simple polygon (Chapter 30), Seidel's randomized algorithm has almost linear running time. The implementation by Narkhede and Manocha is part of the **Graphics Gems** [KHP+13, Part V]. This archive and also Skiena's collection of algorithms [Ski01] contain more specialized code and algorithms for polygons.

Mesh generation is a vast area with numerous applications; see Chapter 29. This is reflected by the fact that there is an abundance of commercial and noncommercial implementations. We mention only a few. From the theoretical point of view the main categories are formed by 2-dimensional triangle meshes, 2-dimensional quadrilateral meshes, 3-dimensional tetrahedral meshes, 3-dimensional cubical (also called hexahedral) meshes, and other structured meshes. A focus on the applications leads to entirely different categories, which is completely ignored here. **Triangle** produces triangle meshes. **QMG** is a program for quadtree/octree 2- and 3-dimensional finite element meshing written by Vavasis [Vav00]. **Trelis** [tre14] can do many different variants of 2- and 3-dimensional meshing; it is a commercial product that is free for scientific use. Note that, depending on the context, triangle or tetrahedra meshes are also called triangulations.

In applications geometric objects are sometimes given as point clouds meant to represent a curve or surface. With the introduction of 3D-scanners and similar devices, appropriate techniques and related software became increasingly important. Obviously, this problem is directly related to mesh generation. **Cocone** by Dey et al. [DGG+02] and **Power Crust** by Amenta, Choi, and Kolluri [ACK02] are designed to produce "watertight" surfaces; see Chapter 35. Geomagic is a company that sells many software products related (not only) to meshing [geo]. For instance, **Geomagic Wrap** generates meshes from 3D-scans.

VisPak by Wismath et al. [W+02] is built on top of **LEDA** and can be used for the generation of visibility graphs of line segments and several kinds of polygons.

Smallest enclosing balls of a point set in arbitrary dimension can be computed with Gärtner's **Miniball** [Gär13].

Software at the junction between convexity and real algebraic geometry is still scarce. A noticeable exception is **Axel** by Mourrain and Wintz [MW]. It is designed for dealing with semi-algebraic sets in a way that is common in geometric modeling.

The computer algebra system **Magma** by Cannon et al. [C+16] has some basic support for real algebraic geometry. Visualization of curves and surfaces can be done with **surf** by Endrass [End10].

LIBRARIES

Most of the above-mentioned software systems for dealing with polytopes and convex hulls are available as libraries of various flavors. **cddlib** [Fuk15] and **lrslib** [Avi15] are the C library versions of Fukuda's **cdd** and Avis's **lrs**, respectively. The **Parma Polyhedra Library** [BHZ16] and **Normaliz** [BSSR16] primarily *are* libraries, written in C++. **polymake** can be used as a C++ library, too. All of the above offer exact convex hull computation and exact linear optimization.

There is a C++ library version of **qhull** [BDH16] that performs convex hulls and Voronoi diagrams in floating-point arithmetic. Moreover, **cddlib** and **polymake** also have a limited support for floating-point arithmetic.

The computation of Voronoi diagrams, arrangements, and related information is a particular strength of CGAL [F⁺16] and LEDA [led]. See Chapter 68.

The Quickhull algorithm (in three dimensions) is implemented as a Java class library [Llo04], and this has been wrapped for JavaScript in the CindyJS project [cin].

There is a parallelized C++ library MinkSum to enumerate the vertices of a Minkowski sum by Weibel [Wei12].

For triangle meshes in $\mathbb{R}^3$ there is the GNU Triangulated Surface Library [gts06] written in C. Its functionality comprises dynamic Delaunay and constrained Delaunay triangulations, robust set operations on surfaces, and surface refinement and coarsening for the control of level-of-detail.

Bhaniramka and Wenger have a set of C++ classes for the construction of isosurface patches in convex polytopes of arbitrary dimension [BW15]. These can be used in marching cubes like algorithms for isosurface construction.

GENERAL PURPOSE SYSTEMS AND THEIR MODULES

Plain Maple [map15] and Mathematica [mat15] only offer 2-dimensional convex hulls and Voronoi diagrams. Higher-dimensional convex hulls can be computed via the Maple package convex [Fra09]. Additionally, the Matlab package GeoCalcLib by Schaich provides an interface to lrs [Sch16].

Mitchell [Mit] has implemented some of his algorithms related to mesh generation in Matlab [mat16]. The finite element meshing program QMG by Vavasis can also be used with Matlab.

Particularly interesting for real algebraic geometry is GloptiPoly [HLL09], a Matlab package by Henrion et al., which, e.g., can find global optima of rational functions.

ADDITIONAL WEB SERVICES

Emiris maintains a Web page [Emi01] with several programs that address problems related to convex hull computations and applications in elimination theory. The web site Polyhedral.info [pol] was already mentioned.

polymake can be tried directly in the web browser from its web page [GJ⁺16]; it offers a front end to a server.

A Web page [Sch] by Schneiders contains a quite comprehensive survey on software related to meshing.

67.1.5 GEOMETRIC DATA STRUCTURES AND SEARCHING

LIBRARIES

Geometric data structures form the core of the C++ libraries CGAL [F⁺16] and LEDA [led]. The algorithms implemented include several different techniques for point location, collision detection, and range searching. See Chapter 68.

As already mentioned above, graph theory plays a role for some of the more advanced geometric algorithms. Several libraries for working with graphs have been developed over the years. It is important to mention in this context the Boost Graph Library [SLL15]. This is part of a general effort to provide free peer-reviewed portable C++ source libraries that extend the STL.

67.1.6 COMPUTATIONAL TECHNIQUES

PARALLELIZATION

One important computational technique that is used in various contexts is parallelization. There are different levels at which this can be employed. At the very low level, modern processors can operate on large data vectors, up to 512 bits, in a single instruction, e.g., via the MMX, SSE, or AVX extensions to the x86 architecture.

At the next level the tasks of an algorithm can be split up and run in parallel on current multi-core machines. A widely used API for this approach is OpenMP [Ope15b], which provides preprocessor statements for easy (shared-memory) parallelization of C, C++, or Fortran code, e.g., parallel loop execution. The OpenMP framework can also be used to some extent on general-purpose graphics processing units (GPGPU).

Parallelization on graphics processing units can provide a large number of processing units, but those are rather limited in the instruction set and memory access. Common frameworks for this are NVIDIA's CUDA [Nvi15] or the Open Computing Language (OpenCL) [Ope15a].

A different approach to achieve massive parallelization is via the Message Passing Interface (MPI), which is implemented for example in OpenMPI [Ope15c]. This allows us to spread the jobs to a large number of nodes connected via low-latency interconnect.

As an example, for the lrs code by Avis [Avi15] there is a wrapper plrs for shared-memory parallelization that works well up to about 16 threads. Moreover, there is also an MPI-based wrapper mplrs that can scale up to 1200 threads [AJ15]. Other libraries that also employ OpenMP are Normaliz [BSSR16] and PHAT [BKR].

67.1.7 APPLICATIONS

Applications of computational geometry are abundant and so are the related software systems. Here we list only very few items that may be of interest to a general audience.

STAND-ALONE SOFTWARE

For linear programming problems, essential choices for algorithms include simplex type algorithms or interior point methods. While commercial solvers tend to offer both, the freely available implementations seem to be restricted to either one. Additionally, there are implementations of a few special algorithms for low dimensions that belong to neither category. Altogether there are a large number of implementations, and we can only present a tiny subset here.

Exact rational linear programming can be done with cdd [Fuk07]. It uses either a dual simplex algorithm or the criss-cross method. An alternative exact linear programming code is lrs [Avi15], which implements a primal simplex algorithm.

SoPlex by Wunderling et al. [W+16] implements the revised simplex algorithm both in primal and dual form; its most recent version supports exact rational arithmetic as well as floating-point arithmetic. It is part of the SCIP Optimization

Suite [sci16]. For an implementation of interior point methods see PCx by Czyzyk et al. [CMWW06]. Since the interior point method relies on numerical algorithms (e.g., Newton iterations) such implementations are always floating-point.

Glop is Google's linear solver, and it is available as open source [glo], but it can also be used via various Google services.

CPLEX [cpl15], Gurobi [gur], and XPress [xpr15] are widespread commercial solvers for linear, integer, and mixed integer programming. Each program offers a wide range of optimization algorithms. However, none of the commercial products can do exact rational linear optimization.

See also Hohmeyer's code linprog [Hoh96] for an implementation of Seidel's algorithm. This is based on randomization, and it takes expected linear time in fixed dimension; see Section 49.4.

There is an abundance of software for solid modeling. In the open source world Blender [ble16] seems to set the standard. For solid modeling with semi-algebraic sets, use Axel [MW].

Another topic with many applications is graph drawing. GraphViz [gra14] is an extensible package that offers tailor-made solutions for a wide range of applications in this area. An alternative is the Open Graph Drawing Framework [ogd15]. Tulip [AB+16] specializes in the visualization of large graphs.

LIBRARIES

cddlib [Fuk15] and lrs offer C libraries for exact LP solving. CPLEX, Gurobi, OSL, PCx, and XPress can also be used as C libraries, while SoPlex has a C++ library version. Other free C libraries for linear and mixed integer programming include GLPK [Mak16] and lpsolve [Ber13].

GDToolkit [gdt07] is a C++ library for graph drawing, which is free for academic use.

In order to meet certain quality criteria, post-processing of mesh data is important. Varylab is a tool for the optimization of polygonal surfaces according to a variety of criteria [RSW].

GENERAL PURPOSE SYSTEMS AND THEIR MODULES

The linear optimization package PCx comes with an interface to Matlab [mat16].

ADDITIONAL WEB SERVICES

The Computational Infrastructure for Operations Research, or COIN-OR for short, is an open source project on optimization software that was initiated by IBM in 2000. Their web site contains lots of useful information and many links to (open source) software for various kinds of optimization [coi].

The geometrica project [B+15] by Boissonnat et al. studies a variety of applications of computational geometry methods.

67.2 FEATURES OF SELECTED SOFTWARE SYSTEMS

All the software packages listed here have been mentioned previously. In many cases, however, we list features not accounted for so far.

Axel [MW] is an algebraic geometric modeler for manipulation and computation with curves, surfaces or volumes described by semi-algebraic representations. These include parametric and implicit representations of geometric objects.

bliss [JK15] is a tool for computing automorphism groups and canonical forms of graphs. It has both a command line user interface as well as C++ and C programming language APIs.

cdd [Fuk07, Fuk15] is a convex hull code based on the double description method that is dual to Fourier-Motzkin elimination. It also implements a dual simplex algorithm and the criss-cross method for linear optimization. `cdd` comes as a stand-alone program; its C library version is called `cddlib`. The user can choose between exact rational arithmetic (based on the `GMP`) or floating-point arithmetic. Can be used via `SageMath`.

Cinderella [RGK16] is a commercial dynamic geometry software with a free version written in `Java`. It supports standard constructions with points, lines, quadrics and more (e.g., physics simulations). The recent `CindyJS` project [cin] aims at porting `Cinderella`'s functionality to JavaScript.

Cocone [DGG+02] is a set of programs related to the reconstruction of surfaces from point clouds in $\mathbb{R}^3$ via discrete approximation to the medial axis transform: `Tight Cocone` produces "watertight" surfaces from arbitrary input, while `Cocone/SuperCocone` is responsible for detecting the surface's boundary. `Geomview` output. Based on `CGAL` and `LEDA`. Not available for commercial use.

Computational Geometry in C [O'R98, O'R00] is a collection of C and `Java` programs including 2- and 3-dimensional convex hull codes, Delaunay triangulations, and segment intersection.

Geomview [Geo14] is a tool for interactive visualization. It can display objects in hyperbolic and spherical space as well as Euclidean space. `Geomview` comes with several external modules for specific visualization purposes. The user can write additional external modules in C. `Geomview` can be used as a visualization back end, e.g., for `Maple` [map15] and `Mathematica` [mat15]. The extension package `Maniview` can visualize 3-manifolds.

GloptiPoly [HLL09] can solve or approximate the Generalized Problem of Moments (GPM), an infinite-dimensional optimization problem that can be viewed as an extension of the classical problem of moments. This allows one to attack a wide range of optimization problems, including finding optima of multi-variate rational functions.

GraphViz [gra14] is a package with various graph layout tools. This includes hierarchical layouts and spring embedders. The system comes with a customizable graphical interface. Also runs on Windows.

GUDHI [BGJV] is a generic C++ library for computational topology and topological data analysis. Features include simplicial and cubical complexes, alpha complexes, and the computation of persistent (co-)homology.

hull [Cla] computes the convex hull of a point set in general position. The program can also compute Delaunay triangulations, alpha shapes, and volumes of Voronoi regions. The program uses exact machine floating-point arithmetic, and it signals overflow. `Geomview` output.

jReality [WGB$^+$09] is a Java-based, full-featured 3D scene graph package designed for 3D visualization and specialized in mathematical visualization. It provides several backends, including a JOGL one for Java-based OpenGL rendering.

LattE integrale [BBDL$^+$15] is software for counting and enumerating lattice points in polyhedra. Its most recent extension is a hybrid C++ and Maple implementation for computing the top coefficients of weighted Ehrhart quasipolynomials.

lrs [Avi15] is a convex hull code based on the reverse search algorithm due to Avis and Fukuda [AF92]. Exact rational, e.g. via the GMP, or floating-point arithmetic. In addition to convex hull computations, lrs can do linear optimization (via a primal simplex algorithm), volume computation, Voronoi diagrams, and triangulations. Moreover, lrs can, in advance, provide estimates for the running time and output size. It also comes as a C library and with wrappers for parallelization.

nauty [MP16] can compute a permutation group representation of the automorphism group of a given finite graph. As one interesting application this gives rise to an effective method for deciding whether two graphs are isomorphic or not. Such an isomorphism test can be performed directly.

Normaliz [BSSR16] is a tool for computations in affine monoids, vector configurations, lattice polytopes, and rational cones. Some algorithms are parallelized via the OpenMP protocol.

Parma Polyhedra Library (PPL) [BHZ16] provides numerical abstractions for applications to the analysis and verification of complex systems. Especially interesting for computational geometry are data types and algorithms for convex polyhedra, which may even be half-open. Fast implementation of the double description method for convex hulls. Can be used via SageMath.

PHAT [BKR] is a software library that provides methods for computing the persistence pairs of a filtered cell complex represented by an ordered boundary matrix with mod 2 coefficients. Uses OpenMP parallelization.

PolyLib [Loe10] is a library of polyhedral functions. Allows for basic geometric operations on parametrized polyhedra. As a key feature PolyLib can compute Ehrhart polynomials, which permits counting the number of integer points in a given polytope.

polymake [GJ$^+$16] is a system for examining the geometrical and combinatorial properties of polytopes. It offers convex hull computation, standard constructions, and visualization. Some of the functionality relies—via interfaces—on external programs including cdd, Geomview, jReality, lrs, nauty, PPL, and vinci. STL-compatible C++ library; computations in exact rational arithmetic based on GMP. Separate modules for simplicial complexes (with homology computation and intersection forms of 4-manifolds), matroids, and other objects.

Regina [BBP$^+$14] deals with manifolds, mostly in dimension 3, and its strength lies in normal surface theory. This way, e.g., it can be checked if a given 3-manifold satisfies the Haken property. Other features include homology and homotopy computations or Turaev–Viro invariants.

qhull [BDH16] computes convex hulls, Delaunay triangulations, Voronoi diagrams, farthest-site Delaunay triangulations, and farthest-site Voronoi diagrams. The algorithm implemented is Quickhull [BDH96]. qhull uses floating-point arithmetic only, but the authors incorporated several heuristics to improve the quality of the output. This is discussed in detail on a special Web page [BDH01] in `qhull`'s documentation; it is an important source for everyone interested in using or implementing computational geometry software based on floating-point arithmetic.

SageMath [S$^+$16] is a very comprehensive general mathematics software system, which builds on top of many existing open-source packages. These can be accessed through a common, Python-based language. Their mission statement reads as follows: "Creating a viable free open source alternative to `Magma`, `Maple`, `Mathematica` and `Matlab`."

Snappy [CDW] is a program for studying the topology and geometry of 3-manifolds, with a focus on hyperbolic structures. Can be used via `SageMath`.

SCIP [sci16] can do mixed integer programming (MIP) and mixed integer nonlinear programming (MINLP). Moreover, `SCIP` is also a framework for constraint integer programming and branch-cut-and-price. The larger `SCIP Optimization Suite` contains the `SoPlex` solver [W$^+$16] for (exact and floating-point) linear programs and more.

TOPCOM [Ram14] is a package for examining point configurations via oriented matroids. The main purpose is to investigate the set of all triangulations of a given point configuration. Symmetric point configurations can be treated more efficiently if the user provides information about automorphisms. `TOPCOM` can check whether a given triangulation is regular.

Trelis [tre14] is a commercial meshing tool for surfaces and 3-dimensional objects to be used in finite element analysis. Mesh generation algorithms include quadrilateral and triangular paving, 2- and 3-dimensional mapping, hex sweeping and multi-sweeping, and others. This replaces/contains the former `CUBIT`.

vinci [BEF03] can be seen as an experimental framework for comparing volume computation algorithms. Exact and floating-point arithmetic. Implemented are Cohen & Hickey-triangulations [CH79], Delaunay triangulations (via `cdd` or `qhull`), and others.

REFERENCES

[AB$^+$16]   D. Auber, M. Bertrand, et al. `Tulip`, version 4.8.1. `http://tulip.labri.fr/`, 2016.

[ABS97]   D. Avis, D. Bremner, and R. Seidel. How good are convex hull algorithms? *Comput. Geom.*, 7:265–301, 1997.

[ACK02]   N. Amenta, S. Choi, and R.K. Kolluri. `Power Crust`, unions of balls, and the medial axis transform, version 1.2. `http://web.cs.ucdavis.edu/~amenta/powercrust.html`, 2002.

[AF92]   D. Avis and K. Fukuda. A pivoting algorithm for convex hulls and vertex enumeration of arrangements and polyhedra. *Discrete Comput. Geom.*, 8:295–313, 1992.

[AGH$^+$15]   B. Assarf, E. Gawrilow, K. Herr, M. Joswig, B. Lorenz, A. Paffenholz, and T. Rehn. Computing convex hulls and counting integer points with `polymake`. Preprint, arXiv:1408.4653v2, 2015.

[AJ15]   D. Avis and C. Jordan. `mplrs`: A scalable parallel vertex/facet enumeration code. Preprint, arXiv:1511.06487, 2015.

[Avi15]   D. Avis. `lrs`, `lrslib`, version 6.1. `http://cgm.cs.mcgill.ca/~avis/C/lrs.html`, 2015.

[B⁺15]   J.-D. Boissonnat et al. Geometrica — geometric computing. https://team.inria.fr/geometrica/, 2015.

[BBDL⁺15]   V. Baldoni, N. Berline, J.A. De Loera, B. Dutra, M. Köppe, S. Moreinis, G. Pinto, M. Vergne, and J. Wu. A user's guide for LattE integrale v1.7.3. http://www.math.ucdavis.edu/~latte/, 2015.

[BBP⁺14]   B.A. Burton, R. Budney, W. Pettersson, et al. Regina: Software for 3-manifold topology and normal surface theory. http://regina.sourceforge.net/, 2014.

[BDH96]   C.B. Barber, D.P. Dobkin, and H. Huhdanpaa. The quickhull algorithm for convex hulls. *ACM Trans. Math. Software*, 22:469–483, 1996.

[BDH01]   C.B. Barber, D.P. Dobkin, and H.T. Huhdanpaa. Imprecision in qhull. http://www.qhull.org/html/qh-impre.htm, 2001.

[BDH16]   C.B. Barber, D.P. Dobkin, and H.T. Huhdanpaa. qhull, version 2015.2. http://www.qhull.org/, 2016.

[BEF03]   B. Büeler, A. Enge, and K. Fukuda. vinci, version 1.0.5. http://www.math.u-bordeaux1.fr/~aenge/index.php?category=software&page=vinci, 2003.

[Beh07]   M. Behle. azove — another zero one vertex enumeration tool. https://people.mpi-inf.mpg.de/~behle/azove.html, 2007.

[Ber13]   M. Berkelaar. lpsolve, version 5.5. http://lpsolve.sourceforge.net/, 2013.

[BG09]   W. Bruns and J. Gubeladze. *Polytopes, Rings, and K-theory*. Springer Monographs in Mathematics, Springer, Dordrecht, 2009.

[BGJV]   J.-D. Boissonnat, M. Glisse, C. Jamin, and R. Vincent. GUDHI (Geometric Understanding in Higher Dimensions). http://gudhi.gforge.inria.fr/.

[BHZ16]   R. Bagnara, P.M. Hill, and E. Zaffanella. The Parma Polyhedra Library, version 1.2. http://bugseng.com/products/ppl/, 2016.

[BIS16]   W. Bruns, B. Ichim, and C. Söger. The power of pyramid decomposition in Normaliz. *J. Symbolic Comput.*, 74:513–536, 2016.

[BKR]   U. Bauer, M. Kerber, and J. Reininghaus. PHAT (Persistent Homology Algorithm Toolbox), version 1.4.0. https://bitbucket.org/phat-code/phat.

[ble16]   Blender v2.77a. https://www.blender.org/, 2016.

[BR07]   M. Beck and S. Robins. *Computing the Continuous Discretely. Integer-Point Enumeration in Polyhedra*. Undergraduate Texts in Mathematics, Springer, New York, 2007.

[BSSR16]   W. Bruns, R. Sieg, C. Söger, and T. Römer. Normaliz. Algorithms for rational cones and affine monoids, version 3.1.1. http://www.math.uos.de/normaliz, 2016.

[BW15]   P. Bhaniramka and R. Wenger. IJK: Isosurface jeneration kode. http://web.cse.ohio-state.edu/research/graphics/isotable, 2015.

[C⁺16]   J. Cannon et al. Magma, version 2.21-12. http://magma.maths.usyd.edu.au/magma, 2016.

[Cab07]   Cabrilog. Cabri 3D. http://www.cabri.com/, 2007.

[CDW]   M. Culler, N.M. Dunfield, and J.R. Weeks. SnapPy, a computer program for studying the topology of 3-manifolds. http://snappy.computop.org.

[CH79]   J. Cohen and T. Hickey. Two algorithms for determining volumes of convex polyhedra. *J. Assoc. Comput. Mach.*, 26:401–414, 1979.

[cho]   Computational Homology Project. http://chomp.rutgers.edu.

[cin]   CindyJS. http://cindyjs.org.

[CL09]   T. Christof and A. Löbel. Porta, version 1.4.1. http://porta.zib.de/, 2009.

[Cla]   K.L. Clarkson. hull, Version 1.0. http://www.netlib.org/voronoi/hull.html.

[CMS93]   K.L. Clarkson, K. Mehlhorn, and R. Seidel. Four results on randomized incremental constructions. *Comput. Geom.*, 3:185–212, 1993.

[CMWW06]   J. Czyzyk, S. Mehrotra, M. Wagner, and S. Wright. PCx, version 1.1. http://pages.cs.wisc.edu/~swright/PCx/, 2006.

[coi]   COmputational INfrastructure for Operations Research. International Business Machines, http://www.coin-or.org/.

[cpl15]   IBM ILOG CPLEX Optimization Studio, Version 12.6.2. IBM Corporation, http://www-01.ibm.com/software/commerce/optimization/cplex-optimizer/, 2015.

[dgd]   DGD gallery. http://gallery.discretization.de.

[DGG+02]   T. Dey, J. Giesen, S. Goswami, J. Hudson, and W. Zhao. Cocone softwares. http://web.cse.ohio-state.edu/~tamaldey/cocone.html, 2002.

[DHSW03]   J.-G. Dumas, F. Heckenbach, D. Saunders, and V. Welker. Computing simplicial homology based on efficient Smith Normal Form algorithms. In M. Joswig and N. Takayama, editors, *Algebra, Geometry, and Software Systems*, pages 177–206, Springer, Berlin, 2003.

[doc]   docker. http://docker.com.

[Dun]   N.M. Dunfield. The CompuTop.org Software Archive. http://computop.org/.

[Ede87]   H. Edelsbrunner. *Algorithms in Combinatorial Geometry*. Springer-Verlag, Berlin, 1987.

[Emi01]   I.Z. Emiris. Computational geometry. http://www-sop.inria.fr/galaad/logiciels/emiris/soft_geo.html, 2001.

[End10]   S. Endrass. surf, version 1.0.6. http://surf.sourceforge.net, 2010.

[Epp03]   D. Eppstein. Polyominoes and other animals. http://www.ics.uci.edu/~eppstein/junkyard/polyomino.html, 2003.

[Epp12]   D. Eppstein. The Geometry Junkyard. http://www.ics.uci.edu/~eppstein/junkyard, 2012.

[ES16]   F. Effenberger and J. Spreer. simpcomp — a gap package, version 2.1.6. https://github.com/simpcomp-team/simpcomp, 2016.

[F+16]   A. Fabri et al. CGAL, version 4.8. http://www.cgal.org, 2016.

[For01]   S.J. Fortune. voronoi. http://cm.bell-labs.com/who/sjf, 2001.

[Fra09]   M. Franz. convex — a maple package for convex geometry, version 1.1.3. http://www-home.math.uwo.ca/~mfranz/convex/, 2009.

[Fuk07]   K. Fukuda. cdd+, version 0.77a. http://www.inf.ethz.ch/personal/fukudak/cdd_home/, 2007.

[Fuk15]   K. Fukuda. cddlib, version 0.94h. http://www.inf.ethz.ch/personal/fukudak/cdd_home/, 2015.

[Gär13]   B. Gärtner. Miniball, version 3.0. http://www.inf.ethz.ch/personal/gaertner/miniball.html, 2013.

[gdt07]   GDToolkit, version 4.0. Dipartimento di Informatica e Automazione, Università di Roma Tre, Rome, Italy, http://www.dia.uniroma3.it/~gdt, 2007.

[geo]   Geomagic. http://www.geomagic.com/.

[Geo13]   The Geometer's Sketchpad. McGraw-Hill Education, `http://www.keycurriculum.com/`, 2013.

[Geo14]   `Geomview`, version 1.9.5. The Geometry Center, `http://www.geomview.org`, 2014.

[Geo15]   `GeoGebra`. International GeoGebra Institute, `http://www.geogebra.org`, 2015.

[GJ+16]   E. Gawrilow, M. Joswig, et al. `polymake`, Version 3.0. `http://polymake.org`, 2016.

[glo]   `Glop`. `https://developers.google.com/optimization/lp`.

[gmp15]   GNU multiprecision arithmetic library, Version 6.1.0. `http://www.gmplib.org`, 2015.

[gra14]   `GraphViz`, version 2.38. AT&T Lab – Research, `http://www.graphviz.org/`, 2014.

[gts06]   GNU triangulated surface library, version 0.7.6. `http://gts.sourceforge.net`, 2006.

[gur]   `Gurobi` optimizer v6.5. `http://www.gurobi.com`.

[HLL09]   D. Henrion, J.-B. Lasserre, and J. Löfberg. `GloptiPoly` 3: moments, optimization and semidefinite programming. *Optim. Methods Softw.*, 24:761–779, 2009.

[Hoh96]   M. Hohmeyer. `linprog`. `http://www3.cs.stonybrook.edu/~algorith/implement/linprog/implement.shtml`, 1996.

[ima]   `Imaginary` – open mathematics. `http://imaginary.org`.

[isl16]   Integer set library, version 0.17.1. `http://isl.gforge.inria.fr/`, 2016.

[Jen]   A.N. Jensen. `Gfan` 0.5, a software system for Gröbner fans and tropical varieties. `http://home.imf.au.dk/jensen/software/gfan/gfan.html`.

[JK15]   T. Junttila and P. Kaski. `bliss`: A tool for computing automorphism groups and canonical labelings of graphs, version 0.73. `http://www.tcs.hut.fi/Software/bliss/`, 2015.

[JT13]   M. Joswig and T. Theobald. *Polyhedral and algebraic methods in computational geometry*. Universitext, Springer, London, 2013.

[jul16]   JuliaGeometry — computational geometry with Julia. `https://github.com/JuliaGeometry`, 2016.

[KHP+13]   D. Kirk, P. Heckbert, A. Paeth, et al. Graphics gems. `http://www.graphicsgems.org/`, 2013.

[KP02]   V. Kaibel and M.E. Pfetsch. Computing the face lattice of a polytope from its vertex-facet incidences. *Comput. Geom.*, 23:281–290, 2002.

[led]   `LEDA`, version 6.3. Algorithmic Solution Software GmbH, `http://www.algorithmic-solutions.com/enleda.htm`.

[Lis98]   D. Lischinski. `cdt`. `http://www.cs.huji.ac.il/~danix/code/cdt.tar.gz`, 1998.

[Llo04]   J.E. Lloyd. Class quickhull3d. `https://quickhull3d.github.io/quickhull3d/apidocs/com/github/quickhull3d/QuickHull3D.html`, 2004.

[Loe10]   V. Loechner. `PolyLib` — a library of polyhedral functions, version 5.22.5. `http://icps.u-strasbg.fr/PolyLib`, 2010.

[Mak16]   A. Makhorin. GNU linear programming kit, version 4.60. `http://www.gnu.org/software/glpk/glpk.html`, 2016.

[map15]   `Maple` 2015. Waterloo Maple, Inc., `http://www.maplesoft.com/products/maple/`, 2015.

[mat15]   `Mathematica`, version 10. Wolfram Research, Inc., `http://www.wolfram.com/mathematica/`, 2015.

[mat16]   Matlab, R2016a.   The Mathworks, Inc., `http://www.mathworks.com/products/matlab`, 2016.

[Mit]   S.A. Mitchell. Computational geometry triangulation results. `http://www.sandia.gov/~samitch/csstuff/csguide.html`.

[Mor]   D. Morozov. `Dionysus`. `http://www.mrzv.org/software/dionysus/`.

[MP16]   B. McKay and A. Piperno. `nauty`, version 2.6r5. `http://pallini.di.uniroma1.it/`, 2016.

[mpf16]   GNU multiple precision floating-point reliably, version 3.1.4. `http://www.mpfr.org`, 2016.

[mpi15]   Multiple Precision Integers and Rationals, version 2.7.2. `http://www.mpir.org`, 2015.

[Müc95]   E.P. Mücke. `Detri`, version 2.6a. `http://www.geom.uiuc.edu/software/cglist/GeomDir`, 1995.

[MW]   B. Mourrain and J. Wintz. `Axel` – algebraic geometric modeling. `http://axel.inria.fr`.

[Nvi15]   Nvidia Corporation. `CUDA`. `http://www.nvidia.com/object/cuda_home.html`, 2015.

[ogd15]   `OGDF`: Open graph drawing framework. `http://www.ogdf.net`, 2015.

[Ope15a]   Open compute language (OpenCL). Khronos Group, `http://www.khronos.org/opencl/`, 2015.

[Ope15b]   Open multi-processing (OpenMP). OpenMP Architecture Review Board, `http://www.openmp.org/`, 2015.

[Ope15c]   Open MPI. `http://www.open-mpi.org/`, 2015.

[O'R98]   J. O'Rourke. *Computational Geometry in C*, second edition. Cambridge University Press, Cambridge, 1998.

[O'R00]   J. O'Rourke. Computational geometry in C. `http://cs.smith.edu/~orourke/books/ftp.html`, 2000.

[pol]   Polyhedral.info. `http://polyhedral.info/software`.

[PR03]   J. Pfeifle and J. Rambau. Computing triangulations using oriented matroids. In M. Joswig and N. Takayama, editors, *Algebra, Geometry, and Software Systems*, pages 49–75, Springer, Berlin, 2003.

[Ram14]   J. Rambau. `TOPCOM`, version 0.17.5. `http://www.rambau.wm.uni-bayreuth.de/TOPCOM`, 2014.

[RGK16]   J. Richter-Gebert and U.H. Kortenkamp. `Cinderella`, version 2.9. `http://www.cinderella.de`, 2016.

[RSW]   T. Roerig, S. Sechelmann, and M. Wahren. `VaryLab`: discrete surface optimization. `http://www.varylab.com`.

[S⁺16]   W.A. Stein et al. Sage Mathematics Software, version 7.1. The Sage Development Team, `http://www.sagemath.org`, 2016.

[Sch]   R. Schneiders. Software: list of public domain and commercial mesh generators. `http://www.robertschneiders.de/meshgeneration/software.html`.

[Sch16]   R. Schaich. `GeoCalcLib`. `http://worc4021.github.io/`, 2016.

[sci16]   SCIP optimization suite 3.2.1. `http://scip.zib.de/`, 2016.

[She05]   J.R. Shewchuk. `Triangle`, version 1.6. `http://www.cs.cmu.edu/~quake/triangle.html`, 2005.

[Ski01]   S.S. Skiena. The Stony Brook Algorithm Repository. http://www3.cs.stonybrook.edu/~algorith/index.html, 2001.

[SLL15]   J. Siek, L.-Q. Lee, and A. Lumsdaine. The boost graph library (bgl), version 1.60.0. http://www.boost.org/libs/graph/doc/index.html, 2015.

[Ste09]   K. Stephenson. CirclePack, version 2.0. http://www.math.utk.edu/~kens/CirclePack, 2009.

[tre14]   Trelis advanced meshing software. http://www.csimsoft.com/trelis.jsp, 2014.

[TVJA14]   A. Tausz, M. Vejdemo-Johansson, and H. Adams. JavaPlex: A research software package for persistent (co)homology. In *Proc. 4th Int. Cong. Math. Software*, vol. 8592 of *LNCS*, pages 129–136, Springer, Berlin, 2014.

[Vav00]   S.A. Vavasis. QMG, version 2.0, patch 2. http://www.cs.cornell.edu/Info/People/vavasis/qmg-home.html, 2000.

[Ver16]   S. Verdoolaege. barvinok, version 0.39. http://barvinok.gforge.inria.fr/, 2016.

[W⁺02]   S.K. Wismath et al. VisPak, version 2.0. http://www.cs.uleth.ca/~wismath/vis.html, 2002.

[W⁺16]   R. Wunderling et al. The sequential object-oriented simplex class library, version 2.2.1. http://soplex.zib.de/, 2016.

[Wee00]   J. Weeks. SnapPea, version 3.0d3. http://www.geometrygames.org/SnapPea, 2000.

[Wei12]   C. Weibel. Minksum, version 1.7. https://sites.google.com/site/christopheweibel/research/minksum, 2012.

[WGB⁺09]   S. Weißmann, C. Gunn, P. Brinkmann, T. Hoffmann, and U. Pinkall. jReality: A java library for real-time interactive 3d graphics and audio. In *Proc. 17th ACM Internat. Conf. Multimedia*, pages 927–928, 2009.

[Wol16]   WolframAlpha. Wolfram Alpha LLC, http://www.wolframalpha.com/, 2016.

[xpr15]   FICO xpress optimization suite. FICO, http://www.fico.com/en/products/fico-xpress-optimization-suite, 2015.

[YD10]   C. Yap and Z. Du. Core Library, version 2.1. http://cs.nyu.edu/exact/core_pages, 2010.

68 TWO COMPUTATIONAL GEOMETRY LIBRARIES: LEDA AND CGAL

Michael Hoffmann, Lutz Kettner, and Stefan Näher

INTRODUCTION

Over the past decades, two major software libraries that support a wide range of geometric computing have been developed: LEDA, the **L**ibrary of **E**fficient **D**ata Types and **A**lgorithms, and CGAL, the **C**omputational **G**eometry **A**lgorithms **L**ibrary. We start with an introduction of common aspects of both libraries and major differences. We continue with sections that describe each library in detail.

Both libraries are written in C++. LEDA is based on the object-oriented paradigm and CGAL is based on the generic programming paradigm. They provide a collection of flexible, efficient, and correct software components for computational geometry. Users should be able to easily include existing functionality into their programs. Additionally, both libraries have been designed to serve as platforms for the implementation of new algorithms.

Correctness is of crucial importance for a library, even more so in the case of geometric algorithms where correctness is harder to achieve than in other areas of software construction. Two well-known reasons are the *exact arithmetic assumption* and the *nondegeneracy assumption* that are often used in geometric algorithms. However, both assumptions usually do not hold: floating-point arithmetic is not exact and inputs are frequently degenerate. See Chapter 45 for details.

EXACT ARITHMETIC

There are basically two scientific approaches to the exact arithmetic problem. One can either design new algorithms that can cope with inexact arithmetic or one can use exact arithmetic. Instead of requiring the arithmetic itself to be exact, one can guarantee correct computations if the so-called *geometric primitives* are exact. So, for instance, the predicate for testing whether three points are collinear must always give the right answer. Such an exact primitive can be efficiently implemented using floating-point filters or lazy evaluation techniques.

This approach is known as the exact geometric computing paradigm and both libraries, LEDA and CGAL, adhere to this paradigm. However, they also offer straight floating-point implementations.

DEGENERACY HANDLING

An elegant (theoretical) approach to the degeneracy problem is **symbolic perturbation**. But this method of forcing input data into general position can cause serious problems in practice. In many cases, it increases the complexity of (intermediate) results considerably; and furthermore, the final limit process turns out to be difficult in particular in the presence of combinatorial structures. For this reason, both libraries follow a different approach. They cope with degeneracies directly by treating the degenerate case as the "normal" case. This approach proved to be effective for many geometric problems.

However, symbolic perturbation is used in some places. For example, in CGAL the 3D Delaunay triangulation uses it to realize consistent point insert and removal functions in the degenerate case of more than four points on a sphere [DT11].

GEOMETRIC PROGRAMMING

Both CGAL and LEDA advocate *geometric programming*. This is a style of higher-level programming that deals with geometric objects and their corresponding primitives rather than working directly on coordinates or numerical representations. In this way, for instance, the machinery for the exact arithmetic can be encapsulated in the implementation of the geometric primitives.

COMMON ROOTS AND DIFFERENCES

LEDA is a general-purpose library of algorithms and data structures, whereas CGAL is focused on geometry. They have a different look and feel and different design principles, but they are compatible with each other and can be used together. A LEDA user can benefit from more geometry algorithms in CGAL, and a CGAL user can benefit from the exact number types and graph algorithms in LEDA, as will be detailed in the individual sections on LEDA and CGAL.

CGAL started six years after LEDA. CGAL learned from the successful decisions and know-how in LEDA (also supported by the fact that LEDA's founding institute is a partner in developing CGAL). The later start allowed CGAL to rely on better C++ language support, e.g., with templates and traits classes, which led the developers to adopt the *generic programming paradigm* and shift the design focus more toward flexibility.

Successful spin-off companies have been created around both LEDA[1] and CGAL[2]. After an initial free licensing for academic institutions, all LEDA licenses are now fee-based. In contrast, CGAL is freely available under the GPL/LGPL [GPL07] since release 4.0 (March 2012). Users who consider the open source license to be too restrictive can also obtain a commercial license.

GLOSSARY

Exact arithmetic: The foundation layer of the *exact computation paradigm* in computational geometry software that builds correct software layer by layer. Exact arithmetic can be as simple as a built-in integer type as long as its precision is not exceeded or can involve more complex number types representing expression DAGs, such as, `leda::real` from LEDA [BFMS00] or `Expr` from CORE [KLPY99].

---

[1] Algorithmic Solutions Software GmbH <www.algorithmic-solutions.com>.
[2] GeometryFactory Sarl <www.geometryfactory.com>.

Floating-point filter: A technique that speeds up exact computations for common easy cases; a fast floating-point interval arithmetic is used unless the error intervals overlap, in which case the computation is repeated with exact arithmetic.

Coordinate representation: Cartesian and homogeneous coordinates are supported by both libraries. Homogeneous coordinates are used to optimize exact rational arithmetic with a common denominator, not for projective geometry.

Geometric object: The atomic part of a geometric kernel. Examples are points, segments, lines, and circles in 2D, planes, tetrahedra, and spheres in 3D, and hyperplanes in dD. The corresponding data types have value semantics; variants with and without reference-counted representations exist.

Predicate: Geometric primitive returning a value from a finite domain that expresses a geometric property of the arguments (geometric objects), for example, CGAL::do_intersect(p,q) returning a Boolean or leda::orientation(p,q,r) returning the sign of the area of the triangle (p,q,r). A ***filtered predicate*** uses a floating-point filter to speed up computations.

Construction: Geometric primitive constructing a new object, such as the point of intersection of two non-parallel straight lines.

Geometric kernel: The collection of geometric objects together with the related predicates and constructions. A ***filtered kernel*** uses filtered predicates.

Program checkers: Technique for writing programs that check the correctness of their output. A checker for a program computing a function f takes an instance x and an output y. It returns true if $y = f(x)$, and false otherwise.

68.1 LEDA

LEDA aims at being a comprehensive software platform for the entire area of combinatorial and geometric computing. It provides a sizable collection of data types and algorithms. This collection includes most of the data types and algorithms described in the textbooks of the area ([AHU74, CLR90, Kin90, Meh84, NH93, O'R94, Tar83, Sed91, Wyk88, Woo93]).

A large number of academic and industrial projects from almost every area of combinatorial and geometric computing have been enabled by LEDA. Examples are graph drawing, algorithm visualization, geographic information systems, location problems, visibility algorithms, DNA sequencing, dynamic graph algorithms, map labeling, covering problems, railway optimization, route planning, computational biology and many more. See <www.algorithmic-solutions.com/leda/projects/index.html> for a list of industrial projects based on LEDA.

The LEDA project was started in the fall of 1988 by Kurt Mehlhorn and Stefan Näher. The first six months were devoted to the specification of different data types and selecting the implementation language. At that time the item concept arose as an abstraction of the notion "pointer into a data structure." Items provide direct and efficient access to data and are similar to iterators in the standard template library. The item concept worked successfully for all test cases and is now used for most data types in LEDA. Concurrently with searching for the correct specifications, existing programming languages were investigated for their suitability as an implementation platform. The language had to support abstract data types and type parameters (genericity) and should be widely available. Based on the experiences with different example programs, C++ was selected because of its flexibility, expressive power, and availability.

We next discuss some of the general aspects of the LEDA system.

EASE OF USE

The LEDA library is easy to use. In fact, only a small fraction of the users are algorithm experts and many users are not even computer scientists. For these users the broad scope of the library, its ease of use, and the correctness and efficiency of the algorithms in the library are crucial. The LEDA manual [MNSU] gives precise and readable specifications for the data types and algorithms mentioned above. The specifications are short (typically not more than a page), general (so as to allow several implementations) and abstract (so as to hide all details of the implementation).

EXTENSIBILITY

Combinatorial and geometric computing is a diverse area and hence it is impossible for a library to provide ready-made solutions for all application problems. For this reason it is important that LEDA is easily extensible and can be used as a platform for further software development. In many cases LEDA programs are very close to the typical textbook presentation of the underlying algorithms. The goal is the equation: *Algorithm + LEDA = Program.*

LEDA *extension packages* (LEPs) extend LEDA into particular application domains and areas of algorithmics not covered by the core system. LEDA extension packages satisfy requirements, which guarantee compatibility with the LEDA philosophy. LEPs have a LEDA-style documentation, they are implemented as platform independent as possible, and the installation process permits a close integration into the LEDA core library. Currently, the following LEPs are available: PQ-trees, dynamic graph algorithms, a homogeneous d-dimensional geometry kernel, and a library for graph drawing.

CORRECTNESS

Geometric algorithms are frequently formulated under two unrealistic assumptions: computers are assumed to use exact real arithmetic (in the sense of mathematics) and inputs are assumed to be in general position. The naive use of floating-point arithmetic as an approximation to exact real arithmetic very rarely leads to correct implementations. In a sequence of papers [BMS94b, See94, MN94b, BMS94a, FGK+00], these degeneracy and precision issues were investigated and LEDA was extended based on this theoretical work. It now provides exact geometric kernels for 2D and higher-dimensional computational geometry [MMN+98], and also correct implementations for basic geometric tasks, e.g., 2D convex hulls, Delaunay diagrams, Voronoi diagrams, point location, line segment intersection, and higher-dimensional convex hulls and Delaunay triangulations.

Programming is a notoriously error-prone task; this is even true when programming is interpreted in a narrow sense: translating a (correct) algorithm into a program. The standard way to guard against coding errors is program testing. The program is exercised on inputs for which the output is known by other means, typically as the output of an alternative program for the same task. Program testing has severe limitations. It is usually only performed during the testing phase of a program. Also, it is difficult to determine the "correct" suite of test inputs. Even if appropriate test inputs are known it is usually difficult to determine the correct outputs for these inputs: alternative programs may have different input and output conventions or may be too inefficient to solve the test cases.

Given that program verification—i.e., formal proof of correctness of an implementation—will not be available on a practical scale for some years to come, *program checking* has been proposed as an extension to testing [BK95, BLR93]. The cited papers explored program checking in the area of algebraic, numerical, and combinatorial computing. In [MNS⁺99, MM96, HMN96] program checkers are presented for planarity testing and a variety of geometric tasks. LEDA uses program checkers for many of its implementations. A more general approach (*Certifying Algorithms*) was introduced in [MMNS11].

68.1.1 THE STRUCTURE OF LEDA

LEDA uses templates for the implementation of parameterized data types and for generic algorithms. However, it is not a pure template library and therefore is based on an object code library of precompiled code. Programs using LEDA data types or algorithms have to include the appropriate LEDA header files in their source code and must link to this library (*libleda*). See the LEDA user manual ([MNSU] or the LEDA book ([MN00]) for details.

68.1.2 GEOMETRY KERNELS

LEDA offers kernels for 2D and 3D geometry, a kernel of arbitrary dimension is available as an extension package. In either case there exists a version of the kernel based on floating-point Cartesian coordinates (called *float-kernel*) as well as a kernel based on rational homogeneous coordinates (called *rat-kernel*). All kernels provide a complete collection of geometric objects (points, segments, rays, lines, circles, simplices, polygons, planes, etc.) together with a large set of geometric primitives and predicates (orientation of points, side-of-circle tests, side-of-hyperplane, intersection tests and computation, etc.). For a detailed discussion and the precise specification, see Chapter 9 of the LEDA book ([MN00]). Note that only for the rational kernel, which is based on exact arithmetic and floating-point filters, all operations and primitives are guaranteed to compute the correct result.

68.1.3 DATA STRUCTURES

In addition to the basic kernel data structures LEDA provides many advanced data types for computational geometry. Examples include:

- A general polygon type (`gen_polygon` or `rat_gen_polygon`) with a complete set of Boolean operations. Its implementation is based on efficient and robust

plane sweep algorithms for the construction of the arrangement of a set of
straight line segments (see [MN94a] and [MN00, Ch. 10.7]).

- Two- and higher-dimensional geometric tree structures, such as range, segment, interval and priority search trees.

- Partially and fully persistent search trees.

- Different kinds of geometric graphs (triangulations, Voronoi diagrams, and arrangements).

- A dynamic `point_set` data type supporting update, search, closest point, and different types of range query operations on one single representation based on a dynamic Delaunay triangulation (see [MN00, Ch. 10.6]).

68.1.4 ALGORITHMS

The LEDA project never had the goal of providing a complete collection of the algorithms from computational geometry (nor for other areas of algorithms). Rather, it was designed and implemented to establish a *platform* for combinatorial and geometric computing enabling programmers to implement these algorithms themselves more easily and customized to their particular needs. But of course the library already contains a considerable number of basic geometric algorithms. Here we give a brief overview and refer the reader to the user manual for precise specifications and to Chapter 10 of the LEDA-book ([MN00]) for detailed descriptions and analyses of the corresponding implementations. The current version of LEDA offers different implementation of algorithms for the following 2D geometric problems:

- convex hull algorithms (also 3D)
- halfplane intersection
- (constraint) triangulations
- closest and farthest Delaunay and Voronoi diagrams
- Euclidean minimum spanning trees
- closest pairs
- Boolean operations on generalized polygons
- segment intersection and construction of line arrangements
- Minkowski sums and differences
- nearest neighbors and closest points
- minimum enclosing circles and annuli
- curve reconstruction

68.1.5 VISUALIZATION (GeoWin)

In computational geometry, visualization and animation of programs are important for the understanding, presentation, and debugging of algorithms. Furthermore, the

animation of geometric algorithms is cited as among the strategic research directions
in this area. *GeoWin* [BN02] is a generic tool for the interactive visualization of
geometric algorithms. *GeoWin* is implemented as a C++ data type. Its design
and implementation was influenced by LEDA's graph editor *GraphWin* ([MN00,
Ch. 12]). Both data types support a number of programming styles which have
been shown to be very useful for the visualization and animation of algorithms.
The animations use *smooth transitions* to show the result of geometric algorithms
on dynamic user-manipulated input objects, e.g., the Voronoi diagram of a set of
moving points or the result of a sweep algorithm that is controlled by dragging the
sweep line with the mouse (see Figure 68.1.1).

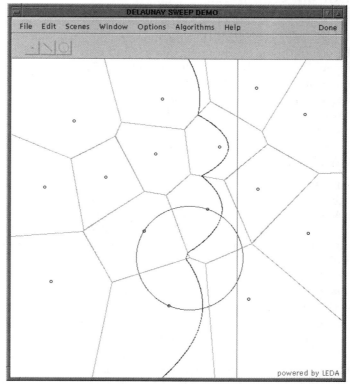

FIGURE 68.1.1
GeoWin animating Fortune's sweep algorithm.

A GeoWin maintains one or more geometric scenes. A geometric *scene* is a
collection of geometric objects of the same type. A collection is simply either a
standard C++ list (STL-list) or a LEDA-list of objects. GeoWin requires that the
objects provide a certain functionality, such as stream input and output, basic geometric transformations, drawing and input in a LEDA window. A precise definition
of the required operations can be found in the manual pages [MNSU]. GeoWin
can be used for any collection of basic geometric objects (geometry kernel) fulfill-
ing these requirements. Currently, it is used to visualize geometric objects and
algorithms from both the CGAL and LEDA libraries.

The visualization of a scene is controlled by a number of attributes, such as color, line width, line style, etc. A scene can be subject to user interaction and it may be defined from other scenes by means of an algorithm (a C++ function). In the latter case the scene (also called the *result scene*) may be recomputed whenever one of the scenes on which it depends is modified. There are three main modes for recomputation: user-driven, continuous, and event-driven.

GeoWin has both an interactive and a programming interface. The interactive interface supports the interactive manipulation of input scenes, the change of geometric attributes, and the selection of scenes for visualization.

68.1.6 PROGRAM EXAMPLES

We now give two programming examples showing how LEDA can be used to implement basic geometric algorithms in an elegant and readable way. The first example is the computation of the *upper convex hull* of a point set in the plane. It uses points and the orientation predicate and lists from the basic library. The second example shows how the LEDA *graph* data type is used to represent triangulations in the implementation of a function that turns an arbitrary triangulation into a Delaunay triangulation by edge flips. It uses points, lists, graphs, and the side-of-circle predicate.

UPPER CONVEX HULL

In our first example we show how to use LEDA for computing the upper convex hull of a given set of points. We assume that we are in LEDA's namespace, otherwise all LEDA names would have to be used with the prefix `leda::`. Function UPPER_HULL takes a list L of rational points (type `rat_point`) as input and returns the list of points of the upper convex hull of L in clockwise ordering from left to right. The algorithm is a variant of Graham's Scan [Gra72].

First we sort L according to the lexicographic ordering of the Cartesian coordinates and remove multiple points. If the list contains not more than two points after this step we stop. Before starting the actual Graham Scan we first skip all initial points lying on or below the line connecting the two extreme points. Then we scan the remaining points from left to right and maintain the upper hull of all points seen so far in a list called *hull*. Note however that the last point of the hull is not stored in this list but in a separate variable p. This makes it easier to access the last two hull points as required by the algorithm. Note also that we use the rightmost point as a sentinel avoiding the special case that *hull* becomes empty.

```
list<rat_point> UPPER_HULL(list<rat_point> L) {
    L.sort();
    L.unique();

    if (L.length() <= 2) return L;

    rat_point p_min = L.front(); // leftmost point
    rat_point p_max = L.back();  // rightmost point

    list<rat_point> hull;        // result list
    hull.append(p_max);          // use rightmost point as sentinel
    hull.append(p_min);          // first hull point
```

```
        // goto first point p above (p_min,p_max)
        while (! L.empty() && ! left_turn(p_min, p_max, L.front())) L.pop();
        if (L.empty()) {                // upper hull consists of only 2 points
            hull.reverse();
            return hull;
        }

        rat_point p = L.pop();          // second (potential) hull point
        rat_point q;
        forall(q,L) {
            while (! right_turn(hull.back(), p, q)) p = hull.pop_back();
            hull.append(p);
            p = q;
        }
        hull.append(p);                 // add last hull point
        hull.pop();                     // remove sentinel
        return hull;
}
```

DELAUNAY FLIPPING

LEDA represents triangulations by bidirected plane graphs (from the graph library)
whose nodes are labeled with points and whose edges may carry additional information, e.g., integer flags indicating the type of edge (hull edge, triangulation edge,
etc.). All edges incident to a node v are ordered in counterclockwise ordering and
every edge has a reversal edge. In this way the faces of the graph represent the
triangles of the triangulation. The graph type offers methods for iterating over the
nodes, edges, and adjacency lists of the graph. In the case of plane graphs there
are also operations for retrieving the reverse edge and for iterating over the edges
of a face. Furthermore, edges can be moved to new nodes. This graph operation is
used in the following program to implement edge flips.

Function DELAUNAY_FLIPPING takes as input an arbitrary triangulation and
turns it into a Delaunay triangulation by the well-known flipping algorithm. This
algorithm performs a sequence of local transformations as shown in Figure 68.1.2 to
establish the Delaunay property: for every triangle the circumscribing circle does
not contain any vertex of the triangulation in its interior. The test whether an
edge has to be flipped or not can be realized by a so-called *side_of_circle* test. This
test takes four points a, b, c, d and decides on which side of the oriented circle
through the first three points a,b, and c the last point d lies. The result is positive
or negative if d lies on the left or on the right side of the circle, respectively, and
the result is zero if all four points lie on one common circle. The algorithm uses a
list of candidates which might have to be flipped (initially all edges). After a flip
the four edges of the corresponding quadrilateral are pushed onto this candidate
list. Note that G[v] returns the position of node v in the triangulation graph G.
A detailed description of the algorithm and its implementation can be found in the
LEDA book ([MN00]).

FIGURE 68.1.2
Flipping to establish the Delaunay property.

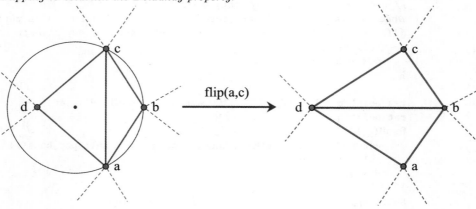

```
void DELAUNAY_FLIPPING(GRAPH<rat_point, int>& G) {
    list<edge> S = G.all_edges();
    while (! S.empty()) {
        edge e = S.pop();
        edge r = G.rev_edge(e);

        edge e1 = G.face_cycle_succ(r);     // e1,e2,e3,e4: edges of quadrilateral
        edge e2 = G.face_cycle_succ(e1);    // with diagonal e
        edge e3 = G.face_cycle_succ(e);
        edge e4 = G.face_cycle_succ(e3);

        rat_point a = G[G.source(e1)];      // a,b,c,d: corners of quadrilateral
        rat_point b = G[G.target(e1)];
        rat_point c = G[G.source(e3)];
        rat_point d = G[G.target(e3)];

        if (side_of_circle(a,b,c,d) > 0) {
            S.push(e1); S.push(e2); S.push(e3); S.push(e4);
            G.move_edge(e,e2,source(e4)); // flip diagonal
            G.move_edge(r,e4,source(e2));
        }
    }
}
```

68.2 CGAL

The development of CGAL, the Computational Geometry Algorithms Library, started in 1995 by a consortium of seven sites from Europe and Israel, funded by European research programs. The central goal of CGAL is to

> *make the large body of geometric algorithms developed in the field of computational geometry available for industrial application.*

The main design goals are correctness, flexibility, efficiency, and ease of use. The focus is on a broad foundation in computational geometry. Important related issues, for example visualization, are supported with standard formats and interfaces.

The first public release 0.9 appeared in June 1997. Since 2009 a biannual release cycle has been established and since 2015 the current development version is publicly available at `<github.com/CGAL>`. About 30 developers work on CGAL, though many of them part-time only. A release provides 4–5 new major features on average, such as new packages or significant updates to existing packages. With release 3.0 in 2003, CGAL became an open source project inviting everybody to join and has since successfully attracted more than 20 feature submissions from outside the founding institutes.

Nowadays CGAL is a standard tool in the area. More than 200 commercial customers work with CGAL and a list of more than 100 projects from diverse areas using CGAL can be found at `<www.cgal.org/projects.html>`. The presentation here is based on CGAL release 4.7 from October 2015, available from CGAL's home page `<www.cgal.org>`.

LIBRARY STRUCTURE

CGAL is structured in three layers: the layer of *algorithms and data structures*, which builds on the *geometric traits* layer with representations for geometric objects and primitive operations on these representations. The geometric traits layer in turn builds on the layer of *arithmetic and algebra* with concepts for algebraic structures and number types modeling these structures. Orthogonally, there is a *support library* layer with geometric object generators, file I/O, visualization, and other nongeometric functionality.

GENERIC PROGRAMMING IDIOMS

Concept: A formal hierarchy of polymorphic abstract requirements on data types.

Model for a concept: A type that fulfils all requirements of that concept and can therefore be used in places where the concept was requested.

Function object: An object that implements a function, e.g., as a C++ class with an `operator()`. It is more efficient and type-safe compared to a C function pointer or object-oriented class hierarchies.

FLEXIBILITY

Following the generic programming paradigm, CGAL is highly *modular* and its different parts are independent of each other. The algorithms and data structures in CGAL are *adaptable* to already existing user code; see the geometric traits class example on page 1821. The library is *extendible*, as users can add implementations in the same style as CGAL. The library is *open* and supports important standards, such as the C++ standard with its standard library, and other established libraries, such as BOOST [Sch14], LEDA, EIGEN [GJ+10], and GMP [Gt14].

CORRECTNESS

CGAL addresses robustness problems by relying on exact arithmetic and explicit degeneracy handling. There is a well-established software process and communication channels set up for the distributed developer community, including a distributed system for revision management, bug and feature tracking. Every night an automatic test-suite is run on all supported platforms and compilers. An editorial board reviews new submissions and supervises design homogeneity.

EASE OF USE

Users with a base knowledge of C++ and generic programming experience a smooth learning curve with CGAL. Many concepts are familiar from the C++ standard library, and the powerful flexibility is often hidden behind sensible defaults. A novice reader should not be discouraged by some of the advanced examples illustrating CGAL's power. CGAL has a uniform design, aims for minimal interfaces, yet rich and complete functionality in computational geometry.

EFFICIENCY

CGAL uses C++ templates to realize most of its flexibility at compile time. That enables flexibility at places normally not considered because of runtime costs, e.g., on the number-type level. Furthermore, choices such as tradeoffs between space and time in some data structures, or between different number types of different power and speed can be made at the application level rather than in the library. This also encourages experimental research.

68.2.1 GEOMETRIC KERNELS

CGAL offers a wide variety of kernels. The *linear* kernels are mostly concerned with linearly bounded objects, such as hyperplanes and simplices. The geometric objects along with some of the available predicates and constructions—classified according to dimension—are summarized in Table 68.2.1.

For circles and spheres an algebraic description involves polynomials of degree two. Therefore, intersections involving circles and spheres do not admit an exact rational representation in general. The *circular 2D kernel* [EKP+04, DFMT02]

TABLE 68.2.1   Linear kernel objects with selected predicates and constructions.

DIM	GEOMETRIC OBJECTS	PREDICATES	CONSTRUCTIONS
all	Point, Vector, Direction, Line, Ray, Segment, Aff_transformation	compare_lexicographically, do_intersect, orientation	intersection, midpoint transform, squared_distance
2	Triangle, Iso_rectangle, Bbox, Circle	collinear, left_turn, side_of_oriented_circle	bbox, centroid, circumcenter, squared_radius, rational_rotation_approximation
3	Plane, Tetrahedron, Triangle, Iso_cuboid, Bbox, Sphere	coplanar, left_turn, side_of_oriented_sphere	bbox, centroid, circumcenter, cross_product, squared_radius
d	Hyperplane, Sphere	side_of_oriented_sphere	center_of_sphere, lift_to_paraboloid

extends the linear kernel to allow for an exact representation of such intersection points, as well as for linear or circular arcs connecting such points. An analogous toolset in 3D is provided by the *spherical 3D kernel* [CCLT09].

PREDEFINED KERNELS

To ease the choice of an appropriate kernel, CGAL offers three predefined linear kernels for dimension two and three to cover the most common tradeoffs between speed and exactness requirements. All three support initial exact constructions from **double** values and guarantee exact predicate evaluation. They vary in their support for exact geometric constructions and exact roots on the number type level. As a rule of thumb, the less exact functionality a kernel provides, the faster it runs.

CGAL::Exact_predicates_inexact_constructions_kernel provides filtered exact predicates, but constructions are performed with **double** and, therefore, subject to possible roundoff errors.

CGAL::Exact_predicates_exact_constructions_kernel also provides filtered exact predicates. Constructions are performed exactly, but using a lazy computation scheme [PF11] as a geometric filter. In this way, costly exact computations are performed only as far as needed, if at all.

CGAL::Exact_predicates_exact_constructions_kernel_with_sqrt achieves exactness via an algebraic number type (currently, either **leda::real** [MS01] or CORE::Expr [KLPY99]).

GENERAL LINEAR KERNELS

The kernels available in CGAL can be classified along the following orthogonal concepts:

Dimension: The dimension of the affine space. The specializations for 2D and 3D offer functionality that is not available in the dD kernel.

Number type: The number type used to store coordinates and coefficients, and to compute the expressions in predicates and constructions.

CGAL distinguishes different concepts to describe the underlying algebraic structure of a number type, along with a concept *RealEmbeddable* to describe an order along the real axis. Most relevant here are the concepts *IntegralDomainWithoutDivision* and *Field*, whose names speak for themselves.

Coordinate representation: The Cartesian representation requires a *field type* (FT) as a number type (a model for *Field* and *RealEmbeddable*). The homogeneous representation requires a *ring type* (RT) as a number type (a model for *IntegralDomainWithoutDivision* and *RealEmbeddable*).

Reference counting: Reference counting is used to optimize copying and assignment of kernel objects. It is recommended for exact number types with larger memory size. The kernel objects have value-semantics and cannot be modified, which simplifies reference counting. The nonreference counted kernels are recommended for small and fast number types, such as the built-in `double`.

The corresponding linear kernels in CGAL are:

`CGAL::Cartesian<FT>`	Cartesian, reference counted, 2D and 3D
`CGAL::Simple_cartesian<FT>`	Cartesian, nonreference counted, 2D and 3D
`CGAL::Homogeneous<RT>`	homogeneous, reference counted, 2D and 3D
`CGAL::Simple_homogeneous<RT>`	homogeneous, nonreference counted, 2D and 3D
`CGAL::Cartesian_d<FT>`	Cartesian, reference counted, d-dimensional
`CGAL::Homogeneous_d<RT>`	homogeneous, reference counted, d-dimensional

FILTERED KERNELS

CGAL offers two general adaptors to create filtered kernels. The first adaptor operates on the kernel level, the second one on the number type level.

(1) `CGAL::Filtered_kernel<K>` is an adaptor to build a new kernel based on a given 2/3D kernel K. All predicates of the new kernel use `double` interval arithmetic as a filter [BBP01] and resort to the original predicates from K only if that filter fails. Selected predicates use a semi-static filter [MP07] in addition. Constructions of the new kernel are those provided by K.

(2) `CGAL::Lazy_exact_nt<NT>` is an adaptor to build a new number type from a given exact number type NT. The new number type uses filters based on interval arithmetic and expression DAGs for evaluation. Only if this filter fails, evaluation is done using NT instead. Due to the exactness of NT, also the new number type admits exact constructions. Specialized predicates avoid the expression DAG construction in some cases.

There is also a dD Version of the Exact_predicates_inexact_constructions_kernel called `CGAL::Epick_d<D>`: a Cartesian, reference counted kernel that supports exact filtered predicates. The parameter D specifies the dimension, which can either be fixed statically at compile-time or dynamically at runtime. Constructions are performed with `double` and, therefore, subject to possible roundoff errors.

68.2.2 GEOMETRIC TRAITS

CGAL follows the generic programming paradigm in the style of the Standard Template Library [Aus98]; algorithms are parameterized with iterator ranges that decouple them from data structures. In addition, CGAL invented the *circulator* concept to accommodate circular structures efficiently, such as the ring of edges around a vertex in planar subdivisions [FGK+00]. Essential for CGAL's flexibility is the separation of algorithms and data structures from the underlying geometric kernel with a *geometric traits class*.

GLOSSARY

Iterator: A concept for a pointer into a linear sequence; exists in different flavors: input, output, forward, bidirectional, and random-access iterator.

Circulator: A concept similar to iterator but for circular sequences.

Range: A pair $[b, e)$ of iterators (or circulators) describing a sequence of items in a half-open interval notation, i.e., starting *with b* and ending *before e*.

Traits class: C++ programming technique [Mye95] to attach additional information to a type or function, e.g., dependent types, functions, and values.

Geometric traits: Traits classes used in CGAL to decouple the algorithms and data structures from the kernel of geometric representations and primitives. Every algorithm and data structure defines a geometric traits concept and the library provides various models. Often the geometric kernel is a valid model.

EXAMPLE OF UPPER CONVEX HULL ALGORITHM

We show two implementations of the upper convex hull algorithm following Andrew's variant of Graham's scan [Gra72, And79] in CGAL. The first implementation makes straightforward use of a sufficient CGAL default kernel and looks similar to textbook presentations or the LEDA example on page 1812.

```
typedef CGAL::Exact_predicates_inexact_constructions_kernel Kernel;
typedef Kernel::Point_2   Point_2;
Kernel kernel;  // our instantiated kernel object

std::list<Point_2> upper_hull( std::list<Point_2> L) {
    L.sort( kernel.less_xy_2_object());
    L.unique();
    if (L.size() <= 2)
        return L;
    Point_2 pmin = L.front();  // leftmost point
    Point_2 pmax = L.back();   // rightmost point
    std::list< Point_2> hull;
    hull.push_back(pmax);       // use rightmost point as sentinel
    hull.push_back(pmin);       // first hull point
    while (!L.empty() && !kernel.left_turn_2_object()(pmin,pmax,L.front()))
        L.pop_front();          // goto first point p above (pmin,pmax)
```

```
        if (L.empty()) {
            hull.reverse();         // fix orientation for this special case
            return hull;
        }
        Point_2 p = L.front();      // keep last point on current hull separately
        L.pop_front();
        for (std::list< Point_2>::iterator i = L.begin(); i != L.end(); ++i) {
            while (! kernel.left_turn_2_object()( hull.back(), *i, p)) {
                p = hull.back();    // remove non-extreme points from current hull
                hull.pop_back();
            }
            hull.push_back(p);      // add new extreme point to current hull
            p = *i;
        }
        hull.push_back(p);          // add last hull point
        hull.pop_front();           // remove sentinel
        return hull;
}
```

The second implementation is more flexible because it separates the algorithm from the geometry, similar to how generic algorithms are written in CGAL. As an algorithmic building block, we place the core of the control flow—the two nested loops at the end—into its own generic function with an interface of bidirectional iterators and a single three-parameter predicate. We eliminate the additional data structure for the hull by reusing the space that becomes available in the original sequence as our stack. The result is thus returned in our original sequence and the function just returns the past-the-end position. Instead of the sentinel, which does not give measurable performance benefits, we explicitly test the boundary case.

```
template <class Iterator, class F>  // bidirectional iterator, function object
Iterator backtrack_remove_if_triple( Iterator first, Iterator beyond, F pred) {
    if (first == beyond)
        return first;
    Iterator i = first, j = first;
    if (++j == beyond)              // i,j mark two elements on the top of the stack
        return j;
    Iterator k = j;                 // k marks the next candidate value in the sequence
    while (++k != beyond) {
        while (pred( *i, *j, *k)) {
            j = i;                  // remove one element from stack, part 1
            if (i == first)         // explicit test for stack underflow
                break;
            --i;                    // remove one element from stack, part 2
        }
        i = j; ++j; *j = *k;        // push next candidate value from k on stack
    }
    return ++j;
}
```

With this generic function, we implement in two lines an algorithm to compute *all* points on the upper convex hull (rather than only the extreme points). All degeneracies are handled correctly. For the sorting, random access iterators are required. Note the geometric traits parameter and how predicates are extracted from the traits class. Any CGAL kernel is a valid model for this traits parameter.

```
template <class Iterator, class Traits>  // random access iterator
Iterator upper_hull( Iterator first, Iterator beyond, Traits traits) {
    std::sort( first, beyond, traits.less_xy_2_object());
    return backtrack_remove_if_triple( first, beyond,
                                       traits.left_turn_2_object());
}
```

EXAMPLE OF A USER TRAITS CLASS

Data structures and algorithms in CGAL can be easily adapted to work on user data types by defining a custom geometric traits class. Consider the following custom point class:

```
struct Point {   // our point type
    double x, y;
    Point( double xx = 0.0, double yy = 0.0) : x(xx), y(yy) {}
    // ... whatever else this class provides ...
};
```

To run the CGAL::ch_graham_andrew convex hull algorithm on points of this type, the CGAL Reference Manual lists as requirements on its traits class a type Point_2, and function objects Equal_2, Less_xy_2, and Left_turn_2. A possible geometric traits class could look like this:

```
struct Geometric_traits {   // traits class for our point type
    typedef double RT;          // ring number type, for random points generator
    typedef Point Point_2;   // our point type
    struct Equal_2 {             // equality comparison
        bool operator()( const Point& p, const Point& q) {
            return (p.x == q.x) && (p.y == q.y);
        }
    };
    struct Less_xy_2 {           // lexicographic order
        bool operator()( const Point& p, const Point& q) {
            return (p.x < q.x) || ((p.x == q.x) && (p.y < q.y));
        }
    };
    struct Left_turn_2 {         // orientation test
        bool operator()( const Point& p, const Point& q, const Point& r) {
            return (q.x-p.x) * (r.y-p.y) > (q.y-p.y) * (r.x-p.x); // inexact!
        }
    };
    // member functions to access function objects, here by default construction
    Equal_2     equal_2_object()    const { return Equal_2(); }
    Less_xy_2   less_xy_2_object()  const { return Less_xy_2(); }
    Left_turn_2 left_turn_2_object() const { return Left_turn_2(); }
};
```

In order to let CGAL know that our traits class belongs to our point class, we specialize CGAL's kernel traits accordingly:

```
namespace CGAL {   // specialization that links our point type with our traits class
    template <> struct Kernel_traits< ::Point> {
        typedef ::Geometric_traits Kernel;
    };
}
```

Now the `CGAL::ch_graham_andrew` function can be used with the custom point class `Point`. The traits class also suffices for the random point generators in CGAL. Here is a program to compute the convex hull of 20 points sampled uniformly from a unit disk.

```
#include <CGAL/ch_graham_andrew.h>
#include <CGAL/point_generators_2.h>

int main() {
    std::vector<Point> points, hull;
    CGAL::Random_points_in_disc_2<Point> rnd_pts( 1.0);
    std::copy_n( rnd_pts, 20, std::back_inserter( points));
    CGAL::ch_graham_andrew( points.begin(), points.end(),
                            std::back_inserter( hull));
    return 0;
}
```

The separation of algorithms and data structures from the geometric kernel provides flexibility, the fingerprint of generic programming. Such flexibility is not only useful when interfacing with custom user data or other libraries but even within CGAL itself. An example is the geometric traits class `CGAL::Projection_traits_xy_3` that models the orthogonal projection of 3D points onto the xy-plane, hence treating them as 2D points. Such a model is useful, for instance, in the context of terrain triangulations in GIS (cf. Chapter 59).

68.2.3 LIBRARY CONTENTS

The library is structured into packages that correspond to different chapters of the reference manual. These packages in turn are grouped thematically.

CONVEX HULL

The *2D convex hull* algorithms return the counterclockwise sequence of extreme points. The *3D convex hull* algorithms return the convex polytope of extreme points or—in degenerate cases—a lower-dimensional simplex. The *d-dimensional* convex hull algorithm is part of the d-dimensional triangulation package. It constructs a simplicial complex from which the facets of the hull can be extracted. See Table 68.2.2 for an overview.

TABLE 68.2.2   Convex hull algorithms on a set of n points with h extreme points.

DIM	MODE	ALGORITHM
2	Static	Bykat, Eddy, and Jarvis march, all in $O(nh)$ time
	Static	Akl & Toussaint, and Graham-Andrew scan, both in $O(n \log n)$ time [Sch99]
	Polygon	Melkman for points of a simple polygon in $O(n)$ time
	Others	lower hull, upper hull, subsequences of the hull, extreme points, convexity test
3	Static	quickhull [BDH96]
	Incremental	randomized incremental construction [CMS93, BMS94b]
	Dynamic	by-product of the dynamic 3D Delaunay triangulation
	Test	convexity test as program checker [MNS$^+$99]
d	Incremental	random. incr. constr. [CMS93, BDH09] in $O(n \log n + n^{\lceil d/2 \rceil})$ expected time

POLYGONS

A *polygon* is a closed chain of edges. CGAL provides a container class for polygons, but all functions are generic with iterators and work on arbitrary sequences of points. The functions available are polygon area, point location, tests for simplicity and convexity of the polygon, and generation of random instances.

CGAL also provides container classes to represent simple *polygons with holes* and collections of such polygons. Moreover, there are generalizations of these classes where the boundary is formed by general x-monotone curves rather than line segments specifically. Besides intersection and containment tests, all of these classes support *regularized Boolean operations*: intersection, union, complement, and difference. The implementation is based on arrangements. Regularization means that lower-dimensional features such as isolated points or antennas are omitted.

Where regularization is undesirable, Nef polygons [Nef78, Bie95] provide an alternative representation. A *Nef polygon* is a point set $P \subseteq \mathbb{R}^2$ generated from a finite number of open halfspaces by set complement and set intersection operations. It is therefore closed under Boolean set operations and topological operations, such as closure, interior, boundary, regularization, etc. It captures features of mixed dimension, e.g., antennas or isolated vertices, open and closed boundaries, and unbounded faces, lines, and rays. The potential unboundedness of Nef polygons is addressed with *infimaximal frames* and an extended kernel [MS03]. The representation is based on the *halfedge data structure* [Ket98] (see below), extended with face loops and isolated vertices [See01].

There are functions to compute the *Minkowski sum* of two simple polygons or an *offset polygon*, i.e., the Minkowski sum of a simple polygon with a disk. In both cases the result may not be simple and is, therefore, represented as a polygon with holes. For offset polygons these are generalized polygons, where edges are line segments or circular arcs. As an alternative to an exact representation of the offset, one can obtain an approximation with a guaranteed error bound, specified as an input. The implementation is based on the arrangement data structure and convex decomposition and convolution methods [Wei06, Wei07, BFH$^+$15].

Polygons can also be *partitioned* into y-monotone polygons or convex polygons.

The y-monotone partitioning is a sweep-line algorithm [BCKO08]. For convex partitioning one algorithm finds the minimum number of convex pieces, and two fast 4-approximation algorithms are given: one is a sweep-line algorithm [Gre83], the other is based on the constrained Delaunay triangulation [HM83].

The *straight skeleton* of a simple polygon [AAAG95] constructs a straight-line graph. The implementation is based on [FO98] with additional vertex events [EE99] and can also be used to compute a corresponding collection of offset polygons.

In *polyline simplification* the goal is to reduce the number of vertices in a collection of polylines while preserving their topology, i.e., not creating new intersections. The input can be a single polygon or an arbitrary set of polylines, which may intersect and even self-intersect. Parameters of the algorithm are: a measure for the simplification cost of removing a single vertex and a stop criterion, such as a lower bound for the number or percentage of vertices remaining or an upper bound for the simplification cost. The algorithm [DDS09] is based on constrained triangulations.

2D visibility [BHH+14] provides algorithms to compute the visibility region of a point in a polygonal domain. The input domain is represented as a bounded face in an arrangement and, in particular, may have holes.

CELL COMPLEXES AND POLYHEDRA

The *halfedge data structure* a.k.a. doubly connected edge list (DCEL) is a general purpose representation for planar structures. It is an edge-based representation with two oppositely directed halfedges per edge [Ket98]. A *polyhedral surface* is a mesh data structure based on the halfedge data structure. It embeds the halfedge data structure in 3D space. The polyhedral surface provides various basic integrity-preserving operations, the "Euler operations" [Ket98]. As an alternative, CGAL provides a *surface mesh* structure that is index rather than pointer based [SB11], which saves memory on 64-bit systems.

Most algorithms operate on models of *MutableFaceGraph*. This concept extends the *IncidenceGraph* concept of the Boost Graph Library (BGL) [SLL02] by a notion of halfedges, faces, and a cyclic order of halfedges around a face. Both polyhedral surface and surface mesh are models of MutableFaceGraph, but also third-party libraries such as OpenMesh [BSBK02] provide such models. Modeling a refinement of IncidenceGraph, these data structures can directly be used with corresponding BGL algorithms, e.g., for computing shortest paths.

A generalization to d-dimensional space is provided by the *combinatorial map* data structure [DT14]. It implements an edge-based representation for an abstract d-dimensional cell complex. The role of halfedges is taken by darts, which represent an edge together with its incident cells of any dimension. The *linear cell complex* adds a linear geometric embedding to the combinatorial structure, where every vertex of the combinatorial map is associated with a point.

3D Nef polyhedra are the natural generalization of Nef polygons to 3D. They provide a boundary representation of objects that can be obtained by Boolean operations on open halfspaces [HKM07]. In addition to Boolean operations, there are also algorithms [Hac09] to construct the *Minkowski sum* of two Nef polyhedra, and to decompose a given Nef polyhedron P into $O(r^2)$ convex pieces, where r is the number of reflex edges in P.

ARRANGEMENTS

Planar *Arrangements* are represented using an extended halfedge data structure that supports inner and outer face boundaries with several connected components. The geometry of edges is specified via a traits class, for which many different models exist, such as line segments, polylines, conic arcs, rational functions [SHRH11], Bézier curves, or general algebraic curves. Edges may be unbounded but are assumed to be x-monotone. In order to support general curves, the corresponding traits classes specify a method to split a given curve into x-monotone pieces. The point location strategy can also be specified via the traits class. Available are, among others, a trapezoidal decomposition using randomized incremental construction [HKH12] and a landmark strategy that uses a set of landmark points to guide the search [HH09]. Other supported operations include vertical ray shooting queries and batched point location using a plane sweep.

Using a sweep in an arrangement, one can also compute all pairwise *intersections for* a set of n *curves* in $O((n + k) \log n)$ time, where k is the number of intersection points of two curves. Upper and lower *envelopes* of x-monotone curves in 2D or surfaces in 3D can be constructed using a divide-and-conquer approach. Finally, *Snap rounding* converts a given arrangement of line segments into a fixed precision representation, while providing some topological guarantees. The package also provides a direct algorithm to construct an iterated snap rounding [HP02]. The book by Fogel, Halperin and Wein [FHW12] gives many examples and applications using CGAL arrangements.

TRIANGULATIONS AND VORONOI DIAGRAMS

Triangulations use a triangle-based data structure in 2D, which is more memory efficient than an edge-based structure. Similarly, the representation in 3D is based on tetrahedra. Both data structures act as container classes with an iterator interface and topologically represent a sphere using a symbolic infinite vertex. The construction is randomized incremental and efficient vertex removal [BDP+02, DT11] is supported. Where not mentioned otherwise explicitly, the following structures are available in both 2D and 3D.

Delaunay triangulations also implicitly represent the dual *Voronoi diagram*. But there is also an adaptor that simulates an explicit representation. Point location is the walk method by default. From 10,000 points on, it is recommended [DPT02] to use the Delaunay hierarchy [Dev02] instead. Batched insertion (including construction from a given range of points) spatially sorts the points in a preprocessing step to reduce the time spent for point location.

Regular triangulations are the dual of power diagrams, the Voronoi diagram of weighted points under the power-distance [ES96]. Both Delaunay and regular triangulations in 3D support parallel computation using a lock data structure.

For a *constrained triangulation* (cf. Chapter 29) one can define a set of constraining segments, which are required edges in the triangulation. Constrained (Delaunay) triangulations are available in 2D only. There is optional support for intersecting constraints, which are handled by subdividing segments at intersection points.

For a *periodic triangulation* the underlying space is a torus rather than a sphere. The space is represented using a half-open cube as an original domain, which contains exactly one representative for each equivalence class of points. Points outside of the original domain are addressed using integral offset vectors. Not all point sets admit a triangulation on the torus, but a grid of a constant number of identical copies of the point set always does [CT09].

An *alpha shape* is a sub-complex of a Delaunay triangulation that contains only those simplices whose empty circumsphere has radius bounded by a given α. Alpha shapes are available both for unweighted and for weighted points under the power distance [EM94].

Apollonius graphs are the dual of *additively weighted* Voronoi diagrams (a.k.a. *Apollonius diagrams*). They are available in 2D only, and support dynamic vertex insertion, deletion, and fast point location [KY02].

A *segment Delaunay graph* is dual to a Voronoi diagram for a set of line segments. It is available in 2D only but for both the Euclidean [Kar04] and the L_∞ metric [CDGP14]. Intersecting segments are supported. Geometric primitives use a combination of geometric and arithmetic filtering.

A *d-dimensional triangulation* is combinatorially represented as an abstract pure simplicial complex without boundary. The data structure supports specification of the dimension at either compile time or runtime. Both general triangulations and Delaunay triangulations are available.

CGAL also provides a generic framework [RKG07] for *kinetic data structures* where points move along polynomial trajectories. Specifically it implements kinetic Delaunay triangulations in 2D and 3D and kinetic regular triangulations in 3D.

MESH GENERATION

In Delaunay refinement meshing, the goal is to obtain a Delaunay triangulation for a given domain that (1) respects certain features and (2) whose simplices satisfy certain shape criteria (e.g., avoid small angles). To this end, Steiner points are added to subdivide constraint features or to destroy bad simplices.

The *3D mesh generator* [JAYB15] computes an isotropic simplicial mesh represented as a subcomplex of a 3D Delaunay triangulation. The concept to describe the input domain is very general: The only requirement is an oracle that can answer certain queries about the domain. For instance, does a query point belong to the domain and—if so—to which part of the domain? Domain models include isosurfaces defined by implicit functions, polyhedral surfaces, and segmented 3D images (for instance, from CT scans). The algorithm can handle lower-dimensional features in the input domain using restricted Delaunay triangulations [BO05] and protecting balls [CDR10]. Several different optimization procedures can optionally be used to remove slivers from the resulting mesh: Lloyd smoothing [DW03], odt-smoothing [ACYD05], a perturber [TSA09] and/or an exuder [CDE+00].

The *3D surface mesher* [RY07] handles an input domain that is a surface in 3D, represented as an oracle. The resulting mesh is represented as a two-dimensional subcomplex of a 3D Delaunay triangulation. Theoretical guarantees [BO05] regarding the topology depend on the local feature size. But the algorithm can also be run to guarantee a manifold when allowed to relax the shape criteria.

A given 2D constrained triangulation can be transformed into a *conforming* Delaunay or conforming Gabriel triangulation using Shewchuk's algorithm [She00]. *3D skin surfaces* provides an algorithm [KV07] to construct the skin surface [Ede99] for a given set of weighted points (i.e., balls) and a given shrink factor.

GEOMETRY PROCESSING

CGAL provides three algorithms to reconstruct a surface from a set of sample points. Poisson reconstruction [KBH06] computes an implicit function from a given set of points with oriented normal vectors. A corresponding surface can then be obtained using the surface mesher. *Scale-space reconstruction* [DMSL11] computes a surface that interpolates the input points by filtering an alpha shape depending on a scale variable. The scale-space method tends to be more robust with respect to outliers and noise. *Advancing Front Surface Reconstruction* [CD04] greedily adds Delaunay triangles subject to topological constraints, so as to maintain an orientable 2-manifold, possibly with boundary.

3D surface subdivision implements four subdivision methods (cf. [WW02]) that operate on a polyhedron: Catmull-Clark, Loop, Doo-Sabin, and $\sqrt{3}$-subdivision. In the opposite direction, *surface mesh simplification* aims to reduce the size of a given mesh while preserving the shape's characteristics as much as possible. The implementation collapses at every step an edge of minimum cost. The cost function appears as a parameter, with the Lindstrom-Turk model [LT99] as a default.

Surface subdivision decomposes a given mesh based on a shape diameter function (SDF) [SSC08], which aims to estimate the local object diameter. The decomposition is computed using a graph cut algorithm [BVZ01] that minimizes an energy function based on the SDF values.

In *surface deformation* we are given a mesh and a subset of vertices that are to be moved to given target positions. The goal is to deform the mesh and maintain its shape subject to these movement constraints. The implementation operates on a polyhedron and uses the as-rigid-as-possible algorithm [SA07] with an alternative energy function [CPSS10].

Planar parameterization operates on a polyhedron and aims to find a one-to-one mapping between the surface of the polyhedron and a planar domain. There are different desiderata regarding this mapping such as a small angle or area distortion or that the planar domain is convex. Several different methods are provided, such as Tutte barycentric mapping [Tut63], discrete conformal map [EDD+95], Floater mean value coordinates [Flo03], discrete authalic [DMA02], and least squares conformal maps [LPRM02], possibly in combination with boundary conditions.

Geodesic *shortest paths* on a *triangulated surface mesh* can be obtained using an algorithm by Xin and Wang [XW09]. *Triangulated surface mesh skeletonization* builds a 1D mean curvature skeleton [TAOZ12] for a given surface mesh. This skeleton is a curvilinear graph that aims to capture the topology of the mesh. Both algorithms work with a model of a generic *FaceListGraph* concept as an input.

Approximation of ridges and umbilics approximately determines regions of extremal curvature [CP05] on a given mesh, interpreted as a discretization of a smooth surface. The algorithm [CP05] needs the *local differential properties* of the mesh to be provided, which can be obtained using a companion package [CP08].

Point set processing provides tools to analyze and process 3D point sets, for instance, compute the average spacing, remove outliers, simplify, upsample, regularize or smooth the point set, or estimate the normals. The *shape* of a given *point set* with unoriented normals can be *detected* using a RANSAC (random sample consensus) algorithm [SWK07]. Supported shapes include plane, sphere, cylinder, cone and torus.

2D placement of streamlines generates streamlines (everywhere tangent to the field) for a given vector field, using a farthest point seeding strategy [MAD05].

OPTIMIZATION

Geometric optimization algorithms in CGAL fall into three categories: *Bounding Volumes, Inscribed Areas*, and *Optimal Distances*; see Table 68.2.3. In addition, *Principal Component Analysis* provides the computation of axis-aligned bounding boxes, centroids, and a linear least squares fitting in 2D and 3D.

As for combinatorial algorithms, there is a *Linear and Quadratic Programming Solver* to compute exact solutions for linear programs and convex quadratic programs. The solver uses a simplex-like algorithm, which is efficient if either the number of variables or the number of constraints is small [GS00]. CGAL also provides generic implementations of *monotone* [AKM$^+$87] and *sorted* [FJ84] *matrix search*.

TABLE 68.2.3   Geometric optimization.

DIM	ALGORITHM
2,3,d	Smallest enclosing disk/sphere of a point set [Wel91, GS98b, Gär99]
2,3,d	Smallest enclosing sphere of a set of spheres [FG04]
2	Smallest enclosing ellipse of a point set [Wel91, GS98a, GS98c]
2	Smallest enclosing rectangle [Tou83], parallelogram [STV$^+$95], and strip [Tou83] of a point set
d	Smallest enclosing spherical annulus of a point set [GS00]
d	Smallest enclosing ellipsoid of a point set (approximation using Khachyian's algorithm [Kha96])
2	Rectangular p-center, $2 \leq p \leq 4$ [Hof05, SW96]
2	Maximum (area and perimeter) inscribed k-gon of a convex polygon [AKM$^+$87]
2	Maximum area axis-aligned empty rectangle [Orl90]
d	Distance between two convex polytopes (given as a convex hull of a point set) [GS00]
3	Width of a point set [GH01]
2	All furthest neighbors for the vertices of a convex polygon [AKM$^+$87]

SPATIAL SEARCHING AND SORTING

Generic *range trees* and *segment trees* [BCKO08] can be interchangeably nested to form higher-dimensional search trees. A related structure in 1D is the *interval skip list*, a fully dynamic data structure to find all intervals that overlap a given point [Han91].

Spatial searching provides various queries based on k-d-trees: k-nearest and k-furthest neighbor searching, incremental nearest and incremental furthest neighbor searching [HS95]. All queries are available as exact and approximate searches. Query items can be points and other spatial objects.

Spatial sorting organizes a sequence of points so as to increase spatial coherence. In combination with carefully dosed randomness this is useful to speed up the localization step of incremental algorithms [ACR03, Buc09].

2D Range and Neighbor Search provides an interface to the dynamic 2D Delaunay triangulation for nearest neighbor, k-nearest neighbors, and range searching in the plane [MN00].

Fast Intersection and Distance Computation in 3D can be done using a hierarchy of axis-aligned bounded boxes stored in an AABB tree. The data structure supports intersection tests, intersection counting, and intersection reporting for a single query object, and computing a closest point for a given query point.

Intersecting Sequences of dD Iso-oriented Boxes efficiently computes all intersecting pairs among a collection of axis-aligned bounded boxes [ZE02]. The algorithm is also useful as a preprocessing heuristic for computing intersections among more general objects.

68.3 SOURCES AND RELATED MATERIAL

FURTHER READING

The LEDA user manual [MNSU] and the LEDA book [MN00] discuss the architecture, the implementation, and the use of the LEDA system.

The most relevant and up-to-date resource about CGAL is its online reference manual at <doc.cgal.org>. The design of CGAL and the reasons to use the C++ language are thoroughly covered in [FGK+00]. Generic programming aspects are discussed in [BKSV00]. The design of the CGAL kernel is presented in [HHK+07, PF11], the d-dimensional kernel in [MMN+98], the circular kernel in [EKP+04, DFMT02], and the spherical kernel in [CCLT09]. Older descriptions of design and motivation are in [Ove96, FGK+96, Vel97]. In particular, precision and robustness aspects are discussed in [Sch96], and the influence of different kernels in [Sch99, BBP01]. The most recent survey about CGAL is [Ber14].

RELATED CHAPTERS

Chapter 26: Convex hull computations
Chapter 27: Voronoi diagrams and Delaunay triangulations

Chapter 28: Arrangements
Chapter 29: Triangulations and mesh generation
Chapter 30: Polygons
Chapter 33: Visibility
Chapter 35: Curve and surface reconstruction
Chapter 38: Point location
Chapter 40: Range searching
Chapter 45: Robust geometric computation
Chapter 49: Linear programming
Chapter 55: Graph drawing
Chapter 56: Splines and geometric modeling
Chapter 59: Geographic information systems
Chapter 67: Software

REFERENCES

[AAAG95]   O. Aichholzer, D. Alberts, F. Aurenhammer, and B. Gärtner. A novel type of skeleton for polygons. *J. Universal Comput. Sci.*, 1:752–761, 1995.

[ACR03]   N. Amenta, S. Choi, and G. Rote. Incremental constructions con BRIO. In *Proc. 19th Sympos. Comput. Geom.*, pages 211–219, ACM Press, 2003.

[ACYD05]   P. Alliez, D. Cohen-Steiner, M. Yvinec, and M. Desbrun. Variational tetrahedral meshing. *ACM Trans. Graph.*, 24:617–625, 2005.

[AHU74]   A.V. Aho, J.E. Hopcroft, and J.D. Ullman. *The Design and Analysis of Computer Algorithms*. Addison-Wesley, Reading, 1974.

[AKM+87]   A. Aggarwal, M.M. Klawe, S. Moran, P.W. Shor, and R. Wilber. Geometric applications of a matrix-searching algorithm. *Algorithmica*, 2:195–208, 1987.

[And79]   A.M. Andrew. Another efficient algorithm for convex hulls in two dimensions. *Inform. Process. Lett.*, 9:216–219, 1979.

[Aus98]   M.H. Austern. *Generic Programming and the STL*. Addison-Wesley, Reading, 1998.

[BBP01]   H. Brönnimann, C. Burnikel, and S. Pion. Interval arithmetic yields efficient dynamic filters for computational geometry. *Discrete Appl. Math.*, 109:25–47, 2001.

[BCKO08]   M. de Berg, O. Cheong, M. van Kreveld, and M. Overmars. *Computational Geometry: Algorithms and Applications*, third edition. Springer, Berlin, 2008.

[BDH96]   C.B. Barber, D.P. Dobkin, and H. Huhdanpaa. The Quickhull algorithm for convex hulls. *ACM Trans. Math. Software*, 22:469–483, 1996.

[BDH09]   J.-D. Boissonnat, O. Devillers, and S. Hornus. Incremental construction of the Delaunay triangulation and the Delaunay graph in medium dimension. In *Proc. 25th Sympos. Comp. Geom.*, pages 208–216, ACM Press, 2009.

[BDP+02]   J.-D. Boissonnat, O. Devillers, S. Pion, M. Teillaud, and M. Yvinec. Triangulations in CGAL. *Comput. Geom.*, 22:5–19, 2002.

[Ber14]   E. Berberich. CGAL—Reliable geometric computing for academia and industry. In *Proc. 4th Internat. Congr. Math. Softw.*, vol. 8592 of *LNCS*, pages 191–197, Springer, Berlin, 2014.

[BFH+15]   A. Baram, E. Fogel, D. Halperin, M. Hemmer, and S. Morr. Exact Minkowski sums of polygons with holes. In *Proc. 23rd Eur. Sympos. Algorithms*, vol. 9294 of *LNCS*, pages 71–82, Springer, Berlin, 2015.

[BFMS00]   C. Burnikel, R. Fleischer, K. Mehlhorn, and S. Schirra. A strong and easily computable separation bound for arithmetic expressions involving radicals. *Algorithmica*, 27:87–99, 2000.

[BHH⁺14]   F. Bungiu, M. Hemmer, J. Hershberger, K. Huang, and A. Kröller. Efficient computation of visibility polygons. Preprint, arXiv:1403.3905, 2014.

[Bie95]   H. Bieri. Nef polyhedra: A brief introduction. In *Geometric Modelling*, vol. 10 of *Computing Supplement*, pages 43–60, Springer, Berlin, 1995.

[BK95]   M. Blum and S. Kannan. Designing programs that check their work. *J. ACM*, 42:269–291, 1995.

[BKSV00]   H. Brönnimann, L. Kettner, S. Schirra, and R.C. Veltkamp. Applications of the generic programming paradigm in the design of CGAL. In *Generic Programming*, vol. 1766 of *LNCS*, pages 206–217, Springer, Berlin, 2000.

[BLR93]   M. Blum, M. Luby, and R. Rubinfeld Self-testing/correcting with applications to numerical problems. *J. Comput. Syst. Sci.*, 47:549–595, 1993.

[BMS94a]   C. Burnikel, K. Mehlhorn, and S. Schirra. How to compute the Voronoi diagram of line segments: Theoretical and experimental results. In *Proc. 2nd Eur. Sympos. Algorithms*, vol. 855 of *LNCS*, pages 227–239, Springer, Berlin, 1994.

[BMS94b]   C. Burnikel, K. Mehlhorn, and S. Schirra. On degeneracy in geometric computations. In *Proc. 5th ACM-SIAM Sympos. Discrete Algorithms*, pages 16–23, 1994.

[BN02]   M. Bäsken and S. Näher. Geowin—A generic tool for interactive visualization of geometric algorithms. In *Software Visualization*, vol. 2269 of *LNCS*, pages 88–100, Springer, Berlin, 2002.

[BO05]   J.-D. Boissonnat and S. Oudot. Provably good sampling and meshing of surfaces. *Graph. Models*, 67:405–451, 2005.

[BSBK02]   M. Botsch, S. Steinberg, S. Bischoff, and L. Kobbelt. OpenMesh—A generic and efficient polygon mesh data structure. In *Proc. 1st OpenSG Symposium*, 2002.

[Buc09]   K. Buchin. Constructing Delaunay triangulations along space-filling curves. In *Proc. 17th Eur. Sympos. Algorithms*, vol. 5757 of *LNCS*, pages 119–130, Springer, Berlin, 2009.

[BVZ01]   Y. Boykov, O. Veksler, and R. Zabih. Fast approximate energy minimization via graph cuts. *IEEE Trans. Pattern Anal. Mach. Intell.*, 23:1222–1239, 2001.

[CCLT09]   P.M.M. de Castro, F. Cazals, S. Loriot, and M. Teillaud. Design of the CGAL 3D spherical kernel and application to arrangements of circles on a sphere. *Comput. Geom.*, 42:536–550, 2009.

[CD04]   D. Cohen-Steiner and F. Da. A greedy Delaunay-based surface reconstruction algorithm. *Visual Comput.*, 20:4–16, 2004.

[CDE⁺00]   S.-W. Cheng, T.K. Dey, H. Edelsbrunner, M.A. Facello, and S.-H. Teng. Sliver exudation. *J. ACM*, 47:883–904, 2000.

[CDGP14]   P. Cheilaris, S.K. Dey, M. Gabrani, and E. Papadopoulou. Implementing the L∞ segment Voronoi diagram in CGAL and applying in VLSI pattern analysis. In *Proc. 4th Internat. Congr. Math. Softw.*, vol. 8592 of *LNCS*, pages 198–205, Springer, Berlin, 2014.

[CDR10]   S.-W. Cheng, T.K. Dey, and E.A. Ramos. Delaunay refinement for piecewise smooth complexes. *Discrete Comput. Geom.*, 43(1):121–166, 2010.

[CLR90]   T.H. Cormen, C.E. Leiserson, R.L. Rivest, and C. Stein. *Introduction to Algorithms*. MIT Press, Cambridge, 1990; third edition 2009.

[CMS93]   K.L. Clarkson, K. Mehlhorn, and R. Seidel. Four results on randomized incremental constructions. *Comput. Geom.*, 3:185–212, 1993.

[CP05]   F. Cazals and M. Pouget. Topology driven algorithms for ridge extraction on meshes. Research Report 5526, INRIA Sophia-Antipolis, 2005.

[CP08]   F. Cazals and M. Pouget. Algorithm 889: Jet_fitting_3—A generic C++ package for estimating the differential properties on sampled surfaces via polynomial fitting. *ACM Trans. Math. Software*, 35:24, 2008.

[CPSS10]   I. Chao, U. Pinkall, P. Sanan, and P. Schröder. A simple geometric model for elastic deformations. *ACM Trans. Graph.*, 29:38, 2010.

[CT09]   M. Caroli and M. Teillaud. Computing 3D periodic triangulations. In *Proc. 17th Eur. Sympos. Algorithms*, vol. 5757 of *LNCS*, pages 37–48, Springer, Berlin, 2009.

[DDS09]   C. Dyken, M. Dæhlen, and T. Sevaldrud. Simultaneous curve simplification. *J. Geogr. Syst.*, 11:273–289, 2009.

[Dev02]   O. Devillers. The Delaunay hierarchy. *Internat. J. Found. Comput. Sci.*, 13:163–180, 2002.

[DFMT02]   O. Devillers, A. Fronville, B. Mourrain, and M. Teillaud. Algebraic methods and arithmetic filtering for exact predicates on circle arcs. *Comput. Geom.*, 22:119–142, 2002.

[DMA02]   M. Desbrun, M. Meyer, and P. Alliez. Intrinsic parameterizations of surface meshes. *Comput. Graph. Forum*, 21:209–218, 2002.

[DMSL11]   J. Digne, J.-M. Morel, C.-M. Souzani, and C. Lartigue. Scale space meshing of raw data point sets. *Comput. Graph. Forum*, 30:1630–1642, 2011.

[DPT02]   O. Devillers, S. Pion, and M. Teillaud. Walking in a triangulation. *Internat. J. Found. Comput. Sci.*, 13:181–199, 2002.

[DT11]   O. Devillers and M. Teillaud. Perturbations for Delaunay and weighted Delaunay 3D triangulations. *Comput. Geom.*, 44:160–168, 2011.

[DT14]   G. Damiand and M. Teillaud. A generic implementation of dD combinatorial maps in CGAL. *Procedia Engineering*, 82:46–58, 2014.

[DW03]   Q. Du and D. Wang. Tetrahedral mesh generation and optimization based on centroidal Voronoi tessellations. *Internat. J. Numer. Methods Eng.*, 56:1355–1373, 2003.

[EDD+95]   M. Eck, T. DeRose, T. Duchamp, H. Hoppe, M. Lounsbery, and W. Stuetzle. Multiresolution analysis of arbitrary meshes. In *Proc. 22nd Conf. Comput. Graph. Interactive Tech*, pages 173–180, ACM Press, 1995.

[Ede99]   H. Edelsbrunner. Deformable smooth surface design. *Discrete Comput. Geom.*, 21:87–115, 1999.

[EE99]   D. Eppstein and J. Erickson. Raising roofs, crashing cycles, and playing pool: Applications of a data structure for finding pairwise interactions. *Discrete Comput. Geom.*, 22:569–592, 1999.

[EKP+04]   I.Z. Emiris, A. Kakargias, S. Pion, M. Teillaud, and E.P. Tsigaridas. Towards and open curved kernel. In *Proc. 20th Sympos. Comput. Geom.*, pages 438–446, ACM Press, 2004.

[EM94]   H. Edelsbrunner and E.P. Mücke. Three-dimensional alpha shapes. *ACM Trans. Graph.*, 13:43–72, 1994.

[ES96]   H. Edelsbrunner and N.R. Shah. Incremental topological flipping works for regular triangulations. *Algorithmica*, 15:223–241, 1996.

[FG04]   K. Fischer and B. Gärtner. The smallest enclosing ball of balls: Combinatorial structure and algorithms. *Internat. J. Comput. Geom. Appl.*, 14:341–378, 2004.

[FGK⁺96]   A. Fabri, G.-J. Giezeman, L. Kettner, S. Schirra, and S. Schönherr. The CGAL kernel: A basis for geometric computation. In *Proc. 1st ACM Workshop Appl. Comput. Geom.*, vol. 1148 of *LNCS*, pages 191–202, Springer, Berlin, 1996.

[FGK⁺00]   A. Fabri, G.-J. Giezeman, L. Kettner, S. Schirra, and S. Schönherr. On the design of CGAL a computational geometry algorithms library. *Softw. Pract. Exp.*, 30:1167–1202, 2000.

[FHW12]   E. Fogel, D. Halperin, and R. Wein. *CGAL Arrangements and Their Applications— A Step-by-Step Guide.* Vol. 7 of *Geometry and Computing*, Springer, Berlin, 2012.

[FJ84]   G.N. Frederickson and D.B. Johnson. Generalized selection and ranking: sorted matrices. *SIAM J. Comput.*, 13:14–30, 1984.

[Flo03]   M.S. Floater. Mean value coordinates. *Comput. Aided Geom. Des.*, 20:19–27, 2003.

[FO98]   P. Felkel and S. Obdržálek. Straight skeleton implementation. In *Proc. 14th Spring Conf. Comput. Graph.*, pages 210–218, 1998.

[Gär99]   B. Gärtner. Fast and robust smallest enclosing balls. In *Proc. 7th Eur. Sympos. Algorithms*, vol. 1643 of *LNCS*, pages 325–338, Springer, Berlin, 1999.

[GH01]   B. Gärtner and T. Herrmann. Computing the width of a point set in 3-space. In *Proc. 13th Canad. Conf. Comput. Geom.*, pages 101–103, 2001.

[GJ⁺10]   G. Guennebaud, B. Jacob, et al. Eigen v3. `http://eigen.tuxfamily.org`, 2010.

[GPL07]   GNU general public license V3+. URL: `http://www.gnu.org/licenses/gpl.html`, June 29, 2007.

[Gra72]   R.L. Graham. An efficient algorithm for determining the convex hull of a finite planar set. *Inform. Process. Lett.*, 1:132–133, 1972.

[Gre83]   D.H. Greene. The decomposition of polygons into convex parts. In *Computational Geomety*, vol. 1 of *Adv. Comput. Res.*, pages 235–259, JAI Press, Greenwich, 1983.

[GS98a]   B. Gärtner and S. Schönherr. Exact primitives for smallest enclosing ellipses. *Inform. Process. Lett.*, 68:33–38, 1998.

[GS98b]   B. Gärtner and S. Schönherr. Smallest enclosing circles—An exact and generic implementation in C++. Technical Report B 98–04, Informatik, Freie Universität Berlin, 1998.

[GS98c]   B. Gärtner and S. Schönherr. Smallest enclosing ellipses—An exact and generic implementation in C++. Technical Report B 98–05, Informatik, Freie Universität Berlin, 1998.

[GS00]   B. Gärtner and S. Schönherr. An efficient, exact, and generic quadratic programming solver for geometric optimization. In *Proc. 16th Sympos. Comput. Geom.*, pages 110–118, ACM Press, 2000.

[Gt14]   T. Granlund and the GMP development team. *The GNU Multiple Precision Arithmetic Library*, 6th edition. Manual, `https://gmplib.org/`, 2014.

[Hac09]   P. Hachenberger. Exact Minkowksi sums of polyhedra and exact and efficient decomposition of polyhedra into convex pieces. *Algorithmica*, 55:329–345, 2009.

[Han91]   E.N. Hanson. The interval skip list: A data structure for finding all intervals that overlap a point. In *Proc. 2nd Workshop Algorithms Data Struct.*, vol. 519 of *LNCS*, pages 153–164, Springer, Berlin, 1991.

[HH09]   I. Haran and D. Halperin. An experimental study of point location in planar arrangements in CGAL. *ACM J. Exp. Algorithms*, 13:3, 2009.

[HHK⁺07]   S. Hert, M. Hoffmann, L. Kettner, S. Pion, and M. Seel. An adaptable and extensible geometry kernel. *Comput. Geom.*, 38:16–36, 2007.

[HKH12]   M. Hemmer, M. Kleinbort, and D. Halperin. Improved implementation of point location in general two-dimensional subdivisions. In *Proc. 20th Eur. Sympos. Algorithms*, vol. 7501 of *LNCS*, pages 611–623, Springer, Berlin, 2012.

[HKM07]   P. Hachenberger, L. Kettner, and K. Mehlhorn. Boolean operations on 3D selective Nef complexes: Data structure, algorithms, optimized implementation and experiments. *Comput. Geom.*, 38:64–99, 2007.

[HM83]   S. Hertel and K. Mehlhorn. Fast triangulation of simple polygons. In *Proc. 4th Internat. Conf. Found. Comput. Theory*, vol. 158 of *LNCS*, pages 207–218, Springer, Berlin, 1983.

[HMN96]   C. Hundack, K. Mehlhorn, and S. Näher. A simple linear time algorithm for identifying Kuratowski subgraphs of non-planar graphs. Unpublished, 1996.

[Hof05]   M. Hoffmann. A simple linear algorithm for computing rectilinear three-centers. *Comput. Geom.*, 31:150–165, 2005.

[HP02]   D. Halperin and E. Packer. Iterated snap rounding. *Comput. Geom.*, 23:209–225, 2002.

[HS95]   G.R. Hjaltason and H. Samet. Ranking in spatial databases. In *Proc. 4th Sympos. Advances Spatial Databases*, vol. 951 of *LNCS*, pages 83–95, Springer, Berlin, 1995.

[JAYB15]   C. Jamin, P. Alliez, M. Yvinec, and J.-D. Boissonnat. CGALmesh: A generic framework for Delaunay mesh generation. *ACM Trans. Math. Softw.*, 41:23, 2015.

[Kar04]   M.I. Karavelas. A robust and efficient implementation for the segment Voronoi diagram. In *Proc. 1st Internat. Sympos. Voronoi Diagrams*, pages 51–62, 2004.

[KBH06]   M.M. Kazhdan, M. Bolitho, and H. Hoppe. Poisson surface reconstruction. In *Proc. 4th Eurographics Sympos. Geom. Process.*, pages 61–70, 2006.

[Ket98]   L. Kettner. Designing a data structure for polyhedral surfaces. In *Proc. 14th Sympos. Comput. Geom.*, pages 146–154, ACM Press, 1998.

[Kha96]   L.G. Khachiyan. Rounding of polytopes in the real number model of computation. *Math. Oper. Res.*, 21:307–320, 1996.

[Kin90]   J.H. Kingston. *Algorithms and Data Structures*. Addison-Wesley, Reading, 1990.

[KLPY99]   V. Karamcheti, C. Li, I. Pechtchanski, and C.K. Yap. A core library for robust numeric and geometric computation. In *Proc. 15th Sympos. Comput. Geom.*, pages 351–359, ACM Press, 1999.

[KV07]   N. Kruithof and G. Vegter. Meshing skin surfaces with certified topology. *Comput. Geom.*, 36:166–182, 2007.

[KY02]   M. Karavelas and M. Yvinec. Dynamic additively weighted Voronoi diagrams in 2D. In *Proc. 10th Eur. Sympos. Algorithms*, vol. 2461 of *LNCS*, pages 586–598, Springer, Berlin, 2002.

[LPRM02]   B. Lévy, S. Petitjean, N. Ray, and J. Maillot. Least squares conformal maps for automatic texture atlas generation. *ACM Trans. Graph.*, 21:362–371, 2002.

[LT99]   P. Lindstrom and G. Turk. Evaluation of memoryless simplification. *IEEE Trans. Vis. Comput. Graph.*, 5:98–115, 1999.

[MAD05]   A. Mebarki, P. Alliez, and O. Devillers. Farthest point seeding for efficient placement of streamlines. In *Proc. 16th IEEE Visualization*, pages 479–486, 2005.

[Meh84]   K. Mehlhorn. *Data Structures and Algorithms 1, 2, and 3*. Springer, Berlin, 1984.

[MM96]   K. Mehlhorn and P. Mutzel. On the embedding phase of the Hopcroft and Tarjan planarity testing algorithm. *Algorithmica*, 16:233–242, 1996.

[MMN+98]   K. Mehlhorn, M. Müller, S. Näher, S. Schirra, M. Seel, C. Uhrig, and J. Ziegler. A computational basis for higher-dimensional computational geometry and applications. *Comput. Geom.*, 10:289–303, 1998.

[MMNS11]   R.M. McConnel, K. Mehlhorn, S. Näher, and P. Schweitzer. Certifying algorithms. *Computer Science Review*, 5:119–161, 2011.

[MN94a]   K. Mehlhorn and S. Näher. Implementation of a sweep line algorithm for the straight line segment intersection problem. Technical Report MPI-I-94-160, Max-Planck-Institut für Informatik, 1994.

[MN94b]   K. Mehlhorn and S. Näher. The implementation of geometric algorithms. In *Proc. 13th IFIP World Computer Congress*, vol. 1, pages 223–231, Elsevier, Amsterdam, 1994.

[MN00]   K. Mehlhorn and S. Näher. *LEDA: A Platform for Combinatorial and Geometric Computing*. Cambridge University Press, 2000.

[MNS+99]   K. Mehlhorn, S. Näher, M. Seel, R. Seidel, T. Schilz, S. Schirra, and C. Uhrig. Checking geometric programs or verification of geometric structures. *Comput. Geom.*, 12:85–103, 1999.

[MNSU]   K. Mehlhorn, S. Näher, M. Seel, and C. Uhrig. The LEDA User Manual. Technical report, Max-Planck-Institut für Informatik. http://www.mpi-sb.mpg.de/LEDA/leda.html.

[MP07]   G. Melquiond and S. Pion. Formally certified floating-point filters for homogeneous geometric predicates. *ITA*, 41:57–69, 2007.

[MS01]   K. Mehlhorn and S. Schirra. Exact computation with leda_real — Theory and geometric applications. In *Symbolic Algebraic Methods and Verification Methods*, pages 163–172, Springer, Vienna, 2001.

[MS03]   K. Mehlhorn and M. Seel. Infimaximal frames: A technique for making lines look like segments. *Internat. J. Comput. Geom. Appl.*, 13:241–255, 2003.

[Mye95]   N.C. Myers. Traits: A new and useful template technique. *C++ Report*, 1995. http://www.cantrip.org/traits.html.

[Nef78]   W. Nef. *Beiträge zur Theorie der Polyeder*. Herbert Lang, Bern, 1978.

[NH93]   J. Nievergelt and K.H. Hinrichs. *Algorithms and Data Structures*. Prentice Hall, Upper Saddle River, 1993.

[O'R94]   J. O'Rourke. *Computational Geometry in C*. Cambridge University Press, 1994.

[Orl90]   M. Orlowski. A new algorithm for the largest empty rectangle problem. *Algorithmica*, 5:65–73, 1990.

[Ove96]   M.H. Overmars. Designing the Computational Geometry Algorithms Library CGAL. In *Proc. 1st ACM Workshop Appl. Comput. Geom.*, vol. 1148 of *LNCS*, pages 53–58, Springer, Berlin, 1996.

[PF11]   S. Pion and A. Fabri. A generic lazy evaluation scheme for exact geometric computations. *Sci. Comput. Program.*, 76(4):307–323, 2011.

[RKG07]   D. Russel, M.I. Karavelas, and L.J. Guibas. A package for exact kinetic data structures and sweepline algorithms. *Comput. Geom.*, 38:111–127, 2007.

[RY07]   L. Rineau and M. Yvinec. A generic software design for Delaunay refinement meshing. *Comput. Geom.*, 38:100–110, 2007.

[SA07]   O. Sorkine and M. Alexa. As-rigid-as-possible surface modeling. In *Proc. 5th Euro-graphics Symp. Geom. Process.*, pages 109–116, 2007.

[SB11]   D. Sieger and M. Botsch. Design, implementation, and evaluation of the sur-face_mesh data structure. In *Proc. 20th Int. Meshing Roundtable*, pages 533–550, Springer, Berlin, 2011.

[Sch96]   S. Schirra. *Designing a Computational Geometry Algorithms Library*. Lecture Notes for Advanced School on Algorithmic Foundations of Geographic Information Systems, CISM, Udine, 1996.

[Sch99]   S. Schirra. A case study on the cost of geometric computing. In *Proc. 1st Workshop Algorithm Eng. Exper.*, vol. 1619 of *LNCS*, pages 156–176, Springer, Berlin, 1999.

[Sch14]   B. Schäling. *The Boost C++ Libraries*, 2nd edition. XML Press, Laguna Hills, 2014.

[Sed91]   R. Sedgewick. *Algorithms*. Addison-Wesley, Reading, 1991.

[See94]   M. Seel. Eine Implementierung abstrakter Voronoidiagramme. Master's thesis, Fachbereich Informatik, Universität des Saarlandes, Saarbrücken, 1994.

[See01]   M. Seel. Implementation of planar Nef polyhedra. Research Report MPI-I-2001-1-003, Max-Planck-Institut für Informatik, 2001.

[She00]   J.R. Shewchuk. Mesh generation for domains with small angles. In *Proc. 16th Sympos. Comput. Geom.*, pages 1–10, ACM Press, 2000.

[SHRH11]   O. Salzman, M. Hemmer, B. Raveh, and D. Halperin. Motion planning via manifold samples. In *Proc. 19th Eur. Sympos. Algorithms*, vol. 6942 of *LNCS*, pages 493–505, Springer, Berlin, 2011.

[SLL02]   J.G. Siek, L.-Q. Lee, and A. Lumsdaine. *The Boost Graph Library—User Guide and Reference Manual*. C++ in-depth series. Prentice Hall, Upper Saddle River, 2002.

[SSC08]   L. Shapira, A. Shamir, and D. Cohen-Or. Consistent mesh partitioning and skele-tonisation using the shape diameter function. *Visual Comput.*, 24:249–259, 2008.

[STV+95]   C. Schwarz, J. Teich, A. Vainshtein, E. Welzl, and B.L. Evans. Minimal enclosing parallelogram with application. In *Proc. 11th Sympos. Comput. Geom.*, pages C34–C35, ACM Press, 1995.

[SW96]   M. Sharir and E. Welzl. Rectilinear and polygonal p-piercing and p-center problems. In *Proc. 12th Sympos. Comput. Geom.*, pages 122–132, ACM Press, 1996.

[SWK07]   R. Schnabel, R. Wahl, and R. Klein. Efficient RANSAC for point-cloud shape detection. *Comput. Graph. Forum*, 26:214–226, 2007.

[TAOZ12]   A. Tagliasacchi, I. Alhashim, M. Olson, and H. Zhang. Mean curvature skeletons. *Comput. Graph. Forum*, 31:1735–1744, 2012.

[Tar83]   R.E. Tarjan. Data structures and network algorithms. In *CBMS-NSF Regional Conference Series in Applied Mathematics*, vol. 44, 1983.

[Tou83]   G.T. Toussaint. Solving geometric problems with the rotating calipers. In *Proc. IEEE MELECON '83*, pages A10.02/1–4, 1983.

[TSA09]   J. Tournois, R. Srinivasan, and P. Alliez. Perturbing slivers in 3D Delaunay meshes. In *Proc. 18th Int. Meshing Roundtable*, pages 157–173, Springer, Berlin, 2009.

[Tut63]   W.T. Tutte. How to draw a graph. *Proc. London Math. Soc.*, 13:743–768, 1963.

[Vel97]   R.C. Veltkamp. Generic programming in CGAL, the computational geometry algo-rithms library. In *Proc. 6th Eurographics Workshop on Programming Paradigms in Graphics*, 1997.

[Wei06]   R. Wein. Exact and efficient construction of planar Minkowski sums using the convolution method. In *Proc. 14th Eur. Sympos. Algorithms*, vol. 4168 of *LNCS*, pages 829–840, Springer, Berlin, 2006.

[Wei07]   R. Wein. Exact and approximate construction of offset polygons. *Comput. Aided Des.*, 39:518–527, 2007.

[Wel91]   E. Welzl. Smallest enclosing disks (balls and ellipsoids). In *New Results and New Trends in Computer Science*, vol. 555 of *LNCS*, pages 359–370, Springer, Berlin, 1991.

[Woo93]   D. Wood. *Data Structures, Algorithms, and Performance*. Addison-Wesley, Reading, 1993.

[WW02]   J. Warren and H. Weimer. *Subdivision Methods for Geometric Design—A Constructive Approach*. Morgan-Kaufmann, San Francisco, 2002.

[Wyk88]   C.J. van Wyk. *Data Structures and C Programs*. Addison-Wesley, Reading, 1988.

[XW09]   S.-Q. Xin and G.-J. Wang. Improving Chen and Han's algorithm on the discrete geodesic problem. *ACM Trans. Graph.*, 28:104:1–104:8, 2009.

[ZE02]   A. Zomorodian and H. Edelsbrunner. Fast software for box intersections. *Internat. J. Comput. Geom. Appl.*, 12:143–172, 2002.

INDEX OF CITED AUTHORS

The name of each author cited in a chapter appears only once with a reference to that chapter; either to its first appearance in the chapter's bibliography, or, if not cited there, to its first appearance in the text of the chapter.

INDEX OF DEFINED TERMS

世界著名数学家 A. W. Tucker 曾指出：

连续数学的成见约束了离散数学的成长，然而将来会在与组织复杂性问题有关的组合数学中有一大挑战. 19 世纪克服了简单性，20 世纪前半叶取得了解决非组织复杂性问题的胜利. 科学必须在两个世纪后半叶学会解决组织复杂性问题，因而组合数学对于人类的未来具有最大的重要性.

说白了就是 21 世纪是组合的世纪.

本书是一部包罗万象的英文版组合学的工具书，中文书名可译为《离散与计算几何手册(第三版)》.

本书的主要作者有 3 位：

雅各布·E. 古德曼(Jacob E. Goodman)，美国数学家，他与 Richard Pollack 是《离散与计算几何》杂志的创始编辑. 2008 年他从纽约大学城市学院退休. 他在代数几何、组合数学、离散几何等领域发表论文 60 余篇，退休后大部分时间都在作曲和创作悬疑小说. 他曾于 1990 年获得 MAA（美国数学协会）颁发的福特奖(Lester R. Ford Award)，他还是 AMS（美国数学学会）的研究员.

约瑟夫·奥罗克(Joseph O'Rourke)，美国数学家，史密斯学院计算机科学和数学教授，统计和数据科学项目主任. 他的研究领域是计算几何，并为几何计算开发算法. 除本书外，他还撰写或与人合著了五本书. 他最近的著作《如何折叠》是为高中生而写的.

乔鲍·D.托特(Csaba D. Tóth),美国数学家,加州州立大学北岭分校数学教授,也是波士顿都市区塔夫茨大学计算机科学的客座教授.其主要研究方向为低维空间的分层细分、拓扑图论和几何优化,发表过关于离散与计算几何的论文90多篇.

正如本书作者在前言中所述:

尽管近年来有关离散几何和计算几何的高质量书籍和期刊激增,但迄今为止,还没有任何一本能够涵盖这两个领域的所有重要内容的参考书籍可供非专业人士和专家共同使用.本书旨在使人们更容易获得这些几何领域中最重要的结果和方法,无论是学术界中数学和计算机科学的研究人员,或者是专业领域——运筹学、分子生物学和机器人学等不同领域中的从业者,都可以从本书中获益.

近年来,作为一个整体的离散数学经历了一次重要的发展,那就是其重要部分离散几何的实质性发展.其发展的部分原因是强大的计算机的出现,以及最近计算几何中相对较新的领域中活动的激增.离散几何和计算几何之间的这种综合是本手册的核心,它们其中一个领域的方法和见解都激发了人们对另一个领域的新理解.

术语"离散几何"曾经主要代表填装、覆盖和平铺等领域,后来逐渐发展到包括组合几何、凸多面体,以及平面和更高维度中点、线、面、圆和其他几何对象的排列等领域.同样,"计算几何"不久前还简单地指几何算法的设计和分析,近年来扩大了范围,现在意味着从计算的角度研究几何问题,也包括计算凸性、计算拓扑,以及涉及排列和多面体的组合复杂性问题.从中我们可以清楚地看出,现在这两个领域之间存在显著的重叠,事实上,这种重叠也成为一种实践,因为数学家和计算机科学家发现他们在研究相同的几何问题,并成功地建立了合作关系.

与此同时,越来越多的领域可以应用这项研究的成果,包括工程、晶体学、计算机辅助设计、制造、运筹学、地理信息系统、机器人、纠错码、断层扫描、几何建模、计算机图形学、组合优化、计算机视觉、模式识别和实体建模.

考虑到这一点,很明显,一本包含离散几何和计算几何最重要结果的手册不仅会使这两个领域的工作人员受益,还会使组合学、图论、几何概率和实代数几何等相关领域的工作人员受益,而且该结果本身的使用者(在工业和学术领域)也会受益.本书旨在填补这一角色.我们相信,对于几何学和几何计算的研究人员以及在工作中使用几何工具的专业人士来说,本书都将成为不可或缺的工作工具.

本书涵盖了离散和计算几何两个领域的广泛主题,还有很多应用领域中的主题,具体包括几何数据结构、多胞腔和多面体、凸包和三角剖分算法、填装和覆盖、沃罗诺伊图式、组合几何问题、计算凸性、最短路径和网络、计算实代数几何、几何排列及其复杂性、几何重构问题、随机化和去随机化技术、射线射击、几何中的并行算法、定向拟阵、计算拓扑、数学规划、运动规划、球填充、计算机图形学、机器人学、晶体学,等等.本书最后一章提供了可用软件的列表.结果以定理、算法和表格的形式呈现,每个技术术语都在术语表中进行了仔细定义,该术语表位于首次使用该术语的章节之前.有许多例子和图表来说明我们所讨论的想法和大量未解决的问题.

本书第 1 版共分为六个部分.前两个部分是关于组合和离散几何以及多胞腔和多面体的,其中涉及基本几何对象,如平面排列、格和凸多面体.下一部分是关于算法和几何复杂性的,将从计算的角度讨论这些基本几何对象.第四部分和第五部分是关于数据结构和计算技术的,讨论了跨越几何对象范围的各种计算方法,例如随机化和去随机化,几何中的并行算法,以及用于搜索和查找点位置的有效数据结构.第六部分在书中所占篇幅最大,所含章节包含了离散和计算几何的 14 个应用领域,包括低维线性规划、组合优化、运动规划、机器人学、计算机图形学、模式识别、图形绘制、样条函数、制造、实体建模、框架刚度、场景分析、纠错码和晶体学.该部分以第 15 章结束,它是与书中涵盖的各个领域相关的可用软件的最新汇编.其后是一个综合索引,包括专有名词以及手册正文中定义的所有术语.

关于参考文献:由于提供手册中所包含的所有数千个结果的完整参考文献是令人望而却步的,所以我们必须在很大程度上限制自己对最重要结果的参考范围,或者是对那些时间太近而没有被包括在早期的调查书籍或文章中的参考文献的引用.对于其余部分,我们提供了有关本书所涵盖主题的易于获得的相关调查的参考资料(带注释),这些调查本身包含了广泛的参考书目.通过这种方式可以帮助那些希望追寻较早结果源头的读者开展工作.

为了方便国内读者使用,我们翻译了本书的目录如下:

组合与离散几何

最近有人曾这样评论趋势专家平克（Daniel H. Pink）的著作《驱动力》，他说："（此书）总结了近50年来有关积极性的几乎所有社会科学研究成果".

如果将社会科学改为离散与计算机几何也同样可以用于评价本书.

平克的书中引用了新闻记者、美国国会女议员卢斯（Clare Booth Luce）告诉肯尼迪总统的一句话："一个伟大的人，就是一句话."这意味着如果一位领导人有着清晰而强烈的目标意识，那么我们用一句话就能总结出他的人生.

平克认为，"一句话"理念对任何人都适用. 因此，不管你是一个国家的总统，还是一个园艺俱乐部的主席，要让自己的人生朝着更伟大的目标进发，就从这个大问题开始："你的那句话是什么？"

这句话必须有排他性，也就是非你莫属，就像只要提到"相对论"，除了爱因斯坦，谁都不敢认领.

如果我们的工作（无论是所有工作还是单项工作）像各种人才计划或各级科技奖励的许多申报书那样，必须用几页纸甚至厚厚的一本才能说清楚，才能有辨识度和标签性，那么就要彻底放弃已有工作得诺贝尔奖的梦想,趁早转向第二件事情:尽快明确"你的那句话是什么?""而今迈步从头越".

毫无疑问,通向"你的那句话"之路,就如唐僧师徒西天取经之路,必然困难重重.唐僧也有一个"一句话"策略,特别值得借鉴.

取经团队之所以最终能取回真经,因素固然很多,但最重要的一点,在我看来,就是唐僧无论何时、无论何地、无论面对谁(从妖怪到国王),都念念不忘一句话:"贫僧是东土大唐差往西天取经者".

仔细琢磨,这么短短一句话却解决了取经团队的根本性问题:我是谁? 从哪里来? 到哪里去? 去干什么?

其实,只要念念不忘这句话并一路向西,不一定是唐僧,换成任何人,都可以带领团队,如乌巢禅师所说,路途虽远,终须有到之日.

如要也用"一句话"总结本书,或许可以说:一书在手,能搜尽收!

刘培志

2022. 12. 9

于哈工大

书　名	出版时间	定　价	编号
数学物理大百科全书.第1卷(英文)	2016－01	418.00	508
数学物理大百科全书.第2卷(英文)	2016－01	408.00	509
数学物理大百科全书.第3卷(英文)	2016－01	396.00	510
数学物理大百科全书.第4卷(英文)	2016－01	408.00	511
数学物理大百科全书.第5卷(英文)	2016－01	368.00	512
zeta函数,q-zeta函数,相伴级数与积分(英文)	2015－08	88.00	513
微分形式:理论与练习(英文)	2015－08	58.00	514
离散与微分包含的逼近和优化(英文)	2015－08	58.00	515
艾伦·图灵:他的工作与影响(英文)	2016－01	98.00	560
测度理论概率导论,第2版(英文)	2016－01	88.00	561
带有潜在故障恢复系统的半马尔柯夫模型控制(英文)	2016－01	98.00	562
数学分析原理(英文)	2016－01	88.00	563
随机偏微分方程的有效动力学(英文)	2016－01	88.00	564
图的谱半径(英文)	2016－01	58.00	565
量子机器学习中数据挖掘的量子计算方法(英文)	2016－01	98.00	566
量子物理的非常规方法(英文)	2016－01	118.00	567
运输过程的统一非局部理论:广义波尔兹曼物理动力学,第2版(英文)	2016－01	198.00	568
量子力学与经典力学之间的联系在原子、分子及电动力学系统建模中的应用(英文)	2016－01	58.00	569
算术域(英文)	2018－01	158.00	821
高等数学竞赛:1962—1991年的米洛克斯·史怀哲竞赛(英文)	2018－01	128.00	822
用数学奥林匹克精神解决数论问题(英文)	2018－01	108.00	823
代数几何(德文)	2018－04	68.00	824
丢番图逼近论(英文)	2018－01	78.00	825
代数几何学基础教程(英文)	2018－01	98.00	826
解析数论入门课程(英文)	2018－01	78.00	827
数论中的丢番图问题(英文)	2018－01	78.00	829
数论(梦幻之旅):第五届中日数论研讨会演讲集(英文)	2018－01	68.00	830
数论新应用(英文)	2018－01	68.00	831
数论(英文)	2018－01	78.00	832

刘培杰数学工作室
已出版（即将出版）图书目录——原版影印

书　名	出版时间	定　价	编号
湍流十讲（英文）	2018—04	108.00	886
无穷维李代数：第 3 版（英文）	2018—04	98.00	887
等值、不变量和对称性（英文）	2018—04	78.00	888
解析数论（英文）	2018—09	78.00	889
《数学原理》的演化：伯特兰·罗素撰写第二版时的 手稿与笔记（英文）	2018—04	108.00	890
哈密尔顿数学论文集（第 4 卷）：几何学、分析学、天文学、 概率和有限差分等（英文）	2019—05	108.00	891
偏微分方程全局吸引子的特性（英文）	2018—09	108.00	979
整函数与下调和函数（英文）	2018—09	118.00	980
幂等分析（英文）	2018—09	118.00	981
李群，离散子群与不变量理论（英文）	2018—09	108.00	982
动力系统与统计力学（英文）	2018—09	118.00	983
表示论与动力系统（英文）	2018—09	118.00	984
分析学练习.第 1 部分（英文）	2021—01	88.00	1247
分析学练习.第 2 部分，非线性分析（英文）	2021—01	88.00	1248
初级统计学：循序渐进的方法：第 10 版（英文）	2019—05	68.00	1067
工程师与科学家微分方程用书：第 4 版（英文）	2019—07	58.00	1068
大学代数与三角学（英文）	2019—06	78.00	1069
培养数学能力的途径（英文）	2019—07	38.00	1070
工程师与科学家统计学：第 4 版（英文）	2019—06	58.00	1071
贸易与经济中的应用统计学：第 6 版（英文）	2019—06	58.00	1072
傅立叶级数和边值问题：第 8 版（英文）	2019—05	48.00	1073
通往天文学的途径：第 5 版（英文）	2019—05	58.00	1074
拉马努金笔记.第 1 卷（英文）	2019—06	165.00	1078
拉马努金笔记.第 2 卷（英文）	2019—06	165.00	1079
拉马努金笔记.第 3 卷（英文）	2019—06	165.00	1080
拉马努金笔记.第 4 卷（英文）	2019—06	165.00	1081
拉马努金笔记.第 5 卷（英文）	2019—06	165.00	1082
拉马努金遗失笔记.第 1 卷（英文）	2019—06	109.00	1083
拉马努金遗失笔记.第 2 卷（英文）	2019—06	109.00	1084
拉马努金遗失笔记.第 3 卷（英文）	2019—06	109.00	1085
拉马努金遗失笔记.第 4 卷（英文）	2019—06	109.00	1086
数论：1976 年纽约洛克菲勒大学数论会议记录（英文）	2020—06	68.00	1145
数论：卡本代尔 1979：1979 年在南伊利诺伊卡本代尔大学 举行的数论会议记录（英文）	2020—06	78.00	1146
数论：诺德韦克豪特 1983：1983 年在诺德韦克豪特举行的 Journees Arithmetiques 数论大会会议记录（英文）	2020—06	68.00	1147
数论：1985—1988 年在纽约城市大学研究生院和大学中心 举办的研讨会（英文）	2020—06	68.00	1148

书　名	出版时间	定　价	编号
数论:1987 年在乌尔姆举行的 Journees Arithmetiques 数论大会会议记录(英文)	2020－06	68.00	1149
数论:马德拉斯 1987:1987 年在马德拉斯安娜大学举行的国际拉马努金百年纪念大会会议记录(英文)	2020－06	68.00	1150
解析数论:1988 年在东京举行的日法研讨会会议记录(英文)	2020－06	68.00	1151
解析数论:2002 年在意大利切特拉罗举行的 C. I. M. E. 暑期班演讲集(英文)	2020－06	68.00	1152
量子世界中的蝴蝶:最迷人的量子分形故事(英文)	2020－06	118.00	1157
走进量子力学(英文)	2020－06	118.00	1158
计算物理学概论(英文)	2020－06	48.00	1159
物质,空间和时间的理论:量子理论(英文)	2020－10	48.00	1160
物质,空间和时间的理论:经典理论(英文)	2020－10	48.00	1161
量子场理论:解释世界的神秘背景(英文)	2020－07	38.00	1162
计算物理学概论(英文)	2020－06	48.00	1163
行星状星云(英文)	2020－10	38.00	1164
基本宇宙学:从亚里士多德的宇宙到大爆炸(英文)	2020－08	58.00	1165
数学磁流体力学(英文)	2020－07	58.00	1166
计算科学:第 1 卷,计算的科学(日文)	2020－07	88.00	1167
计算科学:第 2 卷,计算与宇宙(日文)	2020－07	88.00	1168
计算科学:第 3 卷,计算与物质(日文)	2020－07	88.00	1169
计算科学:第 4 卷,计算与生命(日文)	2020－07	88.00	1170
计算科学:第 5 卷,计算与地球环境(日文)	2020－07	88.00	1171
计算科学:第 6 卷,计算与社会(日文)	2020－07	88.00	1172
计算科学.别卷,超级计算机(日文)	2020－07	88.00	1173
多复变函数论(日文)	2022－06	78.00	1518
复变函数入门(日文)	2022－06	78.00	1523
代数与数论:综合方法(英文)	2020－10	78.00	1185
复分析:现代函数理论第一课(英文)	2020－07	58.00	1186
斐波那契数列和卡特兰数:导论(英文)	2020－10	68.00	1187
组合推理:计数艺术介绍(英文)	2020－07	88.00	1188
二次互反律的傅里叶分析证明(英文)	2020－07	48.00	1189
旋瓦兹分布的希尔伯特变换与应用(英文)	2020－07	58.00	1190
泛函分析:巴拿赫空间理论入门(英文)	2020－07	48.00	1191
卡塔兰数入门(英文)	2019－05	68.00	1060
测度与积分(英文)	2019－04	68.00	1059
组合学手册.第一卷(英文)	2020－06	128.00	1153
＊一代数、局部紧群和巴拿赫＊一代数丛的表示.第一卷,群和代数的基本表示理论(英文)	2020－05	148.00	1154
电磁理论(英文)	2020－08	48.00	1193
连续介质力学中的非线性问题(英文)	2020－09	78.00	1195
多变量数学入门(英文)	2021－05	68.00	1317
偏微分方程入门(英文)	2021－05	88.00	1318
若尔当典范性:理论与实践(英文)	2021－07	68.00	1366
伽罗瓦理论.第 4 版(英文)	2021－08	88.00	1408

刘培杰数学工作室
已出版(即将出版)图书目录——原版影印

书　　　名	出版时间	定　价	编号
典型群,错排与素数(英文)	2020—11	58.00	1204
李代数的表示:通过 gln 进行介绍(英文)	2020—10	38.00	1205
实分析演讲集(英文)	2020—10	38.00	1206
现代分析及其应用的课程(英文)	2020—10	58.00	1207
运动中的抛射物数学(英文)	2020—10	38.00	1208
2—纽结与它们的群(英文)	2020—10	38.00	1209
概率,策略和选择:博弈与选举中的数学(英文)	2020—11	58.00	1210
分析学引论(英文)	2020—11	58.00	1211
量子群:通往流代数的路径(英文)	2020—11	38.00	1212
集合论入门(英文)	2020—10	48.00	1213
酉反射群(英文)	2020—11	58.00	1214
探索数学:吸引人的证明方式(英文)	2020—11	58.00	1215
微分拓扑短期课程(英文)	2020—10	48.00	1216
抽象凸分析(英文)	2020—11	68.00	1222
费马大定理笔记(英文)	2021—03	48.00	1223
高斯与雅可比和(英文)	2021—03	78.00	1224
π 与算术几何平均:关于解析数论和计算复杂性的研究(英文)	2021—01	58.00	1225
复分析入门(英文)	2021—03	48.00	1226
爱德华·卢卡斯与素性测定(英文)	2021—03	78.00	1227
通往凸分析及其应用的简单路径(英文)	2021—01	68.00	1229
微分几何的各个方面.第一卷(英文)	2021—01	58.00	1230
微分几何的各个方面.第二卷(英文)	2020—12	58.00	1231
微分几何的各个方面.第三卷(英文)	2020—12	58.00	1232
沃克流形几何学(英文)	2020—11	58.00	1233
彷射和韦尔几何应用(英文)	2020—12	58.00	1234
双曲几何学的旋转向量空间方法(英文)	2021—02	58.00	1235
积分:分析学的关键(英文)	2020—12	48.00	1236
为有天分的新生准备的分析学基础教材(英文)	2020—11	48.00	1237
数学不等式.第一卷.对称多项式不等式(英文)	2021—03	108.00	1273
数学不等式.第二卷.对称有理不等式与对称无理不等式(英文)	2021—03	108.00	1274
数学不等式.第三卷.循环不等式与非循环不等式(英文)	2021—03	108.00	1275
数学不等式.第四卷.Jensen 不等式的扩展与加细(英文)	2021—03	108.00	1276
数学不等式.第五卷.创建不等式与解不等式的其他方法(英文)	2021—04	108.00	1277

刘培杰数学工作室
已出版(即将出版)图书目录——原版影印

书　名	出版时间	定　价	编号
冯·诺依曼代数中的谱位移函数:半有限冯·诺依曼代数中的谱位移函数与谱流(英文)	2021—06	98.00	1308
链接结构:关于嵌入完全图的直线中链接单形的组合结构(英文)	2021—05	58.00	1309
代数几何方法.第1卷(英文)	2021—06	68.00	1310
代数几何方法.第2卷(英文)	2021—06	68.00	1311
代数几何方法.第3卷(英文)	2021—06	58.00	1312
代数、生物信息和机器人技术的算法问题.第四卷,独立恒等式系统(俄文)	2020—08	118.00	1199
代数、生物信息和机器人技术的算法问题.第五卷,相对覆盖性和独立可拆分恒等式系统(俄文)	2020—08	118.00	1200
代数、生物信息和机器人技术的算法问题.第六卷,恒等式和准恒等式的相等 问题、可推导性和可实现性(俄文)	2020—08	128.00	1201
分数阶微积分的应用:非局部动态过程,分数阶导热系数(俄文)	2021—01	68.00	1241
泛函分析问题与练习:第2版(俄文)	2021—01	98.00	1242
集合论、数学逻辑和算法论问题:第5版(俄文)	2021—01	98.00	1243
微分几何和拓扑短期课程(俄文)	2021—01	98.00	1244
素数规律(俄文)	2021—01	88.00	1245
无穷边值问题解的递减:无界域中的拟线性椭圆和抛物方程(俄文)	2021—01	48.00	1246
微分几何讲义(俄文)	2020—12	98.00	1253
二次型和矩阵(俄文)	2021—01	98.00	1255
积分和级数.第2卷,特殊函数(俄文)	2021—01	168.00	1258
积分和级数.第3卷,特殊函数补充:第2版(俄文)	2021—01	178.00	1264
几何图上的微分方程(俄文)	2021—01	138.00	1259
数论教程:第2版(俄文)	2021—01	98.00	1260
非阿基米德分析及其应用(俄文)	2021—03	98.00	1261
古典群和量子群的压缩(俄文)	2021—03	98.00	1263
数学分析习题集.第3卷,多元函数:第3版(俄文)	2021—03	98.00	1266
数学习题:乌拉尔国立大学数学力学系大学生奥林匹克(俄文)	2021—03	98.00	1267
柯西定理和微分方程的特解(俄文)	2021—03	98.00	1268
组合极值问题及其应用:第3版(俄文)	2021—03	98.00	1269
数学词典(俄文)	2021—01	98.00	1271
确定性混沌分析模型(俄文)	2021—06	168.00	1307
精选初等数学习题和定理.立体几何.第3版(俄文)	2021—03	68.00	1316
微分几何习题:第3版(俄文)	2021—05	98.00	1336
精选初等数学习题和定理.平面几何.第4版(俄文)	2021—05	68.00	1335
曲面理论在欧氏空间 E_n 中的直接表示(俄文)	2022—01	68.00	1444
维纳—霍普夫离散算子和托普利兹算子:某些可数赋范空间中的诺特性和可逆性(俄文)	2022—03	108.00	1496
Maple 中的数论:数论中的计算机计算(俄文)	2022—03	88.00	1497
贝尔曼和克努特问题及其概括:加法运算的复杂性(俄文)	2022—03	138.00	1498

书　名	出版时间	定　价	编号
复分析:共形映射(俄文)	2022—07	48.00	1542
微积分代数样条和多项式及其在数值方法中的应用(俄文)	2022—08	128.00	1543
蒙特卡罗方法中的随机过程和场模型:算法和应用(俄文)	2022—08	88.00	1544
线性椭圆型方程组:论二阶椭圆型方程的迪利克雷问题(俄文)	2022—08	98.00	1561
动态系统解的增长特性:估值、稳定性、应用(俄文)	2022—08	118.00	1565
群的自由积分解:建立和应用(俄文)	2022—08	78.00	1570
混合方程和偏差自变数方程问题:解的存在和唯一性(俄文)	2023—01	78.00	1582
拟度量空间分析:存在和逼近定理(俄文)	2023—01	108.00	1583
二维和三维流形上函数的拓扑性质:函数的拓扑分类(俄文)	2023—03	68.00	1584
齐次马尔科夫过程建模的矩阵方法:此类方法能够用于不同目上的的复杂系统研究、设计和完善(俄文)	2023—03	68.00	1594
狭义相对论与广义相对论:时空与引力导论(英文)	2021—07	88.00	1319
束流物理学和粒子加速器的实践介绍:第2版(英文)	2021—07	88.00	1320
凝聚态物理中的拓扑和微分几何简介(英文)	2021—05	88.00	1321
混沌映射:动力学、分形学和快速涨落(英文)	2021—05	128.00	1322
广义相对论:黑洞、引力波和宇宙学介绍(英文)	2021—06	68.00	1323
现代分析电磁均质化(英文)	2021—06	68.00	1324
为科学家提供的基本流体动力学(英文)	2021—06	88.00	1325
视觉天文学:理解夜空的指南(英文)	2021—06	68.00	1326
物理学中的计算方法(英文)	2021—06	68.00	1327
单星的结构与演化:导论(英文)	2021—06	108.00	1328
超越居里:1903年至1963年物理界四位女性及其著名发现(英文)	2021—06	68.00	1329
范德瓦尔斯流体热力学的进展(英文)	2021—06	68.00	1330
先进的托卡马克稳定性理论(英文)	2021—06	88.00	1331
经典场论导论:基本相互作用的过程(英文)	2021—07	88.00	1332
光致电离量子动力学方法原理(英文)	2021—07	108.00	1333
经典域论和应力:能量张量(英文)	2021—05	88.00	1334
非线性太赫兹光谱的概念与应用(英文)	2021—06	68.00	1337
电磁学中的无穷空间并矢格林函数(英文)	2021—06	88.00	1338
物理科学基础数学.第1卷,齐次边值问题、傅里叶方法和特殊函数(英文)	2021—07	108.00	1339
离散量子力学(英文)	2021—07	68.00	1340
核磁共振的物理学和数学(英文)	2021—07	108.00	1341
分子水平的静电学(英文)	2021—08	68.00	1342
非线性波:理论、计算机模拟、实验(英文)	2021—06	108.00	1343
石墨烯光学:经典问题的电解决方案(英文)	2021—06	68.00	1344
超材料多元宇宙(英文)	2021—07	68.00	1345
银河系外的天体物理学(英文)	2021—07	68.00	1346
原子物理学(英文)	2021—07	68.00	1347
将光打结:将拓扑学应用于光学(英文)	2021—07	68.00	1348
电磁学:问题与解法(英文)	2021—07	88.00	1364
海浪的原理:介绍量子力学的技巧与应用(英文)	2021—07	108.00	1365
多孔介质中的流体:输运与相变(英文)	2021—07	68.00	1372
洛伦兹群的物理学(英文)	2021—08	68.00	1373
物理导论的数学方法和解决方法手册(英文)	2021—08	68.00	1374

书　　名	出版时间	定　价	编号
非线性波数学物理学入门(英文)	2021—08	88.00	1376
波:基本原理和动力学(英文)	2021—07	68.00	1377
光电子量子计量学.第1卷,基础(英文)	2021—07	88.00	1383
光电子量子计量学.第2卷,应用与进展(英文)	2021—07	68.00	1384
复杂流的格子玻尔兹曼建模的工程应用(英文)	2021—08	68.00	1393
电偶极矩挑战(英文)	2021—08	108.00	1394
电动力学:问题与解法(英文)	2021—09	68.00	1395
自由电子激光的经典理论(英文)	2021—09	68.00	1397
曼哈顿计划——核武器物理学简介(英文)	2021—09	68.00	1401
粒子物理学(英文)	2021—09	68.00	1402
引力场中的量子信息(英文)	2021—09	128.00	1403
器件物理学的基本经典力学(英文)	2021—09	68.00	1404
等离子体物理及其空间应用导论.第1卷,基本原理和初步过程(英文)	2021—09	68.00	1405
拓扑与超弦理论焦点问题(英文)	2021—07	58.00	1349
应用数学:理论、方法与实践(英文)	2021—07	78.00	1350
非线性特征值问题:牛顿型方法与非线性瑞利函数(英文)	2021—07	58.00	1351
广义膨胀和齐性:利用齐性构造齐次系统的李雅普诺夫函数和控制律(英文)	2021—06	48.00	1352
解析数论焦点问题(英文)	2021—07	58.00	1353
随机微分方程:动态系统方法(英文)	2021—07	58.00	1354
经典力学与微分几何(英文)	2021—07	58.00	1355
负定相交形式流形上的瞬子模空间几何(英文)	2021—07	68.00	1356
广义卡塔兰轨道分析:广义卡塔兰轨道计算数字的方法(英文)	2021—07	48.00	1367
洛伦兹方法的变分:二维与三维洛伦兹方法(英文)	2021—08	38.00	1378
几何、分析和数论精编(英文)	2021—08	68.00	1380
从一个新角度看数论:通过遗传方法引入现实的概念(英文)	2021—07	58.00	1387
动力系统:短期课程(英文)	2021—08	68.00	1382
几何路径:理论与实践(英文)	2021—08	48.00	1385
论天体力学中某些问题的不可积性(英文)	2021—07	88.00	1396
广义斐波那契数列及其性质(英文)	2021—08	38.00	1386
对称函数和麦克唐纳多项式:余代数结构与 Kawanaka 恒等式(英文)	2021—09	38.00	1400
杰弗里·英格拉姆·泰勒科学论文集:第1卷.固体力学(英文)	2021—05	78.00	1360
杰弗里·英格拉姆·泰勒科学论文集:第2卷.气象学、海洋学和湍流(英文)	2021—05	68.00	1361
杰弗里·英格拉姆·泰勒科学论文集:第3卷.空气动力学以及落弹数和爆炸的力学(英文)	2021—05	68.00	1362
杰弗里·英格拉姆·泰勒科学论文集:第4卷.有关流体力学(英文)	2021—05	58.00	1363

刘培杰数学工作室
已出版（即将出版）图书目录——原版影印

书　名	出版时间	定　价	编号
非局域泛函演化方程:积分与分数阶(英文)	2021－08	48.00	1390
理论工作者的高等微分几何:纤维丛、射流流形和拉格朗日理论(英文)	2021－08	68.00	1391
半线性退化椭圆微分方程:局部定理与整体定理(英文)	2021－07	48.00	1392
非交换几何、规范理论和重整化:一般简介与非交换量子场论的重整化(英文)	2021－09	78.00	1406
数论论文集:拉普拉斯变换和带有数论系数的幂级数(俄文)	2021－09	48.00	1407
挠理论专题:相对极大值,单射与扩充模(英文)	2021－09	88.00	1410
强正则图与欧几里得若尔当代数:非通常关系中的启示(英文)	2021－10	48.00	1411
拉格朗日几何和哈密顿几何:力学的应用(英文)	2021－10	48.00	1412
时滞微分方程与差分方程的振动理论:二阶与三阶(英文)	2021－10	98.00	1417
卷积结构与几何函数理论:用以研究特定几何函数理论方向的分数阶微积分算子与卷积结构(英文)	2021－10	48.00	1418
经典数学物理的历史发展(英文)	2021－10	78.00	1419
扩展线性丢番图问题(英文)	2021－10	38.00	1420
一类混沌动力系统的分歧分析与控制:分歧分析与控制(英文)	2021－11	38.00	1421
伽利略空间和伪伽利略空间中一些特殊曲线的几何性质(英文)	2022－01	68.00	1422
一阶偏微分方程:哈密尔顿－雅可比理论(英文)	2021－11	48.00	1424
各向异性黎曼多面体的反问题:分段光滑的各向异性黎曼多面体反边界谱问题:唯一性(英文)	2021－11	38.00	1425
项目反应理论手册.第一卷,模型(英文)	2021－11	138.00	1431
项目反应理论手册.第二卷,统计工具(英文)	2021－11	118.00	1432
项目反应理论手册.第三卷,应用(英文)	2021－11	138.00	1433
二次无理数:经典数论入门(英文)	2022－05	138.00	1434
数,形与对称性:数论,几何和群论导论(英文)	2022－05	128.00	1435
有限域手册(英文)	2021－11	178.00	1436
计算数论(英文)	2021－11	148.00	1437
拟群与其表示简介(英文)	2021－11	88.00	1438
数论与密码学导论:第二版(英文)	2022－01	148.00	1423

书　名	出版时间	定　价	编号
几何分析中的柯西变换与黎兹变换:解析调和容量和李普希兹调和容量、变化和振荡以及一致可求长性(英文)	2021—12	38.00	1465
近似不动点定理及其应用(英文)	2022—05	28.00	1466
局部域的相关内容解析:对局部域的扩展及其伽罗瓦群的研究(英文)	2022—01	38.00	1467
反问题的二进制恢复方法(英文)	2022—03	28.00	1468
对几何函数中某些类的各个方面的研究:复变量理论(英文)	2022—01	38.00	1469
覆盖、对应和非交换几何(英文)	2022—01	28.00	1470
最优控制理论中的随机线性调节器问题:随机最优线性调节器问题(英文)	2022—01	38.00	1473
正交分解法:涡流流体动力学应用的正交分解法(英文)	2022—01	38.00	1475
芬斯勒几何的某些问题(英文)	2022—03	38.00	1476
受限三体问题(英文)	2022—05	38.00	1477
利用马利亚万微积分进行 Greeks 的计算:连续过程、跳跃过程中的马利亚万微积分和金融领域中的 Greeks(英文)	2022—05	48.00	1478
经典分析和泛函分析的应用:分析学的应用(英文)	2022—03	38.00	1479
特殊芬斯勒空间的探究(英文)	2022—03	48.00	1480
某些图形的施泰纳距离的细谷多项式:细谷多项式与图的维纳指数(英文)	2022—05	38.00	1481
图论问题的遗传算法:在新鲜与模糊的环境中(英文)	2022—05	48.00	1482
多项式映射的渐近簇(英文)	2022—05	38.00	1483
一维系统中的混沌:符号动力学,映射序列,一致收敛和沙可夫斯基定理(英文)	2022—05	38.00	1509
多维边界层流动与传热分析:粘性流体流动的数学建模与分析(英文)	2022—05	38.00	1510
演绎理论物理学的原理:一种基于量子力学波函数的逐次置信估计的一般理论的提议(英文)	2022—05	38.00	1511
R^2 和 R^3 中的仿射弹性曲线:概念和方法(英文)	2022—08	38.00	1512
算术数列中除数函数的分布:基本内容、调查、方法、第二矩、新结果(英文)	2022—05	28.00	1513
抛物型狄拉克算子和薛定谔方程:不定常薛定谔方程的抛物型狄拉克算子及其应用(英文)	2022—07	28.00	1514
黎曼-希尔伯特问题与量子场论:可积重正化、戴森-施温格方程(英文)	2022—08	38.00	1515
代数结构和几何结构的形变理论(英文)	2022—08	48.00	1516
概率结构和模糊结构上的不动点:概率结构和直觉模糊度量空间的不动点定理(英文)	2022—08	38.00	1517

刘培杰数学工作室

已出版(即将出版)图书目录——原版影印

书　名	出版时间	定　价	编号
反若尔当对:简单反若尔当对的自同构(英文)	2022-07	28.00	1533
对某些黎曼-芬斯勒空间变换的研究:芬斯勒几何中的某些变换(英文)	2022-07	38.00	1534
内诣零流形映射的尼尔森数的阿诺索夫关系(英文)	2023-01	38.00	1535
与广义积分变换有关的分数次演算:对分数次演算的研究(英文)	2023-01	48.00	1536
强子的芬斯勒几何和吕拉几何(宇宙学方面):强子结构的芬斯勒几何和吕拉几何(拓扑缺陷)(英文)	2022-08	38.00	1537
一种基于混沌的非线性最优化问题:作业调度问题(英文)	即将出版		1538
广义概率论发展前景:关于趣味数学与置信函数实际应用的一些原创观点(英文)	即将出版		1539
纽结与物理学:第二版(英文)	2022-09	118.00	1547
正交多项式和 q-级数的前沿(英文)	2022-09	98.00	1548
算子理论问题集(英文)	2022-09	108.00	1549
抽象代数:群、环与域的应用导论:第二版(英文)	即将出版		1550
菲尔兹奖得主演讲集:第三版(英文)	2023-01	138.00	1551
多元实函数教程(英文)	2022-09	118.00	1552
球面空间形式群的几何学:第二版(英文)	2022-09	98.00	1566
对称群的表示论(英文)	2023-01	98.00	1585
纽结理论:第二版(英文)	2023-01	88.00	1586
拟群理论的基础与应用(英文)	2023-01	88.00	1587
组合学:第二版(英文)	2023-01	98.00	1588
加性组合学:研究问题手册(英文)	2023-01	68.00	1589
扭曲、平铺与镶嵌:几何折纸中的数学方法(英文)	2023-01	98.00	1590
离散与计算几何手册:第三版(英文)	2023-01	248.00	1591
离散与组合数学手册:第二版(英文)	2023-01	248.00	1592
分析学教程.第1卷,一元实变量函数的微积分分析学介绍(英文)	2023-01	118.00	1595
分析学教程.第2卷,多元函数的微分和积分,向量微积分(英文)	2023-01	118.00	1596
分析学教程.第3卷,测度与积分理论,复变量的复值函数(英文)	2023-01	118.00	1597
分析学教程.第4卷,傅里叶分析,常微分方程,变分法(英文)	2023-01	118.00	1598

联系地址:哈尔滨市南岗区复华四道街 10 号　哈尔滨工业大学出版社刘培杰数学工作室
网　　址:http://lpj.hit.edu.cn/
邮　　编:150006
联系电话:0451-86281378　　13904613167
E-mail:lpj1378@163.com